Windows® 10

ALL-IN-ONE

for dummies®

A Wiley Brand

Windows® 10

ALL-IN-ONE

4th Edition

**by Woody Leonhard
and
Ciprian Adrian Rusen**

A Wiley Brand

Windows® 10 All-in-One For Dummies®, 4th Edition

Published by: **John Wiley & Sons, Inc.,** 111 River Street, Hoboken, NJ 07030-5774, www.wiley.com

Copyright © 2021 by John Wiley & Sons, Inc., Hoboken, New Jersey

Published simultaneously in Canada

For general information on our other products and services, please contact our Customer Care Department within the U.S. at 877-762-2974, outside the U.S. at 317-572-3993, or fax 317-572-4002. For technical support, please visit https://hub.wiley.com/community/support/dummies.

Wiley publishes in a variety of print and electronic formats and by print-on-demand. Some material included with standard print versions of this book may not be included in e-books or in print-on-demand. If this book refers to media such as a CD or DVD that is not included in the version you purchased, you may download this material at http://booksupport.wiley.com. For more information about Wiley products, visit www.wiley.com.

Library of Congress Control Number: 2020948872

ISBN 978-1-119-68057-4 (pbk); ISBN 978-1-119-68059-8 (ebk); ISBN 978-1-119-68058-1 (ebk)

Manufactured in the United States of America

SKY10023288_121520

Contents at a Glance

Table of Contents

Introduction

Windows has a long and glorious history, stretching all the way back to Windows 1.0 in 1985. It was sent to the bit bucket in the sky at the end of 2001. Windows 3.0, starting in 1990, began to fill Microsoft's coffers and 3.1 (1992) is widely held as a landmark achievement in the history of computing. Then came Windows NT — a completely new beast, built to be a server — and the reunification of consumer and server side in Windows 95 (Chicago), Windows 98 (Memphis — the first version to ship with Internet Explorer built-in), 2000, and the much-maligned Windows ME.

With Windows XP (Whistler) in 2001, Microsoft took on the mantle of juggernaut, and blew away everything in its path. Also in 2001, the US Department of Justice sued Microsoft for using its monopoly power to roll over other Internet browsers. Such was the staying power of Windows XP that it was used in a significant number of PCs for many years after its launch.

Windows Vista came along in 2006/2007, but it was upstaged in 2009 by Windows 7 — arguably the most-loved version of Windows. It continues to capture a large share of Windows users and was only recently upstaged by Windows 10.

Then, in 2012, there was Windows 8. Think of Windows 8/8.1 as an extended, bad, no-good, horrible nightmare. Microsoft's woken up now. They fired almost everybody who ran the Windows 8 operation, cleaned out the house, and brought in some truly gifted engineers. Windows 10 is a brand-new day. Whether it's *your* brand-new day is another story.

Windows 10, released in July 2015, looks a little bit like Windows 7 and a little bit like Windows 8.1. It doesn't work like either of them, but for the billion-and-a-half Windows users out there, at least it's recognizable as Windows.

If you haven't yet taken the plunge with Windows 10, I advise you to go slowly. Microsoft is furiously working on extending the product and shoring up problems. The Windows 10 you know today will change in a few months — a new version appears every six months — and you may like the new one better. Before installing Windows 10, I would simply count to ten.

For most Windows 8 and 8.1 users, Windows 10 is a no-brainer. You can kvetch about some problems — the disappearance of Windows Media Center, for example. There are dozens of additional details, but by and large, Windows 10 is what Windows 8.1 should've been.

Windows 7 users did not have as much incentive to move to Windows 10, but there are some good changes. Microsoft effectively ditched Internet Explorer and built a much lighter and more capable browser called Microsoft Edge. Instead of desktop gadgets, which in Windows 7 were held together with baling wire and chewing gum, Windows 10 sports an entire infrastructure for apps (also known as Universal apps). Windows 10 works with all the new hardware, touchscreens, and pens. There's an improved Task Manager, File Explorer, Clipboard, and a dozen other system utilities.

Is that enough to convince Windows 7 users to abandon ship in droves? Probably not. The single biggest allure of Windows 10 for the Windows 7 battle-hardened is that it's clearly the way of the future. Also, since January 2020, Windows 7 no longer receives updates and security patches. That's a major risk for users who want to stay safe on the Internet.

If you want a better Windows, for whatever reason, you'll have to go through Windows 10.

Here's what you should ask yourself before you move from Windows 7 to Windows 10:

» Are you willing to learn a new operating system, with a number of new features that may or may not appeal to you?

» Are you willing to let Microsoft snoop on your actions, more than they did with Windows 7? Microsoft has become more transparent about what it being snooped, and it appears to be roughly on par with Google's snooping and arguably less intrusive than Apple's snooping.

» Are you willing to let Microsoft take control of your machine? The company has already shown that it can take Windows 7 and 8.1 machines to town, with the Get Windows 10 campaign. But in Windows 10, it's considerably more difficult to keep patches at bay.

» Are you willing to ditch a trusted operating system (Windows 7) that is no longer secure because Microsoft has decided to stop supporting it, and deal with Windows 10's annoyance factors for the sake of security?

This isn't the manual Microsoft forgot. This is the manual Microsoft wouldn't dare print. I won't feed you the Microsoft party line or make excuses for pieces of Windows 10 that just don't work: Some of it is junk, some of it is evolving, and some of it is devolving. My job is to take you through the most important parts of Windows 10, give you tips that may or may not involve Microsoft products, point out the rough spots, and guide you around the disasters. Frankly, there are some biggies.

I also look at using non-Microsoft products in a Windows way: iPhones, Androids, Kindles, Gmail and Google apps, Facebook, Twitter, Dropbox, Firefox, Google Chrome, iCloud, and many more. Even though Microsoft competes with just about every one of those products, each has a place in your computing arsenal and ties into Windows 10 in important ways.

I'll save you more than enough money to pay for the book several times over, keep you from pulling out a whole shock of hair, lead you to dozens if not hundreds of "Aha!" moments, and keep you awake in the process. Guaranteed.

About This Book

Windows 10 All-in-One For Dummies, 4th Edition, takes you through the Land of the Dummies — with introductory material and stuff your grandmother can (and should!) understand — and then continues the journey into more advanced areas, where you can truly put Windows to work every day.

I start with the Windows 10 Start menu, and for many of you, that's the only Start you'll ever need. The Start menu coverage here is the best you'll find anywhere because I don't assume that you know Windows 10 and I step you through everything you need to know both with a touchscreen and a mouse.

Then I dig in to the desktop and take you through all the important pieces.

I don't dwell on technical mumbo jumbo, and I keep the baffling jargon to a minimum. At the same time, though, I tackle the tough problems you're likely to encounter, show you the major road signs, and give you lots of help where you need it the most.

Whether you want to get two or more email accounts set up to work simultaneously, turn your tiles a lighter shade of pale, or share photos of your Boykin Spaniel in OneDrive, this is your book. Er, I should say ten books. I've broken out the topics into ten minibooks, so you'll find it easy to hop around to a topic — and a level of coverage — that feels comfortable.

I didn't design this book to be read from front to back. It's a reference. Each chapter and each of its sections are meant to focus on solving a particular problem or describing a specific technique.

Windows 10 All-in-One For Dummies, 4th Edition, should be your reference of first resort, even before you consult Windows Help and Support. There's a big reason why: Windows Help was written by hundreds of people over the course of many years. Some of the material was written ages ago, and it's confusing as all get-out, but it's still in Windows Help for folks who are tackling tough legacy problems. Some of the Help file terminology is inconsistent and downright misleading, largely because the technology has changed so much since some of the articles were written. Finding help in Help frequently boggles my mind: If I don't already know the answer to a question, it's hard to figure out how to coax Help to help. Besides, if you're looking for help on connecting your smartphone to your PC or downloading pictures from your Samsung Galaxy smartphone, Microsoft would rather sell you something different. The proverbial bottom line: I don't duplicate the material in Windows 10 Help and Support, but I point to it if I figure it can help you.

WARNING

A word about Windows 10 versions: Microsoft is trying to sell the world on the idea that Windows 10 runs on everything — desktops, laptops, tablets, assisted reality headsets, huge banks of servers, giant conference room displays, refrigerators, and toasters. While that's literally true — Microsoft can call anything Windows 10 if it wants — for those of us who work on desktops, laptops, and tablets, Windows 10 is Windows 10.

Foolish Assumptions

I don't make many assumptions about you, dear reader, except to acknowledge that you're obviously intelligent, well-informed, discerning, and of impeccable taste. That's why you chose this book, eh?

Okay, okay. The least I can do is butter you up a bit. Here's the straight scoop: If you've never used Windows, bribe your neighbor (or, better, your neighbor's kids) to teach you how to do four things:

>> Play a game with your fingers (if you have a touchscreen) or with a mouse (if you're finger-challenged). Any of the games that ship with Windows 10, or free games in the Microsoft Store, will do. If your neighbor's kids don't have a different recommendation, try the new Microsoft Solitaire Collection.

>> Start File Explorer.

>> Get on the web.

>> Turn Windows 10 off. (Click or tap the Start icon in the lower left of the screen, click the universal on/off button thingy, and then click Shut down.)

That covers it. If you can play a game, you know how to turn on your computer, log in if need be, touch and drag, and tap and hold down. If you run File Explorer, you know how to click a taskbar icon. After you're on the web, well, it's a great starting point for almost anything. And if you know that you need to use the Start menu, you're well on your way to achieving Windows 10 enlightenment.

And that begins with Book 1, Chapter 1.

Icons Used in This Book

Some of the points in *Windows 10 All-in-One For Dummies*, 4th Edition, merit your special attention. I set off those points with icons.

TIP

When I'm jumping up and down on one foot with an idea so absolutely cool that I can't stand it anymore, I stick a tip icon in the margin. You can browse any chapter and hit its highest points by jumping from tip to tip.

ASK WOODY.COM

When you see this icon, you get the real story about Windows 10 — not the stuff that the Microsoft marketing droids want you to hear — and my take on the best way to get Windows 10 to work for you. You find the same take on Microsoft, Windows, and more at my eponymous website, www.AskWoody.com.

REMEMBER

You don't need to memorize the information marked with this icon, but you should try to remember that something special is lurking.

WARNING

Achtung! Cuidado! Thar be tygers here! Anywhere that you see a warning icon, you can be sure that I've been burnt — badly. Mind your fingers. These are really, really mean suckers.

TECHNICAL STUFF

Okay, so I'm a geek. I admit it. Sure, I love to poke fun at geeks. But I'm a modern, New Age, sensitive guy, in touch with my inner geekiness. Sometimes, I just can't help but let it out, ya know? That's where the technical stuff icon comes in. If you get all tied up in knots about techie-type stuff, pass these paragraphs by. (For the record, I managed to write this entire book without telling you that an IPv4 address consists of a unique 32-bit combination of network ID and host ID, expressed as a set of four decimal numbers with each octet separated by periods. See? I can restrain myself sometimes.)

Beyond the Book

When I wrote the 4th edition of this book, I covered the Windows 10 May 2020 update, version 2004. Microsoft promises to keep Windows 10 updated twice a year. For details about significant updates or changes that occur between editions of this book, go to www.dummies.com, search for *Windows 10 All-in-One For Dummies*, and open the Download tab on this book's dedicated page.

In addition, the cheat sheet for this book has handy Windows shortcuts and tips on other cool features worth checking out. To get to the cheat sheet, go to www. dummies.com, and then type *Windows 10 All-in-One For Dummies Cheat Sheet* in the search box.

Where to Go from Here

That's about it. It's time for you to crack this book open and have at it.

If you haven't yet told Windows 10 to show you filename extensions, flip to Book 3, Chapter 1. If you haven't yet set up the File History feature, go to Book 8, Chapter 1. If you're worried about Microsoft keeping a list of all the searches that you conduct *on your own computer,* check out Book 2, Chapter 5.

**ASK
WOODY.COM**

Don't forget to bookmark two websites: www.AskWoody.com and www. digitalcitizen.life. They will keep you up-to-date on all the Windows 10 stuff you need to know — including notes about this book, the latest Windows bugs and gaffes, patches that are worse than the problems they're supposed to fix, useful tutorials, and much more — and you can submit your most pressing questions for free consultation from The Woodmeister and his merry gang.

See ya! Shoot me mail at woody@AskWoody.com.

Sometimes, it's worth reading the Intro, eh?

1

Starting Windows 10

Contents at a Glance

Chapter **1**

Windows 10 4 N00bs

Don't sweat it. We all started as newbies who didn't know much about technology.

If you've never used an earlier version of Windows, you're in luck! With Windows 10, you don't have to force your fingers to forget so much of what you've learned. This version is different from any Windows that has come before. It's a melding of Windows 7 and Windows 8, tossed into a blender, speed turned up full, poured out on your screen.

If you heard that Windows 8 was a dog, you heard only the printable part of the story. By clumsily forcing a touchscreen approach down the throats of mouse-lovers everywhere, Windows 8 frustrated people who loved touch-based interfaces, drove mouse users nuts, and left everybody — aside from a few diehards — screaming in pain.

Windows 10 brings a kinder, gentler approach for the 1.7 billion or so people who have seen the Windows desktop and know a bit about struggling with it. Yes, Windows 10 exposes you to some smartphone-style tiles that you can touch, but they aren't nearly as intrusive or scary as you think.

Some of you are reading this book because you specifically chose to run Windows 10. Others are here because Windows 10 came preinstalled on a new computer or because your company forced you to upgrade to Windows 10. Some of you are here because you fell victim to Microsoft's much maligned "Get Windows 10" campaign. Whatever the reason, you've ended up on a good operating system, and it should serve you well — if you understand and respect its limitations.

Now you're sitting in front of your computer, and this thing called Windows 10 is staring at you. The screen (see Figure 1-1), which Microsoft calls the lock screen, doesn't say *Windows,* much less *Windows 10.* The lock screen doesn't say much of anything except the current date and time, with maybe a tiny icon or two that shows whether your Internet connection is working. You may also see when the next meeting is scheduled in your Calendar, how many unopened emails await, or whether you should just take the day off because your holdings in AAPL stock soared again.

FIGURE 1-1:
The Windows 10 lock screen. Your picture may differ, but the function stays the same.

3:03

Monday, May 25

Working on Windows 10 All-In-One For Dummies
3:00 PM - 4:00 PM

You may be tempted to sit and admire the gorgeous picture, whatever it may be, but if you swipe up from the bottom, click anywhere on the picture, or press any key, you see the login screen, resembling the one in Figure 1-2. If more than one person is set up to use your computer, you'll see more than one name.

That's the login screen, but it doesn't say *Login* or *Welcome to Win10 Land* or *Howdy* or even *Sit down and get to work, Bucko.* It has names and pictures for only the people who can use the computer. Why do you have to click your name? What if your name isn't there? And why can't you bypass all this garbage, log in, and get your email?

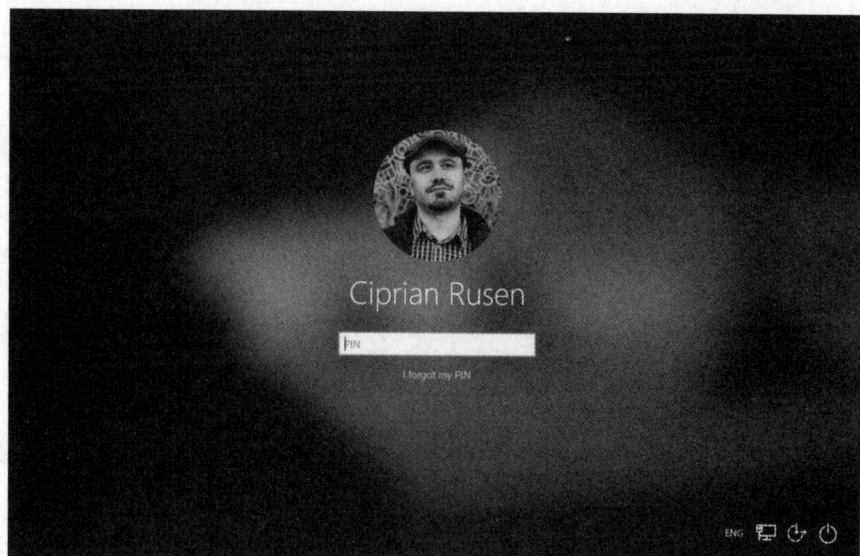

FIGURE 1-2:
The Windows 10
login screen.

Good for you. That's the right attitude.

Windows 10 ranks as the most sophisticated operating system ever made. It cost more money to develop and took more people to build than any previous operating system — ever. So why is it so blasted hard to use? Why doesn't it do what you want it to do the first time? Why do updates constantly break it? For that matter, why do you need it at all?

Someday, I swear, you'll be able to pull a PC out of the box, plug it into the wall, turn it on, and then get your email, look at the news, or connect to Facebook — bang, bang, bang, just like that, in ten seconds flat. In the meantime, those stuck in the early 21st century have to make do with PCs that grow obsolete before you unpack them, software so ornery that you find yourself arguing with it, and Internet connections that involve turtles carrying bits on their backs.

If you aren't comfortable working with Windows and you still worry that you may break something if you click the wrong button, welcome to the club! In this chapter, I present a concise overview of how all this hangs together and what to look for when buying a Windows 10 computer. It may help you understand why and how Windows 10 has limitations. It also may help you communicate with the geeky rescue team that tries to bail you out, whether you rely on the store that sold you the PC, the smelly guy in the apartment downstairs, or your daughter's nerdy classmate.

Hardware and Software

At the most fundamental level, all computer stuff comes in one of two flavors: hardware or software. *Hardware* is anything you can touch — a computer screen, a mouse, a hard drive, a keyboard, a DVD drive (remember those coasters with shiny sides?). *Software* is everything else: the movies you stream on Netflix, the digital pictures of your last vacation, and programs such as Microsoft Office. If you shoot a bunch of pictures, the pictures themselves are just bits — software. But they're probably sitting on some sort of memory card inside your smartphone or camera. That memory card is hardware. Get the difference?

Windows 10 is software. You can't touch it. Your PC, on the other hand, is hardware. Kick the computer screen, and your toe hurts. Drop the big box on the floor, and it smashes into a gazillion pieces. That's hardware.

Chances are good that one of the major PC manufacturers — Lenovo, HP, Dell, Acer, or ASUS, for example — or maybe even Microsoft, with its Surface line, or even Apple, made your hardware. Microsoft, and Microsoft alone, makes Windows 10.

When you bought your computer, you paid for a license to use one copy of Windows on the PC you bought. Its manufacturer paid Microsoft a royalty so it could sell you Windows along with the PC. (That royalty may have been zero dollars, but it's a royalty nonetheless.) You may think that you got Windows from, say, Dell — indeed, you may have to contact Dell for technical support on Windows questions — but Windows came from Microsoft.

If you upgraded from Windows 7 or Windows 8.1 to Windows 10, you might have received a free upgrade license — but it's still a license, whether you paid for it or not. You can't give it away to someone else.

REMEMBER

These days, most software, including Windows 10, asks you to agree to an End User License Agreement (EULA). When you first set up your PC, Windows asked you to click the Accept button to accept a licensing agreement that's long enough to reach the top of the Empire State Building. If you're curious about what agreement you accepted, take a look at the official EULA repository, www.microsoft.com/en-us/Useterms/Retail/Windows/10/UseTerms_Retail_Windows_10_English.htm.

Why Do PCs Have to Run Windows?

Here's the short answer: You don't have to run Windows on your PC.

The PC you have is a dumb box. (You needed me to tell you that, eh?) To get that box to do anything worthwhile, you need a computer program that takes control of the PC and makes it do things. It does things such as show web pages on the screen, respond to mouse clicks or taps, or print résumés. An *operating system* controls the dumb box and makes it do worthwhile things, in ways that mere humans can understand.

Without an operating system, the computer can sit in a corner and display profound messages on the screen, such as *Non-system disk or disk error* or *Insert system disk and press any key when ready.* If you want your computer to do more than that, though, you need an operating system.

ASK
WOODY.COM

Windows is not the only operating system in town. The other big contenders in the PC and PC-like operating system game are Chrome OS, macOS, and Linux:

>> **ChromeOS:** Cheap Chromebooks have long dominated the best-seller lists at many computer retailers — and for good reason. If you want to surf the web, work on email, compose simple documents, or do anything in a browser — which covers a whole lot of ground these days — ChromeOS is all you need. Chromebooks run Google's ChromeOS. They can't run Windows programs such as Office or Photoshop (although they *can* run web-based versions of them, such as Office Online or the Photoshop Express Editor). Despite the limitations, they don't get infected and have few maintenance problems. You can't say the same about Windows 10: That's why you need a thousand-page book to keep it going. Yes, you do need a reliable Internet connection to get the most out of ChromeOS. But some parts of ChromeOS and Google's apps, including Gmail, can work even if you don't have an active Internet connection.

ChromeOS, built on Linux, looks and feels much like the Google Chrome web browser. There are a few minor differences, but in general, you feel like you're working in the Chrome browser. One downside is that ChromeOS, unlike Linux or Windows, can't be installed on any PC you want. It's limited to the devices on which it is sold and preinstalled by their manufacturer. That's why, if you want ChromeOS, you must purchase a Chromebook or Chromebox (the "equivalent" of a desktop PC).

For friends and family who don't have big-time computer needs, I find myself recommending a Chromebook more often than not. It's cheaper, easier for them, and easier for me to help them out.

>> **macOS:** Apple has made great strides running on Intel hardware. If you don't already know how to use Windows or own a Windows computer, it makes sense to consider buying an Apple computer or running macOS or both. Yes, you can build a custom computer and run macOS on it: Check out www.hackintosh.com. But, no, it isn't legal — the macOS End User License Agreement explicitly forbids installation on a non-Apple-branded computer. Also, installing it is not for the faint of heart.

That said, if you buy a Mac — say, a MacBook Air or Pro — it's easy to run Windows 10 on it. Some people feel that the highest quality Windows environment today comes from running Windows 10 on a MacBook, and for years I've run it on my MacBook Pro and Air. All you need is a program called Boot Camp, and that's already installed, free, on the MacBook.

>> **Linux:** The big up-and-coming operating system, which has been up and coming for a couple of decades now, is Linux, which is pronounced "LIN-uchs." It's a viable contender for cheaper PCs and older ones. Linux comes in many names (called distros) and versions. If you want to give it a try, you might want to start with Ubuntu Linux. If you plan to use your PC only to get on the Internet — to surf the web and send emails— Linux can handle all that, with few of the headaches that remain as the hallmark of Windows. By using free programs such as LibreOffice (www.libreoffice.org) and online services such as GSuite and Google Drive (www.drive.google.com), you can even cover the basics in word processing, spreadsheets, presentations, contact managers, calendars, and more. Linux may not support the vast array of hardware that Windows 10 offers — but more than a few wags will tell you that Windows has problems supporting it too.

WINDOWS RT, RIP

Back in the early days of Windows 8, Microsoft developed a different branch of Windows that was christened *Windows RT*. New Windows RT computers at the time were generally small, light, and inexpensive. They had a long battery life and touch-sensitive displays.

Several manufacturers made Windows RT machines, but the only company that sold more than a dumpster full of them was Microsoft. Microsoft's original Surface (later renamed Surface RT) and Surface 2 ran Windows RT — and even they didn't sell worth beans.

The fundamental flaw with Windows RT? It wasn't Windows. You couldn't (and can't) run classic Windows programs on it. You can't upgrade the machine to real Windows. But try explaining that to a garden-variety customer. Microsoft blew it when they gave the new, odd operating system the name *Windows RT*.

The company has essentially orphaned Windows RT. If you own a Windows RT device (most likely a Microsoft Surface or Surface 2), the folks in Redmond provided one last update, called Windows RT 8.1 update 3, which plugs what little they could muster. See www.microsoft.com/surface/en-us/support/install-update-activate/windows-8-1-rt-update-3.

In the tablet sphere, iOS and Android rule, with iOS for iPhones and iPads — all from Apple — and Android for smartphones and tablets from a bewildering number of manufacturers. Windows 10 doesn't exactly compete with any of them. However, Microsoft tried to take on iPad with the now-defunct Windows RT (see the sidebar "Windows RT, RIP") and is trying to dip its billion-dollar toe back in the bare-bones water with Windows 10 S mode and the upcoming Windows 10X.

Yet another branch of Windows is geared toward phones and tablets, especially 8-inch and smaller tablets. Windows 10 Mobile (see the sidebar) owes its pedigree to Windows Phone 8 and Windows RT. At least conceptually (and, in fact, under the hood in no small part), Microsoft has grown Windows Phone up and Windows RT down to meet somewhere in the middle. As we went to press, Windows 10 Mobile was dead. Today, no one creates smartphones with Windows 10 Mobile.

WARNING

Windows 10 in S mode is a relatively confusing development with an unclear future. Designed to compete with ChromeOS and iPads, *S mode* refers to a set of restrictions on "real" Windows 10. Supposedly in an attempt to improve battery life, reduce the chance of getting infected, and simplify your life, the S mode versions of Windows 10 won't run most regular Windows programs. S mode limits users only to apps found in the Microsoft Store. You get Spotify, iTunes, but not Google Chrome or Firefox.

Fortunately, Windows 10 S mode systems can be upgraded so that they're no longer in S mode. For most people who want more than the basics, that's a smart move. If you find that you can't run real Windows programs on your Windows 10 in S mode machine, look into dropping S mode.

What do other people choose? It's hard to measure the percentage of PCs running Windows versus Mac versus Linux. One company, StatCounter (www.statcounter. com), specializes in analyzing the traffic of 3 million sites globally and provides lots of useful statistics based on the data they collect. One stat is tallying how many Windows computers hit those sites, compared to macOS and Linux. Although their data may not be 100 percent representative of real-world market share, it does an excellent job of giving us an idea of operating system penetration. If you look at only desktop operating systems — Windows (on desktops, laptops, 2-in-1s) and macOS X — the numbers in April 2020 (according to StatCounter) break as shown in Figure 1-3. (Linux and ChromeOS, the two bottom lines, have barely more than 1 percent market share, each).

In April 2020, Windows (the top line) had a market share of 76.52 percent of all desktop operating systems, and macOS (the second line from the top) had 18.99 percent. In Microsoft's world, Windows 10 is king with a 73.14 percent market share. Windows 7 is a distant second, with 19.44 percent, and constantly declining, as Microsoft has declared its end of life on January 14, 2020. As of this date, users are no longer receiving support and updates for Windows 7 and are highly encouraged to upgrade to Windows 10.

WINDOWS 10 MOBILE, RIP

Generally, devices with screens smaller than 9 inches ran the other kind of Windows, known as Windows 10 Mobile. Yes, there were devices larger than 9 inches that used to run Windows 10 Mobile and 8-inch devices with the "real" Windows 10. The general argument went like this: If you don't need to use the traditional Windows 7–style desktop, why pay for it? Windows 10 centers on the mouse-friendly desktop. Windows 10 Mobile sticks to the tiled world and is much more finger-friendly.

This book talks about Windows 10, not Windows 10 Mobile. Microsoft gave up and sold its Nokia business in May 2016. Also, the company stopped fixing bugs and providing updates for Windows 10 Mobile in December 2019. Today, no one sells smartphones or tablets with Windows 10 Mobile, and the platform is dead.

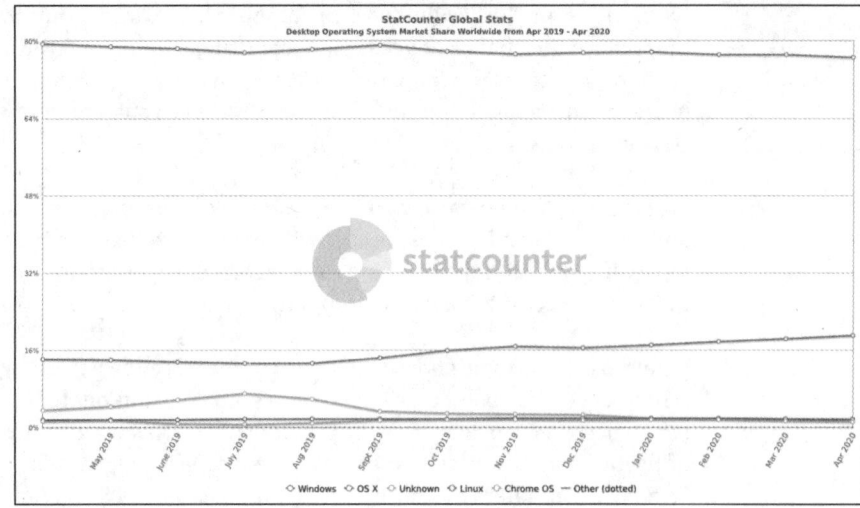

FIGURE 1-3:
The worldwide market share of desktop operating systems — April 2019 – April 2020.

ASK
WOODY.COM

If you look at the bigger picture, including tablets and smartphones, the numbers change dramatically. As of April 2020, StatCounter says that 39.13 percent of all devices on the Internet use Android, while 33.1 percent use Windows. Back in July 2015, Andreesen Horowitz reported that the number of iOS devices (iPhones, iPads) sold per month zoomed ahead of the number of Windows PCs. Mobile operating systems are swallowing the world — and the trend has been in mobile's favor, not Windows. The number of smartphones sold every year exceeded the number of PCs sold in 2011, and the curve has gone steeply in favor of mobile ever since. The number of PCs sold every year peaked in 2014 and has been declining steadily ever since. According to Statista, at the end of 2019, 60 percent of search engine visits in the United States were made from mobile devices. In other countries such as those in Asia, mobile is even more significant because people learn how to access the Internet on mobile devices and not on PCs.

Windows was once the king of the computing hill. Not so anymore. This is good news for you — the Windows customer. Today, Microsoft is branching out to make software for smartphones and tablets of all stripes, and Windows 10 itself works better with whatever tablets and hybrid devices you might use. It's a brave new Windows world.

A Terminology Survival Kit

Some terms pop up so frequently that you'll find it worthwhile to memorize them, or at least understand where they come from. That way, you won't be caught flat-footed when your first-grader comes home and asks whether he can install a Universal app on your computer.

TIP

If you want to drive your techie friends nuts the next time you have a problem with your Windows 10 computer, tell them that the hassles occur when you're "running Microsoft." They won't have any idea whether you mean Windows, Word, Outlook, OneNote, Search, or any of a gazillion other programs. Also, they won't know if you're talking about a Microsoft program on Windows, the Mac, iPad, iPhone, or Android.

Windows 10, the *operating system* (see the preceding section), is a sophisticated computer program. So are computer games, Microsoft Office, Microsoft Word (the word processor part of Office), Google Chrome (the web browser made by Google), those nasty viruses you've heard about, that screen saver with the oh-too-perfect fish bubbling and bumbling about, and others.

An *app* or a *program* or a *desktop app* is *software* (see the earlier "Hardware and Software" section in this chapter) that works on a computer. *App* is modern and cool; *program* is old and boring; *desktop app* or *application* manages to hit both gongs, but they all mean the same thing.

A *Windows app* is a program that, at least in theory, runs on any version of Windows 10. By design, *apps* (which used to be called Universal Windows Platform, or UWP apps) should run on Windows 10 on a desktop, a laptop, and a tablet— and even on an Xbox game console, a giant wall-mounted Surface Hub, a HoloLens augmented reality headset, and possibly Internet of Things tiny computers. They also run on Windows 10 in S mode (see the previous section).

For most people, "Universal" does not mean what they might think it means. Universal Windows apps *don't* work on Windows 8.1 or Windows 7. They don't even run on Windows RT tablets (see the "Windows RT, RIP" sidebar). They're universal only in the sense that they'll run on Windows 10. In theory.

A special kind of program called a *driver* makes specific pieces of hardware work with the operating system. The driver acts like a translator that enables Windows to ask your hardware to do what it wants. Suppose you have a document you want to print. You edit the document in Word, click the Print button, and wait for the document to be printed. Word is an application that asks the operating system to print the document. The operating system takes it and asks the printer driver to print the document. Then the driver translates the document into a language that the printer understands. Finally, the printer prints the document and delivers it to you. Everything inside your computer and all that is connected to it has a driver: The hard disk inside the PC has a driver, the printer has a driver, your mouse has a driver, and Tiger Woods has a driver (several, actually, and he makes a living with them). I wish that everyone was so talented.

Many drivers ship with Windows, even though Microsoft doesn't make them. The hardware manufacturer's responsible for making its hardware work with your Windows PC, and that includes building and fixing the drivers. However, if Microsoft makes your computer, Microsoft's responsible for the drivers, too. Sometimes you can get a driver from the manufacturer that works better than the one that ships with Windows. Also, keep in mind that device manufacturers offer updated drivers on their websites.

When you stick an app or program on your computer — and set it up so that you can use it — you *install* the app or program (or driver).

When you crank up a program — that is, get it going on your computer — you can say you *started* it, *launched* it, *ran* it, or *executed* it. They all mean the same thing.

If the program quits the way it's supposed to, you can say it *stopped, finished, ended, exited,* or *terminated.* Again, all these terms mean the same thing. If the app stops with some sort of weird error message, you can say it *crashed, died, cratered, croaked, went belly up, jumped in the bit bucket,* or *GPFed* (techspeak for "generated a General Protection Fault" — don't ask), or employ any of a dozen colorful but unprintable epithets. If the program just sits there and you can't get it to do anything, no matter how you click your mouse or poke the screen, you can say that it *froze, hung, stopped responding,* or *went into a loop.*

A *bug* is something that doesn't work right. (A bug is not a virus! Viruses work as intended far too often.) US Navy Rear Admiral Grace Hopper — the intellectual guiding force behind the COBOL programming language and one of the pioneers in the history of computing — often repeated the story of a moth being found in a relay of an ancient Mark II computer. The moth was taped into the technician's logbook on September 9, 1947. (See Figure 1-4.)

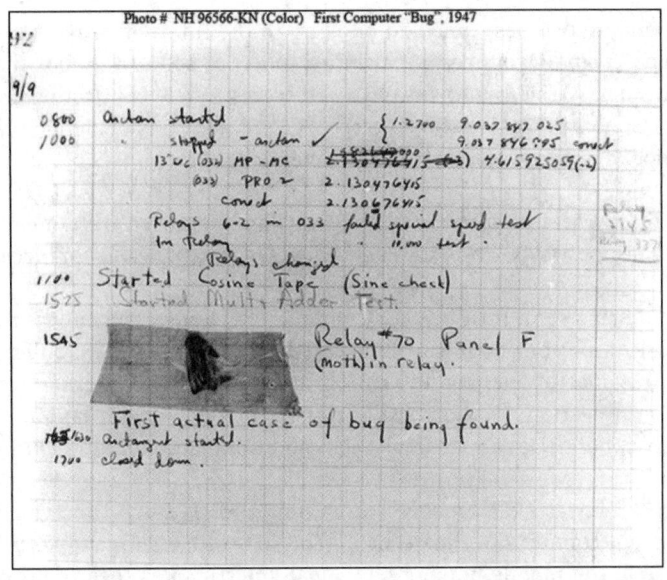

FIGURE 1-4: Admiral Grace Hopper's log of the first actual case of a bug being found.

The people who invented all this terminology think of the Internet as being some great blob in the sky — it's *up*, as in "up in the sky." So, if you send something from your computer to the Internet, you're *uploading*. If you take something off the Internet and put it on your computer, you're *downloading*.

The *cloud* is just a marketing term for the Internet. Saying that you put your data "in the cloud" sounds so much cooler than saying you copied it to storage on the Internet. Programs can run in the cloud — that is, they run on the Internet. Just about everything that has anything to do with computers can be done in the cloud. Just watch your pocketbook.

REMEMBER

If you use *cloud storage*, you're just sticking your data on some company's computers. Put a file in Microsoft OneDrive, and it goes onto one of Microsoft's computers. Put it in Google Drive, and it goes to Google's storage in the sky. Move it to Dropbox, and it's sitting on a Dropbox computer.

When you put computers together, you *network* them, and if your network doesn't use wires, it's called a *Wi-Fi network.* At the heart of a network sits a box, commonly called a *router* or an *access point,* that computers can plug into. If the router has "rabbit ears" on top, for wireless connections, it's usually called a *Wi-Fi router.* Keep in mind that some Wi-Fi routers may not have antennae outside and keep them hidden inside their box. Yes, fine lines of distinction exist among all these terms. No, you don't need to worry about them.

There are two basic ways to hook up to the Internet: *wired* and *wireless.* Wired is easy: You plug your computer into a router or some other box that connects to the Internet. Wireless falls into two categories: Wi-Fi connections, which you find in many homes, coffee shops, airports, and all kinds of public places, and cellular (mobile phone–style) wireless connections.

Cellular Wireless Internet connections are identified with one of the G levels: 2G, 3G, 4G, or maybe even 5G. Each G level is faster than its predecessor.

This part gets a little tricky. If your smartphone can connect to a 3G or 4G network, it may be possible to set it up to behave like a Wi-Fi router: Your laptop talks to the smartphone, and the smartphone talks to the Internet over its 3G or 4G (or 5G) connection. That process is called *tethering* — your laptop is tethered to your smartphone. Not all smartphones can tether, and not all manufacturers allow it.

Special boxes called *mobile hotspot* units work much the same way: The mobile hotspot connects to the 3G or 4G (or 5G) connection, and your laptop gets tethered to the mobile hotspot box. Most smartphones these days can be configured as mobile hotspots.

If you plug your Internet connection into the wall, you have *broadband,* which may run via *fiber* (a cable that uses light waves), *DSL* or *ADSL* (which uses regular old phone lines), *cable* (as in cable TV), or *satellite.* The fiber, DSL, cable, or satellite box is called a *modem,* although it's really a *router.* Although fiber-optic lines are inherently much faster than DSL or cable, individual results can be all over the lot. Ask your neighbors what they're using and then pick the best. If you don't like your current service, vote with your wallet.

ASK
WOODY.COM

Turning to the dark side of the force, Luke, the distinctions among *viruses, worms,* and *Trojans* grow blurrier every day. In general, they're programs that replicate and can be harmful, and the worst ones blend different approaches. *Spyware* gathers information about you and then phones home with all the juicy details. *Adware* gets in your face with dodgy ads, all too frequently installing itself on your computer without your knowledge or consent. *Ransomware* scrambles (or threatens to scramble) your data and demands a payment to keep the data intact. I tend to lump the three together and call them *scumware* or *crapware* or something a bit more descriptive and less printable.

If a bad guy (and they're almost always guys) manages to take over your computer without your knowledge, turning it into a zombie that spews spam by remote control, you're in a *botnet*. (And yes, the term *spam* comes from the immortal *Monty Python* routine that's set in a cafe serving Hormel's SPAM luncheon meat, the chorus bellowing "lovely Spam, wonderful Spam.") Check out Book 9 for details about preventing scumware and the like from messing with you.

The most successful botnets employ *rootkits* — programs that run underneath Windows, evading detection because regular programs can't see them. The number of Windows 10 computers running rootkits is probably two or three or four orders of magnitudes less than the number of zombified Windows XP computers. However, as long as Windows XP computers are out there, botnets will continue to be a major threat to everyone.

ASK
WOODY.COM

This section covers about 90 percent of the buzzwords you hear in common parlance. If you get stuck at a party where the bafflegab is flowing freely, don't hesitate to invent your own words. Nobody will ever know the difference.

What, Exactly, Is the Web?

Years from now, the operating system you use will be largely irrelevant, as will be the speed of your computer, the amount of memory you have, and the number of terabytes of storage that hum in the background. Microsoft will keep milking its cash cow, but the industry will move on. Individuals and businesses will stop shelling out big bucks for Windows and the iron to run it. Instead, the major push will be online. Rather than spend money on PCs that become obsolete the week after you purchase them, folks will spend money on big data pipes: It'll be less about me and more about us. Why? Because so much more is "out there" than "in here." Count on it.

But what is the Internet? This section answers this burning question (if you've asked it). If you don't necessarily wonder about the Internet's place in space and time just yet, you will . . . you will.

REMEMBER

You know those stories about computer jocks who come up with great ideas, develop the ideas in their basements (or garages or dorm rooms), release their products to the public, change the world, and make a gazillion bucks?

This isn't one of them.

The Internet started in the mid-1960s as an academic exercise — primarily with the RAND Corporation, the Massachusetts Institute of Technology (MIT), and the National Physical Laboratory in England — and rapidly evolved into a military project, under the US Department of Defense Advanced Research Project Agency (ARPA), designed to connect research groups working on ARPA projects.

By the end of the 1960s, ARPA had four computers hooked together — at UCLA, SRI (Stanford), UC Santa Barbara, and the University of Utah — using systems developed by BBN Technologies (then named Bolt Beranek and Newman, Inc.). By 1971, it had eighteen. I started using ARPANET in 1975. According to the website www.internetworldstats.com, at the beginning of 2020, the Internet had more than 4.5 billion users worldwide — well over half of the global population.

Today, so many computers are connected directly to the Internet that the Internet's addressing system is running out of numbers, just as your local phone company is running out of telephone numbers. The current numbering system — named IPv4 — can handle about 4 billion addresses. The next version, named IPv6, can handle this number of addresses:

340,000,000,000,000,000,000,000,000,000,000,000,000

That should last for a while, don't you think?

Ever wonder why you rarely see hard statistics about the Internet? I've found two big reasons:

>> Defining terms related to the Internet is devilishly difficult these days. (What do you mean when you say, "*X* number of computers are connected to the Internet"? Is that the number of computers up and running at any given moment? The number of different addresses that are active? The number that could be connected if everybody dialed up at the same time? The number of different computers that are connected in a typical day, or week, or month?)

>> The other reason is that the Internet is growing so fast that any number you publish today will be meaningless tomorrow.

Getting inside the Internet

Some observers claim that the Internet works so well because it was designed to survive a nuclear attack. Not so. The people who built the Internet insist that they weren't nearly as concerned about nukes as they were about making communication among researchers reliable, even when a backhoe severed an underground phone line or one of the key computers ground to a halt.

As far as I'm concerned, the Internet works so well because the engineers who laid the groundwork were utter geniuses. Their original ideas from 60 years ago have been through the wringer a few times, but they're still pretty much intact. Here's what the engineers decided:

>> **No single computer should be in charge.** All the big computers connected directly to the Internet are equal (although, admittedly, some are more equal than others). By and large, computers on the Internet move data around like kids playing hot potato — catch it, figure out where you're going to throw it, and let it fly quickly. They don't need to check with some übercomputer before doing their work; they just catch, look, and throw.

>> **Break the data into fixed-size packets.** No matter how much data you're moving — an email message that just says "Hi" or a full-color, life-size photograph of the Andromeda galaxy — the data is broken into packets. Each packet is routed to the appropriate computer. The receiving computer assembles all the packets and notifies the sending computer that everything came through okay.

>> **Deliver each packet quickly.** If you want to send data from Computer A to Computer B, break the data into packets and route each packet to Computer B by using the fastest connection possible — even if that means some packets go through Bangor and others go through Bangkok.

Taken together, those three rules ensure that the Internet can keep on functioning no matter what happens. If a chipmunk eats through a line, any big computer that's using the gnawed line can start rerouting packets over a different one. If the Cumbersome Computer Company in Cupertino, California, loses power, computers that were sending packets through Cumbersome can switch to other connected computers. It usually works quickly and reliably, although the techniques used internally by the Internet computers get a bit hairy at times.

Big computers are hooked together by high-speed communication lines: the *Internet backbone*. If you want to use the Internet from your business or your house, you must connect to one of the big computers first. Companies that own the big computers — Internet service providers (ISPs) — get to charge you for the privilege of getting on the Internet through their big computers. The ISPs, in turn, pay the companies that own the cables (and satellites) that comprise the Internet backbone for a slice of the backbone.

If all this sounds like a big-fish-eats-smaller-fish-eats-smaller-fish arrangement, that's quite a good analogy.

It's backbone-breaking work, but somebody's gotta do it.

What is the World Wide Web?

People tend to confuse the World Wide Web with the Internet, which is much like confusing the dessert table with the buffet line. I'd be the first to admit that desserts are mighty darn important — life-critical, in fact, if the truth be told. But they aren't the same as the buffet line.

To get to the dessert table, you must stand in the buffet line. To get to the web, you have to be running on the Internet. Make sense?

The World Wide Web owes its existence to Tim Berners-Lee and a few co-conspirators at a research institute named CERN in Geneva, Switzerland. In 1990, Berners-Lee demonstrated a way to store and link information on the Internet so that all you had to do was click to jump from one place — one web page — to another. Nowadays, nobody in his right mind can give a definitive count of the number of pages available, but in 2016, Google reported that it had indexed more than 130 trillion pages. Since then, that number has surely exploded to many hundreds of trillions. Like the Internet itself, the World Wide Web owes much of its success to the brilliance of the people who brought it to life. The following list describes the ground rules:

» Web pages, stored on the Internet, are identified by an address, such as www. dummies.com. The main part of the web page address — dummies.com, for example — is a *domain name*. With rare exceptions, you can open a web page by typing its domain name and pressing Enter. Spelling counts, and under-scores (_) are treated differently from hyphens (-). Being close isn't good enough — there are just too many websites. The part after the dot is the top-level domain. According to VeriSign, in June 2017, approximately 331.9 million domains were on the Internet, with top-level domains such as .com, .net, .org, .info, .biz, or .us. This statistic included all countries and country-specific top-level domains, such as co.uk (the UK equivalent of .com,) .br for Brazil, and .jp for Japan.

» Web pages are written in the funny language HyperText Markup Language (HTML). HTML is sort of a programming language, sort of a formatting language, and sort of a floor wax, all rolled into one. Many products claim to make it easy for novices to create powerful, efficient HTML. Some of those products are getting close.

» To read a web page, you must use a web browser. A *web browser* is a program or desktop app that runs on your computer and is responsible for converting HTML into text that you can read and use. Many people who view web pages use Google's Chrome web browser, although Mozilla Firefox, Opera, and Microsoft's Edge browser in Windows 10 are all contenders. Internet Explorer is still inside Windows 10, but you have to dig deep to find it. (Hint: Click the

Start icon and then open Windows Accessories.) Internet Explorer is no longer actively maintained by Microsoft, and their plan is to convince people to use Microsoft Edge. If you don't stick to the dated and insecure Internet Explorer, any web browser is a good choice.

>> More and more people (including me!) prefer Mozilla Firefox (see www.mozilla.org), Opera (see www.opera.com), or Chrome, from Google (www.google.com/chrome). You may not know that Firefox and Chrome can run right alongside Internet Explorer and Microsoft Edge, with absolutely no confusion between the two. Err, four. In fact, they don't even interact — Microsoft Edge, Firefox, and Google Chrome were designed to operate completely independently, and they do not mess with each other in any way, except when trying to promote themselves over their competitors.

One unwritten rule for the World Wide Web: All web acronyms must be completely, utterly inscrutable. For example, a web address is a *Uniform Resource Locator,* or *URL.* (The techies I know pronounce URL "earl." Those who don't wear white lab coats tend to say "you are ell.") As I said earlier, the HTML acronym means HyperText Markup Language. On the web, a gorgeous, sunny, palm-lined beach with the scent of frangipani wafting through the air would no doubt be called SHS — Smelly Hot Sand. Sheeesh.

ASK
WOODY.COM

The best part of the web is how easily you can jump from one place to another — and how easily you can create web pages with *hot links* (also called *hyperlinks* or just *links)* that transport the viewer wherever the author intends. That's the *H* in HTML and the original reason for creating the web so many years ago.

Who pays for all this stuff?

"Who pays for all this stuff?" is the 64-billion-dollar question, isn't it? The Internet is one of the true bargains of the 21st century. To get online, you probably have to pay AT&T, Comcast, Verizon, Mediacom, Evan, Cable One, CenturyLink, some other cable company, or another ISP a monthly fee. The fee you pay varies depending on the speed you want for your Internet connection and the services bundled with it, such as TV and online streaming subscriptions.

REMEMBER

Microsoft Edge and Internet Explorer are free, sorta, because they come with Windows 10, no matter which version you buy. Firefox is free as a breeze — in fact, it's the poster child for open-source programs: Everything about the browser, even the program code itself, is free. Google Chrome and Opera are free, too. Both Microsoft, with Microsoft Edge and Internet Explorer, and Google, with Chrome, keep tabs on where you go and what you do online — all the better to convince you to click an ad. Firefox collects some data, but its uses are limited. The same with Opera.

Others involved in your security may be selling your personal information. AVG, an antivirus of fame, announced in September 2015 that it would start selling browsing history data to advertisers. Avast — another free antivirus owned by the same company, has similar practices. Your ISP may be selling your data too.

Most websites don't charge a cent. They pay for themselves in any of these ways:

>> **Contract advertising:** Google has made a fortune. In 2019, advertising accounted for $134.81 billion in the company's revenue.

>> **Use display advertising:** Many sites run ads, most commonly from Google, but in some cases, selected from a pool of advertisers. The advertiser pays a bounty for each person who clicks the ad and views its website — a *click-through*.

>> **Use affiliate programs:** Many sites may also participate in a retailer's affiliate program. If a customer clicks through and orders something, the website that originated the transaction receives a percentage of the amount ordered. Amazon is well known for its affiliate program, but many others exist.

>> **Increase a company's visibility:** The website gives you a good excuse to buy more of the company's products. This is why architectural firms show you pictures of their buildings and food companies post recipes.

>> **Reduce a company's operating costs:** Banks and brokerage firms, for example, have websites that routinely handle customer inquiries at a fraction of the cost of H2H (err, human-to-human) interactions.

>> **Draw in new business:** Ask any real estate agent.

Some websites have an entrance fee. For example, if you want to read more than a few articles on *The New York Times* website, you have to part with some substantial coin — $12 for twelve weeks — for their most basic option, the last time I looked. Guess that beats schlepping around a whole lotta paper.

Buying a Windows 10 Computer

Here's how it usually goes: You figure that you need to buy a new PC and spend a couple weeks brushing up on the details — price, storage, size, processor, memory — and doing lots of comparison shopping. You end up at your local Computers Are Us shop, and the guy behind the counter convinces you that the best bargain you'll ever see is sitting right here, right now, and you'd better take it quick before somebody else nabs it.

YOU MAY NOT NEED TO PAY MORE TO GET A CLEAN PC

I hate it when the computer I want comes loaded with all that nice, "free" crapware. I'd seriously consider paying more to get a clean computer.

You don't need an antivirus and Internet security program preinstalled on your new PC. It is going to open and beg for money next month. Windows 10 comes with Windows Security (formerly known as Windows Defender), and it works great — for free.

Browser toolbars? Puh-lease.

You can choose your own Internet service provider. AT&T? Verizon? Who needs you?

And trialware? Whether it's Quicken or any of a zillion other programs, if you must pay for a preinstalled app in three months or six months, you don't want it.

If you're looking for a new computer but can't find an option to buy a PC without all the "extras," look elsewhere. The big PC companies are slowly getting a clue, but until they clean up their act, you may be better served buying from a smaller retailer, who hasn't yet presold every bit that isn't nailed down. Or you can buy direct from Microsoft: Its Surface tablets and laptops are as clean as the driven snow. Pricey, perhaps. But blissfully clean.

Microsoft's online store sells new, clean computers from major manufacturers. Before you spend money on a computer, check to see whether it's available dreck-free (usually at the same price) from the Microsoft Store. Go to www.microsoftstore.com and choose any PC. The ones on offer ship without any of the junk.

If you bought a new computer with all that gunk, you could get rid of it by performing a reset or reinstall. See Book 8, Chapter 2 for details.

Your eyes glaze over as you look at yet another spec sheet and try to figure out one last time whether a RAM is a ROM, whether a solid-state drive is worth the effort, and whether you need a SATA 6 Gbps, or NVMe, or USB 2 or 3 or C. In the end, you figure that the guy behind the counter must know what he's doing, so you plunk down your credit card and pray you got a good deal.

The next Sunday morning, you look at the ads on Newegg (www.Newegg.com) or Best Buy (www.BestBuy.com) or Amazon (www.Amazon.com) and discover you could have bought the same PC for 25 percent less money. The only thing you know for sure is that your PC is hopelessly becoming out of date, and the next time you'll be smarter about the whole process.

If that describes your experiences, relax. It happens to everybody. Take solace in the fact that technology evolves at an incredible pace, and nobody can keep up with it.

ASK
WOODY.COM

Here's everything you need to know about buying a Windows 10 PC:

>> **Decide if you're going to use a touchscreen.** Although a touch-sensitive screen is not a prerequisite for using apps on Windows 10, you'll probably find it easier to use apps with your fingers than with your mouse. Swiping with a finger is easy; swiping with a trackpad works reasonably well, depending on the trackpad; swiping with a mouse is a disaster. However, if you aren't into Windows 10 apps from the Microsoft Store that are optimized for touch, a touchscreen probably isn't worth the additional expense. Experienced, mouse-savvy Windows users often find that using a mouse and a touchscreen at the same time is an ergonomic pain in the arm. Unless you have fingertips the size of pinheads — or you always use a stylus — using classic Windows programs on a touchscreen is an excruciating experience. Best to leave the touching to apps that are demonstrably touch-friendly.

TIP

There's no substitute for trying the hardware on a touch-sensitive Windows 10 computer. Hands and fingers come in all shapes and sizes. What works for size XXL hands with ten thumbs (present company included) may not cut the mustard for svelte hands and fingers experienced at taking cotton balls out of medicine bottles.

See the section "Inside a touch-sensitive tablet" later in this chapter.

>> **Get a screen that's at least 1920 x 1080 pixels — the minimum resolution to play high-definition (1080p) movies.** You probably want to stream movies from Netflix and watch videos on YouTube. To enjoy the experience, do not get stingy when buying the monitor. Make sure that it's Full HD, which means it has 1920 x 1080 pixels in resolution.

>> **If you're going to use the old-fashioned, Windows 7–style desktop, get a high-quality monitor, a solid keyboard, and a mouse that feels comfortable.** Corollary: Don't buy a computer online unless you know for a fact that your fingers are going to like the keyboard, your wrist will tolerate the mouse, and your eyes will fall in love with the monitor.

>> **Go overboard with hard drives.** In the best of all possible worlds, get a computer with a solid-state drive (SSD) for the system drive (the C: drive) plus

a large hard drive for storage, perhaps attached via a USB cable. For the low-down on SSDs, hard drives, backups, and putting them all together, see the upcoming section "Managing disks and drives."

TIP

How much hard drive space do you need? How long is a string? Unless you have an enormous collection of videos, movies, or songs, 1TB (=1,024GB = 1,048,576MB = 1,073,741,824KB = 1,099,511,627,776 bytes, or characters of storage) should suffice. That's big enough to handle about 1,000 broadcast-quality movies. Consider that the printed collection of the US Library of Congress runs about 10TB.

If you're getting a laptop or Ultrabook with an SSD, consider buying an external 1TB or larger drive at the same time. You'll use it. External hard drives are cheap and plug-in easy to use.

Or you can just stick all that extra data in the cloud, with OneDrive, Dropbox, Google Drive, or some competitor. See Book 6, Chapter 1 to get started. For what it's worth, I used Dropbox in every phase of writing this book.

>> **Everything else they try to sell you pales in comparison.**

ASK
WOODY.COM

If you want to spend more money, go for a faster Internet connection and a better chair. You need both items much more than you need a marginally faster, or bigger, computer.

Inside the big box

In this section, I give you just enough information about the inner workings of a desktop or laptop PC that you can figure out what you have to do with Windows 10. In the next section, I talk about touch-enabled tablets, the PCs that respond to touch. Details can change over time, but these are the basics.

The big box that your desktop computer lives in is sometimes called a *CPU*, or *central processing unit* (see Figure 1-5). Right off the bat, you're bound to get confused, unless somebody clues you in on one important detail: The main computer chip inside that big box is also called a CPU. I prefer to call the big box "the PC" because of the naming ambiguity, but you've probably thought of a few better names.

The big box contains many parts and pieces (and no small amount of dust and dirt), but the crucial, central element inside every PC is the motherboard. (You can see a picture of a motherboard here: www.asus.com/Motherboards/PRIME-X570-PRO/).

FIGURE 1-5:
The enduring, traditional big box.

Courtesy of Dell Inc.

You find the following items attached to the motherboard:

» **The processor, or CPU:** This gizmo does the main computing. It's probably from Intel or AMD. Different manufacturers rate their processors in different ways, and it's impossible to compare performance by just looking at the part number. Yes, Intel Core i7 CPUs usually run faster than Core i5s, and Core i3s are the slowest of the three, but there are many nuances. The same is true for AMD's Ryzen 7, Ryzen 5, and Ryzen 3 line-up of processors. Unless you tackle intensive video games, create and edit audio or video files, or recalculate spreadsheets with the national debt, the CPU doesn't count for much. If you're streaming audio and video (say, with YouTube or Netflix) you don't need a fancy processor. If in doubt, check out the reviews at www.tomshardware.com and www.anandtech.com.

» **Memory chips and places to put them:** Memory is measured in megabytes (1MB = 1,024KB = 1,048,576 characters), gigabytes (1GB = 1,024MB), and terabytes (1TB = 1,024GB). Microsoft recommends a minimum of 2GB of RAM. Unless you have an exciting cornfield to watch grow while using Windows 10, aim for 4GB or more. Most computers allow you to add more memory to them. Boosting your computer's memory to 8GB from 2GB makes the machine snappier, especially if you run memory hogs such as Microsoft Office, Photoshop, or Google Chrome. If you leave Outlook open and work with it all day and run almost any other major program at the same time, 8GB is a wise choice. If you're going to do some video editing, gaming, or software development, you probably need more. But for most people, 8GB will run everything well.

» **Video card:** Most motherboards include remarkably good built-in video. If you want more video oomph, you must buy a video card and put it in a card slot. Advanced motherboards have multiple PCI-Express card slots, to allow you to strap together two video cards and speed up video even more. If you want to run a VR or AR headset, such as an Oculus Rift, you're going to need a much more capable video setup. For more information, see the "Screening" section in this chapter.

» **SSD:** Solid-state drives, or SSDs, are fast and cheap storage. You don't have to buy an expensive drive to benefit from tangible speed improvements. If you don't want to wait a long time for your programs to load and don't want Windows 10 to take minutes to boot, buying an SSD is a must. In comparison, hard disks (HDDs) are slow and dated. You should use an HDD for storing your personal files and backing up your data, not for running Windows 10, games, and apps.

» **Card slots (also known as expansion slots):** Laptops have limited (if any) expansion slots on the motherboard. Desktops generally contain several expansion slots. Modern slots come in two flavors: PCI and PCI-Express (also known as PCIe or PCI-E). Many expansion cards, such as video cards, sound cards, and network cards, require PCIe slots. Of course, PCI cards do not fit in PCIe slots, and vice versa. To make things more confusing, PCIe comes in four sizes — literally, the size of the bracket and the number of bumps on the bottoms of the cards is different. The PCIe 1x is smallest, the relatively uncommon PCIe 4x is considerably larger, and PCIe 8x is a bit bigger still. PCIe 16x is just a little bit bigger than an old-fashioned PCI slot. Most video cards these days require a PCIe 16x slot. Or two.

If you're buying a monitor separately from the rest of the system, make sure the monitor takes video input in a form that your PC can produce. See the upcoming section "Screening" for details.

» **USB (Universal Serial Bus) connections:** The USB cable has a flat connector that plugs into your slots. USB 3 is considerably faster than USB 2, and any kind of USB device can plug into a USB 3 slot, whether the device itself supports USB 3 level speeds.

USB Type-C (often called USB C) is a different kind of cable that has a different kind of slot. It has two big advantages: The plug is reversible, making it impossible to plug it in upside-down, and you can run a considerable amount of power through a USB-C, making it a good choice for power supplies. Many laptops these days get charged through a USB C connection.

Make sure you get plenty of USB slots — at least two, preferably four, or more. Pay extra for a USB C slot or two. More details are in the section "Managing disks and drives," later in this chapter.

» **Lots of other stuff:** You never have to play with this other stuff unless you're very unlucky.

Here are a few upgrade dos and don'ts:

>> **Don't** let a salesperson talk you into eviscerating your PC and upgrading the CPU: Intel Core i7 isn't that much faster than Intel Core i5, and a 3.0-GHz PC doesn't run a whole lot faster than a 2.6-GHz PC. The same is true for AMD's Ryzen 7 versus Ryzen 5.

>> When you hit 8GB in main memory, **don't** expect big performance improvements by adding more memory, unless you're running Google Chrome all day with 42 open tabs, or putting together videos.

>> On the other hand, if you have an older video card, **do** consider upgrading it to a faster card, or to one with more memory. Windows 10 takes good advantage of it.

>> Rather than nickel-and-dime yourself to death on little upgrades, **do** wait until you can afford a new PC, and give away your old one.

>> If you can't afford to buy a new PC, and you want more performance, **do** buy a new SSD. Install Windows 10 and all your apps and games on the SSD. No other hardware component delivers bigger performance improvements than the switch from HDD to SSD.

TIP

If you decide to add memory, have the company that sells you the memory install it. The process is simple, quick, and easy — if you know what you're doing. Having the dealer install the memory also puts the monkey on his back if a memory chip doesn't work or a bracket snaps. This is especially true for laptops.

Inside a touch-sensitive tablet

Although tablets have been on the market for more than a decade, they didn't really take off until Apple introduced the iPad in 2010. Since the iPad went ballistic, every Windows hardware manufacturer has been clamoring to join the game. Even Microsoft has entered the computer-manufacturing fray with its line of innovative tablets known as Surface.

The old Windows 7–era tablets generally required a *stylus* (a special kind of pen), and they had truly little software that took advantage of touch input. The iPad changed all that.

ASK
WOODY.COM

The result is a real hodge-podge of Windows tablets and many kinds of 2-in-1s (which have a removable keyboard, as shown in Figure 1-6, and thus transform to a genuine tablet), laptops, and ultrabooks with all sorts of weird hinges, including some that flip around like an orangutan on a swing.

FIGURE 1-6:
Microsoft Surface
Pro tablets
typify the 2-in-1
combination of
removable slates
with tear-away
keyboards.

Courtesy of Microsoft

As sales of Windows 10 machines plummets, the choice has never been broader. All major PC manufacturers now offer traditional laptops as well as some variant on the 2-in-1, many still have desktops, and more than a few even make Chromebooks!

I did most of the touch-sensitive work in this book on an ASUS ZenBook Duo (see Figure 1-7). Its secondary touch-based screen, called ASUS ScreenPad Plus, gives me on-the-go computing that I never experienced with traditional laptops.

FIGURE 1-7:
The ASUS
ZenBook Duo
used to update
this book.

Courtesy of ASUS

With a 10th generation Intel Core i7-10510U processor, 16GB of RAM, and a 512GB solid-state drive, the ASUS ZenBook Duo is the fastest, most capable laptop I've ever owned. It's a lot more powerful than many desktop PCs people buy. Its dual-screen configuration simply blows me away. With it, you can do multitasking that was never possible on a Windows laptop. To make things even better, it has an NVIDIA GeForce MX250 with 2GB of memory that works great for all kinds of professional use cases, including video editing and architecture. It has two USB 3.1 ports, one USB C, an HDMI output for high-definition monitors (or TVs!), and a MicroSD card reader. Another cool feature is the webcam with facial-recognition support, which makes it easy to sign into Windows 10 using your face instead of your password. Don't worry: Your face is not sent to Microsoft and is stored only locally, on your PC. The keyboard is illuminated so that I can see the keys during the night. This feature is useful when I work long hours, and I am often a night-owl when I get to write books like this one.

ASK
WOODY.COM

Of course, that kind of oomph comes at a price. That's the other part — quite possibly the constraining part — of the equation. A couple thousand bucks for a desktop replacement is great, but if you just want a Windows 10 laptop, you can find respectable, traditional Windows 10 laptops (ultrabooks, whatever you want to call them), with or without touchscreens, for a few hundred.

Microsoft's Surface Pro (Figure 1-6) starts at $749 or so, without the keyboard. The Surface Laptop goes for $1000 and up, and it includes the keyboard. The Surface Book, which is both a laptop and a tablet, starts at $1600.

That said, if a Chromebook or an iPad or an Android tablet will do everything you need to do, there's no reason to plunk down lots of money for a Windows 10 tablet, ultrabook, or laptop. None at all.

If you're thinking about buying a Windows 10 tablet, keep these points in mind:

REMEMBER

>> **Focus on weight, heat, and battery life.** Touch-sensitive tablets are meant to be carried, not lugged around like a suitcase. The last thing you need is a box so hot it burns a hole in your pants, or a fan so noisy you can't carry on a conversation during an online meeting.

>> **Make sure you get multi-touch.** Some manufacturers like to skimp and make tablets that respond only to one or two touch points. You need at least four, just to run Windows 10, and ten wouldn't hurt. Throw in some toes and ask for 20 if you want to be ornery about it.

>> **The screen should run at 1920x1080 pixels or better.** Anything smaller will have you squinting to look at the desktop.

>> **Get a solid-state drive.** In addition to making the machine much, much faster, a solid-state drive (SSD) also saves on weight, heat, and battery life. Don't be overly concerned about the amount of storage on a tablet. Many people with Windows 10 tablets end up putting all their data in the cloud with, for example, OneDrive, Google Drive, Dropbox, or Box. See Book 6, Chapter 1.

>> **Try before you buy.** The screen must be sensitive to your big fingers, and look good, too. Not an easy combination. I also have a problem with bouncy keyboards. Better to know about the limitations before you fork over the cash.

>> **Make sure you can return it.** If you have experience with a "real" keyboard and a mouse, you may find that you hate using a tablet to replicate the kinds of things you used to do with a laptop or desktop PC.

As the hardware market matures, you can expect to see many variations on the tablet theme. It isn't all cut and dried.

OLED VERSUS LED

OLED (organic light-emitting diode) screens are found on TVs, computer monitors, laptop screens, tablets, and even smartphones. Their prices are headed down fast. Can or should they supplant LED screens, which have led the computer charge since the turn of the century? That's' a tough question with no easy answer.

First, understand that an LED screen is an LCD screen — an older technology — augmented by backlighting or edge lighting, typically from LEDs or fluorescent lamps. A huge variety of LED screens are available, but most of the screens you see nowadays incorporate IPS (in-plane switching) technology, which boosts color fidelity and viewing angles.

OLED is a horse of a different color. IPS LED pixels (considered far superior to the older TN LED pixels) turn different colors, but they rely on the backlight or sidelight to push the color to your eyes. OLED (pronounced "oh-led") pixels make their own light. If you take an LED screen into a dark room and bring up a black screen, you can see variations in the screen brightness because the backlight intensity changes, if only a little bit. OLED blacks, by (err) contrast, are uniform and thus deeper.

All sorts of new techniques are being thrown at LED, and LED screens are getting better and better. HDR (high dynamic range) improvements, for example, make LED pictures stand out in ways they never could before. Quantum dots improve lighting and color. Many people feel that, at this point, OLEDs have blacker blacks, but the best LEDs produce better bright colors.

The huge difference is in price: OLED screens are still more expensive than LED. The price of OLED is dropping rapidly, though. In addition, OLEDs don't last as long as LEDs — say, a decade with normal use. There's also some concern that OLEDs draw more power — and will burn through a laptop battery — faster than LCDs, but some contest that statement. Much depends on the particular LED and OLED you compare.

Screening

The computer monitor or screen — and LED, LCD, OLED, and plasma TVs — use technology that's quite different from old-fashioned television circuitry from your childhood. A traditional TV scans lines across the screen from left to right, with hundreds of them stacked on top of each other. Colors on each individual line vary all over the place. The almost infinitely variable color on an old-fashioned TV combined with a comparatively small number of lines makes for pleasant, but fuzzy, pictures.

By contrast (pun intended, of course), computer monitors, touch-sensitive tablet screens, and plasma, LED, OLED, and LCD TVs work with dots of light called *pixels.* Each pixel can have a different color, created by tiny, colored gizmos sitting next to each other. As a result, the picture displayed on computer monitors (and plasma and LCD TVs) is much sharper than on conventional TV tubes.

REMEMBER

The more pixels you can cram on a screen — that is, the higher the screen resolution — the more information you can pack on the screen. That's important if you tend to have more than one word-processing document open at a time, for example. At a resolution of 800x600, two open Word documents placed side by side look big and fuzzy, like caterpillars viewed through a dirty magnifying glass. At 1280x1024, those same two documents look sharp, but the text may be so small that you have to squint to read it. If you move up to wide-screen territory — 1920x1080 (full HD), or even 2560x1440 (aka 1440p) — with a good monitor, two documents side-by-side look stunning. Run up to 4K technology at 3840x2160 or better — the resolution available on many premium ultrabooks — and you need a magnifying glass to see the pixels.

A special-purpose computer called a *graphics processing unit (GPU),* stuck on your video card or possibly integrated into the CPU, creates everything that's shown on your computer's screen. The GPU has to juggle all the pixels and all the colors, so if you're a gaming fan, the speed of the video card (and, to a lesser extent, the speed of the monitor) can make the difference between a zapped alien and a lost energy shield. If you want to experience Windows 10 in all its glory, you need a fast GPU with at least 1GB (and preferably 4GB or more) of its own memory.

Computer monitors and tablets are sold by size, measured diagonally (glass only, not the bezel or frame), like TV sets. Just like with TV sets, the only way to pick a good computer screen over a run-of-the-mill one is to compare them side by side or to follow the recommendation of someone who has.

Managing disks and drives

Your PC's memory chips hold information only temporarily: Turn off the electricity, and the contents of main memory go bye-bye. If you want to reuse your work, keeping it around after the plug has been pulled, you have to save it, typically on a hard drive, or possibly in the *cloud* (which means you copy it to a location on the Internet).

The following list describes the most common types of disks and drives:

>> **Hard drive:** The technology's changing rapidly, with traditional hard disk drives (HDDs) now being rapidly replaced by *solid-state drives* (SSDs) with no moving parts, and to a lesser extent *hybrid drives* that bolt together a rotating drive with an SSD. Each technology has benefits and drawbacks. Yes, you can run a regular HDD drive as your C: drive, and it's going to work fine. But tablets, laptops, or desktops with SSD drives run like greased lightning.

The SSD wins as speed king. After you use an SSD as your main system (C:) drive, you'll never go back to a spinning platter, I guarantee. SSDs are great for the main drive, but they may be expensive for storing pictures, movies, and photos. They may supplant the old whirling dervish drive, but price and technical considerations (see the sidebar "Solid-state drives have problems, too") assure that hard drives will still be around. SSDs feature low power consumption and give off less heat. They have no moving parts, so they don't wear out like hard drives. And, if you drop a hard drive and a solid- state drive off the Leaning Tower of Pisa, one of them may survive. Or maybe not.

Hybrid drives combine the benefits and problems of both HDDs and SSDs. Although HDDs have long had *caches* — chunks of memory that hold data before being written to the drive, and after it's read from the drive — hybrid drives have a full SSD to act as a buffer.

If you can stretch the budget, start with an SSD for the system drive, a big hard drive (one that attaches with a USB cable) for storing photos, movies, and music, and get *another* drive (which can be inside your PC, outside attached with a USB cable, or even on a different PC on your network) to run File History (see Book 8, Chapter 1).

TECHNICAL STUFF

If you want full on-the-fly protection against dying hard drives, you can get three hard drives — one SSD, and two hard drives, either inside the box or outside attached with USB or eSATA cables — and run Storage Spaces (see Book 7, Chapter 4).

Ultimately, though, most people opt for a fast SSD for files needed immediately, coupled with cloud storage for the big stuff. Now that Google offers free unlimited photo storage — and with the rise of data streaming instead of purchased CDs — the need for giant hard drives has hit the skids.

For the enthusiast, a three-tier system, with SSDs storing data you need all the time, intermediate backup in the cloud, and multi-terabyte data repositories hanging off your PC, seems to be the way to go. Privacy concerns (and the, uh, intervention of various governments) have people worried about cloud storage. Rightfully so.

» **SD/xD/CF card memory:** Many smaller computers, and some tablets, have built-in SD card readers. (Apple and some Google tablets don't have SD — the companies would rather sell you more on-board memory, at inflated prices!) You probably know Secure Digital (SD) cards best as the kind of memory used in digital cameras, or possibly phones (see Figure 1-8). A microSD card may slip into a hollowed-out card that is shaped like, and functions as, an SD card.

TIP

Even now, long after the demise of floppy disks, many desktop computer cases have drive bays built for them. Why not use the open spot for a multifunction card reader? That way, you can slip a memory card out of your digital camera (or your Dick Tracy wristwatch, for that matter) and transfer files at will. SD card, miniSD, microSD card, xD card, CompactFlash, memory stick — whatever you have — the multifunction readers cost a pittance and read almost everything, including minds.

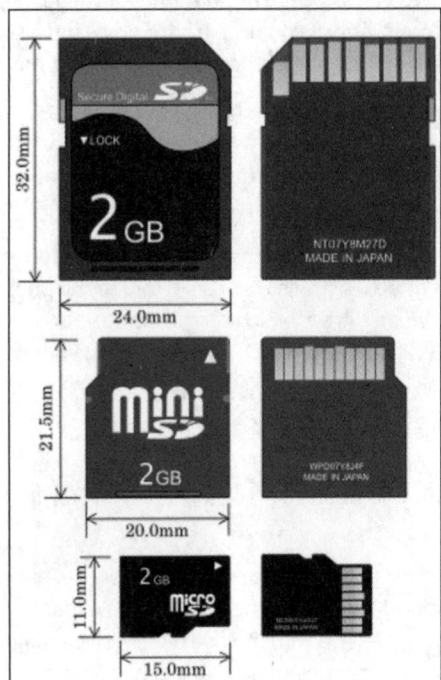

FIGURE 1-8: Comparative sizes of an SD, a miniSD, and a microSD card.

Source: Skcard.svg, Wikimedia

» **CD, DVD, or Blu-ray drive:** Of course, these types of drives work with CDs, DVDs, and the Sony Blu-ray discs, which can be filled with data or contain music or movies. CDs hold about 700MB of data; DVDs hold 4GB, or six times as much as a CD. Dual-layer DVDs (which use two separate layers on top of the disc) hold about 8GB, and Blu-ray discs hold 50GB, or six times as much as a dual-layer DVD.

Fewer and fewer machines these days come with built-in DVD drives: If you want to schlep data from one place to another, a USB drive works fine — and going through the cloud is even easier. For most storage requirements, though, big, cheap USB drives are hard to beat.

» **USB drive or key drive:** Treat it like it's a lollipop: It's half the size of a pack of gum and able to hold an entire PowerPoint presentation or two or six, plus a few full-length movies. Flash memory (also known as a jump drive, thumb drive, or memory stick) should be your first choice for external storage space or for copying files between computers. (See Figure 1-9.) You can even use USB drives on many DVD players and TV set-top boxes.

Pop one of these guys in a USB slot, and suddenly Windows 10 knows it has another drive — except that this one's fast, portable, and incredibly easy to use. It's okay to go for the cheapest flash drives you can find. Some of the features on fancy USB drives are not useful to many users.

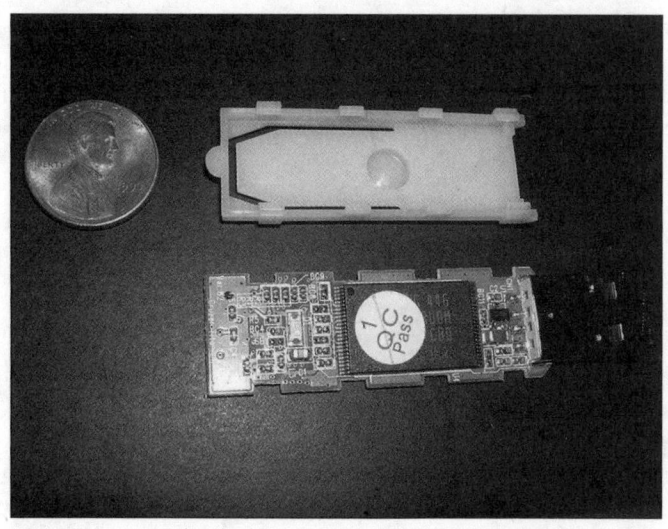

FIGURE 1-9:
The inside of a
USB drive.

Source: Nrbelex, Wikimedia

TIP

What about USB 3? If you have a hard drive that sits outside of your computer — an *external drive* — or a USB drive, it'll run faster if it's designed for USB 3 and attached to a USB 3 connector. Expect performance with USB 3 that's three to five times as fast as USB 2. For most other outside devices, USB 3 is overkill, and USB 2 works just as well.

This list is by no means definitive: New storage options come out every day.

Making PC connections

Your PC connects to the outside world by using a bewildering variety of cables and connectors. I describe the most common in this list:

>> **USB (Universal Serial Bus) cable:** This cable has a flat connector that plugs in to your PC, known as *USB A* (see Figure 1-10). The other end is sometimes shaped like a D (called *USB B*), but smaller devices have tiny terminators (usually called *USB mini* and *USB micro,* each of which can have two different shapes).

USB Connection Type Reference Chart

USB 2.0 A
USB 3.0 A
USB 3.1 A

USB 3.0 B

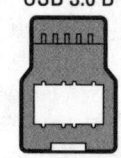

USB 2.0 B

USB 3.0 Micro B

USB 2.0 Mini B
USB 3.0 Mini B

USB 3.0 C
USB 3.1 C
USB 3.1 Gen 2 C

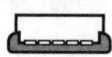

USB 2.0 Micro B

FIGURE 1-10:
The most common USB A, B, C, mini, and micro USB cables.

Source: Amazon Basics

TECHNICAL STUFF

USB 2 connectors work with any device, but hardware — such as a hard drive — that uses USB 3 will be much faster if you use a USB 3 cable and plug it into the back of your computer in a USB 3 port. USB 2 works with USB 3 devices, but you won't get the additional speed. Note that not all PCs, especially older ones, have USB 3 ports.

USB-C is a special kind of USB connection that supports amazingly fast data transmission and high levels of power. You know when you have USB-C because it's impossible to insert the plug upside-down — both sides work equally well. It's becoming the go-to choice for connecting peripherals and, in some cases, power supplies.

ASK WOODY.COM

USB is the connector of choice for just about any kind of hardware — printer, scanner, smartphone, digital camera, portable hard drive, and even the mouse. Apple's iPhones and iPads use a USB connector on one side — to plug in to your computers — but the other side is Thunderbolt (common on Apple devices, not so common on Windows PCs), and doesn't look or act like any other connector.

If you run out of USB connections on the back of your PC, get a USB hub with a separate power supply and plug away.

» **LAN cable:** Also known as a CAT-5, CAT-6, or RJ-45 cable, it's the most common kind of network connector. It looks like an overweight telephone plug (see Figure 1-11). One end plugs in to your PC, typically into a network interface *card* (or *NIC,* pronounced "nick"), a network connector on the motherboard. The other end plugs in to your wireless router (see Figure 1-12) or switch or into a cable modem, DSL box, router, or other Internet connection-sharing device.

» **Keyboard and mouse cable:** Most mice and keyboards (even cordless mice and keyboards) come with USB connectors.

FIGURE 1-11: RJ-45 Ethernet LAN connector.

Source: David Monniaux, Wikimedia

FIGURE 1-12: The back of a wireless router.

» **Bluetooth** is a short-distance wireless connection. Once upon a time, Bluetooth was very finicky and hard to set up. Since the recent adoption of solid standards, Bluetooth's become quite useful. It's now used for connecting all kinds of accessories, including speakers, headsets, mice, and keyboards.

» **DVI-D and HDMI connectors:** Although older monitors still use legacy, 15-pin, HD15 VGA connectors, most monitors and video cards now use the small HDMI connector (see Figure 1-13), which transmits both audio and video over one cable. Some older monitors don't support HDMI but do take a DVI-D digital cable (see Figure 1-14). Newer, premium monitors take advantage of the DisplayPort, which can transmit even more data than HDMI.

TECHNICAL STUFF

Some really old monitors still use the ancient 15-pin VGA connector, the one shaped like a *D*. Avoid VGA if you can. Old-fashioned serial (9-pin) and parallel (25-pin) cables and Centronics printer cables are growing as scarce as hen's teeth. Hey, the hen doesn't need them, either.

FIGURE 1-13:
HDMI has
replaced the old
VGA and DVI-D
video adapters.

Source: D-Kuru, Wikimedia

FIGURE 1-14:
Two different
kinds of DVI-D
cables — they
work well, but
don't carry audio.

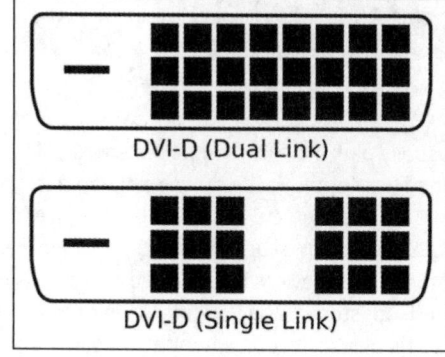

DVI-D (Dual Link)

DVI-D (Single Link)

Source: Hungry Charlie, Wikimedia

Futzing with video, sound, and multitudinous media

Unless you're using a cheap laptop or a tablet, chances are fairly good that you're running Windows 10 on a PC with at least a little oomph in the audio department. In the simplest case, you have to be concerned about four specific sound jacks (or groups of sound jacks) because each one does something different. Your machine may not have all four (are you feeling inadequate yet?), or it may look like a patch board at a Slayer concert, but the basics are still the same.

Here's how the four key jacks are usually marked, although sometimes you must root around in the documentation to find the details (see Figure 1-15):

>> **Line In:** This stereo input jack is usually blue. It feeds a stereo audio signal — generally from an amplified source — into the PC. Use this jack to receive audio output into your computer from an iPad, cable box, TV set, radio, CD player, electric guitar, or other audio-generating box.

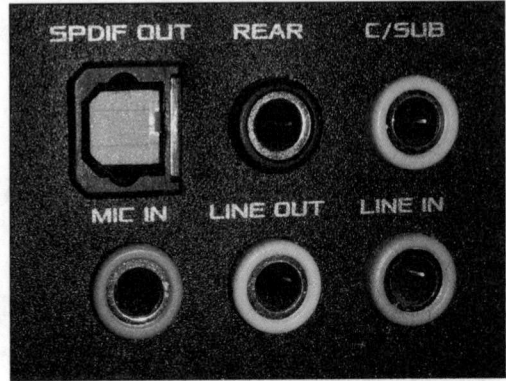

FIGURE 1-15:
The audio jacks
on the back
of a desktop
computer.

>> **Mic In:** This jack is usually pink. It's for unamplified sources, like most microphones or some electric guitars. If you use a cheap microphone for Skype or another VoIP service that lets you talk long distance for free, and the mic doesn't have a USB connector, plug in the microphone here. In a pinch, you can plug any of the Line In devices into the Mic In jack — but you may hear only mono sound, not stereo, and you may have to turn the volume way down to avoid some ugly distortion when the amplifier inside your PC increases the strength of an already-amplified signal.

>> **Line Out:** A stereo output jack, usually lime green, which in many cases can be used for headphones or patched into powered speakers. If you don't have fancy output jacks (such as the Sony-Philips SPDIF), this is the source for the highest-quality sound your computer can produce. If you go for a multi-speaker setup, this is for the front speaker.

>> **Rear Surround Out:** Usually black, this jack isn't used often. It's intended to be used if you have independent, powered rear speakers. Most people with rear speakers use the Line Out connector and plug it into their home theater systems, which then drives the rear speakers; or they use the HDMI cable (see the preceding section) to hook up to their TVs. If your computer can produce full surround sound output, and you have the amplifier to handle it, you'll get much better results using the black audio jack.

ASK
WOODY.COM

Many desktop computers have two more jacks: Orange is a direct feed for your subwoofer, and the gray (or brown) one is for your side speakers. Again, you have to put an amplifier between the jacks and your speakers.

Fortunately, PC-savvy 4-channel amplifiers can handle the lime (front speaker) and black (rear speaker) lines, 6-channel amps may be able to handle all but the gray, and 8-channel amps will take all four: lime (front speaker), orange (subwoofer, or center back), black (rear), and gray (side).

With a sufficiently bottomless budget, you can make your living room sound precisely like the 08R runway at Honolulu International.

Laptops typically have just two jacks, pink for Mic In and lime for Line Out. If you have a headphone with a mic, that's the right combination. It's also common to plug powered external speakers into the lime jack.

Tablets and smartphones usually have an earphone jack, which works just like a lime green Line Out jack. Many new models have ditched the jack and can connect to external sound devices only through Bluetooth. In theory, no cables is a better approach, right?

High-end audio systems may support optical connections. Check both the computer end of the connection and the speaker/receiver end to make sure they'll line up.

TIP

PC manufacturers love to extol the virtues of their advanced sound systems, but the simple fact is that you can hook up a rather plain-vanilla PC to a home stereo and get good-enough sound. Just connect the Line Out jack on the back of your PC to the Aux In jack on your home stereo or entertainment center. *Voilà!*

Ultrabooks and convertibles

While working with Windows 7, I fell in love with an ASUS netbook. Netbooks, which are small laptops, were a popular concept a few years ago, designed to provide the basics people needed from a laptop at an affordable price. Think of them as the precursor to today's Chromebooks.

But then along came the iPad, and at least 80 percent of the reason for using a netbook disappeared. Sales of netbooks have not fared well, and I don't see a comeback any time soon. Tablets blow the doors off netbooks, and 2-in-1s just mopped up the remains.

Ultrabooks are a slightly different story. Intel coined (and trademarked) the term *Ultrabook* and set the specs. For a manufacturer to call its piece of iron an Ultrabook, it must be less than 21mm thick, run for five hours on a battery charge, and resume from hibernation in seven seconds or less. In other words, it must work a lot like an iPad.

Intel threw a $300 million marketing budget at Ultrabooks, but they fizzled. Now the specs seem positively ancient, and the term *Ultrabook* doesn't have the wow factor it once enjoyed.

Right now, I'm having a great time with all the new form factors: I mention the XPS-15, Surface Book 3, and Surface Pro earlier in this chapter. I worked with a trapeze-like machine for a bit, but always worried about snapping the carrier off. There's no one-size-fits-all solution. Now, depending on the situation, I'm just as likely to grab my iPad Pro as I head out the door, or curl back with a Chromebook to watch Netflix. I use Android smartphones and iPhones, too, all the time.

If you're in the market for a new machine, drop by your favorite hardware store and just look around. You might find something different that strikes your fancy. Or you may decide that you just want to stick with a boring desktop machine with a mechanical keyboard and a wide monitor the size of football fields.

Guess what I work on.

What's Wrong with Windows 10?

Microsoft made a lot of mistakes in Windows 10's first year of existence. Chief among them was the widely despised "Get Windows 10" campaign. Combining the worst of an intrusive approach, forced updates, bad interface design, presumptive implementation, and a simple lack of respect for Windows 7 and Windows 8.1

customers, Get Windows 10 (GWX), to me, represents the lowest point in the history of Windows. Microsoft just didn't give a hairy rat's patoutie who they stomped on, as they pushed and pushed and pushed to get everybody on Windows 10. Which is a shame, really, because Windows 10 is a good operating system.

Many people who used to trust Microsoft, lost all trust in the wake of GWX, and it's hard to blame them. I've been writing books about Windows and Office for 25 years, and I think GWX is the most customer-antagonistic effort Microsoft has ever undertaken.

Trust in Microsoft is at the core of what you need to understand about Windows 10. Here is what I feel every Windows 10 customer should know:

» **Forced updates:** Initially, most Windows 10 customers did not have any choice about updates; when Microsoft released a patch, it got applied, unless you went to near-Herculean lengths to block them (see `www.computerworld.com/article/3138088/microsoft-windows/woodys-win10tip-block-forced-win10-updates.html`). I've railed against automatic updating for more than a decade — bad patches have driven many machines and their owners to the brink. Luckily, starting with Windows 10 April 2019 update (codenamed 19H1), all Windows 10 users can pause updates. If you use Windows 10 Home, you can pause them for up to 35 days. If you're using Windows 10 Pro, Education, or Enterprise, you can pause them for up to a year.

» **Privacy concerns:** Microsoft's following the same path blazed by Google and Facebook and, to a lesser extent, Apple and many other tech companies. They're all scraping information about you, snooping on what you're doing, to sell you things. I don't think Microsoft is any worse than the others, but I don't think it's any better either. I talk about reducing the amount of data that Microsoft collects about you in Book 2, Chapter 6. I think that data snooping will be the focus of extensive legislation over the next decade and one of the major battles of our time. The problem, of course, is that the people who control the laws also control the organizations that circumvent the laws.

» **Massive dearth of apps:** A few years ago, apps were a nice part of using an iPhone or iPad. Now, many people rely on them to get their work done and to keep their lives sunny side up. Microsoft missed the ball with UWP apps — they never caught on, and with the demise of a viable Microsoft smartphone ecosystem, developers had little incentive to make UWP apps. That means we get to use two kinds of apps in Windows 10: desktop apps or programs, and Windows 10 apps, which are touch-friendly and similar to past UWP apps. The problem is that users can't tell them apart, and the Microsoft Store in Windows 10 distributes both types of apps. Even Microsoft's Skype team has not decided on what it wants us to use: their Skype Windows 10 app or the classic Skype program. We have both, and most people can't tell which is which, or which is better.

I have learned how to block Microsoft's forced updating and have come to peace with the fact that Microsoft's snooping on me. (Hey, I've used Google's Chrome browser for years, and it's been harvesting data the entire time.) And when I want the convenience of a specific app, I'll pick up my phone, tablet, or Chromebook.

But that's just me. You may have good reason to want to switch to another computing platform. Certainly, Windows 10 is going to give you more headaches and heartaches than the alternatives. But it gives you more opportunities, too.

Welcome to my world!

Chapter **2**

Windows 10 for the Experienced

I f you're among the 1.7 billion or so souls on the planet who have been around the block with Windows 8/8.1, Windows 7, or Windows Vista, you're in for a shock.

On the other hand, if you've been using Windows 10 for a while and want to see what's new, you'll find a few new features and some stuff that's been moved around, but the changes won't be so extreme. This chapter points out the high points.

Although Windows 10 will look relatively familiar to long-time desktop users, the details are different. If you've conquered the Metro side of Windows 8.1 (which is the only side of Windows 8), you're going to be in for a pleasant surprise. And if you're upgrading from one version of Windows 10 to another, the ride may or may not be what you expect.

If You Just Upgraded from Windows 7 or 8.1 to Windows 10

Before digging into an examination of the new nooks and crannies in Windows 10, I'd like to pause for a second and let you know about an option you may or may not have.

If you upgraded from Windows 7 or Windows 8.1 to Windows 10 in the past 30 days, and you don't like Windows 10, you can roll back to your old version. This works for only 30 days because a scheduled program comes in and wipes out the backup after 30 days. But if you're in under the wire and want to roll back, here's how. Note that this technique is only for upgraders; it doesn't apply to new Windows 10 systems or computers in which you installed Windows 10 by wiping out the hard drive. If you love Windows 10 or don't qualify for the rollback, jump down to the next section.

The method for moving back is easy:

1. **Make sure you have your old password.**

   If your original Windows 7 or Windows 8.1 system had login IDs with passwords, you'll need those passwords to log in to the original accounts. If you changed the password while in Windows 10 (local account), you need your old password, not your new one. If you created a new account while in Windows 10, you have to delete it before reverting to the earlier version of Windows.

2. **Make a backup.**

   Before you change any operating system, it's a good idea to make a full system backup. Many people recommend Acronis for the job, but Windows 10 has a good system image program that is identical to the Windows 7 version. However, the program is hard to find. To get to the system image program, in the Windows 10 search box, type **Windows** Backup, press Enter, and click Go to Backup and Restore (Windows 7). Then, click Create a System Image (on the left) and follow the directions.

3. **Run the reset.**

   a. *Click the Start icon and then the Settings icon.*

   b. *Click Update & Security, and then click Recovery.* You see an entry to Go Back to Windows 7 or Go Back to Windows 8.1, depending on the version of Windows from whence you came.

   c. *Click the Get Started button for Back to Windows 7 or Windows 8.1, depending on the version of Windows from whence you came.*

   d. *If asked why you are going back, choose a reason and click or tap Next (see Figure 2-1).*

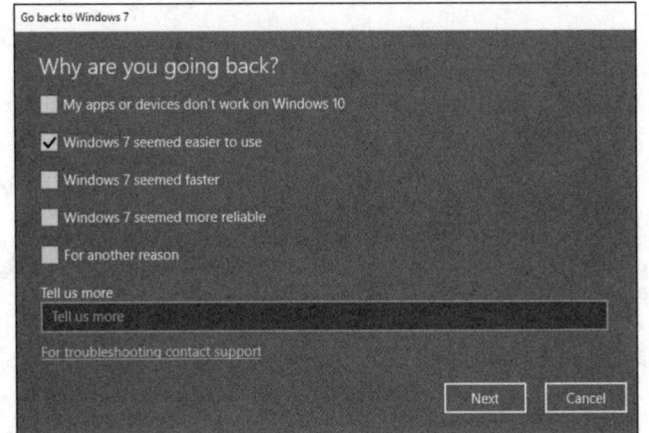

FIGURE 2-1:
When you
roll back to
Windows 7 or 8.1,
you are asked
why you want to
go back.

If you don't see the Get Started button and are using an administrator account, you've likely fallen victim to one of the many gotchas that surround the upgrade. See the next part of this section, but don't get your hopes up.

4. **Revert to Windows 7 or Windows 8.**

Finally, click Go back to Windows 7 or Go back to Windows 8, and your PC reboots and starts the rollback process. During this time, you see the message: "Restore your previous version of Windows." After a while (many minutes, sometimes hours), you arrive back at the Windows 7 (or 8.1) login screen.

5. **Click the user you want to use and enter the password.**

You're ready to go with your old Windows version.

ASK
WOODY.COM

In the Windows 10 May 2020 update, I found that Windows 10 no longer asks users whether they want to keep their files and apps during the rollback. The rollback process automatically restores apps (programs) and settings to their original state and settings (the ones that existed when you upgraded from Windows 7 to Windows 10). Any modifications made to those programs (for example, installing security updates to Office programs) while using Windows 10 are not applied when you return to Windows 7; you have to apply them again.

On the other hand, changes made to your regular files while working in Windows 10 — edits made to Office documents, for example, or to new files created while working with Windows 10 — may or may not make it back to Windows 7. I had no problems with files stored in My Documents; edits made to those documents persisted when Windows 10 rolled back to Windows 7. But files stored in other locations (specifically, in the \Public\Documents folder or on the desktop) didn't always make it back: Sometimes, Word documents created in Windows 10 disappeared when rolling back to Windows 7, even though they were on the desktop or in the Public Documents folder.

One oddity may prove useful: If you upgrade to Windows 10, create or edit documents in a strange location, and then roll back to Windows 7 (or 8.1), those documents may not make the transition. Amazingly, if you then upgrade again to Windows 10, the documents may reappear. You can retrieve the "lost" documents, stick them in a convenient place (such as on a USB drive or in the cloud), roll back to Windows 7, and pull the files back again.

Important lesson: Back up your data files before you revert to an earlier version of Windows.

If you can't get Windows to roll back and detest Windows 10, you're up against a tough choice. The only option I've found that works reliably is to reinstall the original version of Windows from scratch. On some machines, the old recovery partition still exists. You can bring back your old version of Windows by going through the standard recovery partition technique (which varies from manufacturer to manufacturer), commonly called a factory restore. More frequently, you get to start all over with a fresh install of Windows 7 or Windows 8.1.

A Brief History of Windows 10

So you've decided to stick with Windows 10? Good.

Pardon me while I rant for a bit.

Microsoft darn near killed Windows — and most of the PC industry — with the abomination that was Windows 8. Granted, there were other forces at work — the ascendancy of mobile computing, touchscreens, faster cheaper and smaller hardware, better Apple devices, Android, and other competition — but to my mind, the number one factor in the demise of Windows was Windows 8.

We saw PC sales drop. After Windows XP owners replaced their machines in a big wave in late 2014 and early 2015, responding to the end of support for Windows XP, we saw PC sales drop even more. Precipitously. Steve Ballmer confidently predicted that Microsoft would ship 400 million machines with Windows 8 preinstalled in the year that followed Windows 8's release. The actual number was closer to a quarter of that. Normal people like you and me went to great lengths to avoid Windows 8, settling on Windows 7.

Windows 8.1, which arrived a nail-biting year after Windows 8, improved the situation a bit, primarily by not forcing people to boot to the tiled Metro Start screen.

The team inside Microsoft that brought us the wonderful forced Windows 8 Metro experience was also responsible, earlier, for the Office ribbon. Many of us old-timers grumbled about the ribbon, saying Microsoft should at least present an alternative for using the older menu interface. It never happened. Office 2007 shipped with an early ribbon, and subsequent versions have been even more ribbon-ified since. Here's the key point: Office 2007 sold like hotcakes, despite the ribbon, and it's been selling in the multi-billion-dollar range ever since. As a result, the Office interface team figured they knew what consumers wanted, and old-timers were just pounding their canes and waggling toothless gums.

The entire Office 2007 management team was transplanted, almost intact, to the Windows 8 effort. They saw an opportunity to transform the Windows interface, and they took it, over the strenuous objections of many of us in the peanut gallery. I'm convinced they figured it would play out like the Office ribbon. It didn't. Windows 8 is, arguably, the largest software disaster in Microsoft's history.

Essentially all the Windows 8 management team — including some very talented and experienced people — left Microsoft shortly after that operating system shipped. With a thud. Their boss, Steve Ballmer, left Microsoft too. Ballmer's still the largest individual shareholder in Microsoft, with 333,000,000 shares at last count, worth $31 billion and change.

In their place, we're seeing a new generation of managers taking care of Windows. The current head of the Windows effort, Joe Belfiore, oversaw the PC/Tablet/Phone department in the Operating Systems Group at Microsoft. Before Windows 10, he led program management for the Windows Phone team and the effort to create the Metro design language that we hated so much, the disliked Live Tiles, and the much-ignored Cortana.

That said, Microsoft's traditional PC market has sunk into a funk, and it appears to be on a slow ride into the sunset. Or it may just turn belly up and sink, anchored with mounds of iPhones, iPads, MacBooks, Galaxy Tabs, and Chromebooks. Or maybe, just maybe, Windows 10 will breathe some life back into the 35-year-old veteran. Yes, Windows 1.0 shipped in November 1985.

However things play out, at least we have an (admittedly highly modified) Start menu to work with, as shown in Figure 2-2.

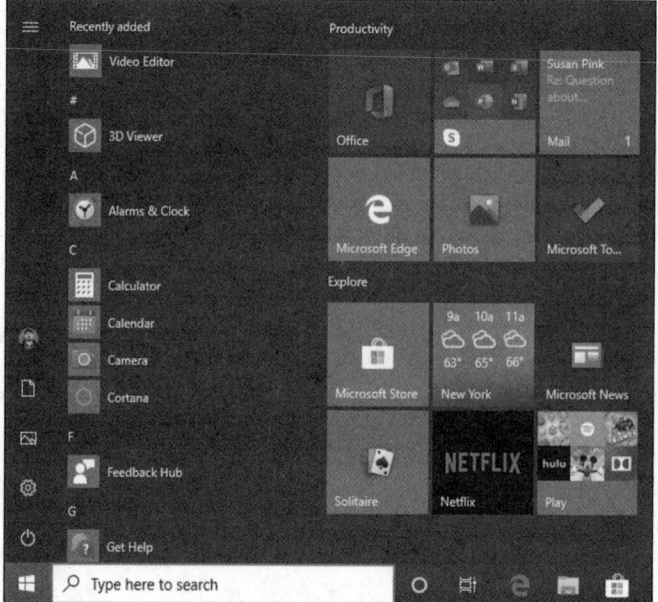

THE "GET WINDOWS 10" DEBACLE

No description of the recent history of Windows, however brief, can gloss over the fear and loathing that Microsoft induced with its Get Windows 10, or GWX, campaign.

The campaign started shortly after the RTM release in July 2015, with a little-noticed program known as KB 3035583. In October 2015, Microsoft started force-updating Windows 7 and Windows 8.1 computers to Windows 10, without the owners' knowledge or consent. A loud scream arose, and, a week after the forced upgrades started, they suddenly stopped. But the GWX campaign continued, showing increasingly persistent ads for Windows 10, all the symptoms of nagware, and even malware. Microsoft proved that it could reach into your Windows 7 machine and start the upgrade to Windows 10, whether you wanted it or not.

The resultant clamor — from an unexpected appearance of a Windows 10 upgrade notification on a weather forecaster's live news show, to Windows experts fretting over their relatives and friends, to more than 1 million posts on a Chinese blog — should have convinced Microsoft to back off. It didn't. If you bump into people who don't trust Windows or Microsoft, they have a good reason.

Exploring the Versions of Windows 10

Microsoft has famously announced that Windows 10 is "the last version of Windows." Which is to say, uh, Windows 10 is anything *but* the last version of Windows.

Instead of continuing Windows version numbers in an obvious way — say, Windows 10, Windows 10.1, Windows 10.2 Service Pack 17, Windows 11, Windows 2019, whatever — Microsoft has developed a new way of naming versions of Windows 10, all to make it look like Windows 10 is an immutable object.

Far from it.

The first version of Windows 10, which didn't have an official name, arrived in July 2015. People are now calling it Windows 10 version 1507 — where *15* stands for *2015* and *07* stands for July. Some people call it Windows 10 RTM, but that's a blasphemous approach because Windows as a Service never reaches Release to Manufacturing status. It's constantly changing. Constantly improving, to hear the marketeers talk about it.

WARNING

In late 2017, Microsoft vowed to turn out a new version (of "the last version" of Windows 10!) every six months. Many people — present company included — think that's crazy because it forces customers to install a new version of Windows every six months, more or less. The six-month horizon gives very little time to create anything new that's worthwhile.

But that's where we stand.

Here are the versions of Windows 10, to date:

- » Version **1507** –RTM released July 29, 2015 — contains the basic elements of Windows 10, few of which worked properly.

- » Version **1511** — Originally Fall update and later November update released November 10, 2015 — became the first stable and generally usable version of Windows 10.

- » Version **1607** — Anniversary update, released August 2, 2016 — spruced up the Start menu and Microsoft Edge, added the Notification (er, Action) Center (Book 2, Chapter 3), started adding features to the Cortana personal assistant, fleshed out a few of the Universal apps (see Book 4), improved Windows Hello to recognize your finger and your face, and added digital ink so you can draw on things.

- Version **1703** — Creators update, released April 11, 2017 — had small improvements for Cortana and the Microsoft Edge browser, a new privacy settings overview, an easier way to control updates (for Win10 Pro only), and lots of stuff for folks who draw in 3D and use virtual/augmented reality.

- Version **1709** — Fall Creators update, released October 17, 2017 — made OneDrive usable again with Files on Demand (Book 6, Chapter 1), and touches up My People (Book 3, Chapter 3), Cortana, and Microsoft Edge.

- Version **1803** — Spring Creators update, released April 10, 2018 — got Dark Mode, another tweak to My People, more Cortana and Edge, and not a whole lot more.

- Version **1809** — October 2018 update, released October 2, 2018 — got the Timeline, an improved clipboard that syncs through the cloud with other Windows 10 PCs you own, improved Search, the Your Phone app (see Book 10, Chapter 2), and Snip & Sketch, a new app for taking screenshots.

- Version **1903** — May 2019 update, released May 21, 2019 — delivered minor improvements to the Start menu, separated Cortana from Search (thank goodness for that), Windows Sandbox (see Book 7, Chapter 2), and a few other quality-of-life improvements.

- Version **1909** — November 2019 update, released November 12, 2019 — improved minor aspects in File Explorer, the Calendar, and notifications. It was one of the most underwhelming updates in terms of new features. Microsoft's focus was mostly on bug fixing and improving reliability.

- Version **2004** — May 2020 update, released May 27, 2020 — gave users more control over their Windows updates (hooray for that), Cortana was separated even more from the operating system, Task Manager was tweaked to show more useful data, Search got faster, and other minor improvements were made.

REMEMBER

Of course, each new version of Windows 10 is "the most secure version ever." That's been a constant claim since Windows 3.0.

You may have a version later than 2004 (type About in the search box and press Enter), but chances are good the new features aren't going to make your life much more interesting.

The Different Kinds of Windows Programs, Er, Apps

Windows 10 runs two very different kinds of programs. Permit me to go back to basics.

Computer programs (call them applications or desktop apps if you want) that you and I know work by interacting with an operating system. Since the dawn of Windows time, give or take a bit, Windows apps have communicated with Windows through a specific set of routines (application program interfaces or APIs) known colloquially and collectively as Win32. With rare exceptions, Windows desktop apps — the kind you use every day — take advantage of Win32 APIs to work with Windows.

In early June 2011, at the D9: The D: All Things Digital conference in California, Steven Sinofsky, and Julie Larson-Green gave their first demo of Windows 8. As part of the demo, they showed off new immersive or Metro apps, that interact with Windows in a different way. They use the newly minted (and still evolving) API set known as Windows Runtime or, more commonly, the WinRT API.

Microsoft started calling the WinRT based apps "immersive" and "full screen." Most of the world settled on Microsoft's internal code name, Metro. Microsoft, however, has since changed the name to Modern UI, then Windows 8, Windows Store App, New User Interface, Microsoft Design Language, Microsoft style design, and more recently Modern and Universal. The preferred terminology used to be *Universal Windows Platform (UWP) app,* although the tech support folks reverted to *Universal app* all the time. In recent times, Microsoft has decided to ditch the UWP platform, and add its functionality to non-UWP software development platforms. Therefore, UWP is now dead. I continue to use the term *Metro* in normal conversation, but in this book, to minimize confusion, I use the terms Windows 10 app or app.

Don't be confused. (Ha!) They all mean the same thing: Those are the names for Universal Windows applications that run with the WinRT API.

Windows 10 (Modern, Metro) apps have many other characteristics: They're sandboxed — stuck inside a software cocoon that isolates the programs so that it's hard to spread infections through them. They can be easily interrupted, so their power consumption can be minimized; if a UWP app hangs, it's almost impossible for it to freeze the machine. But at their heart, Universal Windows Platform apps are written to use the WinRT API.

Windows 8 and 8.1 (and Server 2012) support the WinRT API — Universal apps run on the Metro side of Windows 8, not on the desktop. ARM-based processors also run the WinRT API. You can find ARM architecture processors in many smartphones and tablets. In theory, apps should run on any Windows 10 computer — a desktop, a laptop, a tablet, a phone, a wall-mounted Surface Hub, an Xbox, and even a HoloLens headset. In practice, however, it isn't quite so simple. For example, only the simplest app that works in Windows 10 will run in Windows 8. So "Universal" is something of an aspiration, not a definition.

REMEMBER

In Windows 10, UWP (Modern/Metro/Tiled) apps run in their own boxes, right there on the desktop. Look at the Weather app — formerly a UWP app, now a Windows 10 app — shown in Figure 2-3.

All the other Windows programs — the ones you've known since you were still wet behind the WinEars — are now called desktop apps. Five years ago, you would've just called them programs, but now they have a new name. After all, if Apple can call its programs *apps*, Microsoft can, too. Technically, old-fashioned Windows programs, applications, or desktop apps are built to use the Win32 API.

HOW DID WE GET INTO THIS NICE MESS

Microsoft's been making tablet software for more than a decade, and it never put a dent in the market. Never did get it. Apple started selling tablet software in 2010 and selling tons of it. Boy howdy. Now Microsoft's diving in to get a piece of the touch-enabled action.

There's a big difference in approaches. Apple started with a phone operating system, iOS, and grew it to become the world's best-selling tablet operating system. There's very little difference between iOS 14 on an iPhone and iPadOS 14 on an iPad: Applications written for one device usually work on the other, with a few obvious changes, such as screen size. On the other hand, Apple's computer operating system, macOS, is completely different. It's built and optimized for use with a Mac computer. Apple is slowly changing the apps on both iOS and macOS, so they resemble each other and work together. But the operating systems are fundamentally quite different (even though, yes, iOS did originally start with the Mac OS Darwin foundation).

When Windows 7 was finished, Steve Sinofsky and crew decided to take a fundamentally different tack. Instead of the good people at Microsoft growing their phone software up, they decided to grow their computer operating system down. (The fact that the phone software at that point drew nearly universal scorn could've been part of the reason.) Windows 8 grew out of that decision: There's a touch-friendly part and a mouse-and keyboard-friendly part. The two aren't mutually exclusive: You can use your mouse on the Metro Start screen, and in the Windows 10 apps, you can use your greasy thumb on an old-fashioned desktop app. But the approach is different, the design is different, and the intent is different.

Windows 10 goes back to Windows Start menu roots and tries to grow the same concept down even further, to Windows Phone. Microsoft will be able to say that Windows covers all the bases, from lowly smartphones to gigantic workstations (and server farms, for that matter). The fact that the "Windows" running in each of the device classes is quite different kinda gets swept under the rug.

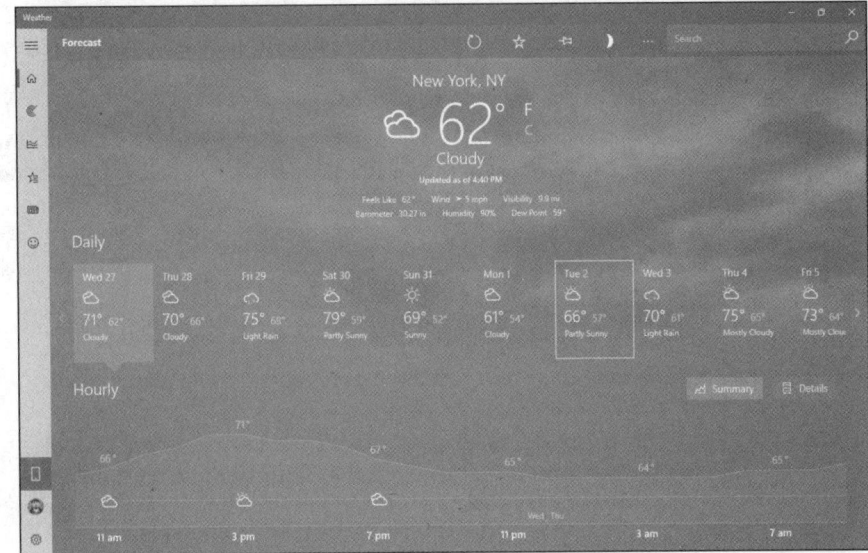

FIGURE 2-3:
The Windows 10
Weather app is
a former UWP
app because it's
based on the
WinRT API. See
the distinctive
design?

ASK
WOODY.COM

Unfortunately, there's a huge difference between Windows 10 apps and desktop apps. For starters:

» **Universal Windows apps — the ones that run on the WinRT API — are on the way out.** Microsoft rebuilt the aging Windows Desktop app Internet Explorer and turned it into the Universal Windows app called Microsoft Edge. Internet Explorer has been gradually marginalized, and now it's dead in a corner. After many years of trying to promote UWP apps, Microsoft has decided that it would stop. Instead, they opened up more and more UWP functionality to non-UWP platforms, including legacy platforms that Microsoft once deprecated, such as Win32, WPF, and WinForms. So what's old is new again.

» **Desktop apps and Universal Windows apps are starting to look the same.** Developers want you to look at their programs and think, "Oh, hey, this is a snappy new version." Also, Microsoft is intertwining the features available in both app platforms, making things blurrier than ever.

» **Universal Windows apps really are better.** Don't shoot me. I'm just the messenger. Now that we can run those newfangled tiled Universal Metro whoozamajiggers in their own resizable windows on the Windows desktop, the underlying new WinRT plumbing beats the pants off Win32. WinRT apps don't bump into each other as much, they (generally) play nice in their own sandboxes, they won't take Windows down with them, and they don't have all the overhead of those buggy Win32 calls.

If you're going to stay with Windows, it's time to get with the system and learn about this new tiled stuff.

Here's a quick guide to what's new — and what's still the same — with some down-and-dirty help for deciding whether you truly need Windows 10.

What's New for the XP Crowd

Time to fess up. You can tell me. I won't rat you out.

If you're an experienced Windows XP user and you're looking at Windows 10, one of two things happened: Either your trusty old XP machine died, and you *had* to get Windows 10 with a new PC, or a friend or family member conned you into looking into Windows 10 to provide tech support.

If you're thinking of making the jump from Windows XP to Windows 10, and you're going to stick with a keyboard (as opposed to going touch-only, or touch-mostly, heaven help ya), you have two big hurdles:

>> Learning the ways of Windows 10 apps (which I outline in the next section, "What's New for Windows 7 Users").

>> Making the transition from XP to Windows 7 because the Windows 10 desktop works much like Windows 7.

Are you sure you want to tackle the learning curve? Er, curves? See the nearby sidebar about switching to a Mac. Or try a Chromebook — see the other sidebar.

That said, if you didn't plunge into the Windows 7 or Vista madness, or the Windows 8/8.1 diversion, and instead sat back and waited for something better to come along, many improvements indeed await in Windows 10.

Improved performance

Windows 10 (and Windows 8 and 7 before it) actually places fewer demands on your PC's hardware. I know that's hard to believe, but as long as you have a fairly powerful video card and 4GB or more of RAM memory, moving from XP to Windows 10 will make your PC run faster.

WOULDN'T IT BE SMARTER TO GET A MAC?

Knowledgeable Windows XP users may find it easier — or at least more rewarding — to jump to a Mac, rather than upgrading to Windows 10. I know that's heretical. Microsoft will never speak to me again. But there's much to be said for making the switch.

Why? XP cognoscenti face a double whammy: learning Windows 7 (for the Windows 10 desktop) and learning how to deal with Metro/Modern Universal Windows apps. If you don't mind paying the higher price — and, yes, Macs are marginally more expensive than PCs, feature-for-feature — Macs have a distinct advantage in being able to work easily in the Apple ecosystem: iPads, iPhones, the App Store, iTunes, iCloud, and Apple TV all work together remarkably well. That's a big advantage held by Apple, where the software, hardware, cloud support, and content all come from the same company. "It just works" may be overblown, but there's more than a nugget of truth in it. Give or take a buggy iOS update.

Yes, Macs have a variant of the Blue Screen of Death. Yes, Macs do get viruses. Yes, Macs have all sorts of problems. Yes, you may have to stand in line at an Apple Store to get help — I guess there's a reason why Microsoft Stores seem so empty.

If you're thinking about switching sides, I bet you'll be surprised at the similarities between macOS and Windows XP.

EXTOLLING THE VIRTUES OF CHROMEBOOKS

If you're looking to buy a new computer, you should definitely consider getting a Chromebook. You know, the machines that Microsoft says "scroogle" you? Yeah. They're amazingly powerful, almost impervious to infections, start on a dime, sip batteries, don't get tied up for hours on end installing upgrades — and they're pretty darn cheap.

How to tell if you're ready for a Chromebook? Try using nothing but the Google Chrome browser on your aging computer for a bit. If you can do everything that you need to do with the Chrome browser, you're automatically ready for a Chromebook. Even if you can't, chances are pretty good that what you need is available in Chromebook land. No, you won't find Photoshop, but you will find plenty of cheap photo-editing packages. No, you won't find the full-blown Office suite, but you can use Office Online. I've moved almost everything to Google Docs and Sheets and rarely turn back to the big guns.

Chromebooks are a breath of fresh air if you don't absolutely need any Windows-based programs. I use mine all the time, and suggest you try it, too.

If you don't have a powerful video card, and you're running a desktop system, you can get one for less than $100, and extra memory costs a pittance. I've upgraded dozens of PCs from XP to Windows 10, and the performance improvement is quite noticeable. You, laptop users, aren't so lucky because the graphics card is usually soldered in.

Better video

Windows 10 doesn't sport the Aero interface made popular in Vista and Windows 7, but some of the Aero improvements persist. The new Windows 10 *reveal* feature lights up items as you hover your cursor over them if that sort of thing appeals.

TIP

The *Snap Assist* feature in Windows 10 lets you drag a window to an edge of the screen and have it automatically resize to half-screen size — a boon to anyone with a wide screen. Sounds like a parlor trick, but it's a capability I use many times every day. You can even snap to the four corners of the screen, and the desktop shows you which open programs can be clicked to fill in the open spot (see Figure 2-4).

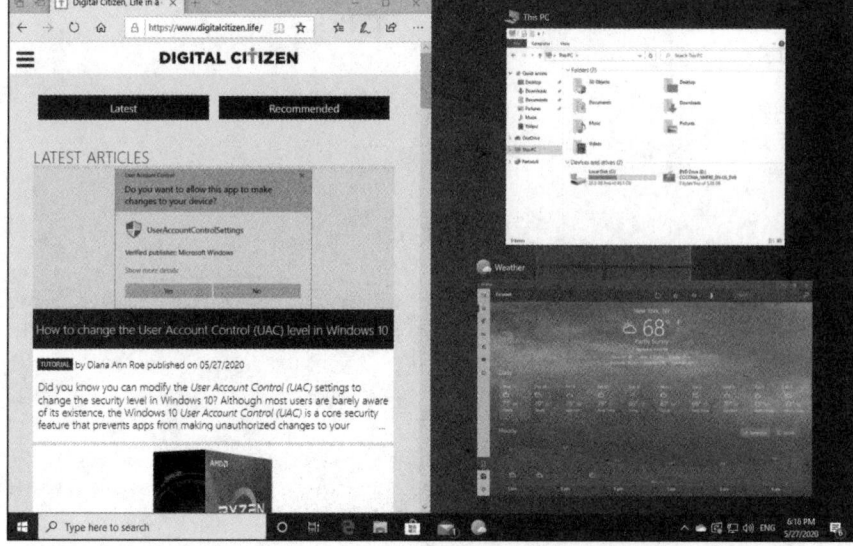

FIGURE 2-4:
Drag a window to the edge or a corner, and the other available windows appear, ready for you to click into place.

Windows 10's desktop shows you thumbnails of running programs when you hover your mouse cursor over a program on the taskbar (see Figure 2-4).

Video efficiency is also substantially improved: If you have a video that drips and drops in Windows XP, the same video running on the same hardware may go straight through in Windows 10.

A genuinely better browser is emerging

Internet Explorer lives in Windows 10, but it's buried deep. If you're lucky, you'll never see it when you use Windows 10. Internet Explorer is old and buggy, and Microsoft has stopped developing it. It became a bloated slug with incredibly stupid and infection-prone "features": ActiveX, COM extensions, custom crap-filled toolbars, and don't get me started on Silverlight. It deserves to die if only in retaliation for all the infections it's brought to millions of machines.

In its place, the new, light, standards-happy, fast Microsoft Edge is everything Internet Explorer should have been, without the legacy garbage. Microsoft built Edge from the ground up as a Windows 10 app that runs on the desktop in its own resizable window. It's a poster boy for the new apps that are coming down the pike. It took Microsoft forever to build, but the final result is well worth the effort.

Unfortunately, Microsoft Edge is still an unfinished work. Few people use it because it lacks many important browser features. The situation's slowly improving, and Microsoft has just launched a revamped version based on the same rendering engine as Google Chrome. Unfortunately, this new version is not built into Windows 10 yet. You have to download it from www.microsoft.com/en-us/edge. Edge might well be ready for prime time at some point.

If you live in fear of Internet Explorer getting you infected and/or hate the massive patches that used to appear every month, Microsoft Edge will be a refreshing change.

Cortana

Apple has Siri. Google has Google Assistant. Amazon has Alexa. Microsoft has Cortana, the Redmond version of an AI-based personal assistant, shown in Figure 2-5. Unlike Siri and Google Assistant, though, Cortana used to take over the Windows 10 search function, so it had a larger potential footprint than its AI cousins.

Cortana never took off, and it was used a lot less than Siri or Google Assistant. Because of that, Microsoft decided to decouple it from the rest of Windows 10, and as of the May 2020 update, it is a separate entity. It no longer takes over Windows 10's search, and you can ignore it if you want. However, if you do enable it, it sits in the background, listening for your commands.

I tell you much more about Cortana in this book — she has a chapter all to herself, Book 3, Chapter 5 — but I'll drop a little tidbit here, tailored for those Windows XP fans among you who may just be a bit intimidated by a talking helper-droid.

You see, Cortana has a history.

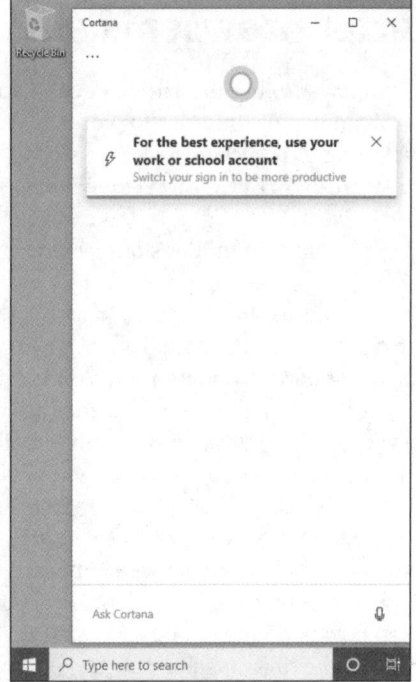

FIGURE 2-5:
Cortana sits,
listening, and
watching, waiting
to help you. That
should either
make you
skeptical or
scared — or
a little of both.

Back in 2001, Microsoft released a game called Halo: Combat Evolved. In Halo: CE, you, the player, take the role of the Master Chief, a kinda-human kinda-cyber soldier known as Master Chief Petty Officer John-117. Cortana is part of you, an artificial intelligence that's built into a neural implant in your body armor. After saving Captain Keyes, Cortana and the Master Chief go into a map room called the Silent Cartographer, and . . . well, you get the idea. Cortana is smooth and creepy and omniscient, just like the Windows 10 character.

ASK
WOODY.COM

Right now, depending on how you measure, Cortana is likely the least intelligent of the assistants, with Google Assistant on top, and Siri and Alexa vying for second place. That may change over time. In fact, someday Cortana may scan this paragraph and call me to task for my impertinence — bad blot on my record, served up to our robotic overlords.

Other improvements

Many other features — not as sexy as Cortana but every bit as useful — put Windows 10 head and shoulders above Windows XP. The standout features include:

>> **The taskbar:** I know many XP users swear by the old Quick Launch toolbar, but the taskbar, after you get to know it, runs rings around its predecessor. Just one example is shown in Figure 2-3 earlier in this chapter.

>> **A backup worthy of the name:** Backup was a cruel joke in Windows XP. Windows 7 did it better, but Windows 10 makes backup truly easy, particularly with File History (see Book 8, Chapter 1).

>> **A less-infested notification area:** Windows XP let any program and its brother put an icon in the notification area near the system clock. Windows 10 severely limits the number of icons that appear and gives you a spot to click if you really want to see them all. Besides, notifications are supposed to go in the Action pane on the right. See Book 2, Chapter 3.

>> **Second monitor support:** Although some video card manufacturers managed to jury-rig multiple monitor support into the Windows XP drivers, Windows 10 makes using multiple monitors one-click easy.

>> **Easy wireless networking:** All sorts of traps and gotchas live in the Windows XP wireless programs. Windows 10 does it much, much better.

>> **Search:** In Windows XP, searching for anything other than a filename involved an enormous kludge of an add-on that sucked up computer cycles and overwhelmed your machine. In Windows 10, search is part of Windows itself, and it works quickly.

On the security front, Windows 10 is light years ahead of Windows XP. From protection against rootkits to browser hardening, and a million points in between, XP is a security disaster — Microsoft no longer supports it — while Windows 10 is relatively (not completely) impenetrable.

Although Windows 10 isn't the Windows XP of your dreams, it's remarkably easy to use and has all sorts of compelling new features.

What's New for Windows 7 Users

Three years after Windows 10 hit the ether, Windows 7 was still riding strong. Depending on how you count and whose numbers you believe, at the three-year mark, Windows 7 was still driving about half of all Windows computers in the world. That's staying power, and it's worthy of your respect.

Nonetheless, Windows 7 is clearly on the way down, and Windows 10 is on the way up. One big reason for that is that Microsoft has stopped providing updates and support for Windows 7. That's as it should be, nostalgia notwithstanding.

Don't be worried. Anything that works with Windows 7, 8, or 8.1 — and almost everything from Vista — will work in Windows 10. Programs, hardware, drivers, utilities — just about anything.

That's a remarkable achievement, particularly because your Windows Desktop apps/Legacy programs (there's that *L* word again) have to peacefully coexist with the WinRT API-based Windows/Universal/Modern/Metro apps.

Windows 10 does have lots going for it. Let me skip lightly through the major changes between Windows 7 and Windows 10.

Getting the hang of the new Start menu

By now, you've no doubt seen the tiles on the right of the Start menu (refer to Figure 2-6).

FIGURE 2-6:
The Windows 10
desktop and
Start menu.

If you're coming to Windows 10 from Windows 7 — without taking a detour through Windows 8 — those tiles are likely to represent your greatest conceptual hurdle. They're different, but in many ways they're familiar.

ASK
WOODY.COM

Do you remember gadgets in Windows 7? See Figure 2-7. They actually started in Windows Vista. Many people (who finally found them) put tiles for clocks on their desktops. I also used to use the CPU gadget and on some machines the Weather gadget.

In Windows 10, you have a layout that's more or less similar to Windows 7, but it has fantastically good gadgets. Microsoft rebuilt all the plumbing in Windows to support these really good gadgets. Those updated, enormously powerful gadgets are now called Windows 10 apps.

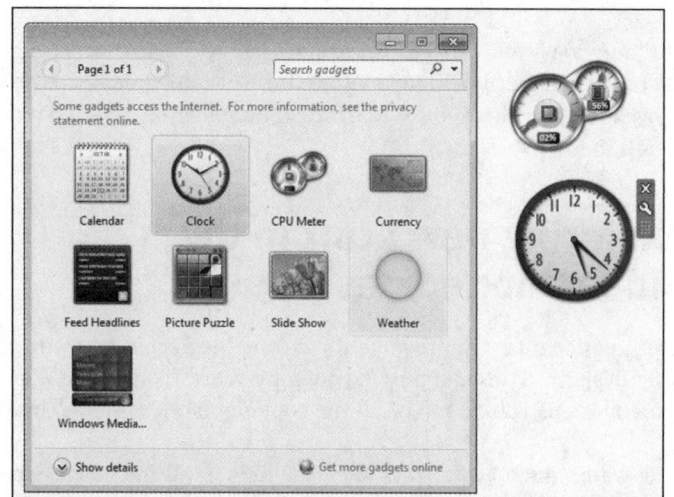

FIGURE 2-7:
Windows 7 gadgets — at least from the interface point of view — work much like the new Universal Windows app tiles.

The new gadgets/apps run in resizable windows on the desktop. They can do phenomenal things. In fact, Microsoft Edge is quite superior to Internet Explorer, even if it doesn't yet have all the bells and whistles. Edge, which runs as a gadget/ Windows 10 app, has become the new default browser.

Tiles for these gadgets/apps appear to the right of the list of programs in the Windows 10 Start menu.

REMEMBER

Here's the big picture, from the Windows 7 perspective: Windows 10 has a desktop, and it's more or less analogous to the desktop in Windows 7. It doesn't have a Windows 8/8.1–style Metro view. Doesn't need a Metro view: The gadgets (or Metro apps or Windows 10 apps) now behave themselves and run in resizable windows on the desktop.

In Windows 10, you can switch from a finger-friendly view of the desktop to a mouse-friendly view and back. The finger-friendly view — called Tablet mode — has larger app tiles, opens the apps at full-screen, and hides most of the text. It takes three clicks to change modes. Or you can plug or unplug your keyboard on a 2-in-1 such as the Surface Pro, and Windows 10 will ask if you want to switch modes.

Here's the ace in the hole: Programmers who write programs for Windows 10 app can have their gadgets run, with a varying amount of modification, on Windows 10 for PCs, Windows 10 for tablets without a keyboard, Windows 10 running on mobile-phone-like ARM chips (primarily from Qualcomm) and even Xbox One. At least, that's the theory. It remains to be seen how it works in practice.

The only way you can get these new gadgets/Windows 10 apps is through the Microsoft (formerly Windows) Store, so — again, at least in theory — they should be well-vetted, checked for malware, and generally in good shape before you can install them.

Exploring new stuff in the old-fashioned desktop

You'll notice many improvements to long-neglected portions of the Windows 7-style desktop. For example, if you copy more than one file at a time, Windows actually keeps you on top of all the copying in one window. Imagine that.

A new and much better *Task Manager* rolls in all the usage reporting that's been scattered in different corners of Windows (see Figure 2-8). The new Task Manager even gives you hooks to look at programs that start automatically, and to stop them if you like. Some serious chops. See Book 8, Chapter 4.

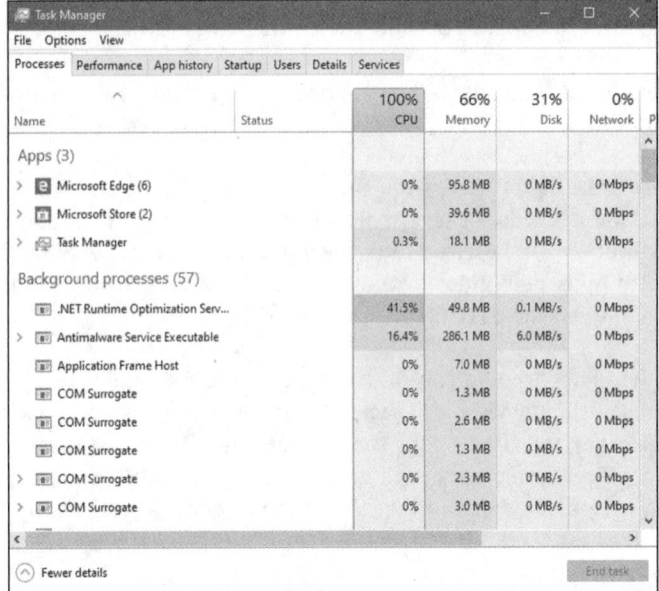

FIGURE 2-8:
The new and greatly improved Task Manager.

ASK
WOODY.COM

File Explorer (formerly known as Windows Explorer) takes on a new face and loses some of its annoying bad habits. You may or may not like the new Explorer, but at least Windows 10 brings back the up arrow to move up one folder — a feature that last appeared in Windows XP. That one feature, all by itself, makes me feel good about the new File Explorer. Explorer also now offers native support for ISO files. About time.

Taking a cue from iPad, Windows 10 also offers a one-stop system restore capability. With *Reset This PC,* you can remove apps and settings and keep personal files, or wipe everything and reset Windows 10 to its factory defaults and files. It's like it pulls in a brand-spanking-new version of Windows 10. See Book 8, Chapter 2 for more details.

TIP

Storage Spaces requires at least two available hard drives — not including the one you use to boot the PC. If you can afford the disk space, Windows 10 can give you a fully redundant, hot backup of everything, all the time. If a hard drive dies, you disconnect the dead one, slip in a new one, grab a cup of coffee, and you're up and running as if nothing happened. If you run out of disk space, stick another drive in the PC or attach it with a USB cable, and Windows figures it all out. It's a magical capability that debuted in Windows Home Server, now made more robust. See Book 7, Chapter 4 for more on Storage Spaces.

Backup gets a major boost with an Apple Time Machine work-alike called *File History.* You may not realize it, but Windows 7 had the capability to restore previous versions of your data files. Windows 10 offers the same functionality, but in a much nicer package — so you're more likely to discover that it's there. See Book 8, Chapter 1. Unfortunately, Windows 10 drops the capability to create whole-disk ghost backups — you need to buy a third-party program such as Acronis if a full backup is in your future.

Power options have changed significantly. Again. The new options allow Windows to restart itself much faster than ever before.

TECHNICAL
STUFF

If you ever wanted to run a Virtual Machine inside Windows, Microsoft has made *Hyper-V* available, free. It's a rather esoteric capability that can come in very handy if you need to run two different copies of an operating system on one machine. You must be running a 64-bit version of Windows 10 Pro (or Enterprise), with at least 4GB of RAM. See Book 8, Chapter 4.

What's New for Windows 8 and 8.1 Users

You're joking, right?

Windows 10 is a no-brainer if you already have Windows 8 or Windows 8.1.

Okay, I'll backtrack a bit. If you're a big fan of the tiled Metro side of Windows 8 or 8.1, you probably won't be happy with Windows 10, at least at first. There's no Charms bar, the taskbar always takes up part of the screen, Metro apps aren't completely immersive because they have title bars, and the full-screen tablet mode in Windows 10 isn't exactly comparable to the Metro side of Windows 8.

But if you use a mouse, even a little bit, or the desktop side of Windows 8/8.1, there's absolutely no question in my mind that you'll be happier with Windows 10.

Here's what you'll find when shifting from Win8 to Win10:

>> The Start menu — need I say more?

>> Big new features (detailed in the next section), along with a bunch of small tweaks really make life easier. Even in tablet mode, you'll find all sorts of things to love about Windows 10.

>> Windows 10 apps are updated and greatly improved, although Windows 10 has only a few more apps than Windows 8.

>> OneDrive is built-in and it works better. You don't need to install a separate app.

Windows 10 is, in many ways, what Windows 8 should've been. If Microsoft had been listening to its experienced Windows customers, Windows 8 never would've seen the light of day.

What's New for All of Windows

Permit me to take you on a whirlwind tour of the most important new features in Windows 10 — of which there are many.

The Start menu

Unless you've been living on an alternate Windows desktop, you know that Windows 10 sports a new Start menu, with shortcuts on the left, a list of all your apps and programs in the middle, and Windows 8–style tiles on the right.

Figures 2-2 and 2-6 earlier in this chapter show the Start menu. In Figure 2-9, I show you the Start menu with the phone-dialer style index; you get to it by clicking the Start icon and then clicking one of the headings for the app groups (A, B, and so on).

You have very few customizing options for the Start menu — for example, you can't drag entries onto the Most Used list in the top left, or drag items from the list on the left and turn them into tiles on the right. Tiles on the right can be resized to small (one-quarter the size of a medium tile), medium, wide (two single-size slots, as with the Store and Mail tiles in the screenshot), and large (twice the size of wide). You can click and drag, group and ungroup tiles on the right, and give groups custom names.

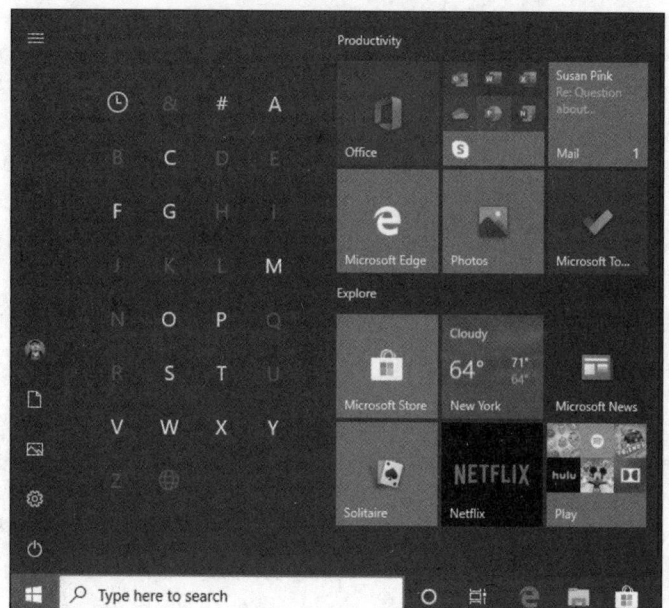

FIGURE 2-9:
The Start menu, with the index that lets you jump to apps quickly.

TIP

You can resize the Start menu, within certain rigid limits. You can adjust it vertically in small increments, but trying to drag things the other way is limited to big swaths of tiles: Groups of tiles remain three wide, and you can add or remove only entire columns. You can drag tiles from the right side of the Start screen onto the desktop for easy access.

Although it's possible to manually remove all the tiles on the right (right-click each, Unpin from Start), the big area for tiles doesn't shrink beyond one column.

In tablet mode, Start looks quite different, although many of the options are the same. See Figure 2-10.

I talk about personalizing the Start menu in Book 3, Chapter 2 and working with tablet mode in Book 3, Chapter 3.

Microsoft Edge

Microsoft Edge (Figure 2-11) finally sheds the albatross that is Internet Explorer. Edge is a stripped-down, consciously standards-compliant, screamingly fast shell of a browser, ready to take on just about any website anywhere. Microsoft Edge may see Microsoft taking back the mindshare it's been steadily losing on the browser front for the past decade or so. As of this writing, though, Google's Chrome rules the roost.

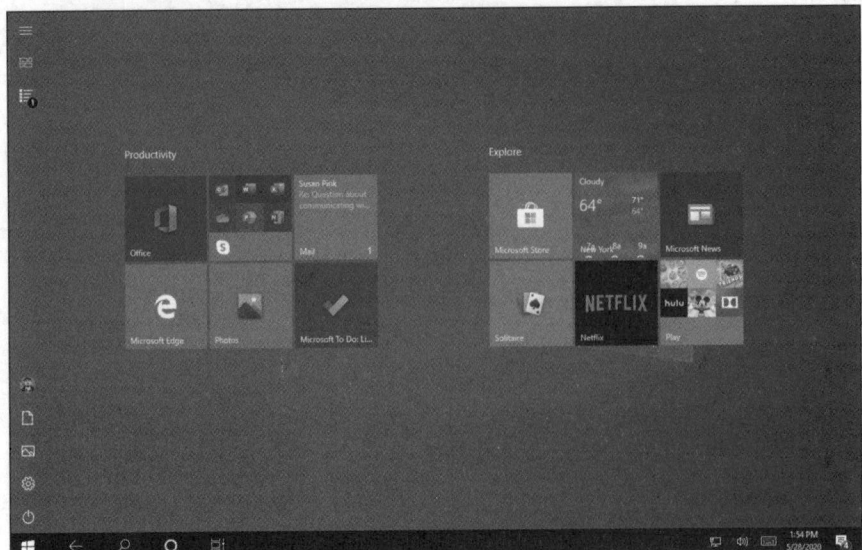

FIGURE 2-10:
The Start menu in tablet mode.

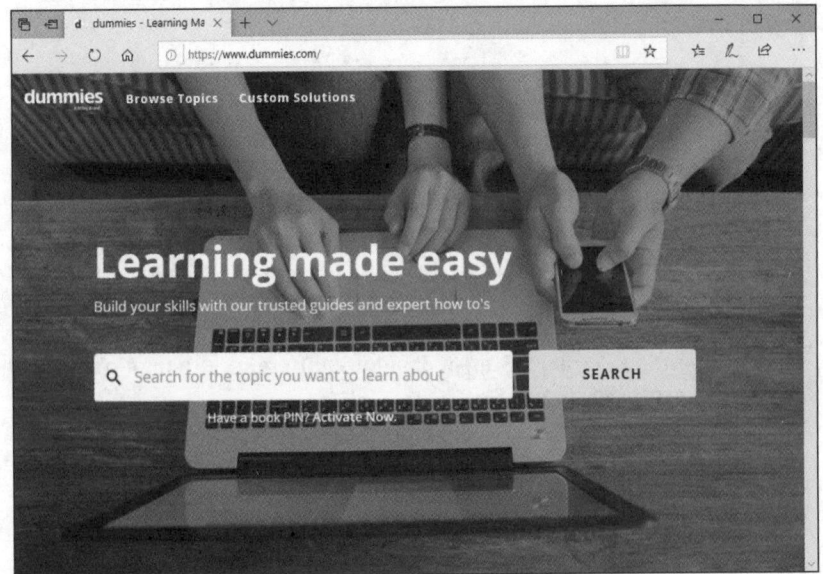

FIGURE 2-11:
Microsoft Edge finally lets you cut the Internet Explorer cord.

ASK
WOODY.COM

Microsoft Edge replaces Internet Explorer, which still lurks in Windows 10, but it's buried in the Start ➪ Windows Accessories list. Microsoft Edge is, however, the default web browser, with its own tile on the right side of the Start menu and its own icon on the taskbar. Internet Explorer continues to use the old Trident rendering engine, while Edge has the newer engine of its own. That makes it faster, lighter, and much more capable of playing nicely with websites designed for Firefox and Google Chrome.

Edge is a Windows 10 app (formerly Universal app, formerly Metro app) that runs inside its own window on the desktop, like every other WinRT API-based Universal Windows app. In contrast, Internet Explorer is an old-fashioned desktop app, and the difference is like a Tesla 3 versus a 1958 Edsel.

Adobe Flash Player is turned off by default for enhanced security; there's a reading view as well, which helps on smaller screens. Click the OneNote icon in the upper right, and all the OneNote markup tools become available. And you can Print as PDF.

REMEMBER

Where Internet Explorer was frequently infected by wayward Flash programs and bad PDF files, Edge is relatively immune. And all the flotsam that came along with IE — the ancient (and penetrable) COM extensions, wacko custom toolbars, even Silverlight — are suddenly legacy and rapidly headed to a well-deserved stint in the bit bucket.

On the other hand, Microsoft Edge has a new version that is not yet built into Windows 10. This new version is based on the same rendering engine as Google Chrome and has support for Google Chrome-like extensions, which play in their own sandboxes, staying isolated. Instead of the spaghetti mess with IE add-ons, we finally have some Microsoft-sponsored order. You can download it and try it at `www.microsoft.com/en-us/edge`.

Microsoft Edge uses Cortana for voice assistance and search capabilities. I talk about Edge in Book 5, Chapter 1.

Search

Search used to be intertwined with Cortana, making it bloated and slow in the initial versions of Windows 10. Also, Search collected a lot of data about what people do on their Windows 10 PCs. As of the May 2020 update, Search has detached itself from Cortana and received many improvements. But as always with Microsoft, people had to hate it first before Microsoft listened and made it better.

You can use Search to start apps using only the keyboard (geeks love that). You also get fast access to Windows 10 settings, your documents, photos, and emails, and even websites. As you would expect, Windows 10 Search is integrated with Bing, not Google, and your web searches are used to make Bing better. As shown in Figure 2-12, Search is used also to provide you with news (getting the latest headlines about coronavirus was not something I loved) and ads (promoting the new Chrome-based version of Microsoft Edge and the like).

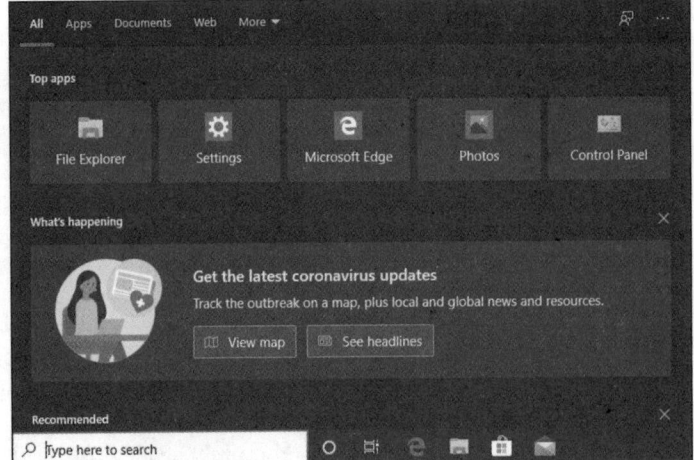

FIGURE 2-12:
Search helps you
find what you are
looking for, but
also displays ads
and the latest
news.

Leaving all these minor annoyances aside, I do like the new Search a lot. The indexing of files works better than ever, it eats up fewer system resources, and search results are returned faster than ever. And Search is well integrated with OneDrive, SharePoint, and Outlook, so finding your stuff in the cloud is easy, as long as you use a Microsoft, Work, or School account with Windows 10. Two other cool feats are that you can tell Windows 10 what folders to exclude from Search so that it doesn't bother indexing them, and have it respect your power mode settings when using Windows 10 on a laptop or tablet. Goodbye Windows 10 Search draining my battery faster than it should!

Cortana

Although Apple partisans will give you a zillion reasons why Siri rules and Googlies swear the superiority of Google Assistant, Cortana partisans think Microsoft rules the AI roost, of course. Unlike Siri and Google Assistant, though, Cortana used to take over the Windows search function. As of the May 2020 update, that is no longer the case, and Cortana has a box of her own, isolated from the rest of the operating system. You can see her in Figure 2-13. She now behaves more like a chat app and can take both voice and text commands from you.

Cortana works only when connected to the Internet and is severely limited unless you use a Microsoft account. You can control some aspects of Cortana's inquisitiveness by clicking the hamburger icon in the upper-left corner and going to Settings. For example, you can select how you want to talk to her (through typing, speaking, or both), the permissions you give her, and your privacy settings.

Frequently overlooked in Cortana discussions is the fact that everything you search for through Cortana goes to Microsoft's giant database in the sky.

REMEMBER

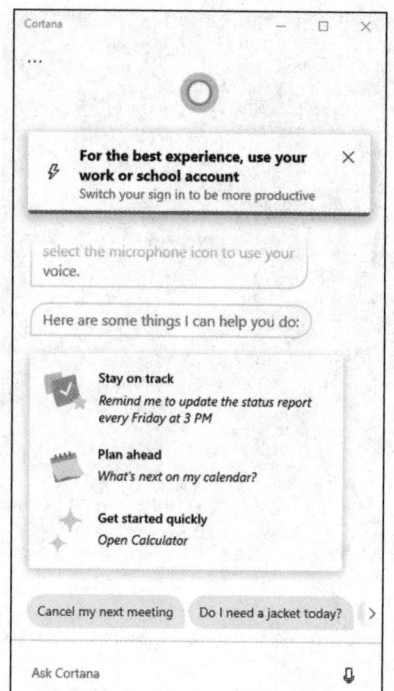

FIGURE 2-13: Cortana knows all, sees all if you enable her.

Cortana improves as it gathers more information about you — yes, by logging what you do. But it also improves as Microsoft hones its artificial intelligence know-how, on the back end. One interesting move on Microsoft's part, and an admittance that they have lost the first round of the virtual assistant battle, is that Cortana is now integrated with Amazon's Alexa. Amazon and Microsoft partnered up in August 2018 to make Cortana available through Amazon Echo devices and Alexa available through Windows 10. Cortana is going to be able to start Alexa, and take Alexa commands, and vice-versa. In theory, it sounds great, but this partnership is in its early stages of development, without much fuss going around.

In actual use, there's no question that Google's AI is superior to all the others, with Siri and Alexa each occupying different niches. Cortana's well adapted to Windows 10, but she isn't all that smart. I talk about Cortana in Book 3, Chapter 5.

Virtual desktops and task view

Windows has had virtual (or multiple) desktops since Windows XP, but before Windows 10, you had to install a third-party app — or something like Sysinternals desktop, from Microsoft — to get them to work. Windows 10 implements virtual desktops (Figure 2-14) in a way that is useful.

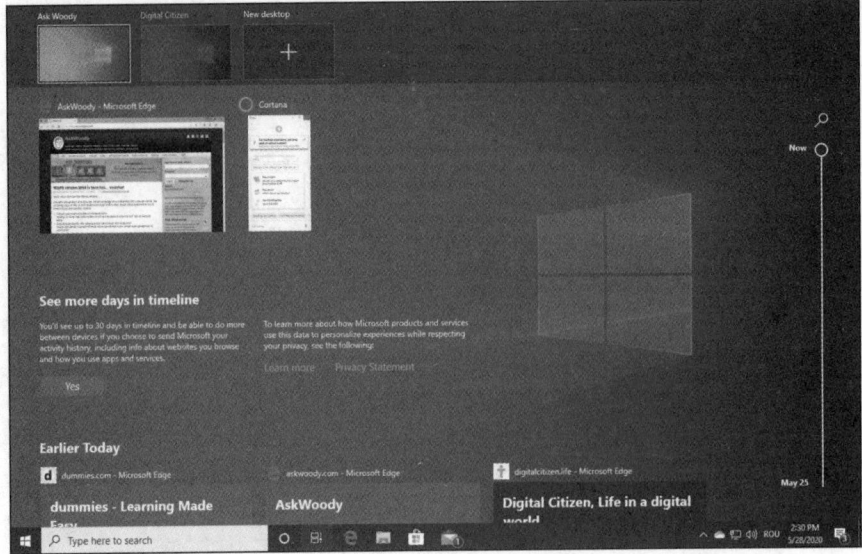

FIGURE 2-14:
Task view (shown
here on top with
the new Timeline
feature below)
displays all the
multiple desktops
you've set up.

**ASK
WOODY.COM**

Don't let the terminology freak you out: Virtual desktops are just multiple desktops and vice versa. If you want to sound cool, you can talk about optimizing your virtual desktops, but people in the know will realize you're just flipping between multiple desktops.

Multiple desktops are handy if you tend to multitask. You can set up one desktop to handle your mail, calendar, and day-to-day stuff, and another desktop for your latest project or projects. Got a crunch project? Fire up a new desktop. It's a great way to put a meta-structure on the work you do every day.

To start a new desktop, press Win+Ctrl+D. To see all available desktops, plus your Timeline, click the Task View icon to the right of the Windows 10 Search bar. Windows can be moved between desktops by right-clicking and choosing Move To. Alt+Tab still rotates among all running windows. Clicking an icon in the taskbar brings up the associated program, regardless of which desktop it's on.

Another nice feature introduced in the May 2020 update is that you can name virtual desktops anyway you want to help you keep track of which is which. It took Microsoft a long time to realize that this tiny improvement makes a world of difference.

Security improvements

I'm told that Pliny the Elder once described the alarm system of ancient Rome by saying, "Even when the dogs sleep, the goose watches."

By that standard, Windows 10 has been goosed.

With Windows 8, Microsoft somehow found a new backbone — or decided that it can fend off antitrust actions — and baked full antivirus, antispyware, antiscumstuff protection into Windows itself. Windows 10 continues to use exactly the same protection as Windows 8/8.1.

Although the 'Softies resurrected an old name for the service — *Windows Defender* — and then changed it to *Windows Security*, the antivirus protection inside Windows 10 is second to none. In Windows 10, Windows Security gives you the following layers of security: antivirus protection, ransomware protection, firewall protection against network and Internet attacks, reputation-based protection (for apps, files, and websites), exploit protection, and parental controls. All this is free!

TECHNICAL
STUFF

Microsoft is also encouraging hardware manufacturers to use a boot-up process called *UEFI*, as a replacement to the decades-old BIOS. UEFI isn't exactly a Windows 10 feature, but it's a requirement for all PCs that carry the Windows 10 (or Windows 8) logo. UEFI can help protect you from rootkits by requiring digital signatures on any operating system that gets loaded. See Book 9, Chapter 3.

Game mode and Xbox

Gaming is a big deal in Windows 10, and Microsoft wants its operating system to be the best choice for gamers. To cater to the needs of gamers, Windows 10 has a *game mode* that starts automatically when it detects that you're playing something. You can also start it manually.

Game mode prioritizes the processor and graphics card resources to your game. It also stops Windows Update from installing driver updates or showing update notifications during your play. Another useful feature is that it stops all notifications from all apps so that they don't interfere with your game.

Another feature is the Xbox game bar, which has been improved with each new Windows 10 update. With it, you can take screenshots while you play and record videos of your gameplay. You can also use it to quickly adjust the audio and voice settings — useful when you play online with others and have to coordinate with them. The Xbox game bar also shows you the performance of your computer (processor, RAM, and graphics card resource consumption) and allows you to chat and interact with your friends on Xbox (see Figure 2-15).

TIP

Press Win+G to display the Xbox game bar at any time, including when you're not playing. Familiarize yourself with all the buttons and features, so that you can use it productively while you play games.

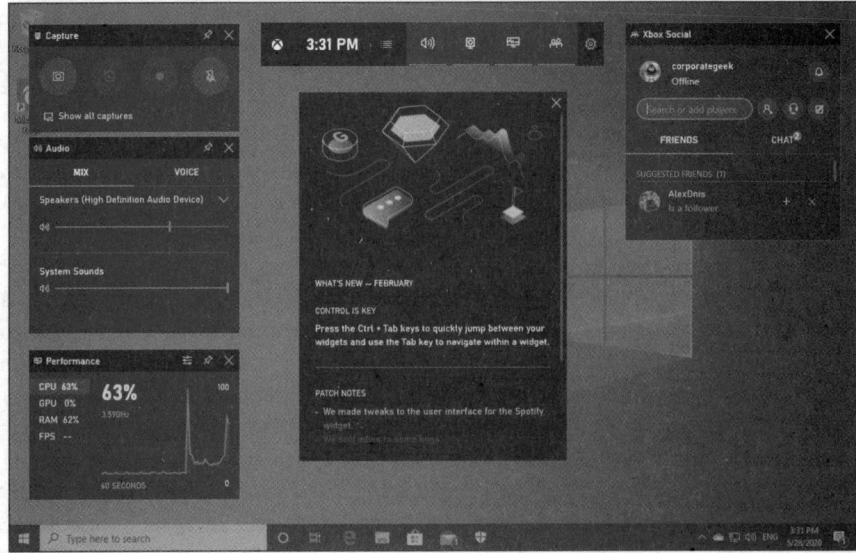

FIGURE 2-15:
The Xbox game
bar has many
features useful to
gamers.

In the Windows 10 May 2020 update, Microsoft also introduced DirectX 12 Ultimate — their best features and tools for game developers, so they can build the best-looking games for Windows 10 and Xbox One. It is yet another effort on Microsoft's part to unify the gaming experience between the Xbox console and Windows 10 PCs. With it, Microsoft has the same common graphics API and graphics architecture for both PCs and consoles.

I discuss Windows 10 gaming more in Book 5, Chapter 4.

Other Windows 10 apps

Microsoft has given most of its built-in apps a much-needed makeover.

Mail and Calendar, unlike their Windows 8.1 analogs, actually work. You don't need to feel like the 90-pound weakling on the beach if you crank them up. I use Gmail and Google Calendar, but the new Windows 10 Mail app is definitely a contender. I talk about Mail and Calendar (which are really one app with two different viewpoints) in Book 4, Chapter 1.

People is a derivative of the Windows Phone People Sense app. It doesn't do much, but it may be useful to some users. If you hate it, you can disable it easily. I talk about People in Book 4, Chapter 2.

Groove Music and Movies & TV have replaced the useless Windows 8.1 Xbox Music and Xbox Video apps. They're surprisingly capable and tie into Microsoft's streaming service. (It took Microsoft only half a decade to put together a decent streaming service.) The Groove Music Pass — a monthly subscription offering that works with Groove Music — has been discontinued, replaced with Spotify and a new Microsoft/Spotify détente. Look at Book 4, Chapter 5, for more.

The new **Photos** app used to be a dud but is evolving slowly. The latest iteration is reasonably decent and sometimes useful. Basic users might not need anything else, but advanced users may be disappointed. I talk about the Photos app in Book 4, Chapter 3.

The **Weather** app shows more weather and less sappy background than its Windows 8.1 counterpart. I cover it along with the other Bing apps — News, Money, Sports — in Book 5, Chapter 3.

Even the **Microsoft Store** is better than it used to be — damning with faint praise, for sure. The best part about it is that it now includes both Windows 10 apps and desktop apps, alongside other content such as games or movies. For details on actual improvements, see Book 5, Chapter 3.

What you lose

Although Microsoft hasn't talked much about it, the fact is that all the old Windows Live programs are disappearing. Windows Live is, in fact, dead. Windows 8 killed it, and Windows 10 drove a stake through its heart. If you use any of the Windows Live apps in Windows 7 (or Vista or XP, for that matter), your old Live apps are still available, but it doesn't look like Microsoft is going to do much with them. They certainly aren't getting any support.

WARNING

Why? The Windows 10 Universal/Metro apps cover many of the Live bases. Consider these:

» **Windows Live ID** (formerly known as Microsoft Wallet, Microsoft Passport, .NET Passport, and Microsoft Passport Network), which now operates from the Windows Live Account site (confused yet?), is rebranded Microsoft Your Account and referred to informally as your *Microsoft Account.*

» **Windows Live OneDrive** has already turned into just plain *OneDrive.* Parts of Ray Ozzie's Windows Live Mesh — formerly Live Mesh, Windows Live Sync, and Windows Live FolderShare — have been folded into OneDrive, although Microsoft has squashed PC-to-PC sync; the only way to synchronize files is through the OneDrive cloud. It appears as if Mesh has met its match.

» **Windows Live Mail** has officially fallen out of favor, with Microsoft announcing that it won't support WLM with any Microsoft accounts. Expect Microsoft to push the new *Universal Windows Mail* as a core Windows communications app. Ditto for Windows Live Calendar.

>> **Windows Live Contacts** is now the *Windows 10 People app.*

>> **Windows Live Photo Gallery** morphed into the *Windows 10 Photos* app.

>> **Windows Live Messenger** is dead. It's been replaced by Skype — or Facebook, or any of a zillion competitors. I use Line, but that's a story for Book 5, Chapter 2.

It's not just the Windows Live apps that are dying. Some of the old Windows programs — **Media Center** being a good example — are just dead. Homegroups got canned, with Microsoft hoping you'll use OneDrive instead. The old Windows 7 Backup is still there, buried under layers of clicks, but Microsoft would clearly prefer if you didn't bother with old-fashioned backup and used OneDrive instead.

Some people feel that losing **Adobe Reader** (and other browser add-ins) in Microsoft Edge is a bad thing. I disagree strongly. Reader (and Flash, which is insulated in Microsoft Edge) have brought on more pain and misery — and hijacked systems — than they're worth. Microsoft's own ActiveX technology, which won't run on Edge, is another malware magnet that deserves to die, as do browser helper objects, home page hijackers, custom toolbars, and much more. You can run all those add-ins in the Legacy desktop version of Internet Explorer if you absolutely must.

Some other odd missing pieces include the following:

WARNING

>> **ClearType** doesn't run on the Windows 10 apps' interface, at all. It's still on the old-fashioned desktop, but your Windows 10/Universal/Metro apps can't use it.

Note that this is different from Microsoft's ClearType HD technology, a marketing term for the monitors on Microsoft Surface tablets. I have no idea why Microsoft used the same term for both.

>> **Flip 3D** is gone. Little more than a parlor trick, and rarely used, the Windows key+Tab used to show a 3D rendering of all running programs and flip among them. Stick a fork in it. Now it cycles among desktops.

Do You Need Windows 10?

ASK WOODY.COM

With the drubbing I gave Windows 8 and Windows 8.1 in the press — and in my *For Dummies* books — you might think that I'd come down hard on Windows 10.

Nope.

I've been using Windows 10 in various stages for more than five years now, and I still love it. This is from a guy who works in front of a monitor about 16 hours a day, 7 days a week (at least during the book-writing season). I use a mouse or trackpad, and I'm proud of it. Windows 10, to my mind, is a great operating system, and it's a big improvement over Windows 8. I know, damning with faint praise again.

If you use a keyboard and a mouse with Windows 8 or 8.1, you need Windows 10. It's that simple.

Switching over to touch computing isn't quite so clear-cut. I have a couple of touch tablets, and I review dozens more, and for simple demands — mail, web, media playing, TV casting — I still prefer Chrome OS, the driving force behind Chromebooks. It's simpler, less prone to infuriating screw-ups, less prone to infection, and less demanding for patches.

On the other hand, if you need one of the (many!) Windows 10 apps or Windows desktop apps that don't run on Chrome OS, and you have a touch-first environment, Windows 10 ain't a bad choice.

ASK WOODY.COM

One thing's for sure. This isn't recycled old Windows 8 garbage. With Windows 10, Microsoft has taken a bold step in the right direction — one that accommodates both old desktop fogies like me and the more mobile newcomers (like me, too, I guess).

I haven't felt this good about a Microsoft product since the original release of Windows 7. I just wish Microsoft hadn't pushed so hard with the Get Windows 10 campaign. It still leaves a bad taste in my mouth after all these years.

Chapter **3**

Which Version?

P ermit me to dispel two rumors, right off the bat. Windows 10 isn't exactly free. And it isn't the last version of Windows.

ASK WOODY.COM

You probably heard either or both of those rumors from well-regarded mainstream publications, and what you heard was wrong.

Here are the facts:

» From July 29, 2015 (when Win10 RTM was released) to July 29, 2016, you could upgrade from a genuine copy of Windows 7 or Windows 8.1 to Windows 10 for free. At the time this book went to press, you can't, although hope springs eternal. For the latest info on free or reduced-price upgrades, drop by www. AskWoody.com.

If you're building a new PC, you have to buy Windows 10. And if you buy a new PC with Windows 10 preinstalled, the PC manufacturer (probably) paid for Windows 10.

» Microsoft may drop the numbering system, in which case Windows 10 would be simply Windows, but there will always be version numbers associated with each release. I tell you how to find yours in this chapter. The number 10 is, was, and always will be a marketing fantasy.

If you haven't yet bought a copy of Windows, you can save yourself some headaches and more than a few bucks by buying the right version the first time. And if you're struggling with the 32-bit versus 64-bit debate, illumination — and possibly some help — is at hand.

Counting the Editions

Windows 10 appears in six different major editions, uncounted numbers of minor editions, and three of the major editions are available in 32-bit and 64-bit incarnations. That makes nine different editions of Windows to choose from. Not counting the kinda-sorta Windows 10 editions for ARM chips (such as Qualcomm's), phones (Mobile), Xbox, HoloLens, refrigerators, and bumper cars.

Fortunately, most people need to concern themselves with only two editions, and you can probably quickly winnow the list to one. Contemplating the 32-bit conundrum may exercise a few extra gray cells, but with a little help, you can probably figure it out easily.

In a nutshell, the four Windows 10 editions (and targeted customer bases) look like this:

REMEMBER

>> **Windows 10 Home (initially named Windows 10)** — the version you probably want — works great unless you specifically need one of the features in Windows 10 Pro. A big bonus for many of you: This version makes all the myriad Windows languages — 96 of them, from Afrikaans to Yoruba — available to anyone with a normal, everyday copy of Windows, at no extra cost. Its biggest downside is that it allows you to postpone updates only up to 35 days.

>> **Windows 10 Pro** includes everything in Windows 10 Home plus the capability to attach the computer to a corporate domain network; the Encrypting File System and BitLocker (see the "Encrypting File System and BitLocker" sidebar later in this chapter) for scrambling your hard drive's data; Hyper-V for running virtual machines; and the software necessary for your computer to act as a Remote Desktop host — the "puppet" in an RD session. A big plus is that it allows users to postpone updates up to a year.

>> **Windows 10 Enterprise** is available only to companies that buy into Microsoft's Volume License program — the (expensive) volume licensing plan that buys licenses to every modern Windows version. Enterprise offers a handful of additional features, but they don't matter unless you're going to buy a handful of licenses or more.

>> **Windows 10 Education** looks just like Windows 10 Enterprise, but it's available only to schools, through a program called Academic Volume Licensing.

Those four editions run only on Intel and AMD processors. They're traditional Windows.

You'll hear increasingly about Windows 10 editions designed for ARM chips — computer chips originally designed for smartphones. In theory, those editions will work exactly the same way as their Intel/AMD brethren but can't run desktop apps unless they are emulated.

In addition, just to make your life more complicated, many of these editions of Windows can run in S mode. Microsoft's peddling S mode as an alternative to Chromebooks — stripped-down, fast starting, battery friendly, and somewhat impervious to infection.

WHAT HAPPENED TO WINDOWS PHONE?

Windows Phone turned into a multibillion-dollar tragedy that sent tens of thousands of people to the unemployment line and put a major drain on Finland's economy.

Finnish company Nokia pioneered the Windows phone and sold millions of them. Nokia sales started drifting off, and Microsoft was faced with a big choice: Either prop up Nokia or lose its only major outlet for the Windows Phone software. Long story short, Microsoft sent one of its execs to lead Nokia, ultimately buying Nokia in April 2014 for $7.2 billion. Three months later, Microsoft announced it was laying off 18,000 Nokia employees.

Fifteen months later, the Nokia phone business had crashed and Microsoft wrote off $7.6 billion in acquisition costs. In May 2016, after several more rounds of layoffs and write-offs, Microsoft announced it was selling the Nokia brand and its smartphones to Foxconn, the company best known for manufacturing computers throughout Asia.

Microsoft sold the remnants of the Nokia brand for $350 million, and Foxconn immediately announced plans to sell Android phones. Windows Phone was rebranded into Windows 10 Mobile but that did not help. In December 2019, Windows 10 Mobile had reached the end of its life.

TECHNICAL STUFF

Windows Vista and Windows 7 both had Ultimate editions, which included absolutely everything. Windows 10 doesn't work that way. If you want the whole enchilada, you have to pay for volume licensing.

Windows Media Center — the Windows XP–era way to turn a PC into a set-top box — is no longer available in any version of Windows 10. Do yourself a favor and buy a Chromecast, or use your cable company's DVR if you really have to record TV.

Any edition of Windows 10 running in S mode runs only apps. That bears repeating: **S mode doesn't run old-fashioned Windows programs.** S mode is restricted to running just Windows 10 apps in the Microsoft Store. You can have Windows 10 Home, Pro, Enterprise, and Education, all running in S mode.

This book covers Windows 10 Home and Windows 10 Pro. Most of the content is applicable also to Windows 10 Enterprise and Windows 10 Education. Only a little bit of the content applies to Windows 10 in S mode.

TIP

Before you tear your hair out trying to determine whether you bought the right version or which edition you should buy your great-aunt Ethel, rest assured that choosing the right version is much simpler than it first appears. Flip to "Narrowing the choices," later in this chapter. If you're considering buying a cheap version now and maybe upgrading later, I suggest that you first read the next section, "Buying the right version the first time," before you make up your mind.

Buying the right version the first time

What if you aim too low? What if you buy Windows 10 and decide later that you really want Windows 10 Pro? Be of good cheer. Switching versions isn't as tough as you think.

Microsoft chose the feature sets assigned to each Windows version with one specific goal in mind: Maximize Microsoft's profits. If you want to move from Windows 10 Home to Windows 10 Pro (the only upgrade available to individuals), you need to buy the Windows 10 Pro Pack. To buy an upgrade, choose the Start icon, the Settings icon, Update & Security, Activation, and then choose Go to Store.

Similarly, moving from Windows 10 in S mode to just plain Windows 10 requires only a trip to the Microsoft Store.

Upgrading is easy and cheap, but not as cheap as buying the version you want the first time. That's also why it's important for your financial health to get the right version from the get-go.

Narrowing the choices

You can dismiss three regular Windows editions and both Windows Mobile editions out of hand:

» **Any Windows 10 version in S mode** may work for a little while, but I'll bet you bucks to buckaroos that you'll get tired of it shortly. S mode is great in schools and places where admins want absolute control. It's onerous for people who have a choice. You can't even run the Google Chrome browser in S mode. Plan on ditching it as soon as you can.

» **Windows 10 Enterprise** is an option only if you want to pay through the nose for five or more Windows licenses, through the Volume Licensing program. Microsoft may change its mind — either lower the price for small bunches of licenses and/or make the Enterprise version available to individuals — but as of this writing, Enterprise is out of the picture for most of you. There are some tricks, but in general they aren't worth the hassle.

» **Windows 10 Education,** similarly, can be purchased only in large quantities. If you're a student, faculty member, or staff member at a licensed school, you must contact the IT department to get set up.

ENCRYPTING FILE SYSTEM AND BITLOCKER

Encrypting File System (EFS) is a method for encrypting individual files or groups of files on a hard drive. EFS starts after Windows boots: It runs as a program under Windows, which means it can leave traces of itself and the data that's being encrypted in temporary Windows places that may be sniffed by exploit programs. The Windows directory isn't encrypted by EFS, so bad guys (and girls!) who can get access to the directory can hammer it with brute-force password attacks. Widely available tools can crack EFS if the cracker can reboot the, uh, crackee's computer. Thus, for example, EFS can't protect the hard drive on a stolen laptop/notebook. Windows has supported EFS since the halcyon days of Windows 2000.

BitLocker was introduced in Windows Vista and has been improved since. BitLocker runs *underneath* Windows: It starts before Windows starts. The Windows partition on a BitLocker-protected drive is completely encrypted, so bad guys who try to get to the file system can't find it.

EFS and BitLocker are complementary technologies: BitLocker provides coarse, all-or-nothing protection for an entire drive. EFS lets you scramble specific files or groups of files. Used together, they can be mighty hard to crack.

BitLocker To Go provides BitLocker-style protection to removable drives, including USB drives. You should use it when storing important data on your USB drives.

That leaves you with Windows 10 Home, unless you have a crying need to do one of the following:

>> **Connect to a corporate network.** If your company doesn't give you a copy of Windows 10 Enterprise, you need to spend the extra bucks and buy Windows 10 Pro.

>> **Play the role of the puppet — the *host* — in a Remote Desktop interaction.** If you're stuck with Remote Desktop, you must buy Windows 10 Pro.

Note that you can use Remote Assistance, any time, on any Windows PC, any version. (See Book 7, Chapter 2.) This Windows 10 Pro restriction is specifically for Remote Desktop, which is commonly used inside companies but not that much by other types of users.

TIP

Many businesspeople find that TeamViewer, a free alternative to Remote Desktop, does everything they need and that Remote Desktop amounts to overkill. TeamViewer lets you access and control your home or office PC from any place that has an Internet connection. Look at its website, www. teamviewer.com.

» **Provide added security to protect your data from prying eyes or to keep your notebook's data safe even if it's stolen.** Start by determining whether you need Encrypting File System (EFS), BitLocker, or both (see the "Encrypting File System and BitLocker" sidebar). Win10 Pro has EFS and BitLocker — with BitLocker To Go tossed in for a bit o' lagniappe.

» **Run Hyper-V.** Some people can benefit from running virtual machines inside Windows 10. If you absolutely must get an old Windows XP program to cooperate, for example, running Hyper-V with a licensed copy of Windows XP may be the best choice. For most people, VMs are an interesting toy, but not much more.

» **Postpone Windows 10 updates up to a year.** Unfortunately, Microsoft has dropped the ball on quality when it comes to Windows 10 updates. Each month there's news of a buggy update that wrecks people's computers. Windows 10 Pro gives people the option to postpone and control updates in a way that Windows 10 Home doesn't.

Choosing 32 Bit versus 64 Bit

If you've settled on, oh, Windows 10 as your operating system of choice, you aren't off the hook yet. You need to decide whether you want the 32-bit flavor or the 64-bit flavor of Windows 10 Home. Similarly, Windows 10 Pro and Enterprise are available in a 32-bit model and a 64-bit model.

Although the 32-bit and 64-bit flavors of Windows look and act the same on the surface, down in the bowels of Windows, they work quite differently. Which should you get? The question no doubt seems a bit esoteric, but just about every new PC nowadays uses the 64-bit version of Windows 10 for good reasons:

» **Performance:** The 32-bit flavor of Windows — the flavor that everyone was using a few years ago— has a limit on the amount of memory that Windows can use. Give or take a nip here and a tuck there, 32-bit Windows machines can see, at most, 3.4 or 3.5 gigabytes (GB) of memory. You can stick 4GB of memory into your computer, but in the 32-bit world, anything beyond 3.5GB is simply out of reach. It just sits there, unused. That's why you see 32-bit Windows only on tiny, cheap tablets and mobile devices.

The 64-bit flavor of Windows opens your computer's memory, so Windows can see and use more than 4GB — much more, in fact. With many desktop apps, such as the Google Chrome browser, acting like resource hogs, you'll want 4GB or more on any PC.

ASK
WOODY.COM

REMEMBER

Although lots of technical mumbo jumbo is involved, the simple fact is that programs are getting too big, and Windows as we know it is running out of room. Although Windows 10 can fake it by shuffling data on and off your hard drive, doing so slows your computer significantly.

>> **Security:** Security is one more good reason for running a 64-bit flavor of Windows. Microsoft enforced strict security constraints on drivers that support hardware in 64-bit machines — constraints that just couldn't be enforced in the older, more lax (and more compatible!) 32-bit environment.

WARNING

And that leads to the primary problem with 64-bit Windows: drivers. Some people have older hardware that doesn't work in any 64-bit flavor of Windows. Their hardware isn't supported. Hardware manufacturers sometimes decide that it isn't worth the money to build a solid 64-bit savvy driver, to make the old hardware work with the new operating system. You, as a customer, get the short end of the stick.

Application programs are a different story altogether. The 64-bit version of Office 2010 was notorious for causing all sorts of headaches. You were better off running 32-bit Office 2010, even on a 64-bit system (yes, 32-bit programs run just fine on a 64-bit system, by and large). Office 2016 and 2019 don't have the 64-bit shakes; they work fine on either 32-bit or 64-bit Windows. Some programs can't take advantage of the 64-bit breathing room. It's not all sweetness and light.

Now that you know the pros and cons, you have one more thing to take into consideration: What does your PC support? To run 64-bit Windows 10, your computer must support 64-bit operations. If you bought your computer any time after 2005 or so, you're fine — virtually all the PCs sold since then can handle 64-bit. But if you have an older PC, here's an easy way to see whether your current computer can handle 64 bits: Go to Steve Gibson's SecurAble site, at www.grc.com/securable.htm. Follow the instructions to download and run the SecurAble program. If your computer can handle 64-bit operations, SecurAble tells you.

ASK WOODY.COM

If you have older hardware — printers, scanners, USB modems, and the like — that you want to use with your Windows 10 computer, do yourself a favor and stick with 32-bit Windows. It's unlikely that you'll start feeling the constraints of 32 bits until your current PC is long past its prime. On the other hand, if you're starting with completely new hardware — or hardware that you bought in the past five or six years — and you plan to run your current PC for a long, long time, 64-bit Windows makes lots of sense. You may end up cursing me when an obscure driver goes bump in the night. But in the long run, you'll be better prepared for the future.

Which Version of Windows Are You Running?

You may be curious to know which version of Windows you're running on your current machine. Here's the easy way to tell:

» If your Start screen resembles the one in Figure 3-1, you have some version of Windows 8, 8.1, RT, or RT 8.1. Swipe from the right or hover your mouse cursor in the lower-right corner, and then choose Change PC Settings. Click or tap PC and Devices, then PC Info. You get a report like the one in Figure 3-2.

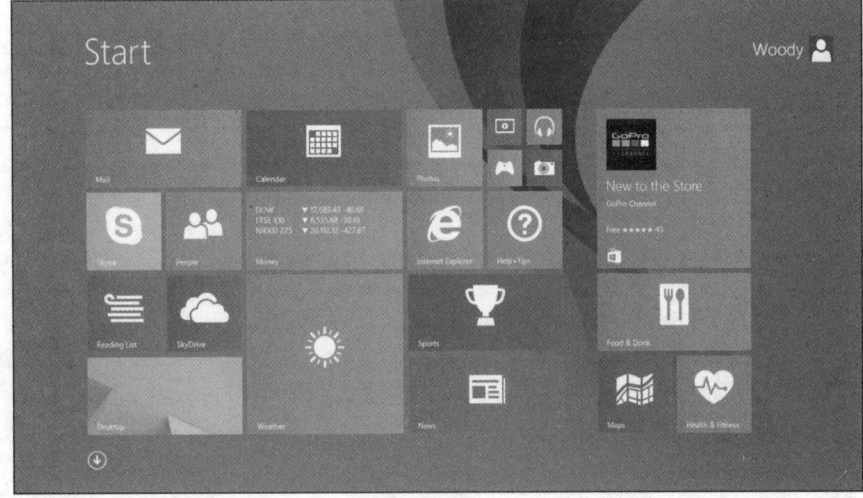

FIGURE 3-1: A Start screen like this is a dead giveaway for 8, 8.1, or RT.

» If you have a desktop like the one in Figure 3-3, you're running some version of Windows 7. Click the Start icon in the lower-left corner, then Control Panel ⇨ System ⇨ Security. Under System, click View Amount of RAM and Processor Speed. You see a report like the one in Figure 3-4.

» If your desktop doesn't look like Figure 3-1 or Figure 3-3, you're running Windows Vista or XP. Click the Start icon in the lower-left corner, then click Control Panel ⇨ System ⇨ Security. Under System, click View Amount of RAM and Processor Speed.

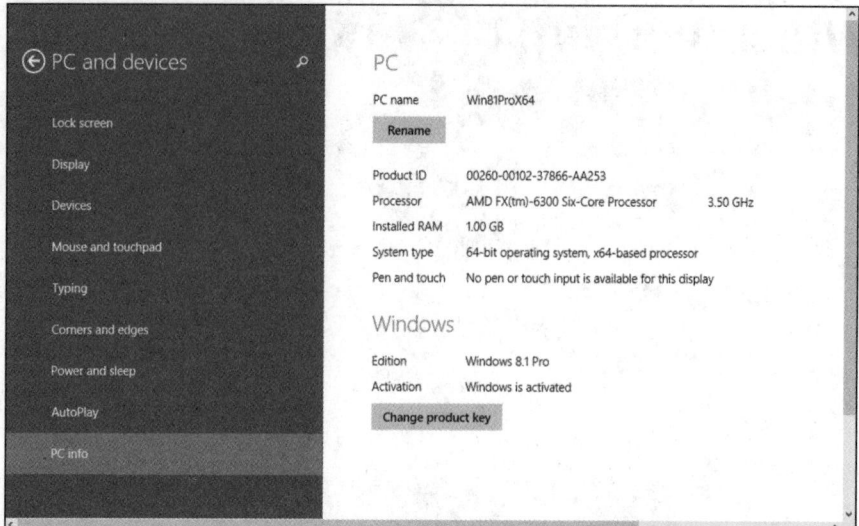

FIGURE 3-2:
This machine
runs 64-bit
Windows 8.1 Pro.

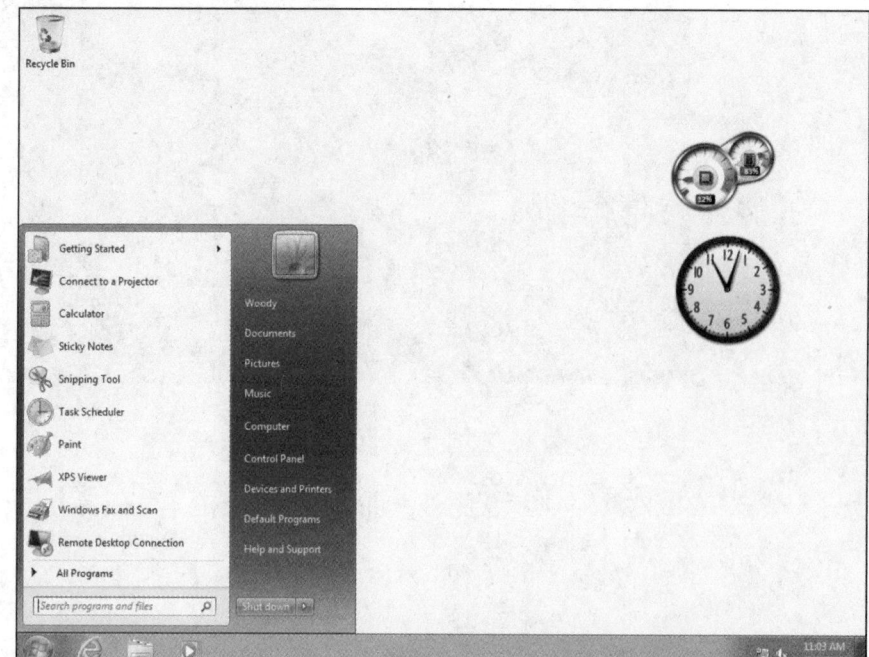

FIGURE 3-3:
Here's a telltale
desktop in
Windows 7.

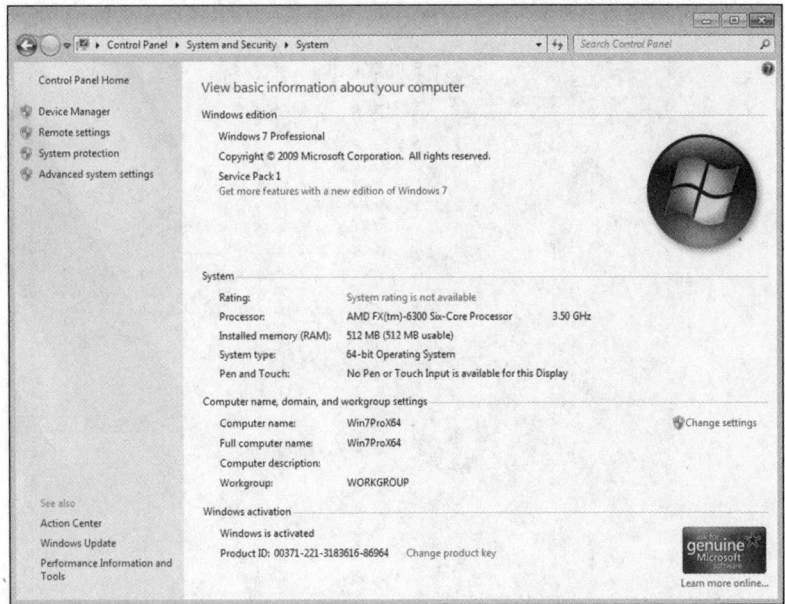

FIGURE 3-4:
This is Windows 7 Pro Service Pack 1, 64-bit.

If you have a 64-bit system installed already, you should upgrade to a 64-bit version of Windows 10. If you currently have a 32-bit system, check Steve Gibson's site, as mentioned in the preceding section.

But if you have a different-looking screen, chances are very good you already have Windows 10. Here's how to see which version you have:

1. In the Search box in the taskbar, type about.

Search results are immediately shown, and at the top of the stack you should see something like *About Your PC*.

2. Press Enter or click About Your PC.

You see an About window like the one in Figure 3-5.

3. On the right, scroll down until you can see Device Specifications.

To the right of the System Type heading, you see whether you have a 32-bit or 64-bit version of Windows.

4. Scroll down farther until you can see Windows Specifications.

To the right of the heading version, you'll see your version number and edition.

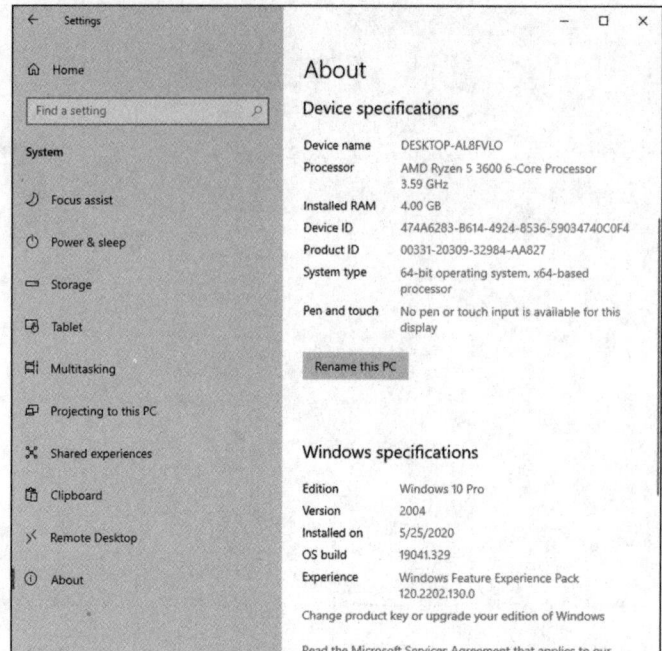

FIGURE 3-5:
Full system
information is in
the About box.

2

Personalizing Windows

Contents at a Glance

Chapter **1**

Getting Around in Windows

Ready to get your feet wet, but not yet up to a full plunge?

Good. You're in the right place for a dip-your-toes-in kind of experience. Nothing tough in this chapter, just a bit of windows cruising. Lay of the land kind of stuff.

If you're an experienced Windows 7 user, you'll find parts of Windows 10 that look a bit familiar and parts that look like they were ripped from an iPhone. If you're an experienced Windows 8.1 user, I salute you and your stamina, and I welcome you to a kinder, gentler version of Windows.

REMEMBER

Former Microsoft General Manager and Distinguished Engineer Hal Berenson said it best: "Consumers increasingly reject the old experiences in both their personal and work lives. For the 20-something-and-under crowd, the current Windows desktop experience is about as attractive as the thought of visiting a 19th-century dentist."

Windows 10 looks a little bit like that 19th-century dentist's office, but underneath it's gone through radical transformations.

ASK
WOODY.COM

I figure that 90 percent of the stuff that most of the people do with a computer runs fine on a tablet or a Chromebook. So why put up with all the hassles of running Windows on a piece of iron that weighs more than your refrigerator, and breaks down a lot more often? Maybe you're addicted to blue screens and frozen mice. Or maybe you're ready to leave it all behind and tap your way to something new. But if you're still solidly stuck in the Windows column, this chapter's for you.

In this chapter, I show you what's to like about both the old-fashioned side of Windows and the new app side, how to get around if you're new to Windows, and if you're an experienced Windows hand, how to reconcile your old finger memory with the new interface. It isn't as hard as you think.

Really.

I also show you how to be input-agnostic — how to use either your fingers, or a pen, or your fork, er, mouse to get around the screen. And I give you a few not-at-all-obvious tips about how to get the most out of your consorting with the beast.

Windows' New Beginnings

ASK
WOODY.COM

The way I look at it, most people starting with Windows 10 start in one of five groups, with the largest percentage in the first group:

>> Somewhat experienced at some version of Windows and primarily comfortable with a mouse and keyboard. (More than 1.5 billion people have used Windows.)

>> Experienced at Windows but want to learn touch input.

>> Windows 8 refugee who's hoping and praying Windows 10 isn't so disorienting.

>> New to Windows, prefer to use touch.

>> New to Windows and want to visit the 19th-century dentist's office to see what all the screaming's about.

TECHNICAL
STUFF

If you fall into that final group, you need to learn to use the antique interface apparatus known as a mouse and keyboard. I'm reminded of Scotty on the Enterprise picking up a mouse and saying, "Computer! Computer! Hello computer . . ." When Scotty's reminded to use the keyboard, he says, "Keyboard. How quaint." At least he didn't say, "Hey, Cortana!"

TAP OR CLICK, PAPER OR PLASTIC?

Lots of people have asked me whether I'm serious about tapping on a Windows machine. Yes, I am, and I hope you will be, too.

I tried the old stylus Windows interface, back when the luggable Windows tablets first appeared, in the Windows XP days. I hated it. I still hate it. I hated it so much that when I saw someone using an iPad, all I could think was, "Oh, that must suck." (Remember, *suck* is a technical term.)

An hour later, I tried an iPad, and suddenly using a finger was fine. More than fine, it was tremendous. When my then-18-month-old son spent a few hours playing on the iPad and started using the interface like a virtuoso, I was hooked. The tap-and-swipe interface is astonishingly easy to learn, use, and remember.

Windows 10's tap interface isn't as elegant as the iPad's. Sorry, but it's true. The main difference is that Windows has to accommodate lots of things that the iPad just doesn't do — right-click comes immediately to mind (although tablets have tap-and-hold to simulate right-click; the iPhone even has 3D touch, which goes way beyond clicking). But for many, many things that I do every day — web surfing, quickly checking email, scrolling through Twitter, catching up on Facebook, reading the news, looking at the stock market, and on and on — the touch interface is vastly superior to a mouse and keyboard. At least, it is to me.

That said, yes, you can get used to a tablet without a mouse and keyboard.

As I'm writing this book, I have three computers on my desk. One's a traditional desktop running Windows 10, and one's running the latest beta test version of Windows 10. The third is a Win10 tablet with a portable keyboard — a Surface Pro. When I want to look up something quickly, guess which one I use? Bzzzzzt. Wrong. I pick up my Nexus phone — or my iPad. "OK, Google, where is Timbuktu?" "Navigate to Costco." "Call the Recreation Center."

So this section offers a whirlwind tour of your new Windows 10 home that helps you start clicking and tapping your way around.

A tale of two homes

As you undoubtedly know by now, Windows 10's Start menu has two faces. They're designed to work together. You can be the judge of how well they live up to the design.

On the left side of the Start menu (see Figure 1-1), you see the Start menu that's supposed to look like the Windows 7 (and Windows Vista and Windows XP) Start menu. On the right side of the Start menu, you see a bunch of tiles, some of which have useful information on them.

FIGURE 1-1:
The Windows
10 Start menu
as seen on a
1920 x 1080 (HD)
monitor.

Although the left side of the Start menu is supposed to bring back warm, comforting memories of Windows 7 (and Windows XP), underneath the surface, the left part of the Start menu has almost nothing in common with Windows 7 and earlier Start menus. The old Start menu has been ripped out and replaced with this Windows 10-style list of links and, on the far left, a set of shortcut icons.

 See the funny icon in the upper-left corner (and shown in the margin)? For the mathematicians in the crowd, it looks just like an equivalence sign. In the computer world, that's known as signaling a hamburger menu (see the nearby sidebar).

TIP The new stab at a Start menu is both good and bad. As you'll see, the left side of the Start menu is a wimpy thing, built according to inflexible rules. If you gnawed away at the Windows 7 Start menu back in the day, you'll find that there's very little meat to the new Start menu. Conversely, the Windows 10 Start menu doesn't get screwed up as easily — or as completely — as the Windows 7 Start menu.

On the right side of the Start menu, you see a vast sea of tiles. Unlike the tiles on your iPhone or iPad or Galaxy, these tiles have some smarts: If prodded, they will tell you things that you might want to know, without opening up the associated app. In this screenshot, you can see a bit of the weather, a news story, a photo, a preview of an email message, and a little peek at the calendar. You also see lots and lots of ads. That's the Windows tile shtick, and it's apparent here in all its glory.

THE HISTORY OF THE HAMBURGER ICON

There have been many harsh words about the lowly hamburger. On the one hand, the icon doesn't really say anything. On the other hand, so many systems and programs now use the icon that it's close to being universal. Even cross-platform.

Ends up that the hamburger icon (like so many things we take for granted today) was designed at the Xerox Palo Alto Research Center, PARC, for use on the first graphical computer, the Xerox Star. Norm Cox designed it at PARC, and you can see its first appearance at https://vimeo.com/61556918. Software designer Geoff Alday contacted Cox, and this is what he said:

"I designed that symbol many years ago as a 'container' for contextual menu choices. It would be somewhat equivalent to the context menu we use today when clicking over objects with the right mouse button. Its graphic design was meant to be very 'road sign' simple, functionally memorable, and mimic the look of the resulting displayed menu list. With so few pixels to work with, it had to be very distinct, yet simple . . . we used to tell potential users that the image was an 'air vent' to keep the window cool. It usually got a chuckle, and made the mark much more memorable."

That's why, 30 years later, Windows 10 uses the hamburger icon that, when clicked, opens a contextual menu with options that differ based on your context and the app you use.

Whether you like having your news boiled down into a sentence fragment, that's for you to decide.

Unlike the left side of the Start menu, the right side with the tiles can get gloriously screwed up. You can stretch and move and group and ungroup until you're blue screened in the face.

ASK
WOODY.COM

I tend to think of the tiles on the right side of the Start menu as the next generation of Windows 7 Gadgets. If you ever used Gadgets, you know that they were small programs that displayed useful information on their faces. Microsoft banned them before releasing Windows 8, primarily because they raised all sorts of security problems.

Windows 10 Start menu tiles don't have the security problems. And the infrastructure that has replaced the Gadget mentality has taken Windows 10 to an entirely new level.

Getting Around in Windows

Switching to tablet mode and back

Get your computer going. Go ahead. I'll wait.

You're looking at the old-fashioned Windows desktop, right? (If you have a mouse and Windows sensed it, you're looking at the desktop. If your machine is only touch, you may be in tablet mode already.)

Time to take a walk on the wild side. Let's flip over to tablet mode. Way down in the lower-right corner, to the right of the date and time, click the Action/Notification Area icon. (Microsoft calls it an Action Area, but every other computer on the planet calls it a Notification area.) At the bottom, in the upper-left of the hive of shortcut icons, click Tablet Mode.

This (see Figure 1-2) is where the finger pickers live. They can tap and swipe and pinch and nudge to their heart's content.

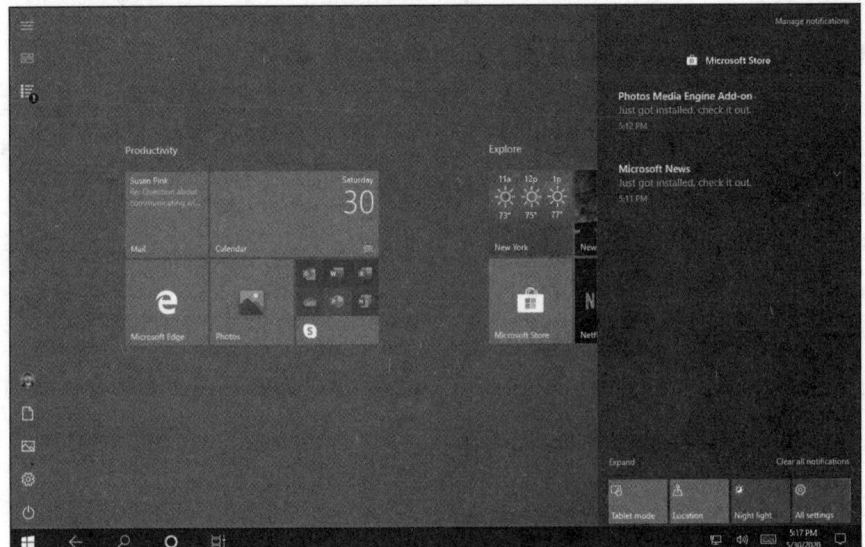

FIGURE 1-2:
Tablet mode, a good place for touch-first types.

Wait. Don't panic.

To get back to normal (I call it *desktop mode*), click or tap the Action/Notification Area icon in the lower-right corner, and click Tablet Mode once again. Like Dorothy tapping her heels together three times, you go back to where there's no place like Home.

REMEMBER

That brings you back to Figure 1-1. Which is probably where you wanted to be.

Although tablet mode is designed for people who want to use a touchscreen, not a mouse, there's no law that says you're stuck in one persona or the other. You can flip back and forth between regular mouse-first mode and tablet mode any time.

Navigating around the Desktop

Whether you use a mouse, a trackpad, or your finger, the desktop rules as your number-one point of entry into the beast itself.

Here's a guided tour of your PC, which you can perform with a mouse, a finger, or even a stylus, your choice:

1. **Click or tap the Start icon.**

You see the Start menu (refer to Figure 1-1).

2. **Tap or click the tile on the right marked Mail.**

You may have to Add an Email Account, but sooner or later, Microsoft's Windows 10 Mail app appears, as in Figure 1-3.

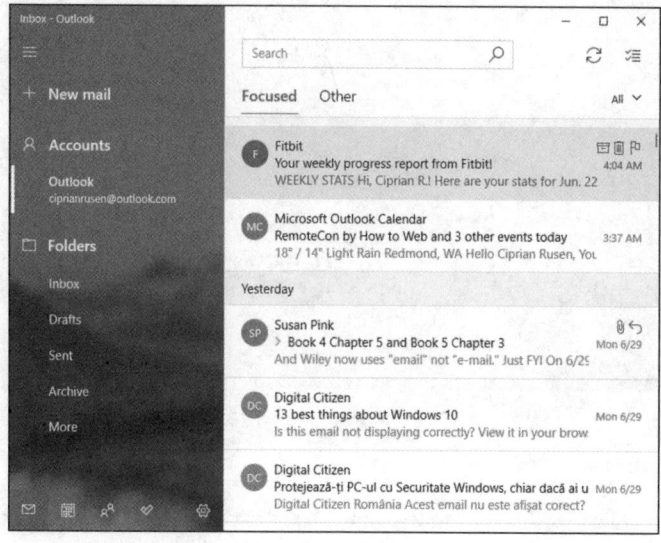

FIGURE 1-3:
The Mail app is indicative of the new Windows 10 apps.

3. **Take a close look at the Mail app window.**

Like other app windows, the Mail window can be resized by moving your mouse cursor over an edge and dragging. You can move the whole window by clicking the title bar and dragging. You can minimize the window — make it float down, to the taskbar — by clicking the horizontal line in the upper-right corner. And, finally, you can close the app by clicking the X in the upper right.

That may seem pretty trivial if you're from the Windows 7 side of the reality divide. But for Windows 8/8.1 veterans, the capability to move a Metro app window around is a Real Big Deal.

4. **At the bottom in the taskbar, to the right of the Search box and Cortana icon, click the Task View icon (shown in the margin).**

The desktop turns gray, and your Mail window shrinks a bit. Your Timeline— a kind of reminder of what you were doing once upon a time — appears, spread across the page. A New Desktop icon, shaped like a + sign, is at the top.

5. **Click the + (New Desktop) icon.**

Windows 10 creates a new, empty desktop, and shows it to you in task view. See Figure 1-4. Note how the Mail app shows up on Desktop 1, and Desktop 2 is blank except for your wallpaper.

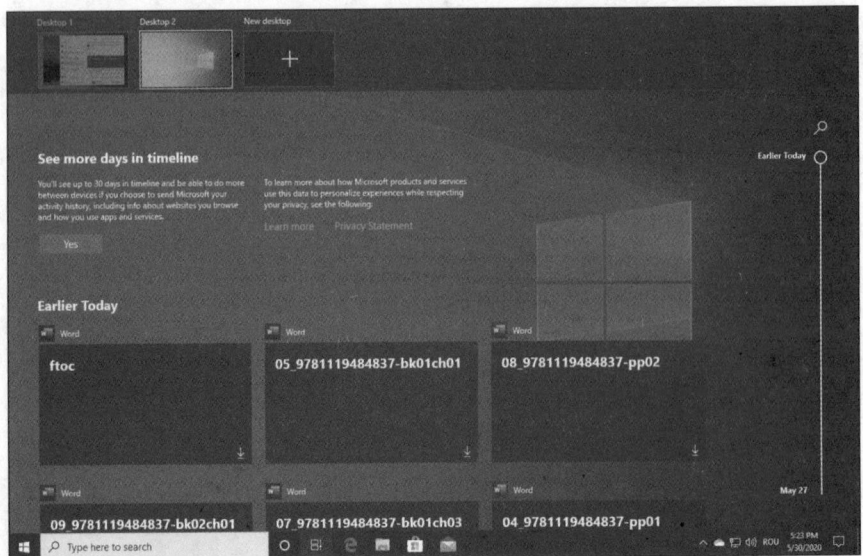

FIGURE 1-4:
Windows 10 lets you create as many desktops as you like.

6. **Click Desktop 2, on the right. Then click or tap Start, and choose the Weather app. Finally, click the Task View icon.**

Windows 10 pops back into task view, showing the Mail app running on Desktop 1 and the Weather app running on Desktop 2. In addition, the background for Desktop 2 has darkened, and you can see a slimmed-down version of the Weather app on the Desktop 2 desktop. See Figure 1-5.

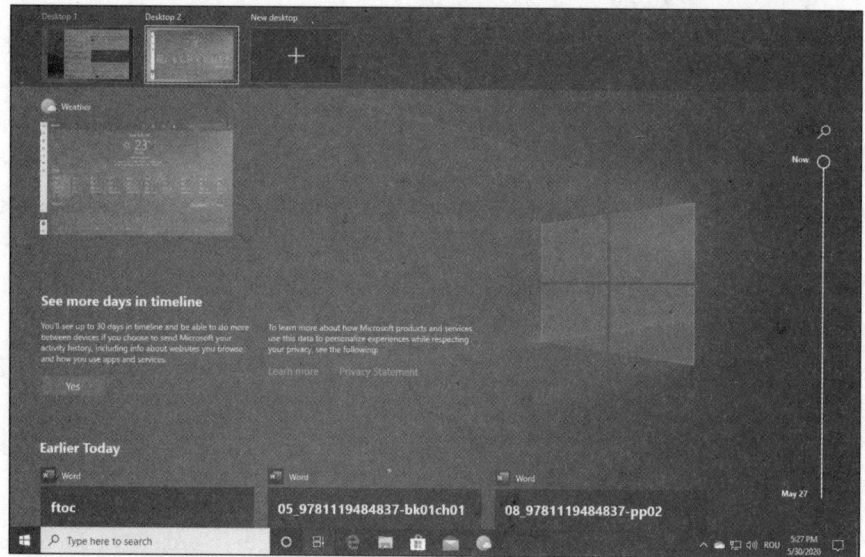

FIGURE 1-5:
Two desktops,
each with
different
programs
running.

7. **From the screen shown in Figure 1-5, right-click (or tap and hold down on) the running Weather app and choose Move to, Desktop 1. Then hover the mouse over the Desktop 1 thumbnail at the top, without clicking it.**

You've just successfully created a second desktop, and then moved a running application from one desktop to another. The results should look like Figure 1-6. That's a quick introduction to the Timeline, task view, and multiple desktops.

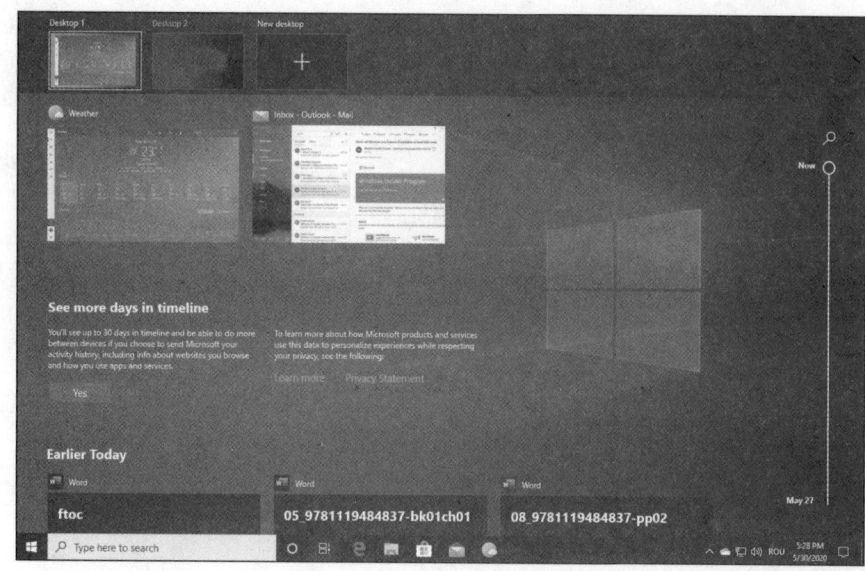

FIGURE 1-6:
Both of the
apps are running
happily on
Desktop 1.

Getting Around in
Windows

8. **Click the X button in the top-right corner of the Mail and Weather apps. Then click or tap the Start icon again.**

Windows 10 brings up a more-or-less alphabetized view of all your apps, in the second column of the Start menu.

9. **Scroll down to Windows Accessories, click the down arrow to its right, and then on Paint.**

The Paint app appears, just like in the good old days, as shown in Figure 1-7.

Note that the Start menu's apps list has a few collections of programs, like Windows Accessories. When you install new programs, they may build drop-down menus on the All Apps list, as you see with Windows Accessories, but far more commonly they just get dumped in the list. That gives you lots of stuff to scroll through.

Also note that the running app — Paint — has an icon down on the taskbar (in this case, in the middle) and shown in the margin. When you close Paint, the icon disappears. If you want to keep Paint on the taskbar, right-click the icon and choose Pin This Program to Taskbar. That'll save you a scroll-scroll-scroll trip through All Apps the next time you want to run Paint.

FIGURE 1-7:
The Paint app in Windows 10.

10. **Click the X button in the top-right of Paint. Again, click or tap Start. This time click or tap one of the alphabetizing indexes.**

For example, click the A above Alarms & Clock. Windows shows you a telephone-like index for all your apps entries, as you can see in Figure 1-8. If you were to click, say, W, you would be immediately transported to the W part of the All Apps list.

FIGURE 1-8:
The Start apps list has an index.

11. **Let's take a quick look at the other notable new Windows 10 apps: At the bottom of the screen, on the taskbar, click or tap the Edge icon (shown in the margin).**

The icon looks like a rolling wave. You're transported into Microsoft Edge, the new Internet browser from the folks in Redmond. See Figure 1-9.

Internet Explorer is still around if you really must use it: Just look in Start apps under Windows Accessories, not far from where you found Paint in Step 9.

REMEMBER

Microsoft has abandoned Internet Explorer. That's good because Internet Explorer has turned into a bloated, buggy, sinking piece of scrap. (You knew that already if you read any of my previous Windows *All-in-One For Dummies* books.) With Microsoft Edge, there's a chance that the 'Softies may actually stand a chance of one day competing against Google's Chrome browser and Mozilla Firefox — both of which I still recommend.

Play with Microsoft Edge for a bit — type something up in the address bar. Click or tap the + sign at the top, and add a new tab. Click some links. See how it works like a browser? Edge is actually a reasonably good browser, although it lacks some key features. See Book 5, Chapter 1 for much more info.

Getting Around in Windows

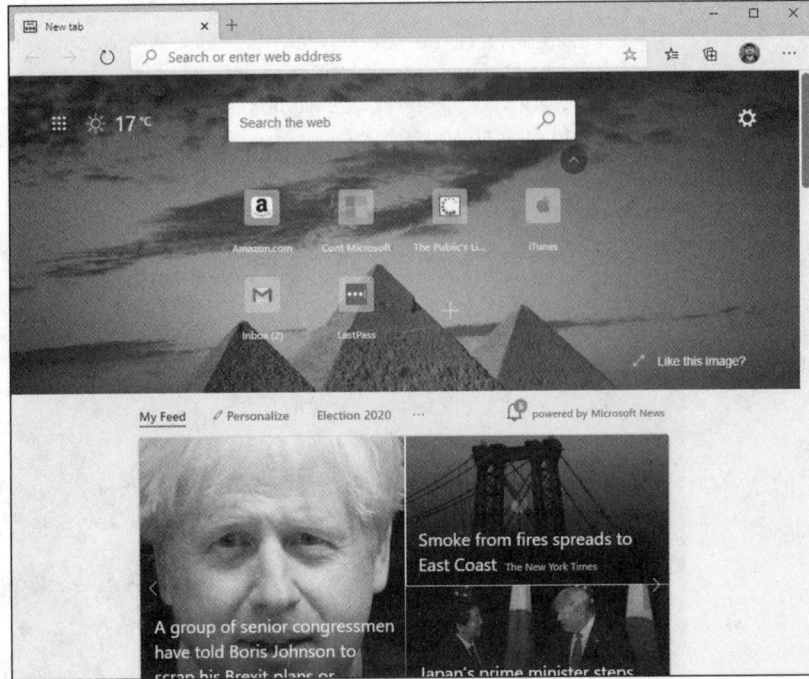

FIGURE 1-9:
Microsoft Edge
takes you straight
to adville; do not
pass go.

12. **If you haven't yet started with Cortana, give her a try. Click her icon near the Search box on the taskbar.**

    Cortana goes through some setup steps, which I describe in Book 3, Chapter 5. If you already have a Microsoft account, it's easy to get set up. (Note that you do need a Microsoft account to get personal information out of Cortana — that's how she/he/it stores your data for later retrieval.)

**TECHNICAL
STUFF**

    Cortana is now an independent app, but more than anything, Cortana is an extension of Bing, Microsoft's search engine. Anything you type in Cortana search — any sweet nothing you whisper in her ear — is destined for Bing.

13. **Click inside the Ask Cortana search box, and type** Tell me a joke**. Or click the microphone icon (Speak to Cortana) and say "Tell me a joke."**

    I won't attest to her sense of humor (see Figure 1-10), but Cortana has certainly been trained well. If you'd like more interesting things to ask Cortana, hop over to Google (sorry) and search for *Cortana questions*.

    That completes the canned tour of Windows 10 highlights. There's much, much more to discover — I only scratched a thin layer of epidermis.

    Take a breather.

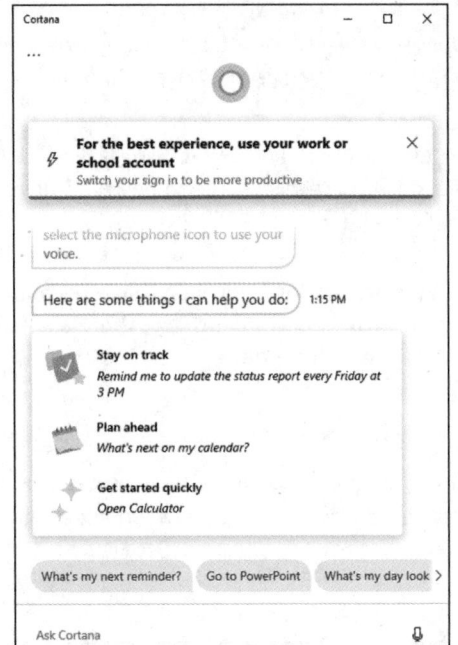

FIGURE 1-10:
Hey, Cortana
(pause, pause).
Tell me a joke!

Keying Keyboard Shortcuts

Windows 10 has about a hundred zillion — no, a googolplex — of keyboard shortcuts.

I don't use very many of them. They make my brain hurt.

REMEMBER

Here are the keyboard shortcuts that everyone should know. They've been around for a long, long time:

>> **Ctrl+C** copies whatever you've selected and puts it on the Clipboard. On a touchscreen, you can do the same thing in most applications by tapping and holding down, and then choosing Copy.

>> **Ctrl+X** does the same thing but removes the selected items — a cut. Again, you can tap and hold down, and Cut should appear in the menu.

>> **Ctrl+V** pastes whatever is in the Clipboard to the current cursor location. Tap and hold down usually works.

>> **Ctrl+A** selects everything, although sometimes it's hard to tell what "everything" means — different applications handle Ctrl+A differently. Tap and hold down usually works here, too.

>> **Ctrl+Z** usually undoes whatever you just did. Few touch-enabled apps have a tap-and-hold-down alternative; you usually have to find Undo on a ribbon or menu.

>> When you're typing, **Ctrl+B, Ctrl+I,** and **Ctrl+U** usually flip your text over to Bold, Italic, or Underline, respectively. Hit the same key combination again, and you flip back to normal.

REMEMBER

In addition to all the key combinations you may have encountered in Windows versions since the dawn of 19th–century dentistry, there's a healthy crop of new combinations. These are the important ones:

>> The **Windows key** brings up the Start menu.

>> **Alt+Tab** cycles through all running Windows programs, one by one — and each running Legacy desktop app is treated as a running program. (Windows key+Tab treats the entire desktop as one app.) See Figure 1-11.

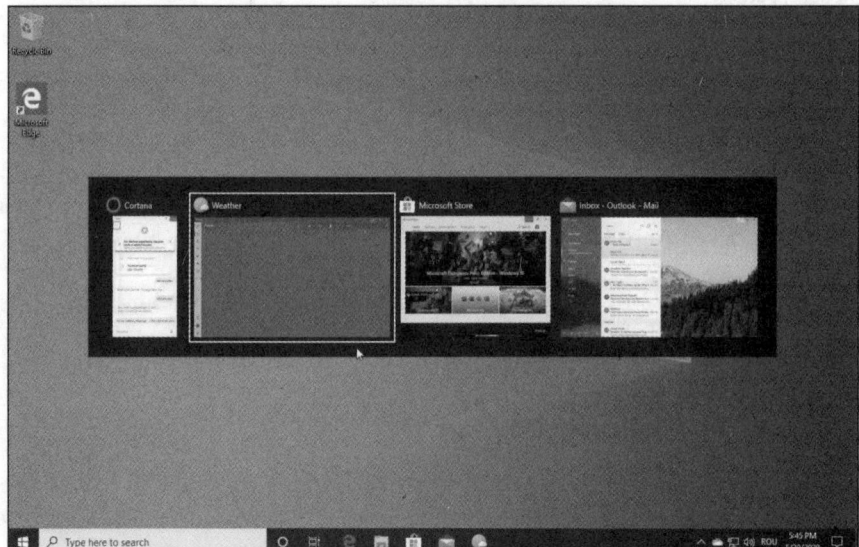

FIGURE 1-11:
Alt+Tab cycles through all running apps.

» **Ctrl+Alt+Del** — the old Vulcan three-finger salute — brings up a screen that lets you choose to lock your PC (flip to Book 2, Chapter 2), switch the user (see Book 2, Chapter 4), sign out, or run the new and much improved Task Manager (see Book 8, Chapter 4).

You can also right-click the Start icon or press Windows key+X to bring up the Power User menu shown in Figure 1-12.

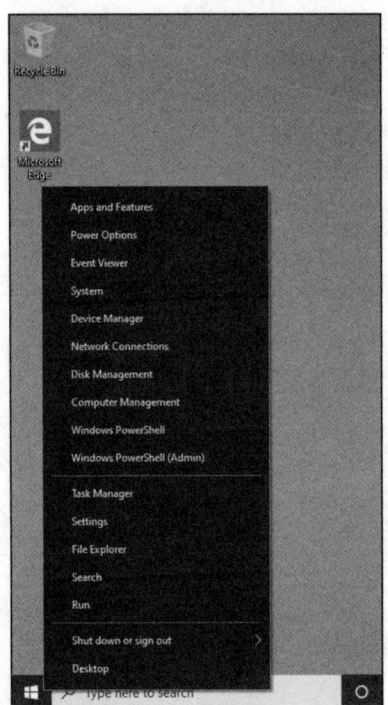

FIGURE 1-12:
The Win-X, or Power User, menu can get you into the innards of Windows 10.

And finally, the trick I know you'll use over and over. Starting with Windows 10 Fall Creators update, version 1709, there's a new built-in emoji keyboard. Simply click wherever you want to type an emoji, hold down the Windows key, and press the period. See Figure 1-13.

Who says Windows 10 isn't as cool as your smartphone? Only took Microsoft a decade or so.

Getting Around in Windows

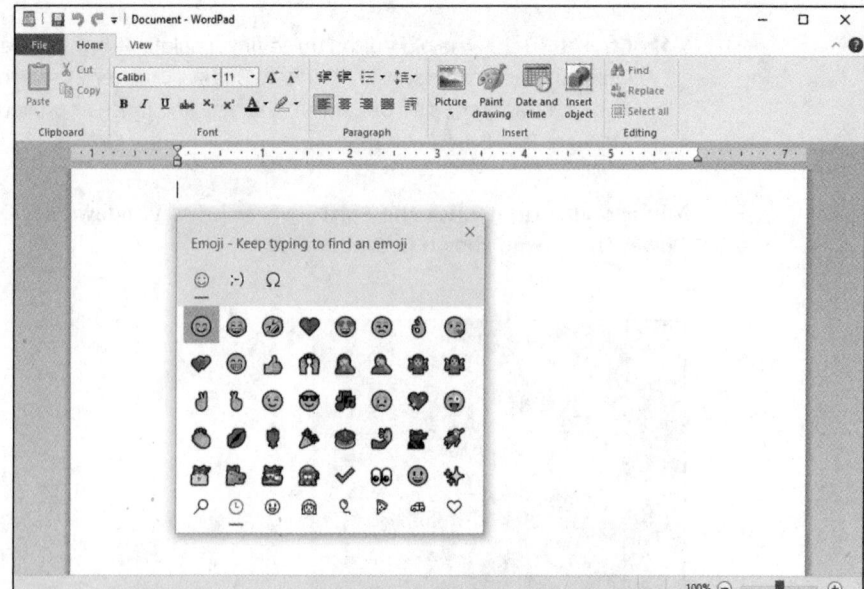

FIGURE 1-13:
Emojis are —
finally! — just
a keyboard
command away.

Chapter **2**

Changing the Lock and Login Screens

Windows 10 presents three hurdles for you to clear before you can get down to work (or play, or whatever):

» You have to get past the *lock screen.* That's a first-level hurdle so your computer doesn't accidentally get started, like the lock screen on a smartphone or an iPad.

» If more than one person — one *account* — is set up on the computer, you have to choose which person will log in. I go into detail about setting up user accounts in Book 2, Chapter 4.

» If a password's associated with the account, you must type it into the computer. Windows allows different kinds of passwords, which are particularly helpful if you're working on a touch-only tablet or a tiny screen like a telephone's. But the idea's the same: Unless you specifically set up an account without a password, you need to confirm your identity.

Only after clearing those three hurdles are you granted access to the desktop and, from there, to everything Windows 10 has to offer. In the sections that follow, you find out how you can customize the lock screen and the login methods to suit yourself.

WHAT'S NEW IN WINDOWS HELLO

Windows Hello gives an additional method for confirming your identity. Windows Hello uses biometric authentication — scanning your face or fingerprint or one day scanning your iris — as a much more secure method than passwords.

The camera version of Windows Hello technology has not hit the mainstream — Microsoft's Surface devices can log you in by recognizing your face, and a few high-end laptops also have the capability, but it's not common. Even expensive laptops with built-in cameras frequently skip that part of Windows Hello. There isn't that much demand — and many people get freaked out knowing their computer is watching, to log them in.

The fingerprint version of Windows Hello has become fairly common — many Windows computers with a recent fingerprint sensor use Hello.

These are the best-known laptops that support Windows Hello facial recognition: Alienware 15, Dell Inspiron 15 5570, ASUS ZenBook Duo, HP Spectre x360, Lenovo Miix 720, Samsung Notebook 9 Pro 15, and all Microsoft Surface family models after Surface Pro 4 and Surface Book 2, including the Surface Laptop. You can buy an add-on RealSense camera that'll support Windows Hello, but it's expensive.

Many devices support fingerprint recognition, but the specific kind of recognition demanded by Hello, once again, isn't common.

Only time will tell if Hello is reliable enough (and the hardware cheap enough!) to make a dent in the market.

If you have a pre-Windows 10 camera or fingerprint reader, chances are very good it won't work with Windows Hello. Many more details are in Microsoft's lengthy Passport guide, at https://technet.microsoft.com/en-us/itpro/windows/keep-secure/microsoft-passport-guide. (Passport is an on-again, off-again Microsoft brand that's being folded into Windows Hello.)

Working with the Lock Screen

The very first time you start Windows, and anytime you shut it down, restart, or let the machine go idle for long enough, you're greeted with the lock screen, such as the one in Figure 2-1.

FIGURE 2-1:
The Windows 10
lock screen.

6:27

Saturday, May 30

You can get through the lock screen by doing any of the following:

» Swiping up with your finger, if you have a touch-sensitive display.

» Clicking with your mouse.

» Pressing any key on your keyboard.

You aren't stuck with the lock screen Microsoft gives you. You can customize your picture and the little icons (or *badges*). The following sections explain how.

Using your own picture

Changing the picture for your lock screen is easy. (See the nearby sidebar "Individualized lock screens" for details about the difference between your lock screen and the system's lock screen.) Customizing the picture is a favorite trick at Windows 10 demos, so you know it must be easy, right? Here's how:

1. **Click or tap the Start icon, the Settings icon, and then Personalization.**

2. **On the left, choose Lock Screen.**

   The lock screen's Preview window appears.

3. **From the Background drop-down list, first try Windows Spotlight, if it's available (see Figure 2-2).**

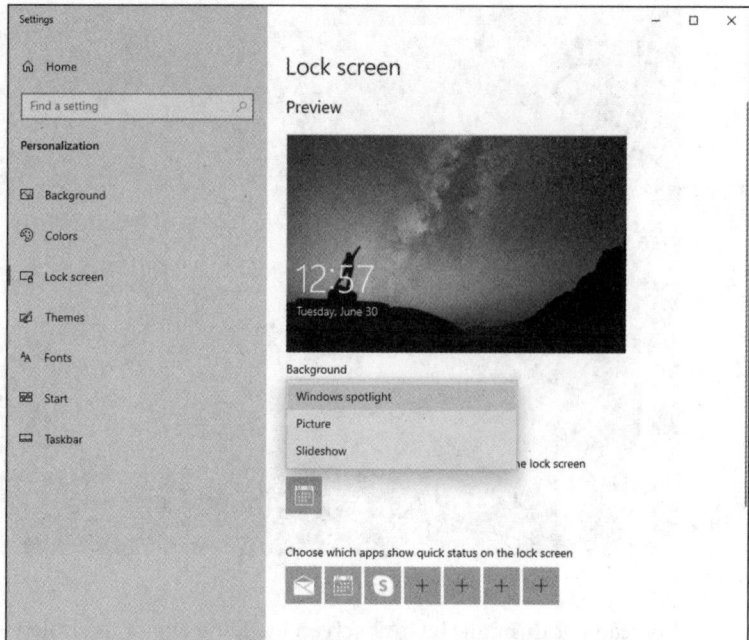

FIGURE 2-2:
Change your lock
screen here.

Windows Spotlight images come directly from Microsoft — more specifically, from Bing — and change frequently. Microsoft reserves the right to put advertising on Windows Spotlight screens, ostensibly to tell you about features in Windows 10 that you haven't used yet. Remains to be seen whether other, uh, partners can purchase spots on the screen.

4. **From the Background drop-down list, choose Picture.**

   This selection (see Figure 2-3) lets you choose which picture will appear. If you like one of the pictures on offer, click it. If you'd rather find your own picture, click Browse.

   You can decide whether you want your chosen picture to be overlaid with "fun facts, tips, tricks, and more on your lock screen." Oh goodie.

5. **If you find a picture you want, click it. If not, choose Slideshow in the Background drop-down box.**

   This option ties into the Albums in the Windows 10 Photos app (see Book 4, Chapter 3), or you can choose to turn a folder of pictures into a slideshow. If you decide to go with a slideshow, click the Advanced Slideshow Settings link to set whether the slideshow can be pulled from your camera roll, whether the chosen pictures have to be large enough to fit your screen, and several additional choices.

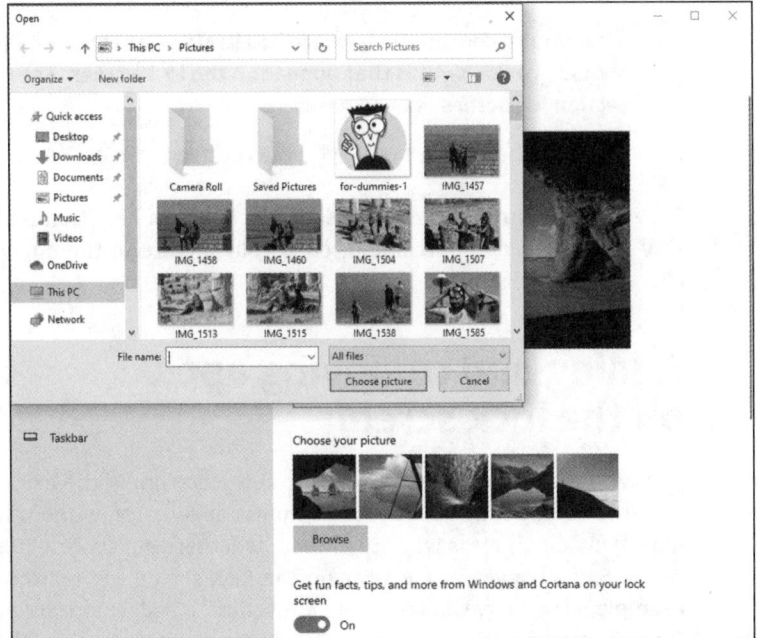

FIGURE 2-3:
Choose your own picture, with or without Microsoft advertising.

INDIVIDUALIZED LOCK SCREENS

If you read the Microsoft help documentation, you may think that Windows 10 keeps one lock screen for all users, but it doesn't. Instead, it has a lock screen for each individual user and one more lock screen for the system as a whole.

If you're using the system and you lock it — say, tap your picture on the Start menu and choose Lock — Windows 10 shows your personal lock screen, with the badges you've chosen. If you swipe or drag to lift that lock screen, you're immediately asked to provide your password. There's no intervening step to ask which user should log in.

If, instead of locking the system when you leave it, you tap your picture and choose Sign Out, Windows 10 behaves quite differently. It shows the system's lock screen, with the system's badges. Your lock screen and badges are nowhere to be seen. If you drag or swipe to go through the lock screen, you're asked to choose which user will log in.

Bottom line: If you change your lock screen using the techniques in this chapter, you change only *your* lock screen. Windows' idea of a lock screen stays the same.

6. **After you've chosen the background itself, you can specify what apps should provide details that appear on the lock screen. See the next section for details about *Badges*.**

You're finished. There's no Apply or OK button to tap or click.

Test to make sure that your personal lock screen has been updated. The easiest way is to go to the Start menu, click your picture in the upper-left corner, and choose Lock or Sign Out.

Adding and removing apps on the lock screen

Badges are the little icons that appear at the bottom of the lock screen. They exist to tell you something about your computer at a glance, without having to log in — how many email messages are unread, whether your battery needs charging, and so on. Some badges just appear on the lock screen, no matter what you do. For example, if you have an Internet connection, a badge appears on the lock screen. If you're using a tablet or laptop, the battery status appears; there's nothing you can do about it.

Mostly, though, Windows 10 lets you pick and choose quick status badges that are important to you. The question I most often hear about badges is, "Why not just choose them all?"

Good question. The programs that support the badges update their information periodically — every 15 minutes, in some cases. If you have a badge on your lock screen, the lock screen app that controls the badge has to wake up every so often, so it can retrieve the data and put it on the lock screen. Putting everything on the lock screen drains your computer's battery.

ASK
WOODY.COM

Corollary: If your computer has a short battery life, whittle your needs down as much as you can, and get rid of every quick status badge you don't absolutely need. But if your computer is plugged in to the wall, put all the badges you like on the lock screen.

Here's how to pick and choose your quick status badges:

1. Click or tap Start and then the Settings icon.

2. Choose Personalization. On the left, choose Lock Screen. On the right, scroll down.

TIP

At the bottom of the Lock screen settings are two rows of gray icons. You can see them in Figure 2-4.

The first icon points to a specially anointed app that shows detailed status information on the lock screen. You get only one. In Figure 2-1, I have the Calendar, which is the default choice.

The detailed status app has to be specially designed to display the large block of information shown in Figure 2-1.

3. **Tap or click the detailed status icon and choose which display badge you want to appear in that slot on the lock screen.**

TIP

Apps must be specially designed to display the badge information on the lock screen. You're given a choice of all the apps that have registered with Windows 10 as being capable of displaying a quick status badge on the lock screen. As you add more apps, some of them appear spontaneously on this list.

The second row, of seven icons, corresponds to seven badge locations at the bottom of the lock screen. They appear in order from left to right, starting below the time. In theory (although this doesn't always work), you can choose which badges appear, and where they appear, in order from left to right.

4. **Click each of the seven gray icons in turn and choose an app to show its status on the lock screen (see Figure 2-4).**

If you choose None, the gray icon gets a plus (+) sign, indicating that it isn't being used. No badge appears in the corresponding slot on the lock screen.

The quick status apps have to be built specifically to show their badges on the lock screen.

You're finished. There's no Apply or OK button to tap or click.

Go back out to the lock screen — click or tap the Start icon, choose your picture at the top, choose Lock — and see whether you like the changes. If you don't like what you see or you're worried about unnecessarily draining your battery with all the fluff, start over at Step 1.

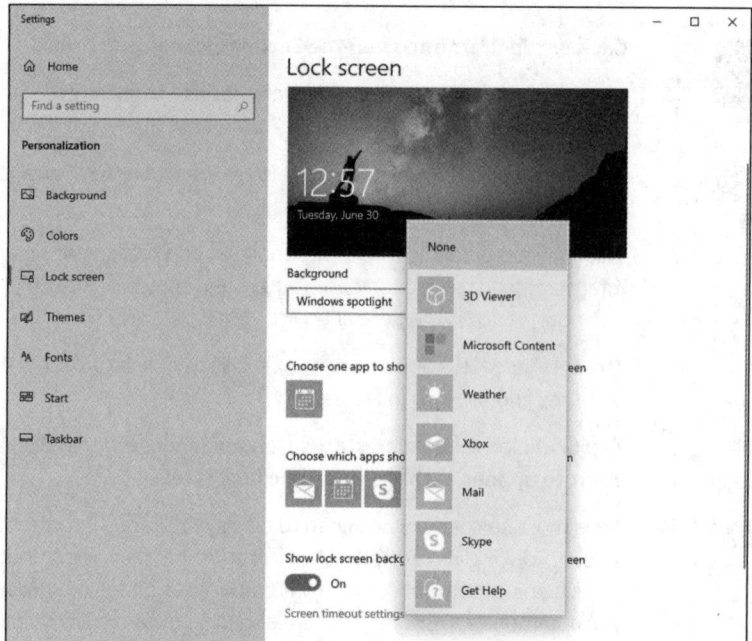

FIGURE 2-4:
Choose which
apps' badges
appear on the
lock screen.

Logging On Uniquely

In this section, I step you through setting up picture passwords and PINs, tell you how to show your face to Windows Hello, and give you a little hint about how you can bypass login completely, if you aren't overly concerned about other people snooping around on your PC. Yes, it can be done, quite easily.

Using a picture password

If you follow the instructions in Book 2, Chapter 4, set up an account, and the account has an everyday, ordinary password, you can use a picture password.

It's easy.

A *picture password* consists of two parts: First, you choose a picture — any picture — and then you tell Windows that you're going to draw on that picture in a particular way, such as taps, clicks, circles, and straight lines, with a finger or a mouse. The next time you want to log in to Windows, you can either type your password or you can repeat the series of clicks, taps, circles, and straight lines.

For example, suppose you have pictures of you and your friends, as shown in Figure 2-5, and you want your picture password to consist of tapping the forehead of the person in the middle, then the man on the left, and then the woman on the right, in that order.

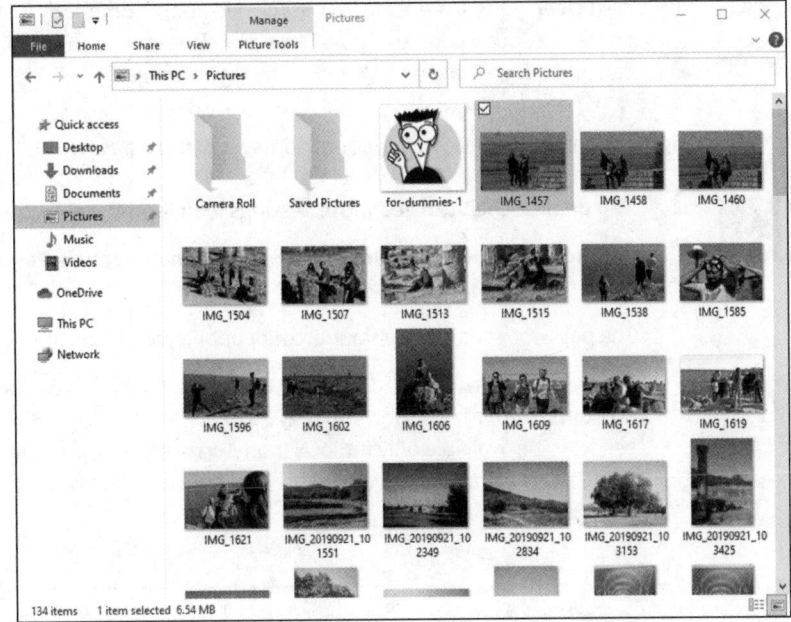

FIGURE 2-5:
Some of your photos from the Pictures folder will make a great picture password.

That picture password is simple, fast, and not easy to guess.

Everybody I know who has a chance to switch to a picture password or PIN loves it. Whether you're working with a mouse or a stubby finger, a few taps or slides are so much easier than trying to remember and type a17LetterP@ssw0rd.

Microsoft has a few suggestions for making your picture password hard to crack. These include the following:

>> **Start with a picture that has lots of interesting points.** If you have just one or two interesting locations in the photo, you don't have very many points to choose from.

>> **Don't use just taps (or clicks).** Mix things up. Use a tap, a circle, and a line, for example, in any sequence you can easily remember.

>> **Don't always move from left to right.** Lines can go right to left, or top to bottom. Circles can go clockwise or counterclockwise.

REMEMBER

>> **Don't let anybody watch you sign in.** Picture passwords are worse than keyboard passwords, in some respects, because the picture password appears on the screen as you're drawing it.

>> **Clean your screen.** Really devious souls may be able to figure out that trail of oil and grime is from your repeated use of the same picture password. If you can't clean your screen and you're worried about somebody following the grime trail, put a couple of gratuitous smudges on the screen. I'm sure you can find a 2-year-old who would be happy to oblige.

ASK WOODY.COM

Here's how to change your account to use a picture password:

1. **Tap or click the Start icon, the Settings icon, and then Accounts.**

2. **On the left, choose Sign-In Options and then on Picture Password, on the right.**

   The password settings for your account appear, as shown in Figure 2-6.

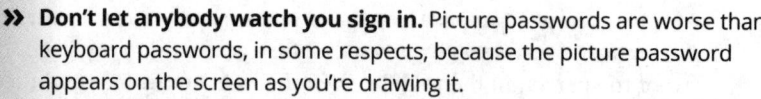
WARNING

   The picture password used to be available for all types of Windows 10 accounts. As of the Windows 10 May 2020 update, this type of password seems to be available only for local (non-Microsoft) accounts. If your account is associated with an email, you might not see Picture Password in your sign-in options.

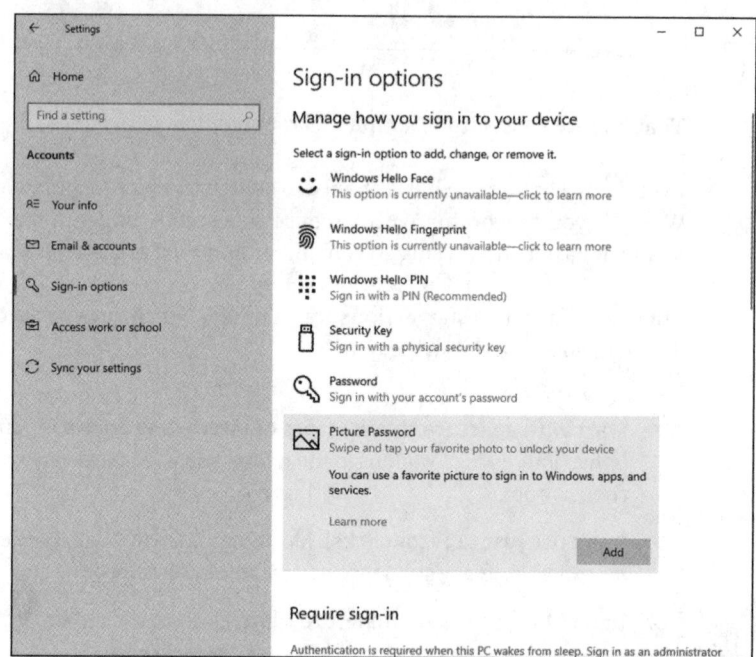

FIGURE 2-6:
Your account's sign-in options.

3. **Under Picture Password, tap or click Add.**

   If your account doesn't yet have a password, you're prompted to provide one. If you do have a password, Windows 10 asks you to enter it.

   You must have a typed password — the password can't be blank — or Windows 10 will just log you in without any password, either typed or picture.

4. **Type your password, and then tap or click OK.**

   Windows 10 asks you to choose a picture.

5. **Tap or click Choose Picture, find a picture (remember, with ten or more interesting points), and tap or click Open.**

   Your picture appears in a cropping bucket. The picture must conform to an odd shape, or it won't fit on the login screen.

6. **Slide the picture around to crop it the way you want. Then tap or click Use This Picture.**

   Windows 10 invites you to set up your gestures, as shown in Figure 2-7.

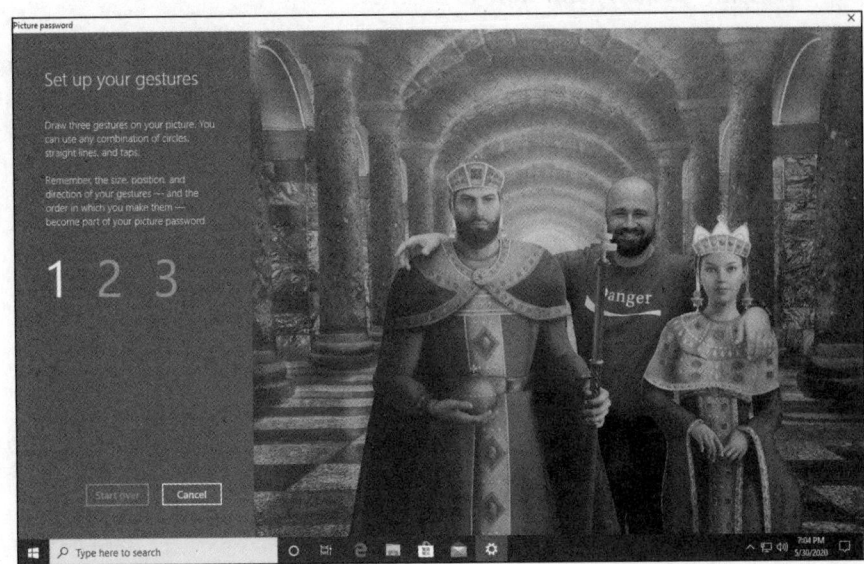

FIGURE 2-7:
Here's where you draw your three taps/clicks, lines, and circles.

7. **Trace out the gestures exactly as you want them.**

   Make sure the gestures are in the correct order and that each of the three consists of a click, a line, or a circle.

   Windows 10 then asks you to repeat your gestures. This is where you get to see how sensitive the gesture-tracking method can be.

8. **Repeat the gestures. When you get them to match (which isn't necessarily easy!), tap or click Finish.**

   Your new picture password is ready.

9. **Go to the Start menu, tap your picture, and choose Lock. Then click anywhere on the lock screen, and make sure you can replicate your gestures.**

REMEMBER

If you can't get the picture password to work, you can always use your regular typed password.

Creating a PIN

Everybody has PIN codes for ATM cards, telephones, just about everything.

WARNING

Reusing PIN codes on multiple devices (and credit cards) is dangerous — somebody looks over your shoulder, watches you type your Windows 10 PIN, and then lifts your wallet. Such nefarious folks can have a good time, unless the PINs are different. Word to the wise, eh?

PINs have lots of advantages over passwords and picture passwords. They're short and easy to remember. Fast. Technically, though, the best thing about a PIN is that it's stored on your computer — it's tied to that one computer, and you don't have to worry about it getting stored in some hacked database or stolen with your credit card numbers. In recent versions of Windows 10, the PIN is part of Windows Hello — Microsoft's service for secure authentication options, which improves with each major update. More on that in the next section of this chapter. For now, realize that creating a PIN is easy. Here's how to do it:

1. **Tap or click Start, the Settings icon, and then Accounts.**

2. **On the left, choose Sign-In Options.**

   The password settings for your account appear (refer to Figure 2-6).

3. **Click or tap Windows Hello PIN and then Add.**

   Windows 10 asks you to verify your password — it must be your typed password; a picture password won't do.

4. **Type your password, and tap or click OK.**

   Windows 10 gives you a chance to type your PIN, as in Figure 2-8, and then retype it to confirm it. *Note:* Most ATM PINs are four digits, but you can go longer, if you want — Windows 10 can handle just about any PIN you can throw at it.

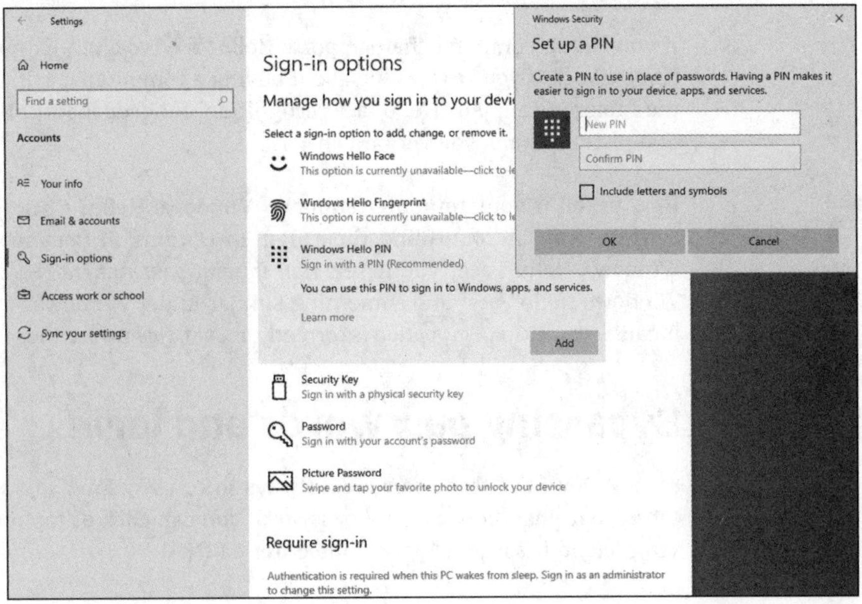

FIGURE 2-8:
Creating a PIN is easy.

5. **Type your PIN, confirm it, and tap or click OK.**

You can log in with your PIN.

Windows Hello

In a nutshell, Windows Hello offers biometric authentication — way beyond a password or a PIN. The Windows Hello technology includes fingerprint, face (and, soon, iris) recognition with a specially designed camera or fingerprint reader or both.

Microsoft is gradually implementing fingerprint recognition with older finger scanners as well. But the hallmark Hello scan for your shiny face is limited to fancy cameras, included with only a limited number of computers (see the "What's New in Windows Hello" sidebar at the start of this chapter).

ASK WOODY.COM

Frankly, I'm not a big fan of Windows Hello face recognition. I use it on the Surface Book and sometimes on the Surface Pro, but it isn't my cup of tea. Why? Many times, I sit in front of a PC and don't want to log in. Heresy, I know. But if I put my face anywhere near the Surface Pro when it's turned on, I'm caught like a deer in headlights — bang, there, I'm logged in. If I want to log in to a different account, I have to manually log out and then beat Hello to the punch, which is surprisingly difficult.

If you have a computer that supports Hello face recognition, give it a try, and see if you like it. If you're thinking about buying a computer specifically because it has the camera to support Hello face recognition, fuhgeddaboutit. I'll stick with a PIN or a picture, thank you very much.

How to tell if your computer supports Windows Hello? Click the Start icon, the Settings icon, Account, and then Sign-In Options. If your machine can handle Windows Hello Face, you'll see *Sign in with your camera (Recommended),* under Windows Hello Face, and something similar under Windows Hello Fingerprint. If it can't, you'll see *This option is currently unavailable* under those sign-in options.

Bypassing passwords and login

So now you have three convenient ways to tell Windows 10 your password: You can type it, just like a normal password; you can click or tap on a picture; or you can pretend it's a smartphone and enter a PIN.

But what if you don't want a password? What if your computer is secure enough — it's sitting in your house, it's in your safe deposit box, it's dangling from a vine over a pot of boiling oil — and you just don't want to be bothered with typing or tapping a password?

TIP

As long as you have a Local account, it's easy. Just remove your password. Turn it into a blank. Follow the steps in Book 2, Chapter 4 to change your password but leave the New Password field blank.

Microsoft accounts can't have blank passwords. But local accounts can.

If you have a blank password, when you click your username on the login screen, Windows 10 ushers you to the desktop.

If only one user is on the PC and that user has a blank password, just getting past the lock screen takes you to the desktop.

If you have a Microsoft account, you have to use your password (picture, PIN, Hello, whatever) once each time you reboot. If you don't want to be bothered after that, see the Require Sign-In drop-down choice at the top of the Sign-In options screen. Click to change the answer to "If you've been away, when should Windows require you to sign in again?" to Never.

Chapter **3**

Working with the Action/ Notification Center

I f you've ever used a moderately sentient smartphone or tablet, you already know about the notification center. Different devices do it differently, but the general idea is that the device watches and gathers notifications — little warning messages or status reports — that are sent to you. The smartphone or tablet gathers all the notifications and puts them in one place, where you can look at them and decide what to do from there.

In Windows 7, notifications just kind of flew by, and there weren't many of them. In Windows 8 and 8.1, you typically see many more notifications (I'm looking at you, Gmail running in Chrome), but they still fly by. There's no way in Windows 8 or 8.1 to look at old notifications. After they're off the screen — frequently for just a few seconds — that's it. And when they pile up, they can pile up and up and up and up, taking over the right edge of your screen.

**ASK
WOODY.COM**

Finally, with Windows 10, we have a place where the operating system collects and displays all the notifications. Or at least some of them. You know, like smartphones have had for a decade or so.

What Is the Action Center?

Unfortunately, this new locus for machine notifications isn't called a notification center, as it's called in almost every operating system, in almost every language, on earth. That name's taken. So we get a strange name for a common sight: It's officially called the *action center*, although everyone I know slips from time to time and calls it the notification center.

This isn't the Windows solution center (born in Windows 7, primarily for security stuff), nor is it the Windows 7 or Windows 8 action center (see Figure 3-1), which includes lots of system-related stuff, but no program notifications.

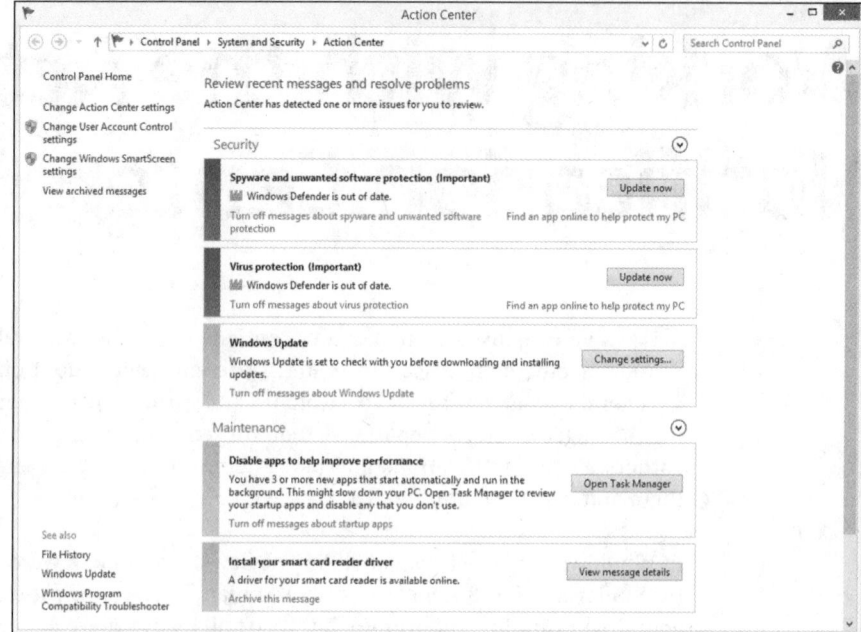

FIGURE 3-1:
The Windows 7/8/8.1 action center is not to be confused with the Windows 10 action center.

Instead, the Windows 10 action center is, well, a real notification center. Click the icon down in the lower right of the screen (and shown in the margin) and you can see the action in Figure 3-2.

At the top, Windows 10 gathers some (but by no means all) of the various programs' notifications. At the bottom, you have a bunch of links to various settings. These links are called *quick actions* and can be personalized from a list or predefined items.

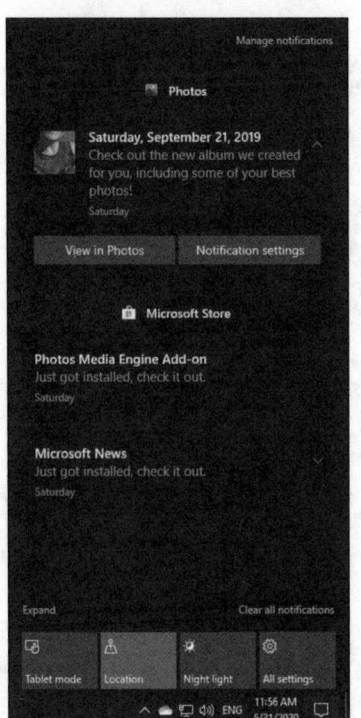

FIGURE 3-2:
The Windows
10 notific. . . err,
action center.

What, Exactly, Is a Notification?

ASK WOODY.COM

Historically, Windows allowed all sorts of notifications: blinking taskbar tiles, balloon messages over the system time (in the lower-right corner), dire-looking icons in the system notification area (near the system time), or dialog boxes that appear out of nowhere, sometimes taking over your computer. Then came Windows 8, and the powers that be started looking down on programs that jilted and cavorted, whittled and wheezed. People who write the programs have gradually become more disciplined.

Except for Scottrade, Figure 3-3, which locks out the screen, but that's another modal dialog story.

These new, politically correct notifications — the things that can happen when Windows 10 or one of its programs wants your attention — fall into three broad categories:

>> They can put rectangular notices, usually gray, in the upper-right edge of your screen, with a few lines of text. Typically, the notifications say things such as *Tap* or *Choose what happens when you insert a USB drive,* or *Turn sharing on or off.*

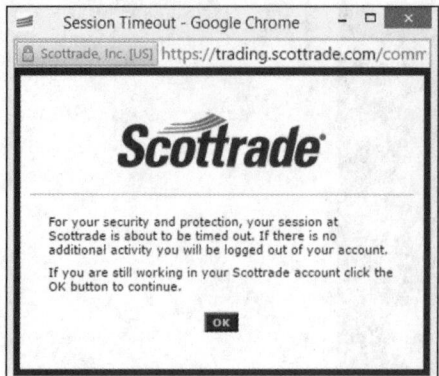

FIGURE 3-3:
Scottrade's
notifications,
generated by its
web app, lock
up the system
entirely. Those
are not nice
notifications.

These notifications are called *toaster notifications* (or sometimes just *toast*), and they're a core part of the new Universal face of Windows 10. It's a fabulous name because they pop up, just like toast, but on their sides, and then they disappear.

>> They can show toaster notifications on the lock screen. This is considered more dire than simply showing the notifications on tiled apps or the desktop. Why? Because the apps that create lock screen notifications may need to run, even when Windows 10 is sleeping. And that leads to battery drainage.

>> They can play sounds. Don't get me started.

Everyone has picked up on Windows 10's notification system. Even websites can now send you notifications when new content is posted. When an app sends you a notification, it appears briefly in the bottom-right corner of the screen. In Figure 3-4, you can see a notification about an email I've received.

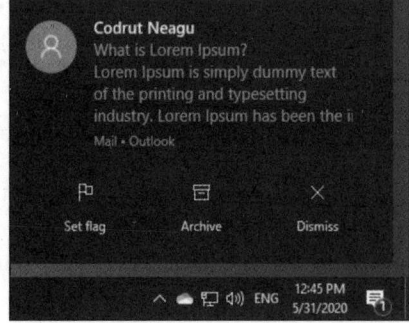

FIGURE 3-4:
A notification
for an email I
received.

When you click the notification — if you're fast enough — it opens the email for you, and puts it in a separate window, ready for you to reply. That's exactly the kind of response you should expect from your notifications — click them, and they do something appropriate. Some notifications display contextual options. For example, notifications from the Mail app have options for flagging, archiving, or dismissing an email you've received.

These notifications can be sent by apps, websites (through the web browser, when you've approved the receiving of notifications), and Windows 10 itself. As time goes by, more and more apps will be integrated with Windows 10's action center.

Working with Notifications

In earlier versions of Windows 10, clicking a notification rarely accomplished anything useful. Due to user feedback, Microsoft has improved the way the action center is organized and displays information. In recent Windows 10 versions, the system is usable and useful.

**ASK
WOODY.COM**

With the Anniversary update of Windows 10, version 1607, the Windows 10 action center became just a little more useful, primarily because you can now click a notification and expect something worthwhile to happen. In the November 2019 update of Windows 10, Microsoft finally allowed users to set the kind of notifications they get from their apps.

If you find that a particular program is generating notifications that you don't want to see, Windows 10 lets you disable all notifications rather easily, or you can pick and choose which apps can send notifications and which just have to stifle their utterances.

Here's how to disable notifications:

1. **Tap or click the Start icon, the Settings icon, and then System.**

Or you can get into Settings from the bottom of the action center (refer to the All Settings icon near the bottom right of Figure 3-2).

2. **On the left, choose Notifications & Actions.**

You see the Notifications & Actions pane shown in Figure 3-5.

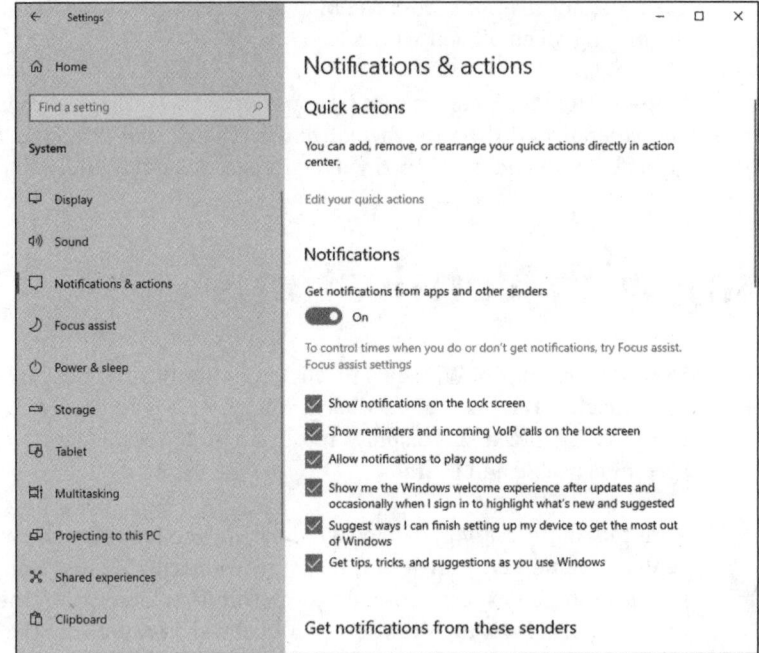

FIGURE 3-5:
Make changes
to how Windows
10 displays
notifications.

3. **Turn off all notifications by scrolling down and finding the slider marked Get Notifications from Apps and Other Senders and then sliding it to Off.**

4. **If you would like to silence just one app, scroll down farther (see Figure 3-6), find the app, and move its slider to Off.**

You're finished. There's no Apply or OK button to tap or click.

TIP

At Step 4, if you click or tap the name of an app instead of moving its slider, you get access to options for controlling how its notifications are displayed. For example, you can disable the sound played for each notification, change the number of notifications visible in the action center for that app, set its priority, and more.

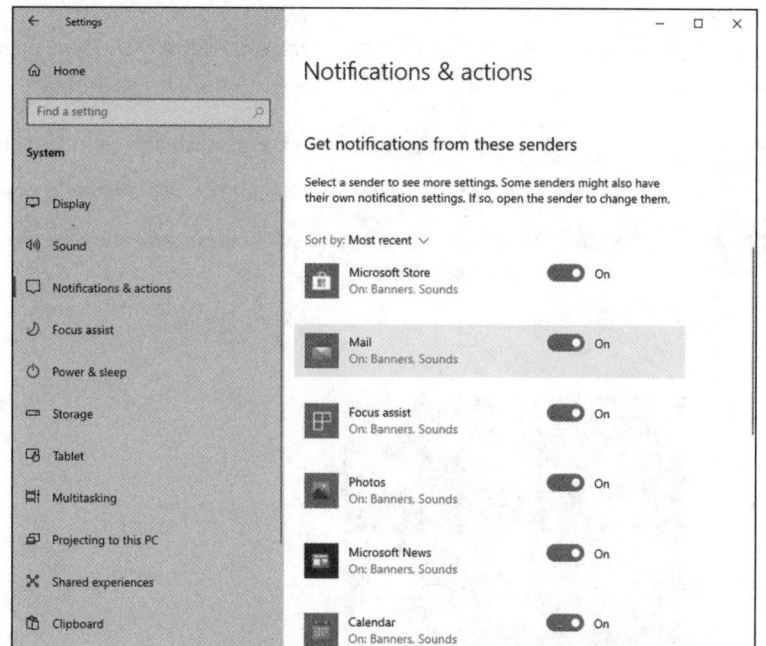

FIGURE 3-6:
You can silence notifications from individual apps.

Working with Settings Shortcuts

The action center contains a gob (that's a technical term) of shortcuts at the bottom of the Notifications pane. In Figure 3-2, I count four of them. If the computer used a Wi-Fi connection, there would have been additional shortcuts for Wi-Fi, Bluetooth, rotation lock, battery saver, and brightness. The maximum number of shortcuts available varies depending on whether you're using a desktop PC, laptop, tablet, or 2-in-1; its hardware configuration; and the apps installed.

Quick actions mimic what you would find on a smartphone — airplane mode is an obvious analog — all readily accessible from the right side of the screen. In many cases, a quick action displays a Settings page, where you can change the individual setting, displays a pane on the right side of the screen (Connect, Project), or toggles a specific setting in, uh, Settings. The Screen Snip quick action opens the Snip & Sketch app, which you can use to take screenshots.

ASK
WOODY.COM

You can think of quick actions as handy shortcuts to frequently adjusted settings, or you can look at them as a testimony to the diverse way Windows 10 has settings scattered all over Hades's half acre. You decide. Too bad you can't add your own quick actions. The action center be a convenient place to stick your own favorite programs.

You have some — but not much — control over which icons appear at the bottom of the pane. Here's how to exert as much influence as you can:

1. **Tap or click the Start icon, the Settings icon, and then System.**

   Or you can get into Settings from the bottom of the action center.

2. **On the left, choose Notifications & Actions, and then on the right click the Edit Your Quick Actions link.**

   The Notifications pane appears (refer to Figure 3-7).

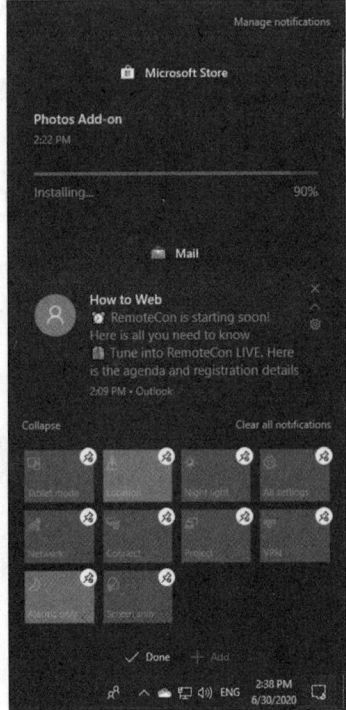

FIGURE 3-7:
Editing the list
of quick actions
available in the
action center.

3. **To rearrange the icons for quick actions, just click and drag.**

   You can't drag a quick action off the grid.

4. **To choose additional quick action icons, click or tap the Add+ button, and make a selection from the list that appears.**

5. **To remove a quick action icon, click or tap its pin on the top-right.**

6. **When you've set things the way you want them, tap Done.**

Table 3-1 explains what each of the configurable quick action icons does.

TABLE 3-1   **Some Quick Action Icon Results in the Action Center**

Click This Icon	And This Happens
Tablet Mode	Flips the computer to tablet mode.
Brightness	Adjusts the screen brightness to the level you want.
Connect	Searches for wireless display and audio devices — Miracast in particular.
All Settings	Takes you to the Settings app.
Battery Saver	Cycles between two battery saver modes, dimming the display. It doesn't work if the machine is plugged in.
VPN	Displays the Settings app's Network & Internet section on VPN, where you can add a new VPN connection or connect to an existing one.
Bluetooth	Turns Bluetooth on and off.
Rotation Lock	Prevents the screen from rotating from portrait to landscape and vice versa.
Wi-Fi	Turns Wi-Fi on and off. There's no provision to select a Wi-Fi connection.
Location	Turns the location setting on and off in the Settings app's Privacy, Location pane.
Night Light	Enables the night light, which filters the blue light emitted by the screen. Useful when working during the night.
Focus Assist	Turns focus assist on or off. When turned on, all notifications are blocked.
Screen Snip	Starts the Snip & Sketch app, for taking a quick screenshot.
Airplane Mode	Turns all wireless communication on and off. See the Settings app's Network & Internet, Airplane Mode setting.
Project	Projects the image on your screen to an external display or a projector.

Chapter **4**

Controlling Users

icrosoft reports that 70 percent of all Windows PCs have just one user account. That's a startling figure. It means that 70 percent of all Windows PCs run at the most permissive security level, all the time. It means that, on a large portion of all Windows PCs, little Billy can install Internet Antivirus 2011 — a notorious piece of scumware — and have it bring down the whole family with a couple of simple clicks. "Sorry, Dad, but it's an antivirus program, and it said that we really need to install it, and it's just $49.95 for a three-month subscription. I thought you said that antivirus was good. They wouldn't lie about stuff like that, would they?"

ASK
WOODY.COM

Although it's undoubtedly true that many PCs are each used by just one person, I think it's highly likely that people don't set up multiple user accounts on their PCs because they're intimidated. Not to worry. I take you through the ins and outs.

Even if you're the only person who ever uses your PC, you may want to create a second account — another user, as it were — even if the second user is just you. (As Pogo said, "We have met the enemy, and he is us.") Then again, you may not. And therein lies this chapter's story.

REMEMBER

If you're running Windows 10 Enterprise or Windows 10 Pro and your PC is connected to a big corporate network (in the parlance, a *domain*), you have little or no control over who can log in to your computer and what a logged-in user can do after she's on the machine. That's a Good Thing, at least in theory: Your company's network administrator gets to worry about all the security issues, relieving you of the hassles of figuring out whether the guy down the hall should be able to look at payroll records or the company Christmas card list. But it can also be a pain in the neck, especially if you have to install a program, like, right now, and you don't have a user account with sufficient capabilities. If your computer is attached to a domain, your only choice is to convince (or bribe) the network admin to let you in.

The nostrums in this chapter apply only to PCs connected to small networks or to stand-alone PCs. If you're on a big network, you must pay homage to the network gods. Pizza, beer, and a smile can help.

ASK
WOODY.COM

Windows 10 has two separate locations that control user accounts. If you want to do only some simple stuff — create a new account, change the password, or switch to a picture password, say — you can do it all on the touch-friendly Settings side of Windows 10. On the other hand, if you want to do something more challenging — set the User Account Control trigger levels, for example — you must work with the old-fashioned Windows 7–style Control Panel. I show you how to use both in this chapter.

User Account Control is a security topic, only tangentially related to user accounts. I talk about it in Book 9, Chapter 3.

Why You Need Separate User Accounts

Windows 10 assumes that, sooner or later, more than one person will want to work on your PC. All sorts of problems crop up when several people share a PC. I set up my screen just right, with all my icons right where I can find them, and then my son comes along and plasters the desktop with a shot of Alpha Centauri. He puts together a killer Taylor Swift playlist and "accidentally" deletes my Grateful Dead playlist in the process.

It's worse than sharing a TV remote.

Windows 10 helps keep peace in the family — and in the office — by requiring people to log in. The process of *logging in* (also called *signing in*) lets the operating system keep track of each person's settings: You tell Windows 10 who you are, and it lets you play in your own sandbox.

REMEMBER

Having personal settings that are activated whenever you log in to Windows 10 doesn't create heavy-duty security. Unless your PC is a slave to a big Active Directory domain network, your settings can get clobbered and your files deleted, if someone else with access to your computer or your network tries hard enough. But as long as you're reasonably careful and follow the advice in this chapter, Windows security works surprisingly well.

WARNING

If someone else can put his hands on your computer, it isn't your computer anymore. That can be a real problem if someone swipes your laptop, if the cleaning staff uses your PC after hours, or if a snoop breaks into your study. Unless you use BitLocker (in Windows 10 Pro), anybody who can restart your PC can look at, modify, or delete your files or stick a virus on the PC. How? In many cases, a miscreant can bypass Windows 10 directly and start your PC with another operating system. With BitLocker out of the picture, compromising a PC doesn't take much work.

Choosing Account Types

When dealing with user accounts, you bump into one existential fact of Windows life over and over again: The type of account you use puts severe limitations on what you can do.

Unless you're hooked up to a big corporate network, user accounts can generally be divided into two groups: the haves and the have-nots. (Users attached to corporate domains are assigned accounts that can exist anywhere on the have-to-have-not spectrum.) The have accounts are *administrator* accounts. The have-nots are *standard* accounts. That's it. Standard. Kinda makes your toes curl just to think about it.

What's a standard account?

A person running with a standard account can do only, uh, standard tasks:

>> Run programs installed on the computer, including programs on USB/key drives.

>> Use hardware already installed on the computer.

>> Create, view, save, modify, and use documents, pictures, and sounds in the Documents, Pictures, or Music folders as well as in the PC's Public folders.

>> Change his password or switch back and forth between requiring and not requiring a password for his account. He can also add a picture or PIN password. If your computer is sufficiently enabled, he can also use Windows Hello to set up a camera, fingerprint, or retina scan. Just like in the movies.

>> Switch between an offline (local) account and a Microsoft account. I talk about both in the next section of this chapter.

>> Change the picture that appears next to his name on the Welcome screen and on the left side of the Start menu, change the desktop wallpaper, resize the Windows toolbar, add items to the old-fashioned desktop toolbar and Start menu, and make other small changes that don't affect other user accounts.

In most cases, a standard user can change systemwide settings, install programs, and the like, but only if he can provide the username and password of an administrator account.

If you're running with a standard account, you can't even change the time on the clock. It's quite limited.

There's also a special, limited version of the standard account called a *child account*. As the name implies, child accounts can be controlled and monitored by those with standard and administrator accounts. See the sidebar on child accounts.

CHILD ACCOUNTS

Microsoft provides a quick-and-dirty way to set up child accounts as part of the account creation process. Child accounts are like standard accounts, but they're automatically set up with child protection enabled — someone with an administrator account can control which websites the child accounts can access, what time of day the accounts can be used, and the total amount of time the accounts are used in a day.

It's all done on the web — the controls aren't in Windows 10 itself, they're in a website maintained by Microsoft. There's a small charge for each child account that you set up. Note that laws in various places — including COPPA in the US — require that an account for anyone under 13 has to be associated with a guardian who controls a child. There's a 44-page synopsis of the COPPA regulations at www.ftc.gov/system/files/2012-31341.pdf. Easy reading for a parent wanting to set up an account for the kids.

Full instructions for bringing a Windows 7 or 8 child account into Windows 10 are at http://windows.microsoft.com/en-us/windows-10/set-up-family-after-upgrade.

What's an administrator account?

REMEMBER

People using administrator accounts can change almost anything, anywhere, at any time. However, certain folders remain off limits, even to administrator accounts, and you must jump through some difficult hoops to work around the restrictions. People using administrator accounts can even change other offline/ local accounts' passwords — a good thing to remember if you ever forget your password.

If you start Windows 10 with a standard account and you accidentally run a virus, a worm, or some other piece of bad computer code, the damage is usually limited: The malware can delete or scramble files in your Documents folder, and probably in the Public folders, but that's about the extent of the damage. Usually. Unless it's exceedingly clever, the virus can't install itself into the computer, so it can't run repeatedly, and it may not be able to replicate. Poor virus.

Someone with an administrator account can get into all the files owned by other users: If you thought that attaching a password to your account and putting a top-secret spreadsheet in your Documents folder would keep it away from prying eyes, you're in for a rude surprise. Anybody who can get into your machine with an administrator account can look at it. Standard users, on the other hand, are effectively limited to looking only at their own files.

Choosing between standard and administrator accounts

REMEMBER

The first account on a new PC is always an administrator account. If you bought your PC with Windows 10 preinstalled, the account that you have — the one you probably set up shortly after you took the computer out of the box — is an administrator account. If you installed Windows 10 on a PC, the account you set up during the installation is an administrator account.

When you create new accounts, on the other hand, they always start out as standard accounts. That's as it should be.

Administrator accounts and standard accounts aren't set in concrete. In fact, Windows 10 helps you shape-shift between the two as circumstances dictate:

>> If you're using a standard account and try to do something that requires an administrator account, Windows 10 prompts you to provide the administrator account's name and password or PIN (see Figure 4-1).

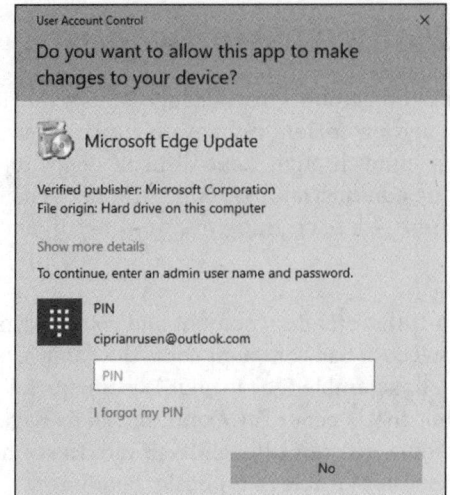

FIGURE 4-1:
Windows asks
permission
before
performing
administrative
actions.

WARNING

If the person using the standard account selects an administrator account without a password, simply clicking the Yes button allows the program to run — one more reason why you need passwords on all your administrator accounts, eh?

>> Even if you're using an administrator account, Windows 10 normally runs as though you had a standard account, in some cases adding an extra hurdle when you try to run a program that can make substantial changes to your PC — and *substantial* is quite a subjective term. You have to clear the same kind of hurdle if you try to access folders that aren't explicitly shared (see Figure 4-2). That extra hurdle helps prevent destructive programs from sneaking into your computer and running with your administrator account, doing their damage without your knowledge or permission.

TIP

Some experts recommend that you use a standard account for daily activities and switch to an administrator account only when you need to install software or hardware or access files outside the usual shared areas. Most experts ignore their own advice: It's the old do-as-I-say-not-as-I-do syndrome.

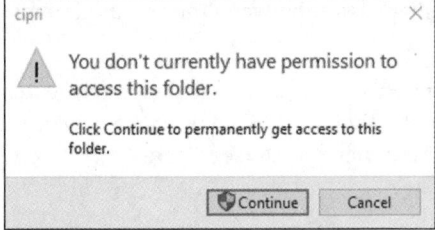

FIGURE 4-2:
Windows lays
down a challenge
before you dive in
to another user's
folder.

I used to recommend that people follow the lead of the do-as-I-say crowd and simply set up every knowledgeable user with an administrator account. Times change and Windows has changed: You rarely need an administrator account to accomplish just about anything in "normal" day-to-day use. (One exception: You can add new users only if you're using an administrator account.) For that reason, I've concluded that you should save that one administrator account for a rainy day, and set up standard accounts for yourself and anyone else who uses the PC. Run with a standard account, and I bet you seldom notice the difference.

What's Good and Bad about Microsoft Accounts

In addition to administrator and standard accounts (and child accounts, which are a subset of standard accounts), Microsoft also has another pair of account types, *Microsoft accounts,* and *offline* (formerly known as *local*) *accounts.* You can have an administrator account that's a Microsoft account or a standard account that's a Microsoft account or an administrator account that's an offline/local account, and so on. If you aren't confused, you obviously don't understand. Heh heh heh.

The basic differentiation goes like this:

>> **Microsoft accounts** are registered with Microsoft. Most people use their @ hotmail.com or @live.com or @outlook.com email addresses. Still, you can register any email address at all as a Microsoft account (details in the next chapter). Microsoft accounts must have a password.

When you log in to Windows 10 with a Microsoft account, Windows goes out to Microsoft's computer in the clouds and verifies your password, and then pulls down many of your major Windows 10 settings and transfers them to the PC you just logged in to. You can control which settings get synced in the Settings app (Start➪Settings➪Accounts➪ Sync Your Settings) — see Figure 4-3.

If you change, say, your background, the next time you log in to Windows 10 — from any machine, anywhere in the world — you see the new background. More than that, if the Microsoft account is set up to do so, you can get immediate access to all your music, email, OneDrive storage, and other Windows 10 features without logging in again.

>> **Offline** (or **Local**) **accounts** are regular, old-fashioned accounts that exist only on this PC. They don't save or retrieve your settings from Microsoft's computers. Offline/Local accounts may or may not have a password.

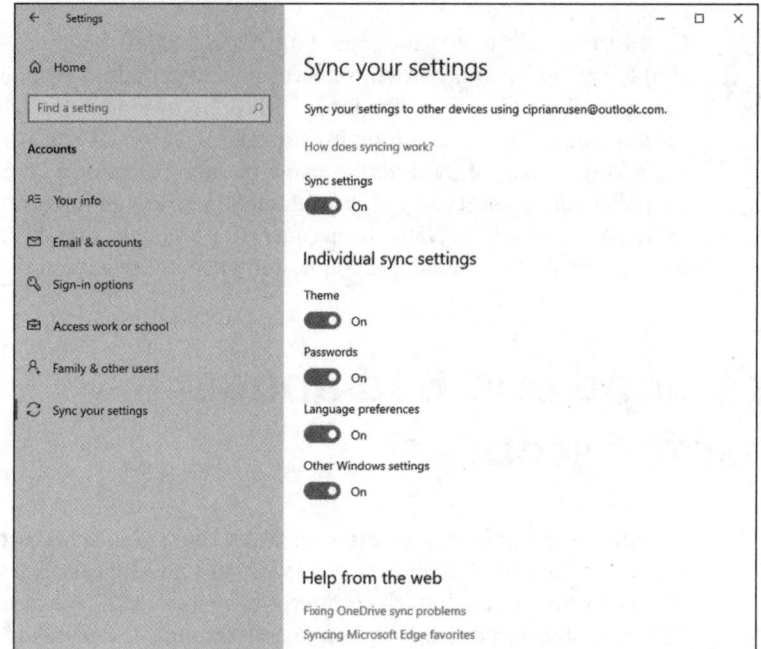

FIGURE 4-3:
Control which
Windows 10
settings get
synced across
your Microsoft
account.

REMEMBER

On a single PC, administrator accounts can add new users, delete existing users, or change the password of any offline/local account on the computer. They can't change the password of any Microsoft accounts.

WARNING

As you may imagine, privacy is among the several considerations for both kinds of accounts. I go into the details in Book 2, Chapter 5.

Microsoft accounts are undeniably more convenient than offline/local accounts. Sign in to Windows 10 with your Microsoft account, and many of your apps will just realize who you are, pull in your email, sync your storage, and much more. On the other hand, using a Microsoft account means that Microsoft has a log of many of your interactions with your machine — when you signed in, how you used the Microsoft apps (including Edge), Bing search results, and so on. The Microsoft account login also lets Microsoft associate your account with a specific electronic address and IP address (see Book 2, Chapter 5).

Is the added convenience worth the erosion in privacy? Only you can decide.

Adding Users

After you log in to an administrator account, you can add more users quite easily. Here's how:

1. **Click or tap the Start button and then the Settings shortcut.**

2. **On the Settings window, click or tap Accounts.**

The Accounts screen appears, as shown in Figure 4-4.

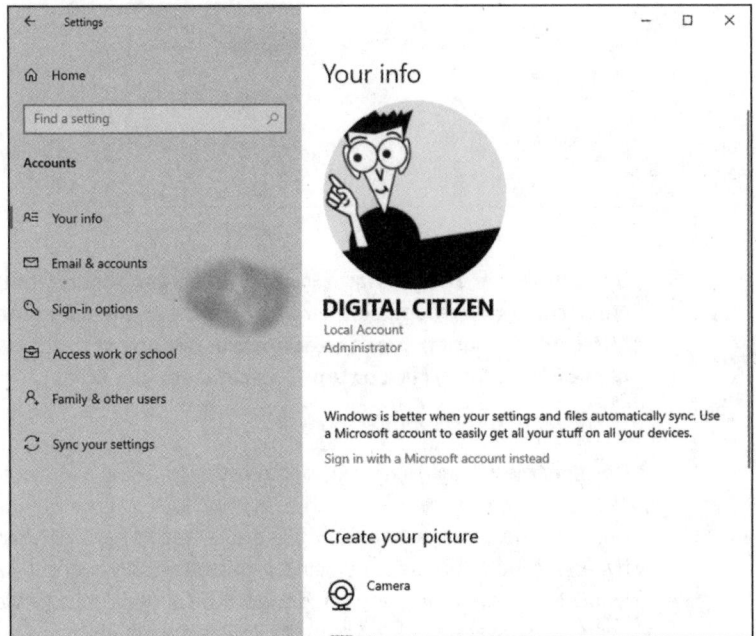

FIGURE 4-4:
Accounts settings.

3. **On the left, click or tap Family & Other Users and then choose one of the following:**

- **Add a Family Member:** Choose this if you want to control the account with Parental Controls. This option is available only if your use a Microsoft account to add the new user.

- **Add Someone Else to This PC:** The someone else could well be a family member — you just don't get easy access to Parental Controls for the new account.

You see the challenging How Will This Person Sign In? dialog box, as shown in Figure 4-5.

FIGURE 4-5:
Microsoft wants
you to set up
a Microsoft
account.

4. **If the new user already has a Microsoft account (or an @hotmail.com or @live.com or @outlook.com email address — which are automatically Microsoft accounts), type the address in the box at the top and then tap or click Next. Then click or tap Finish and you are done.**

   Windows 10 sets up your account.

   Don't get me wrong. There are good reasons for using a Microsoft account — a Microsoft account makes it much easier and faster to retrieve your mail and calendar entries, for example, or use the Microsoft Store or Music or Videos, bypassing individual account logins. It'll automatically connect you to your OneDrive account. Only you can decide if the added convenience is worth the decreased privacy. Book 2, Chapter 5 covers the details.

5. **On the other hand, if you're skeptical about using a Microsoft account, click or tap the link at the bottom that says Take Your Microsoft Account and Shove It (otherwise known as I Don't Have This Person's Sign-In Information).**

   Windows 10 gives you yet another opportunity to set up a Microsoft account, as shown in Figure 4-6.

6. **At the bottom, click or tap Add a User without a Microsoft Account. Sheesh.**

   Windows 10 (finally!) asks you about an offline/local account name and password. See Figure 4-7.

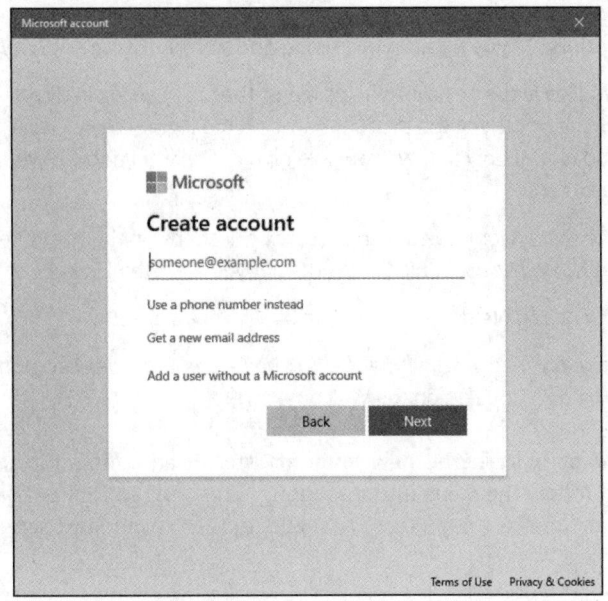

FIGURE 4-6:
Here's the second
time Microsoft
asks whether
you want to set
up a Microsoft
account.

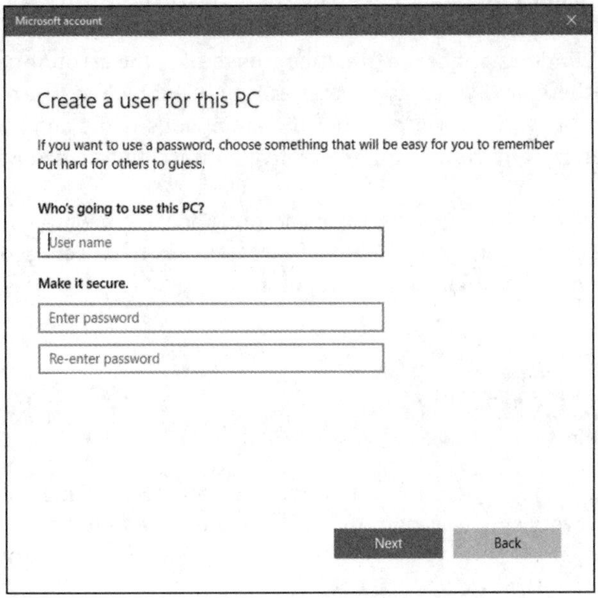

FIGURE 4-7:
Now you get to
the "adding a new
account" part.

7. **In the Who's Going to Use This PC? field, type a name for the new account.**

   You can give a new account just about any name you like: first name, last name, nickname, titles, abbreviations . . . No sweat, as long as you don't use the characters / \ [ ] " ; : | < > + = , ? or *.

8. **(Optional) Type a password twice and answer three security questions.**

   If you leave the password fields blank, the user can log in directly by simply tapping or clicking the account name on the login screen. Usually, that isn't a good idea, if only to thwart people who casually get ahold of your machine for a minute.

   Note that anyone can see the security questions on the computer, so avoid that NSFW (Not Suitable For Work) hint you were thinking about.

9. **Click or tap Next.**

   You're finished. Rocket science. You have a new standard account, and its name now appears on the Welcome screen.

If you want to turn the new account into an administrator account or a child account, follow the steps in the section, "Changing Accounts," later in this chapter. To add an account picture for the login screen and Start screen, flip to Book 3, Chapter 2.

This topic is more than a bit confusing, but you aren't allowed to create a new account named Administrator. There's a good reason why Windows 10 prevents you from making a new account with that name: You already have one. Even though Windows 10 goes to great lengths to hide the account named Administrator, it's there, and you may encounter it one night when you're exploring a blind alley. For now, don't worry about the ambiguous name and the ghostly appearance. Just refrain from trying to create a new account named Administrator.

Just because you have a Microsoft account doesn't mean you can log in to any computer anywhere. Your Microsoft account has to be set up on a specific computer before you can use that computer.

Changing Accounts

If you have an administrator account, you can reach in and change almost every detail of every single account on the computer — except one.

Changing other users' settings

In general, changing other users' settings is easy if you have an administrator account. To change an account from a standard account to an administrator account:

1. Click or tap the Start button and then the Settings shortcut.

2. On the Settings window, click or tap Accounts. On the left, choose Family & Other Users.

A list of all the accounts on the computer appears.

3. Click or tap on the account you want to change.

For example, in Figure 4-8, I chose to change my offline/local account called Digital Citizen.

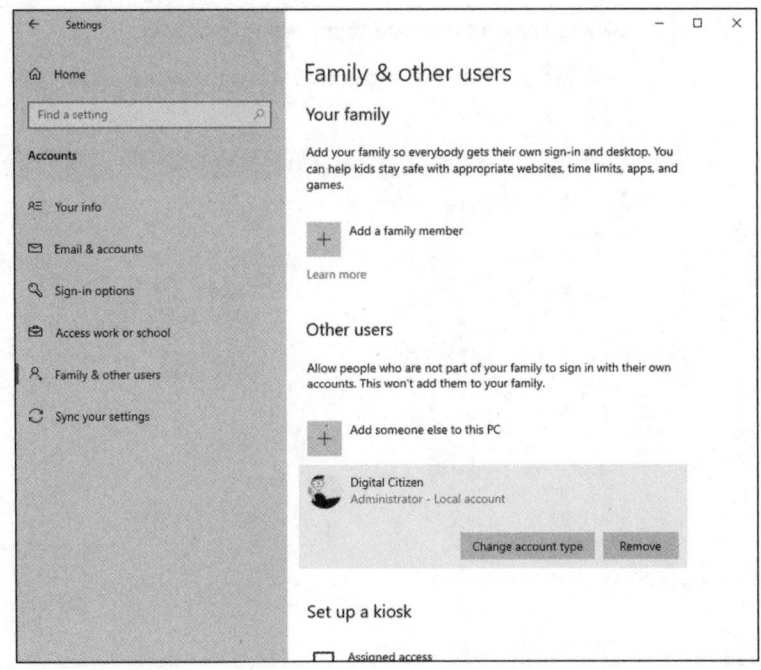

FIGURE 4-8: Choose the account you want to change from standard to administrator, or vice versa.

4. Click or tap the Change Account Type button below the selected account.

Windows 10 responds with the option to change from standard user to administrator account and back.

5. Select the new account type, and click or tap OK.

The account's type changes immediately.

Controlling Users

For other kinds of account changes, you need to venture into the old-fashioned Control Panel applet. Here's how:

1. **In the Windows 10 Search box, type** Control Panel. **In the list of search results, choose Control Panel.**

   The old-fashioned Control Panel appears.

2. **Choose User Accounts, then User Accounts again. Click Manage Another Account.**

   A list of all the accounts on the computer appears.

3. **Click or tap on the account you want to change.**

   Windows 10 immediately presents you with several options (see Figure 4-9).

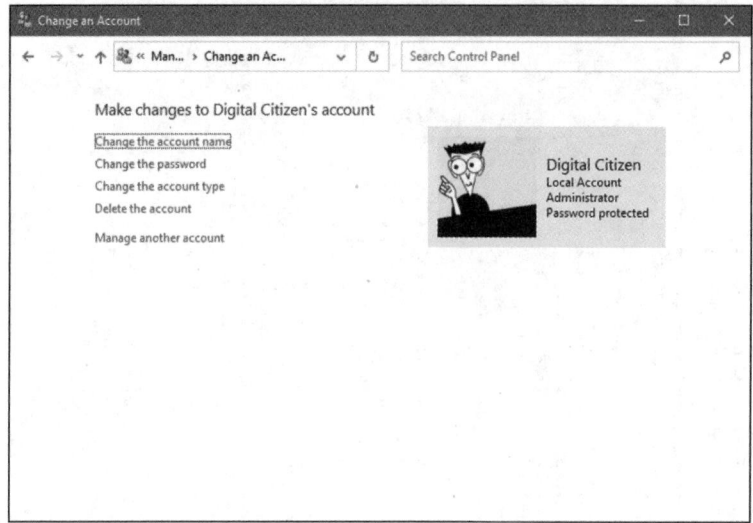

FIGURE 4-9:
Maintain another
user's account.

Here's what the options entail:

» **Change the Account Name:** This option appears only for offline/local accounts. (It'd be kind of difficult if Windows 10 let you change someone's Microsoft account, eh?) Selecting this option modifies the name displayed on the login screen and at the top of the Start menu while leaving all other settings intact. Use this option if you want to change only the name on the account — for example, if Little Bill wants to be called Sir William.

» **Create/Change a Password:** Again, this appears only for offline/local accounts. (Create appears if the account doesn't have a password; Change appears if the account already has a password.) If you create a password for the chosen user, Windows 10 requires a password to crank up that user account. You can't get past the Login screen (using that account) without it. This setting is weird because you can change it for other people: You can force Bill to use a password when none was required before, you can change Bill's password, or you can even make it blank.

If you change someone's password, do her a big favor and tell her how to create a Password Reset Disk. See Book 3, Chapter 6.

REMEMBER

Passwords are cAse SenSitive — you must enter the password, with upper-case and lowercase letters, precisely the way it was originally typed. If you can't get the computer to recognize your password, make sure that the Caps Lock setting is off. That's the number-one source of login frustration.

ASK WOODY.COM

Much has been written about the importance of choosing a secure password, mixing uppercase and lowercase letters with punctuation marks, ensuring that you have a long password, blah blah blah. I have only two admonitions: First, don't write your password on a yellow sticky note attached to your monitor; second, don't use the easily guessed passwords that the Conficker worm employed to crack millions of systems (see Table 4-1, at the end of this list). Good advice from a friend: Create a simple sentence you can remember, and swap out some letters for numbers (G00dGr1efTerry), or think of a sentence and use only the first letters! (toasaoutfl!) Of course, using a picture password or PIN (or even a Hello mugshot, a fingerprint, or an iris scan) makes even more sense.

» **Change the Account Type:** You can use this option to change accounts from administrator to standard and back again. The implications are somewhat complex; I talk about them in the section "Choosing Account Types," earlier in this chapter.

» **Delete the Account:** Deep-six the account, if you're that bold (or mad, in all senses of the term). If you're deleting a Windows 10 account, the account itself still lives — it just won't be permitted to log in to this computer. Windows offers to keep copies of the deleted account's Documents folder and desktop, but warns you quite sternly and correctly that if you snuff the account, you rip out all the email messages, Internet Favorites, and other settings that belong to the user — definitely not a good way to make friends. Oh, and you can't delete your own account, of course, so this option won't appear if your PC has only one account.

» **Manage Another Account:** Displays the list of accounts so you can choose another user and modify the user's account using the options just described.

TABLE 4-1   # Most Frequently Used Passwords*

Most Frequently Used Passwords					
000	0000	00000	0000000	00000000	0987654321
111	1111	11111	111111	1111111	11111111
123	123123	12321	123321	1234	12345
123456	1234567	12345678	123456789	1234567890	1234abcd
1234qwer	123abc	123asd	123qwe	1q2w3e	222
2222	22222	222222	2222222	22222222	321
333	3333	33333	333333	3333333	33333333
4321	444	4444	44444	444444	4444444
44444444	54321	555	5555	55555	555555
5555555	55555555	654321	666	6666	66666
666666	6666666	66666666	7654321	777	7777
77777	777777	7777777	77777777	87654321	888
8888	88888	888888	8888888	88888888	987654321
999	9999	99999	999999	9999999	99999999
a1b2c3	aaa	aaaa	aaaaa	abc123	academia
access	account	Admin	admin	admin1	admin12
admin123	adminadmin	administrator	anything	asddsa	asdfgh
asdsa	asdzxc	backup	boss123	business	campus
changeme	cluster	codename	codeword	coffee	computer
controller	cookie	customer	database	default	desktop
domain	example	exchange	explorer	file	files
foo	foobar	foofoo	forever	freedom	f**k
games	home	home123	ihavenopass	Internet	internet
intranet	job	killer	letitbe	letmein	login
Login	lotus	love123	manager	market	money
monitor	mypass	mypassword	mypc123	nimda	nobody
nopass	nopassword	nothing	office	oracle	owner
pass	pass1	pass12	pass123	passwd	password

Most Frequently Used Passwords					
Password	password1	password12	password123	private	public
pw123	q1w2e3	qazwsx	qazwsxedc	qqq	qqqq
qqqqq	qwe123	qweasd	qweasdzxc	qweewq	qwerty
qwewq	root	root123	rootroot	sample	secret
secure	security	server	shadow	share	sql
student	super	superuser	supervisor	system	temp
temp123	temporary	temptemp	test	test123	testtest
unknown	web	windows	work	work123	xxx
xxxx	xxxxx	zxccxz	zxcvb	zxcvbn	zxcxz
zzz	zzzz	Zzzzz			

* From the Conficker worm, Bowdlerized with an asterisk (*) as a fig leaf

Changing your own settings

Changing your own account is just a little different from changing other users' accounts. Follow these steps:

1. Bring up the Control Panel.

To do so, down in the Windows 10 Search box, type **Control Panel**. Then, up at the top, choose Control Panel.

2. In the upper right, choose User Accounts, then User Accounts again.

Windows 10 offers you the chance to change your own account. If you want to change your password, picture, or family settings, you get bounced out to the Settings app. And from there, if you have a Microsoft account, you can link it to your local account.

Most of the options for your own account mirror those of other users' accounts, as described in the preceding section. If you have the only administrator account on the PC, you can't delete your own account and you can't turn yourself into a standard user. Makes sense: Every PC must have at least one user with an administrator account. If Windows 10 lost all its administrators, no one would be around to add users or change existing ones, much less to install programs or hardware, right?

Switching Users

Windows 10 allows you to have more than one person logged in to a PC simultaneously. That's convenient if, say, you're working on the family PC and checking Billy's homework when you hear the cat screaming bloody murder in the kitchen and your wife wants to put digital pictures from the family vacation on OneDrive while you run off to check the microwave.

The capability to have more than one user logged in to a PC simultaneously is *fast user switching*, and it has advantages and disadvantages:

>> **On the plus side:** Fast user switching lets you keep all your programs going while somebody else pops on to the machine for a quick jaunt on the keyboard. When she's done, she can log off, and you can pick up precisely where you left off before you got bumped.

>> **On the minus side:** All idle programs left sitting around by the inactive (bumped) user can bog things down for the active user, although the effect isn't drastic. You can avoid the overhead by logging off before the new user logs in.

To switch users, click the Start button, click or tap your picture, and choose either the name of the user you want to switch to or Sign Out. If you choose the latter, you're taken to the sign-in screen, where you can choose from any user on the computer.

The Changing Environment

Windows Hello represents a big step forward in Windows 10 login capabilities. Instead of one sudden "Hello, Johnny!" login experience, Microsoft is keeping all the login possibilities you've known for years, while rolling out the various pieces of Windows Hello features over time.

The initial release of Windows 10, in July 2015, included Windows Hello facial identification for the small subset of Windows 10 users who have special cameras. Microsoft also built some fingerprint reading smarts into Windows Hello.

The November 2015 release of Windows 10 added features for corporate machines and Passport for Work features, including an Azure AD cloud interface. The July (actually August) 2016 Anniversary update, version 1607, added some basic iris scanning capabilities. The Anniversary update also dropped the Passport terminology, but not the Passport functionality, for consumers and work.

Subsequent versions of Windows 10 added more features to Windows Hello. For example, Microsoft giving more support to the Trusted Platform Module (TPM) chip that's appearing inside all new Windows 10 machines. A thorough discussion of TPM is on the How-To Geek site, at `www.howtogeek.com/237232/what-is-a-tpm-and-why-does-windows-need-one-for-disk-encryption/`.

Chapter **5**

Microsoft Account: To Sync or Not to Sync?

Microsoft has been trying to get people to sign up for company-branded accounts for a long time.

ASK WOODY.COM

In 1997, Microsoft bought Hotmail and took over the issuance of @hotmail.com email addresses. Even though Hotmail's gone through a bunch of name changes — MSN Hotmail, Windows Live Hotmail, and now Outlook.com, among others — the original @hotmail.com email addresses still work, and have worked, through thick and thin.

Twenty years after its inception, that old @hotmail.com ID still works the same as it ever did — except now it's called a *Microsoft account*. If you picked up an @msn.com ID, @live.com ID, Xbox ID, Skype ID, or @outlook.com ID along the way, it's now a Microsoft account as well.

ASK
WOODY.COM

In this chapter, I show you exactly what's involved with a Microsoft account, show you why it can be useful, explore the dark underbelly of Microsoft accountability, and give you a trick for acquiring a Microsoft account that won't compromise much of anything.

What, Exactly, Is a Microsoft Account?

Now that Microsoft has finally settled on a name for its ID — at least, this month — permit me to dispel some of the myths about Microsoft accounts.

An email address that ends with @hotmail.com, @msn.com, @live.com, or @out-look.com is, *ipso facto*, a Microsoft account. The same is true for Hotmail and Live and Outlook.com accounts in any country, such as @hotmail.co.uk. You don't have to use your Microsoft account. Ever. But you do have one.

REMEMBER

Many people don't realize that *any* email address can be a Microsoft account. You need only to register that email address with Microsoft; I show you how in the section "Setting Up a Microsoft Account" later in this chapter.

In the context of Windows 10, the Microsoft account takes on a new dimension. When you set up an account to log in to Windows, it can either be a Microsoft account or an *offline* (also called a *local*) *account.* The key differences:

» Microsoft accounts are always email addresses, and they must be registered with Microsoft. As I explain in Book 2, Chapter 4, when you log in to Windows 10 with a Microsoft account, the operating system automatically syncs some settings — Windows settings like your picture and backgrounds, Microsoft Edge history and favorites, and others — so if you change something on one machine and log in with the same Microsoft account on another, the changes go with you.

In addition, a Microsoft account gives you something of a one-stop log in to Internet-based Microsoft services. For example, if you have a OneDrive account, logging in to Windows 10 with a Microsoft account automatically hitches you up to your OneDrive files.

WARNING

» Until the Windows 10 May 2020 update, if you logged in to Windows 10 with a Microsoft account, and didn't modify Cortana's behavior, Microsoft tracked every search you made *on your computer.* I'm not talking about a web search. I'm talking about when you searched through your documents or email messages, right there on your machine. If you turned on "Hey, Cortana" recognition, Cortana also listened to everything you said, all the time. However, with Windows 10 version 2004 (May 2020 update), Cortana has been turned into a standalone app and is no longer tied into Windows 10. This is good news because Cortana no longer tracks you as aggressively as it did in the past. I talk about Cortana in Book 3, Chapter 5.

>> Offline/local accounts can be just about any name or combination of characters. If you sign in with an offline/local account, Microsoft can't sync anything on different machines. Sign in with an offline/local account, and you have to sign in to your OneDrive account separately. Windows 10 will remember your settings — your backgrounds, passwords, favorites, and the like — but they won't be moved to other PCs when you log in.

So, for example, phineasfarquahrt@hotmail.com is a Microsoft account. Because it's an @hotmail.com Hotmail email address, it's already registered with Microsoft. I can create a user on a Windows 10 machine with the name phineasfarquahrt@hotmail.com, and Windows will recognize that as a Microsoft account.

On the other hand, I can set up an account on a Windows PC that's called, oh, *Woody Leonhard.* It's an offline/local account. Because Microsoft accounts have to be email addresses (you see why in the section "Setting Up a Microsoft Account"), the Woody Leonhard account has to be an offline/local account.

When you set up a brand-new Windows 10 PC, you must enter an account, and it can be either a Microsoft account or an offline/local account. Microsoft stacks the deck and makes you tap or click all over heaven's half acre to avoid using a Microsoft account. When you add a new account, Microsoft nudges you to use a Microsoft account. Still, it will begrudgingly accept an offline/local account (see Book 2, Chapter 4).

Deciding Whether You Want a Microsoft Account

If Microsoft tracks a Microsoft account, you may ask, why in the world would I want to sign on to Windows 10 with a Microsoft account?

Good question, grasshopper.

Signing on to Windows 10 with a Microsoft account brings a host of benefits. In particular:

>> **Some of your Windows 10 settings will travel with you.** Your user picture, desktop, browser favorites, and other similar settings will find you no matter which PC you log in to.

I find this helpful in some ways, and annoying in others. For example, I have a big-screen Windows 10 desktop PC and a little Windows 10 tablet. If I put a whole bunch of shortcuts on the desktop, they look horrible on the tablet.

Your tiled Windows 10 apps (Universal apps) — the ones that came with Windows 10 or that you downloaded from the Microsoft Store — revert to their last state. So if you're on a killer winning streak with a Solitaire game, that'll go with you to any PC you log in to. Your Microsoft Edge open tabs travel. Settings for the Windows 10 Weather app travel. Even apps *that Microsoft doesn't make* may have their settings moved from machine to machine.

>> **Sign-in credentials for programs and websites travel.** If you rely on Microsoft Edge to keep sites' login credentials, those will find you if you switch machines.

>> **You will be automatically signed in to Windows 10 apps and services** that use the Microsoft account (or Windows Live ID). Mail, Calendar, OneDrive, Skype, and the Microsoft website all fall into that category.

ASK
WOODY.COM

Don't be overly cynical. In some sense, Microsoft dangles these carrots to convince you to sign up for, and use, a Microsoft account. But in another sense, the simple fact is that none of these features would be possible if it weren't for some sort of ID that's maintained by Microsoft.

I use a Microsoft account on my main machine. However, I employ a little trick — creating a new Microsoft account and only using it to sign in to Windows 10 — which I describe in the next section on setting up a new account.

WARNING

That's the carrot. Here's the stick. If you sign in with a Microsoft account, the company has a record of every time you've signed on to every PC you use with that account. More than that, when you crank up Microsoft Edge (or Internet Explorer), you're logged in with your Microsoft account — which means that Microsoft can, at least theoretically, keep records about all your browsing (except, presumably, InPrivate browsing). Bing gets to jot down your Microsoft account every time you search through it. Microsoft gets data on any music you view in the Windows 10 Music app. Your stock interests are logged in the Windows 10 Money app. Even the weather you request ends up in Microsoft's giant database. And if you use Cortana, everything you ask her ends up in Microsoft's big database chock full of your history.

Perhaps it's true that you have no privacy and should get over it. The fact is that most people don't care. My attitude toward data scraping and Windows 10 snooping has changed over the years. I talk about my begrudging conversion in the first part of the next chapter, Book 2, Chapter 6.

WHAT IF MY HOTMAIL OR OUTLOOK.COM ACCOUNT IS HIJACKED?

So you set up a Hotmail account or Outlook.com for logging on to your Windows PC, and all of a sudden the account gets hijacked. Some cretin gets into the account online and changes the password. The next time you try to log in to your Windows 10 PC, what happens?

It's not far-fetched: I get complaints almost every day from people who have been locked out of their Hotmail/Outlook.com accounts.

If you use a Hotmail ID, a Windows Live account, or an Outlook.com account for your Microsoft account and your Hotmail/Outlook.com account gets hijacked and the password changed, Windows 10 lets you log in to your PC, but when you do, you get the notice *You're signed in to this PC with your old password. Sign in again with your current password, or reset it.* If you then try to reset your password, you can't — clicking or tapping the Reset link doesn't do anything.

The increased use of two-factor authentication — where Microsoft sends you a text message on your smartphone, and you must respond before the password gets changed — has improved the situation. But many people don't use 2FA.

Once it's changed, until you can come up with your Hotmail/Outlook.com account's password, you're put in a reduced functionality mode that's like logging on with an offline/local account. As long as you can remember your old password — the last one you used to log in to this machine — you can continue to log in. But ultimately, you're going to want your Windows 10 login account back!

To get your account back, you need to contact the people at Microsoft and convince them that you're the rightful owner. If you set up your Hotmail/Outlook.com account recently, chances are at least fair that you have an alternate email address or phone number designated for just such an emergency, so-called 2FA or two-factor authentication, described in a sidebar later in this chapter. Microsoft started asking for that specific information on sign-up a couple of years ago. Go to `http://account.live.com/resetpassword.aspx`, and have a Microsoft rep contact you.

Setting Up a Microsoft Account

Just to make life a little more complicated, shortly before Microsoft released Windows 8, it suddenly decided to kill off the name Hotmail and replace it with Outlook.com. I talk about the reasons why — basically, Hotmail was losing market share, and Microsoft needed to get it back — in Book 10, Chapter 4.

For purposes of this chapter, a Hotmail or Outlook.com account, a Live.com account, Xbox LIVE account, OneDrive account, Skype account, MSN account, Microsoft Passport account, or a Windows 10 Mobile account are all interchangeable: They're email addresses that have already been automatically signed up as Microsoft accounts. I tend to refer to them collectively as Hotmail accounts because, well, most Microsoft accounts have been Hotmail accounts for the past two decades or so. Old habits die hard.

If you don't have a Microsoft account, the way I see it, you have three choices for setting one up:

>> **You can use an existing email address.** But if you do that, Microsoft will be able to put that email address in its database. It can cross-reference the address to many things you do with Windows 10. (Can you tell my tinfoil hat is showing?)

>> **You can use (or set up) a Hotmail/Live/Xbox/OneDrive/Skype/Outlook. com account.** If you already have one, Microsoft tracks it already — Microsoft knows when you receive and send email, for example. But that's true of any online email program, including Gmail and Yahoo! Mail. Using a Hotmail/ Outlook.com account to log in to Windows 10, though, means that Microsoft can track additional information and associate it with your Hotmail/Outlook. com account — the times you log in to Windows 10, locations, and so on. You may be okay with that, or you may not want Microsoft to be able to track that kind of additional information.

>> **You can create a bogus new Hotmail/Outlook.com account and use it only to log in to Windows 10.** It's free and easy, and if you use it wisely, nobody will ever know the difference. The only downsides: If you use Hotmail/ Outlook.com, you have to tell the Windows 10 Mail app to look in your other inbox; your existing Hotmail/Outlook.com contacts won't get carried over into the tiled People automatically; and Skype will want to work with your new, bogus ID — although you can change it.

I love to use bogus Outlook.com accounts. I keep in mind that every time I use Microsoft Edge, having signed in to Windows 10 with a Microsoft account, that Microsoft will dump all my browsing history in its coffers.

So, of course, I use Firefox or Google Chrome when I want to use the Internet. Google keeps Chrome data, but it doesn't have Microsoft's database of logged in Windows 10 users, and Firefox isn't beholden to anybody.

Search engines, of course, are a different story entirely. Bing/Microsoft and Google keep track of everything you send their way.

Setting up a Hotmail/Outlook.com account

Here's how to set up a new Hotmail/Outlook.com account:

1. **Using your favorite web browser, go to** www.Outlook.com.

The main screen, laden with ads that go on forever, lets you sign in or create a free account.

2. **If you aren't automatically signed in, tap or click Create Free Account.**

You see the Create Account form, as shown in Figure 5-1.

FIGURE 5-1:
Sign up for an
anonymous
Hotmail/Outlook.
com ID.

3. **Type an account name and press Next.**

If someone already has the email address you entered, type another and click Next again.

4. **Type the password you want to use, deselect the box that tells Microsoft to send you spam, and click Next.**

5. **Fill out a fanciful first and last name, and then click Next.**

6. **Give Microsoft your country (which they can find anyway by looking at your Internet IP address) and fill out a birthdate. Click Next.**

If your birthdate indicates that you're less than 18 years old, you may have problems using the account.

7. **Type the CAPTCHA codes, if you can figure them out, and then click Next.**

If you passed the IQ test, Outlook.com loads your inbox and Microsoft's welcome message, as in Figure 5-2. That's it.

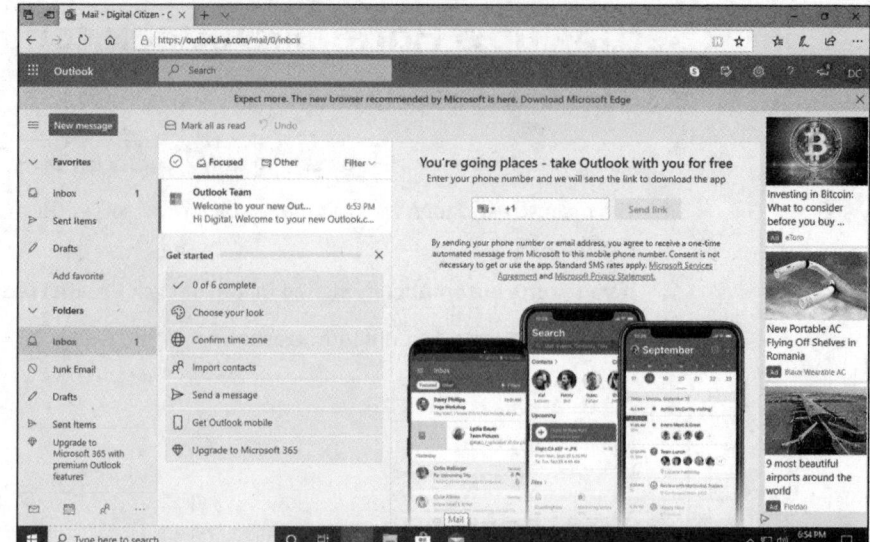

FIGURE 5-2:
Your new
Microsoft account
(née Windows
Live ID, Hotmail
account, MSN
account, Outlook.
com account,
Xbox Live
account) is alive
and working.

You can now use your new Outlook/Hotmail account as a Windows 10 login ID. You can use it for email, Skype, Xbox . . . just about anything from Microsoft.

Making any email address a Microsoft account

You must follow a different procedure to turn any email address into a Microsoft account. The steps are simple, as long as you can retrieve email sent to the address:

1. **Using your favorite web browser, go to** `signup.live.com`.

   You see the Create Account message, as shown in Figure 5-3, where you can create a Microsoft account without a Microsoft email address.

2. **Type your email address from Gmail, Yahoo! Mail, or some other place. Then click Next.**

3. **Enter the password you want to use and click Next.**

   REMEMBER

   Note that the password you provide here is for your Microsoft account. It is *not* your email password. The password you enter here will be the password you need to use to log in to Windows 10 or any website that requires a Microsoft account. Most experts advise you not to reuse your email password as your Microsoft account password.

4. **Fill out a fanciful first and last name, and then click Next.**

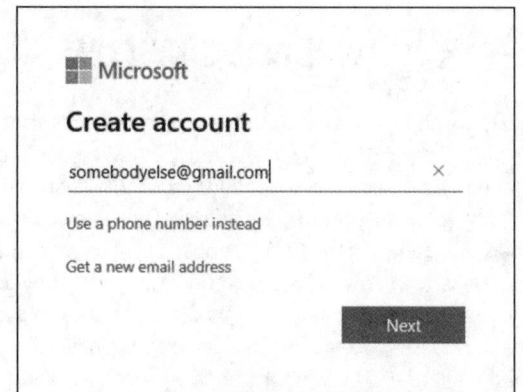

FIGURE 5-3:
Creating a
Microsoft account
with an email
from another
company.

5. **Give Microsoft your country and fill out a birthdate. Click Next.**

   Microsoft sends a 4-digit verification code to your email address. If you do not see the email, check your Spam/Junk folder.

6. **Type the verification code, deselect the box that allows Microsoft to send you emails, and press Next.**

7. **Type the CAPTCHA code and then tap or click Next.**

   Your Microsoft account page is loaded, as in Figure 5-4.

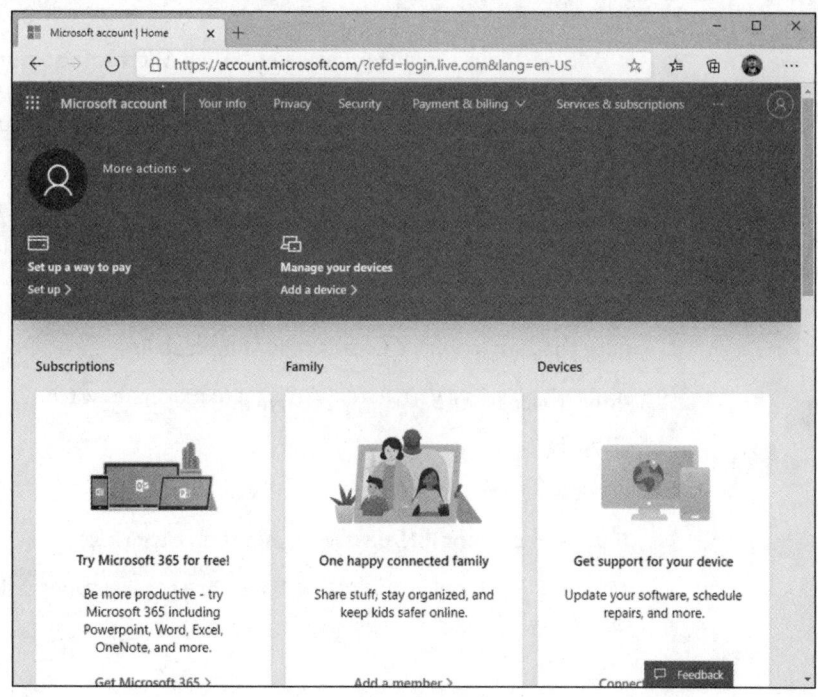

FIGURE 5-4:
Your new
Microsoft account
is alive, with an
email that is not
from Microsoft.

TWO-FACTOR AUTHENTICATION

Microsoft has been developing and expanding a feature called *two-factor authentication*. Details vary, but it's a good choice. Usually, when you log in with your Microsoft account using a machine that hasn't been explicitly identified (by you) as being an acceptable computer, Microsoft issues a challenge to verify that you are who you say you are. Usually, the authentication comes in the form of an SMS sent to your smartphone, or an email sent to your registered email address. Benefits are pretty obvious: Somebody may be able to steal your password, but it's rare that they get both your password and your computer (which bypasses two-factor authentication entirely), and almost impossible to get both your password and your smartphone — or access to your email address.

Most people are leery about giving their phone numbers to Microsoft. Hey, it took me almost a decade before I learned to stop worrying and love the bomb. I've found, though, that Microsoft doesn't use my phone number for nefarious purposes. And that smartphone-based two-factor authentication works great, even if I do mumble from time to time about it being so slow. You should try apps such as Microsoft Authenticator or Google Authenticator., which are available for Android and iPhone. I prefer the Microsoft app.

Stop Using Your Microsoft Account

So you've read about the differences between a Microsoft account and an offline/local account, and you've decided that you just don't want to keep feeding Microsoft information. You want to move to an offline/local account. Fortunately, that's pretty easy:

1. **Click the Start button, then the Settings shortcut, and then Accounts. On the left, choose Your Info.**

   You see the account settings for your account.

2. **Click the Sign In with a Local Account Instead link.**

   Windows 10 asks if you're sure you want to make the switch.

3. **Click Next.**

   Windows 10 asks you to enter your PIN or password.

4. **Type your current PIN/password and then click Next.**

   Windows 10 presents you with the Switch to a Local Account dialog shown in Figure 5-5.

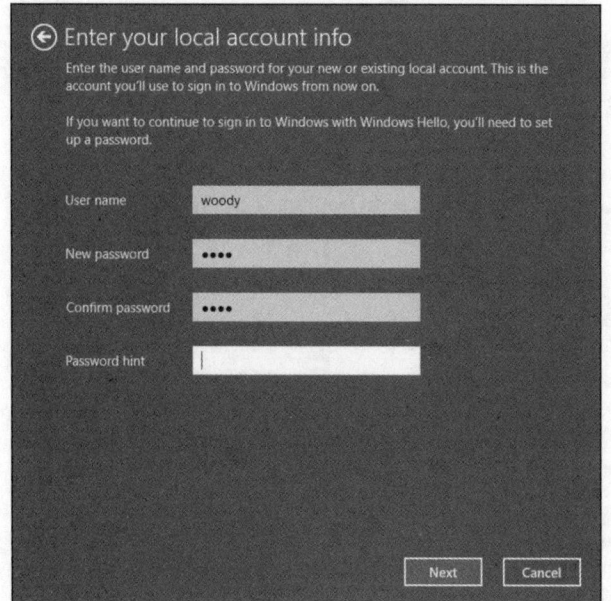

FIGURE 5-5:
Type the offline/
local account, its
password, and
the password
hint.

5. **Enter the offline/local account to use in place of your Microsoft account, the password you want, and a password hint. Then click Next.**

   Windows 10 warns you to make sure you've saved your work — it's about to restart — and to ensure that you know your new password.

6. **Click Sign Out and Finish.**

   Windows 10 signs you out and displays the lock screen. Now you can sign in with your offline/local account.

Note that your old Windows 10 account is no longer valid for signing in to this computer. Instead, you sign in only through the offline/local account. If you want to switch back, click the Start button, the Settings shortcut, and then Accounts. Click the Sign In with a Microsoft Account Instead link. Then, go through the hoops of adding the details of your Microsoft account. Your old Microsoft account reappears.

Taking Care of Your Microsoft Account

If you ever want to change any of the details in your Microsoft account, it's easy — if you know where to go.

For reasons understood only by Microsoft, to maintain your Microsoft account, go to `https://account.microsoft.com`. Sign in, and you see full account information, as shown previously in Figure 5-4.

To change any of the information for your account, or the password, tap or click the related link below the item you want to change.

Controlling Sync

If you don't explicitly change anything, logging on to Windows 10 with a Microsoft account syncs some settings across all the PCs that you use.

You can tell Microsoft that you don't want to sync specific items. Here's how:

1. **Click the Start button, the Settings shortcut, and then Accounts.**

2. **On the left, click or tap Sync Your Settings.**

   The Sync Your Settings screen appears, as shown in Figure 5-6.

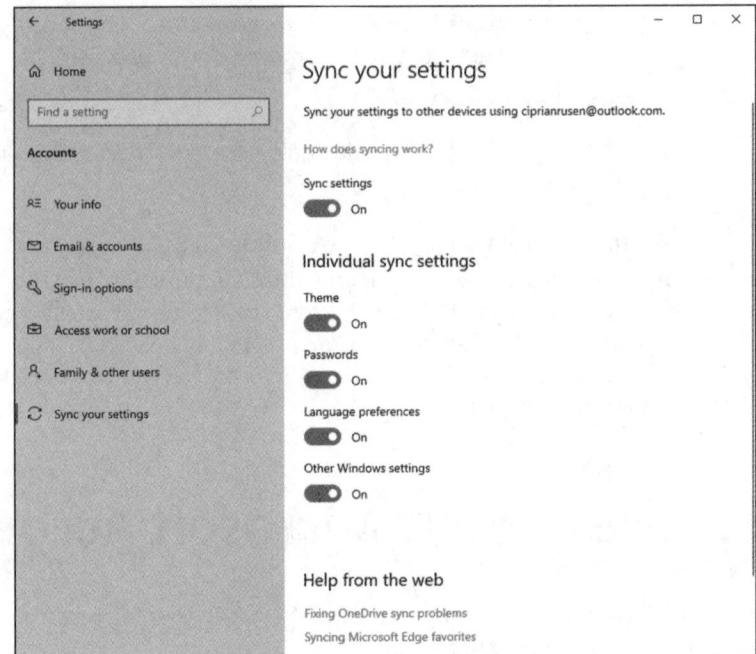

FIGURE 5-6:
Control the way Microsoft accounts sync here.

3. **Follow the list in Table 5-1 to choose whether you want to sync specific items.**

REMEMBER

Sync happens only when you log in with the same Microsoft account on two different PCs.

You're finished. No need to tap or click OK or Apply. The changes take effect with your next login.

TABLE 5-1 ## Sync Settings

Setting	What It Controls
Sync Settings	This is an overall off switch. If you don't want to sync anything, turn this off.
Theme	Your user picture, Start menu tiles, color, background, and desktop settings.
Passwords	Potentially sensitive information, including login credentials for Windows 10 apps and some website passwords.
Language preferences	The input method (language/keyboard) in effect.
Other Windows settings	All ease-of-access settings, your mouse settings, the list of installed apps, shared printers, and so on.

Chapter **6**

Privacy Control

"The best minds of my generation are thinking about how to make people click ads. That sucks."

— Jeff Hammerbacher, early Facebook employee

ASK
WOODY.COM

When you work with "free" services — search engines such as Google and Bing (which is Microsoft); social networks such as Facebook, Pinterest, and LinkedIn (Microsoft); online storage services such as OneDrive (Microsoft) and Google Drive; free email services such as Gmail (Google), Hotmail/Outlook.com (Microsoft), and Yahoo! Mail; even the "free" versions of Windows 10 — these services may not charge you anything, but they're hardly free. You pay with your privacy. Every time you go to one of those sites or use one of those products, with a few noteworthy exceptions, you leave a trail that companies are eager to exploit, primarily for advertising.

The exceptions? Google doesn't scan activity for any paid account, or any educational account. (They've been sued up the wazoo.) Apple swears it doesn't wallow in the data grabbing cesspool. Microsoft loves to say it doesn't scan the contents of Outlook.com/Hotmail messages. There are lots of if's, and's, but's, and nuances. But by and large, if it's free, you're the product, not the customer.

There's a reason why you buy something on, say, Alibaba, and then find ads for Alibaba appearing on all sorts of websites. One of the big advertising conglomerates has your number. Maybe just your IP address. Maybe a planted cookie. But they've connected enough dots to know that, whatever site you happen to be on at the moment, you once bought something on Alibaba.

Now, even when you log in to Windows 10, if you opt to use a Microsoft account, you leave another footprint in the sand. (I talk about Microsoft accounts in Book 2, Chapter 5.)

This isn't horrible. Necessarily. It isn't illegal — although laws in different countries differ widely, and lawsuits are reshaping the picture even as we speak. In most cases, anyway. The advertisers view it as a chance to direct advertising at you that's likely to generate a response. In some respects, it's like a billboard for a cold Pepsi on a hot freeway or an ad for beer on Super Bowl Sunday.

In other respects, though, logging your activity online is something entirely different.

I talk about privacy in general in Book 9, Chapter 1, and the browser Do Not Track flag (which may or may not do what you think it should do) in Book 5, Chapter 1. In this chapter, I want to give you an overview of privacy settings — and some privacy shenanigans — specifically inside Windows 10.

Why You Should Be Concerned

As time goes by, people are becoming more and more aware of how their privacy is being eroded by using the Internet. Some people aren't particularly concerned. Others get paranoid to the point of blocking anything that has a remote chance of tracking them. Chances are pretty good you're somewhere between the two poles.

ASK WOODY.COM

Windows 10 users need to understand that this version of Windows, *much* more than any version of Windows before, pulls in data from all over the web. Every time you elect to connect to a service, you're connecting the dots for Microsoft's data-collection routines. And if you use a Microsoft account, Microsoft's dot connector is even more productive.

I'm not implying that Microsoft is trying to steal your data or somehow use your identity for illegal purposes. It isn't. At this point, Microsoft mostly wants to identify your buying patterns and your interests, so it can serve you ads that you will click, for products that you will buy. The Google shtick. That's where the money is.

WARNING

Although Google freely admits that it scans inbound and outbound Gmail email, on free accounts, all the better to generate ads that you will click, Microsoft insists that it doesn't — ergo, the infamous Scroogled ads, wherein the pot and kettle somehow tie it on. Don't be fooled. Microsoft *does* scan Hotmail/Outlook.com mail and Windows 10 Mail app messages that you receive with Windows 10 — for spam detection, if nothing else. Whether Microsoft will start keeping track of detailed information about your messages in the future is hard to say.

Here's how the services stack up when it comes to privacy (or the lack thereof):

TIP

>> **Google:** Without a doubt, Google has the largest collection of data. You leave tracks on the Google databases every time you use Google to search for a website. That's true of every search engine (except www.DuckDuckGo.com), not just Google, but Google has 90 percent or more of the search engine market worldwide. You also hand Google web-surfing information if you sign in to your Google Chrome browser (so it can keep track of your bookmarks for you) or if you sign in to Google itself (for example, to use GSuite or Google Drive). The native Android browser ties into Google, too, and using an Android smartphone or tablet also sends tons of data to Google.

Google also owns *DoubleClick,* the best-known, third-party cookie generator on the web. Any time you go to a site with a DoubleClick ad — most popular sites have them — a little log about your visit finds its way into Google's database.

ASK WOODY.COM

Google's scanning policies changed significantly in late 2014. As of mid-2015, Google no longer scans email, or the contents of Google Drive files, for paid accounts, Academic accounts, or non-profit accounts. If you have a free Google account, you should expect that Google will sift through your mail and files, looking for information that can convince you to click on an ad.

>> **Facebook:** Although Facebook may not have the largest collection of data, it's the most detailed. People who sign up for Facebook tend to give away lots of information. When you connect your Microsoft account in Windows 10 to Facebook — for example, add your Facebook Friends to your Ultimate People app list (unless Facebook has shut Microsoft out this week, which happens from time to time) — some data that you allow to be shared on Facebook is accessible to Microsoft. That's why it's important to lock down your Facebook account (see Book 6, Chapter 2).

REMEMBER

Every time you go to a website with a Facebook Like icon, that fact is tucked away in Facebook's databases. If you're logged in to Facebook at the time you hit a site with a Like icon, your Facebook ID is transmitted, along with an indication of which site you're looking at, to the Facebook databases. As of this writing, Microsoft can't get into the Facebook database — which is truly one of the crown jewels of the Facebook empire — although it can pull a list of your Friends, if you allow it.

>> **Microsoft:** Microsoft's Internet access database may not be as big as Google's, or as detailed as Facebook's, but the 'Softies are trying to get there too. One of the ways they're catching up is by encouraging you to use a Microsoft account. The other is to create all these connections to other data-collecting agencies inside Windows 10. Then there's Bing, which logs what you're looking at just like a Google Search does.

Windows 10 is light-years ahead of earlier versions of Windows when it comes to harvesting your data. Or perhaps I should say it's light-years behind earlier versions of Windows when it comes to protecting your privacy.

WARNING

Until the May 2020 update, the single biggest leaker in Windows 10 was Cortana's Smart Search feature — which was smart for Microsoft's data collection efforts. Unless you went to great lengths to trim back Cortana's snooping, Microsoft (through Bing) kept a list of all the terms you searched for *on your computer.* Because Cortana's Smart Search was enabled by default when you installed Windows 10, chances were good that Microsoft was collecting information about every single search you made for your documents, pictures, email, and so on. However, the situation changed recently. Since the May 2020 update, Cortana is wholly decoupled from Windows 10 and tracks data only when you're specifically using her. Local Windows 10 searches now happen separately, and that data remains stored on your PC. A considerable improvement, if you ask me! I talk about Cortana in Book 3, Chapter 5.

For an ongoing, authoritative discussion of privacy issues, look at the Electronic Frontier Foundation's Defending Your Rights in the Digital World page at www. eff.org/issues/privacy.

Privacy Manifesto

Privacy had become such a huge issue with Windows 10 that many folks did not install it, just because they figured Windows 10 was sending all their private information to Microsoft. In one sense, that was true — Windows 10 snooped in ways no previous version of Windows ever dared. In another sense, though, increased snooping is a sign of changing times. And I'm convinced that Microsoft is no worse than most of the alternatives. Also, their recent updates to Windows 10 have provided more transparency about the data collected and sent to their company's servers.

The point is that you, the Windows 10 user, need to understand what's going on — and you need to make decisions accordingly.

Like it or not, times have changed, and attitudes toward snooping have changed along with them.

The past: Watson to WER

Back in the distant past, the Windows 3.0 beta (in 1989-1990) included a program called Dr. Watson, which responded to Windows crashes by gathering all the data it could find and packaging it as a text file (drwtsn32.log). Dr. Watson was also smart enough to generate a core dump, which could be fed into a debugger on a diagnostic machine.

Dr. Watson worked offline. If you wanted to send your text log file or core dump to somebody, that was up to you. Dr. Watson was highly successful, leading to the identification and eradication of thousands of bugs (most, it must be said, in non-Microsoft drivers).

Around the time of Windows XP, Dr. Watson turned into the Problem Reports and Solutions program, which became part of the broader Windows Error Reporting (WER) system built into Windows XP and then enhanced for Vista, Windows 7, and Windows 8. WER differs from Dr. Watson in many respects, not the least of which is an optional automated upload to Microsoft's servers.

The folks who wrote WER, and those who poured through the dumps, knew full well that sensitive information might be transmitted as part of the WER collection. That's why the good doctor asked for permission before sending the info on to Microsoft's servers.

WER was a resounding success. Steve Ballmer says that WER let the Windows team fix 29 percent of all Windows XP errors in Service Pack 1. More than half of all Office XP bugs were squashed in Office XP SP1, thanks to WER. WER became the envy of the operating system software class, propelling many doctoral theses.

Frighteningly, WER data wasn't encrypted before transmission until March 2014. If you had a crash before then and WER kicked in and delivered it to Microsoft, anybody snooping on your Internet connection could see the contents of the report. There have also been allegations that the NSA hooked into WER reports.

Customer Experience Improvement Program

While Watson and WER concentrated on crash reports, an independent force arose in the Windows camp. Borrowing on the Business School buzz phrase "customer experience," Microsoft's Customer Experience Improvement Program (CEIP) gathers a wide array of information about your computer and how you use it, and then shuttles it all off to Microsoft. Historically, when Microsofties used the term *telemetry*, they were referring specifically to CEIP data. That's changing as more telemetry becomes accessible.

CEIP (known internally in Microsoft as SQM, or Software Quality Management) started with MSN Messenger, moving rapidly to Office 2003, and then to Windows Vista and Windows Media Player. It's been part of Windows and Office ever since. When you install any of those programs, Microsoft activates CEIP by default, although you can opt out.

Feedback & Diagnostics tab and DiagTrack

One part WER, one part CEIP, Windows 10 brings all the snooping together under the Feedback & Diagnostics tab. Telemetry in Windows 10 includes data uploaded by the Connected User Experience and Telemetry component, also known as Universal Telemetry Client, with a service application name of DiagTrack.

Microsoft has a detailed description of its telemetry collection policy in a TechNet post by Brian Lich at https://technet.microsoft.com/en-us/itpro/windows/manage/configure-windows-telemetry-in-your-organization. Lich includes an informative diagram that explains Microsoft's conceptual levels of telemetry. See Figure 6-1.

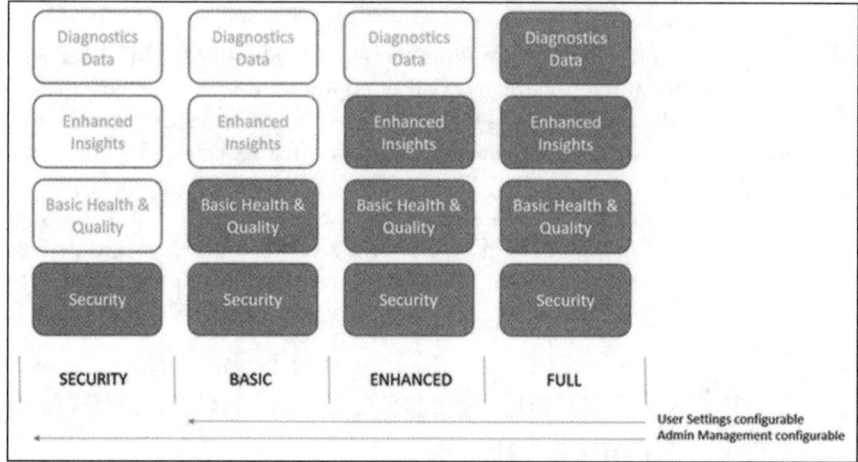

FIGURE 6-1: Microsoft's explanation of stock telemetry levels.

It's far from a definitive list of what data gets sent to Microsoft. Still, the diagram should give you a basic understanding.

To see what you're up against, click the Start button, the Settings icon, and then Privacy. On the left, choose Diagnostics & Feedback. You see the dialog shown in Figure 6-2.

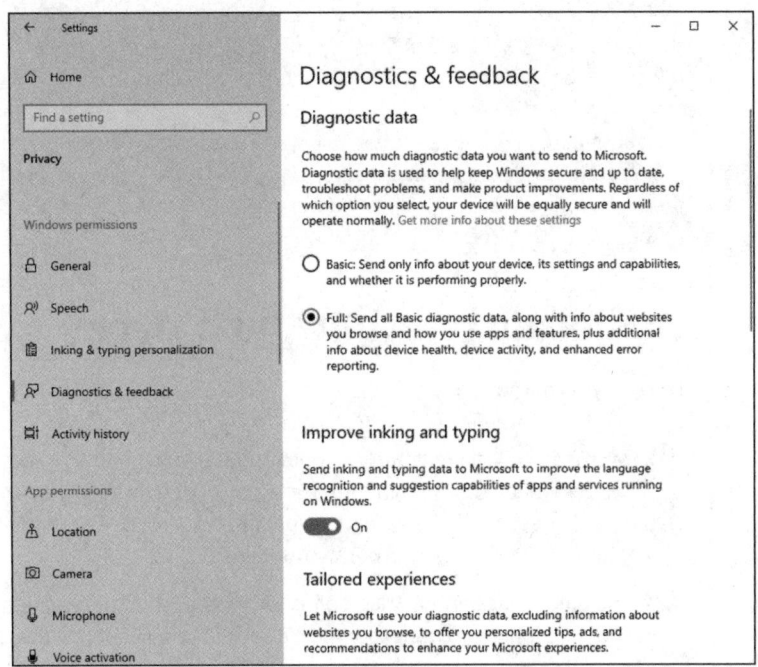

FIGURE 6-2:
The old crash reporting and CEIP settings have a new guise.

The Diagnostic Data setting is one of the key methods you have to reduce — but not eliminate — the Windows 10 telemetry sent from your PC to Microsoft. If you're concerned about sending Microsoft your usage information, select Required Diagnostic Data in the Diagnostic Data section shown in Figure 6-2 and see the nearby sidebar. Then work through the detailed list in "Minimizing Privacy Intrusion" at the end of this chapter.

ASK
WOODY.COM

Although the Settings app only offers two telemetry settings — Basic (or Required) and Full (or Optional) — Windows 10 supports four settings. You can get to the other two (called Security [Enterprise Only] and Enhanced) only if you run the Group Policy Editor. If you don't know about the Group Policy Editor, you're best off sticking with Basic.

Denial ain't nuthin' but a river

Here's what I know:

>> Microsoft collects telemetry — data about your use of Windows 10 — no matter what. You can minimize the amount of data collected (the Basic setting, described in the nearby sidebar,) but you can't stop the flow unless you're connected to a corporate domain.

>> The data being sent to Microsoft is encrypted. That means anyone who's snooping on your connection won't be able to pull out any useful information. As of the April 2018 update (Windows 10 version 1803), Microsoft offers a free app named Diagnostic Data Viewer, which allows people to view what data is sent to Microsoft from their Windows 10 PC.

There's a larger picture. Windows 10, like the rest of the industry, is evolving. I've seen no indication that Microsoft is any worse than, say, Google — and Apple likely undertakes similar data stockpiling. So do Facebook and dozens, if not thousands, of lesser snoopers.

TIP

To enable the Diagnostic Data Viewer and use it to see the data sent to Microsoft, click Start ⇨ Setting ⇨ Privacy. On the left, choose Diagnostics & Feedback. On the right, scroll down to View Diagnostic Data and set its switch to On. Then you can click the Open Diagnostic Data Viewer button.

WARNING

If you want to minimize the identifiable data harvested from you and don't feel comfortable with the fact that Microsoft collects data about you, best to switch to Linux. Then, avoid Google Chrome and use Firefox, use DuckDuckGo instead of Google Search, and always run a VPN (see Book 9, Chapter 4).

'Course, you'd also have to avoid using a smartphone — or even a landline for that matter — and pay with cash or Bitcoin only. You'd also need to avoid walking in public, given the current state of facial recognition, and hope you never end up in a hospital!

ASK WOODY.COM

The question is how comfortable you feel entrusting all these companies — not just Microsoft — with your data. And heaven help you if you live in a house that has a smart electric meter.

I think that data privacy will be one of the foremost legal questions of the next decade. We already have some data protection regulations in place for health records and credit records, but they don't apply in this case. Unless people give up — which may be a reasonable reaction — I predict large-scale problems.

Knowing What Connections Windows Prefers

If you use Windows 10, you're not on a level playing field. Microsoft plays favorites with some online companies and shuns others as much as it possibly can. Cases in point:

>> **Microsoft owns part of Facebook.** You see Facebook here and there in Windows 10. There's a reason why: Microsoft owns a 1.6-percent share of Facebook (at the time of this writing, anyway). Facebook is ambivalent about Microsoft, at best, and as of mid-2015, some open warfare had started. Hard to say how it will play out.

It isn't clear whether Microsoft and Facebook share any data about individual users. But that's a possibility, if not now, at some point in the undefined future.

>> **Microsoft doesn't play well with Google.** Windows 10 has some hooks into Google, but invariably they exist to pull your personal information out of Google (for example, Contacts) and put it in Microsoft's databases. When you see a ready-made connector in Windows 10's Mail app to add a Gmail account — so you can retrieve your Gmail messages in Microsoft's Mail app — there's an ulterior motive.

>> **Microsoft gives lip service to Apple.** There's no love lost between the companies. Microsoft makes software for Mac and iPad platforms (for example, Office for iPad is a treat, OneNote runs on the iPad, and Office has been on the Mac for longer than it's been on Windows!). Apple still makes software for Windows (such as iTunes, Safari, and QuickTime). But they're both fiercely guarding their turf. Don't expect to see any sharing of user information.

>> **Microsoft once tried to buy Yahoo!, which owns Flickr.** Microsoft has hired a boatload of talented people from Yahoo!. Microsoft also still has strong contractual ties to Yahoo!, particularly for running advertising on its search engine, although that could change.

And of course, you know that Microsoft also owns Skype, Hotmail/Outlook.com, Xbox, and OneDrive, right?

Your information — aggregated, personally identifiable, vaguely anonymous, or whatever — can be drawn from any of those sources and mashed up with the data that Microsoft has in its databases. No wonder data mining is a big topic on the Redmond campus.

Controlling Location Tracking

Just as in Windows 8 of yore, Windows 10 has *location tracking.* You must tell Windows 10 and specific applications that it's okay to track your location, but if you do, those apps — and Windows itself — know where you are.

ASK
WOODY.COM

Location tracking isn't a bad technology. Like any technology, it can be used for good or not-so-good purposes, and your opinion about what's good may differ from others'. That's what makes a horse race. And a lawsuit or two.

Location tracking isn't just one technology. It's several.

If your PC has a *GPS* (Global Positioning System) chip (see Figure 6-3) — they're common in tablets but unusual in notebooks and rare in desktops — and the GPS is turned on, and you've authorized a Windows 10 app to see your location, the app can identify your PC's location within a few feet.

HOW APPLE'S LOCATION TRACKING RANKLED

In April 2011, two researchers — Alasdair Allan and Pete Warden — found that iPads and iPhones with GPS systems were keeping track of location and time data, inside the devices, even if the user explicitly disallowed location tracking. They discovered a log file inside every iPad and iPhone running iOS 4 that included detailed information about location and time since 2010.

They also found that the file was being backed up when the iPhone or iPad was backed up, and the data inside the file wasn't encrypted or protected in any way, and a copy was kept on any computer you synced with the iPhone or iPad.

When confronted with the discovery, Apple at first denied it, and then said that "Apple is not tracking the location of your iPhone. Apple has never done so and has no plans to ever do so" — effectively confirming the researchers' discoveries. As details emerged, Apple claimed it was storing the information to make the location programs work better, but it wasn't being used in, or passed to, any location tracking programs.

In May 2011, Apple released iOS 4.3.3, which no longer kept the data. A series of lawsuits and a class action suit followed in the United States, ending with the court granting Apple's motion to dismiss the case. In Korea, the Communications Commission fined Apple about $3,000 for its transgressions.

Location tracking in tablets is a relatively new phenomenon, and it's bound to have some bugs. With a little luck, the bugs — and gaffes — won't be as bad as Apple's.

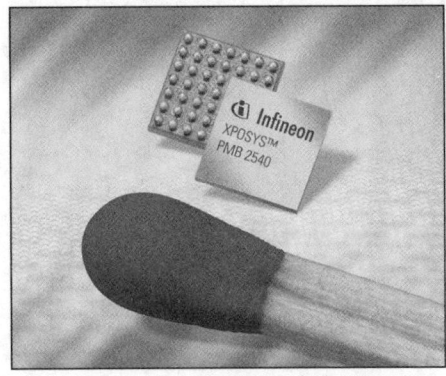

FIGURE 6-3:
GPS chips
turn tiny.

Source: Infineon press release

GPS is a satellite-based method for pinpointing your location. Currently, two commercial satellite clusters are commonly used — GPS (United States, two dozen satellites) and GLONASS (Russia, three dozen satellites). They travel in specific orbits around the earth (see Figure 6-4); the orbits aren't geosynchronous, but they're good enough to cover every patch of land on earth. The GPS chip locates four or more satellites and calculates your location based on the distance to each.

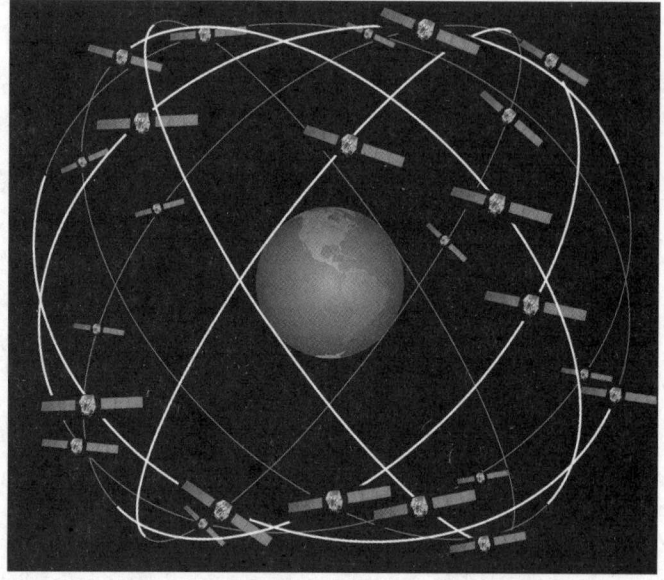

FIGURE 6-4: Carefully crafted orbits ensure that a GPS chip can almost always find four satellites.

Source: HEPL, Stanford University

TRACKING YOUR SHOTS

Any time you put a GPS system and a camera together, you have the potential for lots of embarrassment. Why? Many GPS-enabled cameras — including notably the ones in many phones and tablets — brand the photo with an exact location. If you snap a shot from your tablet and upload it to Facebook, Flickr, or any of a thousand photo-friendly sites, the photo may have your exact location embedded in the file, for anyone to see.

Law enforcement has used this approach to find suspects. The US military warns active duty personnel to turn off their GPSs to avoid disclosing locations. Even some anonymous celebrities have been outed by their cameras and phones. Be careful.

If your Windows 10 PC doesn't have a GPS chip, or it isn't turned on, but you do allow apps to track your location, the best Windows 10 can do is to approximate where your Internet connection is coming from, based on your IP address (a number that uniquely identifies your computer's connection to the Internet). And in many cases, that can be miles away from where you're actually sitting.

When you start a Windows 10 app that wants to use your location, you may see a message asking for your permission to track it, as in the Maps app shown in Figure 6-5.

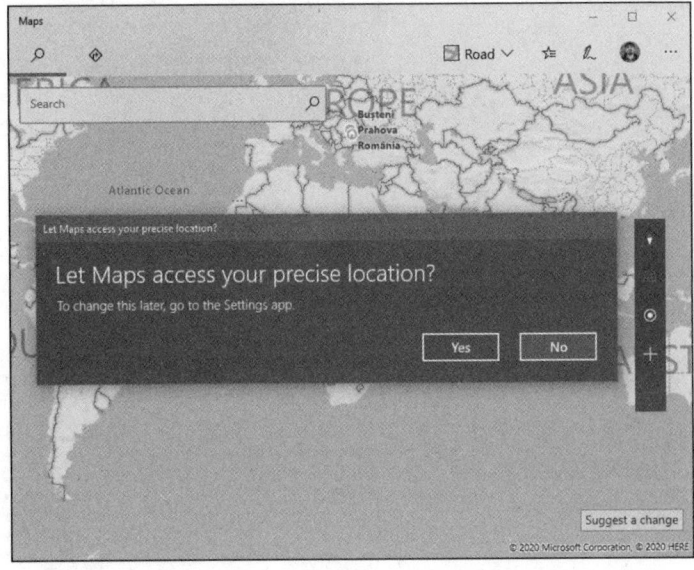

FIGURE 6-5: Windows 10's Maps app wants you to reveal your location.

If you've already turned on location services, each time you add another app that wants to use your location, you see a notification that says, "Let Windows 10 app access your precise location?" You can respond either Yes or No. The following sections explain how you can control location tracking in Windows 10.

Blocking all location tracking

To keep Windows 10 from using your location in *any* app — even if you've already turned on location use in some apps — follow these steps:

1. **Click or tap the Start button and then the Settings icon.**

2. **Click or tap Privacy. On the left, choose Location.**

   The Location Privacy screen appears, as shown in Figure 6-6.

3. To turn off location tracking — even if you've already given your permission to various apps to track your location — click the Change button in the Allow Access to Location on This Device section, and set it to Off.

That's all it takes.

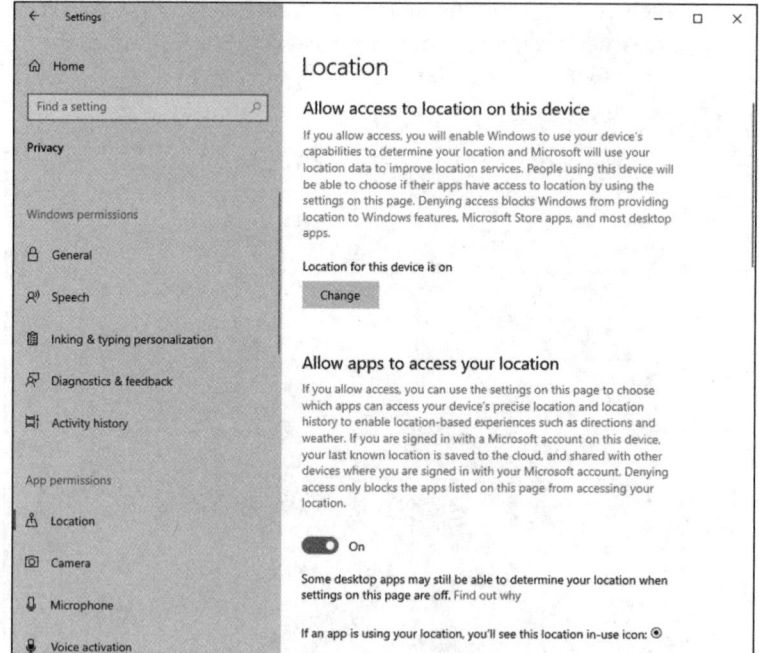

FIGURE 6-6:
Click the Change button to shut off the switch for location tracking.

Blocking location tracking in an app

If you've given an app permission to use your location, but want to turn it off, without throwing the big Off switch described in the preceding steps, here's how to do it:

1. Click or tap the Start button and then the Settings icon.

2. Click or tap Privacy. On the left, choose Location.

3. Scroll down until you find the app you want to cut off.

In Figure 6-7, I looked for the Weather app.

4. On the right, slide the app's Location slider to Off.

The app loses its permission to access your location.

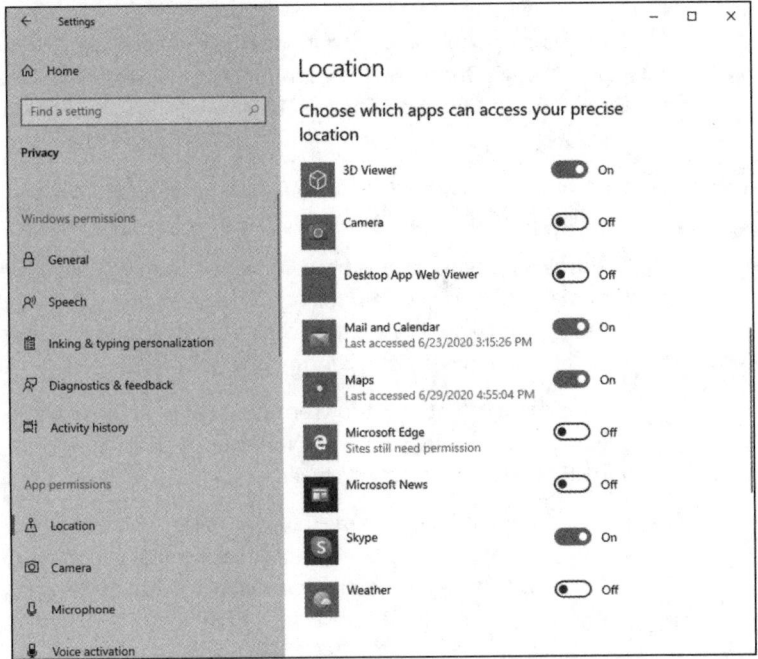

FIGURE 6-7:
You can turn off location tracking for individual apps, as well.

 Some apps keep a history of your locations or searches that may pertain to your location. If you want to verify that's been deleted, too, bring up the app, click or tap the hamburger icon in the upper-left corner (and shown in the margin) and choose Settings. The Settings pane appears on the right. In most cases, you can choose Options and then click the link that says Clear Searches.

Minimizing Privacy Intrusion

Although it's true that using Windows 10 exposes you to many more privacy concerns than any previous version of Windows, you can reduce the amount of data kept about you by following a few simple rules:

>> If you want to log in to Windows 10 using a Microsoft account — and there are many good reasons for doing so — consider setting up a Microsoft account that you use only for logging in to Windows 10 (and possibly for OneDrive, Xbox, and/or Skype). See Book 2, Chapter 5, for details.

>> **Don't use the Windows 10 apps for Mail, People, Calendar, Skype, or OneDrive.** If you have a Hotmail/Outlook.com or Gmail account, don't access them through Windows 10's Mail app; go to your browser (Firefox?), and log in to Hotmail/Outlook.com or Gmail. If you keep a separate Microsoft account

for logging in to Windows 10 only, use the web interface for OneDrive — by going through OneDrive. Run your Contacts, Calendar, and Messaging through Hotmail/Outlook.com or Gmail as well. It isn't as snazzy as using the Windows 10 apps, but it works almost as well.

I use Gmail. If Google wants to bombard me with ads, so be it: I don't buy anything from the ads anyway.

>> **Always use private browsing.** In Microsoft Edge, it's called *InPrivate;* Google Chrome calls it *Incognito;* Firefox says *Private Browsing.* Turning on this mode keeps your browser from leaving cookies around, and it wipes out download lists, caches, browser history, forms, and passwords.

Realize, though, that your browser still leaves crumbs wherever it goes: If you use Google to look up something, for example, Google still has a record of your IP address and what you typed.

Private browsing isn't the same thing as Do Not Track. In fact, as of this writing, Do Not Track is a largely futile request that you make to the websites you visit, asking them to refrain from keeping track of you and your information. For details, see Book 5, Chapter 1.

>> **If you use Office, turn off telemetry in it.** In any Office 2016 program, choose File ⇨ Options ⇨ Trust Center. Click the Trust Center Settings option, and then on the left choose Privacy Options. Deselect the box marked "Send us information about your use and performance of Office software to help improve your Microsoft experience."

In addition to rolling the Diagnostics Data setting to Basic, as described at the beginning of this chapter, you can clamp down further on your privacy settings by going through the lengthy list compiled by Martin Brinkmann at ghacks. net. Go to www.ghacks.net/2015/07/30/windows-10-and-privacy/. Martin literally wrote the book on Windows 10 privacy, an e-book called *The Complete Windows 10 Privacy Guide,* which is available from Amazon at www.amazon.com/ Complete-Windows-10-Privacy-Guide/dp/1978104723.

3

Working on the Desktop

Contents at a Glance

Chapter **1**

Running Your Desktop from Start to Finish

This chapter explains how to find your way around the Windows 10 interface. If you're an old hand at Windows, you know most of this stuff — such as mousing and interacting with dialog boxes — but I bet some of it will come as a surprise, particularly if you've never taken advantage of Windows libraries or if Windows 8/8.1's Metro side tied you in knots. You know who you are.

Most of all, you need to understand that you don't have to accept all the default settings. Windows 10 was designed to sell more copies of Windows 10. Much of that folderol just gets in the way. What's best for Microsoft isn't necessarily best for you, and a few quick clicks can help make your PC more usable, and more . . . yours.

TIP

If you're looking for information on customizing the Windows 10 Start menu and the taskbar, skip ahead to Book 3, Chapter 2. To look at personalizing the desktop (and tablet mode), read Book 3, Chapter 3.

Tripping through Windows 10's Three Personas

As soon as you *log in* to the computer (that's what it's called when you click your name), you're greeted with an enormous expanse of near-nothingness, cleverly painted with a pretty picture. Your computer manufacturer might have chosen the picture for you, or you might see the default Windows 10 wallpaper.

Your Windows destiny, such as it is, unfolds on the computer's screen.

When you crank up Windows 10, it can take on one of three personas. They're pretty easy to discern, if you follow these guidelines:

» **Windows desktop:** Almost everybody starts with the desktop. It has a Start button in the lower-left corner, more icons along the taskbar at the bottom, and larger icons (possibly just the Recycle Bin and Microsoft Edge) on top of the desktop. The picture on the desktop could look like just about anything.

If you click the Start button in the lower-left corner (and shown in the margin), you see a Start menu on the left and a whole bunch of tiles on the right, as in Figure 1-1. That's what I think of as regular Windows 10. Your background picture will no doubt differ, as will the contents of the Start menu on the left and probably the Start tiles on the right.

FIGURE 1-1: The Start Menu from Windows 10 opens when you click the Start button.

If you look at the row of little icons on the far left, the most important one to remember is the one shaped like a gear (and shown in the margin). It's just above the on-off switch icon. Others have used the gear for Settings for decades, and Microsoft has finally caught on. You now find Settings in many parts of Windows 10 behind an eight-spoked icon.

>> **Full-screen start:** If you've been playing around with your computer, or someone else has done it for you, you may arrive in full-screen start, shown in Figure 1-2.

If you're in full-screen start, I recommend that you get out of it for now, while you're still getting your bearings. To do so, click the hamburger icon, the Settings icon (both shown in the margin), Personalization, Start. On the right, slide the switch marked Use Start Full Screen to Off. That will put you back in Figure 1-1, where you use the Start menu.

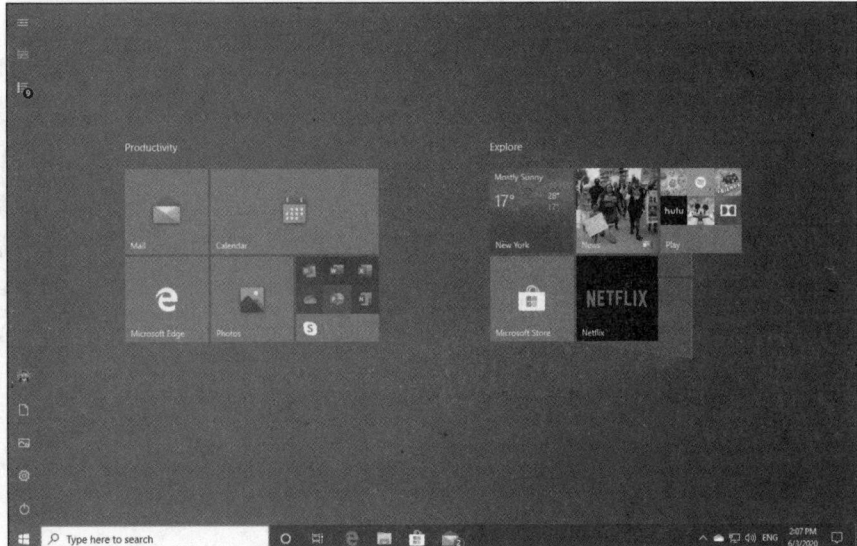

FIGURE 1-2:
The full-screen start. If you see this, drop back to regular Start before you try to change anything.

>> **Tablet mode:** The third possibility is that you started in tablet mode, shown in Figure 1-3. The differences between full-screen start and tablet mode are subtle, but you can see major differences in the taskbar at the bottom. Full-screen start has a big search box to the right of the Start button, but tablet mode has a back arrow.

If you're going to use Windows 10 primarily with your pinkies, instead of a mouse, tablet mode is a good way to, uh, start. I talk about tablet mode extensively in its own section of this chapter. If you find yourself in tablet mode and want to get back to a mouse-happy desktop, click the Action Center icon, waaaaaay down in the lower-right corner, to the right of the time (see Figure 1-3), and then deselect the Tablet Mode tile.

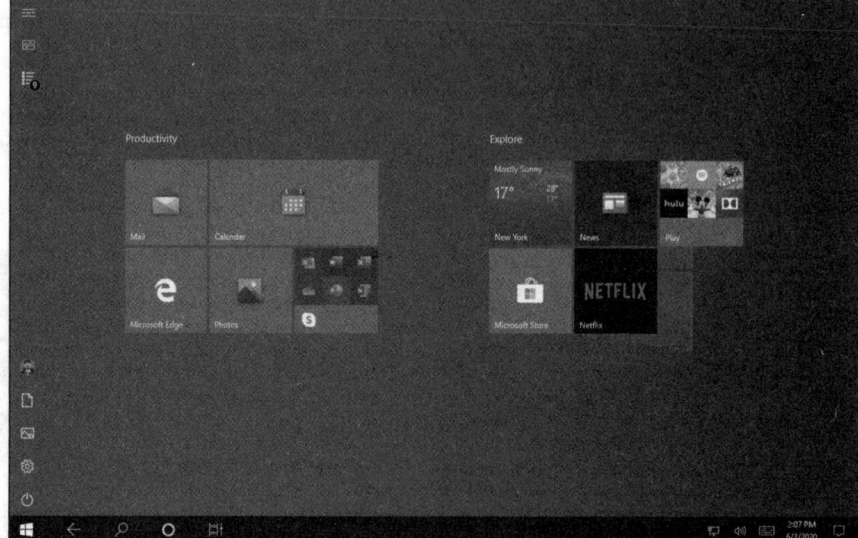

FIGURE 1-3:
Tablet mode is similar to full-screen start but is designed for touch interactions.

Working with the Traditional Desktop

So your main starting screen looks like Figure 1-1, yes? Good. This is where you should start.

The screen that Windows 10 shows you every time you start your computer is the *desktop*, although it doesn't bear much resemblance to a real desktop. Try putting a pencil on it.

ASK
WOODY.COM

I talk about changing and organizing your desktop in Book 3, Chapter 3, but every new Windows 10 user will want to make a few quick changes. That's what you do in this chapter.

TECHNICAL STUFF

The Windows 10 desktop looks simple enough, but don't fool yourself: Under that calm exterior sits one of the most sophisticated computer programs ever created. Hundreds of millions of dollars went into creating the illusion of simplicity — something to remember the next time you feel like kicking your computer and screaming at the Windows 10 gods.

Changing the background

Start taking your destiny into your own hands by changing the wallpaper (err, the *desktop background*). If you bought a new computer with Windows 10 installed, your background text probably says *Dell* or *Vaio* or *Billy Joe Bob's Computer Emporium*, or *Dial 555-3106 for a good time*. Bah. Change your wallpaper by following these steps:

1. **Right-click an empty part of the desktop, or tap and hold down, and then choose Personalize.**

Windows 10 hops to the Settings app's Background pane, shown in Figure 1-4.

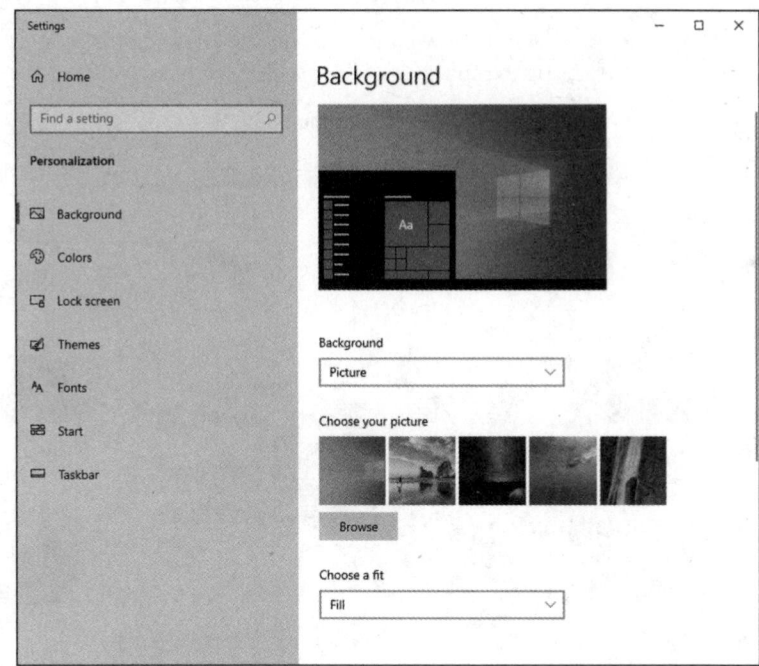

FIGURE 1-4: Choose your desktop background (even a slideshow) here.

2. **Play with the Background drop-down box, and choose a type of background you like.**

   You can choose one of the pictures that Windows 10 offers, a solid color, or a slideshow of what is found inside your Pictures folder.

3. **If you don't see a background that tickles your fancy, or if you want to roll your own backgrounds, click Browse.**

   Windows 10 responds by going into your machine and letting you pick a pic, any pic.

4. **If you find a picture that you like but it looks like a smashed watermelon on your screen or is too small to be visible, in the Choose a Fit drop-down list, tell Windows 10 how to use the picture.**

   These are your options:

   ● Fill the screen. Windows 10 may stretch or crop the image to make this happen.

   ● Fit the image to the available space on the screen.

   ● Stretch the image so that it has the same dimensions as your screen.

   ● Tile the desktop. Windows 10 puts the image on the screen multiple times to fill up the space. See Figure 1-5.

   ● Center the image in the middle of the desktop.

   ● Span the image across multiple displays, if you have two or more.

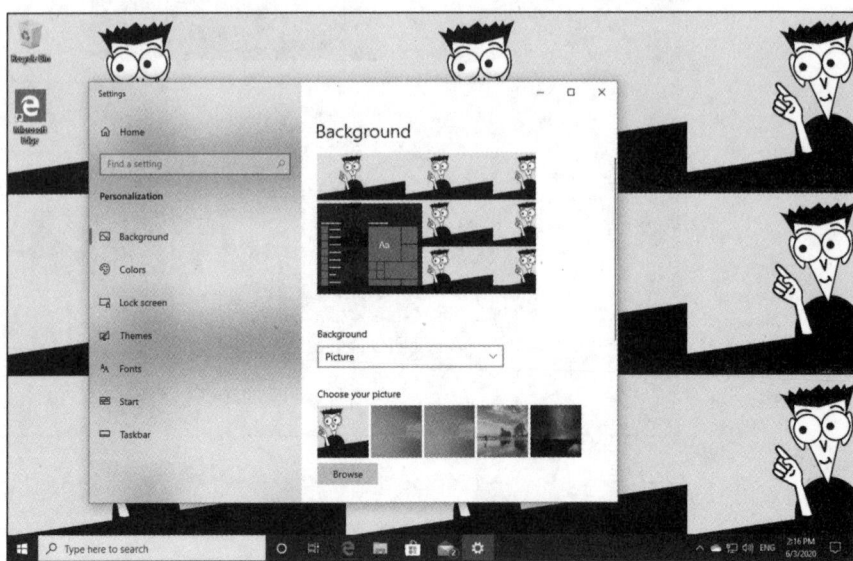

5. **Tap or click the Close (X) button to close the Settings app's Background pane.**

Your new wallpaper settings take effect immediately.

Cleaning up useless icons and programs

If you haven't yet taken control and zapped those obnoxious programs that your PC vendor probably stuck on your machine, now is the time to do it.

TIP

You might think that your brand-spanking-new Windows 10 computer wouldn't have any junk on it. Ha. The people who make and sell computers — all the big-name manufacturers — sell chunks of real estate on your computer, just to turn a profit. Hate to break it to you, but the McAfees and Nortons of the world pay Dell, HP, Sony, Asus and all the others for space. The manufacturers want you to think that they've installed this lovely software for your convenience. Bah. Humbug.

Even Microsoft has taken a dip in the ad-dispersing sewer by cluttering your desktop and Start menu with all sorts of must-have Windows-enhancing products, such as Candy Crush Soda Saga and March of Empires. Ka-ching! Ka-ching!

>> **To get rid of most icons,** simply right-click them and choose Delete or Unpin from Start.

>> **To get rid of the icons' associated programs,** try to remove them the Settings app way first: Click or tap the Start button, the Settings icon, Apps, Apps and Features. See if the program is listed. If so, click or tap it, click Uninstall, and Uninstall one more time, to confirm your action. If you can't find the program in the Settings app, type **Control Panel** in the Windows 10 search box. In the Control Panel, under Programs, click Uninstall a Program. When the Programs and Features window opens, double-click a program to remove it.

ASK WOODY.COM

Unfortunately, many scummy programs don't play by the rules: Either they don't have uninstallers or the uninstaller that appears in the Programs and Features window doesn't get rid of the program entirely. (I won't mention Norton Internet Security by name.) To get rid of the scummy stuff, take a look at Geek Uninstaller. It's at geekuninstaller.com.

Mousing with Your Mouse

For almost everybody, the computer's mouse (or the lowly touchpad) serves as the primary way of interacting with Windows 10. But you already knew that. You can click the left mouse button or the right mouse button, or you can roll the wheel in the middle (if you have one), and the mouse will do different things, depending on where you click or roll. But you already knew that, too.

ASK
WOODY.COM

The Windows 10 Multi-Touch technology and those ever-fancier 11-simultaneous-finger screens let you act like Tom Cruise in *Minority Report,* if you have the bucks for the multiple-finger stuff, the right application software, and the horsepower to drive it. But for those of us who put our gloves on one hand at a time, the mouse remains the input device of choice.

The best way to get the feel for a new mouse? Play one of the games that ships with Windows 10. Choose Start, Microsoft Solitaire Collection, and take it away — just realize that Microsoft will charge you for ad-free versions of their apps. Or hop over to the Microsoft Store for amazing new versions of Minesweeper, Chess Titans, and many others for mouse orienteering. In Figure 1-6, I'm playing a rousing game of traditional Klondike, the program you probably think of when you think "Solitaire."

Try clicking in unlikely places, double-clicking, or right-clicking in new and different ways. Bet you'll discover several wrinkles, even if you're an old hand at the games. (See Book 5, Chapter 4 for more on Windows 10 games.)

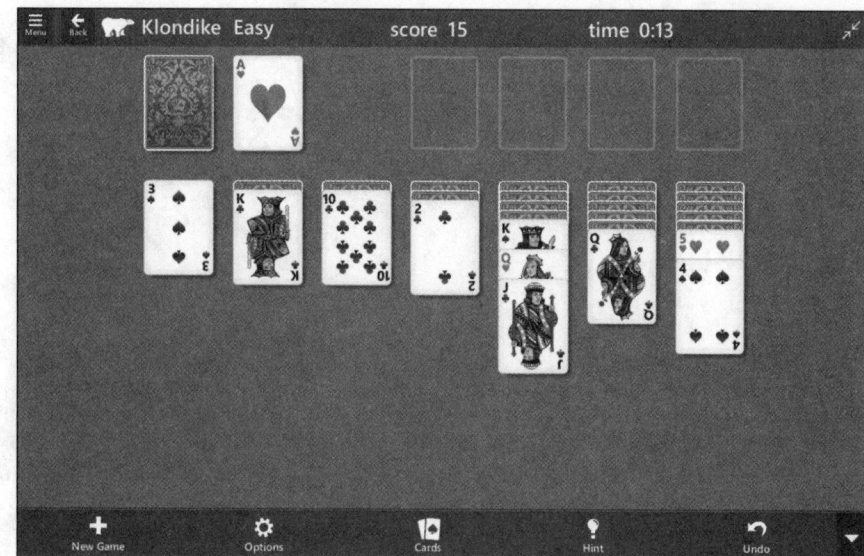

FIGURE 1-6:
The Microsoft Solitaire Collection is great for mouse practice.

TECHNICAL STUFF

Inside the computer, programmers measure the movement of mice in units called *mickeys*. Nope, I'm not making this up. Move your mouse a short distance, and it travels a few mickeys. Move it to Anaheim, and it puts on lots of mickeys.

What's up, dock?

Windows 10 includes several gesture features that can save you lots of time. Foremost among them: a quarter- and half-window docking capability called *snap.*

REMEMBER

If you click the title bar of a window and drag the window a-a-all the way to the left side of the screen, as soon as the mouse hits the edge of the screen, Windows 10 resizes the window so that it occupies the left half of the screen and then docks the window on the far left side. Similarly, *mutatis mutandis,* for the right side. That makes it two-drag easy to put a Word document and a spreadsheet side by side or a list of files from File Explorer alongside your Solitaire game, as shown in Figure 1-7.

A new feature in Windows 10 called Snap Assist makes snapping easier than ever. If you snap one app window to an edge, Windows 10 brings up thumbnails of all the other programs that are running at the time. Click or tap the program, and it occupies the vacant part of the screen, as shown in Figure 1-8.

You can also drag into the corners of the screen and snap four programs into the four corners. (If you're curious, these all are controlled in the Settings app. Tap or click the Start button, the Settings icon, System, Multitasking; the relevant settings are at the top of the pane.)

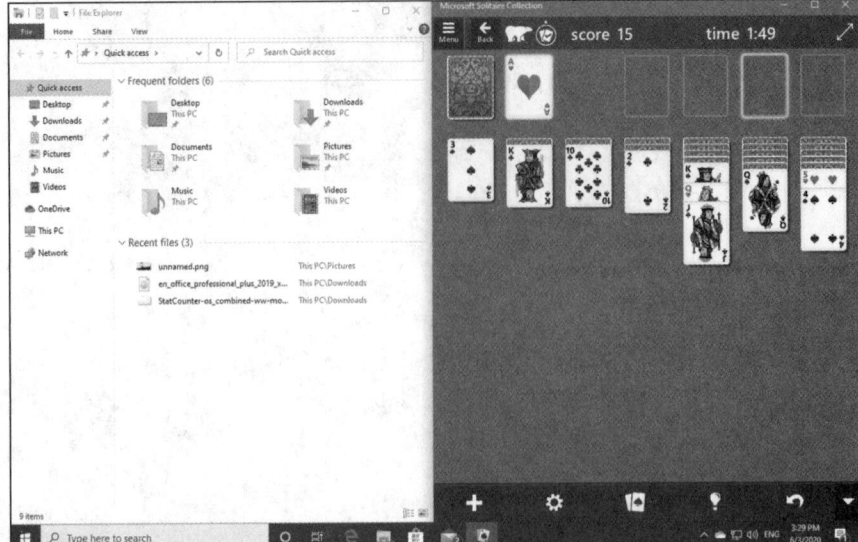

FIGURE 1-7:
Two drags and
you can have
Windows 10
arrange two
programs side
by side.

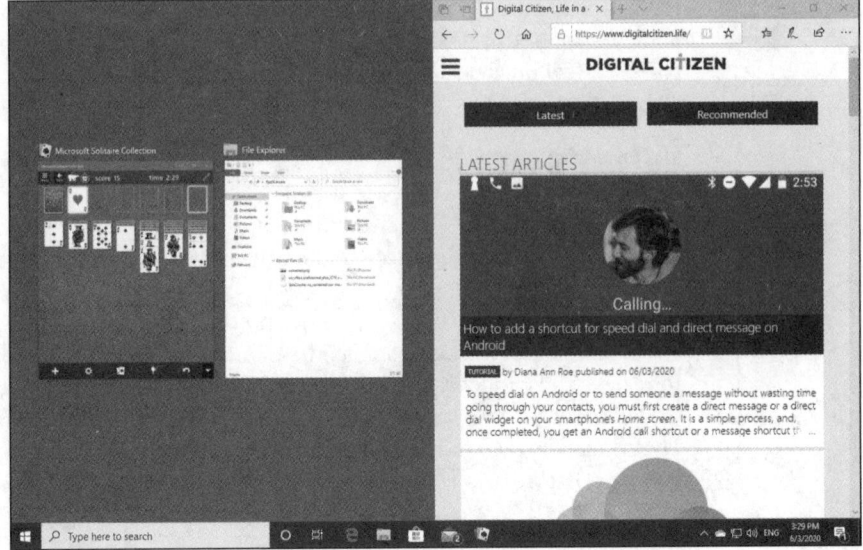

FIGURE 1-8:
Snap Assist
helps you put
two programs
side by side by
offering to snap
the other running
programs.

Those aren't the only navigation tricks. If you drag a window to the top of the screen, it's *maximized*, so it occupies the whole screen. (Yeah, I know: You always did that by double-clicking the title bar.) And, if you click a window's title bar and shake it, all other windows on the screen move out of the way: They *minimize* themselves on the toolbar.

If you have rodentophobia, you can also do the mouse tricks explained in this section by pressing the following key combinations:

>> **Snap left:** Windows key+left arrow

>> **Snap right:** Windows key+right arrow

>> **Maximize:** Windows key+up arrow

Changing the mouse

TIP

If you're left-handed, you can interchange the actions of the left and right mouse buttons — that is, you can tell Windows 10 that it should treat the left mouse button as though it were the right button and treat the right button as though it were the left. The swap comes in handy for some left-handers, but most southpaws I know (including both of my sons) prefer to keep the buttons as is because it's easier to use other computers if your fingers are trained for the standard setting.

TIP

The ClickLock feature can come in handy if you have trouble holding down the left mouse button and moving the mouse at the same time — a common problem for laptop users who have fewer than three hands. When Windows 10 uses ClickLock, you hold down the mouse button for a while (you can tell Windows exactly how long) and the operating system locks the mouse button so that you can concentrate on moving the mouse without having to hold down the button.

To switch left and right mouse buttons or turn on ClickLock, follow these steps:

REMEMBER

1. **Click or tap Start, then the Settings icon, Devices, Mouse (or Touchpad).**

   The Settings icon looks like a gear.

   Windows 10 opens the Mouse (or Touchpad) dialog box, shown in Figure 1-9. *Note:* If you have a sufficiently powerful touchpad, you may be able to adjust settings for single-tap clicking, two-finger drag to scroll, and so on. Specific options vary depending on your brand of touchpad.

2. **If you want to switch the functions of the left and right mouse buttons, change the entry in the Select Your Primary Button box.**

3. **If you want to turn on ClickLock:**

   a. *Tap or click the Additional Mouse Options link, at the bottom.* You get an old-fashioned Control Panel dialog box called Mouse Properties, which you can see in Figure 1-10.

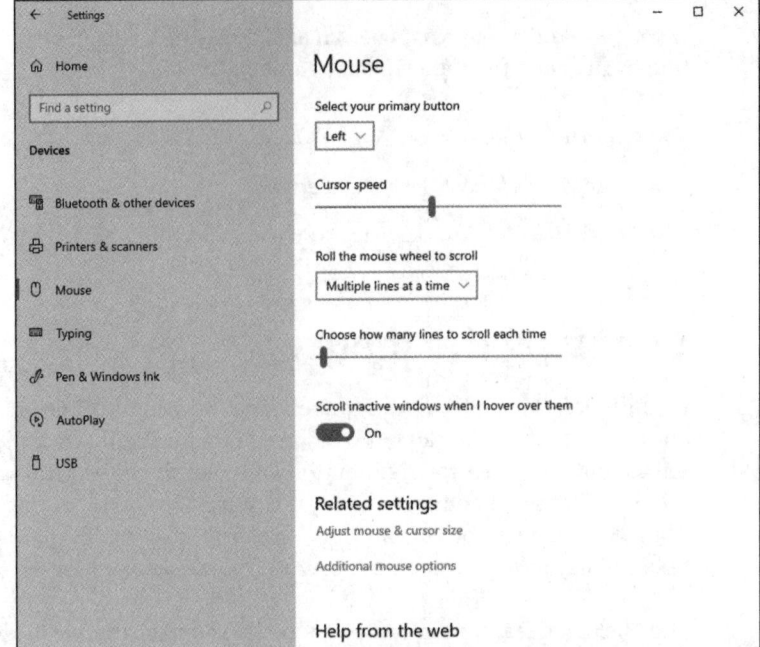

FIGURE 1-9:
Reverse the left and right mouse buttons with one click in the Settings app.

b. *At the bottom of the Mouse Properties dialog box, select the Turn on ClickLock option, and then click OK.*

Although changes made in the Settings app take effect immediately, changes in the old-fashioned Control Panel don't go into effect until you click Apply or OK.

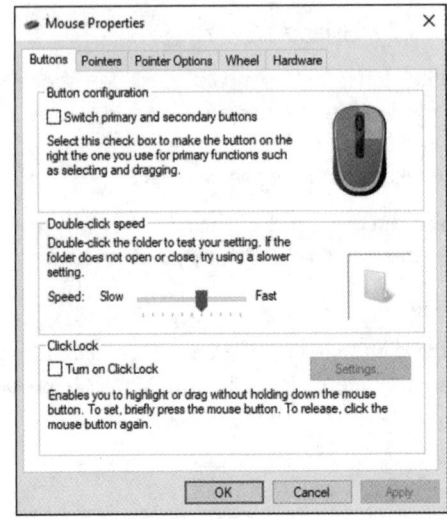

FIGURE 1-10:
This old-fashioned Control Panel dialog box offers the setting for ClickLock.

Starting with the Start button

 Microsoft's subverting of the classic Rolling Stones song "Start Me Up" for Windows 95 advertising might be ancient history now, but the royal road to Windows 10 still starts at the Start button. Click the Start button in the lower-left corner of the screen to open the new Windows 10 Start menu, which looks something like the one shown in Figure 1-11.

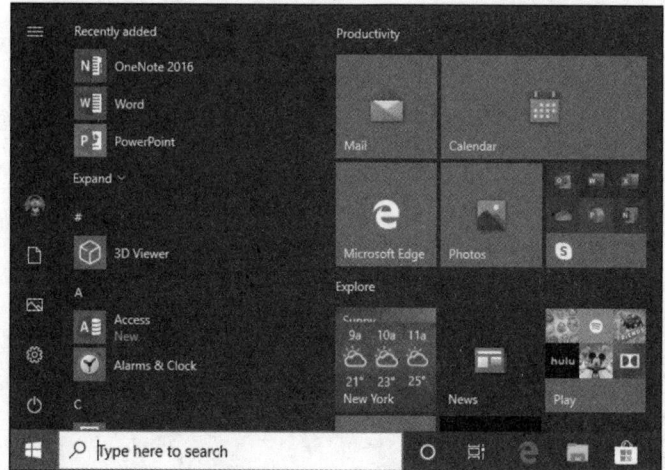

FIGURE 1-11:
The Windows
Start menu can
be customized a
little bit.

The Start menu looks like it's etched in granite, but it isn't. You can change three pieces without digging deep:

>> **To change the name or picture of the current user,** see Book 2, Chapter 2.

>> **To remove a program from the Recently Added or Most Used programs lists,** right-click it, choose More, and then select Remove from This List.

>> **To move a tile on the right or resize one,** just click and drag the tile. You also can right-click (or click and hold down), choose Resize, and then pick a new size — see Book 3, Chapter 2 for details.

TIP

If you bought a new computer with Windows 10 preinstalled, the people who make the computer may have sold one or two or three of the spots on the Start menu. Think of it as an electronic billboard on your desktop. Nope, I'm not exaggerating. I keep expecting to bump into a Windows machine with fly-out Start menu entries that read, oh, "Statistics prove/Near and far/That folks who drive/Like crazy are/ Burma Shave." (See Burma-shave.org/jingles.)

WINDOWS 10 APPS, FORMERLY UWP APPS

Microsoft has always had a hard time with branding — making its technical achievements sparkle and fizz and convey meaning with a name. I think *Windows RT* was the all-time low in Microsoft marketing nomenclature — *Windows Live* sure gave it a run for the money — but that's a bygone.

Near the top of the list (or the bottom, depending on how you stand) of bad branding is the term *Universal Windows app* and/or *Universal Windows Platform app* and/or *UWP app*. At least in theory, Universal Windows apps (sometimes called, confusingly, just *Windows 10 apps* or just *Universal apps* or even *Universal Windows Platform apps*) are computer programs that can, in theory, run on any Windows 10 device, whether it's a desktop, a laptop, a tablet, an Xbox gaming console, a HoloLens headset, or a hearing aid. I wonder what a Blue Screen sounds like.

They can, in theory, run on any device because they make use of a new set of Windows programming interfaces, called the WinRT API. The WinRT API is very different from the old Windows programming interfaces, generally called the Win32 API. The programs you've used on Windows for years run on the Win32 API, and they work on the Windows 10 desktop, much as they always have. But the Universal Windows apps run inside their own little boxes — yep, they look just like Windows programs— and the boxes sit on the desktop.

When you think of Windows versions, the Universal Windows apps run only on Windows 10. In general, they won't run on Windows 8 or 8.1. They definitely can't run on Windows 7 or earlier, because those versions of Windows didn't include the WinRT API.

The WinRT API has all sorts of advantages over the old Win32 API — security, for one, because it's harder to hack a system from inside a WinRT app, but there are lots of additional capabilities that have become more important as we've turned more mobile. The WinRT API reduces battery demand, makes programming easier for touch input and for resizing screens. It keeps programs from clobbering each other. And on and on.

In the ripe old days (circa Windows 8), the programs that used the WinRT API were called Metro apps. When, according to legend, the German supermarket chain Metro threatened to take Microsoft to task (Who is Microsoft to complain? They trademarked *windows*), Microsoft stopped calling Metro apps *Metro* and the result has been pandemonium.

The names used in the interim include Metro, Metro Style, Windows 8 application, Windows Store app (before *Windows Store* morphed into *Microsoft Store*), Windows 8–style user interface app (that really sizzles, doesn't it?), new user interface app, Modern app, and a handful of additional names that aren't entirely printable. Just ask the developers.

Microsoft seems to have dropped the name *program* entirely, no doubt because Apple and Google have apps, not (sniff) programs. Recently, they decided to promote the term Windows 10 apps.

No matter what you call them, Universal Windows apps are clearly the way of the future. The WinRT API has the Win32 API beat in all sorts of ways, except compatibility: Win32 programmers have to learn a completely new way of programming and a new way of thinking, and transferring those tens of billions of lines of code from Win32 to WinRT will take decades. By which time WinRT will be obsolete, no doubt.

The trouble is that UWP apps have not gained a lot of ground with Windows 10 developers, who decided to stick to the old desktop programs. As a result, the future of Windows 10 is Windows 10 apps, which may be UWP apps, desktop apps, or a hybrid of both. Fun, isn't it?

This is how Microsoft is trying to bridge the gap between the two worlds. Welcome to the future.

The right side of the Start menu contains a plethora of tiles. At the beginning, the built-in tiles are all for Windows 10 apps (see the nearby sidebar) from Microsoft itself, plus a peppering of tiles from companies that have emerged on Microsoft's good side. Your computer vendor may have stuck in a couple extras. Ka-ching. And in the normal course of using your computer, you may well put some tiles over there, too.

Here's what you find on the Right Side of the Start Force:

>> The **productivity apps from Microsoft** (Calendar and Mail) are marginally useful, but not likely to draw you away from your current email or calendar program, especially if you use email or a calendar on your smartphone or tablet. Windows 10 used to have a People app here, but it was so bad the powers-that-be got rid of it. See Book 4, Chapters 1 and 2.

>> **Microsoft Edge** may be the most complex Windows 10 app ever written. Microsoft got rid of Internet Explorer and is trying to get people moved over to a modern browser. See Book 5, Chapter 1.

>> **OneNote** is a useful note-taking and clipping app from Microsoft. I use Evernote, but they're directed at two different audiences. See Book 4, Chapter 4.

Also included are a whole bunch of **shovelware apps**, including Groove Music, Movies & TV (Book 4, Chapter 5), Photos (Book 4, Chapter 3), Skype, and Weather, plus an enormous number of apps that invite you to spend more money.

You can modify most of the right side of the Start menu by dragging and dropping tiles, and right-clicking (or tapping and holding down) a tile to resize. There's much more about working with Windows 10 app tiles in Book 3, Chapter 2.

Touching on the Taskbar

Windows 10 sports a highly customizable taskbar at the bottom of the screen (see Figure 1-12). I go into detail in Book 3, Chapter 3.

FIGURE 1-12:
The Windows 10 taskbar lets you pinpoint what's running and jump to the right location quickly.

The taskbar's a wonderfully capable locus for most of the things you want to do, most of the time. For example:

» **Hover your mouse cursor over an icon to see what the program is running.** In Figure 1-12, I hovered my mouse cursor over the Microsoft Edge icon and can see that www.digitalcitizen.life is open.

TECHNICAL STUFF

Some applications, such as File Explorer, show each tab or open document in a separate thumbnail. Clicking a thumbnail brings up the application, along with the chosen tab or document. This nascent feature is implemented unevenly at this point.

>> **Right-click an icon, and you see the application's jump list.** The jump list MAY show an application's most recently opened documents and, for many apps, a list of common tasks and activities. It may show a browser's history list.

TIP

If you click an icon, the program opens, as you would expect. But if you want to open a second copy of a program — say, another copy of Firefox — you can't just click the icon. You must press the Shift key on the keyboard and click the program's icon again. An alternative is to right-click and choose the application's name.

You can move most of the icons around on the taskbar by simply clicking and dragging. (You can't move the Start, Cortana, or the Task View icons.)

REMEMBER

If you want to see all the icons on your desktop and relegate all open windows to shadows of their former selves, click the far right edge of the taskbar.

The Windows 10 taskbar has many tricks up its sleeve, but it has one capability that you may need if screen real estate is at a premium. (Hey, you folks with 30-inch monitors need not apply, okay?)

Auto-Hide lets the taskbar shrink to a thin line until you bump the mouse pointer way down at the bottom of the screen. As soon as the mouse pointer hits bottom, the taskbar pops up. Here's how to teach the taskbar to auto-hide:

1. **Right-click an empty part of the taskbar.**

2. **Click Taskbar Settings.**

    The Taskbar Settings should be visible.

3. **Slide the Automatically Hide the Taskbar in Desktop Mode setting to On.**

    The taskbar holds many surprises. See Book 3, Chapter 3.

Working with Files and Folders

"What's a file?" Man, I wish I had a nickel for every time I've been asked that question.

A file is a, uh, thing. Yeah, that's it. A thing. A thing that has stuff inside it. Why don't you ask me an easier question, like "What is a paragraph?" or "What is the meaning of life, the universe, and everything?"

A *file* is a fundamental chunk of stuff. Like most fundamental chunks of stuff (say, protons, Congressional districts, or ear wax), any attempt at a definitive definition gets in the way of understanding the thing itself. Suffice it to say that a Word document is a file. An Excel workbook is a file. That photograph your cousin emailed you the other day is a file. Every track on the latest Coldplay album is a file, but so is every track on every audio CD ever made. Chris Martin isn't that special.

WARNING

Filenames and folder names can be very long, but they can't contain the following characters:

/ \ : * ? " < > |

KEEPING FOLDERS ORGANIZED

If you set folders up correctly, they can help you keep track of things. If you toss your files around higgledy-piggledy, no system of folders in the world can help. Unfortunately, folders have a fundamental problem. Permit me to illustrate.

Suppose you own a sandwich shop. You take a photograph of the shop. Where do you stick the photo? Which folder should you use? The answer: There's no good answer. You could put the photo in with all your other shop stuff — documents and invoices and payroll records and menus. You could stick the photo in the Pictures folder, or in your OneDrive Pictures folder, which Windows 10 automatically provides. You could put it in the Public or Public Documents or Public Pictures folder so other people using your PC, or other folks connected to your network, can see the photo of the shop. You could create a folder named Photos and file away the picture chronologically (that's what I do), or you could even create a folder named Shop inside the Photos folder and stick the picture in \Photos\Shop.

I stick my photos in the Google Photos app (see Book 4, Chapter 3) and rely on a Google search to find them, but you see the point.

This where-to-file-it-and-where-to-find-it conundrum stands as one of the hairiest problems in all of Windows, and until Windows 7, you had only piecemeal help in keeping things organized. Now, using the Windows 10 libraries, and a Search function that (finally!) works the way you would expect, you stand a fighting chance of finding that long-lost file, especially if you're diligent in assigning tags to pictures and videos. For more info on that, see the sidebar "Creating libraries," later in this chapter.

But if you stick the photo in OneDrive, ay, that's another story entirely. See Book 6, Chapter 1 for the sad story (and sidebar) of smart files.

Files can be huge. They can be tiny. They can even be empty, but don't short-circuit any gray cells on that observation.

Folders hold files and other folders. Folders can be empty. A single folder can hold millions — yes, quite literally millions — of files and other folders.

 To look at the files and folders on your machine that you probably use every day, click or tap the File Explorer icon down in the taskbar (and shown in the margin). A program named File Explorer appears, and it shows you the contents of your frequently used folders (see Figure 1-13).

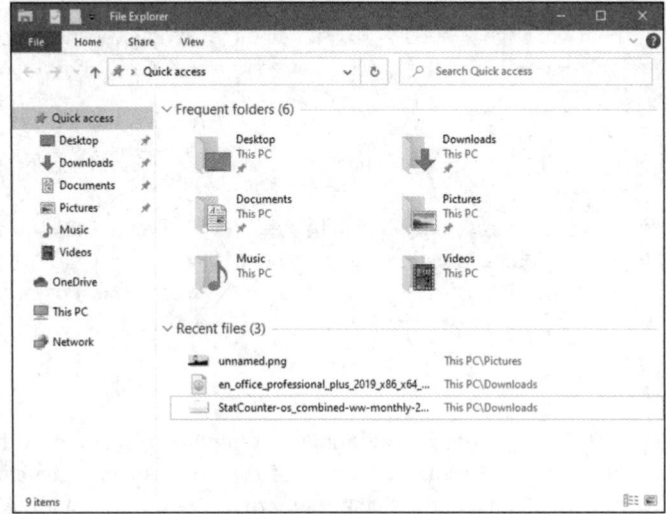

FIGURE 1-13:
The most frequently used folders and recently accessed files, shown by File Explorer.

File Explorer can show you the contents of a hard drive — folders and files — or a thumb drive or a CD/DVD drive. File Explorer can also help you look at other computers on your network, if you have a network.

Using File Explorer

Your PC is a big place, and you can get lost easily. Microsoft has spent hundreds of millions of dollars to make sure that Windows 10 points you in the right direction and keeps you on track through all sorts of activities.

Amazingly, some of it actually works.

If you're going to get any work done, you must interact with Windows 10. If Windows is going to get any work done, it must interact with you. Fair 'nuff.

CREATING LIBRARIES

Windows 7 brought a powerful new concept to the table: libraries. Think of them as easy ways to mash together the contents of many folders: You can work with a collection of folders as easily as you work with just one folder, no matter where the folders live. You can pull together pictures in ten of the folders on your desktop plus the ones in your computer's \Public folder plus the ones on that external 4 terabyte drive and the \Public folder on another computer connected to your network, and treat them all as though they were in the same folder.

Unfortunately, as Microsoft pushed deeper into the cloud and brought OneDrive to the fore, libraries got left behind. In Windows 8 and 8.1, it's hard to find the vestiges and make them work right. Windows 10 continues in the Windows 8/8.1 tradition. Microsoft wants you to stick your data in its cloud — on its computers — not on your piddlin' little PC.

Many people find libraries too difficult. I find working *without* libraries is too difficult.

I refer to libraries occasionally in this chapter, but if you want the whole story, check out Book 7, Chapter 3. Unless you want to put all your data in OneDrive (which isn't a bad idea, really) or Google Drive or Dropbox (my choice for most of my online storage), libraries are a better way to organize your data here on earth.

Microsoft refers to the way Windows interacts with people as the *user experience*. Gad. File Explorer lies at the center of the, er, user experience. When you want to work with Windows 10 — ask it where it stuck your wedding pictures, show it how to mangle your files, or tell it (literally) where to go — you usually use File Explorer.

Navigating

File Explorer helps you get around in the following ways:

>> **Click a folder to see the files you want.** On the left side of the File Explorer window (refer to Figure 1-13), you can click a real folder (such as Desktop or Downloads), a shortcut you dragged to the Quick Access list on the left, other drives on your computer, or other computers on the network. You can also reach into your OneDrive account in the sky, as you can see in Figure 1-14.

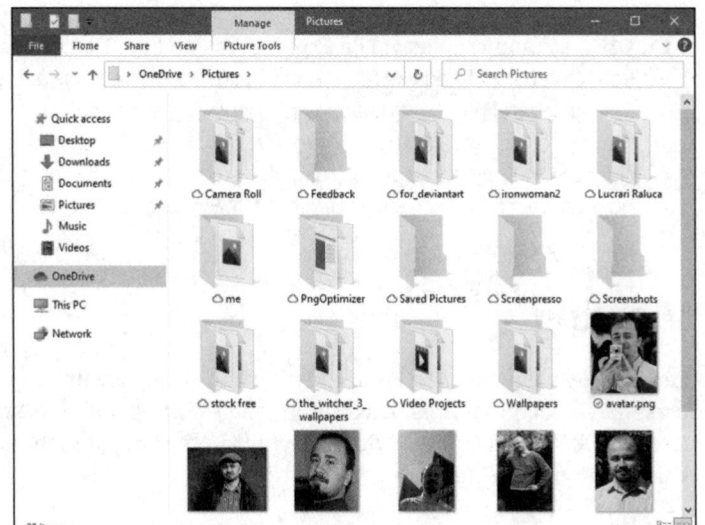

FIGURE 1-14:
File Explorer helps you move around, even into the sky, with OneDrive.

>> **Use the "breadcrumb" navigation bar to move around.** At the top of the File Explorer window (refer to Figure 1-14), you can click the wedges to select from available folders. I clicked OneDrive, and then Pictures, and ended up in the OneDrive Pictures folder, as shown in Figure 1-14. I can click OneDrive in the breadcrumbs on the top or in the column on the left, to get back my OneDrive's home location.

>> **Details appear below.** If you click a file or folder once, details for it (number of items, Sharing state) appear in the Details box at the bottom of the File Explorer window. If you double-click a folder, it becomes the current folder. If you double-click a document, it opens. (For example, if you double-click a Word document, Windows 10 fires up Word, if you have it installed, and has it start with that document open and ready for work.)

>> **Many of the actions you might want to perform on files or folders show up in the command bar at the top.** Most of the other actions you might want to perform are accessible by right-clicking the file or folder.

TIP

>> **To see all options, press Alt.** Depending on how you have it configured, File Explorer may show you an old-fashioned command bar (File, Edit, View, Tools, Help) with dozens of functions tucked away. It will also show you keyboard shortcuts (single letters in small boxes) that you can use to get to the commands from the keyboard. (For example, Ctrl+V displays the View tab.)

>> **Open as many copies of File Explorer as you like.** That can be helpful if you're scatterbrained like I am — er, if you like to multitask and you want to look in several places at once. Simply right-click the File Explorer icon in the taskbar (and shown in the margin) and choose File Explorer. An independent copy of File Explorer appears, ready for your finagling. Or press and hold down the Shift key while you click the File Explorer icon one more time.

Viewing

Large icons view (refer to Figure 1-14) is, at once, visually impressive and cumbersome. If you grow tired of scrolling (and scrolling and scrolling) through those icons, click the View tab and choose Details in the Layout section. You see the succinct list shown in Figure 1-15.

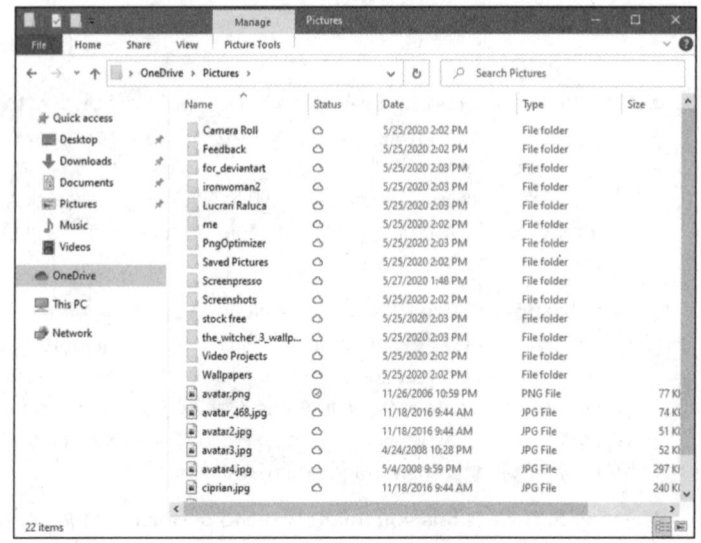

FIGURE 1-15:
Details view has more meat, less sizzle.

Windows 10 offers several picturesque views — dubbed extra-large icons, large icons, medium icons, small icons, and infinitesimal eye-straining icons (okay, I got carried away a bit) — that can come in handy if you're looking through a bunch of pictures. In most other cases, though, the icons only get in the way.

TIP

In details view, you can sort the list of files by clicking a column heading — Name or Date, for example. You can right-click one of the column headings and choose More to change what the view shows (get rid of Type, for example, and replace it with Date Taken).

PREVIEW

Every File Explorer window can show a Preview pane — a strip along the right side of the window that, in many cases, shows a preview of the file you selected.

Some people love the preview feature. Others hate it. A definite speed hit is associated with previewing — you may find yourself twiddling your thumbs as Windows 10 gets its previews going. The best solution is to turn off the preview unless you absolutely need it. And use the right tool for the job — if you're previewing lots of picture files, fire up a Photo app (not necessarily the one in Windows 10; see Book 4, Chapter 3).

You can enable the Preview pane (and all other File Explorer panes) by clicking the View tab and then clicking Preview pane, in the Panes section in the top left.

Creating files and folders

Usually, you create new files and folders when you're using a program. You make new Word documents when you're using Word, say, or come up with a new folder to hold all your offshore banking spreadsheets when you're using Excel. Programs usually have the tools for making new files and folders tucked away in the File, Save and File, Save As dialog boxes. Click around a bit and you'll find them.

But you can also create a new file or folder directly in an existing folder quite easily, without going through the hassle of cranking up a 900-pound gorilla of a program. Follow these steps:

1. **Move to the location where you want to put the new file or folder.**

For example, if you want to stick the new folder Revisionist Techno Grunge in your Documents folder, click the File Explorer icon in the taskbar (and shown in the margin), and on the left, under Quick Access, click Documents.

2. **Right-click a blank spot in your chosen location.**

By "right-click a blank spot," I mean "don't right-click an existing file or folder," okay? If you want the new folder or file to appear on the desktop, right-click an empty spot on the desktop.

3. **Choose New (see Figure 1-16) and then the kind of file you want to create.**

If you want a new folder, choose Folder. Windows 10 creates the new file or folder and leaves it with the name highlighted so that you can rename it by simply typing.

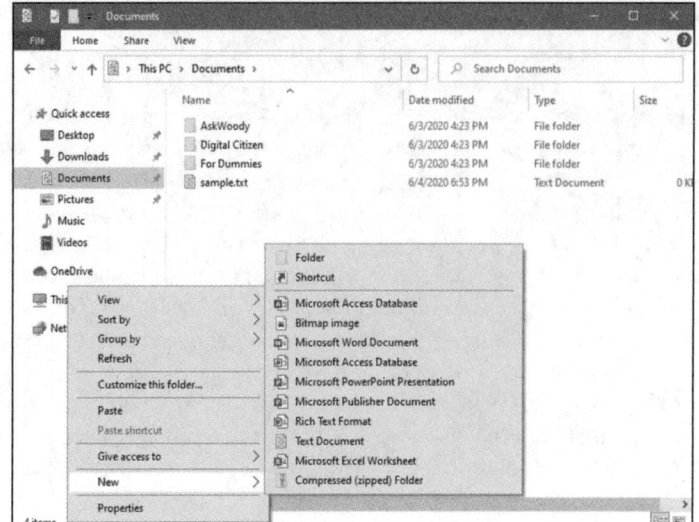

FIGURE 1-16:
Right-click an empty location, and choose New to create a file or folder.

Modifying files and folders

REMEMBER

As long as you have permission (see the section "Sharing folders," later in this chapter), modifying files and folders is easy — rename, delete, move, or copy them — if you remember the trick: right-click (or, for the painfully tap addicted, tap and hold down).

To copy or move more than one file (or folder) at a time, select all the files (or folders) before right-clicking. You can select more than one file using any of these methods:

>> Hold down Ctrl while clicking.

>> Click and drag around the outside of the files and folders to lasso them.

>> Use the Shift key if you want to choose a bunch of contiguous files and folders — ones that are next to each other. Click the first file or folder, hold down Shift, and click the last file or folder.

Showing filename extensions

If you're looking at the Recent files on your computer and you can't see the period and three-letter suffixes of the filenames (such as .txt and .tiff and .jpg) that are visible in Figures 1-13, 1-14, 1-15, 1-16, and most of the rest of this book, don't panic! You need to tell Windows 10 to show them — electronically knock Windows upside the head, if you will.

In my opinion, every single Windows 10 user should force the operating system to show full filenames, including the (usually three-letter) extension at the end of the name.

I've been fighting Microsoft on this topic for many years. Forgive me if I get a little, uh, steamed — yeah, that's the polite way to put it — in the retelling.

Every file has a name. Almost every file has a name that looks more or less like this: Some Name or Another.ext.

The part to the left of the period — Some Name or Another, in this example — generally tells you something about the file, although it can be quite nonsensical or utterly inscrutable, depending on who named the file. The part to the right of the period — ext, in this case — is a filename *extension*, the subject of my diatribe.

Filename extensions have been around since the first PC emerged from the primordial ooze. They were a part of the PC's legacy before anybody ever talked about legacy. Somebody somewhere decided that Windows wouldn't show filename extensions anymore. (My guess is that Bill Gates himself made the decision, about 20 years ago, but it's only a guess.) Filename extensions were considered dangerous: too complicated for the typical user, a bit of technical arcana that novices shouldn't have to sweat.

WARNING

No filename extensions? That's garbage. Pure, unadulterated garbage.

The fact is that nearly all files have names such as Letter to Mom.docx, Financial Projections.xlsx, or ILOVEYOU.vbs. But Windows, with rare exception, shows you only the first part of the filename. It cuts off the filename extension. So you see Letter to Mom, without the .docx (which brands the file as a Word document), Financial Projections, without the .xlsx (a dead giveaway for an Excel spreadsheet), and ILOVEYOU, without the .vbs (which is the filename extension for Visual Basic programs).

I really hate it when Windows hides filename extensions, for four big reasons:

>> **If you can see the filename extension, you can usually figure out which kind of file you have at hand and which program will open it.** People who use Word 2003, for example, may be perplexed to see a .docx filename extension — which is generated by Word 2019 and can't be opened by bone-stock Word 2003.

Legend has it that former Microsoft CEO (and current largest individual stockholder) Steve Ballmer once infected former CEO (and current philanthropist extraordinaire) Bill Gates's Windows PC using a bad email attachment, ILOVEYOU.VBS. If Ballmer had seen the .VBS on the end of the filename, no doubt he would've guessed it was a program — and might've been disinclined to double-click it.

>> **It's almost impossible to get Windows to change filename extensions if you can't see them.** Try it.

>> **Many email programs and spam fighters forbid you from sending or receiving specific kinds of files, based solely on their filename extensions.** That's one of the reasons why your friends might not be able to email certain files to you. Just try emailing an .exe file, no matter what's inside.

>> **You bump into filename extensions anyway.** No matter how hard Microsoft wants to hide filename extensions, they show up everywhere — from the Readme.txt files mentioned repeatedly in the official Microsoft documentation to discussions of .jpg file sizes on Microsoft web pages and a gazillion places in between.

Take off the training wheels, okay? To make Windows show you filename extensions the easy way, follow these steps:

1. **In the taskbar, click the File Explorer icon.**

   File Explorer appears (refer to Figure 1-13).

2. **Click or tap View.**

   You see File Explorer's View ribbon, shown in Figure 1-17.

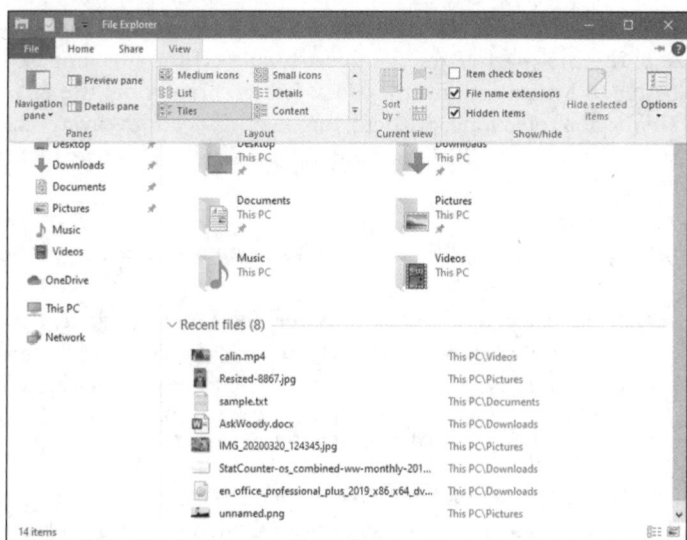

FIGURE 1-17:
Make Windows 10 show you filename extensions.

3. **Click to select the File Name Extensions box, in the Show/Hide section.**

Your changes take place immediately. Look at your unveiled filename extensions.

**ASK
WOODY.COM**

While you're here, you may want to change another setting. If you can avoid the temptation to delete or rename files you don't understand, select the Hidden Items box. That way, Windows 10 shows you all files on your computer, including ones that have been marked as hidden, typically by Microsoft. Sometimes, you need to see all your files, even if Windows wants to hide them from you.

Sharing folders

Sharing is good, right? Your mom taught you to share, didn't she? Everything you need to know about sharing you learned in kindergarten — like how you can share your favorite crayon with your best friend and get back a gnarled blob of stunted wax, covered in mysterious goo.

SHARING AND ONEDRIVE

Microsoft wants you to put all your files in OneDrive. No, they aren't trying to snoop the contents. Microsoft gives away lots of "free" cloud storage in OneDrive because they want you to use (and pay for) other Microsoft products. Microsoft's cost for 5 GB of "free" cloud space is measured in pennies, and it's getting cheaper. Microsoft's income from keeping you in the Microsoft fold — maybe buying a subscription to Microsoft 365 (formerly known as Office 365), say, or clicking an ad in Bing — pays for the free storage and then some.

That's why Windows 10 doesn't put a big emphasis on file sharing, here on earth. This book shows you many ways to share files — libraries, public folders — that Microsoft isn't particularly interested in proliferating. They don't make money and don't lock you into their ecosystem when your files are all down here, out of the cloud.

In some cases, OneDrive is your best choice for storing and sharing files. I use it all the time, although I tend to put my most important files (including all the files used in preparing this book) in Dropbox. For many people who get nosebleeds in the cloud, for a wide variety of reasons, though, keeping your sharing out of Microsoft's cloud makes good sense.

It's your data. You can choose. You can even change your mind if you want. This book has an extensive discussion of OneDrive and sharing in Book 6, Chapter 1. But if you want to keep your data out of Microsoft's cloud and off Microsoft's computers, follow along here to see how it's done the Windows 10 way.

You can put your files in the cloud, and use the features built into all the cloud services for sharing files or folders. OneDrive (see the "Sharing and OneDrive" sidebar), Dropbox, Google Drive, Box, Mega, and the others have different rules. If you want to share from a Windows 10 computer, though, you must follow Microsoft's rules.

Windows 10 supports two very different ways for sharing files and folders:

>> **Move the files or folders you want to share into the \Public folder.**
The \Public folder is kind of a big cookie jar for everybody who uses your PC: Put a file or folder in the \Public folder so all the other people who use your computer can get at it. The \Public folder is available to other people on your network, if you have one and you've told Windows to share its files, but you have little control over who, specifically, can get at the files and folders.

>> **Share individual files or folders, without moving them anywhere.** When you share a file or folder, you can specify exactly who can access the file or folder and whether they can just look at it or change it or delete it.

Using the \Public folder

You might think that simply moving a file or folder to the \Public folder would make it, well, public. At least to a first approximation, that's exactly how things work.

Any file or folder you put in the \Public folder, or any folder inside the \Public folder, can be viewed, changed, or deleted by all the people who are using your computer, regardless of which kind of account they may have and whether they're required to log in to your computer. In addition, anybody who can get into your computer through the network will have unlimited access. The \Public folder is (if you'll pardon a rather stretched analogy) a big cookie jar, open to everybody who is in the kitchen.

(For more details, and important information about Public networks and big-company domains, check out *Networking All-in-One For Dummies,* 7th Edition, by Doug Lowe [Wiley].)

TIP

Follow these easy steps to move a file or folder from one of the built-in personal folders (Desktop, Documents, Downloads, Music, Pictures, or Videos) into its corresponding location in one of the \Public folders:

1. Tap or click the File Explorer icon in the taskbar.

2. Navigate to the file or folder that you want to move into the \Public folder.

In Figure 1-18, I double-clicked the Quick Access Pictures folder to get to my pictures.

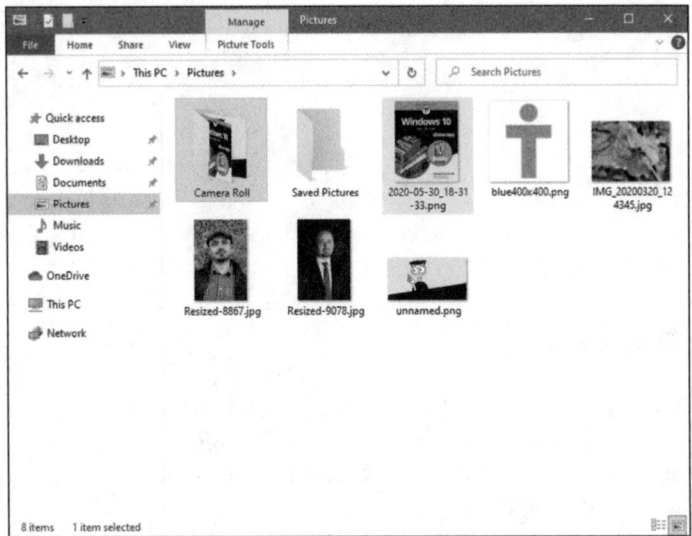

FIGURE 1-18:
Moving a folder
to the \Public
folder is easy,
if you know the
trick.

3. Right-click the folder or file you want to move, and choose Cut.

In this case, I wanted to move the Camera Roll folder, so I cut it.

4. Navigate to the \Public folder where you want to move the folder or file.

This is more difficult than you might think. In general, on the left of File Explorer, click This PC (scroll down if necessary, to see it), then scroll way down and double-click or tap Local Disk (C:). Then double-click Users, then Public. You see the list of Public folders shown in Figure 1-19.

5. Double-click the \Public folder you want to use. Then right-click inside the folder, and choose Paste.

In this case, I double-clicked Public Pictures and pasted the Camera Roll folder into the Public Pictures folder. From that point on, the photos are available to anybody who uses my computer and to people who connect to my computer. (It may also be available to other computers connected to your network, workgroup, or domain, depending on various network settings. See *Networking All-in-One For Dummies* for specific examples.)

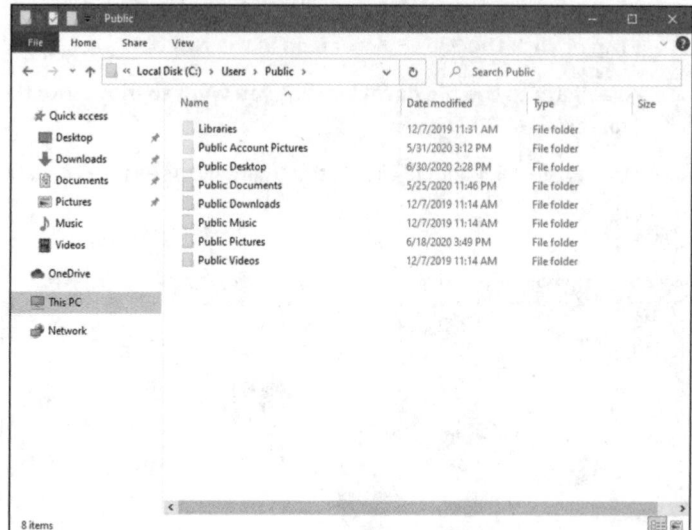

FIGURE 1-19:
Your \Public
folders live here.

Recycling

When you delete a file, it doesn't go to that Big Bit Bucket in the Sky. An intermediate step exists between deletion and the Big Bit Bucket. It's called purgatory — oops. Wait a sec. Wrong book. (*Existentialism For Dummies*, anybody?) Let me try that again. Ahem.

The step between deletion and the Big Bit Bucket is the Recycle Bin.

When you delete a file or folder from your hard drive — whether by selecting it in File Explorer and pressing Delete or by right-clicking and choosing Delete — Windows 10 doesn't actually delete anything. It marks the file or folder as being deleted and displays it in the Recycle Bin. But other than that, it doesn't touch it.

WARNING

Files and folders on USB key drives, SD cards, and network drives don't go into limbo when they're deleted. The Recycle Bin doesn't work on USB key drives, SD cards, or drives attached to other computers on your network. That said, if you accidentally wipe out the data on your USB drive or camera memory card, there is hope. See the discussion of the Recuva program in Book 10, Chapter 5.

To rummage around in the Recycle Bin, and possibly bring a file back to life, follow these steps:

1. Double-click the Recycle Bin icon on the desktop (and shown in the margin).

File Explorer opens the Recycle Bin, shown in Figure 1-20. You may have to click the Recycle Bin Tools tab at the top to bring up the ribbon.

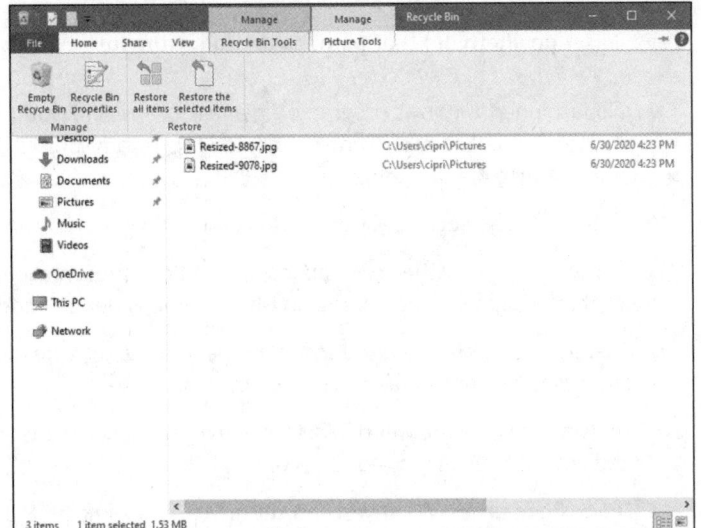

FIGURE 1-20:
Restore files one
at a time or en
masse.

2. **To restore a file or folder (sometimes Windows 10 calls it *undeleting*), click the file or folder, click Recycle Bin Tools on the top, and then click the Restore the Selected Items button.**

TIP

You can select a bunch of files or folders by holding down the Ctrl key as you click.

If you set things up properly, Windows 10 maintains shadow copies of previous versions of many kinds of files. If you can't find what you want in the Recycle Bin, follow the steps in Book 8, Chapter 1 to see whether you can dig something out of the Windows Time Machine.

To reclaim the space that the files and folders in the Recycle Bin are using, click the Empty the Recycle Bin icon. Windows 10 asks whether you really, truly want to get rid of those files permanently. If you say Yes, they're gone.

Creating Shortcuts

Sometimes, life is easier with shortcuts. (As long as the shortcuts work, anyway.) So, too, in the world of Windows, where shortcuts point to things that can be started. You may set up a shortcut to a Word document and put it on your desktop. Double-click the shortcut and Word starts with the document loaded, as if you double-clicked the document in File Explorer.

You can set up shortcuts that point to the following items:

- Old-fashioned Windows programs (er, apps), of any kind (you can put a shortcut for a Windows 10 app on the desktop, by dragging its tile from the Start menu to the desktop).

- Web addresses, such as www.dummies.com

- Documents, spreadsheets, databases, PowerPoint presentations, and anything else that can be started in File Explorer by double-clicking it

- Specific chunks of text (called *scraps*) inside documents, spreadsheets, databases, and presentations, for example

- Folders (including the weird folders inside digital cameras, the Fonts folder, and others that you may not think of)

- Drives (hard drives, CD drives, and key drives, for example)

- Other computers on your network, and drives and folders on those computers, as long they're shared

- Printers (including printers attached to other computers on your network), scanners, cameras, and other pieces of hardware

- Network connections, interface cards, and the like

You have many different ways to create shortcuts. In many cases, you can go into File Explorer, right-click a file, choose Send To, and then choose Desktop (Create Shortcut).

Here's a more general-purpose method that works for, say, websites:

- **Right-click a blank area on the desktop, and choose New ⇨ Shortcut.**

  The Create Shortcut wizard appears, as shown in Figure 1-21.

- **In the box, type the name or location of the program (not Windows 10/ UWP app), file, folder, drive, computer, or Internet address. Click Next.**

  Windows 10 asks you for a name for the shortcut.

- **Give the shortcut a memorable name, and click Finish.**

  Windows 10 places an icon for the program, file, folder, drive, computer, website, document . . . whatever . . . on the desktop.

Anytime I double-click the AskWoody icon on my desktop, the default browser pops up and puts me on the www.AskWoody.com main page.

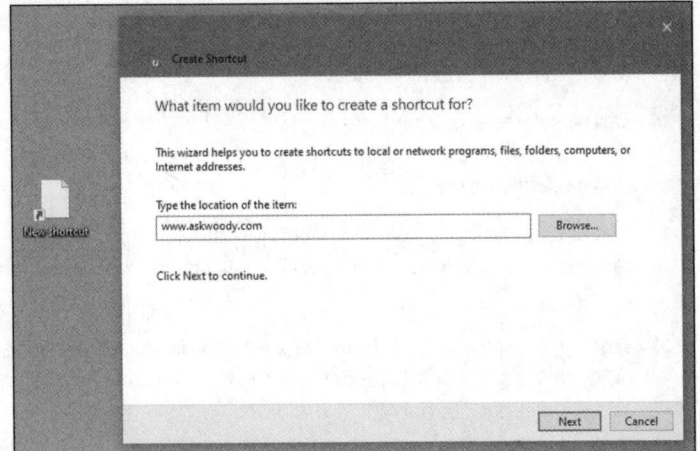

FIGURE 1-21:
Create shortcuts
the old-fashioned
manual way.

You can use a similar procedure for setting up shortcuts to any file, folder, program, or document on your computer or on any networked computer.

**TECHNICAL
STUFF**

Believe it or not, Windows 10 thrives on shortcuts. They're everywhere, lurking just beneath the surface. For example, every single entry on the Start menu is a (cleverly disguised) shortcut. The icons on the taskbar are all shortcuts. Most of File Explorer is based on shortcuts — although they're hidden where you can't reach them. Even the Windows 10 app icons work with shortcuts; they're simply hard to find. So don't be afraid to experiment with shortcuts. In the worst-case scenario, you can always delete them. Doing so gets rid of the shortcut; it doesn't touch the original file it points to.

Keying Keyboard Shortcuts

As I mention in Book 2, Chapter 1, Windows 10 has loads of keyboard shortcuts, but I don't use very many of them.

REMEMBER

Here are the keyboard shortcuts that everyone should know. They've been around for a long, long time:

>> **Ctrl+C** copies whatever you've selected and puts it on the Clipboard. On a touchscreen, you can do the same thing in most applications by tapping and holding down, and then choosing Copy.

>> **Ctrl+X** does the same thing but removes the selected items — a cut. Again, you can tap and hold down, and Cut should appear in the menu.

- » **Ctrl+V** pastes whatever is in the Clipboard to the current cursor location. Tap and hold down usually works.

- » **Ctrl+A** selects everything, although sometimes it's hard to tell what "everything" means — different applications handle Ctrl+A differently. Tap and hold down usually works here, too.

- » **Ctrl+Z** usually undoes whatever you just did. Few touch-enabled apps have a tap-and-hold-down alternative; you usually have to find Undo on a ribbon or menu.

- » When you're typing, **Ctrl+B, Ctrl+I, and Ctrl+U** usually flip your text over to Bold, Italic, or Underline, respectively. Hit the same key combination again, and you flip back to normal.

Sleep: Perchance to Dream

Aye, there's the rub.

Windows 10 has been designed so that it doesn't need to be turned off.

TECHNICAL STUFF

Okay, that's a bit of an overstatement. Sometimes, you have to restart your computer to let patches kick in. Sometimes, you plan to be gone for a week and need to give the beast a blissful rest. But by and large, you don't need to shut off a Windows 10 computer — the power management schemes are very green.

Laptops and tablets are a different story altogether. Most laptops, when they're working properly, will shut themselves off shortly after you fold them together. Many tablets will power off, too. If yours doesn't, you should take the initiative and shut the machine down before stashing it away.

The only power setting most people need to fiddle with is the length of time Windows 10 allows before it turns the screen black. Here's the easy way to adjust your screen blackout time, whether your machine is plugged into the wall or running on battery:

1. Click or tap the Start button and then the Settings icon.

2. Choose System and then Power & Sleep.

Windows brings up the Power & Sleep dialog box shown in Figure 1-22.

3. In the drop-down boxes at the top, choose whatever time you like.

Your changes take effect immediately.

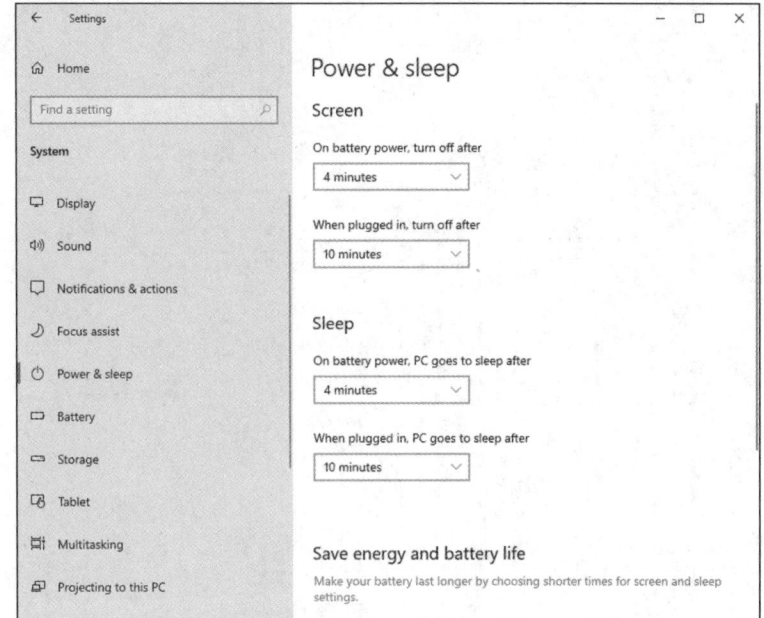

FIGURE 1-22:
Tell your
machine how
long to run off to
never-never-land.

You can click the Additional Power Settings link, if you want to open up the old-fashioned Control Panel pane for power settings.

Although Microsoft has published voluminous details about the power down and power up sequences, I haven't seen any details about how long it takes before your PC actually goes to sleep. In theory, that shouldn't matter too much because the wake-ups are so fast.

**ASK
WOODY.COM**

Microsoft recently published some recommendations that I found fascinating. To truly conserve energy with a desktop computer, be aggressive with the monitor idle time (no longer than two minutes), and make sure that you don't have a screen saver enabled. If you want to conserve energy with a notebook or netbook, your top priority is to reduce the screen brightness!

Chapter **2**

Personalizing the Start Menu

I f you're an experienced Windows user, chances are good that the first time you saw the Windows 8 Start screen, you wondered who put an iPad on it. However, if you're an experienced iPad user, chances are good that the first time you worked with the Windows 8 Start screen, you went screaming for your iPad.

Windows 10 has, I'm convinced, improved upon the Windows 8 experience greatly. If you have a mouse, the Windows 10 Start menu — the screen that almost everybody sees when he or she clicks the Start button, and the screen you'll come back to over and over again — defines and anchors Windows. Like it or not. See Figure 2-1.

My advice, if you don't like those newfangled Start tiles, is to give it a real workout for a month or two. I don't expect that you'll end up singing hosannas about the tiles. But I do expect that you'll warm up to it a little bit — and, like me, you may even miss it when you go back to Windows 7. That goes double if you can use Windows 10 on a touch-friendly tablet.

TIP

In this chapter, I take you through the Start menu, from beginning to end. It's a bit confusing because changes in the desktop's Start menu (refer to Figure 2-1) affects the appearance of the tablet mode Start screen (refer to Figure 2-2).

Hey, if you can get your thumb and all your pinkies on the screen simultaneously, touch has the mouse beat five to one. Sort of.

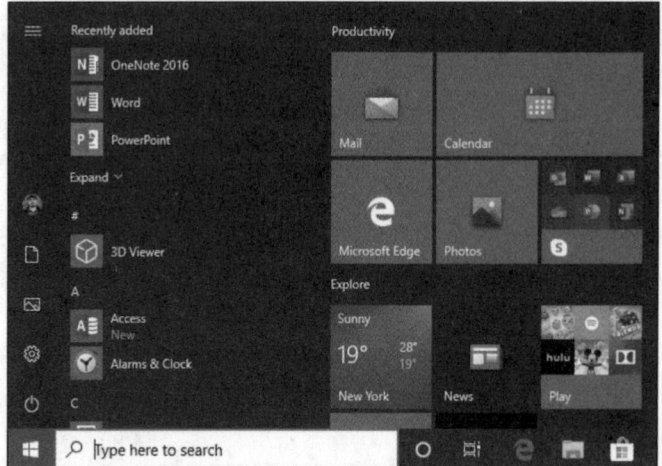

FIGURE 2-1:
The normal mouse-and-keyboard version of the Start menu, in Windows 10.

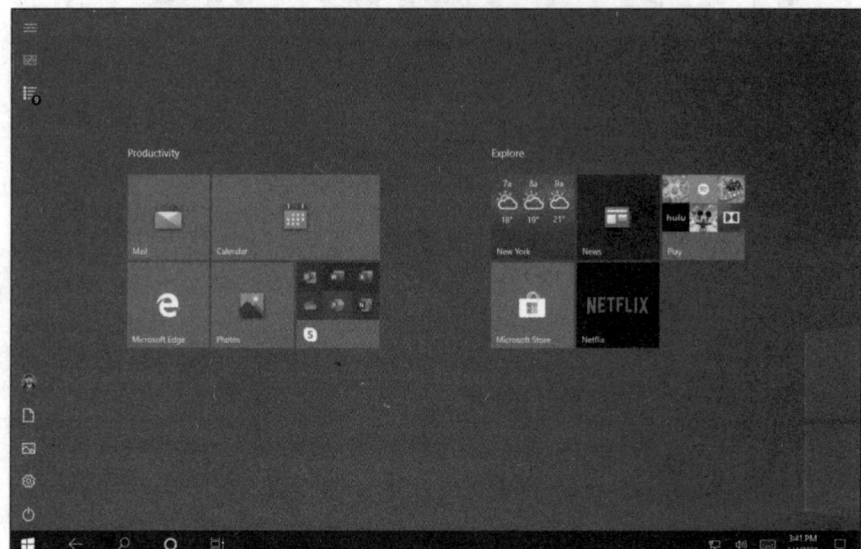

FIGURE 2-2:
The tablet mode Start screen.

Touring the Start Menu

 The very first screen you see when you click or tap the Start button, the Start menu (refer to Figure 2-1), is designed to be at the center of your Windows 10 universe. Don't let the fact that the right side's intentionally made to look like a smartphone screen deter you in the least.

You've probably sworn at the Start menu a few times already, but if you can keep a civil tongue, permit me to expound a bit:

>> The left side of the Start menu (refer to Figure 2-1) consists of a handful of icons that you're likely to use all the time.

If you're in tablet mode — identifiable because you don't see a list of program names (refer to Figure 2-2) — click or tap the three-line (hamburger) icon in the upper-left corner (and shown in the margin). The full left side of the Start menu unfolds, as shown in Figure 2-3.

A third mode, called full-screen start, looks and acts much like tablet mode. It also has a hamburger menu that brings up the left side. See Book 3, Chapter 1 for details.

FIGURE 2-3: In tablet mode, the left side of Start sits under the hamburger icon.

>> *Tiles* (the squares on the right side of the screen) appear in four sizes: large, wide, medium, and small (rocket science). In Figure 2-4, I changed my tiles around a bit to make most advertising tiles small (Candy Crush Soda Saga, Farm Heroes Saga, Disney Magic Kingdoms). Comparing sizes, the Mail tile is medium, Calendar is wide, and Weather looms large. Many tiles that come from Microsoft are *live tiles,* with *active content* (latest news, stock prices, date, temperature, email messages) that changes the face of the tile.

» Tiles are bunched into *groups,* which may or may not have *group names.* Figure 2-4 shows three groups: one marked Productivity, another marked Explore, and a third marked Play. Don't shoot me. Those are the names Microsoft gave them. You can change the group names, as described later in this chapter.

» Somewhere near the middle of the far-left bar, you see your picture. Hover your mouse cursor over the picture and you see either your username or (if you're logged in with a Microsoft account) your full name. For a description of the Microsoft account and the pros and cons of using one, see Book 2, Chapter 5.

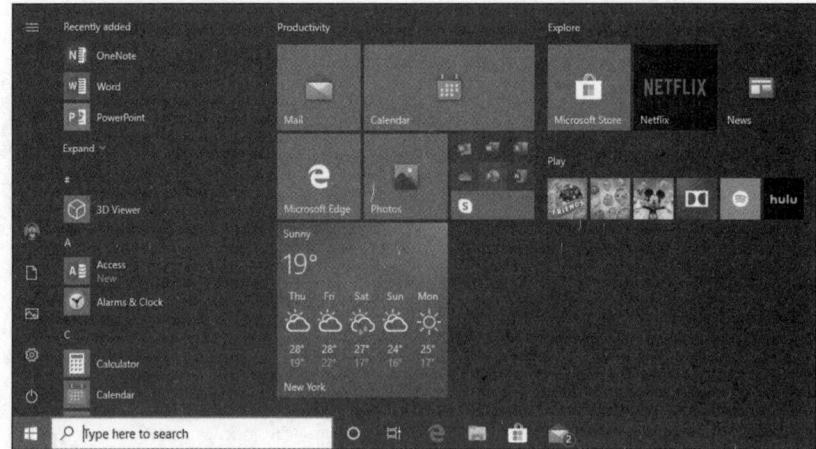

FIGURE 2-4:
After a few swift changes, your Start menu can look like this.

Modifying the Start Menu

Windows 7 has a marvelously malleable Start menu. You can click and drag and poke and rearrange it every which way but loose. I particularly enjoyed setting up nested folders and having them show up as cascading items on the Start menu. But that was then.

The left side of Windows 10's Start menu, by comparison, has a very rigid format that can be changed only in a few specific, preprogrammed ways (see Figure 2-4). Customizing the Start menu in Windows 10 is nothing like customizing it in Windows 7. (And, of course, Windows 8/8.1 didn't *have* a Start menu.)

Changing your picture

I start with an easy change to the Start menu: changing the picture on the far-left edge.

Here's how to change your picture:

1. **Open the Start menu, tap or click your picture, and select Change Account Settings.**

   Windows 10 takes you to a familiar-looking place in the PC Settings hierarchy, as shown in Figure 2-5.

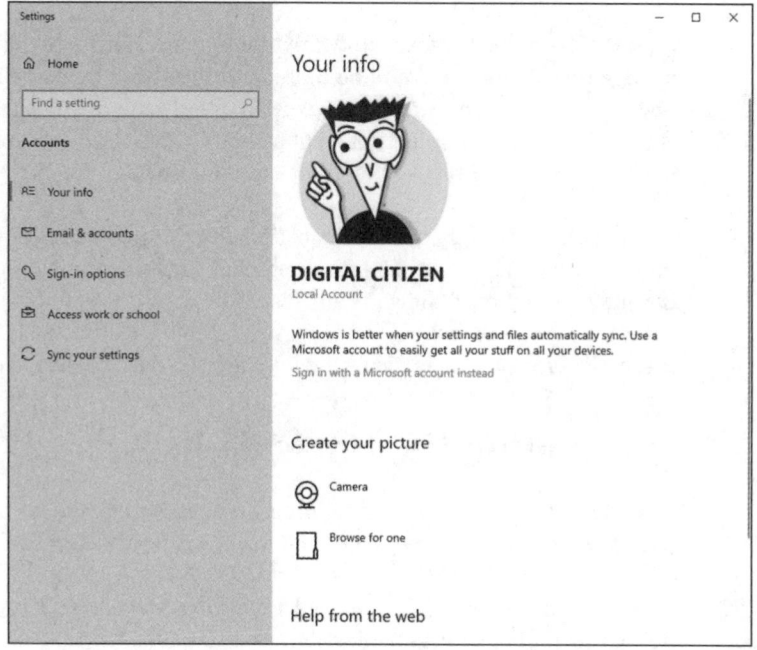

FIGURE 2-5: Change your picture in the Settings app.

2. **If you already have a picture in mind, follow these steps (if you'd rather take a picture, continue to Step 3):**

   a. *At the bottom, choose Browse for One, and navigate to the picture.*

   b. *When you find the picture you want, select it, and tap or click Choose Image.* You return to the PC Settings location shown in Figure 2-5, with your new picture in place.

3. **If you'd rather take a picture with your computer's webcam, then comb your hair, pluck your eyebrows, and tap or click Camera (in that order).**

   In any case, however you create your new picture, it takes effect immediately — no need to click OK or anything of the sort.

ASK
WOODY.COM

Want a weird picture? Any picture you can find on the Internet and download to your computer is fair game — as long as you aren't violating any copyrights.

Manipulating the Most Used section

You would think that the next part of the Start menu — Most Used — would contain links to the apps and locations that you use most often. Ha. Silly mortal.

TECHNICAL
STUFF

Microsoft (and, likely, your hardware manufacturer) salts the list: They put items in there that don't deserve to be there, and they keep items on the list long after they should've disappeared. I've experimented with it for ages, and the list of which items appear on the list, and how rapidly they fall off, seems to be controlled by some sort of counter — a counter that isn't updated correctly all the time.

At this point, the only action I can find that you can perform on the list is to remove a link you don't like. Just right-click (or tap and hold) an entry you don't like and choose Don't Show in This List.

Alternatively, you can get rid of the list entirely. See the next section.

Controlling the left-side lists

Although you can't pin individual programs on the left side of Start, as you could in Windows 7, you do have some high-level say in what appears on the left.

To see the choices on offer, click Start and then the Settings icon (both shown in the margin). Choose Personalization, and then, on the left, choose Start. (Yeah, the sequence starts and ends with Start.) You see the Start menu options shown in Figure 2-6.

Some of those choices are obscure. Here's what they mean:

» **Show More Tiles on Start:** Normally, the tiled area on the right of the Start screen displays columns that are three normal-sized tiles wide. Slide this setting On and the area becomes four columns wide. It takes a little more real estate off your screen, but most people with Full HD monitors can handle four readily.

» **Show App List in Start Menu:** A big part of the reason for having a Start menu is listing all your apps. Leave this on.

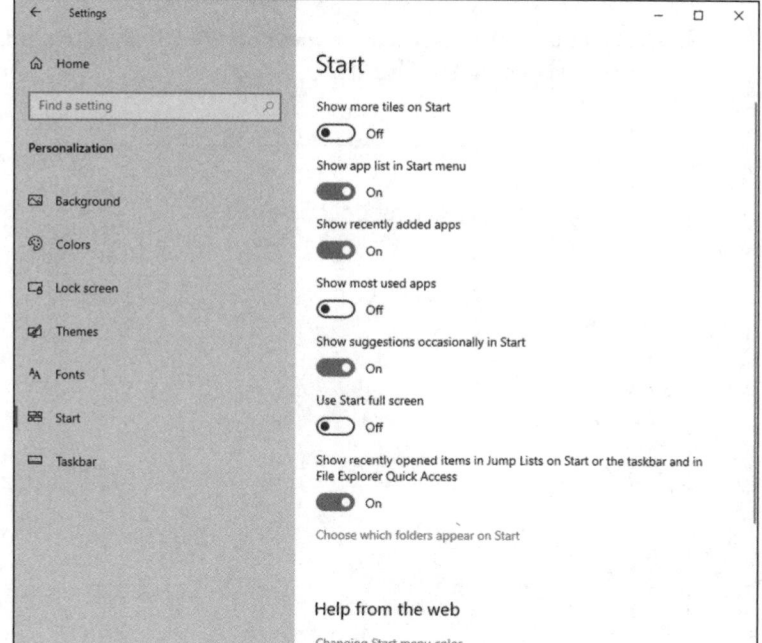

FIGURE 2-6:
You do have
some control
over what
appears on the
left side of the
Start menu.

>> **Show Recently Added Apps:** When you install a new program or app, Start notifies you by putting at the top of the Start apps list a new entry labeled Recently Added. The word *New* also appears under Recently Added apps, in the main apps list. It's an innocuous setting that saves some time, if you can't remember or figure out where your new app falls alphabetically.

>> **Show Most Used Apps:** That's the salted most-recently-used set that I talked about in the preceding section. I find it useful — you may not.

>> **Show Suggestions Occasionally in Start:** One of Microsoft's big advertising "features" in Windows 10 sticks a purposefully chosen app on the left side, in the list of apps. If you ever wondered why Microsoft keeps track of what you do in Windows 10, here's one of the reasons. Microsoft may make money when you click the suggested app, they may put specific apps there to fulfill contractual obligations, or they may use it to nudge you once again to install a Microsoft app. On by default (I wonder why), you can safely turn it off.

>> **Use Start Full Screen:** Full-screen start is a compromise between the regular Start menu and the tablet mode Start screen. It's unlikely you'll want to use it, but I discuss the effect in Book 3, Chapter 1.

>> **Show Recently Opened Items in Jump Lists on Start or the Taskbar and in File Explorer Quick Access:** This option lets you, for example, jump directly to a specific recently opened document when you right-click the Word icon on the taskbar or play a specific video when starting VLC.

Click the Choose Which Folders Appear on Start link, and another set of options appears, as shown in Figure 2-7.

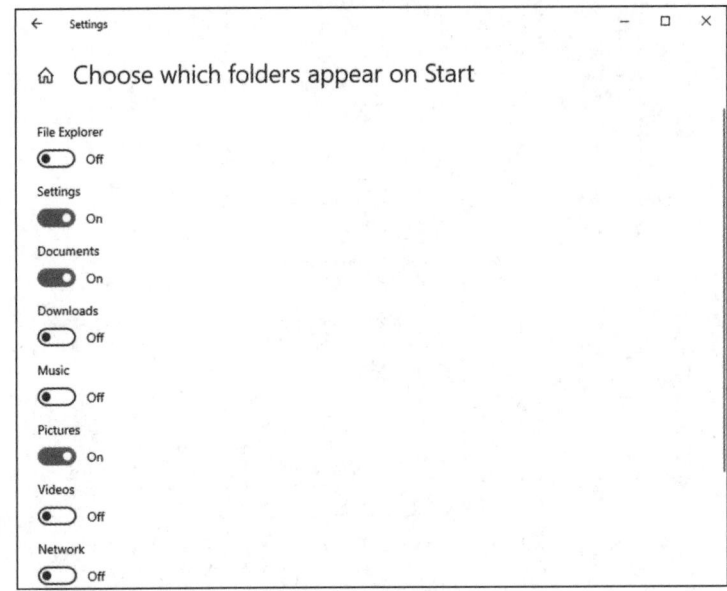

FIGURE 2-7:
You can add a long list of icons to the far left of the Start menu.

Table 2-1 shows you what each of the settings means.

TABLE 2-1   Start Menu Customizing

Choose This	And the Start Menu Starts
File Explorer	File Explorer as usual
Settings	The Settings app
Documents	File Explorer with your Documents folder (not your Documents library) open
Downloads	File Explorer at your personal Downloads folder
Music	File Explorer in your Music folder (not your Music library)
Pictures	File Explorer in your Pictures folder
Videos	File Explorer in your Videos folder
Network	File Explorer with Network selected on the left
Personal Folder	File Explorer at \Users\<yourname>

Circumnavigating the Start apps list

After the Most Used list, the advertising (oops, the "occasionally show suggestions" entry), and the Recently Added list, Windows 10 starts listing all the programs/apps installed on your computer. I call it the *Start apps list* — an alphabetized list of programs installed on your computer. In some cases, the programs are arranged in logical groups (apparently corresponding to instructions in the programs' installer). Most of the time, though, you may spend a while trying to find what you seek.

In Figure 2-8, for example, you can see how the old Control Panel doesn't appear under *C*; it's under *W* for *Windows System*.

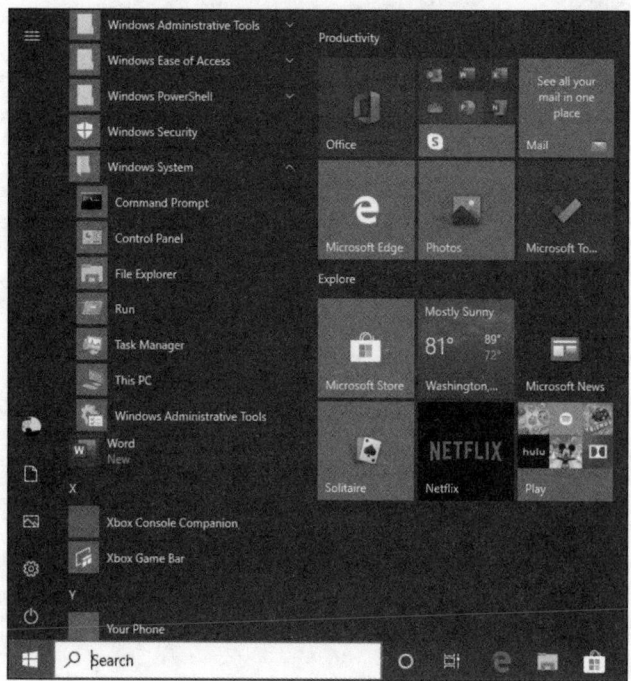

FIGURE 2-8: Looking for Control Panel? Check under Windows System.

Count on all sorts of oddities. With my copy of Office 2016, the link for Word appears under *M* for *Microsoft Office.* If you have Office 365, you'll probably find Word under *W*.

WARNING

There doesn't appear to be any way to rearrange the entries in the Start apps list — another Windows 7 feature that's sorely missed. You can uninstall some of the programs by right-clicking and choosing Uninstall, but there's no way to move the entries around, create new groups or coalesce old ones, rename, or shuffle in any way.

You can, however, click one of the alphabetic headers in the list — such as the X in Figure 2-8 — to bring up an unintelligent phone book, which lets you skip to a specific letter by clicking it. See Figure 2-9.

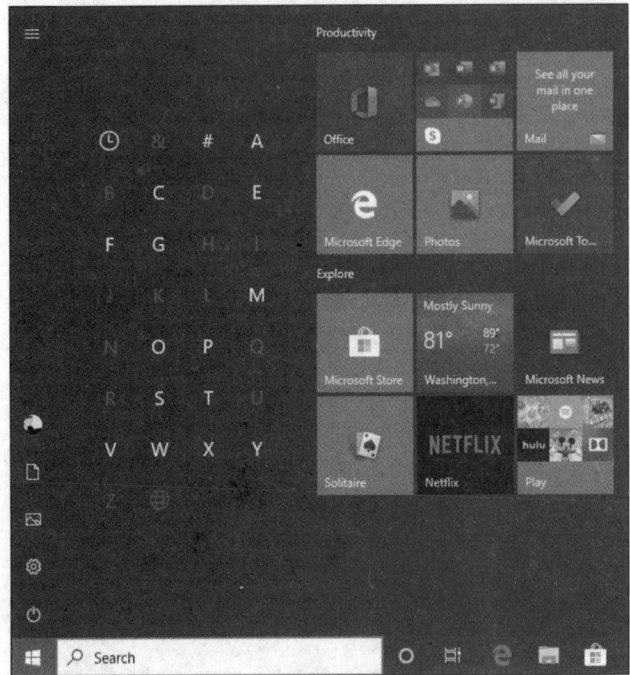

FIGURE 2-9:
This is all the organizing the Start apps list can give.

When you're in the unintelligent phone book mode, simply clicking any app brings back the usual Start menu behavior.

If you right-click (or tap and hold down) one of the apps in the Start apps list, you're usually given two choices:

>> **Pin to Start:** Creates a new tile on the right side of the Start menu that runs the program. (Yeah, I know it's confusing: *Start,* to me, means the left side of the Start menu, and I bet it does to you, too. Still, that's the terminology Microsoft uses.)

>> **Pin to Taskbar:** You have to click More first. This option creates a shortcut on the bottom of the taskbar, which also runs the program.

In some cases, right-clicking a program gives you the option to uninstall the program or run it as if you were an administrator (see Book 2, Chapter 4), or both.

WARNING

Also in some cases, you can click an app in the Start apps list and drag it over to the right, tiled part of the Start menu. I've had problems with that in the past, where the app disappears from the Start apps list and it won't come back. Beware.

Resizing the Start Menu

The Start menu can be resized, either taller and shorter (vertically) or wider and skinnier (horizontally). If you click the upper edge of the Start menu and slide it down, you see something like the screen shown in Figure 2-10.

FIGURE 2-10:
Adjust the Start menu vertically.

In general, you can shorten the Start menu only so most of the most used apps show. Beyond that, it won't shrink. There's also a limit to the height of the Start menu, which varies according to screen size.

Similarly, you can widen the Start menu to the width of two (sometimes more) columns of tiles, or squish it to one column, as you can see in Figure 2-11.

That appears to be the extent of the Start menu shrinking-expanding range. Remember that you can adjust the number of tiles in each column from three to four, using the Show More Tiles setting described earlier in this chapter under "Controlling the left-side lists."

Personalizing the
Start Menu

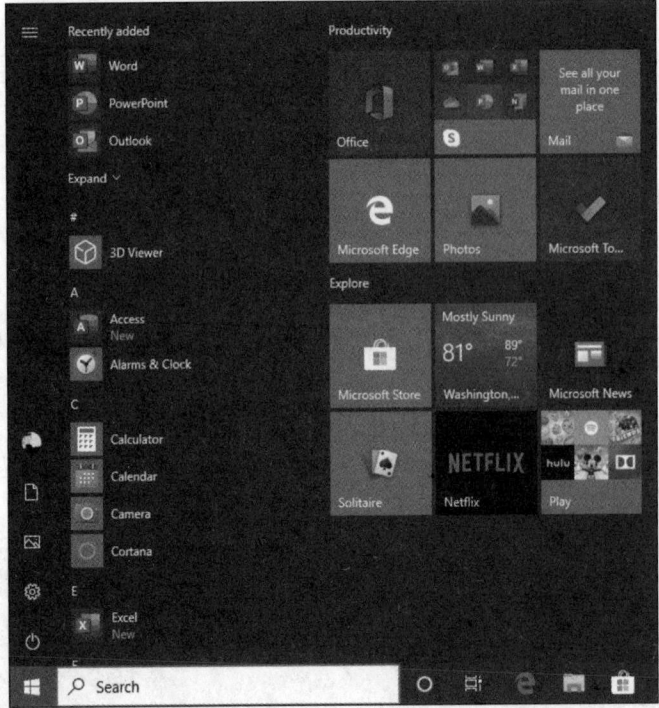

FIGURE 2-11:
Widen or squish
the Start menu
by dragging the
edges.

Changing Tiles on the Start Menu

You can click and drag tiles anyplace you like on the right side of the Start menu. Drag a tile way down to the bottom, and you start a new group. Pin a new program to the Start screen (see the preceding section), and its tile magically appears, probably in a new group made just for that tile.

You can change every tile, too. The actions available depend on what the creator of the tile permits. Here's how to mangle a tile:

1. **Open the Start menu, and right-click (or tap and hold down) the tile you want to change.**

   In Figure 2-12, I right-clicked the Weather tile. A list of actions appears.

2. **Select the desired action, using Table 2-2 as a guide.**

   You can easily delete any tile, and you can resize many of them.

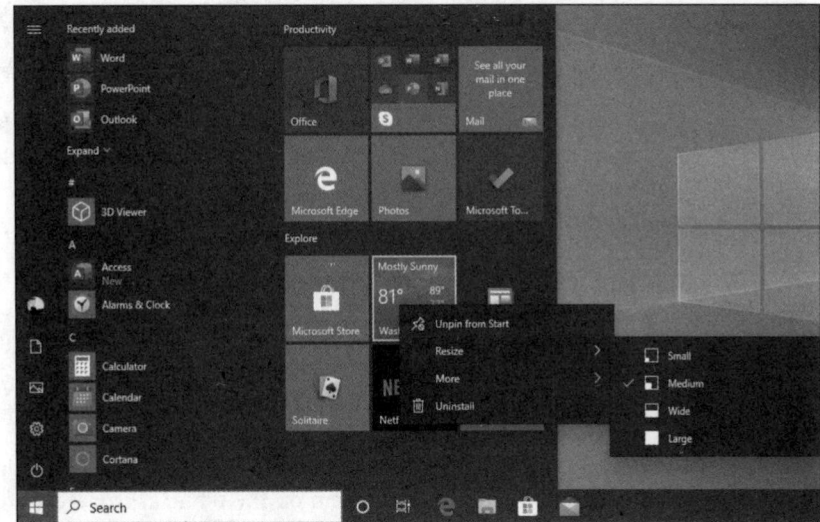

FIGURE 2-12:
You can control
tiles individually.

3. **If you would like to put a name above any of the groups of tiles, simply click and type it in the indicated spot.**

For example, you can change Productivity (at the top of the left column of tiles) to Another Sticky Day in Paradise by clicking (or tapping) Productivity and typing. The changes you make take effect immediately, and they carry through on both the traditional Start menu and over in tablet mode.

TABLE 2-2 **Tile Actions**

Tile Action Name	What the Action Does
Unpin from Start	Removes the tile from the right side of the Start menu. Doesn't affect the app itself. If you later change your mind, you can right-click the app in the Start apps list and choose Pin to Start.
Uninstall	Removes all vestiges of the program. If the program is a Windows 10 app, just confirm the uninstall to remove the app. If the program is a desktop app, the Control Panel's Remove Programs window opens. This option isn't available for programs that come with Windows 10, such as Microsoft Edge, nor is it available for advertising tiles that point to apps you haven't installed yet.
Resize	Makes the tile icon large, wide, medium (the size of the Calendar tile), or small (one-quarter the size of a medium tile).
More/Turn Live Tile Off/On	Stops or starts the animation displayed on the tile. Stopping the active content can help reduce battery drain, but the big benefit is stifling obnoxious flickering tiles — of which there are many.
Pin to taskbar	Creates a shortcut or an icon for the app on the taskbar.

Organizing Your Start Menu Tiles

The beauty of the Start menu tiles is that, within strictly defined limits, you can customize them like crazy. As long as you're happy working with the basic building blocks — four sizes of tiles, and groups — you can slice and dice till the cows come home.

The hard part about corralling the Start menu is figuring out what works best for you.

Add, add, add your tiles

Some people never use the Start menu's tiles. But if you do use them, it's easier to get organized if you put all of them on the table, as it were, before trying to sort them out.

TIP

You don't really need to have *any* tiles in the Start menu. You can right-click and choose Unpin from Start and get rid of every single one. Unfortunately, having done that, you can't make the Start screen narrower, but such is life.

The process for sticking tiles in the Start menu couldn't be simpler, although it may take an hour or two. Click the Start icon (shown in the margin), and go through your apps one by one. Right-click any apps that amuse you and choose Pin to Start. The tile appears on the right.

At the same time, you can also right-click (or tap and hold) and choose to put the app on the taskbar. Or, in most cases, you can drag the app onto the desktop and create a link to the app on the desktop.

TIP

The only significant decision you need to make is whether you want a specific app among the tiles on the Start menu, on the desktop, on the taskbar (see Book 3, Chapter 3), or on all three. As a general rule, I put my most-used apps on the taskbar, put tiles that convey useful information (such as Weather, News, and even Photos — for bringing back memories) on the Start menu, and only rarely stick anything on the desktop.

Before you start working with the tiles on your Start menu, it'll behoove you to go through your Start apps list and pull out the tiles you want or need.

Forming and naming your groups

After you have all your tiles on the right side of the Start menu, it's easy to get the menagerie organized. Try this:

1. **Tap (or click) and drag your tiles so similar tiles are in the same group.**

ASK
WOODY.COM

   For example, if you use Mail, Messaging, People, and Calendar all day long, put them in the same group. If you have Office installed, go through the procedure described in the preceding section to move the tiles you want over to the right side of the Start screen.

   Don't worry just yet if the groups are in the wrong sequence: There are easy ways to move entire groups. Just concentrate on getting your similar tiles into the same group.

TIP

   If you have programs that you look at constantly because they have important information — stock market results, your Spotify music playlist, Skype notifications, or new mail — keep them in one or two groups.

   If you need to create a new group, drag a tile all the way to the bottom. You see a faint vertical bar, which indicates that a new group has just been formed. Drop the tile below the bar.

2. **To give your groups names, click or tap the existing name (which may be Name Group) and type over the name.**

   In Figure 2-13, I put together all the tiles from Microsoft Office 2019, and gave the group the name Microsoft Office.

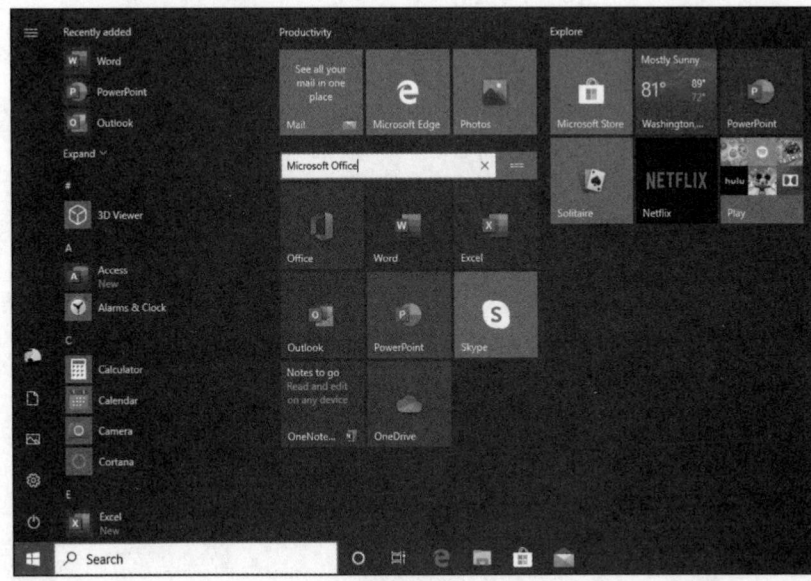

FIGURE 2-13: Here's my homemade collection of Microsoft Office tiles.

3. **To move the group, click or tap the name of the group, and drag it anywhere you like on the right side of the Start menu.**

   I put this group in the upper-left corner. Then I put together another group of the tools I use most often and called the group Tools.

4. **Click or tap and drag, and resize the Start menu if you like.**

Move tiles around any way you like. Don't be bashful! It's your machine. And if you find that you don't like something, change it around a bit and see if you like an alternative.

Chapter **3**

Personalizing the Desktop and Taskbar

t's your desktop. Do with it what you will.

In Book 3, Chapter 2, I talk about gussying up your Start menu — the left side, with icons, the middle, with links, and the right side, with tiles. This chapter looks at the rest of your desktop, what you can do about it, and how you can grab Windows 10 by the throat and shake it up a bit. Player's gotta play, play, play, play, and tweakers gotta tweak, tweak, tweak, tweak.

Shake it up.

With Windows 10's tiles now replacing (and improving upon) Windows 7's gadgets, there are fewer reasons to use the desktop now than ever before. Still, many installers put links for their own programs on the desktop, avoiding Start menu tiles like the plague, and you may have your own reasons for using desktop shortcuts.

No matter what your bias, the taskbar is also an excellent place to put your most heavily used icons.

Decking out the Desktop

The Windows 10 desktop may look simple, but it isn't. In Figure 3-1, for example, you can see the Start menu and the taskbar at the bottom, an icon for the Recycle Bin and one for Microsoft Edge at the top left, a picture file inside the Photos app in the middle, and the action/notification center on the right.

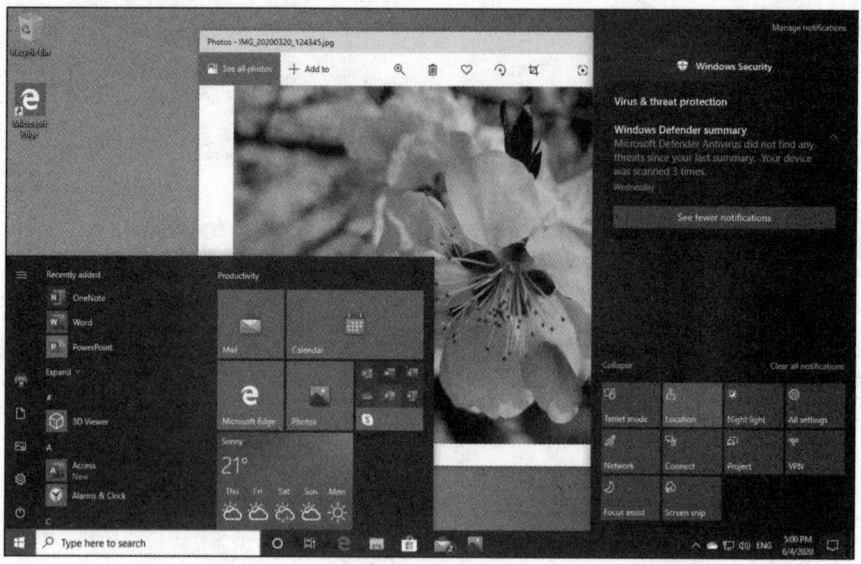

FIGURE 3-1:
The desktop is
a complicated
place.

Underneath everything is a background picture (the Windows 10 wallpaper, in this case). And there is subtle blurring between the windows.

Windows 10 lays down the desktop in layers — and paints the mouse cursor on top of all of them.

You have a handful of options when it comes to making the desktop your kind of place. Let me step you through them.

1. **Click or tap the Start button, the Settings icon, Personalization. On the left, select Background.**

    Windows 10 shows you the Background personalization page.

2. **If you're going to use a picture that stretches all the way across the screen as your background (what we used to call wallpaper), skip to Step 5.**

    If your background doesn't fill up the entire screen, you should first set a background color.

3. **In the Background drop-down box, choose Solid Color.**

   The dialog box shown in Figure 3-2 appears.

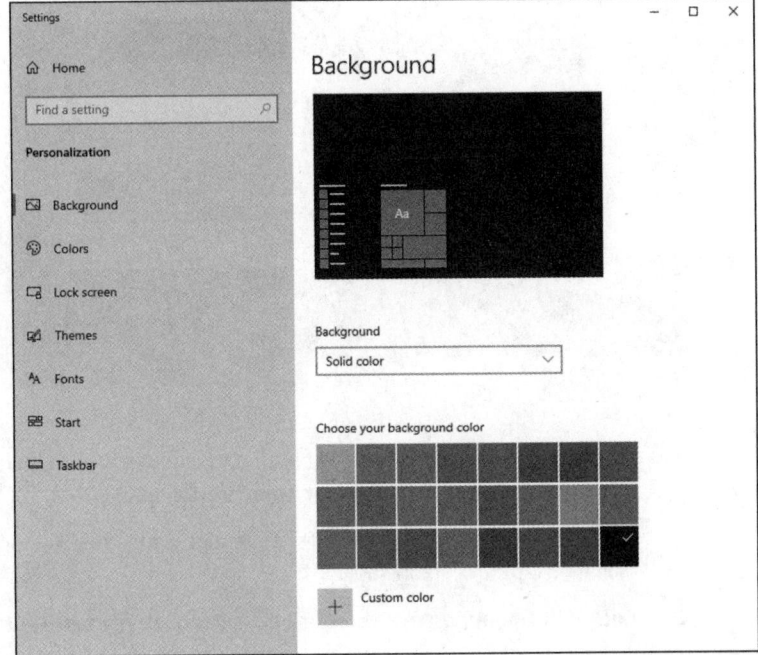

FIGURE 3-2:
If your picture won't fit the entire screen, first set the background color.

4. **Pick a color.**

   At this point you're limited to just the colors that appear in the standard colors box. After you've picked a new color, it should appear in the Preview box and on the screen itself.

5. **If you want to use a picture as your background, in the Background box, choose Picture.**

   That sets up everything to not only pick a pic but also to fit it on the screen, as shown in Figure 3-3.

   TIP

   If you'd rather use a whole bunch of pictures as a slideshow on your Start screen, in the Background box, choose Slideshow. You must have all the pictures in one album or folder; see Book 4, Chapter 3 for a discussion of albums.

6. **Choose a picture from the ones on offer, or click Browse and go out (using File Explorer) to find one you like better.**

   You can use a picture in any common picture file format.

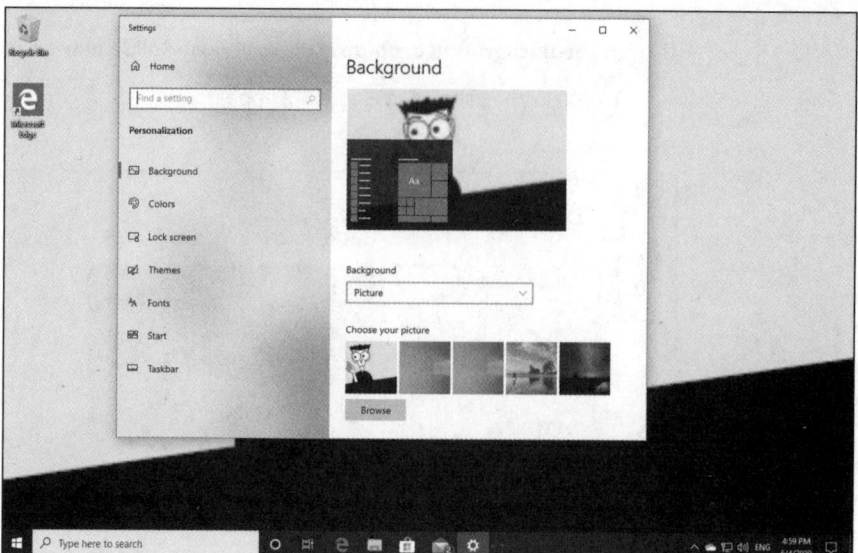

7. **If your picture is too big or too small to fit on the screen, you can tell Windows 10 how to shoehorn it into the available space.**

Use the Choose a Fit drop-down list at the bottom of the Desktop Background dialog box. Details are in Table 3-1.

8. **Click X in the upper-right corner of the Desktop Background dialog box.**

Your changes take effect immediately.

TABLE 3-1 **Picture Position Settings**

Setting	What It Means
Fill	Windows 10 expands the picture to fit the entire screen and then crops the edges. The picture doesn't appear distorted, but the sides or top and bottom may get cut off.
Fit	The screen is letterboxed. Windows 10 makes the picture as big as possible within the confines of the screen and then shows the base color in stripes along the top and bottom (or left and right). No distortion occurs, and you see the entire picture, but you also see ugly strips on two edges.
Stretch	The picture is stretched to fit the screen. Expect distortions.
Tile	The picture is repeated as many times as necessary to fill the screen. If it's too large to fit on the screen, you see the Fill options.
Center	This one is the same as the Fit setting except that the letterboxing goes on all four sides.
Span	Expand the picture to fit as many monitors as are active, left to right.

Windows 10 lets you right-click a picture — a JPG or GIF file — using File Explorer and choose Set as Desktop Background. When you do so, Windows 10 makes a copy of the picture and puts it in the C:\Users\<*username*>\AppData\Roaming\Microsoft\Windows\Recent Items folder and then sets the picture as your background.

You can also control a few aspects of the colors on your desktop, although the pickings are meager, compared to earlier versions of Windows. Here's how to colorize your life:

1. **Click or tap the Start button, the Settings icon, Personalization. On the left, choose Colors.**

   Scroll down to the Choose Your Accent Color section shown in Figure 3-4.

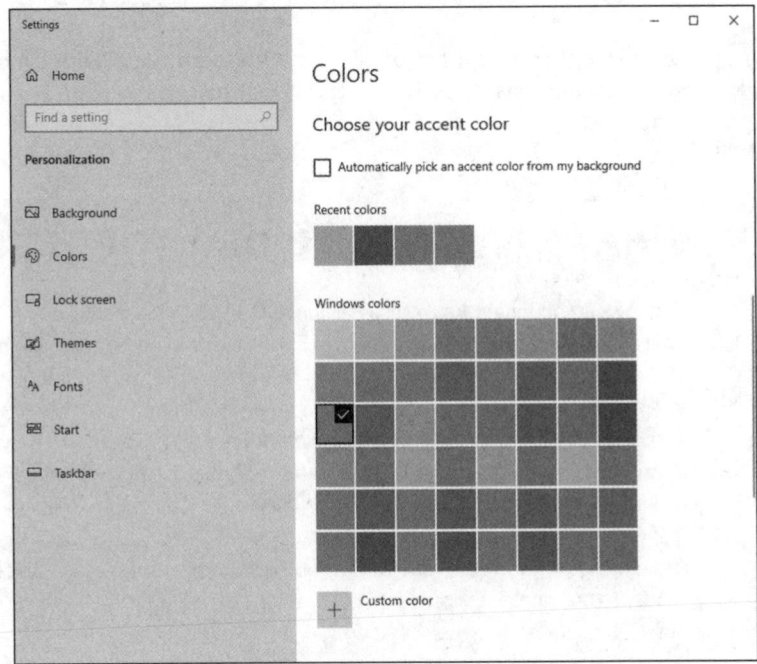

FIGURE 3-4:
Choose a
secondary
color here.

2. **Do one of the following:**

   - **If you want to let Windows choose an accent color for you:** Select the Automatically Pick an Accent Color from My Background option. The accent color will be used sporadically to highlight choices in menus, the background for navigational arrows, and other odd spots.

   - **If you want to choose your own accent color:** Deselect the Automatically Pick an Accent Color from My Background option, and choose from a limited selection of colors.

Personalizing the
Desktop and Taskbar

CHAPTER 3 **Personalizing the Desktop and Taskbar** 245

3. **To have your chosen accent color appear as the background color on the Start menu, on the taskbar, and on the action/notification center pane, select the box for displaying the accent color on those surfaces.**

Usually, Windows 10 uses varying shades of gray for those colors.

4. **To put some transparency and blur on the Start menu, taskbar, and action center, turn on the Transparency Effects slider.**

I rather like the blurring effect.

5. **To make (almost) all the apps appear with white text on a black background, select Dark for Choose Your Default Windows Mode and Choose Your Default App Mode.**

Your changes are visible instantly in Windows 10. I prefer white-on-black (especially for making screenshots), so I set the option to Dark.

**ASK
WOODY.COM**

Of course, I'm still a fan of Windows 7's Aero Glass with its blurred edges and striking contrasts. Yes, I have the visual discernment of a cow. I can live with that. Moo.

WHAT HAPPENED TO DESKTOP THEMES?

Windows 10 no longer has a vestigial link to old-fashioned desktop themes. Themes are collections of the Windows desktop background, window color, sound scheme, and screen saver. At this moment, they don't sit front-and-center in Windows 10 as they did in Windows 7. To find them, go to Settings, Personalization, and then Themes (on the left). You'll see a few themes that Microsoft offers for Windows 10. (They aren't many because Microsoft no longer considers them a focus area.) To change the active theme, click the name of the theme you want to use, under Change Theme.

As an upside, Windows 10 also offers two new theme modes: dark mode and light mode. To activate either one, from Figure 3-2, click or tap the Choose Your Color drop-down, and choose between Light or Dark.

When choosing Dark, the white background colors used in apps and menus turn to black, the light gray in scrollbars turns to dark-gray, and the black text displayed in apps turns to white. When choosing Light, the background used in apps turns white or gray, the background color used in menus turns to white, and the default text displayed in apps turns to gray or black. Both modes (or themes if you like) are excellent and worth trying. Microsoft has borrowed the concept of dark and light modes from smartphones. Dark mode is all the rage with mobile users and looks good in Windows 10 too.

Resolving Desktop Resolution

The best, biggest monitor in the world "don't mean jack" if you can't see the text on the screen. Windows 10 contains a handful of utilities and settings that can help you whump your monitor upside the head and improve its appearance.

With apologies to Billy Crystal, sometimes it *is* more important to look good than to feel good.

Setting the screen resolution

I don't know how many people ask me how to fix this new monitor they just bought. The screen doesn't look right. Must be that %$#@! Windows, yes? The old monitor looked just fine.

Nine times out of ten, when somebody tells me that a new monitor doesn't look right, I ask whether the person adjusted the screen resolution. Invariably, the answer is no. So here's the quick answer to one of the questions I hear most.

REMEMBER

If you plug in a new monitor (or put together a new computer) and the screen looks fuzzy, the most likely culprit entails a mismatch between the resolution your computer expects and the resolution your monitor wants. To a first approximation, a screen resolution is just the number of dots that appear on the screen, usually expressed as two numbers: 1920x1080, for example. Every flat-panel screen has exactly one resolution that looks right and a zillion other resolutions that make things look like you fused your monitor with the end of a Coke bottle.

Setting the screen resolution is easy:

1. **Right-click any empty place on the desktop, and choose Display Settings.**

   You see the Display dialog box shown in Figure 3-5. (If you have more than one monitor or certain kinds of video cards, you may see multiple monitors in the top box.)

2. **Scroll down and click or tap the Advanced Display Settings link, at the bottom.**

   You see your monitor's desktop resolution in the Active Signal Resolution field (see Figure 3-6).

   The hard part? If you don't see a desktop resolution, you must figure out which resolution your monitor likes — its *native resolution*. Some monitors have the resolution printed on a sticker that may still adhere to the front. (Goo Gone works wonders.) All monitors have their native resolutions listed in the manual. (You do have your monitor's manual, yes? No, I don't either.)

Personalizing the
Desktop and Taskbar

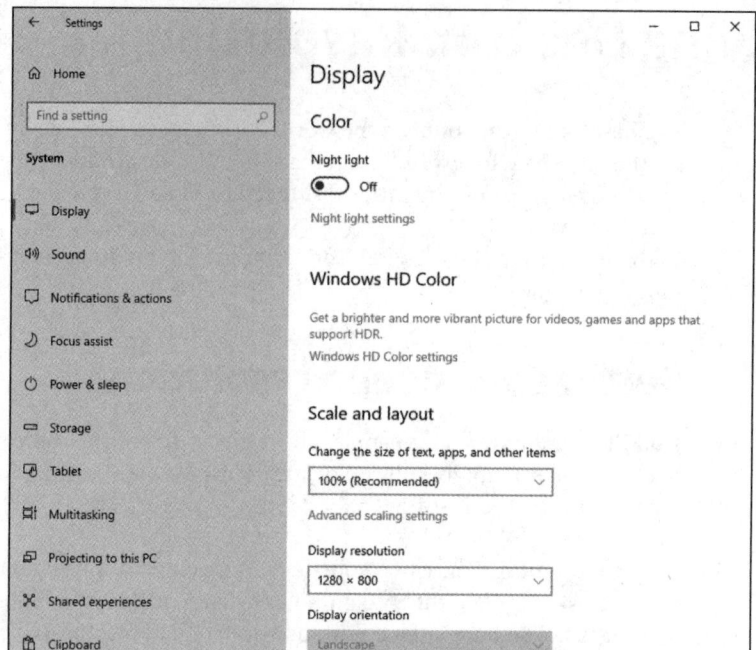

FIGURE 3-5:
Seeking clarity
the Windows
10 way.

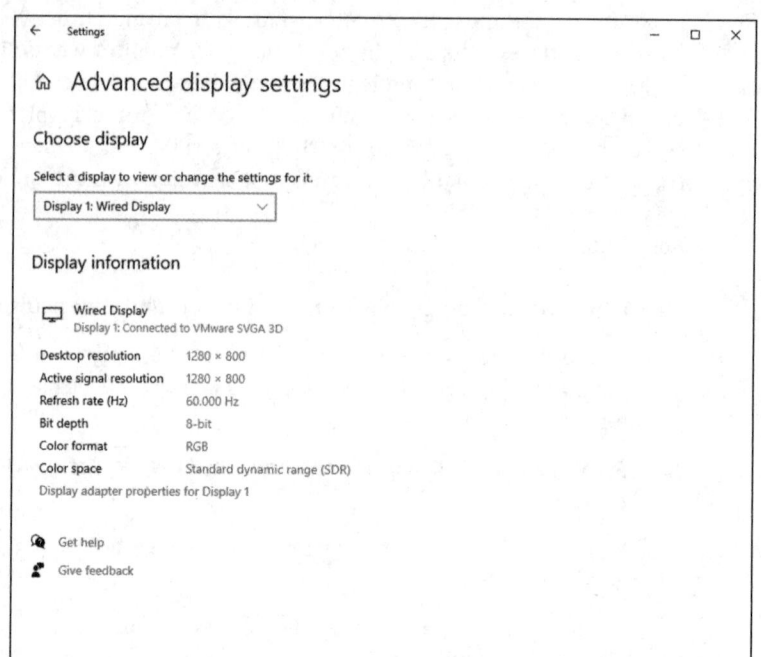

FIGURE 3-6:
See the native
resolution of your
monitor here.

TIP

If you don't know your monitor's native resolution, Google is your friend. Go to www.google.com and type *native resolution* followed by your monitor's model number, which you can (almost) always find engraved in the bezel or stuck on the side. For example, typing *native resolution U3011* immediately finds the native resolution for a Dell U3011 monitor.

If you have, uh, mature eyesight, you may find it helpful to ignore recommendations on tablets and bump up the resolution to make everything larger anyway.

3. **In the upper-left corner, click or tap the back arrow.**

    You return to Figure 3-5. I had you check the resolution first, because if you change it, everything else in this dialog box changes, too.

4. **Click the Display Resolution drop-down box, and choose the resolution you want.**

    Ideally, choose the native resolution of your monitor.

    Everything will become bigger or smaller. The higher the resolution, the smaller everything becomes on the screen. The lower the resolution, the bigger things get.

5. **If you want to lock the orientation of the display — make it portrait all the time, or landscape — change the Display Orientation drop-down box.**

    It's unusual that you want to lock the orientation, but sometimes it happens — like when you're trying to read the news while skinning the cat. I mean the acrobatic maneuver, of course.

    That's all it takes. Your changes take effect immediately.

Changing the size of text, apps, and other items

The problem with high-resolution displays that are Full HD or 4K (everyone wants 4K TVs, smartphones, and so on) is that items get too small on the screen. This makes it difficult to navigate Windows 10, and it puts a strain on your eyes. To keep the native resolution on and make things look bigger than the default, you can set Windows 10 to improve the way it scales the size of text, apps, and other items. Here's how:

1. **Right-click any empty place on the desktop and choose Display Settings.**

    You see the Display dialog box (refer to Figure 3-5).

2. **In the Scale and Layout section on the right, click or tap the drop-down box for Change the Size of Text, Apps, and Other Items.**

You see different scaling options. The default is 100%, and you should increase it to 125%, 150%, or 175%. Experiment with these scaling options and see which one is best. Keep in mind that you'll see different scaling options for different resolutions and monitors, and you may not have all three options available.

If your eyes aren't what they used to be (mine never were), you may want to tell Windows 10 to increase the size of text and other items on the screen. It's just enough boost to help, particularly if you're at an Internet cafe and forgot your glasses.

I strongly recommend that you use this setting with caution. Changing the magnification can cause older programs to go bananas. The overall effect can be chilling. Go slowly, test often, and go back to your default if things don't look or act right.

3. **Close the Settings app, and you are done.**

That's all it takes. Your changes take effect immediately.

Using magnification

If you need more zoom than the font enlarger can offer, you can always use the Ease of Access tool called the Magnifier. As you can see in Figure 3-7, the Magnifier can make everything very big.

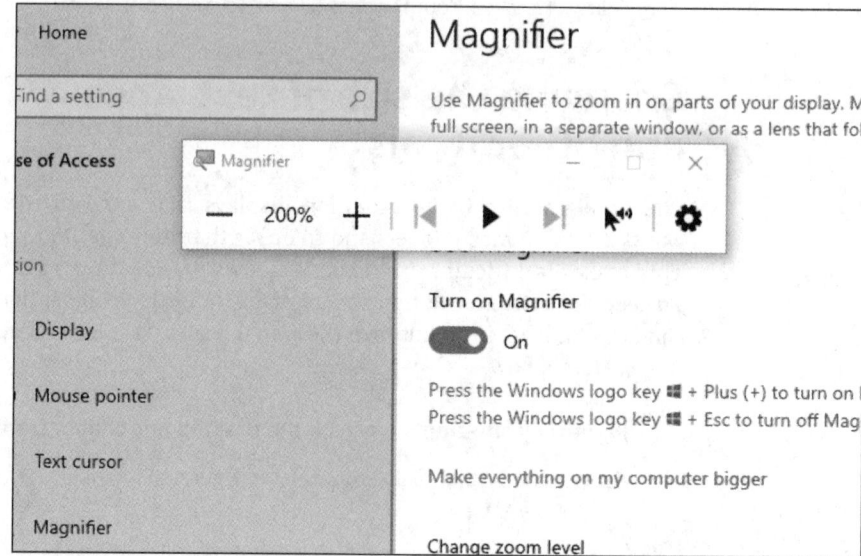

FIGURE 3-7:
The Magnifier can help make everything onscreen really big.

The Magnifier lets you zoom the entire screen by a factor of 200, 300, or 400 — or as high as you like.

REMEMBER

Note that magnifying doesn't increase the quality or resolution of text or pictures. It makes them bigger not finer. That *CSI* "David, can you make the picture sharper?" thing doesn't work with Windows. Sorry, Grissom.

To use the Magnifier, do this:

1. **Click or tap Start, the Settings icon, and Ease of Access. On the left, choose Magnifier.**

2. **Slide the Magnifier setting to On.**

   Everything immediately displays at twice its normal size — 200% in the parlance.

3. **Experiment with moving around. It's odd.**

   Slide your mouse cursor all the way to the left or right to move the screen to the left or right. Same with up and down. This is one situation where a touchscreen really does help.

   A small control shows up with buttons to increase and decrease magnification. (It turns into a magnifying glass icon if you don't use it immediately.)

4. **Scroll down the Magnifier Settings to Change Magnifier View. Click the Choose a View drop-down box and choose Lens.**

   The lens view, shown in Figure 3-8, lets you drag a viewing window across a regular-size screen and magnify what's under the window.

5. **Play with the settings to get the right combination for your eyesight.**

   The settings are sticky, so when you come back to the Magnifier, it'll remember what settings you like best.

6. **To reduce the magnification, press the Windows key and – (minus) repeatedly.**

   That steps you down the magnification levels, until you reach the normal 100% magnification. To turn off magnification, go back to Step 2 and set the Magnifier slider to Off.

If these nostrums don't do the job, you should take advantage of the Windows 10 high-contrast themes. They use color to make text, in particular, stand out. High-contrast themes are available from the Ease of Access dialog box on the left side.

Personalizing the
Desktop and Taskbar

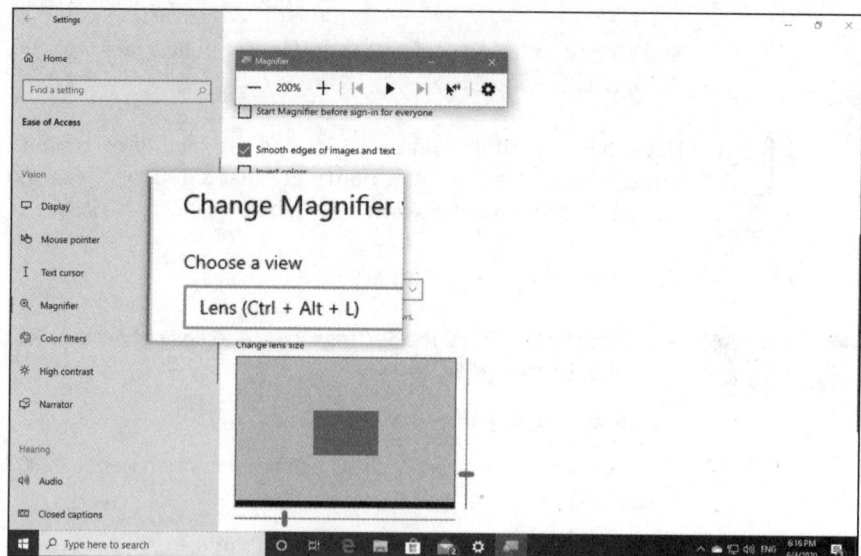

FIGURE 3-8:
The lens view
slides across the
top of a normal-
sized view.

WARNING

If you accidentally hit the Windows key and the + or − key, and your magnification changes mysteriously, now you know the culprit. Go to Ease of Access and turn off Magnifier.

Putting Icons and Shortcuts on the Desktop

Back in the day, if you wanted to get at a program (er, app) quickly, you put a shortcut for it on your desktop. Nowadays, life isn't quite so straightforward. Your choices are many — and that's a good thing.

To access a program/app quickly in Windows 10, you can do any of these:

TIP

>> **Stick a tile on the right side of the Start menu.** This is almost always pretty easy: You find the program (usually by going into the Start ➪ All Apps menu, but also possibly through File Explorer, or maybe there's already a shortcut on your desktop that was put there when the app was installed). Right-click the program, and choose Pin to Start. See Figure 3-9.

>> **Put a link to it in the taskbar.** Using the same technique as with Pin to Start, instead choose More, then Pin to Taskbar. That puts a link to the program in the taskbar, where it's generally available (although, in odd situations — such as tablet mode — it may not be).

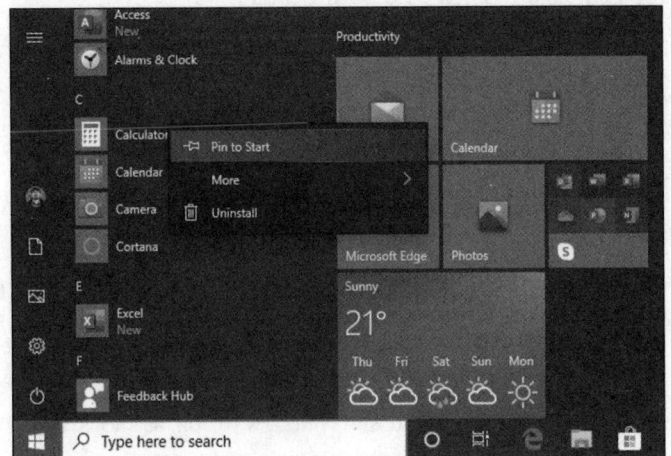

FIGURE 3-9:
It's usually easy to
put a program on
the right side of
the Start menu.

>> **Use Windows 10 search and type the name of the program.** When the program's name is displayed, click or tap it or press the Enter key.

If you've considered adding the program to the Start menu's tiles and putting it on the taskbar, and both approaches leave you a little bit cold, then it's not hard to stick a shortcut to the program on your desktop.

The wonder of desktop shortcuts: You can put many things on the desktop that you just can't get hornswaggled into the Start menu or the taskbar.

Creating shortcuts

Back in Book 3, Chapter 1, I showed you how to put a shortcut to a website on your desktop. Now it's time for the advanced course.

You can set up shortcuts that point to the following items:

>> Old-fashioned Windows programs (desktop apps), of any kind.

>> Web addresses, such as www.dummies.com.

>> Documents, spreadsheets, databases, PowerPoint presentations, pictures, PDF files, and anything else that can be started by double-clicking it.

>> Folders (including the weird folders inside digital cameras, the Fonts folder, and others that you may not think of).

>> Drives (hard drives, CD drives, and key drives, for example).

>> Other computers on your network, and drives and folders on those computers, as long they're shared.

>> Printers (including printers attached to other computers on your network), scanners, cameras, and other pieces of hardware.

>> Network connections, interface cards, and the like.

Here's a whirlwind tour of many different desktop shortcut techniques:

1. **To pin a Windows 10 app (UWP app) to the desktop, find the app in the Start ⇨ All Apps list, click the link, and drag it to the desktop.**

   That creates a shortcut to the Windows 10 app, as shown with the Calendar app in Figure 3-10.

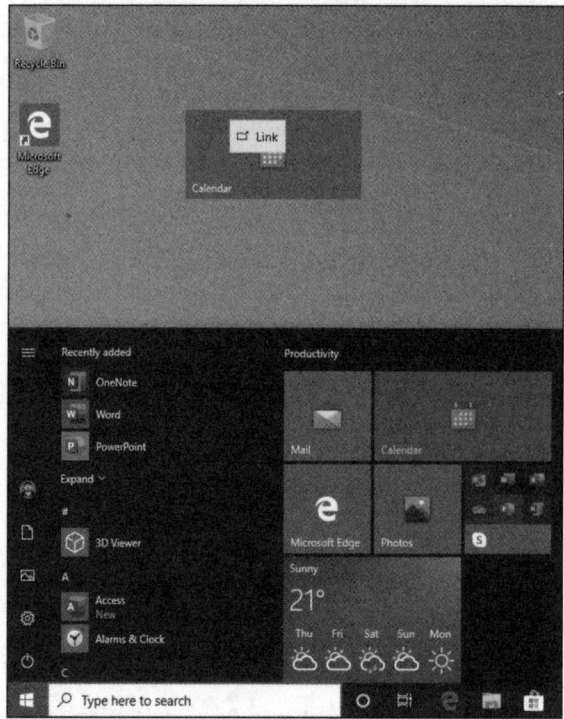

FIGURE 3-10:
Drag a Windows 10 app to the desktop to create a shortcut there.

2. **To create a shortcut to a document (such as a Word file you open over and over), file or folder:**

   a. *Use File Explorer to go to the document, file, or folder.*

   b. *Right-click it, choose Send To, and then choose Desktop (Create Shortcut).*

3. **To create a desktop shortcut for a drive, or another computer on your network (even in a homegroup):**

   a. Use File Explorer to navigate to the drive or computer.

   b. Right-click the folder or drive and choose Create Shortcut.

   c. When Windows says that it can't create a shortcut here, and asks whether you want to place it on the desktop instead, click Yes.

   In Figure 3-11, shortcuts to Calendar, Microsoft Edge, my C: drive, and a document on OneDrive (AskWoody.docx) are all set up and ready to click.

FIGURE 3-11: Shortcuts are easy to set up, if you work through File Explorer.

Arranging icons on the desktop

If you bought a PC with Windows 10 preloaded, you probably have so many icons on the desktop that you can't see straight. That desktop real estate is expensive, and the manufacturers receive a pretty penny for dangling the right icons in your face.

ASK
WOODY.COM

Know what? You can delete all of them, without feeling the least bit guilty. The worst you'll do is delete a shortcut to a manufacturer's tech support program, and if you need to get to the program, the tech support rep can tell you how to find it. The only icon you need is the Recycle Bin, and you can bring that back pretty easily (see the nearby sidebar).

Windows 10 gives you several simple tools for arranging icons on your desktop. If you right-click any empty part of the desktop, you see that you can do the following:

>> **Sort:** Choose Sort By, and then choose an option to sort icons by name, size, type (folders, documents, and shortcuts, for example), or the date on which the icon was last modified. See Figure 3-12.

>> **Arrange:** Right-click an empty place on the desktop, and choose View, Auto Arrange Icons. That is, have Windows 10 arrange them in an orderly fashion, with the first icon in the upper-left corner, the second one directly below the first one, the third one below it, and so on.

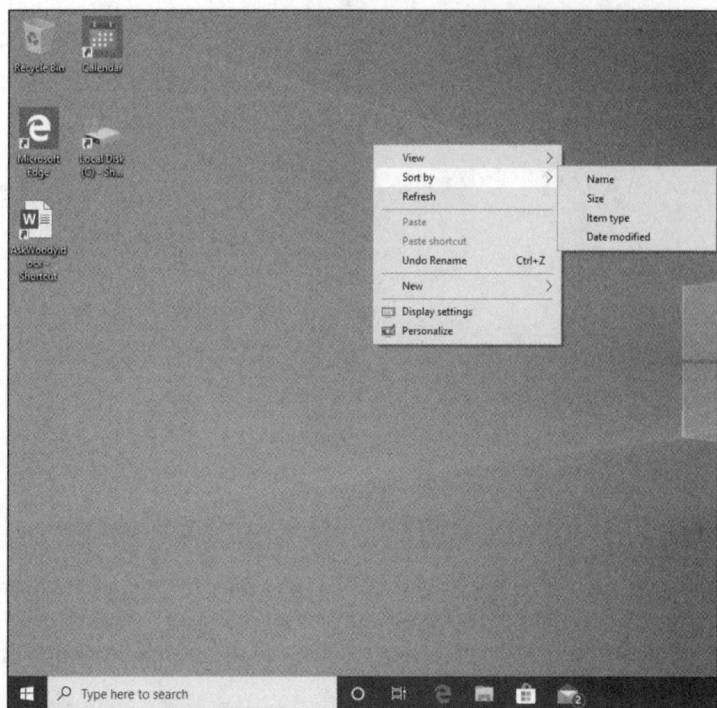

FIGURE 3-12:
Sort all the icons on your desktop with a few clicks.

>> **Align to a grid:** Choose View, Align Icons to Grid. If you don't want to have icons arranged automatically, at least you can choose Align Icons to Grid so that you can see all the icons without one appearing directly on top of the other.

>> **Hide:** You can even choose View, Show Desktop Icons to deselect the Show Desktop Icons option. Your icons disappear — but that kind of defeats the purpose of icons, doesn't it?

TIP

>> **Delete:** In general, you can remove an icon from the Windows 10 desktop by right-clicking it and choosing Delete or by clicking it once and pressing the Delete key.

The appearance of some icons is hard wired: If you put a Word document on your desktop, for example, the document inherits the icon — the picture — of its associated application, Word. The same goes for Excel worksheets, text documents, and recorded audio files. Icons for pictures look like the picture, more or less, if you squint hard.

Icons for shortcuts, however, you can change at will. Follow these steps to change an icon — that is, the picture — on a shortcut:

1. **Right-click the shortcut, and choose Properties.**

2. **In the Properties dialog box, click the Change Icon button.**

3. **Pick an icon from the offered list, or click the Browse button and go looking for icons.**

   Windows 10 abounds with icons. See Table 3-2 for some likely hunting grounds.

RESTORING THE RECYCLE BIN ICON

Sooner or later, it happens to almost everyone. You delete the Recycle Bin icon, and you're not sure how to get it back.

Relax. It isn't that hard . . . if you know the trick.

In the Windows 10 search box type (precisely) **desktop icon settings**. Click or tap the first search result shown, which should be named Themes and Related Settings. Click the Desktop Icon Settings link on the far right. In the Desktop Icon Settings window, select the box for Recycle Bin, and click or tap OK.

You're welcome.

4. **Click the OK button twice.**

   Windows 10 changes the icon permanently (or at least until you change it again).

TABLE 3-2

Where to Find Icons

Contents	File
Windows 10, 8.1, 8, 7, and Vista icons	C:\Windows\system32\imageres.dll
Everything	C:\Windows\System32\shell32.dll
Computers	C:\Windows\explorer.exe
Household	C:\Windows\System32\pifmgr.dll
Folders	C:\Windows\System32\syncui.dll
Old programs (Quattro Pro, anybody?)	C:\Windows\System32\moricons.dll

TIP

Lots and lots of icons are available on the Internet. Use your favorite search engine to search for the term *free Windows icons.* If you go out looking for icons, be painfully aware that many of them come with *crapware wrappers* — programs that install themselves on your machine when all you wanted was an icon. Be careful.

Tricking out the Taskbar

Microsoft developers working on the Windows 7 taskbar gave it a secret internal project name: the Superbar. Although one might debate how much of the Super in the bar arrived compliments of Mac OS, there's no doubt that the Windows 10 taskbar is a key tool for anyone who uses the desktop. Now that you can pin Windows 10 apps on the taskbar, it's become productivity central for many of us.

The Windows Super, uh, taskbar appears at the bottom of the screen, as in Figure 3-13.

If you hover your mouse cursor over an icon, and the icon is associated with a program that's running, you see a thumbnail of what it's doing. For example, in Figure 3-13, Microsoft Edge is running, and the thumbnail gives you a preview of what's on offer.

Anatomy of the taskbar

The taskbar consists of two kinds of icons:

>> **Pinned icons:** Windows 10 ships with eight icons on the taskbar, one for Start, one for Search, another for Cortana, one for task view (and the Timeline), and one each for Microsoft Edge, File Explorer, the Microsoft Store, and Mail. You can see them at the bottom in Figure 3-13. If you install a program and tell the installer to put an icon on the taskbar, an icon for the program appears on the taskbar. You can also pin programs of your choice on the taskbar.

TECHNICAL STUFF

Some older programs have installers that offer to attach themselves to the Quick Launch toolbar. It's a Windows XP–era thing. If you agree to put the icon on the Quick Launch toolbar, the icon for the program gets put on the far-more-upscale taskbar.

>> **Icons associated with running desktop programs:** Every time a program starts, an icon for the program appears on the taskbar. If you run three copies of the program, only one icon shows up. When the program stops, the icon disappears.

You can tell which icons represent running programs: Windows 10 puts an almost imperceptible line under the icon for any running program. If you have more than one copy of the program running, you see more than one line underneath. It's subtle. In Figure 3-13, Microsoft Edge has a line under the icon.

Jumping

If you right-click any icon in the taskbar or tap and hold down, whether or not the icon is pinned, you see a bunch of links called a *jump list,* as shown in Figure 3-14.

The contents of the jump list vary depending on the program that's running, but the bottom pane of every jump list contains the name of the program and the entry Unpin from Taskbar (or conversely, Pin to Taskbar, if the program is running but hasn't been pinned).

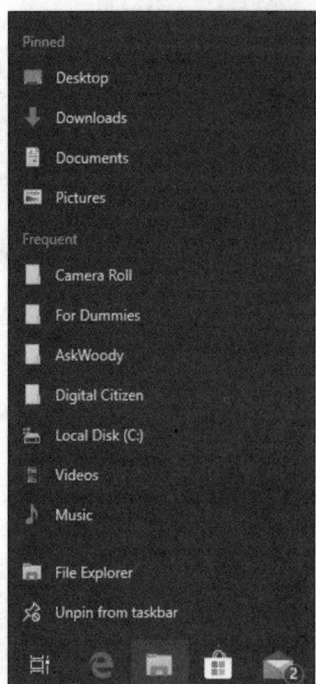

Jump lists were new in Windows 7, and they haven't taken off universally. Implementation of jump lists ranges from downright obsessive (such as Microsoft Edge) to completely lackadaisical (including most applications that aren't made by Microsoft).

Here are the jump list basics:

>> **Jump lists may show your frequent folders or files or recently opened file history.** For example, the File Explorer jump list (shown in Figure 3-14) shows you the same Frequent list that appears inside the app. The Paint jump list (shown in Figure 3-15) shows you the Recent files list found inside the app.

>> **It's generally easy to pin an item to the jump list.** When you pin an item, it sticks to a program's jump list whether or not that item is open. To pin an item, run your mouse out to the right of the item you want to pin and click the stickpin. That puts the item in a separate pane at the top of the jump list.

The jump list has one not-so-obvious use. It lets you open a second copy of the same program. Suppose you want to copy a handful of albums from the music library to your thumbdrive on F:. You start by clicking the File Explorer icon in the taskbar, and on the left, click Music Library. Cool.

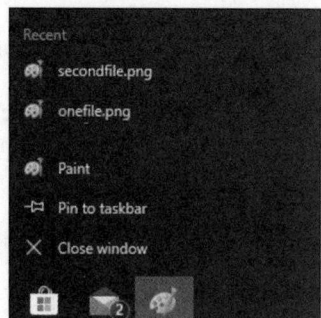

FIGURE 3-15:
Lowly Paint's
jump list shows
recently opened
documents.

You can do the copy-and-paste thang — select an album, press Ctrl+C to copy, use the list on the left of File Explorer to navigate to F:, and then press Ctrl+V to paste. But if you're going to copy many albums, it's much faster and easier to open a second copy of File Explorer and navigate to F: in that second window. Then you can click and drag albums from the Music folder to the F: folder.

To open a second copy of a running program (File Explorer, in this example), you have two choices:

- Hold down the Shift key, and click the icon.

- Right-click the icon (or tap and hold down, perhaps with a nudge upward), and choose the program's name.

In either case, Windows 10 starts a fresh copy of the program.

Changing the taskbar

The taskbar rates as one of the few parts of Windows 10 that is highly malleable. You can modify it till the cows come home:

>> **Pin any program** on the taskbar by right-clicking the program and choosing Pin to Taskbar. Yes, you can right-click the icon of a running program on the taskbar.

>> **Move a pinned icon** by clicking and dragging it. Easy. You know, the way it's supposed to be. You can even drag an icon that isn't pinned into the middle of the pinned icons. When the program associated with the icon stops, the icon disappears, and all pinned icons move back into place.

>> **Unpin any pinned program** by right-clicking it and choosing Unpin from Taskbar. Rocket science.

Unfortunately, with a few exceptions, you can't turn individual documents or folders into icons on the taskbar. But you can pin a folder to the File Explorer jump list, and you can pin a document to the jump list for whichever application is associated with the document. For example, you can pin a song to the jump list for Windows Media Player.

Here's how to pin a folder or document to its associated icon on the taskbar:

1. **Navigate to the folder or document that you want to pin.**

You can use File Explorer to go to the file or folder or you can make a shortcut to the file or folder.

2. **Drag the folder or document (or shortcut) to the taskbar.**

Windows 10 tells you where it will pin the folder, text file, document, or shortcut, as shown in Figure 3-16. For example, if you are dragging a .docx file, Windows 10 will let you pin it to WordPad, Word, File Explorer, or any program that can open a .docx file. If you're dragging a .txt file, Windows 10 lets you pin it to Notepad.

3. **Release the mouse button.**

That's all it takes.

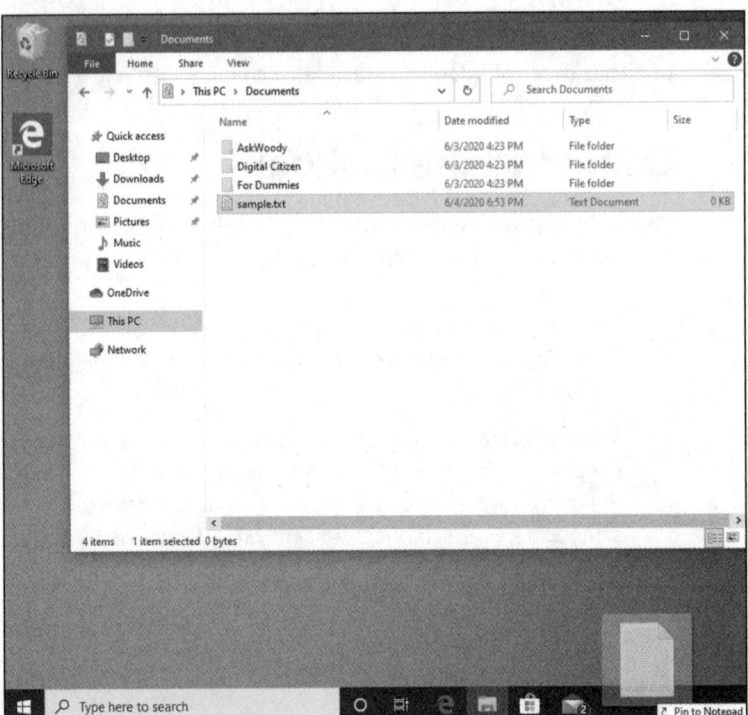

FIGURE 3-16:
Drag a file or folder to pin it to a taskbar icon.

TIP

A little-known side effect: If you pin a file to a program on the taskbar, the program itself also gets pinned to the taskbar, if it wasn't already.

Working with the taskbar

I've discovered a few tricks with the taskbar that you may find worthwhile:

>> Sometimes, you want to shut down all (or most) running programs, and you don't want Windows 10 to do it for you. It's easy to see what's running by looking at the underline under the icon, if your eyesight and your monitor are good enough (refer to Figure 3-13). To close down all instances of a particular program, right-click its icon and choose Close Window or Close All Windows.

TIP

>> Sometimes, if a program is frozen and won't shut itself down, forcing the matter through the taskbar is the easiest way to dislodge it.

>> The terminology is a bit screwy here. Normally, you would choose Exit the Program, Choose File, Exit, Click the Red X, or some such. When you're working with the taskbar, you choose Close Window or Close All Windows from the choices that pop up when you right-click the icon on the taskbar. Different words, same meaning.

TIP

If you move your mouse to the lower-right corner and then click, Windows 10 minimizes all open windows. Click again, and Windows 10 brings back all minimized windows. You can also right-click and choose Peek at Desktop or Show Desktop.

Chapter **4**

Internet Explorer, Chrome, and Firefox

For hundreds of millions of people, the web and Internet Explorer (IE) were synonyms. It's fair to say that IE has done more to extend the reach of PC users than any other product — enabling people from all walks of life, in all corners of the globe, to see what a fascinating world we live in.

At the same time, Internet Explorer has become an object of attack by spammers, scammers, thieves, and other lowlifes. As the Internet's lowest (or is it greatest?) common denominator, IE drew lots of unwanted attention. This has changed. Microsoft is no longer actively developing Internet Explorer, and instead has switched to Microsoft Edge. Users have switched, too, mostly to Google Chrome.

IE usage rose rapidly from its release in 1995, taking half the browser market share by 1998. Usage of IE peaked in 2002–2003, with roughly 90 percent of all browser use worldwide. By early 2018, IE was down to about 12 percent of desktop computer use. (See the sidebar "The history of Internet Explorer.")

By 2015, IE had clearly lost its decade-long supreme position in the web browser pecking order, with strong competition from Firefox and Chrome. By mid-2016, Chrome outflanked IE, and the trend is now clear: IE is a dead horse. Everybody's giving up on it — even Microsoft.

That's the story for desktop and laptop browsers. When you take mobile browsers (browsers used from smartphones and tablets) into account, mobile is taking over the world. In May 2015, Google reported that more than half of all Google searches were from mobile phones. Just phones. By the time you read this, mobile will likely capture 60 to 70 percent of Google searches. Clearly, the future of web browsing looks mobile, with IE (which doesn't run on mobile) rapidly fading into the sunset.

THE HISTORY OF INTERNET EXPLORER

More than any other product, Internet Explorer reflects the odd and tortured Microsoft approach to the web. After largely ignoring the Internet for many years, Microsoft released the first version of Internet Explorer in 1995, as an add-on to Windows. In 1996, Microsoft built Internet Explorer version 3 into Windows itself, violating antitrust laws and using monopolistic tactics to overwhelm Netscape Navigator.

Having illegally pummeled its competitor in the marketplace, Microsoft made almost no improvements to Internet Explorer between August 2001 and August 2006 — an eternity in Internet time. IE became the single largest conduit for malware in the history of computing, with major security patches (sometimes several) appearing almost every month.

And then there was Firefox. Dave Hyatt, Blake Ross (who was a sophomore at Stanford at the time), and hundreds of volunteers took on the IE behemoth, producing a fast, small, free alternative that quickly grabbed a significant share of the browser market. Microsoft responded by incorporating many Firefox features into Internet Explorer.

Although Google did provide most of the money that originally drove Firefox's development, the Googlies decided to make their own browser, with a different slant. First released in late 2008, Chrome has grown to the point that Chrome and Firefox ran neck-and-neck in web utilization statistics, with IE on a downward trend below the 50-percent line. In 2020, Google Chrome became the dominant web browser, leading a market share of more than 60 percent.

With Windows 10's release, Microsoft didn't deprecate Internet Explorer as much as throwing it in a bottle of formaldehyde. You can still use IE all you want (it's under Start ⇨ Windows Accessories), but Microsoft would much rather you use Edge. Which is good, because I would much rather you use Edge, too.

This chapter looks at desktop browsers: Internet Explorer, Firefox, and Google Chrome, each with its own strengths and weaknesses.

TIP

If you're looking for Edge, Microsoft's long-overdue replacement for Internet Explorer, you're in the wrong place. Edge is a Universal Windows Platform (UWP) app, one that lives on the new WinRT-based Universal/Modern/Metro side of the street. For that reason, I talk about it in Chapter 1 of the book that deals with the other side, Book 5.

If you're using Windows 10 in S mode, you don't have any choice: Your only browser is Microsoft Edge. You can skip this chapter entirely.

This chapter looks at what's out there for the old-fashioned desktop, helps you choose one (or two or three) desktop browsers for your everyday use, shows you how to customize your chosen browser, and then offers all sorts of important advice about using the web.

Which Browser Is Best?

I must hear the question, "Which browser is best?" a dozen times a week.

The short answer: It depends.

The long answer: It depends on lots of things. But one thing we know for sure. Microsoft itself doesn't think about Internet Explorer. The old guy's been given the boot, tucked away in an obscure corner where you can conjure him up if you insist. Microsoft's money (and talent) is on Edge.

ASK WOODY.COM

I use Chrome for my day-to-day browser, but increasingly I find myself using the browsers on my Pixel phone and the family iPads. On those devices, I've installed Chrome and use it exclusively. My wife, though, uses the native Safari on her iPhone and iPad. Our Chromebook, which I love with a passion, runs only Chrome (of course).

TIP

I also hear, again and again, the question "How can I make my browser run faster?" The short answer: 99 percent of the time, you can't. The big problem isn't your browser — it's the speed of your Internet connection.

Considering security

Without a doubt, the number-one consideration for any browser user is security. The last thing you need is to get your PC infected with a drive-by attack, where merely looking at an infected web page takes over your computer.

ASK
WOODY.COM

Fortunately, for the first time in many years, if not ever, I feel confident in telling you that many browsers — Chrome, Firefox, Opera, and Microsoft Edge — are excellent choices. None has clear superiority over the others. All are (finally!) secure, as long as you follow a few simple rules.

The days of Microsoft taking all the heat for security holes has passed. Although it's true that there were more frontal assaults on Internet Explorer than on the other two, it's also true that Firefox- and Chrome-specific attacks exist.

TECHNICAL
STUFF

In fact, browsers aren't the major source of attacks any more. Starting in 2007 or so, the bad guys turned their attention away from browsers and went to work on add-ons, specifically Flash and Acrobat PDF Reader, as well as browser toolbars. According to IBM's X-Force Team, the number of browser-attacking exploits has been declining steadily since 2007, with a concomitant rise in infections based on Flash, Reader, Java, toolbars, and other third-party add-ons. Microsoft Edge limits all of them. Score one for the new kid on the block — although Edge remains vulnerable to some of the security holes that dog Internet Explorer.

WARNING

Old versions of Internet Explorer still have major security problems. Microsoft's been actively trying to kill IE 10 for years now. But as long as you stick to the latest browser version, keep your browsers reasonably well updated, and don't install any weird toolbars or other add-ons, your only major points of concern for any of the major browsers are Flash, Reader, and Java. I talk about all three in the following sections.

The place where the latest versions of IE fall down? The infernal parade of patches. Month after month, we're seeing dozens, if not hundreds, of patched parts of IE running out the Automatic update chute. Inevitably one or more of the patch parts causes problems. IE may be the grand old gold standard, but it's on life support.

There's a good case to be made for running Microsoft Edge, and I talk about that in Book 5, Chapter 1.

ASK
WOODY.COM

Both Chrome and Firefox have, in the Windows 8–era past, tried to make a browser that runs well in the new Windows 10 UWP app arena. To date, they haven't had much luck. That may change though, and if it does, I will keep you up to date on www.AskWoody.com.

IE, Firefox, and Chrome aren't the only games in the Windows desktop app milieu. Some people swear by Safari (which is the Apple browser); others go for Opera. I don't like Safari (although it does sync bookmarks on Apple devices), but I do like Opera. I have my hands full just juggling the other three.

Looking at privacy

Privacy is one area that differentiates the Big Three. As best I can tell, nobody knows for sure how much data about your browsing proclivities is kept by the browser manufacturers, but this much seems likely:

- » If you turn on the Suggested Sites feature or SmartScreen Filter in Internet Explorer (see the section on Internet Explorer), IE sends your browsing history to Microsoft, where it is saved and analyzed.

- » Google keeps information about where you go with Chrome. Get over it.

- » Although Firefox is capable of keeping track of where you're going with your browser, Firefox is the least likely of the Big Three to keep or use the data. Why? Because, in direct contrast to both Microsoft and Google, Firefox doesn't have anything to sell you.

In general, the browser manufacturers can't track you directly, as an individual; they can track only your IP address (see the sidebar, "What's an IP address?").

But both Microsoft and Google mash together information that they get from multiple sources. As Microsoft puts it in the Internet Explorer Privacy Statement:

> In order to offer you a more consistent and personalized experience in your interactions with Microsoft, information collected through one Microsoft service may be combined with information obtained through other Microsoft services. We may also supplement the information we collect with information obtained from other companies.

Funny that the statement doesn't mention targeted advertising.

Google does the same thing: It actively collects information about you from every interaction you have with a Google product or location, including the search site and the browser. Google also gets info when you visit a page with a Google ad.

If privacy is very important to you, Firefox is your best choice of the major browsers. No question.

WHAT'S AN IP ADDRESS?

When you're connected to the Internet interacting with a website, the website must be able to find you. Instead of using names (Billy Bob's broken-down ThinkPad), the Internet uses numbers, such as 207.46.232.182, something like a telephone number (that's one of Microsoft's addresses). When you go to a website, you leave behind your IP address. That's the only way the website has to get back to you. Nothing nefarious about it: That's the way the Internet works.

Although your IP address doesn't identify you, uniquely, the IP address for most computers with broadband connections rarely changes. Your IP address changes if you turn off your router and turn it back on again, but for most people, most of the time, the IP address stays constant.

The IP address actually identifies the physical box that's attached to the Internet. For homes and businesses with a network, the address is associated with the router, not individual computers on the Internet. If you're using a mobile (3G, 4G, or 5G) connection, the IP address is associated with your mobile phone provider's equipment, not yours. In some developing countries, the whole country has a handful of IP addresses, and connections inside the country are handled as if they were on an internal network.

Picking a browser

With all the pros and cons, which browser should you choose?

For everyday browsing, I'd say stick with one of the major web browsers. Although each version of each browser is different, a few generalities about the different browsers seem to hold true:

>> **Microsoft Edge** is Microsoft's new kid on the block. The latest version (which must be downloaded from www.microsoft.com/en-us/edge) shares the same rendering engine with Google Chrome and can use Chrome add-ons.

>> **Firefox** has the most extensions, and some of them are quite worthwhile. Ghostery, for example, shows every tracking cookie on every web page (available for Chrome, too); DownThemAll! can download every link on a page and manage them all; IE Tab brings IE compatibility to most ancient web pages (also on Chrome); NoScript blocks Flash and Java unless you unleash them on a specific site. Firefox is also the least likely to sprout privacy problems (see the preceding section).

>> **Chrome** has built-in support for both PDF reading and Flash, and the Java programming language. It can handle all three without relying on the Flash,

Reader, or Java plug-ins, which are historically riddled with security holes. Chrome has also been a pioneer in new features and standards adoption and will take your settings along with you as you move from PC to PC. Increasingly, Chrome is taking top prize for number and quality of extensions. Why? The folks at Chrome have devoted a ton of programming skill and talent to making Chrome extensions rock-solid. As the book went to press, there were strong rumors that Microsoft would announce full Chrome extension support in Edge (plus or minus a few minor changes). There are also rumors that Chrome will start running the full Android app menagerie on an upcoming version of Chrome. Think of that: Any Android app could run on Chrome and on Chromebooks. It's a brave new world, and Chrome is in the lead.

» **Internet Explorer** holds the title for most compatible with ancient websites. Unfortunately, this compatibility comes at a cost: You may have to install programs (such as ActiveX controls) that might have security holes. IE also has a few features that some people find useful, such as the capability to pin websites to the Windows taskbar.

The choice isn't an either/or one. You can easily run Edge, Internet Explorer, Firefox, and Chrome side by side. Here's what I do:

» Most of the time, I run **Chrome.** Yes, I know that Google looks at everything I do while in the clutches of Chrome, but so be it. I particularly like the bookmarking interface — and the bookmarks travel with me, wherever I go, because I'm signed in. Chrome's capability to keep track of where I am and what I'm doing comes in handy when I switch from the desktop to my phone or an iPad: Chrome works the same way wherever I am, whatever I'm doing. And because I'm signed in, it's my number-one favorite digital assistant.

» I have a specific set of tabs open in **Firefox,** all day, every day. I keep two browsers open simultaneously to help me concentrate on Windows updating — that's in Firefox — and all of its nuances, while also working on everything else in Chrome. I like Firefox, with NoScript turned on and Ghostery sniffing out the frighteningly large number of cookies watching me. I don't block cookies with Ghostery, although I can. Mostly I want to see how much sites have sold out, reducing my privacy for their profits.

» I move to **Edge** on the rare occasion when I want a third browser window open, or when I'm testing something. Edge has nice rendering — pages show up better in Edge — and it's fast.

» And **Internet Explorer** is always ready, standing by in case I hit an older web page that doesn't work right in Edge, Firefox, or Chrome. Yes, there are a few — I won't mention my bank by name. Instead of switching Firefox over to the IE Tab add-on, I just jump the monkey and go to IE.

Setting a browser as your default

When you get Windows 10, Microsoft Edge is set up as the default browser: Click a web link in a document, for example, and Edge jumps up to load the web page.

Both Firefox and Chrome offer to become your default web browser, as soon as you install them. Internet Explorer has the option, but it isn't so in your face. They also have a check box that basically tells them to quit asking. I always select that box.

WHAT IS DO NOT TRACK?

Microsoft made a huge step in the direction of helping to protect consumer privacy back when Windows 8 hit. Yes, *that* Microsoft. It turned on Do Not Track by default during Windows 8 setup, in both the desktop and tiled versions of Internet Explorer 11.

Unfortunately, Microsoft was backed into a corner when the folks who promulgated DNT specifically said that a browser can't turn it on by default. Thus, in Windows 10, both IE and Edge don't have DNT enabled by default.

What's DNT? Good question.

Whenever you go to a website, your browser leaves certain fingerprints at each site you visit: the name of your browser, your operating system, your IP address, time zone, screen size, whether cookies are enabled, the address of the last website you visited, that kind of thing. I'm not talking about cookies. I'm talking about data that's inside the header at the beginning of the interaction with every web page. Even if you go incognito (in Chrome), private (in Firefox), or InPrivate (in Internet Explorer or Edge), your browser still sends all that information to every site, every time you visit.

The Do Not Track proposal would assign one more bit in the header that says, "The person using this browser requests that you not track anything he's doing." DNT was originally developed by Firefox. You can turn on DNT in any recent version of Firefox by clicking the Firefox button, Options, Privacy, and selecting the Tell Web Sites I Do Not Want to Be Tracked check box.

As with everything Internet-related, DNT isn't cut and dried. There are lots and lots of nuances. First and foremost, it's entirely voluntary: Websites can ignore the DNT bit if the site's programmers want to. Second, the precise definition of *track* can get a little squishy. Third, there's no possible way to enforce the DNT settings — no way to tell which of the dozens of billions of websites now readily accessible even claim to have a DNT policy, much less implement it. The advertising industry and the privacy partisans have yet to agree on anything, much less a DNT proposal. Still, it's a start in the right direction.

It's easy to change your default browser. Here's how:

1. **Click Start, then click the Settings icon (which looks like a wheel).**

2. **Click Apps. Then, on the left, choose Default Apps.**

You see the default apps shown in Figure 4-1.

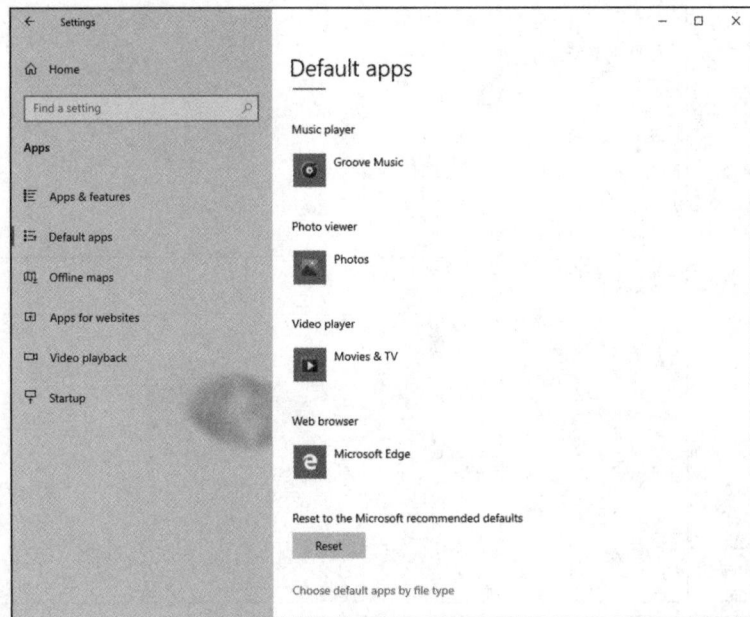

FIGURE 4-1:
Set your default
browser here.

3. **On the right, scroll down to the Web browser entry; chances are you'll see Microsoft Edge. Click Microsoft Edge.**

You see a list of all browsers currently installed on your computer.

4. **Choose the browser that you want to turn into your default browser. Then switch anyway if Microsoft tries to convince you to continue using Edge.**

This tells Windows 10 to associate with the browser almost all filename extensions that the browser can handle.

WARNING

5. **Don't trust Microsoft's re-assigning your browser defaults? Good. Down at the bottom, click the link that says Set Defaults by App.**

You see all the apps installed in Windows 10.

6. **Scroll down until you find Google Chrome or the browser that you want as the default. Click its name, and then click Manage.**

You see the list shown in Figure 4-2. Whoa! When Windows 10 sets defaults for Google Chrome, it doesn't shuffle PDF files to Chrome. Instead, it keeps PDF files inside Microsoft Edge. A little bit of dirty pool here.

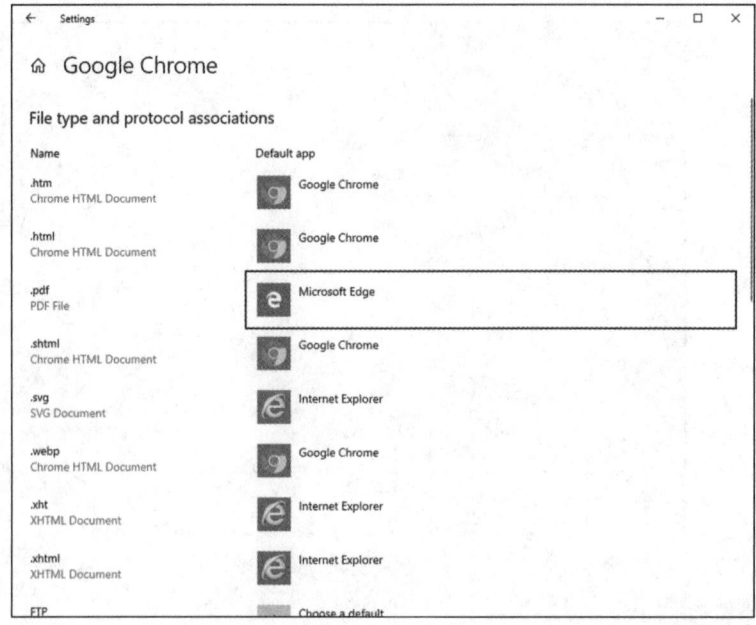

FIGURE 4-2: Chrome can handle these kinds of files and protocols.

7. **Click the name of the default app next to .pdf (Microsoft Edge in Figure 4-2) and choose Google Chrome (or the browser you prefer).**

Your chosen program (in this case, Chrome) becomes the default for that particular kind of file. If you change the PDF box over to Chrome, for example, double-clicking a PDF file will open it in Chrome — not in Microsoft Edge.

8. **Click the X button in the top-right corner of the Settings window.**

Your settings are applied instantly.

Using Internet Explorer on the Desktop

Internet Explorer 11 on the Windows 10 desktop (see Figure 4-3) is similar to — almost indistinguishable from — Internet Explorer 11 on Windows 7 or Windows 8. It has the old, familiar interface. It runs all the add-ons you've come to know and love and distrust. Internet Explorer 11 gives you some of the things you expect from a modern browser — except an extensive library of customized add-ons — and is big, fat, slow, and curiously buggy. Any way you look at it, Microsoft isn't giving IE any love these days. It's definitely on the way out. Which isn't necessarily a bad thing, for you and me.

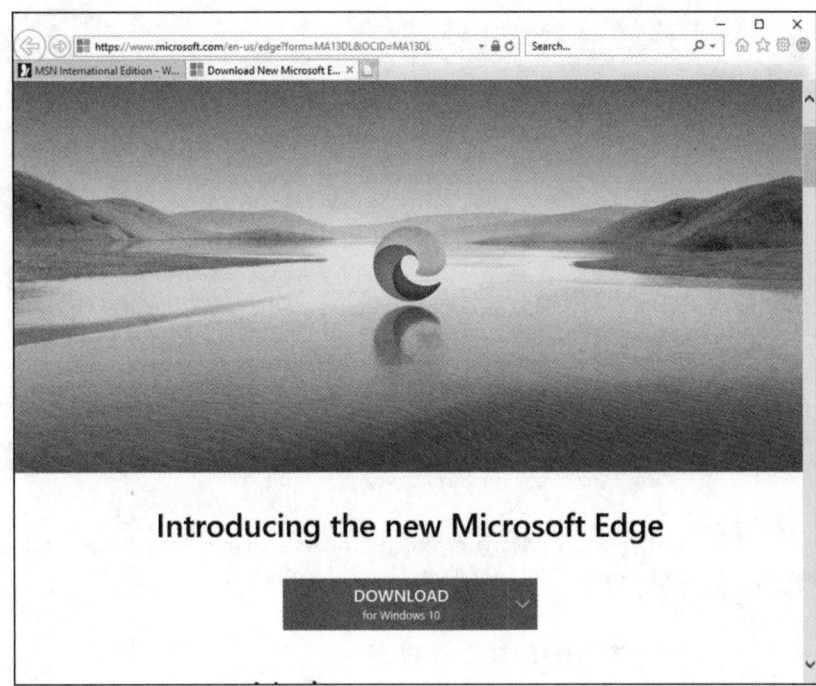

FIGURE 4-3:
Sign of the times: Internet Explorer 11's new welcome page invites you to switch to Edge.

Navigating in IE

One great thing about Internet Explorer is that you can be an absolute no-clue beginner, and with just a few hints about tools and so on, you can find your way around the web like a pro. A big part of the reason why: Hundreds of millions of people, if not more than a billion, have already used IE. For many, *IE* is synonymous with *web*. And that's kind of sad.

Figure 4-4 gives you a diagram of the basic layout of the Internet Explorer window. You get back and forward buttons, an address bar, search box (magnifying glass) and refresh (circle) icons, and icons for home page(s), favorites/history (star), settings (gear), and a "let us know what you think" (happy face).

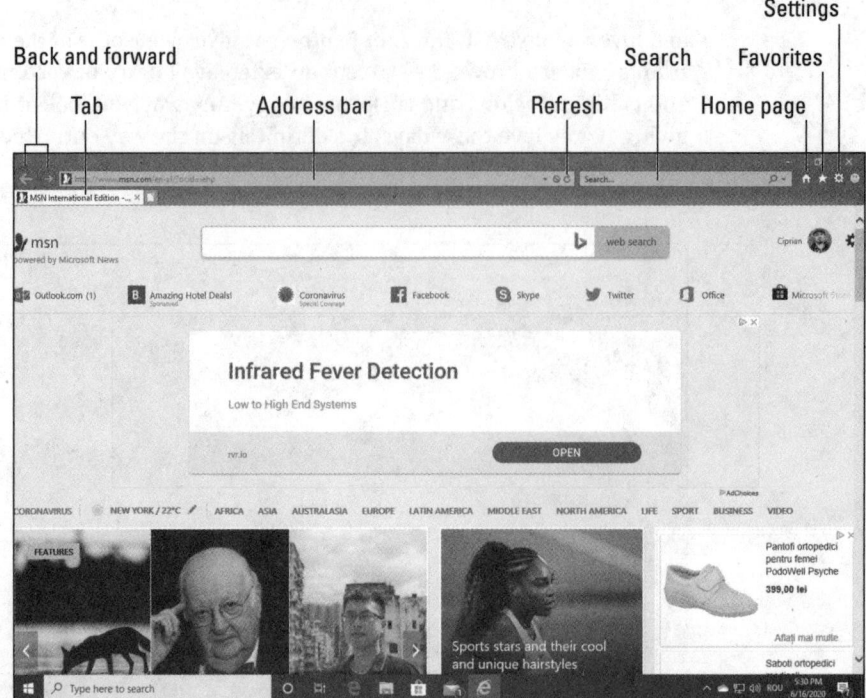

Settings
Back and forward   Search   Favorites
Tab   Address bar   Refresh   Home page

FIGURE 4-4:
The IE
window includes
everything you
need to work on
the web.

And if you're starting on Microsoft's msn.com website, you get ads, ads, ads and — golly! — more ads.

Don't work too hard

A handful of Internet Explorer tricks can make all the difference in your productivity and sanity. Every IE user should know these shortcuts:

>> **You rarely need to type *www* in the address bar at the beginning of an address and you never need to type *http://.*** People who build websites these days are almost always savvy enough to let you drop the use of the *www* at the beginning of the website's name. Unless the site you're headed to was last updated in the late 17th century, you can probably get there by simply typing the name of the site, as long as you include the part at the end. So you can type `http://www.dummies.com` if you want to, but typing *dummies.com* works just as well.

REMEMBER

>> **IE automatically sticks *http://www.* onto the front of an address you type and *.com* on the end if you press Ctrl+Enter.** So if you want to go to the site `http://www.dummies.com`, you need to type only *dummies* in the address bar and press Ctrl+Enter.

>> **With a few exceptions, address capitalization doesn't matter.** Typing either *AskWoody.com* or *askwoody.com* gets you to my website — as does *asKwoodY.cOm*. On the other hand, hyphens (-) and underscores (_) aren't interchangeable: some-site.com and some_site.com would be two different sites if they were the real deals. Similarly, the number 0 isn't the same as the letter O, the number 1 isn't a letter l, and radishes aren't the same as turnips. Or so my niece tells me.

ASK WOODY.COM

The exceptions? Web addresses from one of the thousands of websites that now have shortened URLs. Go to `https://bitly.com` for example, or `https://goo.gl`, feed it an URL that's a gazillion letters long, click a button, and you get back something that looks like this: goo.gl/XY2Am. In those kinds of addresses — shortened ones — capitalization *does* matter.

TIP

While we're on the topic of working too hard, keeping track of passwords rates as the single biggest pain in the neck in any browser. You have passwords for, what, a hundred different sites? If you haven't yet discovered LastPass (or RoboForm or 1Password or KeePass), get to Book 10, Chapter 5, and check it out.

Moving around the main desktop window

As you can see, IE packs lots of possibilities into that small space. The items you use most often are described in this list:

>> **Backward and forward arrows:** Go to the previously displayed page; hold down to see a list of all previous pages.

>> **Address bar:** This enables you to type the web address of a page that you want to move to directly. You can also type search terms here; click the spyglass or press Enter, and IE looks them up using your default search engine.

>> **Refresh:** If you think the page has changed, tap or click this circle arrow icon to have IE retrieve it for you again.

>> **Tab:** You can have many pages open at a time, one on each tab. To create a new tab, click the small, gray blank tab on the right.

>> **Home page:** This replaces the current tab with the tab(s) on your home page(s).

>> **Favorites icon:** This lets you set, go to, and organize favorite websites, as well as look at your browsing history.

>> **Settings:** This eight-spoke wheel takes you under the covers to change the way IE behaves. Or misbehaves.

If you want to see the old-fashioned toolbar menus (File, Edit, View, and all the others) in Internet Explorer, press Alt. Yep, that's how you get to IE's inner workings.

Tinkering with tabs

Tabs offer you a chance to bring up multiple web pages without opening multiple copies of IE. They're a major navigational aid because it's easy to switch among tabs. If you've never used browser tabs, you may wonder what all the fuss is about. It doesn't seem like there's much difference between opening another window and adding a tab (see Figure 4-5). But after you get the hang of it, tabs can help you organize pages and quickly jump to the one you want.

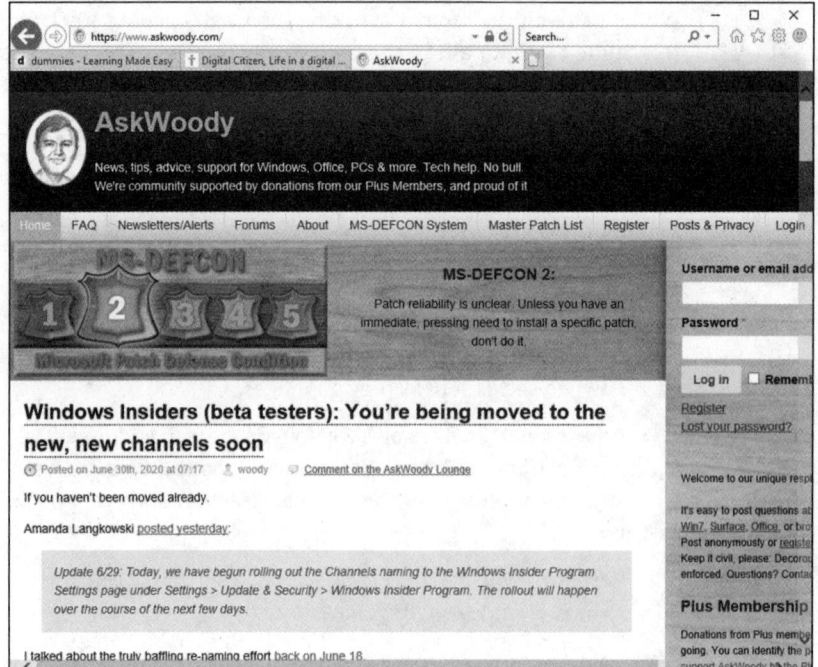

FIGURE 4-5:
If you've never used tabs, you're in for a treat.

You can add a new tab to IE in any of these four ways:

>> Click the gray box to the right of the rightmost tab. That starts a blank new tab, and away you go.

>> Ctrl+click a link to open the linked page in a new tab.

>> Press Ctrl+T to start a new tab. When the tab is open, you get to navigate manually, just as you would in any other browser window.

>> Right-click a link, and choose Open in New Tab.

In addition, the web page you're looking at may specify that any links on the page are to open in a new tab, instead of overwriting the current one.

**ASK
WOODY.COM**

Why do I like tabs? I can set up a single window with a bunch of related tabs and then bookmark the whole shebang. That makes it one-click easy to open all my favorite news sites, research sites, or financial sites. While my browser's out loading pages, I can go do something else and return to the tabbed window when everything's loaded and ready to go.

TIP

You can reorganize the order of tabs by simply clicking a tab and dragging it to a different location.

Using the address bar

No doubt you're familiar with basic browser functions, or you can guess when you know what the controls mean. But you may not know about some of these finer points:

>> When you type on the address bar, IE looks at what you're typing and tries to match it with the list of sites it has in your history list and in your favorites. Sometimes, you can get the right address (URL) by typing something related to the site. Watch as you type and see what IE comes up with.

TIP

If you turn on Bing Suggestions (sometimes called Suggested Sites), IE sends all your keystrokes to Mother Microsoft and has Bing try to guess what you're looking for. Depending on how you feel about privacy, that idea may or may not be a good one. See the section later in this chapter.

>> Click a link, and the web page decides whether you move to the new page in the current browser tab or a new tab appears with the clicked page loaded. Many people don't realize that the web page makes the decision about following the link in the same tab or creating a new one. You can override the web page's setting.

• Shift+click, and a new browser window always opens with the clicked page loaded.

• Ctrl+click, and the clicked page appears on a new tab in the current browser window. Similarly, if you type in the search bar and press Ctrl+Enter, the results appear in a new tab.

>> Even if the web page hijacks your backward and forward arrows, you can always move backward (or forward) by clicking and holding down the directional arrow, and choosing the page you want.

TIP

You can bring up a history of all the pages you visited in the past few weeks by pressing Ctrl+H, as shown in Figure 4-6.

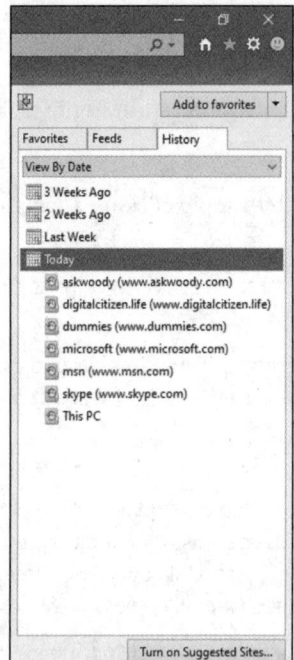

FIGURE 4-6:
Bring up the
browsing history
with Ctrl+H.

To search for a particular word or phrase on a page, press Ctrl+F. Force your browser to refresh a web page (retrieve the latest version, even if a version is stored locally) by pressing F5. If you need to make sure that you have the latest version, even if the timestamps may be screwed up, press Ctrl+F5.

Saving space, losing time

Increasing or decreasing the number of days of browsing history that IE stores doesn't have much effect on the amount of data stored on the hard drive: Even a hyperactive surfer will have a hard time cranking up a History folder that's much larger than 1MB. By contrast, temporary Internet files on your computer can take up 10, 50, or even 100 times that much space.

Those temporary Internet files exist only to speed your Internet access: When IE hits a web page that it has seen before, if a copy of the page's contents appears in the Temporary Internet Files folder, IE grabs the stuff on the hard drive rather than wait for a download. That can make a huge difference in IE's responsiveness, particularly if you have a slow Internet connection, but the speed comes at a price: 250MB, if you haven't changed it.

To clear out the IE temporary Internet files, follow these simple steps:

1. **Start Internet Explorer.**

2. **Click the Settings icon and choose Internet Options.**

   The Settings icon is in the upper right. The Internet Options dialog box appears.

3. **On the General tab, under Browsing History, click the Delete button.**

   You see the Delete Browsing History dialog box shown in Figure 4-7.

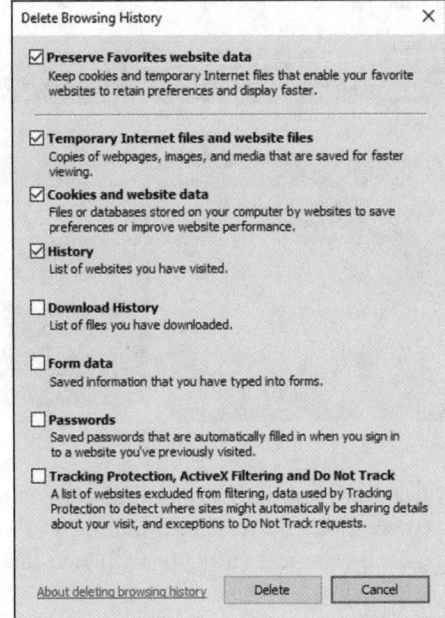

FIGURE 4-7:
You have full control over what kinds of browsing history gets deleted.

4. **Choose the kinds of data you want to delete, and click Delete; then click OK to close the Internet Options dialog box.**

   You won't hurt anything, but revisited web pages take longer to appear. For advice about cookies, see the next section.

Changing the home page

Every time you start the desktop version of IE, it whirrs, and after a relatively brief moment (how brief depends primarily on the speed of your Internet connection), a web page appears. The information that page contains depends on whether your computer is set up to begin with a specific page known as a *home page.*

Microsoft sets up www.msn.com as the IE home page (see Figure 4-8) by default — a page best known for its, uh, quirky choice of news items and phenomenally high density of ads, including Microsoft's own ads. Many PC manufacturers set the Internet Explorer home page to display something related to their systems.

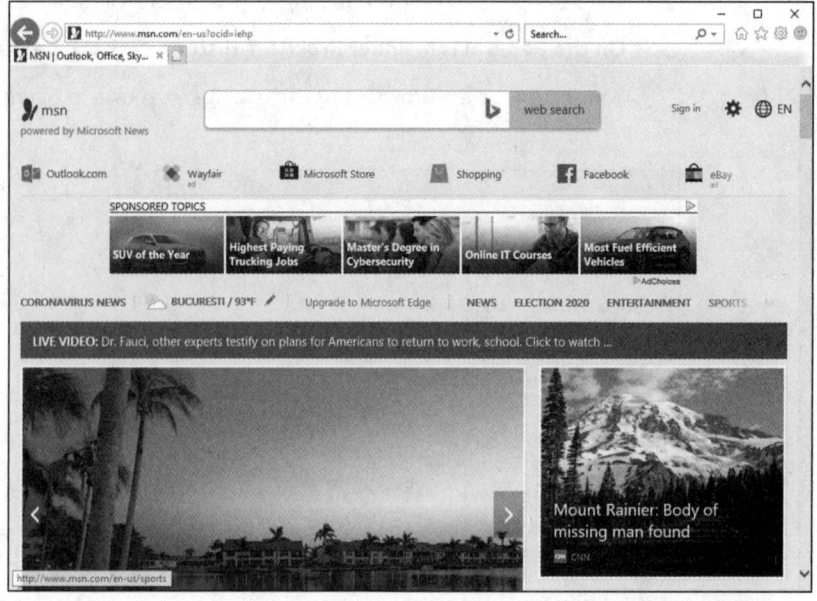

FIGURE 4-8:
If msn.com is your favorite page on the web, you may want to consider a prefrontal lobotomy.

If the ditzy, ad-laden MSN home page leaves you wondering whether P.T. Barnum still designs web pages (there's one born every minute), or if your PC manufacturer's idea of a good home page doesn't quite jibe with your tastes, you can easily change the home page. Here's how:

1. **Start IE.**

2. **Navigate to the page or pages you want to use for a home page.**

   You can bring up as many pages as you like on separate tabs. All the tabs will become your home page. See the previous section, "Tinkering with tabs," if you're not sure how to use tabs.

3. **Tap or click the Setting icon, choose Internet Options, and click the General tab.**

You see the Home Page settings shown in Figure 4-9.

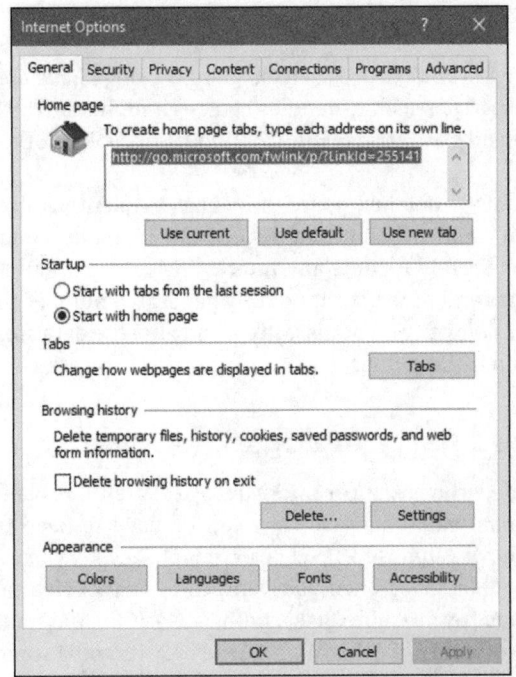

FIGURE 4-9:
Set the home
page(s) here.

4. **At the top, make sure you have the list of all the tabs you would like to open as your home page, and then tap or click the button marked Use Current.**

If you choose Use New Tab, IE starts with no new page at all. That can be considerably faster than starting with a real home page.

TIP

5. **Click OK.**

Every time IE runs, it brings up the tabs you selected.

Dealing with cookies

A *cookie*, as you probably know, is a text file that a website stores on your computer. The website can put information inside its own cookie (say, the time and date of your last visit or the page you were last viewing or your account number). At least in theory, a website can look at and change only its own cookies: The cookie provides a means for an individual website to store information on your computer and to retrieve it later, using your browser.

In general, that's A Good Thing. Cookies can minimize the amount of futzing around that you need to do on a site. For example, shopping cart/checkout sites need cookies.

Of course, nothing ever goes precisely as planned. Bugs have appeared in the way Internet Explorer, in particular, handles cookies and, historically, it's been possible for rogue websites to retrieve information from cookies other than their own.

ASK
WOODY.COM

Because of ongoing problems, sound and fury frequently raised by people who don't understand, and concomitant legislation in many countries, first-party cookies these days rarely include any interesting information. Mostly, they store innocuous settings and perhaps a randomly generated number that's used to track a customer in the company's database. To a bad guy, the data stored in most cookies varies between banal and useless.

What's a third-party cookie?

By contrast, *third-party cookies* (or *tracking cookies*) aren't as bland. They have significant commercial value because they can be used to keep track of your web surfing. Here's how: Suppose ZDNet (www.zdnet.com), which is owned by CBS, sells an ad to DoubleClick. When you venture to any ZDNet page (they all have tiny, one-pixel ads from DoubleClick), both ZDNet/CBS and DoubleClick can stick cookies on your computer. ZDNet can retrieve only its cookie, and DoubleClick can retrieve only its cookie. Cool. DoubleClick may keep information about you visiting a ZDNet site that talks about, oh, an Android phone.

Now suppose that DealTime (www.dealtime.com) sells an ad to DoubleClick. You go to any page on DealTime (they also have tiny 1-pixel DoubleClick ads on every page), and both DealTime and DoubleClick can look at their own cookies. Deal-Time may be smart enough to ask DoubleClick whether you've been looking at Android phones and then offer you a bargain tailored to your recent surfing. Or an insurance company may discover that you've been looking at information pages about the heartbreak of psoriasis. Or a car company may find out you're very interested in its latest Stutzmobile.

Multiply that little example by 10, 100, or 100,000, and you begin to see how third-party cookies can be used to collect a whole lot of information about you and your surfing habits. There's nothing illegal or immoral about it. But some people (present company certainly included) find it disconcerting. Oh, you know that Google owns DoubleClick, yes?

**ASK
WOODY.COM**

I don't get too worked up about cookies these days. If you've ever worked with them programmatically, you're probably at the yawning stage too. But the potential is there for them to become pernicious.

Deleting cookies

Cookies don't have anything to do with spam — you receive the same junk email even if you tell your computer to reject every cookie that darkens your door. Cookies don't spy on your PC, go sniffing for bank accounts, or keep a log of those . . . ahem . . . artistic websites you visit. They do serve a useful purpose, but like so many other concepts in the computer industry, cookies are exploited by a few companies in questionable ways. I talk about cookies extensively in Book 9, Chapter 1. If you're worried about cookies and want to know what's really happening, that's a great place to start.

To delete all cookies in Internet Explorer, follow the instructions in the earlier section "Saving space, losing time" to bring up the Delete Browsing History dialog box (refer to Figure 4-7). Make sure you select Cookies and Website Data, and click Delete. IE deletes all your cookies.

**ASK
WOODY.COM**

Internet Explorer has a mechanism for blocking third-party cookies, but I don't think it works very well. It's based on an old standard known as P3P, which is actually used by about a dozen websites based in Lower Slobovokia — and that's about it. Even some of Microsoft's own sites don't use P3P. I talk about the problems with IE's third-party cookie blocking in one of my *Info-World* Tech Watch articles, at `www.infoworld.com/t/internet-privacy/ googles-cookie-runaround-in-ie-not-big-deal-186889`.

Why you should stop using IE

I highly recommend that you stop using Internet Explorer. If you have to use it, do so only on websites that won't work without it. For everything else, there's Google Chrome, Firefox, Opera, Microsoft Edge, and others — all modern browsers that are a lot more secure and feature-packed. Today, even Microsoft tells you to ditch IE and go for greener pastures, preferably Microsoft Edge. If I haven't convinced you so far to ditch IE, here's one final try:

>> Microsoft keeps IE in Windows 10 only for businesses with ancient web services that haven't been replaced yet. They actively developed IE for years, and this browser only gets security fixes for its never-ending list of vulnerabilities.

>> IE no longer offers support for modern web standards, and most websites look and work poorly when using this browser. Make an experiment and visit

the same news website in Chrome, Edge, and IE, side by side. You'll immediately notice the differences.

>> Internet Explorer no longer has an ecosystem of useful add-ons. Even less-known browsers such as Opera or Vivaldi (I'm sure this is the first time you've heard about them) offer a lot more useful features and extensions.

>> If you like your privacy and security, IE is the last browser you should use. The tools and techniques to hack and steal data from IE have been fine-tuned for decades, and it's the most vulnerable browser you could use.

Customizing Firefox

Firefox is a great browser that respects your privacy more than others. If you don't want Google or Microsoft knowing what you do online (or at least knowing less than what they know when you use their browsers), you should switch to Mozilla Firefox.

ASK WOODY.COM

I use Chrome and Firefox. I've recommended Firefox in my books for years and have recently switched to primarily using Chrome. Debating the relative merits of web browsers soon degenerates to a fight over the number of angels that can stand on the head of a pin. Suffice it to say that I feel Firefox has more options, although both Firefox and Chrome have started grabbing system resources like they own the place. I also like the fact that Firefox has no vested interest in keeping track of what I'm doing.

REMEMBER

I don't mean to imply that Firefox is perfect. It isn't. The Firefox team releases security patches too, just like Edge and Chrome teams, and you need to make sure you keep Firefox updated. But I think you'll enjoy using Firefox more than Internet Explorer. I also would bet that you hit far fewer in-the-wild security problems with the Fox.

Installing Firefox

Installing Firefox can't be simpler. You don't need to disable Internet Explorer, pat your head and rub your belly, or jump through any other hoops (although clicking your heels and repeating "There's no place like home" may help). Just follow these steps:

1. Using any convenient browser (even Edge or IE), go to www.firefox.com and follow the instructions to download and run the installer for the latest version of Firefox.

In Figure 4-10, I pull up Firefox by using Edge.

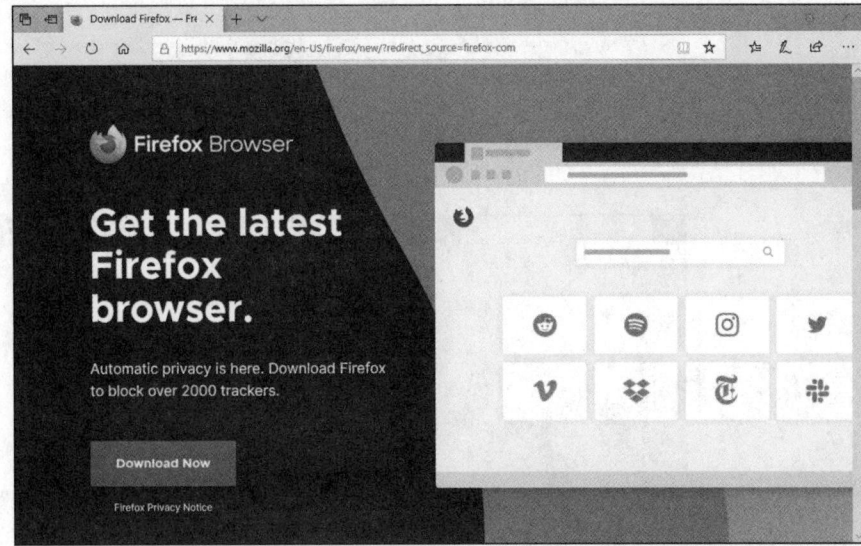

FIGURE 4-10:
You can install
Firefox from
Edge or any other
browser.

2. **Do the following:**

   a. *Click the big Download Now button.*

   b. *Chances are good that you'll need to click Save and then Run to get the installer going.*

   c. *Give the User Account Control message box a Yes.*

   d. *Wait for Firefox to install.*

   You'll likely end up staring at the Firefox main screen, shown in Figure 4-11.

3. **Choose whether you want to sign in to sync with your Firefox account.**

   One of the nicest features in most modern browsers is their capability to sync bookmarks, history, and other settings. With Firefox, you can sync across Windows, Android, or iOS versions of the browser. All it takes is a (gulp) Firefox account.

4. **To get a Firefox account:**

   a. *Log in with a valid email address (it can be a throwaway free address, as long as you can retrieve mail from that address).*

   b. *Provide a password (which does not need to be the same as your email password).*

   c. *Type an age (creativity counts).*

   d. *Select what you want to sync.*

Any time you want to sync your settings with another copy of Firefox, follow the same procedure and sign in with the same Firefox Account. You can use multiple Firefox accounts, if you want to sync groups of machines in different ways.

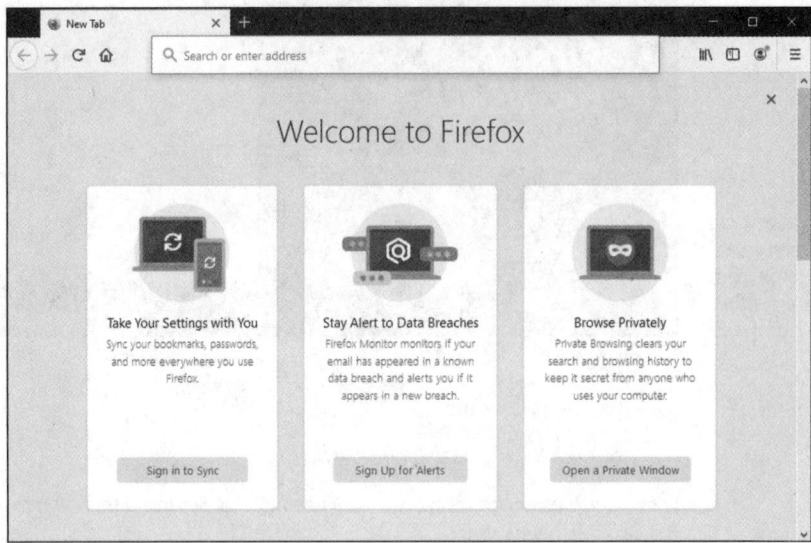

FIGURE 4-11:
Firefox is up and running.

ASK
WOODY.COM

You may be asked whether you want to make Firefox your default browser. I click Not Now, because I'd rather make Chrome my default browser.

All the tricks I mention in the previous IE section called "Don't work too hard," also work in Firefox. You never need to type *http://*, almost never need to type *www*, and typing something like *dummies* followed by a Ctrl+Enter puts you spot-on for `www.dummies.com`.

Setting a home page in Firefox is similar to setting one in IE. To get to the right place, click the Firefox hamburger (three lines) menu, in the upper-right corner, to bring up the Settings menu shown in Figure 4-12. Choose Options. Home page settings are on the General tab.

Browsing privately in Firefox

Firefox has a private browsing feature similar to IE's InPrivate browsing. Firefox's version is called, er, private browsing. (Hey, Firefox invented it!)

FIGURE 4-12:
The Firefox
settings
(hamburger)
menu.

To start a private browsing session, click the hamburger icon in the upper right and choose New Private Window.

Some people prefer to always work in private browsing mode. There's much to be said for that approach, although you won't get the advantages of having cookies hanging around. Staying in private browsing mode is easy to do in Firefox. Here's how:

1. **Start Firefox. Click the hamburger icon in the upper right, and choose Options. On the left, click Privacy & Security.**

2. **Scroll down to the History section. In the Firefox Will drop-down box, choose Use Custom Settings for History.**

   You see the options shown in Figure 4-13.

3. **Select the Always Use Private Browsing Mode option.**

   You may have to restart Firefox after turning on this option.

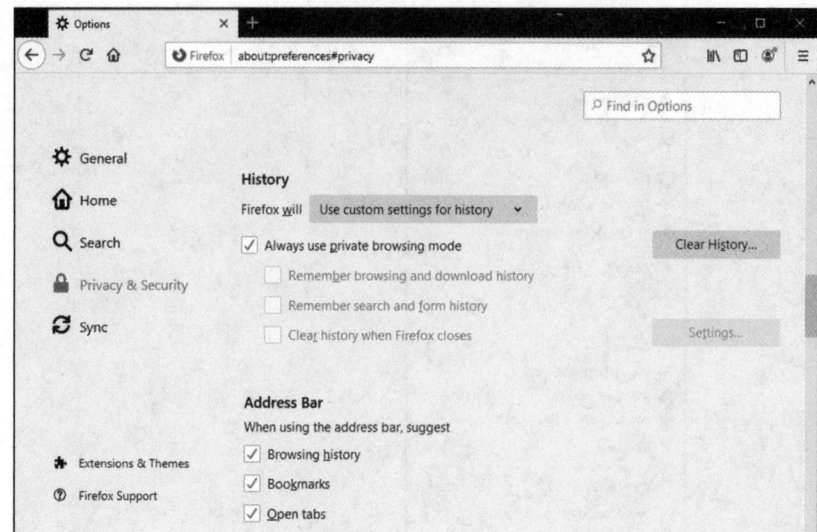

FIGURE 4-13:
It's easy to have
Firefox always
start in private
browsing mode.

4. **If you want to turn on Do Not Track (see the "What Is Do Not Track?" sidebar, earlier in the chapter), scroll down a bit, and select the Always check box in the Enhanced Tracking Protection section.**

   Admittedly, DNT doesn't do much, but it doesn't hurt and may block a few sites.

5. **Close Options by clicking X on its tab in Firefox.**

The next time you start Firefox, it'll be in private browsing mode. If you ever want to drop back into regular mode, click the hamburger icon and follow the above steps, choosing Firefox Will: Remember History.

Bookmarking with the Fox

Firefox handles bookmarks differently from Internet Explorer. (In IE, they're called favorites. Same thing.)

The easiest way to understand Firefox bookmarks? Start with the Unsorted Bookmarks folder. Go to the site you want to bookmark, and tap or click the Bookmark icon (the big star) to the right of the address bar. This step bookmarks the page and puts the bookmark in a type of "all others" folder named Other Bookmarks.

Now follow these steps to assign a tag to your bookmark and then stick your bookmark in a place where you can find it later:

1. **Hold down the Ctrl key and type** B.

   Firefox displays all bookmarks in the left section. Ones that haven't yet been assigned to a folder appear at the bottom, in the list marked Other Bookmarks, as in Figure 4-14.

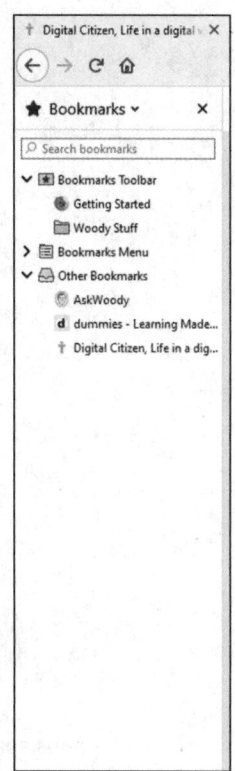

FIGURE 4-14:
Edit your raw, unsorted bookmarks.

2. **To add a new top-level entry on the Bookmarks toolbar, right-click Bookmarks Toolbar and choose New Folder. Type a name, and click Add.**

   The new folder appears both in the Bookmarks Toolbar list (on the left) and on the toolbar itself.

3. **If you want to move a bookmarked item to the toolbar, click and drag it to the corresponding location in the Bookmarks Toolbar list.**

   In Figure 4-15, I created a folder called Woody Stuff and dragged AskWoody, Dummies, and Digital Citizen into the folder. Woody Stuff appears up at the top, on the Bookmarks toolbar.

If you create a folder, you can leave it in the Other Bookmarks folder, but if you want to make it more readily accessible from the bookmarks toolbar, click and drag the new folder in the bookmarks sidebar so the folder appears under the Bookmarks Toolbar folder.

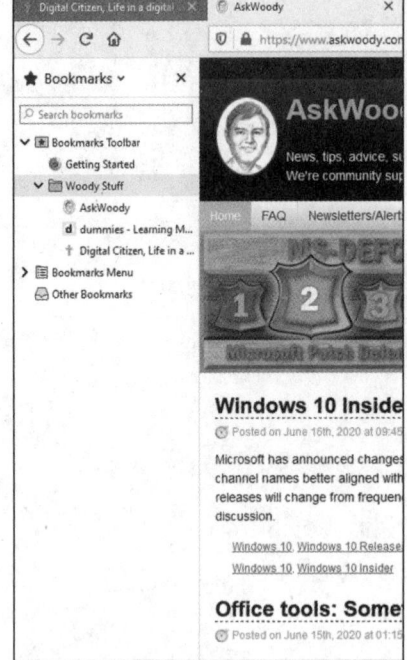

FIGURE 4-15:
The Woody Stuff folder is under the Bookmarks Toolbar folder, so it also appears up on the Bookmarks toolbar.

The bookmarks toolbar is convenient, but it takes up precious space on the screen. Many people prefer to work with the Bookmark icon, to the right of the address bar. Optionally, you can usually click and drag bookmarks into different (existing) folders.

After the folder has been created (and, optionally, located on the Bookmarks menu or the bookmarks toolbar), you can place any bookmark in the folder by double-clicking the bookmark star.

Changing the default search engine

Firefox used to put its searches through Yahoo! Search. In recent versions, it uses Google by default. What is Yahoo! Search, you ask? Good question. As the book went to press, Yahoo! Search (as, indeed, all of Yahoo!) is in a state of rapid flux. Back in more deterministic times, Yahoo! Search was just a front for Microsoft's Bing; a search made through Yahoo! Search drew its answers from Bing, and Yahoo! paid Microsoft big time, pulling through Microsoft's advertising and spitting it out in

Firefox. In late 2015, Yahoo! renegotiated its agreement with Microsoft, and as of this writing, up to 49 percent of all responses come from Google, including Google ads. What will the future bring, er, bring, especially with Yahoo! in its currently precarious financial position? Who knows. Stay tuned.

It's hard to change the default search engine in most browsers. Not so in Firefox. Here's how you do it:

1. Click the Settings icon (the three-line hamburger icon) in the upper right. Choose Options.

2. On the left, choose Search.

3. In the Default Search Engine drop-down list (see Figure 4-16), simply choose your preferred search engine.

As we went to press, Firefox offered Google, Bing, Amazon, DuckDuckGo, Twitter, and Wikipedia search engines. Remember that DuckDuckGo — the icon with a duck on it — doesn't track your searches and doesn't sell your data to advertisers.

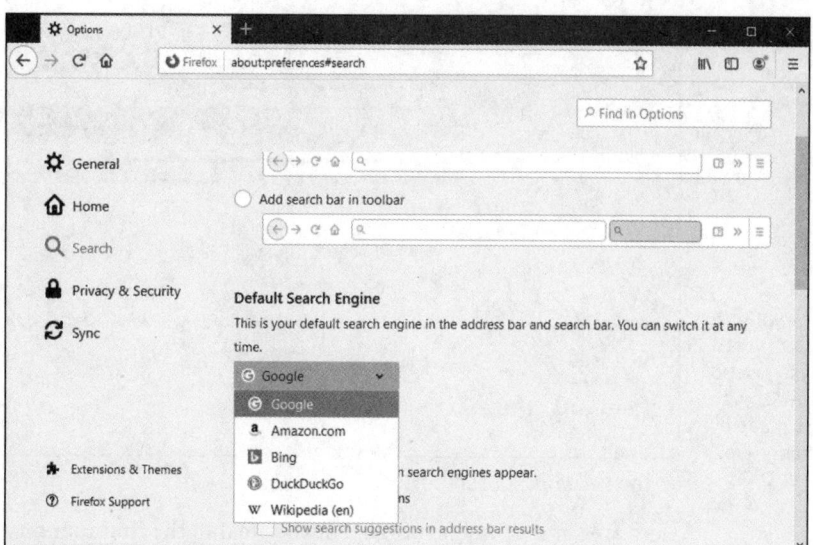

FIGURE 4-16: Firefox makes it easy to switch search engines.

That's all it takes. Whichever search engine you choose becomes your default, and it'll stay that way until you change it.

Firefox's competitors could learn a thing or three.

Adding Firefox's best add-ons

One of the best reasons for choosing Firefox over IE and Edge is the incredible abundance of add-ons. If you can think of something to do with a browser, chances are good there's already an add-on that'll do it.

An enormous cottage industry has grown up around Firefox. The Firefox people made it relatively easy to extend the browser itself. As a result, tens of thousands of add-ons cover an enormous range of capabilities.

To search for add-ons, click the hamburger icon in the upper right and choose Add-Ons. You can search for recommended add-ons by using the search box in Figure 4-17, or you can use Google to look at tons of free add-ons.

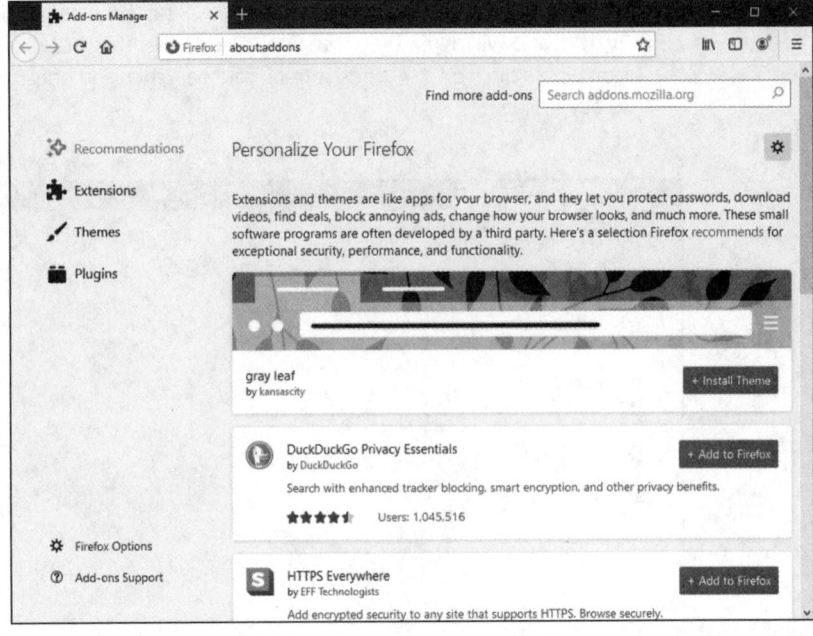

FIGURE 4-17:
Firefox makes it easy to extend the browser with add-ons made by other groups.

Here are some of my favorites. I always install the first four on any Firefox system I come into contact with:

ASK
WOODY.COM

>> **NoScript** lets you shut down all active content — Java, JavaScript, Flash, and more — either individually or for a site as a whole. Some sites don't work with JavaScript turned off, but NoScript gives you a fighting chance to pick and choose the scripts you want. Between JavaScript and Flash blocking, NoScript significantly reduces your exposure to online malware.

>> **Ghostery** keeps an eye on sites that are watching you. It tells you when sites contain web beacons or third-party cookies that can be used to track your surfing habits. I don't use Ghostery to stop cookies, but I do use it to watch who's watching me.

>> **AdBlock Plus** blocks ads. (What did you expect?) It blocks lots of ads — so much so that you may want to pull it back a bit. That's easy too. See a demo at adblockplus.org/en.

REMEMBER

>> **DownThemAll!** scrapes all downloadable files on a web page and presents them so you can choose which files to download. Click Start, and they all come loading down.

>> **Greasemonkey** adds a customizable scripting language to Firefox. After you install Greasemonkey, you can download scripts from https://greasyfork. org/en that perform an enormous variety of tasks, from tweet assistance to downloading Flickr files.

>> **Open in IE** embeds Internet Explorer inside Firefox. If you hit a site that absolutely won't work with Firefox, right-click the link, choose Tools and then choose Open This Link in IE, and Internet Explorer takes over a tab inside Firefox.

>> **eBay for Firefox** watches your trades while you're doing something else. It's from eBay.

>> **Video DownloadHelper** makes it easy to download videos from the web. **Easy YouTube Video Downloader** does the same thing, but it's specialized for YouTube.

>> **Linky** lets you open all links or images on a page, all at once, either on separate tabs or in separate windows. It's a helpful adjunct to Google image search.

To install the latest edition of any of these add-ons, go to Add-Ons Manager (click the hamburger icon, then Add-Ons) and search for the add-on's name. Each add-on's page has download and installation instructions — usually just a click or two and a possible restart of Firefox.

Optimizing Google Chrome

Google Chrome has several advantages over IE and Firefox. Foremost among them: world-class sandboxing of Flash, Java, and PDF support, which greatly reduces the chances of getting stung by the largest source of infections these days. IE and Firefox have both added similar protection, but Chrome was first and, I think, best.

As for Edge . . . it looks like Edge is going to beat Chrome at the sandboxing game, but it's still too early to tell. The bad guys are smart and getting smarter. Edge is still the new kid on the block. Time will tell.

That said, the biggest disadvantage is Google's (readily admitted!) tendency to keep track of where you've been, as an adjunct to its advertising program. If you install Chrome, sign in with your Google account, and start browsing, Google knows all, sees all, saves all — unless you turn on Incognito (private) browsing.

ASK
WOODY.COM

The second major disadvantage? Chrome's a resource hog. If you only open, oh, 10 tabs, Chrome's great. But if you open 20 or 30 at a time — I'll confess I'm among the guilty — Chrome can bog things down significantly.

Installing Chrome

Installing Google Chrome is like falling off a log:

1. With any browser, go to www.google.com/chrome/.

You probably see a big blue button that says Download Chrome.

2. Click the button to download.

3. Click Run, or Save and then Run, depending on what browser you're using to download Chrome.

The installer takes a minute or two, and then asks you to choose a default browser.

4. Give the User Account Control message box a Yes.

Chrome installs itself automatically and then loads, as shown in Figure 4-18.

5. If you want your Chrome settings to follow you onto any computer, tablet, or phone:

a. Sign in with a Google ID, such as a Gmail address.

b. Click the hamburger icon and choose Settings.

c. Click Turn on Sync, and then sign in with your Google account.

Syncing across many kinds of devices is one of the best parts about Chrome. But I'm ever mindful of the fact that Google keeps tabs on everywhere I go and uses the accumulated information to dish up ads designed to convince me to click.

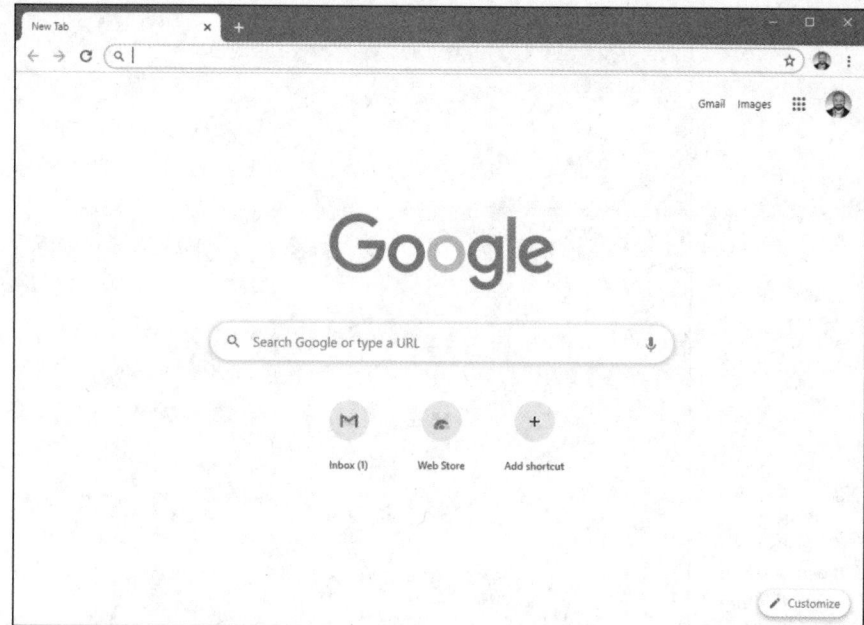

FIGURE 4-18:
Google Chrome has all the usual controls, easily available.

Navigating in Chrome

Navigation in Chrome is very similar to that in Firefox, except there's no search bar. Chrome doesn't need one: You just type in the address bar. (Google calls it an *omnibar*, which is cool because they came up with the idea.)

All the tricks I mention in the earlier IE section called "Don't work too hard" also perform in Chrome. You never need to type *http://*, almost never need to type *www*, and typing something like *dummies* followed by a Ctrl+Enter puts you directly into www.dummies.com.

The default home page in Chrome is a little different from both IE and Firefox. Chrome displays its New Tab page, which has an icon for Google Apps (the tic-tac-toe icon in the upper right, next to Images). On the main part of the New Tab page, you see thumbnails for the Chrome Web Store and other pages that you've frequently visited. The New Tab page adds more entries as you use the browser. Click the Apps icon to bring up Maps, Gmail, Google Drive, YouTube, Gmail, Meet, and more (see Figure 4-19).

If you want to change the home page in Chrome, navigate to the page(s) you want to use. Click the three-dot icon in the upper right, and choose Settings. A new tab opens with various Chrome settings. Under the heading On Startup, select the Open a Specific Page or Set of Pages option. At the bottom, tap or click Use Current Pages. You see a list like the one in Figure 4-20. Verify that you have the right pages, and tap or click OK.

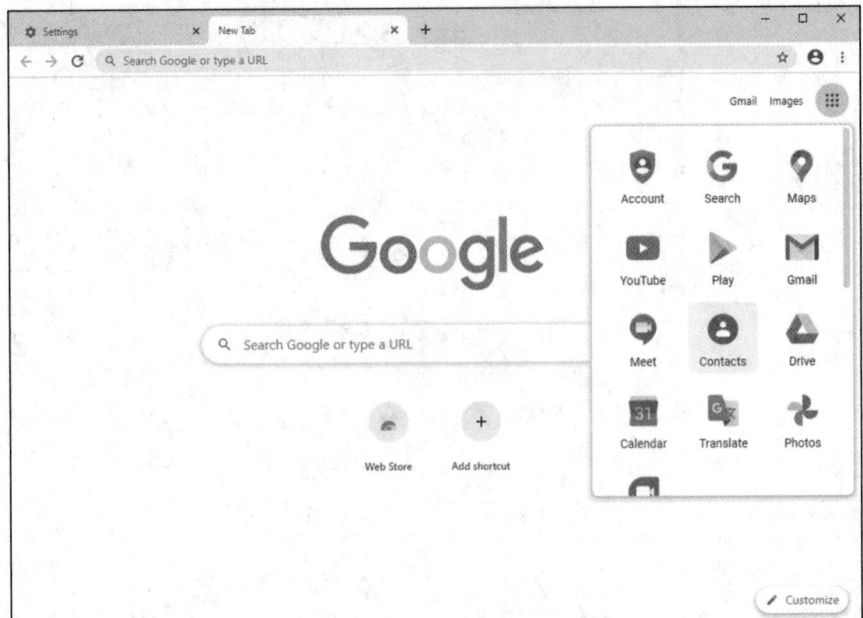

FIGURE 4-19:
The New Tab page in Chrome includes an Apps icon.

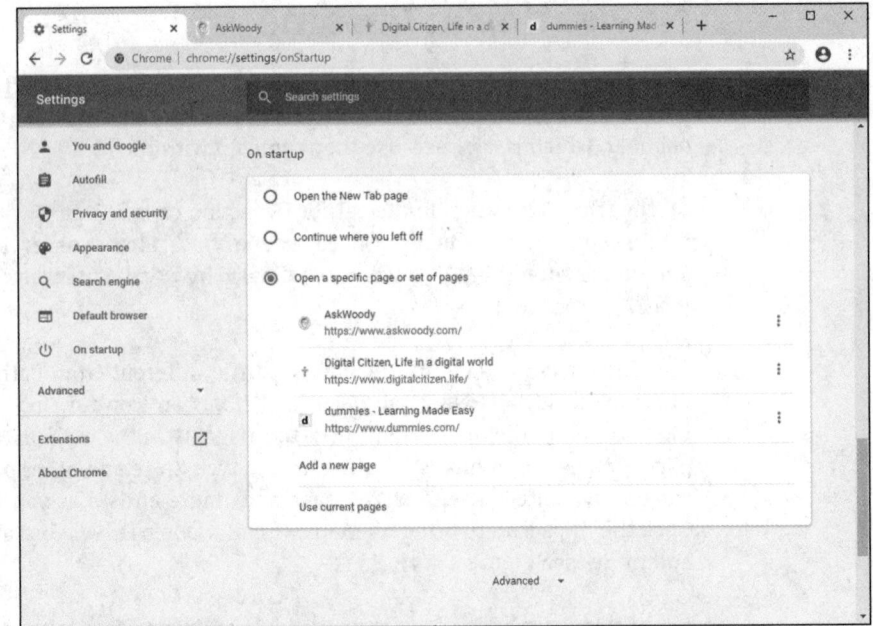

FIGURE 4-20:
It's easy to set the home page(s) for Chrome.

Like IE, Firefox, and Edge, if you signed in to Chrome using a Google ID (such as a Gmail email address), changing the home page(s) here will change your Chrome home pages on all the computers — whether they're on PCs, tablets, smartphones — anywhere you go. Your add-ons and favorites travel with you, too.

The following Chrome features are helpful as you move around the web using Chrome:

>> **The default search engine:** The default search engine setting is on the same Settings tab shown in Figure 4-20. Bing is one of the listed options, but you can add just about any search engine.

>> **Private browsing:** Chrome's version of InPrivate Browsing is called Incognito. To start a new Incognito window, click the hamburger icon and choose New Incognito Window.

ASK
WOODY.COM

>> **Bookmarks:** I find Chrome's bookmarks capability much easier to use than Firefox's. To see why, go to a web page that you'd like to bookmark, and click the Bookmark (star) icon, on the right. If you want to rearrange your bookmark folders, click the vertical three-dot icon, choose Bookmarks, then Bookmark Manager, and you can work with a full, hierarchical organization of folders, as in Figure 4-21.

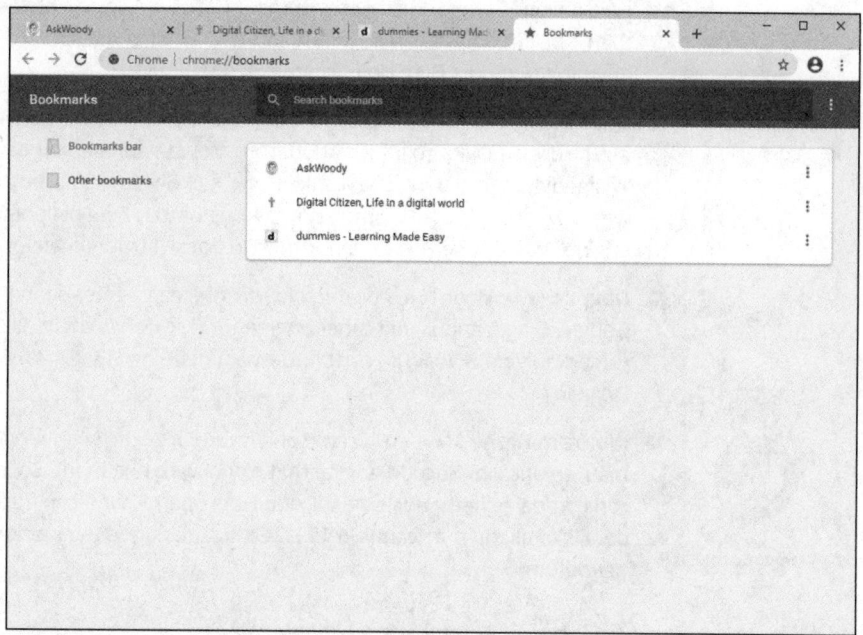

FIGURE 4-21:
Chrome
bookmarks are
simple and easy
to organize.

Chrome Extensions, once a small subset of all great add-ons, now rate as absolutely first-class. Click the three-dot icon in the upper right, and choose More Tools, Extensions. If you're looking for great extensions, try TooManyTabs and the newer OneTab to let you consolidate your tabs, reducing the amount of screen real estate consumed and also cutting back on Chrome's infamous memory hogging. I use LastPass for Chrome all the time.

Searching on the Web

Internet searching can be a lonely business. You're out there, on the Internet range, with nothing but gleaming banner ads and text links to guide you. What happens when you want to find information on a specific subject, but you're not sure where to start? What if Google leads you on a wild goose chase? What if the Microsoft Bing decision engine takes the wrong turn?

Google's good. It's the search engine I use every day. But there are some decent alternatives, several of which can help in specific situations. For example:

>> Microsoft's **Bing** (www.bing.com) isn't all that bad, and it's getting better. It remains to be seen if Bing can come up with any really compelling reasons to switch from Google. Microsoft's dumping a ton of money into search — more than a billion dollars a year, at last count — and I'm not sure it's come up with anything that puts Bing clearly in the lead.

>> **DuckDuckGo** (www.duckduckgo.com) is an up-and-comer that I find fascinating. It relies heavily on information from crowd-sourced sources, including Wikipedia. At this point, the results DuckDuckGo delivers aren't as close to what I want as Google's, but they're getting better. One big point in this search engine's favor: Like Firefox, DuckDuckGo doesn't track what you do.

>> **Dogpile** (www.dogpile.com), an old favorite, aggregates search results from Google, Bing, Yahoo!, and other engines and smashes them all together in a remarkably quick way. If I can't find what I need on Google, I frequently turn to Dogpile.

>> **Wolfram Alpha** (www.wolframalpha.com) isn't exactly a search engine. It's a mathematical deduction engine that works with text input. So, for example, it can compare methanol, ethanol, and isopropanol. Or it can describe to you details of all the hurricanes in 1991. Or it can analyze the motion of a double pendulum.

But I find myself going back to Google.

Google has gone from one of the most admired companies on the web to one of the most criticized — on topics ranging from copyright infringement to pornography to privacy and censorship — and the PageRank system has been demonized in terms rarely heard since the Spanish Inquisition. Few people now believe that PageRank objectively rates the importance of a web page; millions of dollars and thousands of person-months have been spent trying to jigger the results. Like it or not, Google just works. The Google spiders (the programs that search for information), which crawl all over the web, night and day, looking for pages, have indexed billions of pages, feeding hundreds of millions of searches a day. Other search engines have spiders too, but Google's outspiders them all.

REMEMBER

As this book went to press, Google (and its parent company, Alphabet) was worth about $730 billion, the verb *google* had been embraced by prestigious dictionaries, the company was taking on Microsoft *mano a mano* in many areas, and many other search engines offered decent alternatives to the once almighty Google. "OK, Google" has entered my lexicon for querying my phone about anything under the sun. Self-driving cars, robots, fiber optic cable, Internet from hot air balloons, play Go — even a run at "curing" death — are now part of the Google fold. Everything's changing rapidly, and that's good news for us consumers.

In this section, I show you several kinds of searches you can perform with Google (and the other search engines). No matter what you're looking for, a search engine can find it!

Finding what you're looking for

Google has turned into the 800-pound gorilla of the searching world. I know people who can't even find AOL unless they go through Google. True fact.

The more you know about Google, the better it can serve you. Getting to know Google inside and out has the potential to save you more time than just about anything in Windows 10 proper. If you can learn to search for answers quickly and thoroughly — and cut through the garbage on the web just as quickly and thoroughly — you can't help but save time in everything you do.

**ASK
WOODY.COM**

You can save yourself lots of time and frustration if you plot out your search before your fingers hit the keyboard.

Obviously, you should choose your search terms precisely. Pick words that will appear on any page that matches what you're looking for: Don't use *Compaq* when you want *Compaq S710*.

Beyond the obvious, the Google search engine has certain peculiarities you can exploit. These peculiarities hold true whether you're using Google in your browser's search bar or you venture directly to www.google.com:

>> **Capitalization doesn't matter.** Search for *diving phuket* or *diving Phuket* — either search returns the same results.

>> **The first words you use have more weight than the latter words.** If you look for *phuket diving,* you see a different list than the one for *diving phuket.* The former list emphasizes websites about Phuket that include a mention of diving; the latter includes diving pages that mention Phuket.

>> **Google first shows you only those pages that include all the search terms.** The simplest way to narrow a search that returns too many results is to add more specific words to the end of your search term. For example, if *phuket diving* returns too many pages, try *phuket diving beginners.* In programmer's parlance, the terms are ANDed.

>> **If you type more than ten words, Google ignores the ones after the tenth.**

>> **You can use OR** to tell Google that you want the search to include two or more terms — but you have to capitalize OR. For example, *phuket OR samui OR similans diving* returns diving pages that focus on Phuket, Samui, or the Similans.

>> **If you want to limit the search to a specific phrase, use quotes.** For example, *diving phuket "day trip"* is more limiting than *diving phuket day trip* because in the former, the precise phrase *day trip* has to appear on the page.

>> **Exclude pages from the results by putting a space and then a hyphen in front of the words you don't want.** For example, if you want to find pages about diving in Phuket but you don't want to associate with lowly snorkelers, try *diving phuket -snorkeling.*

>> **You can combine search tricks.** If you're looking for overnight diving, try *diving phuket -"day trip"* to find the best results.

>> **Google supports wildcard searches** in quite a limited way: The asterisk (*) stands for a single word. If you're accustomed to searches in, say, Word or Windows 10, the * generally indicates a sequence of characters, but in Google it stands for only an entire word. You may search for *div** and expect to find both diver and diving, but Google won't match on either. Conversely, if you search for, oh, *email * * wellsfargo.com,* you find lots of email addresses. (The second * matches the at sign [@] in an address. Try it.)

TIP

If you use Google to search for answers to computer questions, take advantage of any precise numbers or messages you can find. For example, Googling *computer won't start* doesn't get you anywhere; but *two beeps on startup* may. *Can't install* won't get you anywhere. *Install error 800F9004* turns up wonders. If you're trying to track down a Windows error message, use Google to look for the precise message. Write it down, if you have to.

Using Advanced Search

Didn't find the results you need? Use Google Advanced Search. There's a trick.

If you need to narrow your searches — in other words, if you want Google to do the sifting rather than do it yourself — you should get acquainted with Google's Advanced Search capabilities. Here's a whirlwind tour:

1. **Run your search; if it doesn't have what you want, click Settings and choose Advanced Search.**

REMEMBER

Settings is located in the upper-right corner *of the search results page* — it's not part of your browser, it's actually on the search results page.

Google brings up its Advanced Search page (see Figure 4-22).

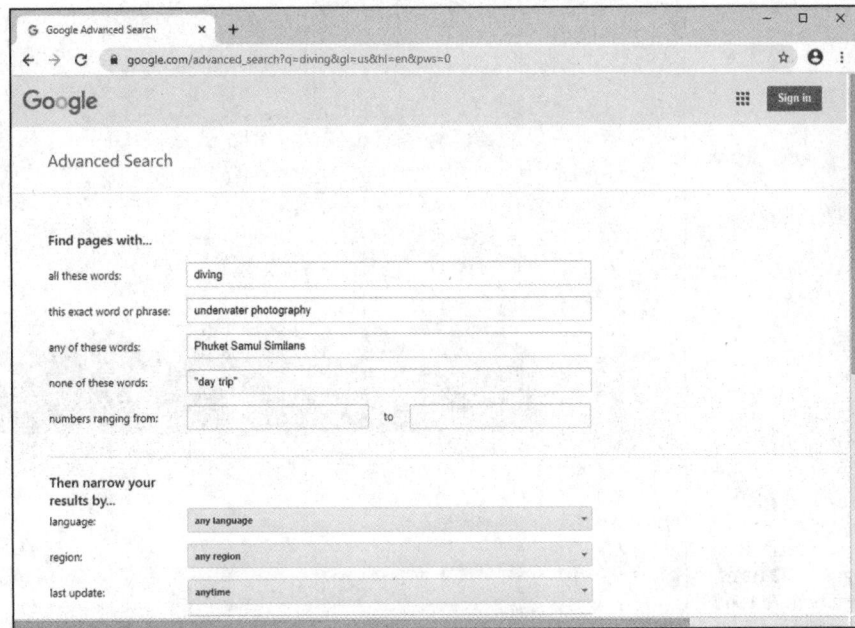

FIGURE 4-22: Advanced Search lets you narrow your Google search quickly and easily.

2. **Fill in the top part of the page with your search terms.**

   In Figure 4-22, I ask for sites that include the word *diving* and the exact phrase *underwater photography*. I also want to exclude the phrase *day trip* and return pages pertaining only to Phuket, Samui, or the Similans.

REMEMBER

   Anything you can do in the top part of this page can also be done by using the shorthand tricks mentioned in the preceding section. If you find yourself using the top part of the page frequently, save yourself some time and brush up on the tricks (such as typing **OR**, -, "") that I mention in the earlier section, "Finding what you're looking for."

3. **In the bottom part of the Advanced Search page, further refine your search by matching on the identified source language of the page (not always accurate); a specific filename extension (such as .pdf or .doc); or the domain name, such as** www.dummies.com.

   You can also click the link at the bottom to limit the search to pages stamped with specific dates (notoriously unreliable), pages with specific licensing allowances (not widely implemented), and ranges of numbers.

4. **Press Enter.**

   The results of your advanced search appear in a standard Google search results window (see Figure 4-23).

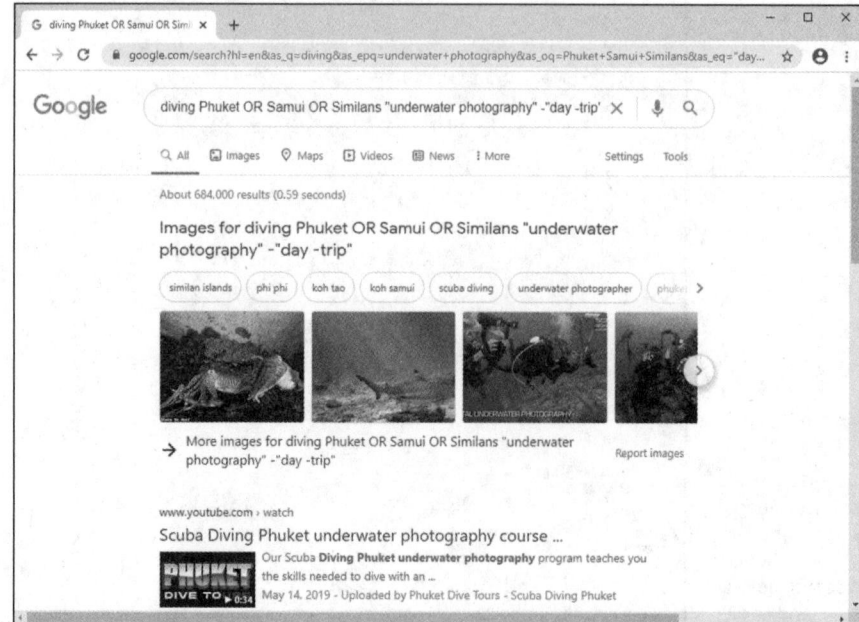

FIGURE 4-23:
Running the stringent search specified in Figure 4-22 turns up hundreds of thousands of hits.

You can find more details about Google Advanced Search on the Google Advanced Search page, `www.google.com/help/refinesearch.html`.

Pulling out Google parlor tricks

Google has many tricks up its sleeve, some of which you may find useful — even if it's just to win a bet at a party. For example:

» To find the status of your UPS, FedEx, or USPS delivery, just type the package number (digits only) in the Google search box.

» The search box is a stock ticker. Type a symbol such as **MSFT, GOOG,** or **AAPL.**

» To use Google as a calculator, just type the equation in the Google search box. For example, to find the answer to 1,234 × 5,678, type **1234*5678** in the search box and press Enter. Or to find the answer to 3 divided by pi, type **3/pi.** No, Google doesn't solve partial differentials or simultaneous equations — yet. For that, check out Wolfram Alpha.

» Google has a built-in units converter. The word *in* triggers the converter. Try **10 meters in feet** or **350 degrees F in centigrade** (or **350 f in c**) or **20 dollars in baht** or (believe me, this is impressive) or **1.29 euros per liter in dollars per gallon.**

» To find a list of alternative (and frequently interesting) definitions for a word, type **define,** as in **define booty.**

» You can see movie reviews and local showtimes by typing **movie** and then the name of the movie, such as **movie star wars 7.**

» Try quick questions for quick facts. For example, try **height of mt everest** or **length of mississippi river** or **currency in singapore.**

If you click the microphone icon in the search bar, or if you're using a smartphone and start by saying "OK Google," all these tricks work under voice command, as well. "OK, Google" (pause) "when was the end of the Cretaceous period?"

Referring to Internet Reference Tools

**ASK
WOODY.COM**

I get questions all the time from people who want to know about specific tools for the Internet. Here are my choices for the tools that everyone needs.

Internet speed test

Everybody, but everybody, needs (or wants) to measure her Internet speed from time to time. The sites I use these days for testing is `www.dslreports.com/speedtest` and `www.testmy.net`.

A million different speed tests are available on the Internet, and 2 million different opinions about various tools' accuracy, reliability, replicability, and other measurements. I used to run speed tests at Speakeasy, but then found that my ISP was caching the data — in fact, caching all the data from a OOKLA-based test — so the results I saw were just local; they didn't reflect long-distance speeds. So I moved to DSLReports, with its tests that can't be cached, and haven't looked back.

I later added `www.testmy.net` because the reports appear valid — and the site has automatic testing, so I can run tests every hour for days on end.

DNSstuff

Ever wonder whether the website BillyJoeBobsPhishery.com belongs to BillyJoeBob? Head over to `www.dnsstuff.com` (see Figure 4-24) and find out.

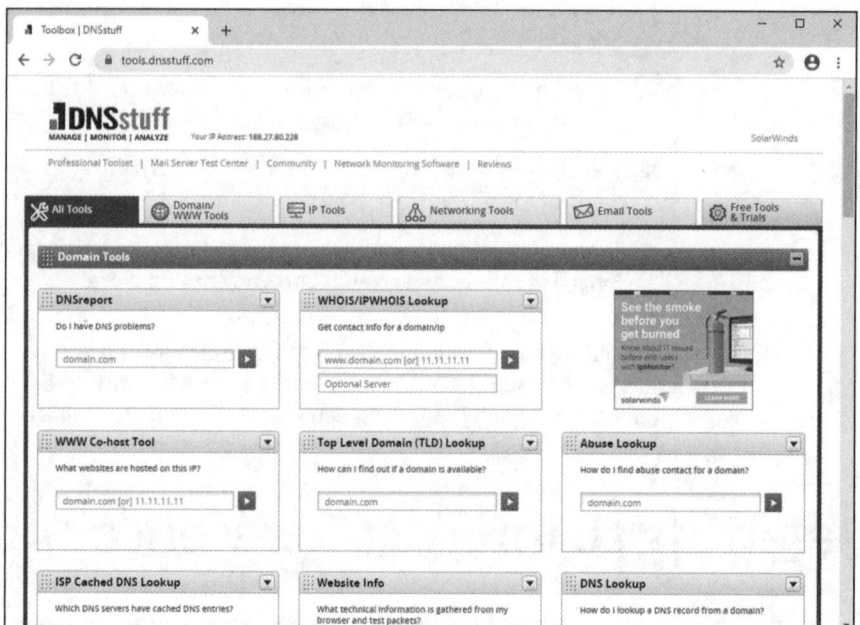

FIGURE 4-24: DNSstuff offers a wide array of web- and Net-related tools.

You give DNSstuff a domain name, and the site divulges all the public records about the site, commonly known as a whois: who owns the site (or at least who registered it), where the rascals are located, and whom to contact — although you must register a valid email address to get all the info.

DNSstuff also tells you the official abuse contact for a particular site (useful if you want to lodge a complaint about junk mail or scams), whether a specific site is listed on one of the major spam databases, and much more.

Monitis Traceroute

So where's the hang-up? When the Internet slows down, you probably want to know where it's getting bogged down. Not that it will do you much good, but you may be able to complain to your ISP.

My favorite tool for tracing Internet packets is the free product Traceroute at this website: `gsuite.tools/traceroute`. When you run Traceroute, you feed it a target location — a web address to use as a destination for your packets (see Figure 4-25). As soon as you enter a target, Traceroute runs out to the target and keeps track of all *hops* — the discrete jumps from location to location — along the way. It also measures the speed of each hop.

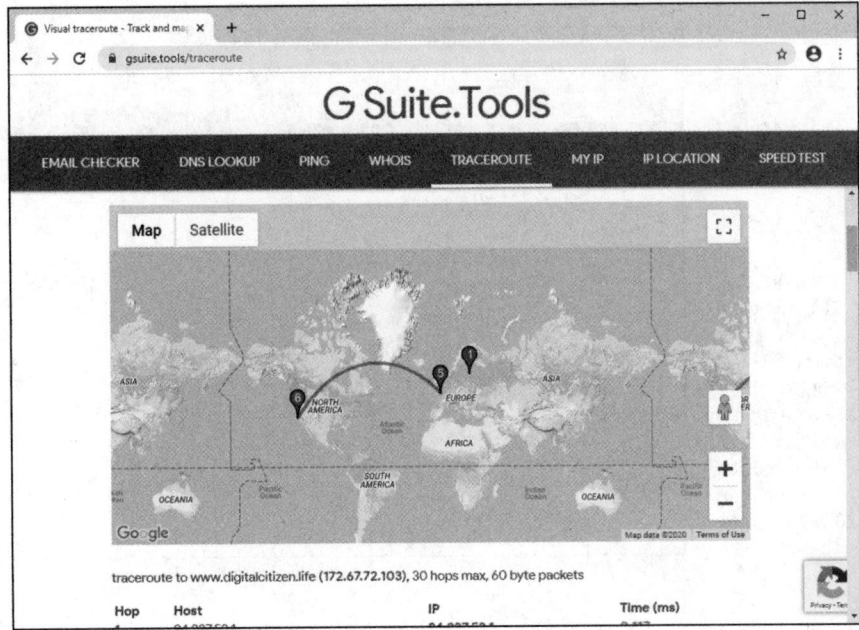

FIGURE 4-25:
Why is the
Internet so slow?
Traceroute
pinpoints pileups.

Down for everyone or just me?

So you try and try and can't get through to Wikipedia, or Outlook has the hiccups: The browser keeps coming back and says it's timed out, or it just sits there and does nothing.

It's time to haul out the big guns. Hop over to www.downforeveryoneorjustme. com (no, I don't make this stuff up), and type the address of the site that isn't responding. The computer on the other end checks to see whether the site you requested is still alive. Cool.

The Wayback Machine

He said, she said. We said, they said. Web pages come and go, but sometimes you just have to see what a page looked like last week or last year. No problem, Sherman: Just set the Wayback Machine for November 29, 1975. (That's the day Bill Gates first used the name Micro Soft.)

If you're a Mr. Peabody look-alike and you want to know what a specific web page really said in the foggy past, head to the Internet Archive at www.archive.org, where the Wayback (or is it WABAC?) Machine has more than 85 billion web pages archived and indexed for your entertainment (see Figure 4-26).

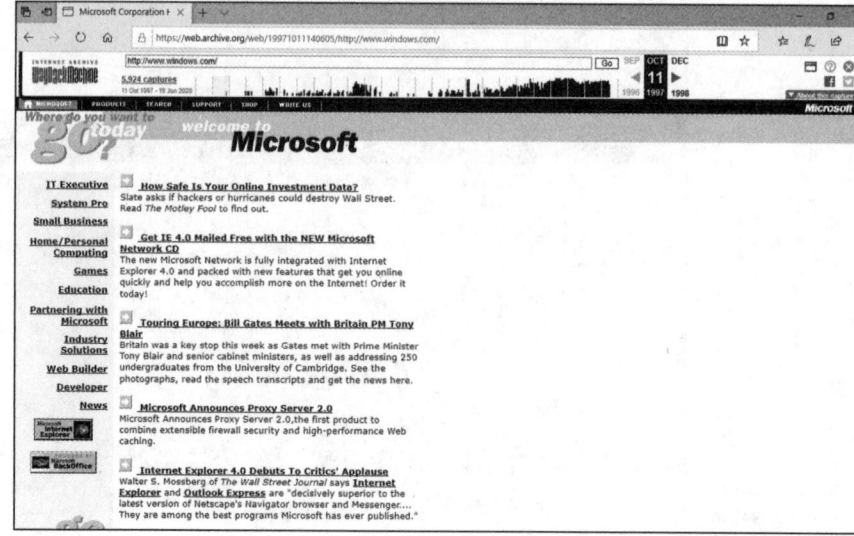

FIGURE 4-26: Everything old is new again with the Archive. org Wayback Machine. This is what windows. com looked like more than 20 years ago on October 11, 1997.

Chapter **5**

Hey, Cortana!

"**H**ey, Cortana!"

"Yes, Boss."

"Get me a tall skinny latte."

"I'll bring it to you. Okay to charge your Amex four dollars and thirty-seven cents?"

Cortana isn't quite there yet. That was in my dream last night. In fact, if you try to order a tall skinny latte in Windows 10, you get the response in Figure 5-1.

Cortana's good, but she isn't *that* good.

Initially launched on Windows Phone, Cortana is Microsoft's digital-assistant answer to Apple's Siri and Google's OK Google.

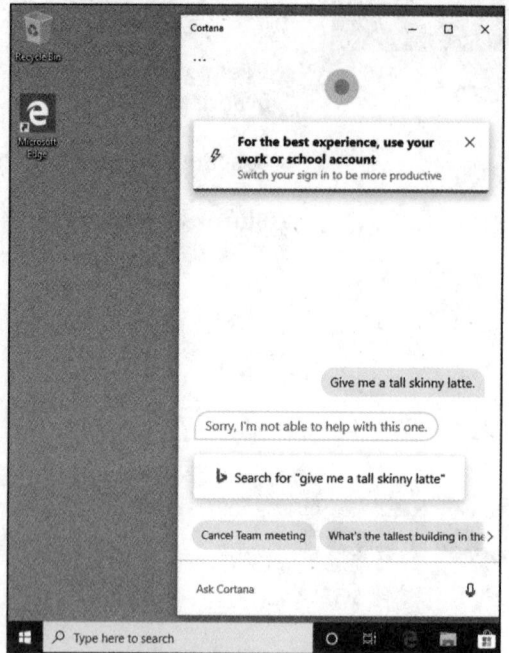

FIGURE 5-1:
Baristas don't
have to worry
about their
jobs. Yet.

Cortana, however, is a bit more refined: She (and I'll relentlessly refer to her as *she*; please forgive the anthropomorphism!) used to be tied into all Windows 10 searches. That situation was both good and bad, as I discuss later in this chapter. Since the May 2020 update, she's an independent app, no longer tied to the inner workings of Windows 10, and that's great.

She's also lots of fun.

GETTING THE *HEY* OUTTA *HEY, CORTANA*

A few years ago, Microsoft announced that it would make Cortana respond to, simply, "Cortana" — instead of requiring the no-doubt more formal, and oh! so old-fashioned "Hey, Cortana."

With Alexa, Siri, Bixby (that's from Samsung) and various other single-name assistants on the rise, Cortana has finally dropped the "Hey," too, starting with the May 2020 update.

The Cortana Backstory

There aren't many parts of Windows 10 that have a backstory, so indulge me for a minute here.

Cortana is a fully developed artificial intelligence character from the video game series Halo. She lives (or whatever AIs do) 500 years in the future. In the Halo series, she morphs/melds into Master Chief Petty Officer John-117 and, in that position, tries to keep Halo installations from popping up all over the galaxy. Halo installations, of course, destroy all sentient life.

Cortana chose John-117, not the other way around. She was supposed to be the resident AI on a ship, temporarily, but plans changed, and she ended up the permanent AI, apparently because of the deviousness of a Colonel Ackerson. It's not nice to fool Cortana, so she hacked into Ackerson's system and blackmailed him.

If that sounds a little bit like the kind of life you lead, well, you're ready for Cortana.

IF YOU DON'T PAY FOR IT, YOU'RE THE PRODUCT

Cortana, as a flagship product in Windows 10, has lots to like — it's smart and getting smarter by the day, and it can help in a zillion ways with some real intelligence.

But Cortana, like other virtual assistants, comes at a price. The price is your privacy.

To use Cortana in anything but a very stripped-down mode, you must provide a Microsoft account. And after you've paired Cortana with your Microsoft account, everything you do with her is logged in Microsoft's database.

Some people don't mind; they figure the benefits of Cortana justify parting with all that personal information. In fact, in a very real sense, Cortana *can't do her job* unless she can see your email, check your calendar, and keep track of what you see and hear and search for. It's a two-way street.

In this chapter, I step you through ways to minimize Cortana's acquisitiveness.

In the end, the choice is yours, but be very much aware of the deal you're making when you invite Cortana into your machine.

Make Cortana Respond to "Hey, Cortana"

Up until the Windows 10 May 2020 update, you could make Cortana respond to "Hey, Cortana" even though she didn't do that by default. Instead, you had to click the Type Here to Search box to the right of the Start icon. Cortana responded with a notice that she could do much more, as shown in Figure 5-2. Hard to tell if that's a brag or a fact.

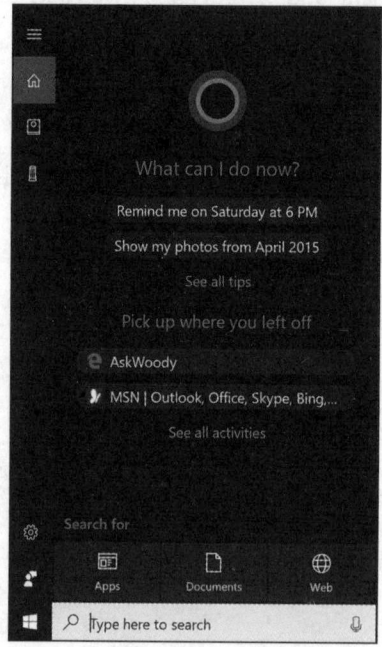

FIGURE 5-2:
Start by letting
Cortana in the
door.

If you run a version of Windows 10 older than the May 2020 update, here's a quick run-through of how to make Cortana respond to "Hey, Cortana":

1. **Open Cortana with a click in the Type Here Search box, and then click the Settings (gear) icon.**

Windows 10 brings up the Settings, Talk to Cortana pane, as shown in Figure 5-3.

2. **If you'd like to let a 500-years-in-the-future AI listen to everything you say, attempting to parse the words "Hey, Cortana," move the Let Cortana Respond to "Hey Cortana" slider to On.**

Cortana works whether or not you turn on the voice prompt. It's kind of the face of Windows' built-in search routines.

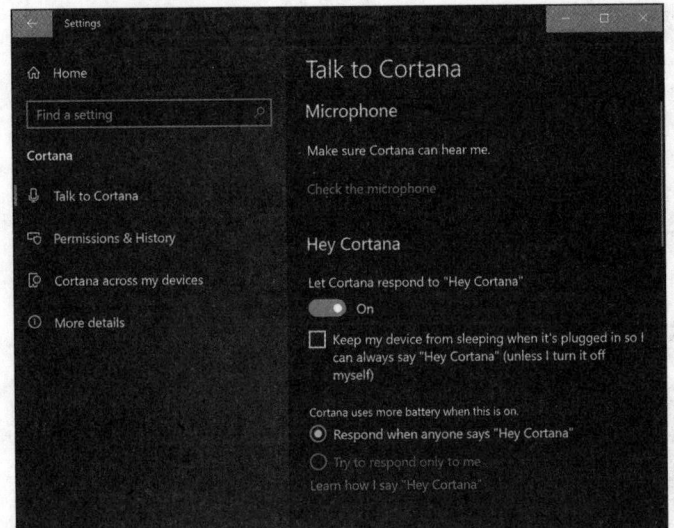

FIGURE 5-3:
Time to fish or
cut bait.

3. **If you want Cortana to listen, even when your machine is locked (typically when the cover is closed), do one of the following:**

- Select the box that says Spy Away O Mighty One.

- Select the box below the Hey Cortana slider that says Keep My Device from
  Sleeping When It's Plugged in So I Can Always Say "Hey Cortana."

4. **If you've logged on to Windows 10 with a Microsoft Account, Cortana can see
all your Microsoft-stored data. If you want to give Cortana the capability to
thumb through that data even when the machine is locked, scroll down and
select the box next to Let Cortana Access My Calendar, Email, Messages, and
Power BI Data When My Device Is Locked.**

But wait! There's more!

5. **Back in the main Cortana screen (refer to Figure 5-2), click the notebook
icon on the left bar of the Cortana pop-up.**

A bunch of customizing options appear, as in Figure 5-4.

6. **Click the Edit button to the right of your name, the one that looks like a
partial figure 8 with a pencil next to it. Then click Change my name.**

7. **Type the name you want Cortana to use for you.**

I'll be first to pay obeisance to our AI overlords when the time comes, but for
now I'll just have her call me Boss (see Figure 5-5).

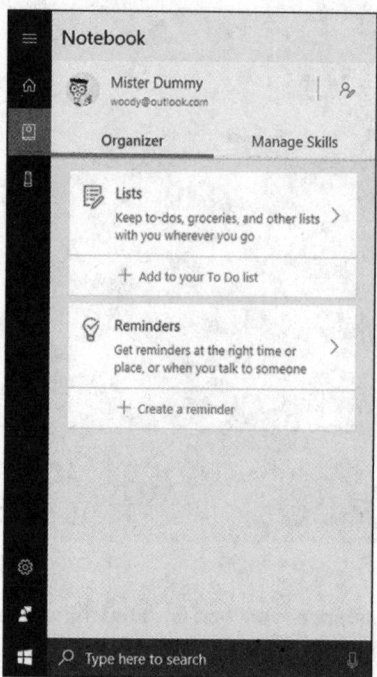

FIGURE 5-4:
Start the customizing.

FIGURE 5-5:
Even more settings.

8. **Click Enter. Then click Hear How I'll Say It.**

Cortana then tries to pronounce your name. If her pronunciation sounds good, click Sounds Good. If she messes up, click That's Wrong and teach Cortana how to say your name by speaking into the mic.

9. **Say "Hey, Cortana!"**

If that doesn't rouse the old biddy, click the microphone icon in the Type Here to Search box and then ask your question. I tried, "What is the sound of one hand clapping?" The impertinent minus-500-year-old told me, "It may be the same as the sound of a tree falling in an empty forest." Which is a fairly good answer, come to think of it.

After the first time or two, Cortana gets the idea that she's supposed to be listening for the sound of your voice.

10. **Practice asking all sorts of questions. Test a bit.**

You'll find that you need to pause slightly after saying "Hey, Cortana." For example, if I say "Hey, Cortana" (pause a second) "How is the weather," I get a response like the one in Figure 5-6.

As of the May 2020 update, Cortana is a separate app that responds to "Cortana" – a shorter version of the past "Hey, Cortana." You can also click her icon, next to the Windows 10 search box. Unlike in the past, there's no need to configure her to respond. Personally, I like the new approach better.

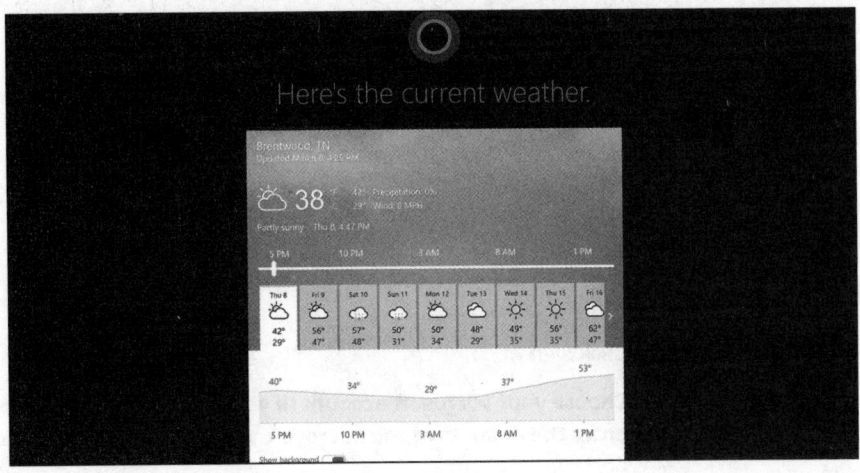

FIGURE 5-6:
Cortana's great at telling you the MSN weather forecast.

Setting up Cortana

Cortana's getting better but, in many cases, he/she/it still isn't as good as the Google Assistant, Alexa, or Siri (see nearby sidebar). As time goes by, we're assured, she'll get better and better at bringing up information that interests you, collating things like your flight schedules, warning about appointments, and on and on. If you're using the Windows 10 May 2020 update or newer, you need to set up the Cortana app and add a Microsoft account to it. Here's how:

1. **Click the Cortana icon near Type Here to Search.**

   Windows 10 brings up the Cortana app, as shown in Figure 5-7.

FIGURE 5-7:
Time to sign in with a Microsoft account.

2. **Click Sign In.**

3. **Choose your Microsoft account (if you're using one in Windows 10) or enter the defaults of the Microsoft account that you want to use with Cortana.**

4. **Click Continue.**

   Cortana is ready to go, and asks how she can help today, as shown in Figure 5-8.

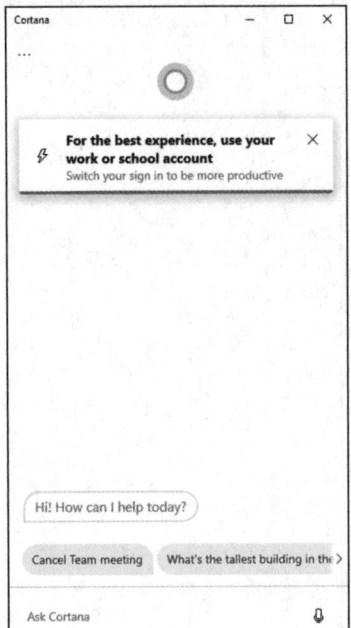

FIGURE 5-8:
Cortana is
ready to go.

Using Cortana Settings

To access Cortana's settings, click the three dots in the top-left corner, and then click Settings. You can set how you want to talk to Cortana (through typing, voice, or both) and change her permissions to access the microphone and speech (you want to give her these permissions if you want to talk to her instead of typing). The most interesting part is the Privacy section of Settings, where you can revoke her permission to access your data (calendar, contacts, email, and so on) or clear your chat history with Cortana. You can also access the Microsoft Privacy dashboard, where Microsoft lists all the data about you stored in their cloud. See Cortana's privacy settings in Figure 5-9.

Keep in mind that turning off pieces of Cortana (say, her capability to keep track of your calendar details) deletes everything Cortana knows on this device but won't delete anything from Microsoft's servers. Give Microsoft an A for full disclosure, but a D for how deep you must dig to find it.

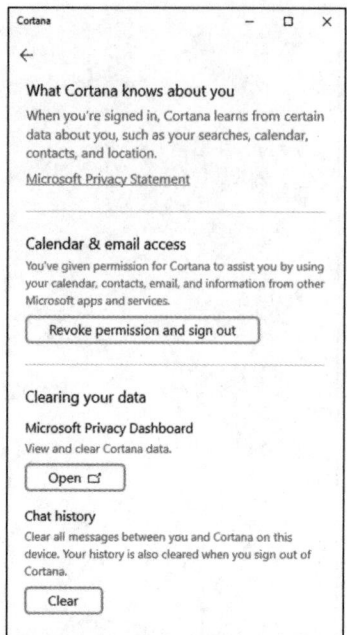

FIGURE 5-9:
Cortana's
settings lead
to interesting
places.

The big Cortana off switch is in her settings: Go to the Privacy settings as explained earlier, and then click the Remove Permission and Sign Out button. Confirm your choice and Cortana is turned off. If you want to see the data that Microsoft has stored about you while using Windows 10 and Cortana, click the Open button under Microsoft Privacy Dashboard in the same Privacy settings. Sign in with your Microsoft account again if you must. You get to the Privacy web page for your Microsoft account — in other words, the place Microsoft uses to store all sorts of nifty things about you, as shown in Figure 5-10.

You can see all the things that Microsoft has stored about you: your browsing history, search history, location activity, voice activity (from using Cortana), media activity, product and service activity, product and service performance data, Cortana's notebook (from previous Windows 10 versions when Cortana was embedded in the operating system), and LinkedIn data.

Spend some time looking through the types of information collected about you and click Clear at will. Unfortunately, you can't see the details. But at least you can delete wide swaths of history from this web page.

If you want to tweak and improve your privacy in Windows 10, click Start, the Settings icon, and Privacy. Then on the left, choose General. You see the Settings app's Privacy settings page, shown in Figure 5-11.

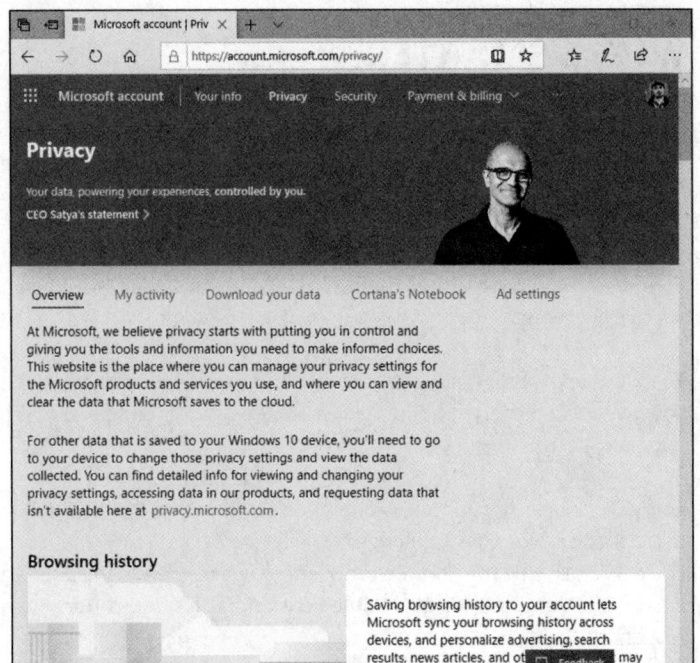

FIGURE 5-10:
Getting closer to
the sanctorum
of your details
inside Microsoft.

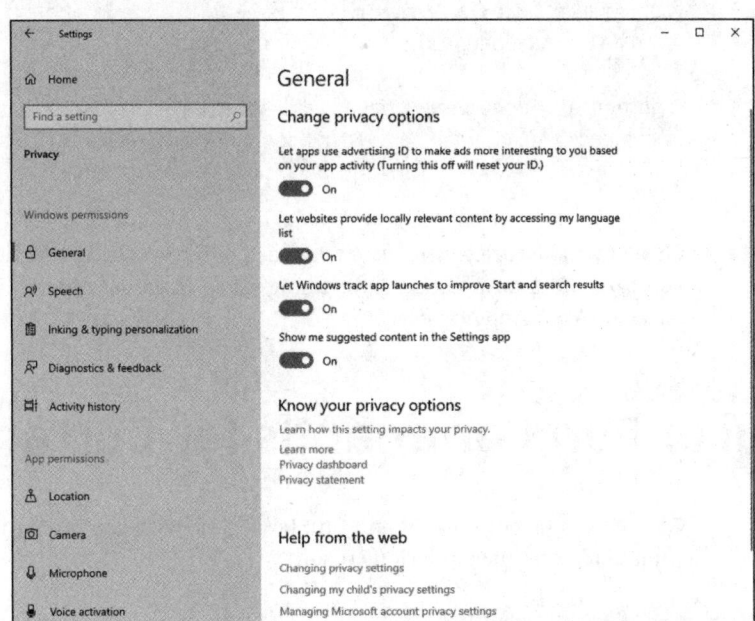

FIGURE 5-11:
These privacy
settings are
inside the
Settings app.

Check to make sure you're okay with each of these settings. Most of all, make sure you turn off Let Apps Use Advertising ID to Make Ads More Interesting to You Based on Your App Activity.

Useful or Fun Commands for Cortana

Cortana can be both useful and fun. Let's start with the fun part, and share some commands that have hilarious results:

>> Tell me a joke!

>> What's your favorite song?

>> Recite Shakespeare!

>> Can you talk like a pirate?

>> Tell me about Halo!

>> Make an impression!

>> What is the meaning of life?

>> Do you know Alexa?

>> What do you think about Google?

>> Testing!

>> Tell me an animal fact

>> Can I borrow some money? :)

Leaving the jokes aside, Cortana can be a productive assistant too. Here's some of the stuff you can ask her to do:

>> Check the weather: "What's the weather in Tokyo?" "Is it going to rain tomorrow?"

>> Ask questions: "What's the tallest building in the world?" "What's the value of Apple stock?" "What's the bitcoin exchange rate?"

>> Check the news: "Show me the latest news." "Show me the latest news in Europe."

>> Do math: "What is 16 multiplied by 25?" "What is 100 by 25?"

>> Make conversions: "How many meters in a mile?" "How many liters in a gallon?"

>> Define words: "What's an epiphany?" "Define philosophy."

>> Translate specific words: "Translate something to Japanese."

Chapter **6**

Maintaining Your System

Windows 10 is a computer program, not a Cracker Jack toy, and it will have problems. The trick lies in making sure that *you* don't have problems too.

This operating system is notorious for crashing and freezing, making it impossible to start the computer or garbling things so badly that you'd think the screen went through a garbage disposal. This situation is especially true when Microsoft launches a major Windows 10 update, which brings not only new features, apps, and improvements but also new bugs and problems. Microsoft has poured lots of time, effort, and money into teaching Windows 10 how to heal itself. You can take advantage of all that work — if you know where to find it.

Chapter 6 is devoted to the topic of how to keep Windows 10 alive and well. In this chapter, I introduce you to the basic ideas and get you started with some of the parts of Windows that you can use in many ways. If you log in to Windows 10 with a local account (as opposed to a Microsoft account, which is always an email address), I also want to cajole you into creating password reset security questions, which may well save your tail someday.

You're welcome.

Rolling Back with the Three Rs

Rollback, restore, reset. System repair. Start fresh. The terminology stinks. Bear with me.

WHEN FRESH START IS BETTER

Microsoft had an all-encompassing option called Fresh Start. It was available for Windows 10 versions before the May 2020 update (version 2004). For the May 2020 update and after, the Fresh Start functionality has been moved to Reset This PC, available in the Settings app. To access it, click the Start button, the Settings icon, and then Update & Security, followed by Recovery. Under Reset This PC, click or tap Get Started. Then choose between Keep My Files and Remove Everything.

Few people realize that your PC manufacturer has a say in what Reset's "Remove Everything" means. Most hardware manufacturers have the command jury-rigged to put their crapware back on your PC. If you run Reset with Remove Everything on those systems, you don't get a clean copy of Windows 10. You get the factory settings version. Yes, you get the original manufacturer's drivers, but you also get the manufacturer's garbageware.

To use Fresh Start, select Keep My Files, choose Cloud or Local, change your settings, and set Restore Preinstalled Apps? to No. With this option, Microsoft is making it easier to create a non-bloatware-addled system. They're circumventing the hardware manufacturers in the process. It's hard to say if this will tick off the few hardware companies that aren't already alienated by Microsoft's Surface sales. After all, the Dells and Lenovos and HPEs of the world make most of their PC-sales profit by sharecropping out screen real estate on new PCs and selling it to the highest bidder. It remains to be seen if the Fresh Start option will eat into their revenue.

There's one clear winner in all this: users. Not only do we get an easy option to nuke all the junk on new PCs, but we also have a way to wipe a Win10 PC clean without going through the hassles of booting to an installation drive. Thank you, Microsoft.

Windows 10 has three very different technologies for pulling you out of a tough spot. I liken them to a little hiccup, the WABAC Machine, a brain transplant, and global thermonuclear war.

>> **Go back to the previous version of Windows 10:** If you figure you just got hit with a bad update to Windows 10, it's now easy to roll back to the previous version, or build. The most recent build, which may include bad device drivers or other tweaks that make things go bump in the night, return to their previous incarnation. That's a hiccup, in the grand scheme of themes.

>> **Restore with a Restore Point:** Like Rocky and Bullwinkle's WABAC Machine (thank you, Mr. Peabody), setting and using *restore points* provides a relatively simple way to switch your PC's internal settings to an earlier, and presumably happier state, should something go awry.

Unlike in earlier versions of Windows, restore points aren't set automatically. And in Windows 10, getting to the rollback settings is hard. Not to worry. The mechanism is still intact and useful. Details are in this chapter.

TIP

Restore points aren't intended to restore earlier versions of files that you work with — that's the function of File History (sometimes called the Windows version of the Mac's Time Machine). I talk about File History extensively in Book 8, Chapter 1.

**ASK
WOODY.COM**

>> **Reset This PC:** In my experience, a Windows 10 reset with the Keep My Files option works almost all the time. It's light years ahead of System Repair, safe mode, and recovery mode, and should be your fix-it method of second resort, after you try using a restore point. If refresh doesn't work, you're in a world of hurt. Search online for instructions on manually booting into safe mode and running a recovery. Good luck.

If a reset with the Keep My Files option doesn't work and you don't mind losing all your data and installed programs, or if you want to wipe your computer clean before you sell it or give it away, you want to run a reset, choosing Remove Everything. Global thermonuclear war.

Most of the time, you run a reset with the Keep My Files option when your computer starts acting flakey. You run a reset to wipe the whole system when you're going to sell your PC. But either or both — or using restore points — may be offered as options when your computer won't boot right. I go into detail on restore points and reset in Book 8, Chapter 2.

REMEMBER

Creating Password Reset Questions

If you have a Microsoft account, the only way to reset the password is online. Go to `https://account.live.com/resetpassword.aspx`, and follow the instructions.

ASK WOODY.COM

If you have a local account (not a Microsoft account), Microsoft doesn't store your password on its computers. (If you need a refresher on the different types of accounts, flip to Book 2, Chapter 4.) Prior to Windows 10 version 1803, Microsoft had a set series of steps that would allow you to create a password reset disk — basically a simple file that would unlock your PC, should you forget the local account's password. As of version 1803, those days are gone. Microsoft has not only done away with the password reset disk, they now specifically acknowledge that they can't and won't help you get your local account password back, should you lose it.

If you forget your local account password, you're out of luck. Windows 1803 and later won't let you in. Your only option is to reinstall Windows, which you need about as much as an IRS audit.

The one exception? If you have the presence of mind to set up three specific password challenge questions before you forget your password — and you can remember the answers to all three of those questions — Windows 10 will let you in.

WARNING

If you use a local account, I can't emphasize enough how important it is to establish your three password challenge questions, particularly if your PC has only one administrator account, and it's a local account. I get mail practically every day from people who have forgotten their passwords and can't get in. This one simple trick, which takes a couple of minutes, will save you untold grief should you forget that lousy password!

REMEMBER

Here's the basic idea: You log into Windows 10, using any kind of password — typed, PIN, or picture. You find the magic location to update your security questions, and then fill in answers to three questions that you choose (from a very small set). When you forget your password, Windows 10 will prompt you to answer those questions. Say the magic words and click your heels three times. Bingo, you're in!

It doesn't matter if somebody has changed your password without your knowing. The password challenge questions let you in, *no matter what the password may be.* As long as you have a local account, you're in like Flynn.

Establishing password security questions

If you have a password-protected local account, follow these steps to set up the magical three questions that will let you back into your account, should you ever get locked out:

1. Log in to Windows 10 using your local account.

It doesn't matter what kind of password you use.

2. Click Start ⇨ Settings. Click the Accounts icon.

3. On the left, choose Sign-In Options.

You get the Sign-In Options page.

4. On the right, click Password and then the link to Update Your Security Questions.

Windows 10 asks you to enter your password, and shows you drop-down boxes for three security questions, as shown in Figure 6-1.

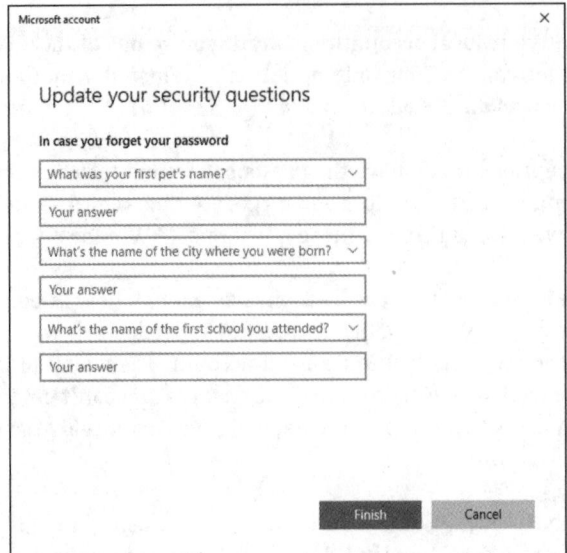

FIGURE 6-1:
Windows 10 has
a bare-bones
security
questions list.

5. **Choose the best questions you can find in each of the three boxes and type the answers.**

   Only six questions are available, and they're the same questions in all three drop-down boxes. Thus, you're forced to enter responses to three of the built-in questions — one in each box. You can't enter your own questions.

   The answers are case-sensitive so, for example, *Dummies* is not the same as *dummies.* The fact that they're case-sensitive may make you change a question.

6. **If you're concerned that you won't remember the precise answers — you'll need to type the answers exactly — make a note for yourself on your phone, or someplace safe.**

7. **Click Finish.**

   Or X out of the Sign-In Options page, but your answers are not going to be saved.

Once again: The password security questions are only for logging in to your PC with your local account. They don't work for Microsoft accounts.

REMEMBER

Using password recovery questions

So, you followed the steps in the preceding section and set up the challenge questions for your local account's password. The time comes when you forget your password. Here's how to get in:

1. **On the login screen, type an incorrect password and click OK.**

   You see a Reset Password prompt, as shown in Figure 6-2.

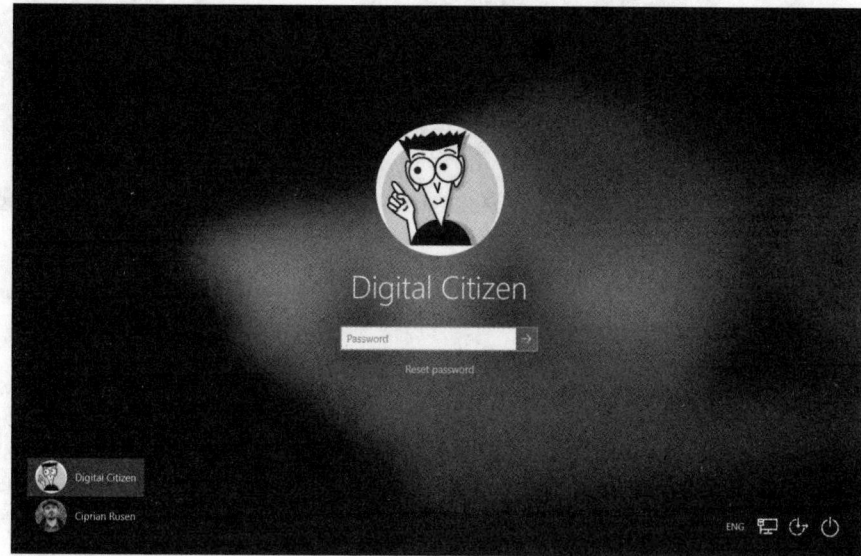

FIGURE 6-2:
If you can't remember your password, type a bad one. You see this screen.

2. **Click Reset Password.**

   You're prompted to enter the answers to your three security questions.

3. **Type answers to all three questions, and then press Enter or click the right arrow next to the bottom answer.**

   Windows 10 immediately prompts you to reset your password, as shown in Figure 6-3.

Don't lose the answers to those questions, okay?

To reiterate: If you have a local account, the ability to answer those three questions will get you into the machine, regardless of the original password.

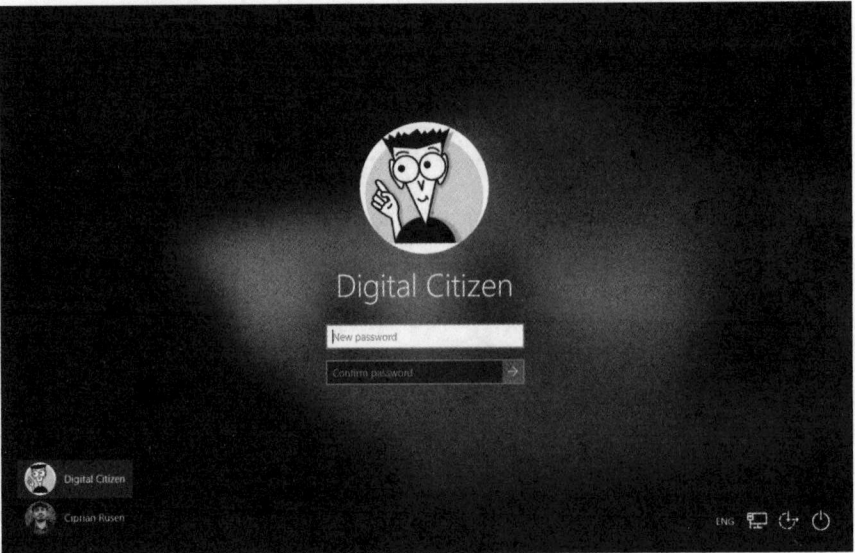

FIGURE 6-3:
Windows 10
forces you to
create a new
password
and hint.

Making Windows Update Work

Microsoft has never been particularly good at patching Windows, but we've hit new lows with Windows 10. I constantly get questions about messed up Windows 10 Cumulative updates — why does my machine say it can't install updates, why does it roll back to an earlier version, what does it take to get Windows 10 to get its act in order?

Microsoft knows that folks are having problems and, much to their credit, they've devised a web page that helps step you through your updating problems and — maybe — find a solution.

The Windows Update troubleshooter, KB 10164, is useful if you're having trouble updating Windows 10. It traces through all the obvious steps for getting Windows Update smacked upside the head as well as some lesser-known approaches.

To use it, go to `https://support.microsoft.com/en-us/help/4089834/windows-10-troubleshoot-problems-updating` and step through the advice. Microsoft claims, "The steps provided here should fix any errors that come up during the Windows Update process," but that's a bit facile for an extremely complex problem. Still, the Knowledge Base article hits all the high points.

Maintaining Drives

Rotating drives (hard drives, CDs, DVDs, even those ancient floppies if you can still find one, and other types of storage media) seem to cause more computer problems than all other infuriating PC parts combined. Why? They move. And unlike other parts of computers that are designed to move (printer rollers and keyboard springs and mouse balls, for example), they move quickly and with ultrafine precision, day in and day out.

E pur, si muove

That's what Galileo said in 1633, after being forced during the Inquisition to recant his beliefs about the earth moving around the sun. "And yet it moves" — and that's the crux of the problem.

As with any other moving mechanical contraption, an ounce of drive prevention is worth ten tons of cure. WD-40 may cure other moving mechanical contraptions, but WD-40 is *not* recommended for PCs. Duct tape and baling wire are another consideration altogether.

USB key drives and solid-state drives (SSD) are a whole different kettle of fish. SSD manufacturers typically offer diagnostic and health maintenance tools to keep their products in top shape, but they contain no moving parts, and thus aren't subject to the vagaries associated with moving drives. I talk about SSDs later in this chapter.

ASK
WOODY.COM

If you're looking for help installing a new hard drive, you're in the wrong place. I talk about adding new drives and getting Windows 10 to recognize them in Book 8, Chapter 4.

What is formatting?

Drives try to pack lots of data into a small space, and because of that, they need to be calibrated. That's where formatting comes in.

When you format a drive, you calibrate it: You mark it with guideposts that tell the PC where to store data and how to retrieve it. Every hard drive (and floppy disk, for that matter) must be formatted before it can be used. The manufacturer probably formatted your drive before you got it. That's comforting because every time a drive is reformatted, everything on the drive is tossed out, completely and (almost) irretrievably. Everything.

TIP

You can format or reformat any hard drive other than the one that contains Windows 10 by starting File Explorer (in the taskbar) and scrolling down to This PC. Then right-click the hard drive and choose Format. You can also format rewritable CDs, DVDs, USB (key) flash drives, and SD or other removable memory cards — delete all the data on them — by following the same approach. To reformat the drive that contains Windows, you must reinstall Windows.

Introducing hard-drive-maintenance tools

Hard drives die at the worst possible moments. A hard drive that's starting to act flaky can display all sorts of strange symptoms: everything from long, long pauses when you're trying to open a file to completely inexplicable crashes and other errors in Windows 10 itself.

Windows 10 comes with a grab bag of utilities designed to help you keep your hard drives in top shape.

>> **Storage Spaces:** The best, most comprehensive of the bunch is Storage Spaces (see Book 7, Chapter 4), which keeps duplicate copies of every file in hot standby, should a hard drive break down. But to use Storage Spaces effectively, you need at least three hard drives and twice as much hard drive space as you have data. Not everyone can afford that. Not everyone wants to dig into the nitty-gritty.

>> **Basic utilities:** Three simple utilities stand out as effective ways to care for your hard drives, and one of them runs automatically once a week. You should get to know Check Disk, Disk Cleanup, and Disk Defragmenter because they all come in handy at the right times.

   You must be a designated administrator (see the section on using account types in Book 2, Chapter 4) to get these utilities to work. I explain how to use Check Disk and Disk Defragmenter in the following two sections.

TIP

>> **Task Scheduler:** If you're really short on disk space, you can use the Task Scheduler to periodically remove temporary files that you don't need by scheduling runs of the Disk Cleanup utility. Task Scheduler has other uses, but most Windows users never really need it.

Running an error check

If a drive starts acting weird (for example, you see error messages when trying to open a file, or Windows 10 crashes in unpredictable ways, or a simple file copy takes hours instead of minutes), run the Windows error-checking routines.

TECHNICAL STUFF

If you're an old hand at Windows (or an even older hand at DOS), you probably recognize the following steps as the venerable CHKDSK routine, in somewhat fancier clothing.

Follow these steps to run Check Disk:

1. Bring up the drive you want to check in File Explorer:

a. *Click the File Explorer icon.*

b. *On the left, choose This PC.*

c. *Right-click the drive that's giving you problems and choose Properties.*

You see the Local Disk Properties dialog box.

2. On the Tools tab, click the Check button, as shown in Figure 6-4.

Windows 10 may tell you that you don't need to scan the drive, because it hasn't found any errors on the drive. If you're skeptical, though, go right ahead.

3. Tap or click Scan Drive.

Windows 10 tells you about any problems it encounters and asks for your permission to fix them.

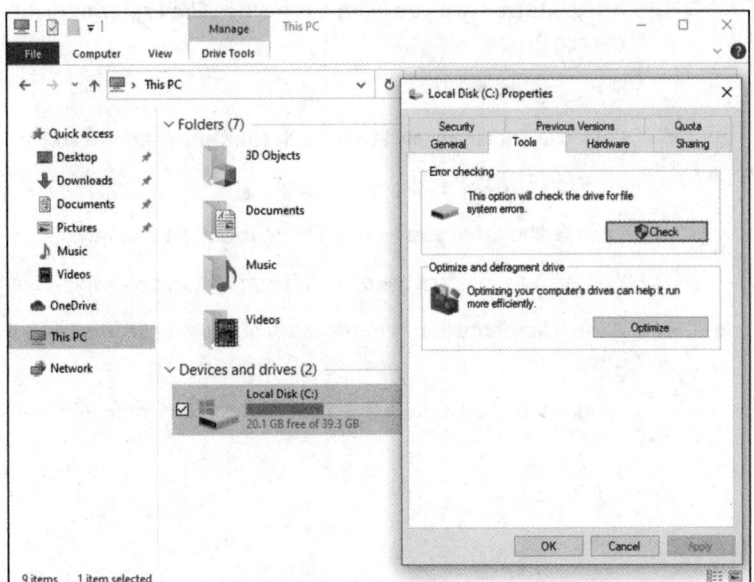

FIGURE 6-4:
Run a check disk.

Defragmenting a drive

Once upon a time, defragmenting your hard drive — instructing Windows to rearrange files on a hard drive so that the various parts of a file all sit next to one another — rated as a Real Big Deal. Windows didn't help automate running defrags, so few people bothered. As a result, drives started to look like patchwork quilts with pieces of files stored higgledy-piggledy. On the rare occasion that a Windows user ran the defragmenter, bringing all the pieces together could take hours — and the resulting system speed-up rarely raised any eyebrows, much less rocketed Windows fans into hyperthreaded bliss.

Windows 7 changed that quietly scheduling a disk defragmentation to run every week. Windows 10 continues in that proud tradition. To get defragmented, you don't need to touch a thing.

TECHNICAL STUFF

Windows 10 doesn't run automatic defrags on SSDs, which is to say, flash memory drives that don't have any moving parts. SSDs don't need defragmentation. They also have a finite lifespan, so there's no need to overwork the drives with a senseless exercise in futility.

If you're curious about how your computer's doing in the defragmentation department, you can see the Defragmenter report this way:

1. **Bring up the drive you want to check in File Explorer. Right-click it and choose Properties.**

    (Refer to Figure 6-4.)

2. **Click the Tools tab, and then click the Optimize button.**

    The Optimize Drives dialog appears.

3. **Choose the drive you want to look at and click Analyze.**

    You see how much of the drive is fragmented, as shown in Figure 6-5.

4. **If the fragmentation is more than, oh, 20% or so, click the Optimize button.**

    Windows 10 runs a defragmentation and optimization re-shuffling.

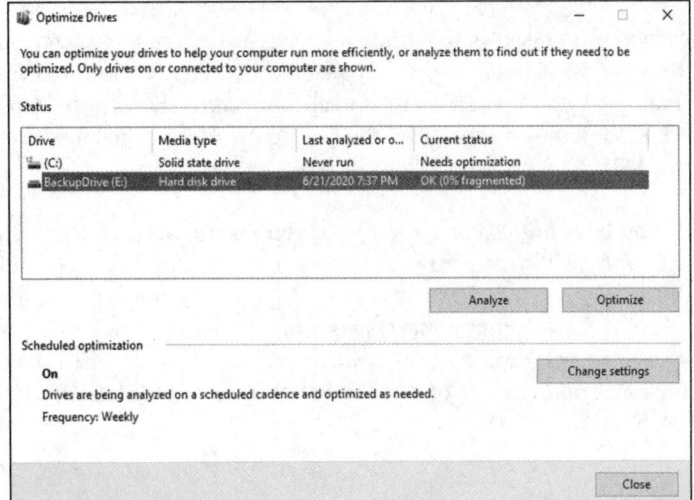

FIGURE 6-5:
Here's a full report of defragmenting activities.

Maintaining Solid-State Drives

Solid-state drives (SSDs) are a completely different breed of cat. You don't want to run a Check Disk on them, even if you can, because the results aren't conclusive, and you'd end up overworking the SSDs. You certainly don't want to run a defrag because the drives are (depending on how you look at it) already defragmented and/or horrendously fragmented and there's no reason to change.

Most SSDs these days are made from NAND flash memory, which is memory that doesn't lose its settings when the power's turned off. Although an SSD may fit into a hard drive slot and behave much like a regular hard drive, the technology's completely different.

ASK
WOODY.COM

While the jury's still out on whether SSDs are *much* more reliable than hard disk drives (HDDs), just about everyone agrees they are reliable. And there's absolutely no doubt that they're enormously faster. Change your C: drive over from a spinning platter to an SSD and strap on your seat belt, Nelly.

WARNING

SSDs have controllers that handle everything. Data isn't stored on SSDs the same way it's stored on HDDs, and many purpose-built hard-drive tools don't work at all on SSDs. The controller must take on all the housekeeping that just comes naturally with HDDs. For example, if you want to erase an HDD, you can format it or just delete all the files on it. If you want to erase an SSD, you should use the manufacturer's utilities, or data can be left behind. See the *Computerworld* article at `www.computerworld.com/article/2506511/solid-state-drives/can-data-stored-on-an-ssd-be-secured-.html`.

Windows 10 disables the utilities known as Defrag, Superfetch, and ReadyBoost on SSDs — you should never see Windows 10 offer to run a Defrag on an SSD, for example — and Windows 10 startup works directly with the hardware during boot. That simultaneously makes the boot go faster and reduces unnecessary wear on the SSD.

TIP

If you have an SSD or get an SSD, you should drop by the manufacturer's website and pick up any utilities it may have for the care and feeding of the furious little buggers. Windows 10 does a particularly good job of looking after them, but the manufacturer may have a few tricks up its sleeve. Intel's SSD Toolbox at `www.intel.com/content/www/us/en/support/articles/000006395/memory-and-storage.html` is one of the better-known utility packs, but you should only use it on Intel SSDs.

Zipping and Compressing

Windows 10 supports two different kinds of file compression. The distinction is confusing but important, so bear with me.

File compression reduces the size of a file by cleverly taking out parts of the contents of the file that aren't needed, storing only the minimum amount of information necessary to reconstitute the file — extract it — into its full original form. A certain amount of overhead is involved because the computer must take the time to squeeze extraneous information out of a file before storing it, and then the computer takes more time to restore the file to its original state when someone needs the file. But compression can reduce file sizes enormously. A compressed file often takes up half its original space — even less, in many cases.

TECHNICAL STUFF

How does compression work? That depends on the compression method you use. In one kind of compression, known as Huffman encoding, letters that occur frequently in a file (say, the letter *e* in a word-processing document) are massaged so that they take up only a little bit of room in the file, whereas letters that occur less frequently (say, *x*) are allowed to occupy lots of space. Rather than allocate eight 1s and 0s for every letter in a document, for example, some letters may take up only two 1s and 0s, and others can take up 15. The net result, overall, is a big reduction in file size. It's complicated, and the mathematics involved get quite interesting.

Following are the two Windows 10 file compression techniques:

>> Files can be compressed and placed in a *compressed (zipped) folder*. The icon for a zipped folder, appropriately, has a zipper on it.

>> Folders or even entire drives can be compressed by using the built-in compression capabilities of the Windows file system (NTFS).

Here's where things get complicated.

NT file system (NTFS) compression is built into the file system: You can use it only on NTFS drives, and the compression doesn't persist when you move (or copy) the file off the drive. Think of NTFS compression as a capability inherent to the hard drive itself. That isn't really the case — Windows 10 does all the sleight-of-hand behind the scenes — but the concept can help you remember the limitations and quirks of NTFS compression.

Although Microsoft would have you believe that compressed (zipped) folder compression is based on folders, it isn't. A compressed (zipped) folder is really a file — *not* a folder — but it's a special kind of file, called a zip file. If you ever encountered zip files on the Internet (they have a .zip filename extension and are read and created directly in File Explorer), you know exactly what I'm talking about. Zip files contain one or more compressed files, and they use the most common kind of compression found on the Internet. Think of compressed (zipped) folders as being zip files, and if you have even a nodding acquaintance with zips, you'll immediately understand the limitations and quirks of compressed (zipped) folders. Microsoft calls them folders because that's supposed to be easier for users to understand. You be the judge.

TIP

If you have Windows 10 show you filename extensions (see my rant about that topic in the section on showing filename extensions in Book 3, Chapter 1), you see immediately that compressed (zipped) folders are, in fact, simple zip files.

Zipping is very common, particularly because it reduces the amount of data that needs to be transported from here to there. NTFS compression isn't nearly as common. It's more difficult, and hard drives have become so cheap there's rarely any need for most people to use it.

Table 6-1 shows a quick comparison of NTFS compression and zip compression.

WARNING

If you try to compress the drive that contains Windows 10 itself (normally your C: drive), you can't compress the files that are in use by Windows.

TABLE 6-1   **NTFS Compression versus Compressed (Zipped) Folder Compression**

NTFS	Zip
Think of NTFS compression as a feature of the hard drive itself.	Zip technology works on any file, regardless of where it is stored.
The minute you move an NTFS-compressed file off an NTFS drive (by, say, sending a file as an email attachment), the file is uncompressed, automatically, and you can't do anything about it: You'll send a big, uncompressed file.	You can move a compressed (zipped) folder (it's a zip file, with a .zip filename extension) anywhere, and it stays compressed. If you send a zip file as an email attachment, it goes over the Internet as a compressed file. The person who receives the file can view it directly in Windows or use a product such as WinZip to see it.
Lots of overhead is associated with NTFS compression. Windows must compress and decompress those files on the fly, and that sucks up processing power.	Very little overhead is associated with zip files. Many programs (for example, antivirus programs) read zip files directly.
NTFS compression is helpful if you're running out of room on an NTFS-formatted drive.	Compressed (zipped) folders (that is to say, zip files) are in a near-universal form that can be used just about anywhere.
You must be using an administrator account to use NTFS compression.	You can create, copy, or move zip files just like any other files, with the same security restrictions.
You can use NTFS compression on entire drives, folders, or single files. They cannot be password protected.	You can zip files, folders, or (rarely) drives, and they can be password protected.

Compressing with NTFS

To use NTFS compression on an entire drive, follow these steps:

1. **Make sure you're using an administrator account.**

   See Book 2, Chapter 4.

2. **Bring up File Explorer by clicking its icon. On the left, choose This PC.**

3. **On the right, right-click or tap and hold down on the drive you want to compress. Choose Properties, and click the General tab.**

4. **Select the Compress This Drive to Save Disk Space check box, as shown in Figure 6-6. Then click OK.**

   Windows 10 asks you to confirm that you want to compress the entire drive. It takes some time to compress the drive; in some cases, the estimated time is measured in days. Good luck.

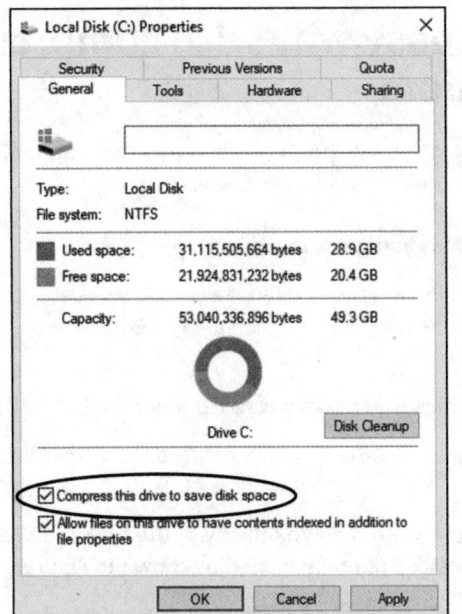

The dialog box shows:

Local Disk (C:) Properties

Security | Previous Versions | Quota
General | Tools | Hardware | Sharing

Type: Local Disk
File system: NTFS

Used space: 31,115,505,664 bytes — 28.9 GB
Free space: 21,924,831,232 bytes — 20.4 GB

Capacity: 53,040,336,896 bytes — 49.3 GB

Drive C: [Disk Cleanup]

☑ Compress this drive to save disk space
☑ Allow files on this drive to have contents indexed in addition to file properties

[OK] [Cancel] [Apply]

To use NTFS compression on a folder, follow these steps:

1. **Make sure you're using an administrator account.**

   See Book 2, Chapter 4.

2. **Bring up File Explorer by clicking its icon. On the left, choose This PC.**

3. **On the right, navigate to the folder you want to compress. Then, right-click (or tap and hold down on) its name.**

4. **In the menu that opens, choose Properties. Then, click the Advanced button.**

5. **Select the Compress Contents to Save Disk Space check box. Then click OK, twice.**

   Windows 10 asks you to confirm that you want to compress the folder. Unless the folder is enormous, it should compress in a few minutes.

TIP

To uncompress a folder, reopen the Advanced Properties dialog box (right-click the folder, choose Properties, and click the Advanced button) and deselect the Compress Contents to Save Disk Space check box.

Zipping the easy way with compressed (zipped) folders

The easiest way to create a zip file, er, a compressed (zipped) folder, is with a simple right-click (or tap and hold). Here's how:

1. **Navigate to the files you want to zip.**

   Usually you find them using File Explorer, although there are other ways. For File Explorer, click the Start icon and then the File Explorer icon. On the left, choose This PC.

2. **Select the file or files that you want to zip together.**

   You can tap and hold down, or Ctrl+click to select individual files, or Shift+click to select a bunch.

3. **Right-click (or tap and hold down on) any of the selected files, and choose Send To ⇨ Compressed (Zipped) Folder, as shown in Figure 6-7.**

   Windows 10 responds by creating a new zip file with a .zip filename extension and placing copies of the selected files inside the new zip folder. You can rename it if you want by typing a new name and pressing Enter. File Explorer selects the file and shows a Compressed Folder Tools context tab. Double-click the new file (er, folder) and you see something like Figure 6-8.

   The new zip file is just like any other file: You can rename it, copy it, move it, delete it, send it as an email attachment, save it on the Internet, or do anything else to it that you can do to a file. That's because it *is* a file.

4. **To add another file to your compressed (zipped) folder, simply drag it onto the zipped folder icon.**

5. **To copy a file from your zip file (uh, folder), double-click the zipped folder icon and treat the file the same way you would treat any regular file.**

6. **To copy all files out of your zip file (folder), click the Extract All button on the File Explorer ribbon.**

   From there, you can choose the location.

TIP

By default, the Extract All icon recommends that you extract all the compressed files into a new folder with the same name as the zip file, which confuses the living bewilickers out of everybody. Unless you give the extracted folder a different name from the original compressed (zipped) folder, you end up with two folders with precisely the same name sitting on your desktop. Do yourself a huge favor and feed the wizard a different folder name while you're extracting the files.

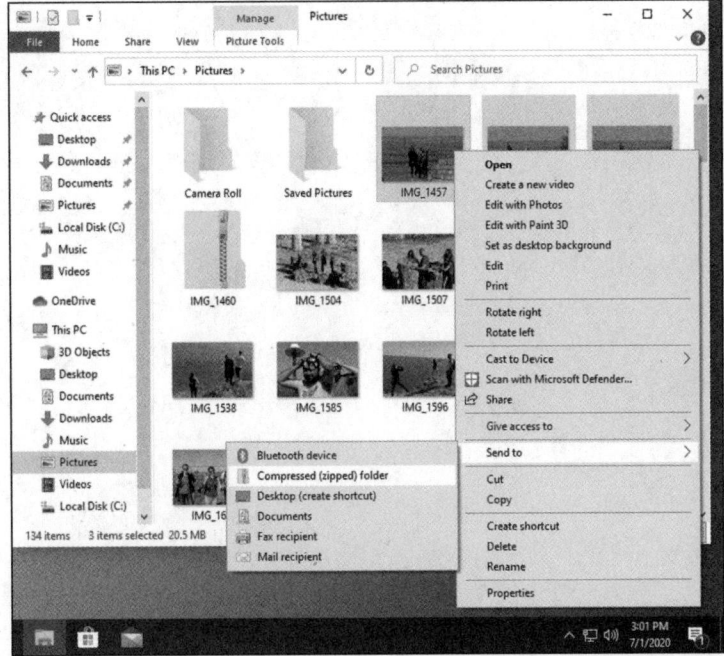

FIGURE 6-7:
Select the files
that you want to
put in a zip file
and right-click
to display
this menu.

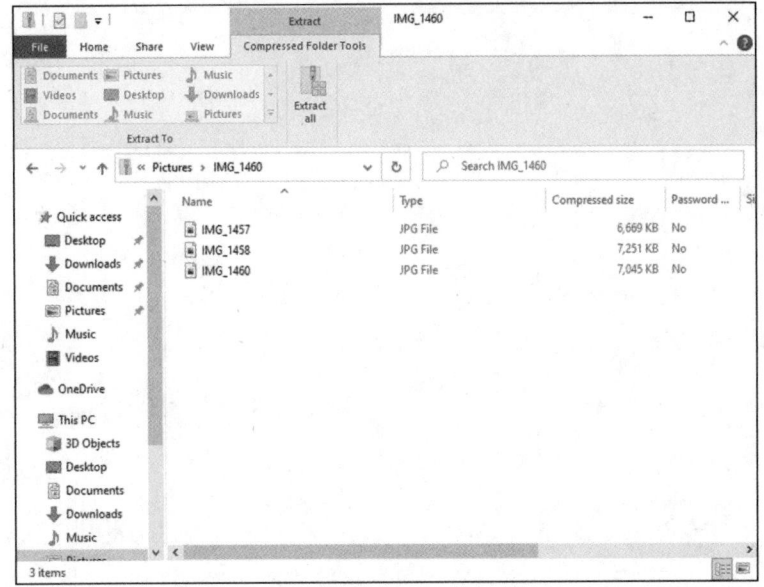

FIGURE 6-8:
Click a zip, and
you get a
context tab for
Compressed
Folder Tools.

Maintaining Your System

4

Using the Built-in Windows 10 Apps

Contents at a Glance

Chapter **1**

Using the Mail and Calendar Apps

The whole productivity app situation — Mail, Calendar, and People — has gone through enormous change since the days of Windows 7. In the, ahem, good old days, Mail, Calendar, and People were basically just one app — very similar to the current situation in Office, where Outlook covers all the bases. That single app, confusingly, was called Windows Live Mail, even though it handled mail and contacts and calendar. It worked reasonably well, but it was old and clunky, and didn't have many features.

ASK WOODY.COM

In Windows 8, Microsoft turned out three separate Metro tiled apps: Mail, Calendar, and People. In fact, all three were connected, but they each had their own Metro tiles, and each worked more or less independently. Not to put too fine a point on it, but the Windows 8 Metro productivity apps were horrible (as you read in my *Windows 8* and *8.1 All-in-One For Dummies* books). Microsoft promised it would make them better.

When Windows 8 hit, the Metro productivity apps were already second rate. By the time Windows 8.1 faded into the sunset, they were all, at best, third rate, eclipsed by Gmail (see Book 10, Chapter 3) and various iThings. Even Microsoft itself had run rings around the apps it shipped in Windows 8.1, with Hotmail, then renamed to Outlook.com. (I talk about Outlook.com in Book 10, Chapter 4.)

THE MANY FACES OF MAIL

Like Gaul, all of email is divided into three parts.

- **Email programs** — commonly called *email clients, email readers,* or *mail user agents* — run on your computer. They reach out to your email, which is stored somewhere on a server (in the cloud, which is to say, on your email company's computer), bring it down to your machine, and help you work on it there, on your machine. Messages get stored on your machine and, optionally, removed from the server when you retrieve them. When you write a message, it too gets stored on your machine, but it also gets sent out, via your email company. Your email client interacts with your email company's computer through strictly defined processes called *protocols.* The most common protocols are POP3 and IMAP. As is the case with most computer acronyms, the names don't really mean anything, although the protocols are quite different.

- **Online email** — most commonly, Gmail, Hotmail/Outlook.com, or Yahoo! Mail, but there are many others — work directly through a web browser, or a program that operates much like a web browser but runs on your computer. You see mail on your computer, but the mail's really stored on your email company's computer. To a first approximation anyway. You can log in to your mail service from any web browser, anywhere in the world, and pick back up right where you left off.

- **Hybrid systems** increasingly combine local mail storage on your machine with online email. Just as email clients are getting more and more online email characteristics, so too are online email systems adopting limited local storage. For example, Gmail — the prototypical online email program — can be set up to store mail on your machine, so you can work on email while away from an Internet connection. The Windows 10 Mail app is a hybrid system, which can be set up to work with any email company's computers.

All the approaches are getting offshoots, as email engulfs mobile devices. Microsoft now has Outlook variants on Windows 10 (with Office 2019 or 365), iPad, Android, and directly through an Internet browser on any kind of machine. The Android, iPad, and browser versions are free for personal use but require Office 365 subscriptions for organizational use and to unlock certain features. With considerable effort from Microsoft, all those variants are starting to look and act like each other.

Google, similarly, has Gmail variants on Android, iPhone, and iPad, although Windows access to Gmail goes through a browser. Unlike Outlook, Gmail has consistently offered the same interface and the same behavior on all its different platforms. The free version of Gmail is identical to the organizational version, but organizations are required to sign up for (and pay for) Google G Suite.

Perhaps surprisingly, thanks to POP3 and IMAP, both Outlook and Gmail work well with just about any email account. You can use @gmail.com email addresses with Outlook and @outlook.com (and @hotmail.com, @msn.com, @live.com, and so on) addresses with Gmail. The people reading your messages will never know that you're consorting with the enemy. See Book 10, Chapter 3.

That's just for email. When you enter the world of SMS (phone messages) and MMS (video/multimedia) instant messaging, life becomes considerably more complex.

In Windows 10, Microsoft threw away the Windows 8 Metro apps. Nobody regrets that less than I do. What has emerged are two apps — one for Mail and Calendar, the other for People — that work the same way, more or less, on Windows 10 PCs, laptops, tablets, and smaller tablets.

The new Windows 10 Mail and Calendar apps are basically two apps, with two tiles, that hook into the same accounts. I talk about the Windows 10 Mail and Calendar app (yes, they're one app, even though there are two tiles on the Start menu) in this chapter. In the next chapter, I step gingerly through the Windows 10 People app.

Choosing a Mail/Calendar App

The Windows 10 Mail app looks like Figure 1-1.

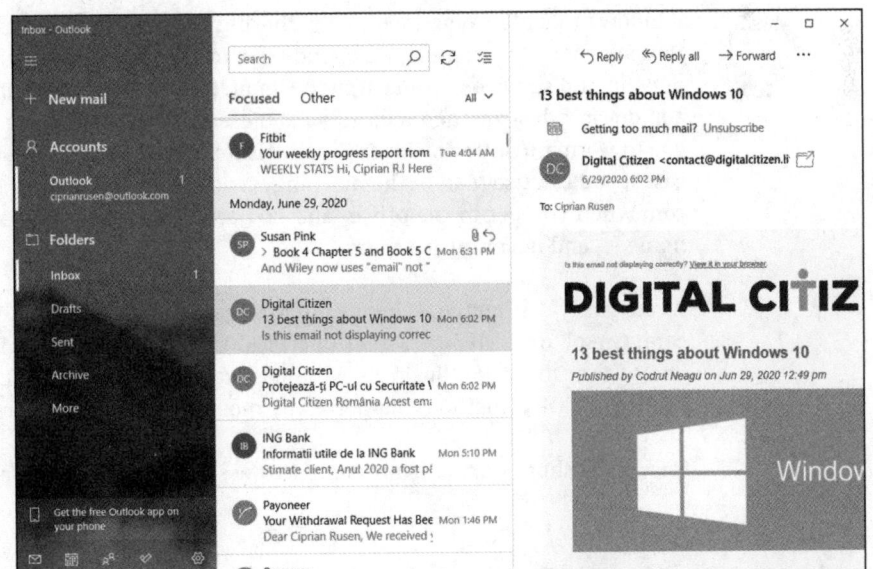

FIGURE 1-1:
Here's a preview of the Windows 10 Mail app.

Your Mail may not quite look the same — the column on the left may not be expanded, the preview pane on the right may not exist. There are lots of differences between tablet mode and regular mode, wide and narrow screens, and portrait and landscape mode.

The Windows 10 Mail has improved enormously since its debut on July 29, 2015, and it keeps getting better. For many, that's damning with faint praise — Mail will never get the development attention lavished on Office 365's Outlook, or even Outlook.com — but there's a steady trickle of improvements.

Is Windows 10's Mail app the right one for you? Good question. Life is full of difficult choices, and I swear Microsoft sits behind about half of them. For me, anyway.

Before you jump into the productivity wallow, think about how you want to handle your mail and calendar.

Comparing email programs

Windows 10 Mail has its benefits, but it may not best suit your needs.

Complicating the situation: Mail isn't an either/or choice. For example, you can set up Hotmail/Outlook.com (see Figure 1-2) or Gmail accounts (see Figure 1-3), and then use either Mail to work with the accounts or the Internet-based interfaces at www.outlook.com and www.gmail.com. In fact, you can jump back and forth between working online at the sites and working on your Windows 10 computer.

ASK
WOODY.COM

Windows 10's Mail functions as a gathering point: It pulls in mail from Hotmail/Outlook.com, for example, and sends out mail through Hotmail/Outlook.com. It pulls in and sends out mail through Gmail. But when it's working right, the Mail app doesn't destroy the mail: All your messages are still sitting there waiting for you in Hotmail/Outlook.com or Gmail. Although there are some subtleties, in most cases, you can use Mail in the morning, switch over to Gmail or Hotmail/Outlook.com when you get to the office, and go back to the tiled Mail app when you get home — and never miss a thing.

As currently configured, Windows 10 Mail can pull in mail from Hotmail/Outlook.com, Gmail, or Exchange Server (a typical situation at a large office or if you use one of the Office 365 business editions), Yahoo! Mail, and AOL Mail, as well as IMAP and POP3 (methods supported by most Internet service providers).

That's the short story. Permit me to throw some complicating factors at you.

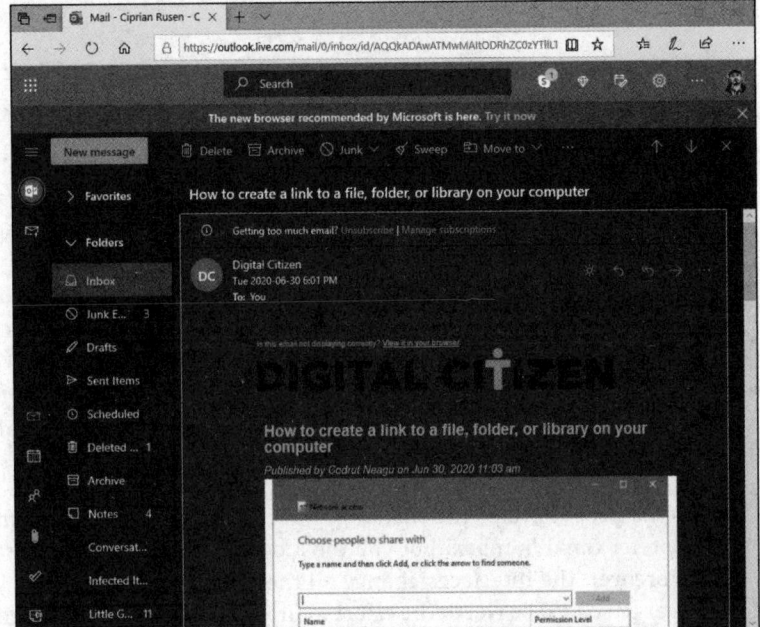

FIGURE 1-2:
Outlook.com
(formerly
Hotmail) —
note the ad on
the top about
Microsoft's new
recommended
browser.

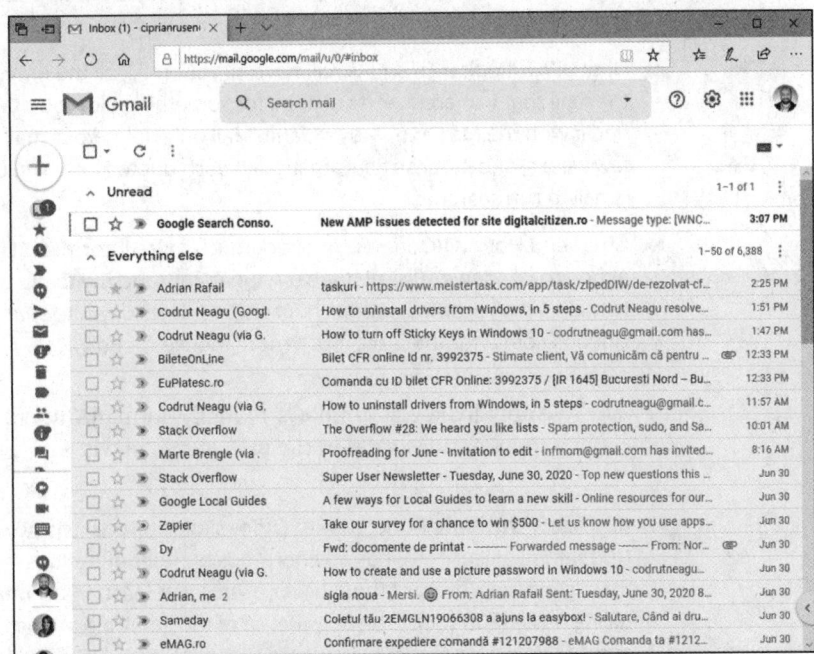

FIGURE 1-3:
I use Gmail as my
email service.

You can add your Hotmail/Outlook.com account to Gmail or add your Gmail account to Hotmail/Outlook.com. In fact, you can add just about any email account to either Hotmail/Outlook.com or Gmail. If you're thinking about moving to Windows 10 Mail just because it can pull in mail from multiple accounts, realize that Gmail (see Book 10, Chapter 3) and Hotmail/Outlook.com (see Book 10, Chapter 4) can do the same thing.

TIP

The main benefit to using Windows 10 Mail rather than Hotmail/Outlook.com or Gmail is that the tiled Mail app stores some of your most recent messages on your computer. (Gmail running on the Google Chrome browser can do the same thing, but you must set it up.) If you can't get to the Internet, you can't download new messages or send responses, but at least Mail can look at your most recent messages.

Some people prefer the Mail app interface over Gmail or Hotmail/Outlook.com. I prefer Gmail's Inbox, but you must decide for yourself. *De gustibus* and all that. Moreover, the interfaces change all the time, so if you haven't looked in the last year or so, it'd be worth the effort to fire up your web browser and have a look-see.

Hotmail/Outlook.com and Gmail are superior to the Mail app in these respects:

REMEMBER

» Hotmail/Outlook.com and Gmail have all your mail, all the time — or at least the mail that you archive. If you look for something old, you may or may not find it with the Mail app — by default, Mail only holds your mail from the past few weeks, and it doesn't automatically reach out to Hotmail/Outlook.com or Gmail to run searches.

» Gmail and Hotmail/Outlook.com pack much more information on the screen. Although Windows 10 Mail has been tuned for touch, with big blocks set aside to make an all-thumbs approach feasible and lots of white space, Hotmail/Outlook.com and Gmail are much more mouse-friendly.

But wait! I've only looked at Windows 10 Mail, Hotmail/Outlook.com, and Gmail. Many, many more options exist in the mail game, to wit:

» **Microsoft Outlook:** Bundled with Office since pterodactyls powered PCs, Outlook (see Figure 1-4) has an enormous number of options — many of them confusing, most of them never used. Or at least, that's what I keep telling myself. Outlook's the Rolls Royce of the email biz, with all the positive and negative connotations.

Among the many, many different versions of Outlook, each has its own foibles. I know people who are still stuck on Office 2007 because it was the last version without the Office ribbon.

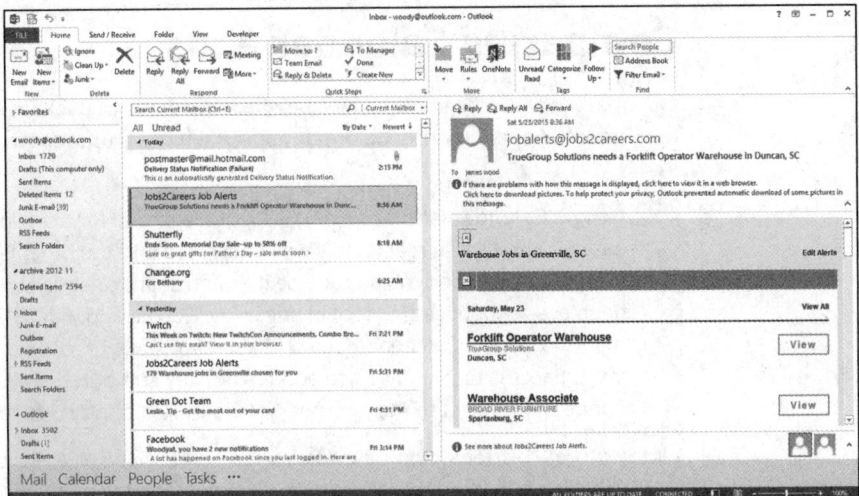

FIGURE 1-4:
Here's Outlook 2013, the way I used to see it.

>> **Outlook.com (was known as Outlook Web App):** It isn't exactly Outlook — at least, not the kind that runs directly on your PC — but Microsoft marketing wants you to believe that it is. Big companies can run their own copies as part of Exchange Server. Most people just log in to www.outlook.com.

>> **Windows Live Mail (WLM):** It's dead for anyone who uses a Microsoft-issued account (@hotmail.com, @outlook.com, @live.com, @msn.com). Microsoft announced in May 2016 that it is no longer supporting WLM. Of course, if you want to use WLM with other email accounts — including, spectacularly, Gmail accounts — it still works the same as always. Microsoft has not only cut the cord with WLM, much to many customers' dismay, it's been given the wet ops treatment. Kinda like a *Game of Thrones* finale.

>> **Free, open-source, inexpensive alternatives:** These include Mozilla Thunderbird, SeaMonkey, Eudora, and many more that have enthusiastic fan bases.

>> **Your Internet service provider (ISP):** It may well have its own email package. My experience with ISP-provided free email hasn't been very positive, but the service generally doesn't hold a candle to Gmail (my favorite), Outlook.com/ Hotmail, Yahoo! Mail, or any of the dozens of competitive email providers. If you use ISP-based email, mail2web (www.mail2web.com) lets you get into just about any mailbox from just about anywhere — if you know the password.

. . . and that's just the Mail app!

ASK WOODY.COM

Most of the time, whether on the road or sitting in my home cave, I use Gmail. It's just so easy to flip out my smartphone or iPad and check on the latest. I gave up on Outlook a couple of years ago and haven't regretted it once.

Using the Mail and Calendar Apps

Comparing Calendar apps

Calendars can also be handled by a bewildering array of packages and sites. Among the hundreds of competing Calendar apps, each has a unique twist. The highlights:

>> **Google Calendar** (see Figure 1-5) is highly regarded for being powerful and easy to use. It's also reasonably well integrated into the other Google Apps, er, Google Drive, although you can use it — and share calendars with other people — without setting foot in any other Google app. Put all your appointments in Google Calendar (`http://calendar.google.com`), and you have instant access to your latest calendar from any computer, tablet, or smartphone that can get to the Internet. See Book 10, Chapter 3 for details.

>> **Hotmail/Outlook.com Calendar,** on the other hand, lives inside Hotmail/Outlook.com. It's reasonably powerful and integrated, and you can share the calendar with your contacts or other people (see Figure 1-6).

>> **Outlook** also does calendars, ten ways from Tuesday, with so many options that it'll bring a tear to your eye. Or maybe that tear is from tearing out your hair.

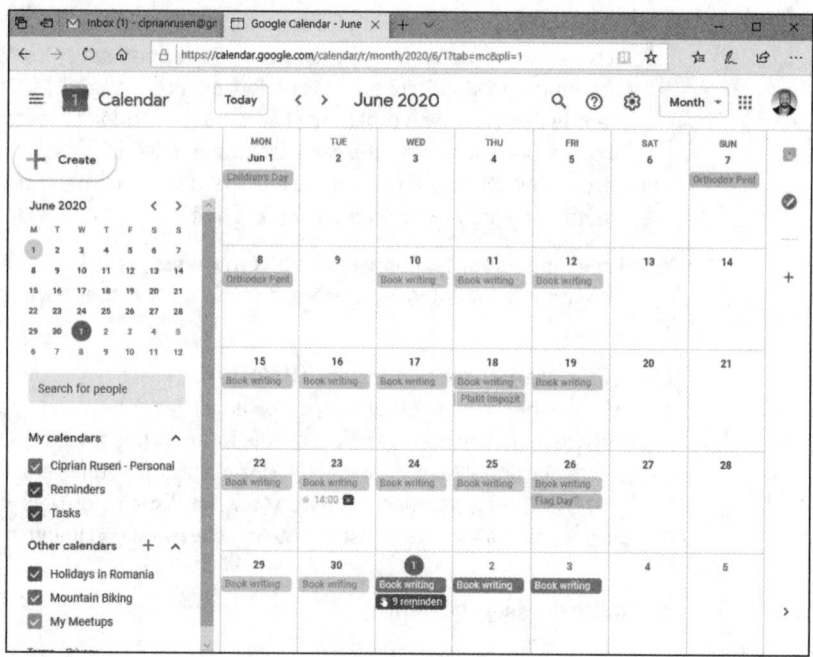

FIGURE 1-5:
I use this Google calendar on many different devices.

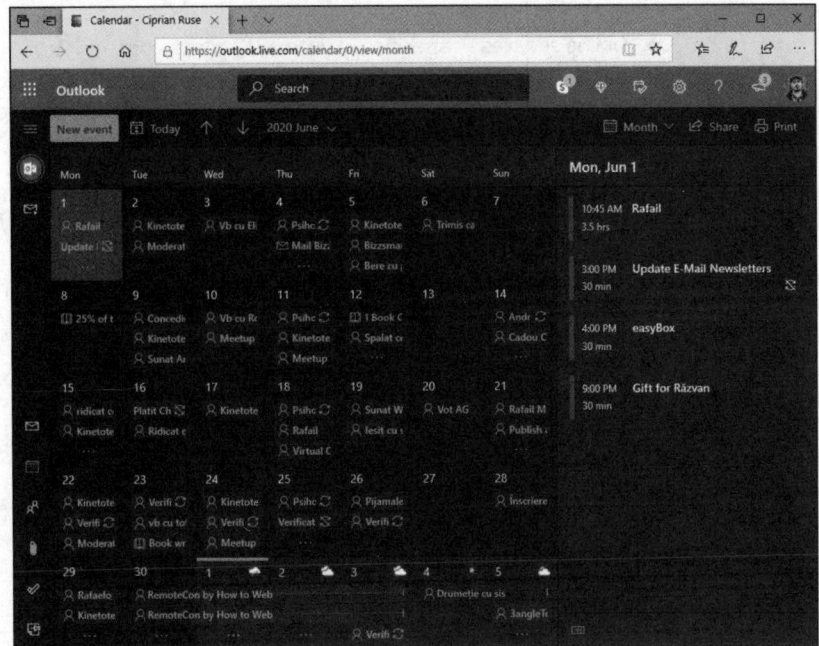

FIGURE 1-6:
The Outlook.com
Calendar has
lots and lots of
options.

TIP

If you want to schedule one conference room in an office with a hundred people, all of whom use Outlook, the Outlook Calendar is definitely the way to go. If you want to keep track of your flight departure times, Aunt Martha's birthday, and the kids' football games, any of the Calendar apps will work fine.

I'm very happy to say that the Windows 10 Calendar app syncs very well with Google Calendar. I use Google Calendar everywhere — Android phone, Android tablet, iPhone, iPad, on the road, in the shower — and with Windows 10, I can finally use Google Calendar on my PCs and laptops.

Choosing the right package

**ASK
WOODY.COM**

So how do you choose a Mail/Calendar program? Tough question, but let me give you a few hints:

TIP

>> The Windows 10 productivity apps — Mail and Calendar — work well enough if your demands aren't great.

But if you have an iPad, consider using the built-in Mail and Calendar apps, or any of a dozen other Apple Apps instead. If you're an Android user, the Google apps work just great.

>> Online services — specifically Hotmail/Outlook.com and Gmail — have many more usable features than either Windows 10 Mail or iPad Mail. If you can rely on your Internet connection, look at both before settling on a specific Mail/Contacts/Calendar program.

>> Gmail and Hotmail/Outlook.com make it easy to use their programs to read ordinary email. I can set up my email account, woody@askwoody.com, to work through Gmail, for example, so mail sent to that email address ends up in Gmail, and if I respond to the message, it appears as if it's coming from woody@askwoody.com, not from Gmail.

TIP

A good compromise is to use either Gmail or Hotmail/Outlook.com most of the time but hook up either iPad Mail, the Gmail app, or Windows 10 Mail (or all three!) to the Gmail or Hotmail/Outlook.com account, so you can grab your iPad when you're headed out the door.

>> Ancient dinosaurs will probably keep using the old, PC-based Office Outlook until its bits rot away. It's ponderous and painful, the embodiment of 19th-century dentist's office chic. But it works. (I can't tell you how happy I am that I finally moved over to Gmail!)

Drilling Down on Windows 10 Mail

The first time you tap or click the Start menu's Mail tile, you're given the chance to Add an Account. If you signed in to Windows 10 with a Microsoft account, you just click a couple of times and end up at the Mail screen, which I show at the beginning of this chapter in Figure 1-1.

REMEMBER

If you signed in to Windows 10 with a local account — one that isn't known to Microsoft (see Book 2, Chapter 5) — or if you say that you want to add an additional email account (click or tap the first screen, on the left where it says Settings, then on the right pane's Manage Accounts, followed by Add Account), the Mail app presents you with the choices shown in Figure 1-7.

Table 1-1 explains the option you should choose, depending on what kind of email provider you have.

People trying to use work email on a computer may have to talk to their IT department for additional configuration options.

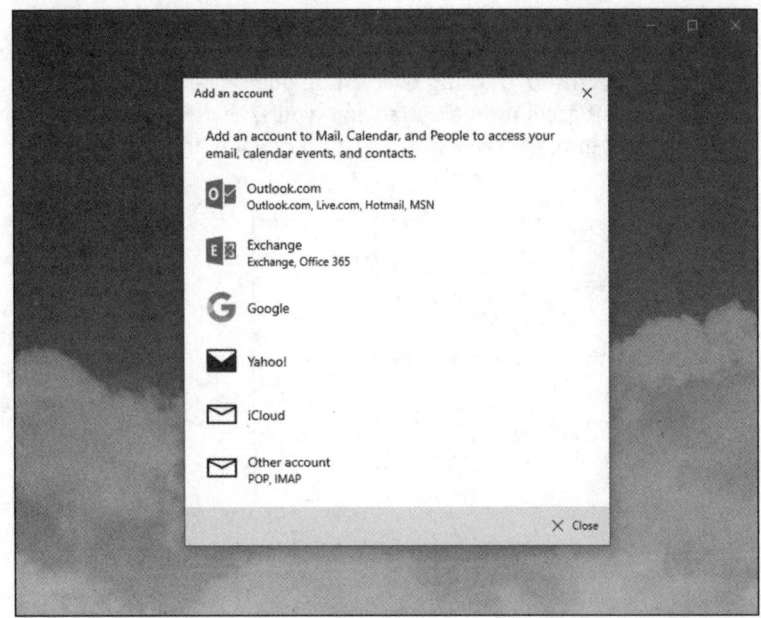

TABLE 1-1

New Mail Account Types

Use This Type	For This Email Service
Outlook.com	If you get mail through Microsoft's servers, your email address looks like something@outlook.com, @live.com, @hotmail.com, or @msn.com.
Exchange	If you get your mail through a company mail server, or if you use Office 365 to handle your mail.
Google	If you have a Google account, most commonly an email address that looks like something@gmail.com, but also if you use Google's servers for email, as you can with GSuite for Business, or if you've just registered your email address with Google and want to retrieve your mail through Google.
Yahoo!	If you get mail from Yahoo!, your email address looks like somebody@yahoo.com.
iCloud	For those from the Apple side of the street, if you have an @icloud.com or an @me.com or @mac.com address.
Other	For any other kind of email address. When you type your email address, Microsoft looks for a bunch of associated information (such as the POP or IMAP server name) in its ginormous database and can almost always set you up with a click or two.
Advanced setup	Use only if you have an Exchange ActiveSync account or if Other fails to find your address, which is rare.

If you signed into Windows 10 with a local account (probably because you don't want Microsoft tracking everything you do — see Book 2, Chapter 4) and add a Microsoft account to the Mail app, you'll be asked if you want to change that local account into a Microsoft account everywhere, per Figure 1-8.

FIGURE 1-8: Unless you want to change your Windows 10 login to a Microsoft account, tell Windows to take a hike.

WARNING

Watch out! Do not click Next. Instead, click Microsoft Apps Only. Otherwise, Windows 10 takes that as permission to switch your local account over to a Microsoft account. There's nothing to see here, Obi-Wan.

By hook or by crook, you end up at the Mail main page, which looks like Figure 1-9. The Mail app pulls in about a month's worth of messages and shows them to you. (Details of the display vary depending on many things, including the width of the screen and whether you're in tablet mode.)

Mail's standard layout has three columns:

>> The left column holds a bunch of icons, which can be hard to decipher.

- The hamburger icon at the top lets you look at all the options (see Figure 1-10).

- The plus icon starts a new message.

- The shadow guy icon — looks like the top two-thirds of an 8-ball — lets you switch among accounts, if you have multiple email accounts.

- The file folder icon lets you switch between your Inbox, Outbox, Archive, and so on.

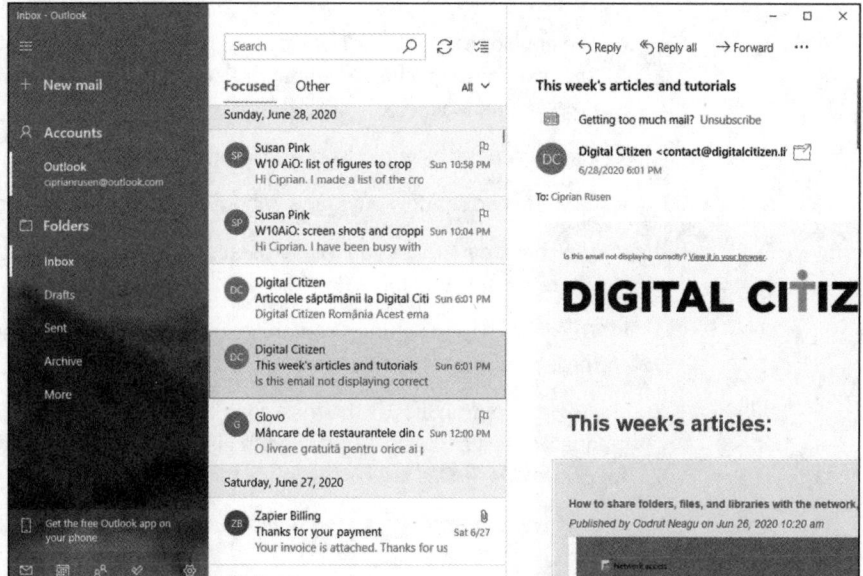

FIGURE 1-9:
A month's worth of messages.

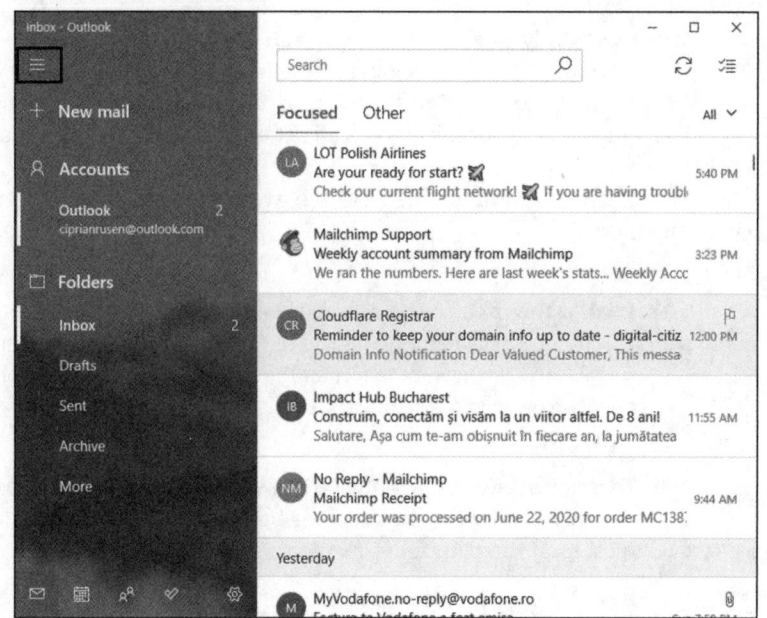

FIGURE 1-10:
The hamburger icon expands to let you choose among the options.

- The envelope icon doesn't do anything. If you're using one of the other Microsoft apps (such as Calendar or People), clicking the envelope icon takes you to Mail.

- Click or tap the calendar icon, and Windows 10 launches the Calendar app.

- Click the two shadow guys, and Windows 10 brings up the People app.

- The checkmark takes you to Microsoft's To Do app (or to Microsoft Store so you can install the app).

- The gear-shaped Settings icon brings up a Settings pane, which I discuss in the later section, "Mail Settings."

» The middle column lists all the messages in the selected folder. If you don't manually select a folder — by using the file folder icon and clicking to pin the specific folder — Mail selects the Inbox for you.

» The right column shows you the selected message.

Creating a new message

When you reply to a message, Mail sets up a typical reply (or a reply to all) in a three-column screen, as shown in Figure 1-11. Similarly, if you tap or click the + icon in the upper left, Mail starts a new, blank message. Whether you reply or start a new message, your message is all set up and ready to go — just start typing.

Here's a quick tour of the features available to you as you create your email message:

» **Format the text:** The new text you type appears in Calibri 11-point type, which is a good all-around middle-of-the-road choice. Don't get me started on Comic Sans. If you want to format the text, just select the text, and click the down arrow next to the underscore icon; you see the formatting options in Figure 1-12.

Those who have a keyboard and know how to use it will be pleased to know that many of the old formatting keyboard shortcuts still work. Here are the most used shortcuts for formatting:

- *Ctrl+B* toggles bold on and off.

- *Ctrl+I* toggles italic on and off.

- *Ctrl+U* toggles underline on and off.

- *Ctrl+Z* undoes the last action.

- *Ctrl+Y* redoes the last undone action.

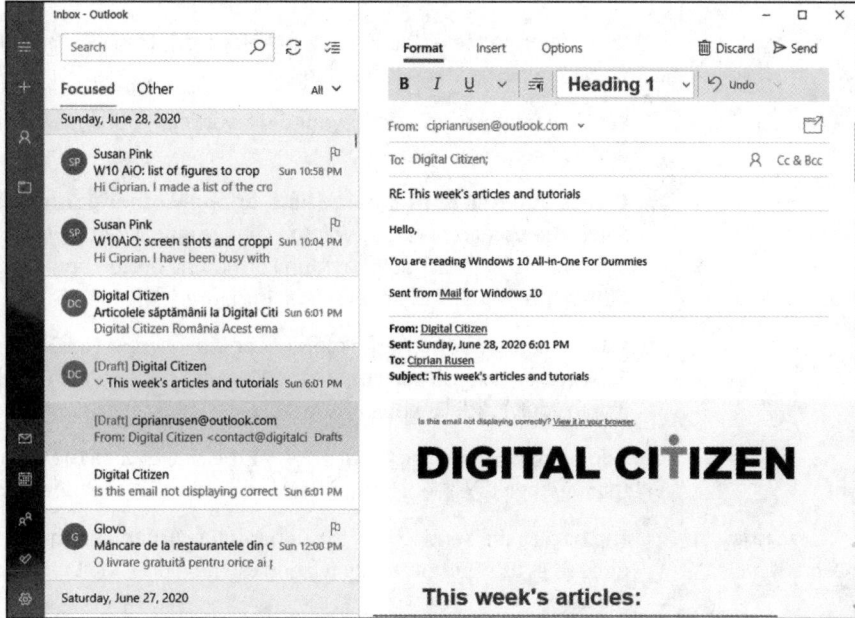

FIGURE 1-11:
When you reply to a message or compose a new message, Mail gives you these options.

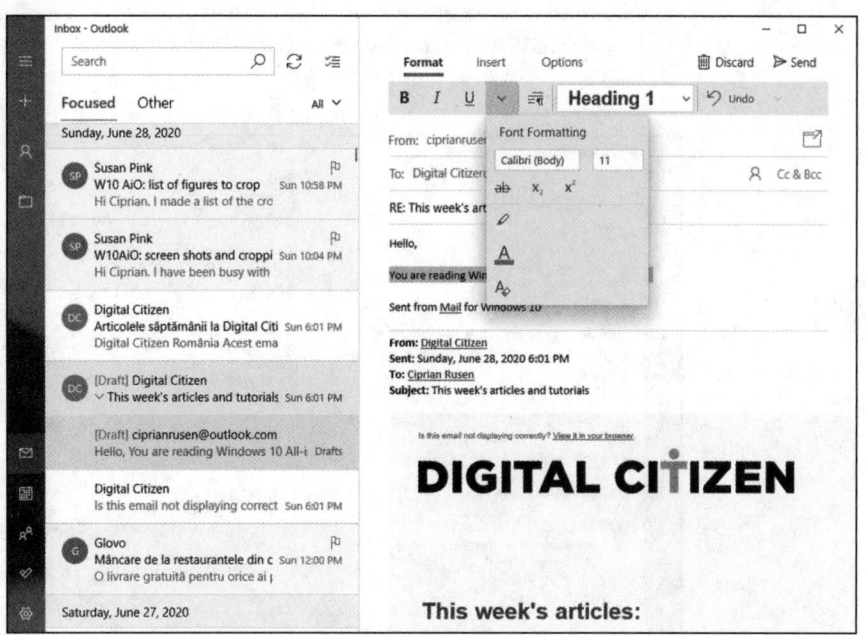

FIGURE 1-12:
Select the text and apply formatting in the usual way.

(In addition to the old stalwarts Ctrl+C for copy, Ctrl+X for cut, and Ctrl+V for paste, of course.)

You'll be happy to know that your old favorite emoticons work, too. Type :-) and you get a smiley face.

>> **Create bulleted or numbered lists, or apply other paragraph formatting:** Select the paragraph(s) you want to change, and click the icon to apply bullets or numbers, or click the Paragraph Formatting button and choose from many other paragraph formats, as shown in Figure 1-13.

>> **Add an attachment:** At the top, click or tap Insert ⇨ Files. You end up in File Explorer, where you can choose the file you want to attach. Click it and (confusingly) click Open.

>> **Add a message priority indicator:** At the top, choose the Options tab and set the message to either High (exclamation point) or Low (down arrow) priority.

REMEMBER

Tap or click the Send icon in the upper-right corner, and the message is queued in the Outbox, ready to send the next time Mail syncs for new messages.

If at any time you don't want to continue, click or tap the Discard trash can button in the upper right. To save a draft, you don't need to do anything: Windows 10 Mail automatically saves everything, all the time.

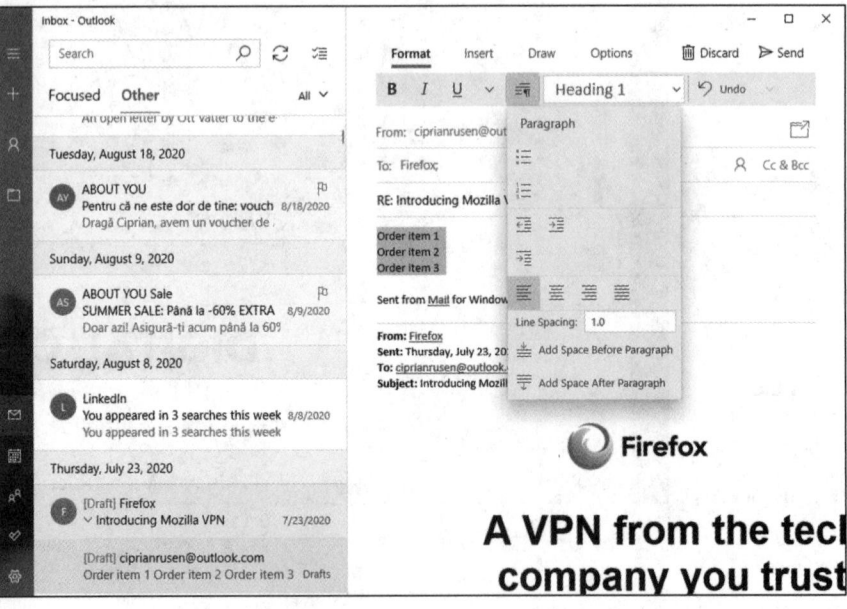

FIGURE 1-13: To create a bulleted list, type the paragraphs, select them, and apply bullets.

The Windows 10 Mail app's editing capabilities are impressive, with many of the features you would expect in Microsoft Word. Styles, tables, fancy formatting, and easy manipulation of in-line pictures top the most-used list. The lack of customized folders counts as a significant shortcoming for many.

Searching for email in the Mail app

Searching for mail is relatively easy if you remember two very important details:

REMEMBER

» **If you have multiple accounts, navigate to the account that you want to search before you perform the search.** If you search while you're looking at the askwoody.com Inbox, for example, you won't find anything in your hotmail.com account.

» **Don't use Cortana.** She isn't up to the challenge. However, the Windows 10 Search might be useful in finding email messages.

To search for email messages:

1. **If you have more than one email account, move to the account you want to search.**

   The easiest way to do that is to click the folder icon on the left and choose whichever account you like.

2. **At the top, above the second column, tap or click the magnifying glass.**

3. **Type your search term, and press Enter or tap the magnifying glass icon again.**

Mail Settings

The Windows 10 Mail app has several worthwhile settings. On the left, at the bottom, tap or click the gear icon (shown in the margin). If the window's wide enough, Settings appears on the right, as shown in Figure 1-14. (If the window isn't wide enough, Settings will tromp over to the left side.)

The next sections tell you what you can do.

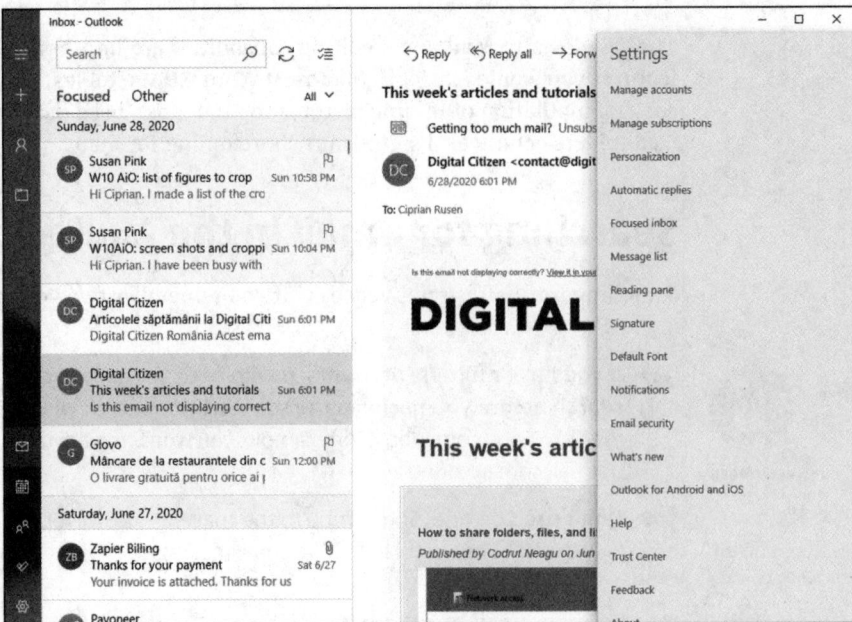

Adding a new email account

The Windows 10 Mail app has built-in smarts for you to connect to any Hotmail/Outlook.com, Gmail, Exchange Server (including Office 365 business edition), AOL, Yahoo!, or IMAP or POP accounts. You can add any number of different types of those accounts — two different Gmail accounts and a few Hotmails — no problem.

To add a new account:

1. **From the Mail app, click or tap the gear icon at the bottom left.**

    The Settings menu appears (refer to Figure 1-14).

2. **Click or tap Manage Accounts ⇨ Add Account.**

    The Add an Account list appears (refer to Figure 1-7).

3. **Tap or click the account type that you want to add.**

    Refer to Table 1-1 for a list of account types. If you click Outlook.com, you're telling Mail that you want to add a Microsoft account, so you see the dialog box shown in Figure 1-15.

FIGURE 1-15:
Enter your
Microsoft email
account.

4. **Enter your email ID and tap or click Next. Then enter your and password and any ancillary information that may be required, and tap or click Sign In.**

    Mail is probably smart enough to look up and find any other information it needs, but you may have to provide additional information (such as a POP3 mail server name) from your email provider.

5. **If Mail presents you with an option to use this account everywhere on your device (refer to Figure 1-8), click or tap Microsoft Apps Only. (Don't click Next.)**

    When Mail comes back, your new account appears under the hamburger icon on the left.

If you want to change the details about your account — in particular, if you don't particularly want to see the name Hotmail, Outlook, or Gmail as an account name — click or tap the Settings icon (shown in the margin), click or tap Manage Accounts, and then tap or click the account you want to change. The Account Settings pane appears, as shown in Figure 1-16. In the top box, you can type a name that will appear in the first column of the Mail main page. If you also want to change the number of days' worth of email downloaded (the default is All Available Mail) or change the sync frequency, click the link marked Change Mailbox Sync Settings.

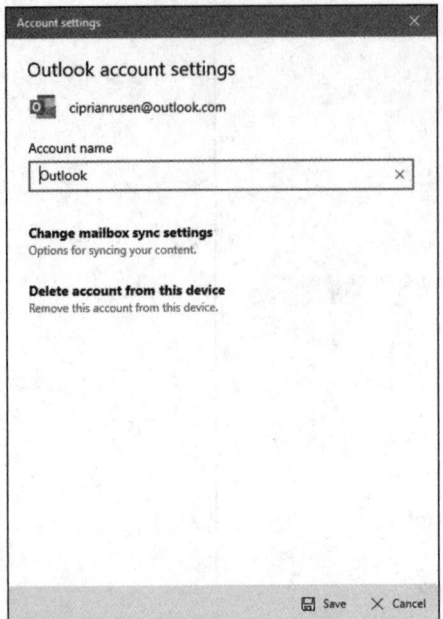

FIGURE 1-16:
Change the
details of an
account.

Setting extra options

The Settings pane has several additional worthwhile options. If you tap or click the Settings icon in the lower-left corner, and then choose Options on the right, you can do these things:

>> **Personalization:** Change the picture that appears in the far-right pane when no mail has been selected or the background for the entire app. You can also enable light or dark mode for the Mail app.

>> **Quick Actions:** Set the response to a swipe from the left or right (set flag, delete, and so on).

>> **Automatic Replies:** Have an account automatically send a response to any received message. (Spammers love this setting because it helps identify active accounts.)

>> **Reading:** Have Mail automatically open the next item when you're finished with the current message.

>> **Signature:** Put a signature (*Sent from Mail for Windows 10* is the default) on all new mail and all responses. Or disable the default signature if you don't want it.

» **Default Font:** Change the default font, size, and formatting for one or all email accounts.

» **Notifications:** Show notifications or play a sound when new mail arrives. Yes, "You've got mail" will work.

Avoiding Calendar App Collisions

The Windows 10 Calendar is relatively straightforward, but the first time you bring up the Calendar app, you may think you're seeing double. Or triple. In Figure 1-17, you can see what I mean.

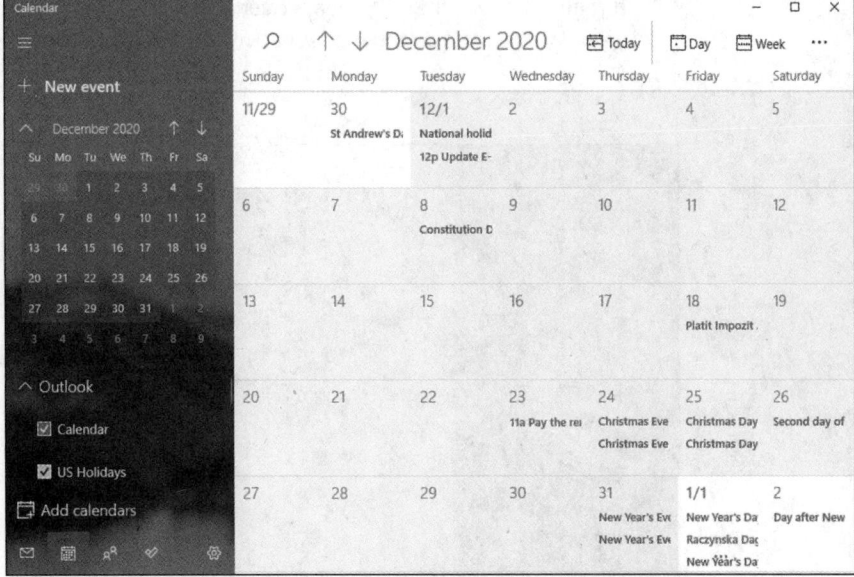

FIGURE 1-17:
Your first time in Win10 Calendar may make your head spin. Note the duplicate entries for Christmas Day, from two different calendars.

Don't panic.

The reason for the duplication? Assuming you have added two or more accounts into Windows 10 Mail, or Calendar, if one or more of the accounts has duplicated entries, the calendars associated with those accounts come along for the ride, and any appointment that appears in both calendars shows up as two stripes on the consolidated calendar.

Fortunately, it's easy to see what's going on and to get rid of the duplicates. Or at least some of the duplicates. Maybe. Here's how to reorganize your Calendar:

1. **From the Start menu, click or tap the Calendar tile to start the Windows 10 Calendar app.**

   You can also click the calendar icon on the bottom of the Mail app.

   If this is the first time you've looked at the Calendar app, it may look like the one in Figure 1-17.

2. **Look at the color-coded selected boxes on the left at the bottom to see whether two or more of your calendars have a source that overlaps. If so, turn off one of the interfering calendars.**

   For example, in Figure 1-17, I have both the Outlook calendar and a US Holidays calendar, but both have entries about holidays in the USA. By simply turning off one of the US Holidays calendars and deselecting the Birthdays calendar, the main calendar goes back to looking somewhat normal, as shown in Figure 1-18.

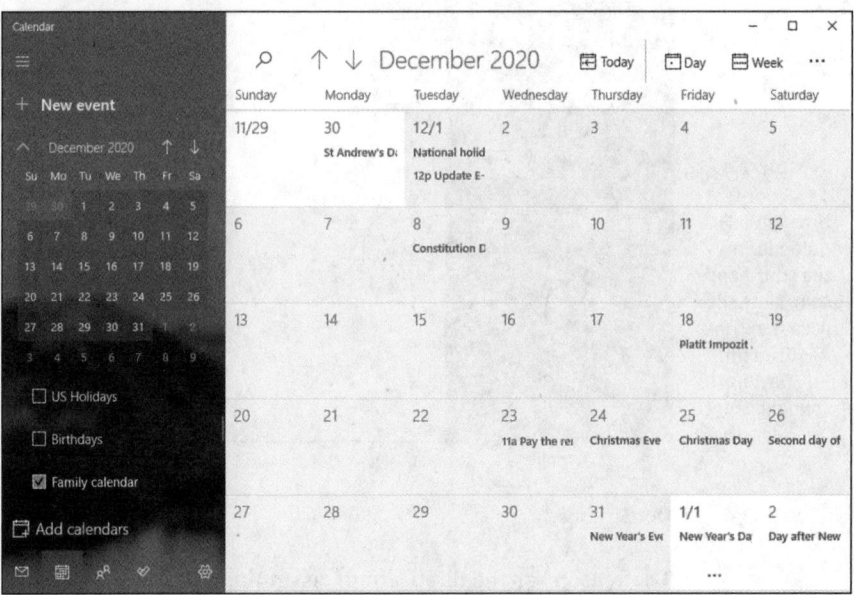

FIGURE 1-18: Getting rid of the Holidays calendar cuts down the clutter.

3. **Go through the calendars, one by one, and set the color-coding for each calendar component to something your eyes can tolerate.**

4. **When you're finished, close the Calendar.**

On the top, you can choose the detail of the calendar you want to see:

>> **Day** brings up an hourly calendar, for two or more days (depending on the number of pixels across your screen).

>> **Week** shows Sunday through Saturday.

>> **Work Week** lists Monday through Friday of the current week only.

>> **Month** brings up one month at a time.

>> **Year** shows a calendar for all the months of the current year.

TIP

In call cases, there are up and down arrows at the top of the screen to move one unit (day, week, month) earlier or later.

Click the hamburger icon to get rid of the left column, and let the calendar take up the entire Calendar window.

Adding Calendar items

To add a new appointment, or other calendar item, tap or click the New event + icon in the upper-left corner. Calendar shows you the Details pane, as shown in Figure 1-19.

Most of the entries are self-explanatory, except these:

REMEMBER

>> **You must choose a calendar — actually, an email account — that will be synchronized with this appointment.** As soon as you enter the appointment, Calendar logs in to the indicated account and adds the appointment to the account's calendar.

>> **You may optionally specify email addresses in the People box.** If you put valid email address(es) in the People box, Calendar automatically generates an email message and sends it to the recipient, asking the recipient to confirm the appointment.

When you finish the appointment, in the upper-left corner, tap or click Save or Send, depending on whether you're setting the appointment or setting it and sending invitations.

ASK WOODY.COM

If you click the icon with two arrows chasing each other in a circle — the Repeat icon — Calendar lets you choose how often to repeat and when to end the repetition.

Using the Mail and Calendar Apps

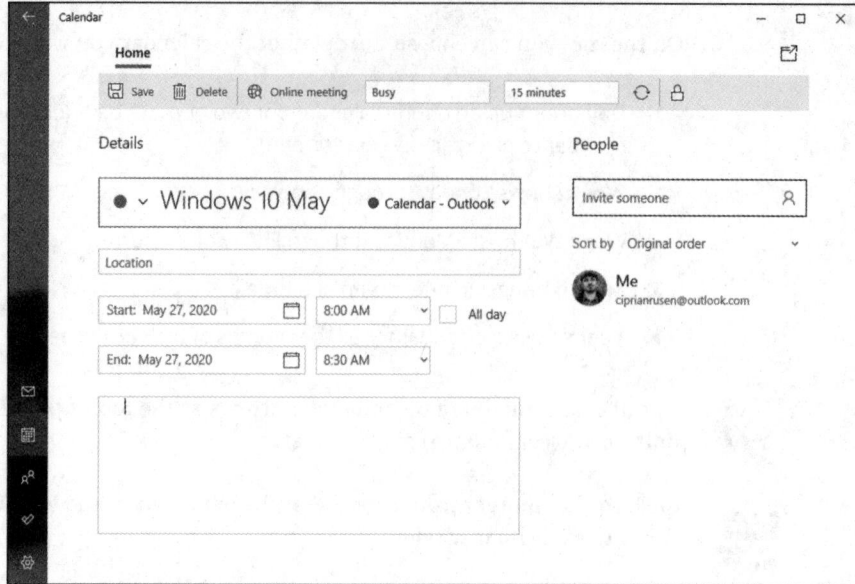

FIGURE 1-19:
Create a new
appointment or
calendar entry.

Struggling with Calendar shortcomings

The Calendar app is a passable calendaring program. It doesn't have any of the goodies you would expect from more advanced calendaring apps, except for toaster-style slide-from-the-right notifications.

On the plus side, you can have Calendar notifications placed on your lock screen. The notifications list individual appointments for the current day. See Book 2, Chapter 2 for details.

**ASK
WOODY.COM**

If you want to look at better calendars (which work from a browser, but not as an independent Windows app), check out these:

>> **Google Calendar** — www.calendar.google.com — is free as a breeze. That's the one I use. When I'm on a Windows 10 tablet, I'll scurry back to the Windows 10 Calendar app, but only to plug in my Google Calendar.

>> For the iPhone and iPad, I use the Google Calendar apps (available in the Apple App Store), but a good friend of mine recommends **Calendars 5** — https://readdle.com/products/calendars5 — which integrates very well indeed with the Apple products.

Beyond Email

It pains me to admit it, but email is changing a lot. Those of us who grew up with email have a hard time accepting it, but in the past week I've used Facebook, Twitter, Slack, WhatsApp, and Line for interactions that are more or less email. The distinction between texting and email is disappearing — in fact, the line between video calls and email is crumbling, voice messages now turn into text entries on my Android smartphone, and it's deucedly difficult to figure out if my wife sent me an email message, an SMS text, a Facebook message, or a Line message. I tend away from video calls because I must comb my hair, but other than that, there are advantages to all the new alternatives.

Don't lock yourself out of the new ways — Facebook, WhatsApp, Twitter, Line, Slack, Snapchat (not just for sexting selfies anymore, in spite of what some politicians think or do; current valuation over $20 billion), Yik Yak (with capitalization over $350 million), Yammer (which Microsoft now owns), Skype (which Microsoft now owns), and many others. Each has a slightly different approach, and in some situations, they're clearly better than good old email.

Chapter **2**

Keeping Track of People

O nce upon a time, contact lists were the meat 'n taters of the PC world. Being able to keep one single list of all your contacts — and keep their addresses, email addresses, and phone numbers all up to date — was one of the most important chores for a burgeoning PC.

**ASK
WOODY.COM**

Those days have long passed. Nowadays, contact lists get gummed up with out-dated entries and useless information. Worse, the contact lists don't talk to each other: My contacts in Facebook, Line, Skype, Gmail, inside my phones, Flickr and Snapchat and Twitter and Pinterest and Outlook just don't talk to each other. Which is all for the better, actually, because if they did start talking to each other, there'd be some really heated arguments and lots of name-calling.

Even if your contacts are better behaved than mine, changing a detail in one place — say, a new email address in Gmail — doesn't ripple to all the lists. Instead, it just means that one of the lists is out of sync with all the others.

I wish I could say that Microsoft has built a better contact list, but they haven't. The Windows 10 People app is a toy app, which may evolve into a superior central repository someday, but I'm not holding my breath.

Microsoft's been working on contact lists since the days of Windows 3.1 and Out-look 4, and none of the lists has worked worth a hill of beans. Don't get me started about changing an email address in Outlook, and not having it updated on the automatic fill-in list for new emails. I don't know how many times I embarrassed myself with that one.

The Contact List in Windows 10

You may want to think of it as the Windows 10 People app (see Figure 2-1), but it's really just Windows 10's contact list. Nothing pretentious about it. In fact, at this point, it isn't even as capable as the contact list in Windows 8.1, which is saying something.

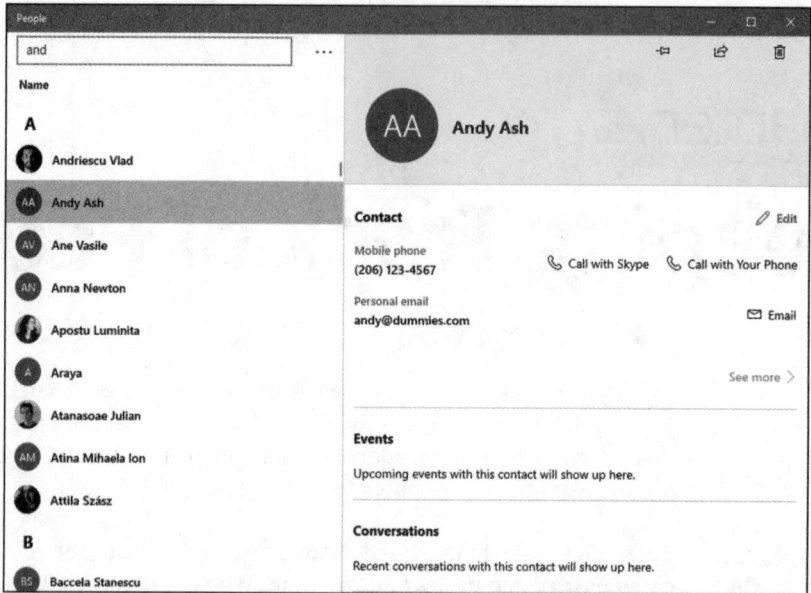

FIGURE 2-1: The Universal People app is a simple contact list.

The Windows 10 People app keeps a list of contacts. If you hook things up right, it'll import contact lists from a variety of sources — the usual email contact lists (Office 365, Exchange, Outlook.com, Gmail.com, iCloud), plus a very few contact list managers available for sale in the Microsoft Store.

As we went to press, Microsoft promised that it was going to build bridges to more apps and sites with contact lists — Twitter, Pinterest, LinkedIn, the various Messenger and chat apps, Facebook, Sina Weibo, and heaven-knows-what-all, but it isn't clear how far Microsoft will get in its quixotic quest.

It would be nice if we could have a Microsoft Contacts Babel Fish, but it's hard to believe the job will ever be done.

Putting Contacts in the People App

If you set up Mail with a Hotmail/Outlook.com, Gmail, or Exchange Server account, all the contacts belonging to that account have already been imported into People. If you set up more than one Hotmail/Outlook.com account, for example, all the contacts in both accounts have been merged and placed in People.

But you aren't even halfway done yet.

Adding accounts to People

REMEMBER

Before you start pulling all your contacts from Hotmail/Outlook.com, Gmail's Contacts, Exchange Server, Office 365, and all the others, realize that there are side effects, not just in establishing Microsoft-controlled links with outside applications, but even inside the core Windows 10 productivity apps, Mail, People, and Calendar.

Before you add an account to People, be aware of the effect that adding that account has in other tiled apps. Here's how connecting the following accounts with People affects other Windows 10 apps:

>> **Google account:** This brings in your Gmail contacts. In addition, it adds your Gmail account to the Windows 10 Mail app.

>> **Hotmail/Outlook.com account:** This brings in your Hotmail/Outlook.com (and Windows Live) contacts and hooks up the email accounts to the Windows 10 Mail app.

>> **Other accounts:** Although you can add other accounts (POP3 and IMAP email accounts) to the Windows 10 People app, as best I can tell doing so does not import anything to People. Rather, it simply adds the connected email account to the Windows 10 Mail app.

Now that you understand the implications, you're ready to add accounts. Here's how to add many/most/all your contacts (you get to choose how many accounts to connect) to the Windows 10 People app:

1. **Bring up the People app from the Start menu by tapping or clicking the People tile, if you have one. Or click Start, scroll down to the Ps, and choose People.**

If you've added only a single email address to Mail, you may see a prompt to add an account. If so, click Add an Account, and skip to Step 3.

2. **From the main People screen (refer to Figure 2-1), click the Settings icon at the top (and shown in the margin).**

You see the Settings pane shown in Figure 2-2.

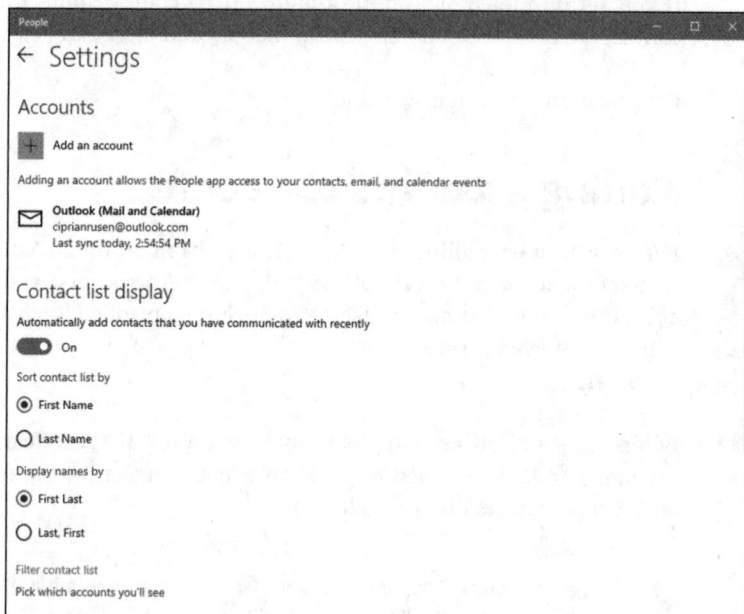

FIGURE 2-2:
Add an account
to your Windows
10 People app.

3. **If you have a contact list with entries that you want to see, pause and think about it a minute.**

TIP

If you have old information in one or more of those accounts, you may want to think carefully about whether including all the contacts in your People list will be more of a pain than it's worth. Modifying existing contacts, People is intensely time-consuming: You must tap or click each contact one by one, review the information about the contact, and modify accordingly. Although Windows 10 People tries to identify duplicate entries — the same people coming from two different sources — and merge the data, it's not good at resolving differences.

4. **If you want to proceed, click or tap Add an Account.**

You see the Add an Account dialog box shown in Figure 2-3. Table 2-1 explains the option you should choose, depending on what kind of email provider you have.

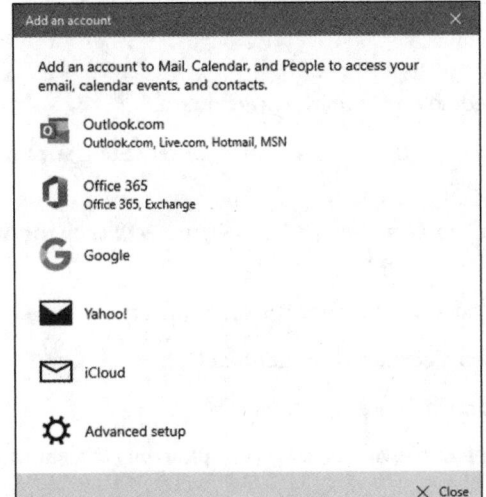

FIGURE 2-3:
The Add an
Account dialog
box looks just like
the analogous
dialog box in the
Mail app.

TABLE 2-1   ## New Mail Account Types

Use This Type	For This Email Service
Outlook.com	If you get mail through Microsoft's servers — your email address looks like something@ outlook.com, @live.com, @hotmail.com, or @msn.com
Office 365	If you get your mail through a company mail server or you use Office 365 to handle your mail
Google	If you have a Google account, most commonly an email address that looks like something@gmail.com, but also if you use Google's servers for email, as you can with Google's G Suite for Business, or if you've just registered your email address with Google and want to retrieve your mail through Google
Yahoo!	Yes, a few people still have @yahoo.com accounts.
iCloud	For those from the Apple side of the street, if you have an @icloud.com or @me.com or @ mac.com address
Advanced setup	Use only if you have an Exchange ActiveSync account, or an IMAP or a POP3 email account

5. **Choose the type of account you have, and follow the directions to add that account's contacts to People.**

You're bound to find many duplicates and lots of mismatched data. Hang in there. There's another trick.

TIP

If you added too many accounts to your Windows 10 People list, there's a way to drop back ten yards and punt — prevent People from showing all the contacts from a specific source — without laboriously deleting individual entries.

Here's how:

1. Bring up the People app from the Start menu.

Tap or click the People tile, if you have one. Or click Start, scroll down to the Ps, and choose People.

2. From the main People screen (refer to Figure 2-1), click the Settings icon, at the bottom (shown in the margin).

The dialog box that was shown in Figure 2-2 appears.

3. At the bottom, click or tap Filter Contact List.

The options shown in Figure 2-4 appear.

4. Select and deselect the boxes, so you display only the contacts that you want to see.

It's easier to scale back duplicates that way — but harder to update older entries.

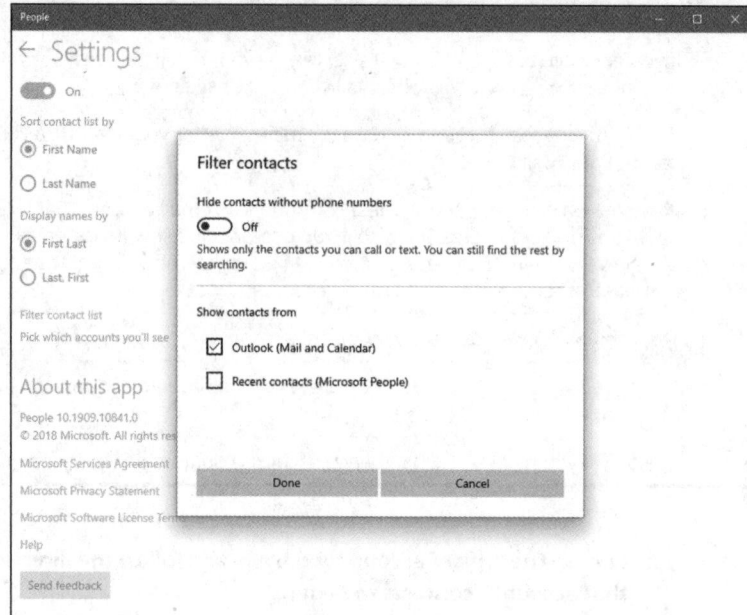

FIGURE 2-4:
Disable all the contacts from a single source.

A little English translation: "Hide contacts without phone numbers — Off" means you want to see all your contacts, whether you have phone numbers for them or not.

SEARCHING FOR PEOPLE

Just to confuse things: Search in People looks only for the beginning of names. If you search for *umm*, you won't find *Dummy*, for example. That's usually not a real big deal, unless you've imported names where both the first and last names have been magically mashed together and stuck in the First Name field.

Editing a contact

If you want to change the information associated with a Windows 10 People person (altogether now: "One eyed, one horned, flying purple people person") — a contact — here's how to do it:

1. **Inside the People app, tap or click a contact's tile.**

The contact details appear, as in Figure 2-5. It's not at all obvious, but you can click the email address and send a message, or click one of the Map links and see the Windows 10 Map app, pointing to the indicated address.

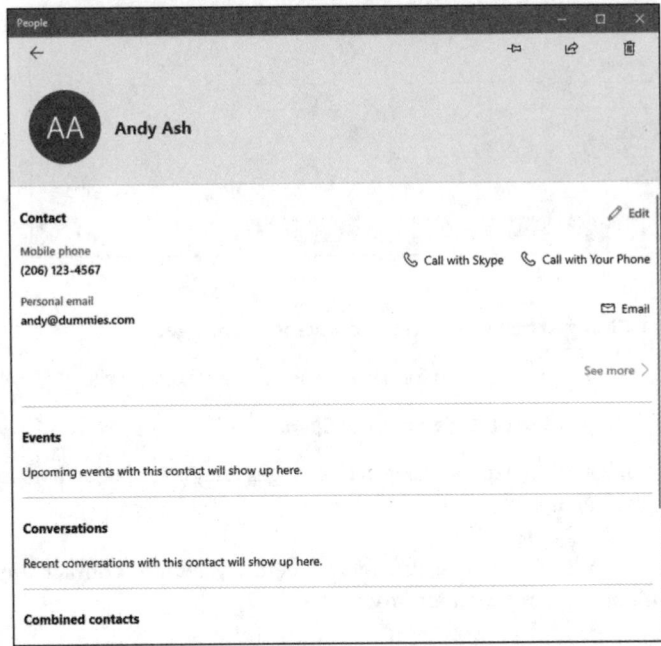

FIGURE 2-5:
The contact info for one of the world's great philanthropists.

2. Click or tap the Edit icon (pencil).

The Edit Outlook Contact pane appears.

3. If you have multiple sources of contacts (say, multiple Outlook accounts, or accounts added in the Windows 10 Mail app), choose which contact you want to edit.

Ultimately, you end up on the editing page, as shown in Figure 2-6.

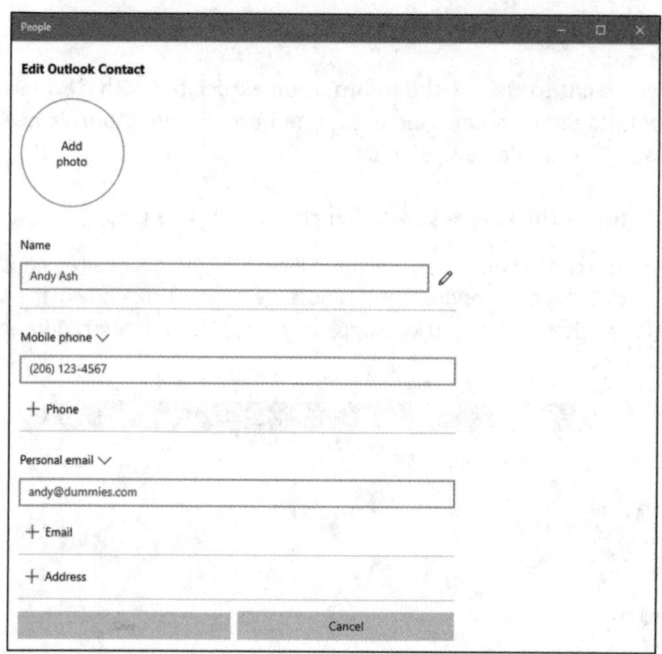

FIGURE 2-6:
Change the
contact's
information here.

4. Change the information you want to change.

See the next section for a list of the different data fields.

5. Very important: Tap or click Save.

WARNING

If you don't explicitly save your changes, they'll disappear, and you won't be warned.

No, it isn't like you're in the 21st century, where contact apps make changes immediately and without prompts.

Adding people in People

Adding a new contact in People isn't difficult, if you can keep in mind one oddity: You add *accounts* via the gear–shaped Settings icon at the top, but to add a *contact*, you use the + (plus sign) icon.

REMEMBER

A People, er, contact doesn't have to be a person. Your local animal shelter is a person, too. Or at least a contact.

Here's how to add a new contact. Keep in mind that People alphabetizes by first name (unless you change the sort order in Settings) or by company name if there is no first or last name.

1. **Start the Windows 10 People app.**

   That puts you on the main screen (refer to Figure 2-1).

2. **Tap or click the + on the left.**

   The new contact dialog box shown in Figure 2-7 appears.

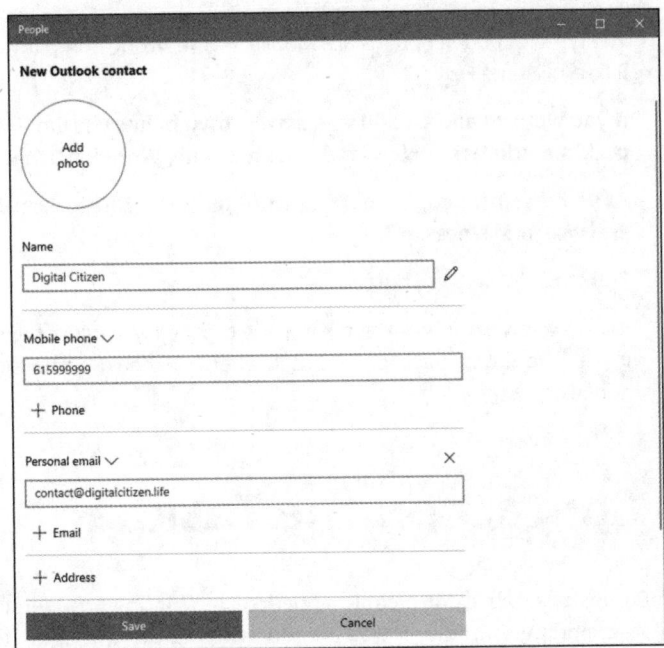

FIGURE 2-7:
Enter your new contact — your new Windows 10 People purpleperson.

3. **Choose the account to which you want to sync this new contact, if you have more accounts in the People app.**

   You can choose from any account identified to the Mail app. When you add a contact to that account, People goes to the account and puts the person in your contact list for that account. Suppose that I add Phineas Farquahrt to my woody@msn.com account. As soon as I'm finished, the People app will log in to my woody@msn.com account and add poor Phineas to my contact list.

4. **Type a first and last name, keeping in mind that People alphabetizes by the first name, by default.**

   For additional name options — phonetic names, middle names, nicknames, title, or suffix — you can tap or click the pencil icon to the right of the Name field.

5. **If you have an email address for the contact, choose what kind of email address and then type the address in the box.**

   The types of email addresses are Personal, Work, and Other.

6. **Similarly, if you have a phone number, choose the type — and type it in the indicated box.**

   The types of phone options are Mobile, Home, Work, Company, Pager, Work Fax, and Home Fax.

7. **If you want to add an address, scroll down below the Email entries, tap or click the Address button, and choose Home, Work, or Other Address.**

8. **As you feel inclined, fill in Other Info, such as Job Title, Significant Other, Website, and Notes.**

9. **Tap or click the Save button.**

   It takes a few seconds —People is going to your mail account and updating it — but you come back to the screen where you see the details of the newly added account.

Putting a Contact on the Taskbar

A new feature called My People appeared in Windows 10 version 1709, the Fall Creators update. Initially, it was pinned by default to the right side of the taskbar. However, in the May 2020 update or newer, My People is hidden. To enable it, right-click anywhere in the empty space of the taskbar and choose Show People on the Taskbar.

My People lets you pin contacts to the taskbar to make it easier to get in touch by email, instant messaging (using Skype or an ever-changing list of other compatible messaging applications), or to just look up details.

Pinning contacts is easy. Here's how:

1. **Make sure the contact you want is in the People app (see the preceding section).**

2. **Click the double-ghost-man icon in the lower right, to the left of the system tray and the time.**

   You see a window like the one in Figure 2-8.

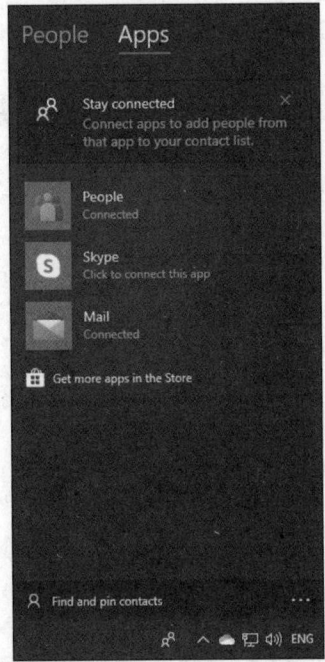

FIGURE 2-8:
Choose a contact to stick on the taskbar.

3. **Click People, and select your contact:**

   - If Windows 10 has guessed accurately and the contact you want is listed, click it.

   - Otherwise, click Find and Pin Contacts, type a search string in the Find and Pin Contacts box at the bottom and click Search. Find your long-lost contact and click it.

   You see a list like the one in Figure 2-9.

4. **If you want to make sure you have the right contact, click the People, or Mail, or Skype icon, or any other icon that belongs to a compatible app.**

In Figure 2-9, I know I have the right contact because I know only one person whose first name is Andy.

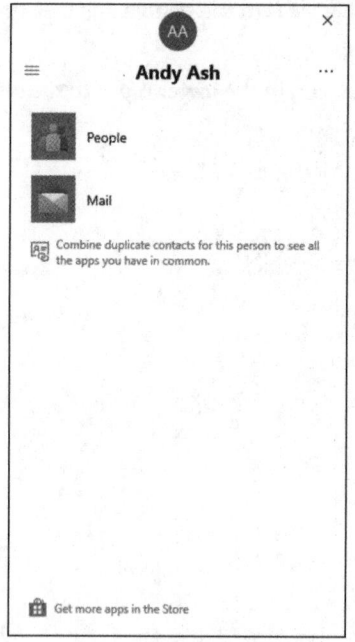

FIGURE 2-9: The chosen contact appears, ready to be pinned.

5. **Here's the odd part. You don't need to do anything more to pin the contact on your taskbar. You can X out of the contact's pane if you like.**

Even if you don't do anything, the new contact appears on the taskbar.

Alternatives to the Win10 People App

If the Windows 10 People app were the only contact app at your disposal, you'd be sitting smack dab in the dumb phone era of the late 1990s. Even Outlook 98 had considerably more sophisticated contact handling than Windows 10 People.

Fortunately, while Microsoft's been playing at contacts, the rest of the world has zoomed right ahead. When you choose a contact app, your top consideration should be whether it runs on all your computers: desktop, laptop, tablet, and smartphone. Windows 10 People doesn't even rate a meh on that scale.

If you're looking for a contact app and you aren't forced into People, try one of these free alternatives:

>> **Sync.me:** Android, iOS, or online, works with Google+, Facebook, and LinkedIn contacts. Features are caller ID (a godsend if you get lots of spammy calls), social syncing, spam protection, world phone book, and reminders. Find it at https://sync.me/.

>> **Contacts +:** Android, iOS, or online, syncs with Facebook, Google+, Twitter, and LinkedIn. It's very visual. Find it at www.contactspls.com/.

>> **Google Contacts:** I use this one on my desktop, laptops, tablets, and smartphones. It works like a champ and ties in to Gmail, which I also use. Find it at https://contacts.google.com.

A whole big world of contact apps is out there. Don't get stuck on one just because it ships with Windows 10.

Chapter **3**

Zooming the Photos App and Beyond

Windows 10's Photos app is meant to be a pleasing, easy way to look at your picture collection, coupled with some easy-to-use photo-editing capabilities. If your expectations go a little bit outside that box, you're going to be disappointed.

In this chapter, you find an introduction to what Photos can and can't do. A quick tour shows you how to navigate around the Photos app. Then I explain how to edit with the simple but surprisingly powerful Photo tools and how to import images from your camera (or smartphone) with Photos. And I show you how to organize pics in your very own Albums.

Finally, if the Photos app doesn't do what you want — and unless your needs are modest, it won't — I talk about the many photo storage and management apps available on the Internet. You may be surprised how much photo moxie is available, free, in the cloud.

Spoiler: With unlimited free (or at least zero-cost) storage online and an amazingly versatile indexing capability, Google Photos deserves your consideration. You may not like the fact that Google will learn where you've been, what you've eaten, and who your friends might be, but the facilities offered are stellar. And free. I talk about the Google Photos app at the end of this chapter.

Discovering What the Windows 10 Photos App Can Do

Photos has a simple layout for viewing your photos. Here's what you get:

>> A central place to view photos from your Windows 10 computer or from your OneDrive account.

>> Help searching for a photo.

>> A way to see your photos organized according to date, by manually built collections called *albums* or by the source folder.

The next section, "Touring Photos," explains the photo sources and how to search or change the display of your photos. The section "Editing Photos" shows how to apply Windows 10 Photos' built-in editing tools to the photo of your choice. The section "Adding Photos" later in this chapter explains how to connect Photos with the web. The section "Working with Albums" shows you how to physically set up albums, and what to do with them. And the last section takes you beyond Windows 10, to explore the amazing tools, available for free online, that will help you store, edit, and distribute your photos, whether you just pass them around the family or crow about them around the world.

The new Windows 10 Photos app is reasonably capable, but it suffers from the same clunky navigation problems that plague all of the Windows side of Windows 10: Try to copy a handful of photos from one folder to another, for example, and your finger could fall off.

Touring Photos

To take a walk around the Photos app:

1. From the Start menu, tap or click the Photos tile.

The main screen of the Photos app appears, displaying the Pictures collection shown in Figure 3-1.

TIP

The collection is a simple, reverse chronological view of all the pictures (and videos) in your computer's Pictures folder, combined with all your pictures and videos in the OneDrive Pictures folder. You can add additional folders, one by one, from the Settings pane (see later in this chapter). If you have a Pictures library (see Book 7, Chapter 3), Windows 10 Photos isn't smart enough to look inside.

TIP

Common point of confusion: Note that pictures outside your Pictures folder aren't included.

By default, Windows 10 Photos automatically enhances the pictures and removes duplicates.

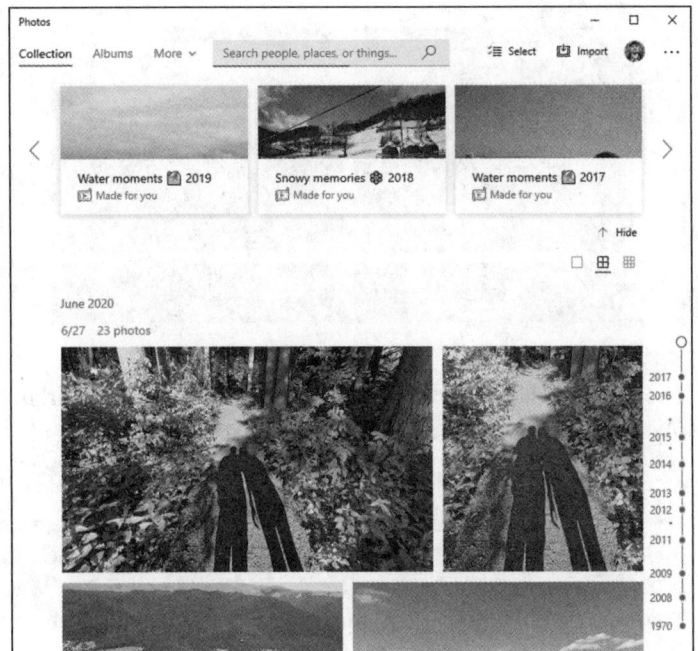

FIGURE 3-1:
The Photos app can bring in files from the Pictures folder on your computer and from OneDrive.

2. **To search for a specific photo, scroll down to find the date the photo was taken, as shown in Figure 3-2.**

You can also search for people, places or things by using the search box at the top.

FIGURE 3-2:
Scroll down to the photo's date.

Photos doesn't have many options, but Microsoft is working on them. While Microsoft's busy trying to write a Windows 10 Photos app, hop over to `http://photos.google.com` and see how a Photos app should work.

TECHNICAL STUFF

The Photos app can display an enormous variety of picture and video formats, including AVI, BMP, GIF (including animated GIFs), JPG, MOV, MP4, MPEG, MPG, PCX, PNG, many kinds of RAW (high-quality photos), TIF, WMF, and WMV files. That covers most picture and movie formats you're likely to encounter.

SEEING VIDEOS AND NETWORK-ATTACHED FOLDERS

In Photos, you see some videos and photos, but not others. The reason has to do with the nuances of how Photos works behind the scenes to show you images. The following points may clear up a few mysteries for you:

- **Photos shows all the picture or video files in your Pictures folder, or in OneDrive.** Although Photos does show videos, the videos need to be in your Pictures folder (not your Videos library) on your computer in order to appear in Photos. Your video files in the Video folder don't appear in Photos at all — odd, but true. Contrariwise, Windows 10 Photos picks up all the videos in OneDrive, in any folder. This is a good place to note that the Windows 10 Video app — called Movies & TV — isn't anything at all like the Photos app. Movies & TV shows a tiny slice of your videos wedged in between mountains of marketing aimed at getting you to rent or buy movies. For more on the Windows 10 Video app, flip to Chapter 5 in this minibook.

- **But . . . if you have a network-attached folder in your Windows 10 Pictures library, Photos won't look at it.** That means you can't put a bunch of photos on a Windows Home Server, a network attached server, or even a different PC in your home network and have the pictures appear in Photos — even if you add the folder to your Pictures library.

Yeah, I know it's ridiculously confusing.

Editing Photos

If you can find a photo you want to edit, in spite of Windows 10 Photos' underwhelming search capabilities, editing it is quite easy — and the tools at hand, while rudimentary, are quite powerful. Here's how:

1. **Navigate to the photo in the Photos app, and click or tap it.**

   An app bar appears at the top of the screen, as in Figure 3-3. You can (from the left): add the photo to an album (see the next section), magnify it, delete it, add it to favorites, rotate it, crop it, search for it online, edit it, share it (if you have any programs that accept shared photos), or print it. If you click the ellipsis in the upper-right corner, you'll see options to start a slideshow, save the photo with a different file name, resize it, copy the photo, open the photo with a different program; set the photo as your lock screen background, tile the photo, and several more.

FIGURE 3-3:
Select a photo
to start working
on it.

2. **Click the Edit & Create (two pencils) icon, and choose Edit.**

Draw, the other option, gives you a typical free-form drawing layer. You have many options: add filters, create adjustments, straighten the photo, rotate it, flip it, and change its aspect ratio, as shown in Figure 3-4.

3. **Click Filters.**

You have a wide variety of options for applying standard (and sometimes fanciful!) enhancement techniques to the picture.

4. **Click Adjustments, to the right of Filters.**

The editing functions shown in Figure 3-5 appear:

- *Light:* Adjust a combination of contrast, exposure, highlights, and shadows.

- *Color:* Adjust saturation, from 0 for grayscale to 100 for ultra-saturated.

- *Clarity:* Increase the outline on automatically chosen edges.

- *Vignette:* Move the slider left to add white to the outer edges or right to add black.

- *Red eye:* Fix problems with red eyes when editing pictures of people.

- *Spot fix:* Click or tap on a spot or blemish to fix it.

- *Undo:* Use the Undo icon liberally.

FIGURE 3-4:
You can select
filters from a
predefined list.

FIGURE 3-5:
Many traditional
photo-adjusting
tools are a swipe
away.

5. **Try the red-eye function by clicking Red Eye and then clicking a specific eye or eyes with the blue dot that appears (see Figure 3-6).**

   The Photos app does a good job of automatically adjusting for red eye.

6. **To blur a specific part of the picture, click the Spot Fix button, and then click the part you want to blur a bit.**

7. **To save your changes, click Save a Copy. To quit without saving, click the X icon in the upper right.**

   You drop out of editing mode.

When you're ready to go back to the collection, click or tap at the top and then click or tap the back arrow.

FIGURE 3-6:
Accurate red-eye correction is just a few clicks away.

Setting Settings

The Windows 10 Photos app has a small group of settings that you may (or may not) find useful.

To see them, return to the Photos app (refer to Figure 3-1), tap or click the ellipsis icon in the upper right, and then choose Settings. You see something similar to Figure 3-7.

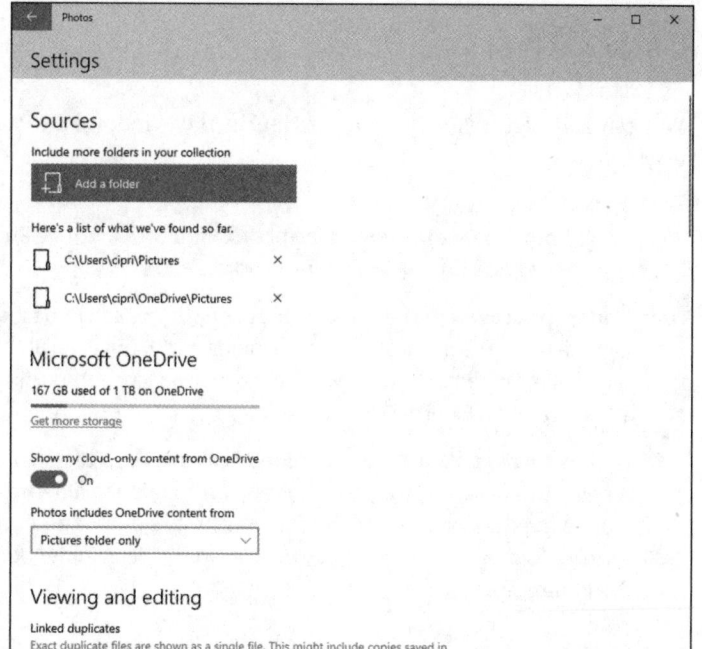

FIGURE 3-7:
These basic
settings may
prove useful.

Don't be too surprised if your Setting panel looks different from this one. Micro-soft changes it all the time.

Not all settings on offer are obvious. Here's the backstory on the ones that aren't so obvious:

>> **Sources:** Click the Add a Folder link to add individual folders to the Windows 10 Photos app's search list. Unfortunately, you can't use the Photos library, but you can pull in folders from anywhere File Explorer can reach.

>> **Microsoft OneDrive:** By default, when the Photos app scans OneDrive, it goes into only your OneDrive Pictures folder, looking for file types that are pictures or videos. Note, in particular, that any kind of graphic file — not only photos and videos — gets picked up. You can tell Photos to pull pictures from All Folders in OneDrive. You can also see a helpful ad for Office 365, which includes additional OneDrive storage space for a price.

>> **Linked Duplicates:** This option is under Viewing and Editing. Slide its switch On to allow Windows 10 Photos to scan your photos and eliminate duplicates, showing you only one copy of each file.

Adding Photos

You can add pictures to your collection in the Windows 10 Photos app in three ways:

>> **Add photos to OneDrive:** Putting photos into the OneDrive Photos folder is a simple drag and drop. File Explorer works great.

>> **Import photos with the Photos Import app:** You can import pictures from a camera or any removable device, including a USB drive, an SD card, or even a big honking external hard drive. See the next section, "Importing Pictures from a Camera or an External Drive," for details.

ASK
WOODY.COM

>> **Add pictures to your Pictures folder:** I call this the old-fashioned way, and it's how I add pictures to the Photos app (in addition to OneDrive). Simply flip over to the desktop and use File Explorer to stick photos and videos in your Pictures folder. Remember that videos in your Videos folder don't show up in the Photos app.

Importing Pictures from a Camera or an External Drive

It's easy to import pictures from a camera or any kind of external data source, including a USB-attached hard drive, an SD card, or a USB-shaped peanut butter stick (if it has photos anyway). Just attach the camera, or plug in the data card or external drive, and wait for File Explorer to recognize it. In File Explorer, click This PC (on the left), and look for the camera or drive. Right-click the camera or drive, choose Import Pictures and Videos, and follow the instructions.

You can even tag your photos as you import them.

If you want to import pictures into your Windows 10 PC from your iPhone, iPad, Android smartphone or tablet, or Windows 10 tablet, you're looking in the wrong place.

Go back to the main Photos app window (Figure 3-1). Then click Import at the top (see Figure 3-8) and choose whether you want to import from a folder or from a USB or other drive, a phone, or a camera.

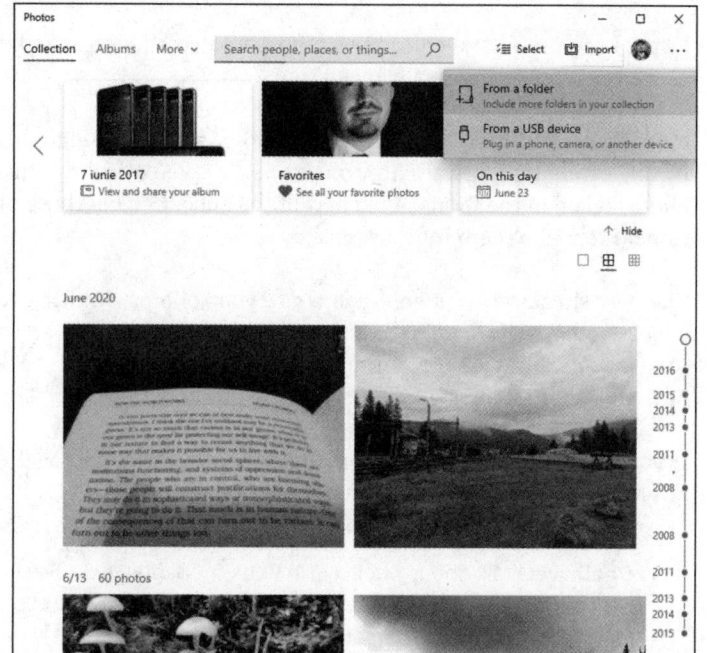

FIGURE 3-8:
To add more folders to the Photos app's trove or import pictures from a mobile device or a USB drive, crank up Import.

Working with Albums

Once you have photos visible in the collection view, Windows 10 works hard at sorting the photos into albums. As of this writing, there's nothing you can do to speed up the process, and it can take many hours for even a small photo collection. You can also create your own album in the Photos app by clicking Albums and then New Album. Then select the pictures you want to include and click or tap Create. Don't forget to give the album a name when you're finished.

After the photos are sorted into albums, there are no tools for rearranging the albums. Look for improvements in the Photos app in the near future (if they aren't there already).

Storing and Managing Photos Online

Hundreds — hundreds — of websites and apps, on all platforms, help you pull your photos or videos from your camera (phone, tablet, phablet, laptop, massive external hard drive, whatever), stick your photos or videos somewhere else (cloud, Windows machine, Mac, network server), automatically sort and categorize them, label or tag them, and edit for common problems (such as red eye or trimming), as well as offer far more advanced traits.

More than a trillion — yes, with a *t* — photos are uploaded to the cloud every year. By one estimate, there are now more photos taken every two minutes than all the photos taken in the 1800s. A big part of the push: Getting those old photos off your camera to make room for new ones.

The best sites and apps help you share your photos, limiting the distribution to people you specify or opening them to everyone. They let you edit with easy-to-use tools. They help you find those old photos that are stuck in weird places. And the very best sites do it all for free or for very little.

If you're looking for a place to put and manage your photos, here are some (but not necessarily the best!) options:

>> **OneDrive:** I talk about OneDrive in Book 6, Chapter 1. Suffice it to say that you can get 5GB of storage free, and 1TB of data for less than $100 a year — with Office thrown into the bargain. You can share that data in many ways. OneDrive has web-based tools for organizing the data, and there are rudimentary tools (outside of Windows) for tagging and searching. At some point, Windows 10 should give us some tools to handle the rest of the typical processes, but as of this writing, Windows 10 isn't helping much.

>> **Flickr** (www.flickr.com): This has long been the photo site of choice for professionals, amateurs, and the completely clueless (see Figure 3-9). Flickr seemed to be in rapid decline, but in 2013, it started fighting back. Under the Yahoo! umbrella now, the phone and tablet apps aren't as capable as its competitors, but it still has excellent editing capabilities.

>> **iCloud** (www.icloud.com): Unlike Windows, the Apple ecosystem has amazing editing and photo-management software. Unfortunately, as I write this, storage in iCloud is relatively expensive and the hooks from iCloud into Windows aren't great. If you use Apple machines, iCloud is an easy (if expensive) choice. If you live and breathe Windows, not so much. Confounding the situation, Apple has two different photo services: iCloud Photos lets you store your last 1,000 photos for free. iCloud Photo library syncs your iPhone and iPad photos to the new Photos app for Mac.

>> **Dropbox** (www.dropbox.com): Dropbox has very good backup capability, called Carousel, with apps that pull photos and videos off your camera (phone, or whatever) and stick the files in Dropbox. Unfortunately, Dropbox doesn't have any photo-editing tools.

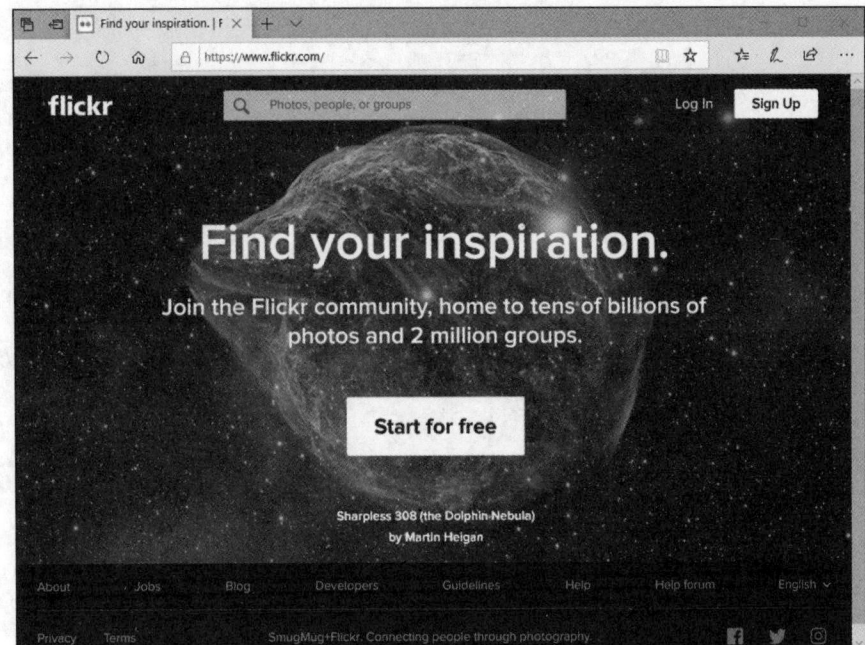

FIGURE 3-9:
Flickr has long been the favorite of photographers, professional and amateur alike.

>> **Amazon** (`www.amazon.com/clouddrive`): Amazon has a photo service that was introduced in late 2014. If you subscribe to Amazon Prime ($99 per year), you get unlimited free photo storage. The service is rudimentary, but if you already belong to Prime, the price is sure hard to beat.

>> **Shutterfly ThisLife** (`www.thislife.com`): As shown in Figure 3-10, Shutterfly grabs photos from everywhere. It can pull photos off your camera and Windows, Mac, iOS, and Android devices, and it can grab your photos on Facebook, Instagram, Twitter, Flickr, Picasa, Tumblr, and SmugMug — all of which are favorite parking places for photos. ThisLife is free for photos, and videos are not expensive.

You should also look at the picture storage and sharing capabilities of Facebook (`www.facebook.com`), which I discuss in Book 6, Chapter 2, and SmugMug (`www.smugmug.com`), which charges $48 per year but gives you unlimited storage. SmugMug (which has absorbed the old site PictureLife) is a good place to go if you're going to want to sell your pictures or turn your pics into T-shirts.

ASK WOODY.COM

Finally — most importantly — Google's Photos (`www.photos.google.com`) app runs rings around anything else ever offered to casual photographers. Keep these caveats in mind: Google scans your photos and uses them to target ads in your direction; if you have very high-quality photos, they undergo some reduction in quality; the editing tools aren't great, but they're adequate.

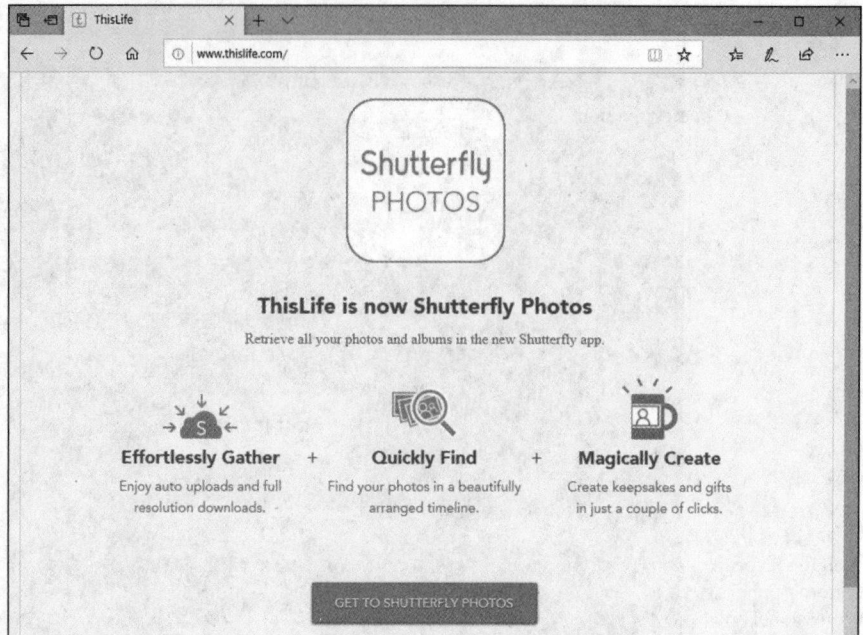

FIGURE 3-10:
ThisLife pulls
pictures from just
about anywhere,
easily, and lets
you control
how they're
distributed.

On the plus side, Google's capability to form albums (see Figure 3-11) — from the kinds of beer you drink to the costumes you wear at Halloween — automatic generation of travelogues and montages, generated animation from groups of stills, capability to group by facial recognition, easy importing from your phone or tablet or camera, and on and on will leave you amazed.

Best of all, it's free. Unlimited storage, unlimited processing, all the time. And it's available in your web browser (Chrome works great), on your iPhone, iPad, Android tablet — just about anywhere except in the Windows 10 app world. Ten thumbs up.

WARNING

Whatever site you use for storage, be aware of the fact that the company running the site has the capability (and likely the permission) to mine your pictures for information that may be used to convince you to click an advertisement. That scanning has grown incredibly sophisticated. For example, in many photos, scanning can identify where you've been and face recognition can tell, with a high degree of accuracy, who's been with you. What you wear, what you eat, how you walk — it's all fair game.

No need to put on a tinfoil hat, but take a look at the description on the Next Web, at `http://thenextweb.com/insider/2016/05/03/photos-next-big-battleground-fight-privacy/`.

FIGURE 3-11:
Google Photos has revolutionized personal photo collecting, storage, sharing, and management. Here's one of many automatically generated albums.

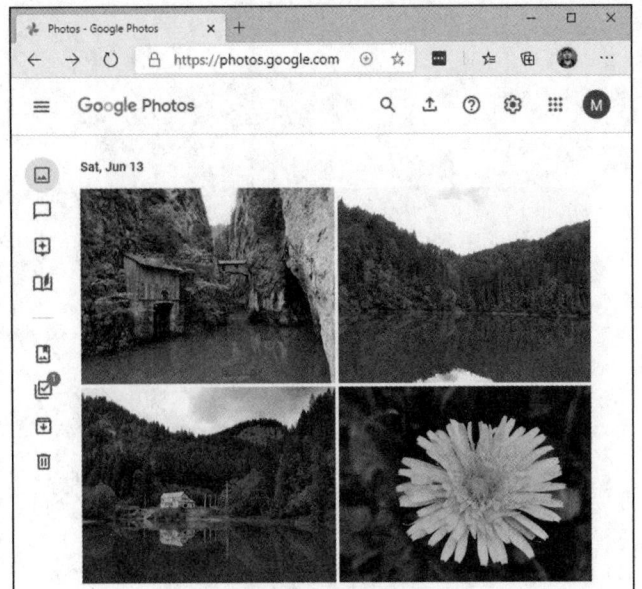

Chapter **4**

Noting OneNote

I f you haven't used OneNote, you've missed out on Microsoft's premier example of a cloud-first, mobile-first application. OneNote started as a piece of Office. It's grown though, so now — particularly in Windows 10 — it's part of Windows itself. It's arguably the most advanced Windows 10 app, although Microsoft Edge fans may beg to differ.

REMEMBER

OneNote isn't Windows-only. Far from it. From early in this decade, it has been available on iPhones, iPads, Android phones and tablets, and other mobile devices. Working with OneDrive (see Book 6, Chapter 1), you can use OneNote to talk to yourself — pass all sorts of things around to your computer(s), your tablet(s), your smartphone(s) — and the OneNote interface makes working with those things surprisingly easy.

To understand OneNote, it helps to understand how it started and grew. It's unique in the Microsoft pantheon.

Believe it or not, OneNote started on the Windows XP Tablet PC, as a program inside Office 2003. All three of the people who actually used XP Tablet PCs with a stylus — a pen — got to struggle with the features, capabilities, and bugs of Microsoft's latest and greatest.

Maybe OneNote's developers thought they had developed a killer app for pens. What they really had was a red herring that took almost a decade to take root. Both the software and hardware to drive it had to stew for a long, long time.

ASK
WOODY.COM

Nowadays, OneNote is a strong product that's valuable for both the touch-and-pen crowd and for those of us who still live in a keyboard-and-mouse world. I don't use OneNote day to day: I'm a longtime Evernote user (`www.EverNote.com`). The features in Evernote don't match up with OneNote, one to one, but if you're not particularly attached to a pen (or even if you are!), you should look at the Evernote alternative.

Getting Started in OneNote with or without a Pen

The nicest part about OneNote is that it's already installed — part and parcel of Windows 10. To get it going, just click Start and then click the OneNote tile. You see a strange welcome like that in Figure 4-1. Click Get Started and OneNote displays your notebook and the default Quick Notes.

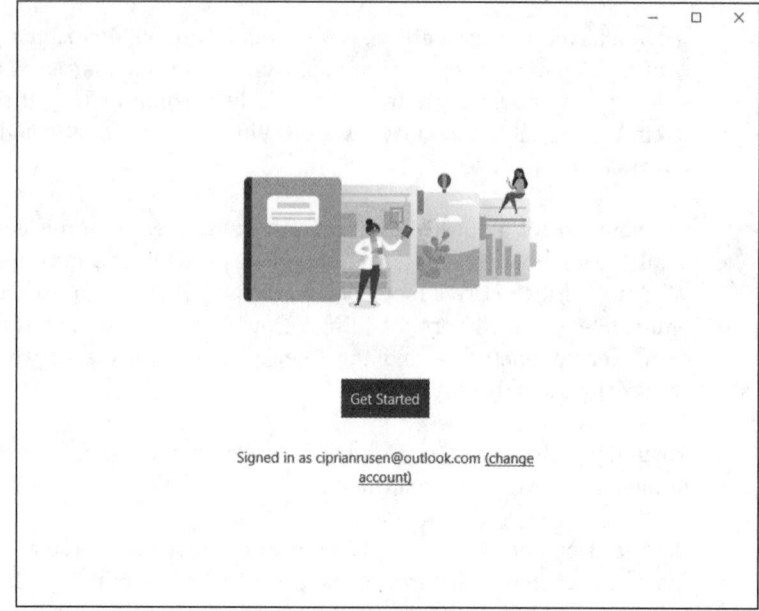

FIGURE 4-1:
If you're already
logged in to
OneDrive,
OneNote is ready
and willing.

Here's a tip for Surface Book and Surface Pro owners who have a sufficiently talented pen and have set up Windows 10 to start by using Hello face recognition (see Book 2, Chapter 4). Once your machine is turned on, you may be able to crank up OneNote by simply clicking the top of the pen. Even if Hello isn't enabled, OneNote is just a click away after you log in to Windows 10. If you're in tablet mode,

OneNote will take up the whole screen. That's a convenient shortcut because it's easy to start Windows 10 on a OneNote page, ready to take notes.

The first time you log into OneNote, sign in with your Microsoft account if asked, and click OK for any additional notifications that may appear.

OneNote works with notebooks, just like Word works with documents, Excel with workbooks, and PowerPoint with presentations. Inside a notebook, there are sections. Within each section, there are pages. And on each page can be . . . many things. Typed notes. Screenshots. Photos. Voice recordings. Marked-up web pages. Tables. Attached files. Web links. Lots and lots of things.

TIP

You can store a notebook just about anywhere. If you store it in some place where others can get to it (OneDrive, or a computer on your home or office network, for example), you can set things up so they can look at and/or modify your notebook.

Try it. I guarantee you'll find OneNote is easy to use.

Setting Up Notebooks, Sections, Pages

Noting OneNote

Here's how to get going with your very own notebook.

1. **Get OneNote fired up by clicking or tapping Start and then the OneNote tile. Or if you have a fancy pen, just click it.**

    If you've used OneNote before, you get a main screen that looks something like Figure 4-2.

2. **Click the downward-pointing arrow, to the right of the Notebook list.**

    You see the Notebook list shown in Figure 4-3.

3. **Click or tap the + Add Notebook link at the bottom of the list of available notebooks.**

    OneNote opens a box, asking you to give your new notebook a name.

4. **Assuming you want to store the new notebook in OneDrive, type a name and press Enter or click Create Notebook.**

    OneNote creates a new notebook and puts a link to it in your OneDrive Documents folder. In Figure 4-4, the new notebook appears as New Notebook (the filename extension is .one, although you can't see it).

REMEMBER

Because of the weird way OneDrive works (see Book 6, Chapter 1), the new notebook may not appear if you look for it in File Explorer's Documents folder. But if you log in to OneDrive (www.onedrive.com), you'll see it.

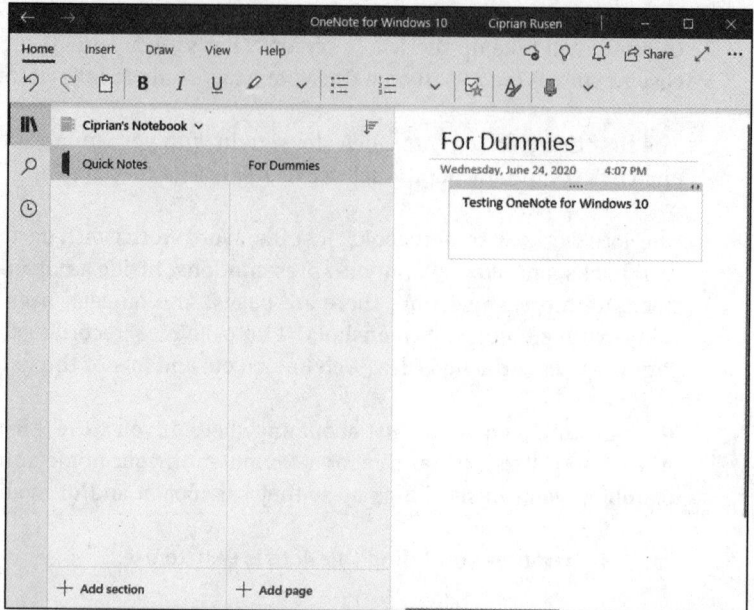

FIGURE 4-2:
OneNote is ready
to get started.

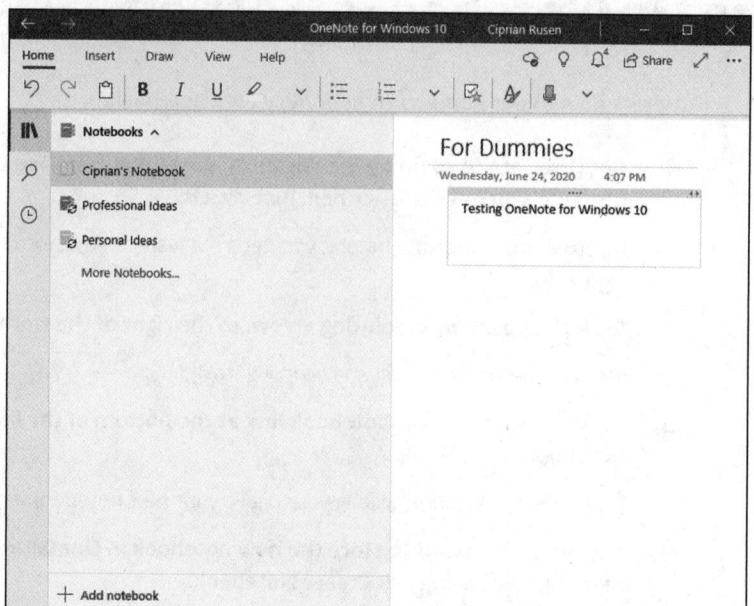

FIGURE 4-3:
You can add a
new notebook
through
the + Add
Notebook link.

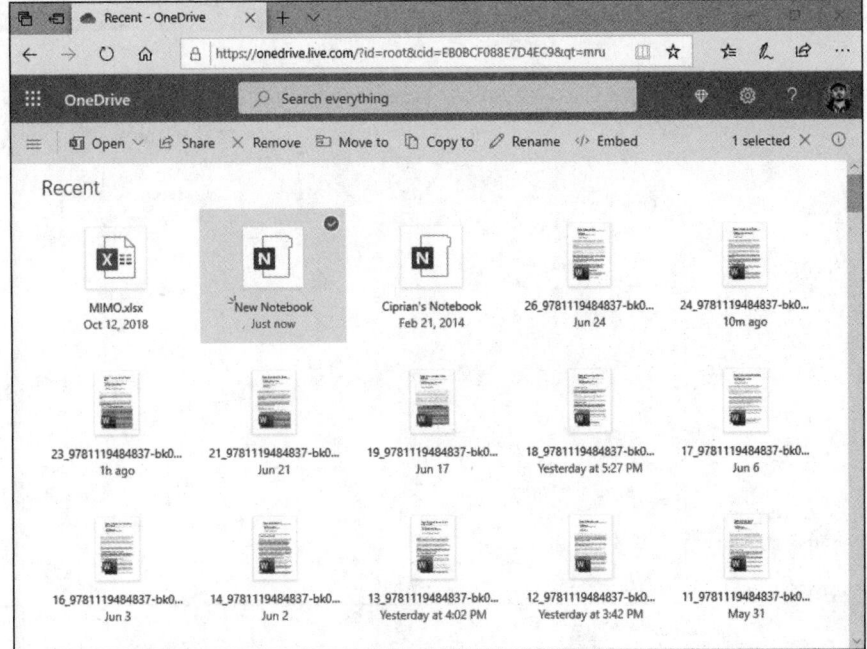

FIGURE 4-4:
The new
notebook really
does get saved to
your OneDrive.

Now that you have a new notebook, let's add a couple of sections. Like adding tabs in a web browser, adding new sections is flat-out simple:

1. **Return to the default notebook.**

   To do so, click the arrow next to the new notebook and choose the default notebook, which should have your name.

2. **On the first tab (where it says Quick Notes in Figure 4-2), right-click or tap and hold down, and then choose Rename Section.**

   Right-clicking also lets you change the section color and do a lot more, as you can see in Figure 4-5.

3. **Type a new name, change the color if you like, and press Enter.**

   The new name appears on the tab.

4. **To add another section, tap or click the + Add Section link at the bottom of the list of sections and type a name.**

   If you've ever worked with tabs in a browser, you already know all you need.

To add pages to a section, right-click (or tap and hold down) + Add Page, at the bottom of the screen. A new, empty, and untitled page appears, ready for you to fill it, as shown in Figure 4-6.

Noting OneNote

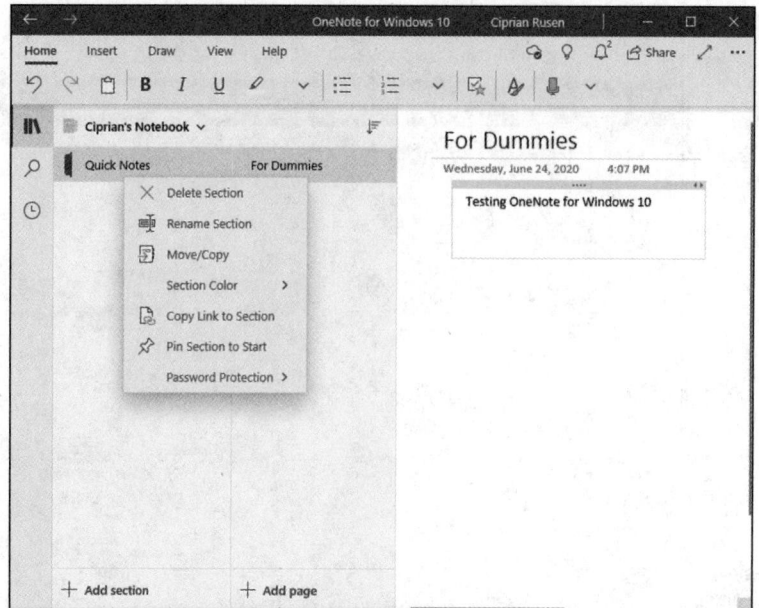

FIGURE 4-5:
Rename a tab —
a section — by
right-clicking.

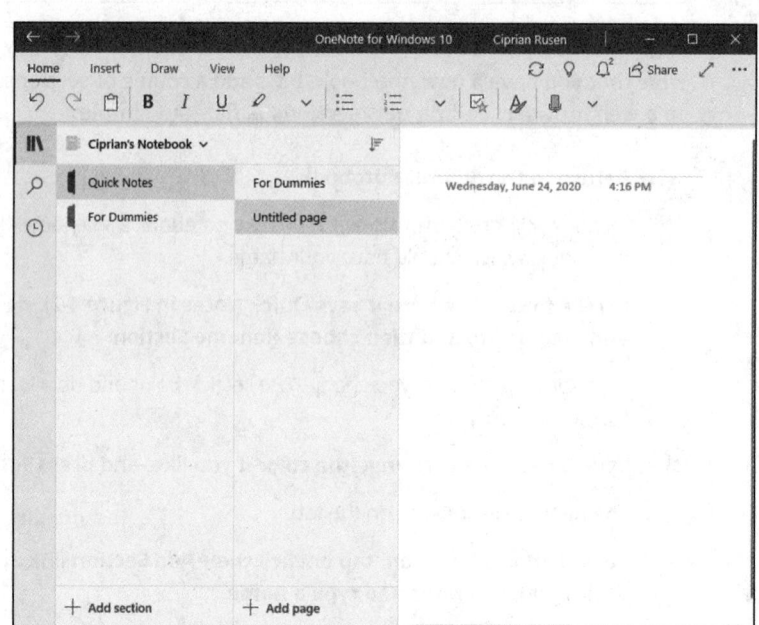

FIGURE 4-6:
Creating a page,
the OneNote way.

OneNote (like all sentient mobile apps) saves everything automatically. You don't have to do a thing.

The typing, formatting, and editing controls at the top work just like you would expect. In Figure 4-7, I typed text into a resizable box by simply typing on the keyboard. Formatting is easy.

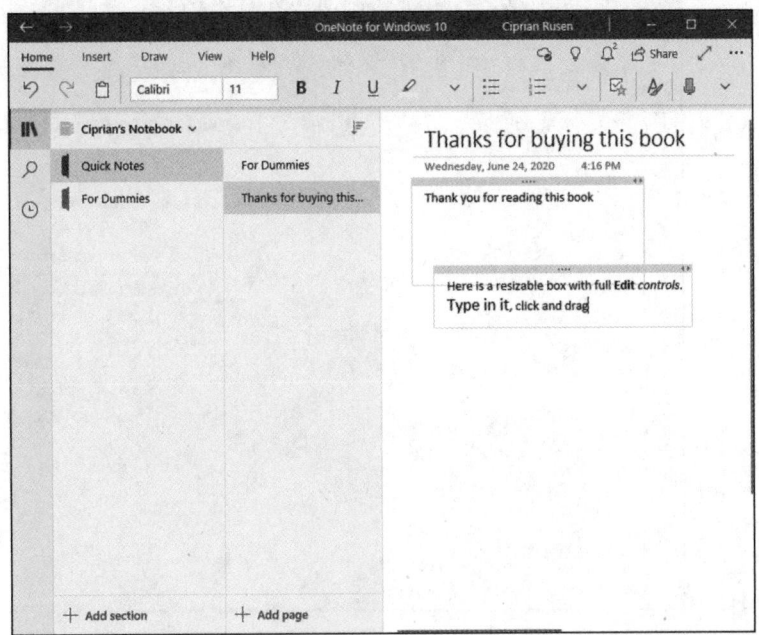

FIGURE 4-7:
Typing in a OneNote page is like falling off a log.

Embellishing on a OneNote Page

You might think that you need a pen in order to draw in OneNote — and, believe me, a good pen helps! — but the fact is that you can doodle with your finger on a touch-sensitive computer, or with a mouse or trackpad if need be. It's just that some pens are sensitive to pressure, so your lines and doodles look much more refined than they do with a mouse.

Microsoft takes a great deal of pride in the way its pen interacts with the Surface Pro machines — click yer Bic (uh, pen), and the computer responds, booting to OneNote in an astonishingly short amount of time. The trick doesn't work with all pens, or all computers, but it's worth a try if you have a pen and a slate: Wait for the computer to go to sleep, and then try clicking any or all of the buttons on your pen.

Here's how to draw on a OneNote page:

1. **Start with whatever page you want to doodle on. Then click the Draw tab at the top.**

   OneNote responds with the tools and palette shown in Figure 4-8.

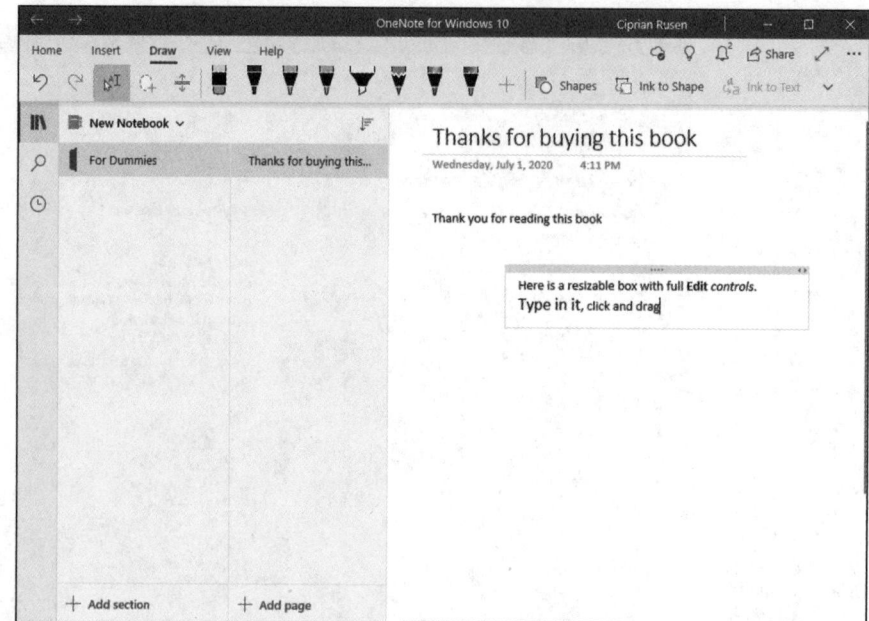

FIGURE 4-8:
Extensive drawing
tools work better
with a pen, but
they'll do okay
with a mouse.

2. **Prepare for drawing:**

   a. *Select a pen — narrow, highlighter, multicolor.*

   b. *Click the down-arrow at the bottom of the pen icon and choose a color.*

   c. *Adjust the thickness of the pen by clicking the plus and minus signs or by choosing a bigger or smaller dot.*

   The cursor turns into a circle.

3. **Draw away.**

   In Figure 4-9, I drew a heckling callout for . . . guess who.

4. **If you don't like what you just drew, press Ctrl+Z.**

   That deletes the drawing you just put on the notebook page and lets you start all over.

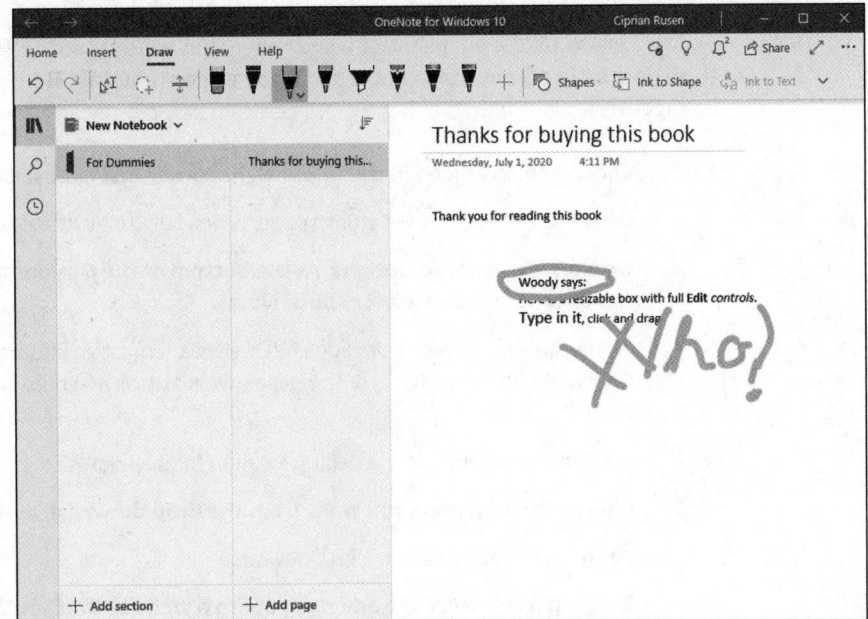

FIGURE 4-9:
Drawing — even
with a mouse —
is easy.

Remember that everything is saved for you automatically; you don't need to do a thing.

**ASK
WOODY.COM**

The icon on the left in Figure 4-8 is a combination select/insert text box control. Use it to select text (say, to apply formatting from the Home tab) or to create a box into which you can type or insert a picture. The behavior is similar to that in Word.

The second icon, which looks like a dotted loop with a +, is a lasso select. Use it to select items to move, copy, or delete as a group. The fourth icon, which looks like an eraser, is an eraser (saints be praised!). If you're an experienced word-processing geek, it's a little difficult to think about the typed text as being just a picture, but you can erase it like a picture. Erase half a letter or right down the middle of a line. Go ahead. OneNote doesn't mind.

The third icon lets you add or delete whitespace — kind of a "move it down" shortcut.

Sending to OneNote

There is one surprising place where OneNote is reasonably well connected: Microsoft Edge.

It's easy to take a snapshot of a web page and send it to OneNote, but you need to do a little prep work to make the transfer go smoothly. Here's how to put it all together:

1. **Click the ellipsis icon in the upper-right corner and choose Settings.**

   OneNote shows you the Settings pane, which I discuss in the next section.

2. **On the right, choose Options. At the bottom of the Options list, click or tap Choose a Notebook for Quick Notes.**

   When you send a page from Edge to OneNote, it must go into a specific notebook. No surprise there. The page will end up as an entire section, with a tab.

   OneNote asks you to pick a default location for new notes.

3. **Choose the notebook you want from the drop-down list, and click OK.**

   You're now ready to share with OneNote.

4. **Bring up Microsoft Edge and navigate to a website that you'd like to save in OneNote.**

   In Figure 4-10, I found a page of interest.

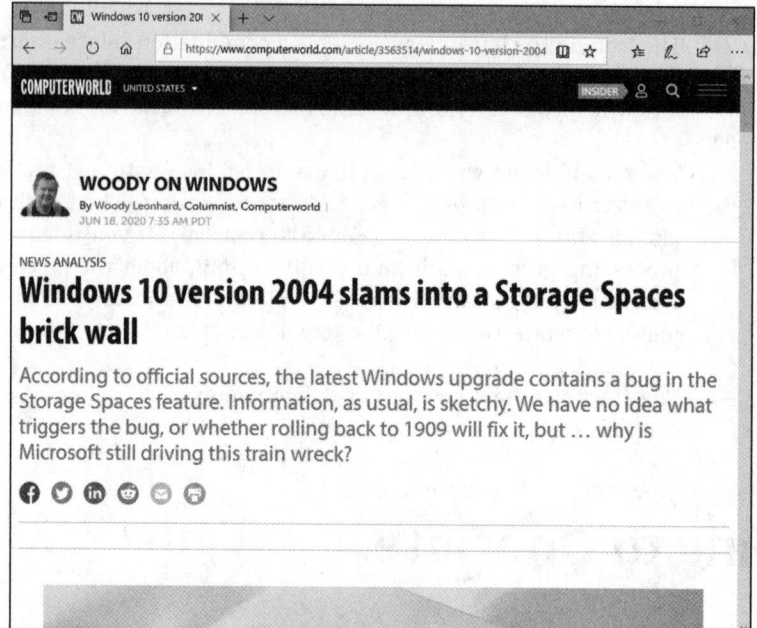

FIGURE 4-10:
Here's the page I want to put in OneNote.

5. **In Microsoft Edge, click or tap the add Notes icon (pen).**

Edge flops over into web note mode, which looks a lot like the web page with OneNote's icons on top.

6. **Make any annotations you like on the web page.**

7. **Click the Save Web Note icon (diskette), choose a recent section, and click Save.**

Save to OneNote is the default. OneNote whirs for a bit and then saves the web page you selected.

8. **Go to OneNote.**

The page is sitting in the Quick Notes tab or section (or whatever section you chose in the Save operation) in the notebook you chose. See Figure 4-11.

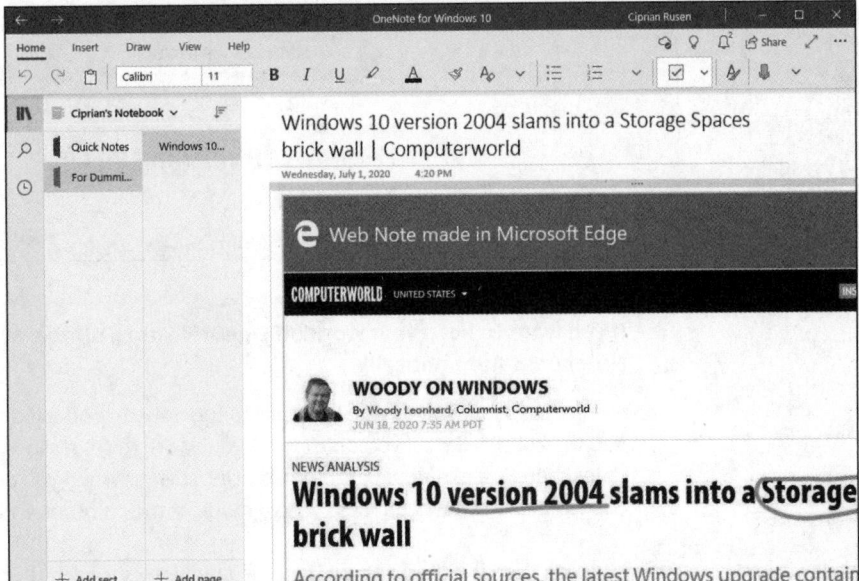

FIGURE 4-11:
OneNote
shares nicely with
Microsoft Edge.

Setting Settings

OneNote has a handful of settings you might want to try some day. Or maybe not. To see them, follow these steps:

1. **Inside OneNote, click the ellipsis icon in the upper-right corner and choose Settings.**

The Settings pane appears on the right.

2. Choose Options.

The other settings aren't very interesting. You end up with the pane shown in Figure 4-12.

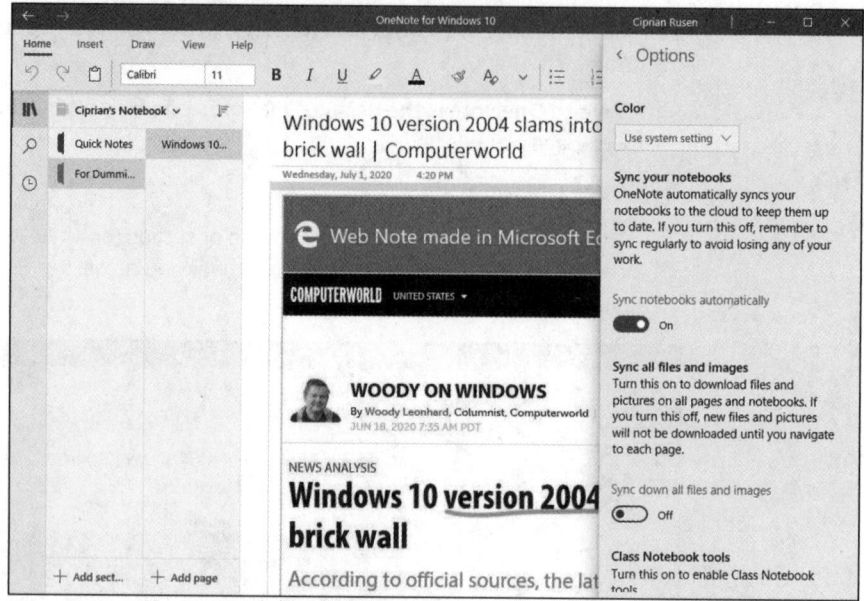

FIGURE 4-12:
A few settings
may prove
worthwhile.

3. If you want to keep your work off OneDrive, turn off the switch for Sync Notebooks Automatically.

It's rare that you would want to return to the not-so-good old days where you had to explicitly Save if you didn't want your work to get trashed. But sometimes there are extenuating circumstances, such as when you don't want your co-workers to see what a mess you've made of the communal notebook.

This chapter just touched the surface of OneNote's capabilities, and you'll find that the app itself has many different guises in many different locations — OneNote online (www.onenote.com) is different from OneNote for the iPad, which is different from OneNote for smartphones, and so on.

Chapter **5**

Maps, Music, Movies — and TV

n this chapter, I cover three Windows 10 apps that are usable, in a pinch, but come nowhere close to other apps that you may have used:

» Windows 10 **Maps,** based on the HERE mapping system (previously owned by Navteq, bought out by Nokia, and now owned by Audi, BMW, and Daimler), provides maps that are competitive in the auto market, perhaps. But the app doesn't hold a candle to the mapping apps from Google and Apple.

» **Groove Music** (yes, it's under *G* in the apps list) has a beautiful interface but almost no brains. A dozen alternative players for Windows (VLC and Microsoft's own Windows Media Player immediately come to mind) and great streaming alternatives (Pandora, Spotify, Deezer) run rings around the Groove.

» The Windows 10 **Media & TV** app may not have a groovy name, but it's in the same dumb-bomb category. Great if you want to buy movies from Microsoft. Need I say more?

With all three apps, Microsoft simply isn't keeping up with the industry at large. Things are moving fast on all three fronts — mapping, audio, and video — and if you're using Windows 10 to tag along, be sure to check your expectations at the door.

Making Maps

Microsoft has had a short and rather tortured history with maps. Microsoft Map-Point emerged from the Expedia Streets and Trips Planner 98, which shipped in Office 97. It was released as a stand-alone product in 2000, and updated many times afterward, finally succumbing to much better mapping products in 2013.

Bing Maps, an outgrowth of MapPoint and MSN Virtual Earth, started in late 2010, and it's still alive. You can see the latest at www.microsoft.com/maps.

ASK
WOODY.COM

The Windows 10 Maps app, on the other hand, draws from one of the seminal sources of map information: HERE, a Nokia brand, which Nokia kept when it sold its much-larger telephone business to Microsoft, then sold to a consortium of auto manufacturers when Nokia had a hard time keeping the doors open. Windows 10 Maps uses the HERE database but superimposes Telenav Scout traffic information.

Remarkably, the Maps app is a for-real Universal Windows app (UWP), which means it runs almost the same way on both Windows 10 and Windows 10 Mobile (the now-dead version of Windows for smartphones). What you see here can be replicated on a smartphone with Windows 10 Mobile (if you can find one), and vice versa.

Basic map functions

If you've ever used Google Maps (I do, every day) or the Apple Map app (my wife does, every day), you already have a basic understanding of the Windows 10 Maps app (which both of us avoid, for reasons that will become obvious).

There are two basic map views:

» **Road** shows a traditional roadmap, at least to a first approximation. See Figure 5-1.

» **Aerial** shows a satellite view of the terrain, augmented by superimposed roads. See Figure 5-2.

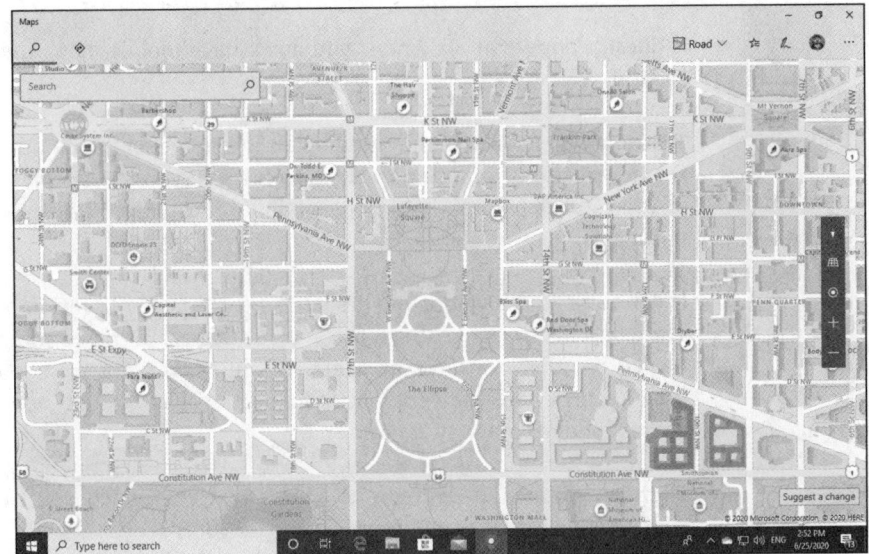

FIGURE 5-1:
The basic road
view calls out the
major landmarks.

FIGURE 5-2:
Aerial view has a
satellite shot with
various notations.

To switch between the two views, click the down arrow at the top right and select the view you want.

TIP

At least in theory, both the road and aerial maps can be superimposed with traffic information, which appears color-coded on the roads. Again, click the icon that looks like a stack of paper and turn Traffic on or off. In my experience, the traffic information is far only occasionally useful. In some cases, for reasons unknown, it doesn't appear at all. And if you don't have mobile data, any traffic data isn't helpful.

Traffic problems are highlighted by an ! (emergency) icon, but in my experience the information connected to these icons is very old — problems cleared up days or even weeks before — and not terribly informative. The times posted are also unreliable.

If you want to see where you are, click the Show My Location button, the bull's-eye icon on the right, above the + and – signs. If you're mobile and have GPS turned on, the location's accurate. If you're working from a computer with a Wi-Fi connection, the best you're going to get is a rough approximation of the nearest phone company router.

There's a rotate-30-degrees-or-so "Tilt" view, which you can enable or disable by clicking the grid icon above the bull's-eye. See Figure 5-3. It's not interesting — places that should have breathtaking elevation differences, as in the area surrounding Homer, Alaska, end up looking like Flatland.

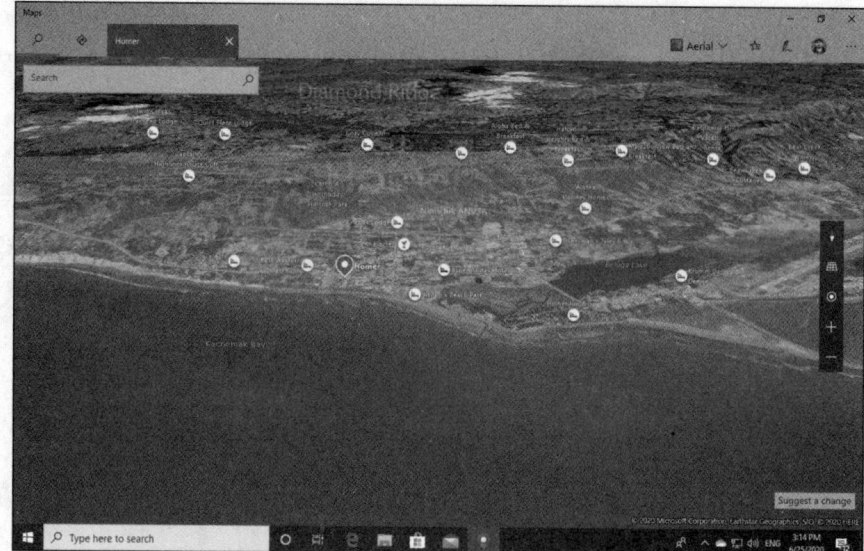

FIGURE 5-3:
Even places with lots of elevation differences look like a 12th-century depiction of a flat earth.

The map has the usual navigation controls: Click and drag to move the map, rotate the mouse button to zoom. With a touchscreen, tap and drag, and pinch or unpinch. For the life of me, though, I couldn't get it to rotate.

Navigating with the Maps app

If you're expecting a Google Maps turn-and-gander experience, you're going to be disappointed.

Type in the search box to bring up a search pane, which includes a list of all the places you've searched for recently, plus some general searches including hotels, coffee (made in Seattle, no doubt), restaurants, shopping, and museums.

If you search for destinations with a qualification (for example, *Coffee near Homer*) using the search box, you get a map with dots on it (see Figure 5-4), which should look familiar to anyone who's used a map search function, a digest of Yelp reviews for the location, plus a link to the review at www.yelp.com.

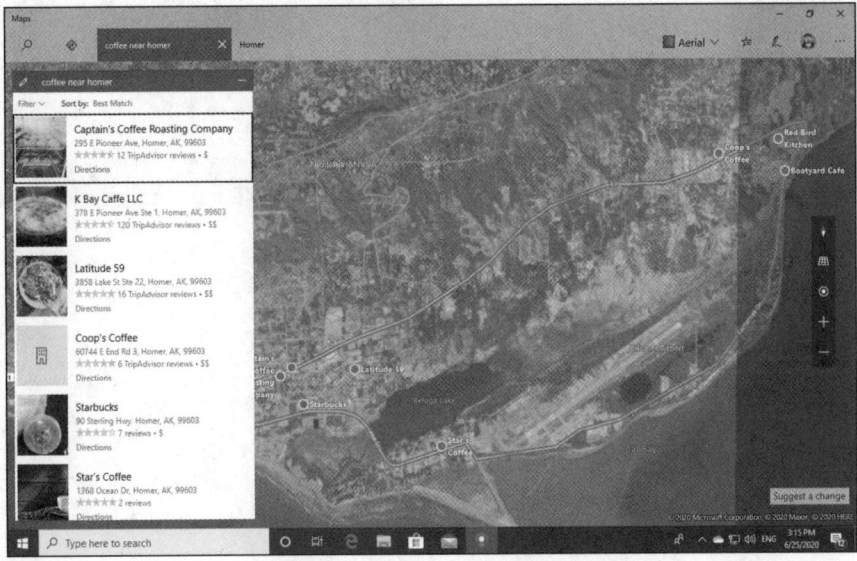

FIGURE 5-4:
Points of interest appear with dots.

Back at the main screen, click the Directions icon (it looks like a right turn sign). You find a reasonably complete direction-navigating feature. Type your From and To locations, and Windows 10 Maps draws you a map with estimated travel times, as shown in Figure 5-5.

From the Directions map, near the estimated time, click the Go button, and you get mapped turn-by-turn instructions, but without any sound. See Figure 5-6.

Maps, Music, Movies — and TV

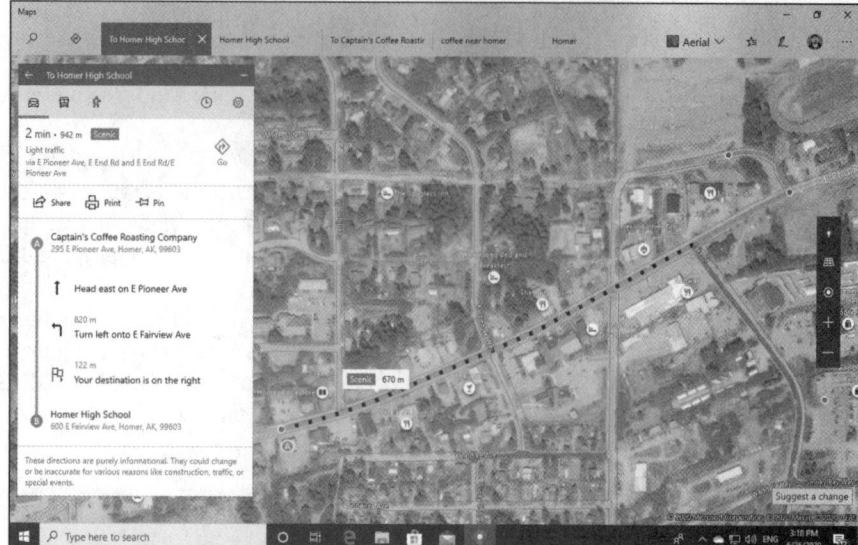

FIGURE 5-5:
Maps provides detailed driving instructions, sometimes with public transport options and walking instructions.

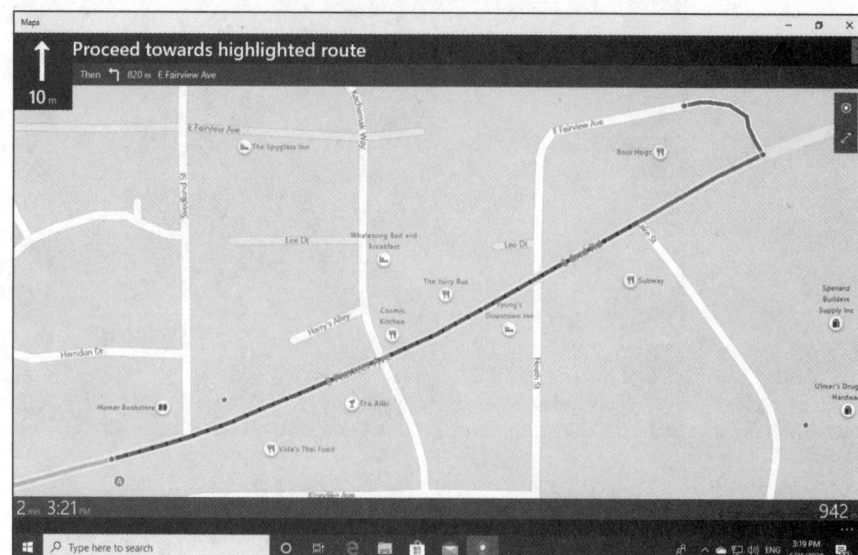

FIGURE 5-6:
Maps offers turn-by-turn instructions, with no voice. Where are you, Cortana?

Taking a map offline

Windows 10 Maps let you download a map and use it even if you aren't connected to the Internet. Here's how to download a map:

1. In the Windows 10 Maps app, click the ellipsis icon, in the upper-right corner, and choose Settings.

2. **At the top, under Offline Maps, click or tap the Choose Maps button.**

You are flipped over to the Settings app, in the Apps/Offline Maps section, as shown in Figure 5-7.

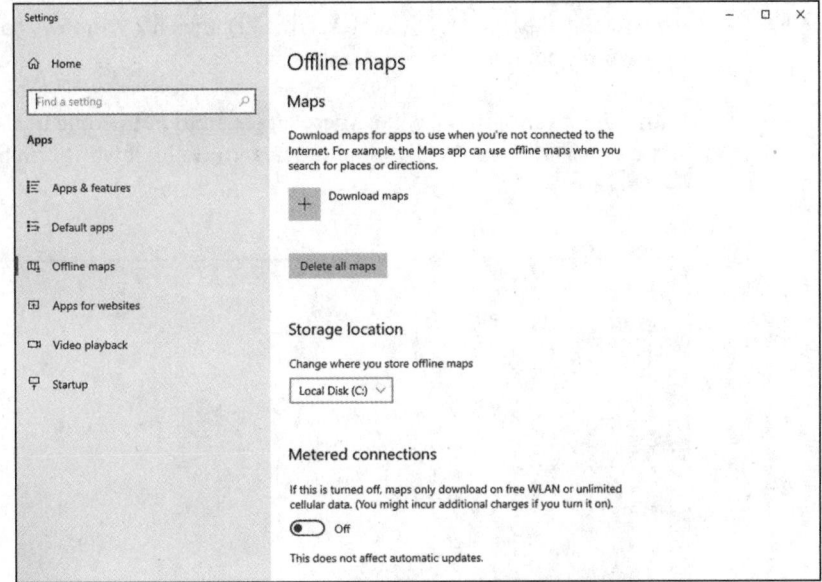

3. **Near the top, click the + icon next to Download Maps.**

4. **Choose a continent, a country, and if necessary, a region or state.**

The downloader has you choose a continent and then a country. If you choose France, Germany, Italy, Russia, China, India, Brazil, Canada, or USA, you are further asked to choose a region or state.

Windows 10 downloads the map and shows you the progress in the Apps/Offline Maps section.

5. **When the download is complete, you can navigate with the stored map, without being connected to the Internet.**

ASK
WOODY.COM

All in all, for day-to-day navigation, the Windows 10 Maps app doesn't even begin to hold a candle to the analogous programs from Google and Apple. You may want to play with it while on your PC, but even then I find the Google website `http://maps.google.com` much easier to use and more thorough.

Maps, Music, Movies —
and TV

Get Your Groove Music On

Microsoft has been trying for years to put together a decent media player — a program that can play songs and videos. For years, we Windows users have had to settle for programs from other companies, notably VLC, ignobly Apple, to get worthwhile players. And I use the terms "iTunes for Windows" and "worthwhile" in the same sentence only under duress.

With Windows 10, it looks like Microsoft tried to put people in charge who understand music and who understand movies. Initially, it was like a breath of fresh air (see Figure 5-8).

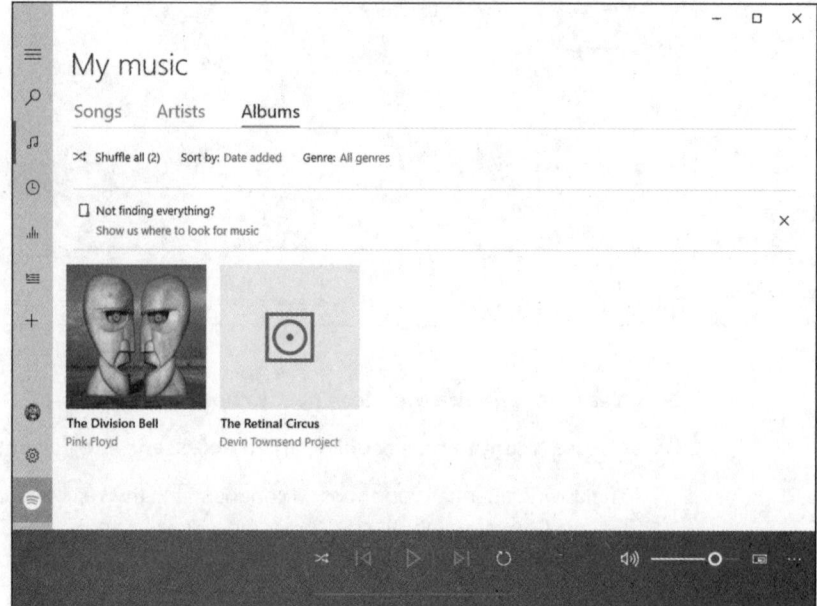

FIGURE 5-8:
Groove Music is like a breath of fresh air.

ASK
WOODY.COM

There was a powerful incentive to do so. Apple has made billions from iTunes. Microsoft wanted a piece of that action and tried to sell both songs and flicks. Unfortunately for them, it didn't work as planned. At the end of 2017, Microsoft has stopped selling music and Groove Music Passes. Since then, Microsoft has partnered with Spotify and are actively promoting Spotify even inside the Groove Music app.

In Windows 8 and 8.1, the Xbox Music app (as it was then called) didn't really want to play your music. It wanted, desperately, to sell you music. With the supremacy of streaming music services — notably Spotify but also Pandora (see the nearby sidebar) — Microsoft tried to sell you a subscription. That's where the future lies for music.

SPOTIFY, PANDORA, AND MICROSOFT

Pandora started the ball rolling, with a subscription-based streaming music service first available in 2000, and then completely redesigned in 2004. If you've never used it, think of Pandora as a smart radio station, able to respond to your likes and dislikes, dishing up songs that match your preferences. Its "discoverability" sets the standard. Pay for your subscription, and the song quality goes up and ads go away. By late 2014, Pandora had grown to 250 million registered users, 80 million active every month, with a profit approaching $1 billion.

Spotify appeared later, in 2008, and it had a different approach: Let people choose the songs they want to hear and make those songs easily available; you listen only to music you know and like. Pandora has a "mere" 1.5 million songs available; Spotify has 30 million or more. Pandora's social interaction offerings are meager. Spotify has tons of features for sharing music and bringing together friends, including collaborative playlists. Spotify has roughly 100 million paying active subscribers.

Where Pandora is like a smart radio station, Spotify is more like a rental service. But as time marches on, the distinction between the two continues to blur, with Spotify offering discoverability aids and Pandora picking up on sharing. And then there's Tidal (big collection) and Amazon Prime streaming (free with an Amazon Prime subscription). The choices, it seems, change every day.

Apple (Beats), Google (Play Music), and Microsoft all jumped into the fray, with variations on the theme, combining elements of Pandora's radio stations with Spotify's pick-and-choose approach. Apple's latest run into the market, called Apple Music, should give the others a run for the money. (Apple Beats is the radio station; Apple Music is the service.)

In 2017, Apple added 6 million paid subscribers. Spotify added about 5.2 million. Apple and Spotify each had somewhere between 35 and 40 million subscribers. Both make well over two billion dollars a year. Microsoft looked at their rivals' successes, rolled over, and gave up. See the final section of this chapter for the Groove Music Pass epitaph.

In 2015, streaming music (that is, services that you pay for, by the month, to serve up music you request) surpassed revenue for digital downloads. In 2015, streaming was a $2.4 billion business. The three primary ways of paying for music — physical format (primarily CDs), digital downloads, and streaming — generated just about the same amount of revenue. Projections for the future put streaming way out in front.

On the TV and movies side, streaming is rolling over the cable and broadcast TV industries as more and more people in the US and Europe "cut the cord." (That is to say, they give up traditional subscription services and get what they want to watch via the Internet.) Streaming video (from Netflix, Amazon Prime Video, and Hulu) took in $6.7 billion or so in 2016, while DVD purchases decline to $5.7 billion, DVD rentals to $2.8 billion, and the video download market picks up the remainders, maybe $4 billion. Video streaming is increasing by leaps and bounds.

With apologies to Robert Zimmerman, "You don't need a weatherman to know which way the wind blows."

ASK
WOODY.COM

Pandora has a Windows 10 app; you can download it from the Microsoft Store. Apple, with Apple Music, wouldn't get caught dead with a Windows 10 app. Spotify avoided the Universal (UWP) treatment until June 2017, when it released a Windows 10 app to great fanfare.

Getting Your Music and Movies into the Apps

If you want to buy your movies or TV shows from Microsoft, the mechanics are easy: Buy them in the Microsoft Store, and they magically appear in the Movies & TV apps.

If you want to buy your music from Microsoft, you're outta luck. The Microsoft Store hasn't carried music since October 2017.

REMEMBER

But what if you already have music and videos, and you want to be able to play them through the Music app and the Movies & TV app? That's a little more complicated — and it can take hours (if not days), depending on the speed of your Internet connection and the state of Microsoft's servers.

The answer is to stick everything in the OneDrive account that you'll use to play the files. Here's a quick course:

1. **Go to a computer that has the music and videos (or has access to the music and videos) that you want to make accessible to the Music and Movies & TV apps.**

   If you already have your music in the iTunes store, or inside Google or Amazon, you may have to copy the files onto your computer (download them). Each vendor has a different way to download its music — and some vendors' plans won't let you download them at all. Make sure you get DRM-free files (see the nearby sidebar).

2. **On that computer, log in to OneDrive** (www.onedrive.live.com) **and use the Microsoft account that you use on the machine where you want the music available.**

   For example, I have a new Windows 10 PC and I use woody@msn.com to log in to Windows on that PC. To transfer files via OneDrive, I'd find a PC that has the music I want and, using a web browser, log in to OneDrive using the account woody@msn.com.

3. **Drag and drop your music from the computer into the OneDrive account's Music folder.**

WARNING

   You won't be able to drag folders into OneDrive. For reasons I can't begin to fathom, Groove Music insists on organizing things according to the details inside the files, and it doesn't want your steeeenkin' folders. So you must reach into each folder, select all the files (Ctrl+A), and drag the files into OneDrive.

   That can take hours, weeks, years, depending on how much music you have.

4. **While you're at it, create a folder called Videos in the OneDrive account, and drag and drop all your video files — MP4s, AVIs, and the like — from the computer into your OneDrive account.**

   With the media files all in OneDrive, you're ready to start.

Of course, if your only music is music you bought from Microsoft (there must be ten of you out there), you don't need to lift a finger.

WHAT IS DRM?

Music and video come in many different formats — think of them as different methods for converting sight and sound into bits. The formats are all different, and translating a video or song from one format to another can really put a crimp on the quality of the recording. Some of the formats put locks on the data, so you can only play or view the file if the creator gives you permission. That's DRM, digital rights management, the scourge of the entertainment industry. In my opinion, anyway.

Back in the dark ages, if you wanted to record music on a computer, you used the MP3 format. It wasn't (and isn't) the fanciest format on the street; it makes files that are bigger than they need to be, and it doesn't support some truly cool capabilities in newer formats (such as Dolby-style 5.1 or 7.1 channel recording). Despite all its shortcomings, MP3 took off and became the universal language of digital music. If you have a device that plays digital music — whether it's an old PC, an ancient portable audio player (they're called "MP3 players" for a reason), a 200GB iPod, a Galactic Zune, or a beat-up 2003 Chevy — it understands MP3.

In the video arena, AVI and MPG file formats play a similar role: They're long-established (okay, old-fashioned) formats. They were invented before anybody thought much about DRM.

AVI, MP3, and MPG files aren't just DRM-free. They're DRM-impossible: The file format doesn't support any attempts to lock you out of your own music or videos. If you buy an MP3 file, for example, you know from the get-go that it doesn't bear any digital rights restrictions — nobody else has control over your music. There are no hidden restrictions, such as limitations on whether you can burn the song on a CD or whether you can play the song on a specific Windows PC.

Apple started out selling DRM-encumbered files in AAC format. But in late 2008, Amazon announced that all its music would be DRM-free. Apple wised up and in early 2009 took DRM off all its new offerings.

DRM-locked music is disappearing. Consumers wised up. Companies that used to peddle locked-up music now sing the praises of DRM-free, with all the fervor of a saved sinner caught with his hand in the till. Yeah, that includes Microsoft, which — for a brief period — sold DRM-enabled WMA audio files.

With a little luck, DRM in the audio world will go the way of the dodo, although you may be stuck with DRM-laden dreck that you got suckered into paying for months or years ago.

Running around the Groove Music App

With your music sitting inside your computer's Music folder, or your OneDrive's Music folder, you're ready to crank it up to 11. Here's how.

1. **Tap or click Start ⇨ Groove Music.**

Groove may hastily assemble your tunes or albums, but sooner or later you see the Groove Music window (refer to Figure 5-8). As you can see, Music takes a brave stab at finding and organizing your music, but it doesn't always get the details right. Note that you're in albums mode, as shown by the third item up in the menu.

2. **If Groove Music didn't find all the music on your machine:**

a. *Click or tap the Show Us Where to Look for Music link.* Groove Music opens a hokey touch-centric file picker window, as shown in Figure 5-9. *Hint:* If you have music in the Public Music folder, look for c:\Users\Public\Music. No, the Groove Music app isn't smart enough to recognize your Music library or the Public Music folder.

b. *When you've added all the folders Groove missed, click Done.*

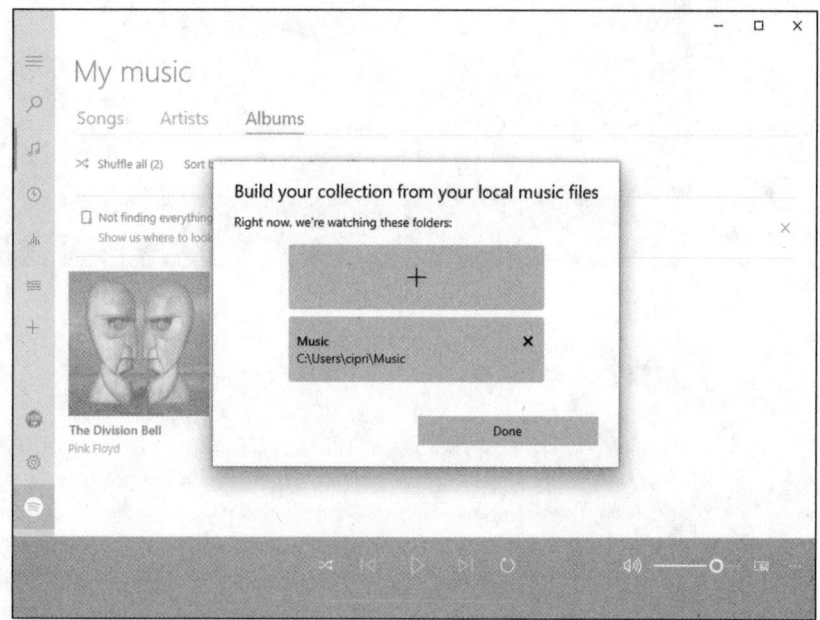

FIGURE 5-9: The clunky interface for adding more folders to your music collection.

Maps, Music, Movies — and TV

3. **Double-click one of the artificially assembled albums to play all the songs in the album.**

You hear the music and see the playlist, as in Figure 5-10.

4. **Use the playback controls — play, pause, change volume, repeat, fast forward, and so on, at the bottom of the screen — exactly as you would expect.**

After you have the music in the machine, and the app knocked upside the head so it can find the music, the rest is easy.

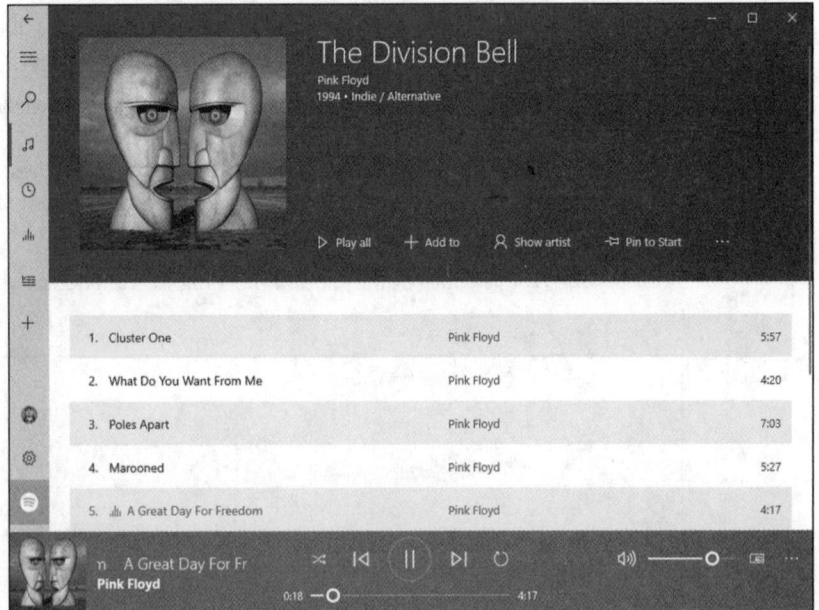

FIGURE 5-10: The playlist constructed from Pink Floyd's *The Division Bell* album.

WHAT'S A PLAYLIST?

Showing your age, aren't you? A *playlist* is a list of songs (or videos) that you want to treat as a group. In the normal course of events, you play a playlist from beginning to end, regardless of where the tracks came from. If you want to stick a rousing rendition of *Who Let the Dogs Out* in between Beethoven's Fifth first movement *Allegro con brio* and its second *Andante con moto*, you just make a playlist and play it. Slice and dice.

Advanced music management programs give you lots and lots of tools for building, modifying, and managing playlists. Groove Music, not so much.

Finding music and playlists

In the next section, I talk about buying new music, but if you already have music in your machine, and you can't find it, you have several options. Here's how to do it to it:

1. **In Groove Music, click the hamburger icon, at the upper left.**

   That gives you the list of actions shown in Figure 5-11.

2. **To see a list of all the artists in your collection, click My Music, and then click Artists.**

3. **To see a list of all the songs in your collection (sorted by the date added to your collection), click My Music, and then click Songs.**

   The list can be very, very long — and not very informative. You can sort the list by song name, artist, or album.

4. **To see what's playing right now — the current playlist — click Now Playing.**

5. **To create a playlist, click Playlists, and then click the + icon to the right. When prompted, type a name for the playlist and click Create Playlist.**

   You can then add tracks to the list.

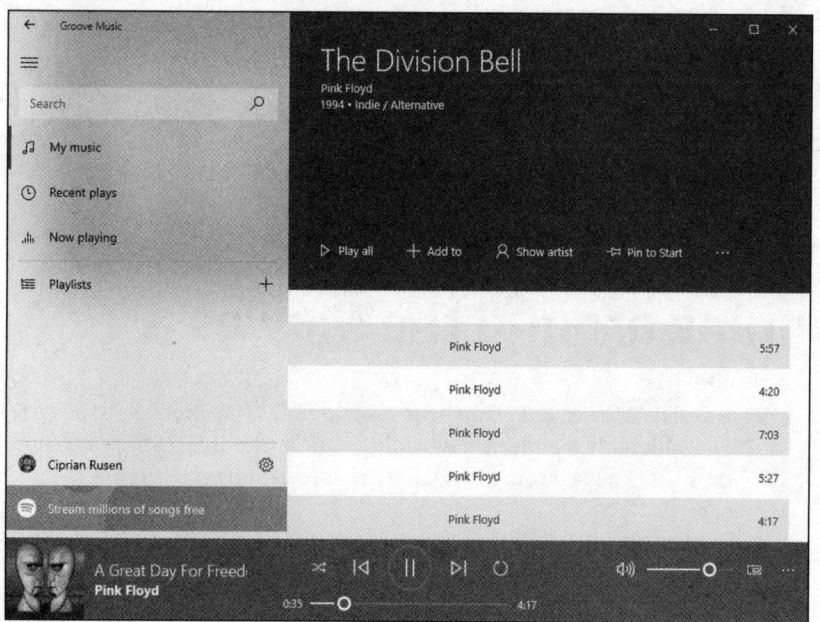

FIGURE 5-11:
Slice and dice your music and playlists here.

ASK
WOODY.COM

You can add an individual song from a playlist by right-clicking the song, choosing Add To, and then selecting the playlist. To remove a song from a playlist, right-click it and choose Delete. Other than that, few tools for maintaining playlists are available.

Running around the Movies & TV App

The Movies & TV app behaves much like the Groove Music app, although it's considerably pushier about selling stuff. Click or tap Start and choose the Movies & TV tile. You see something like the screen shown in Figure 5-12.

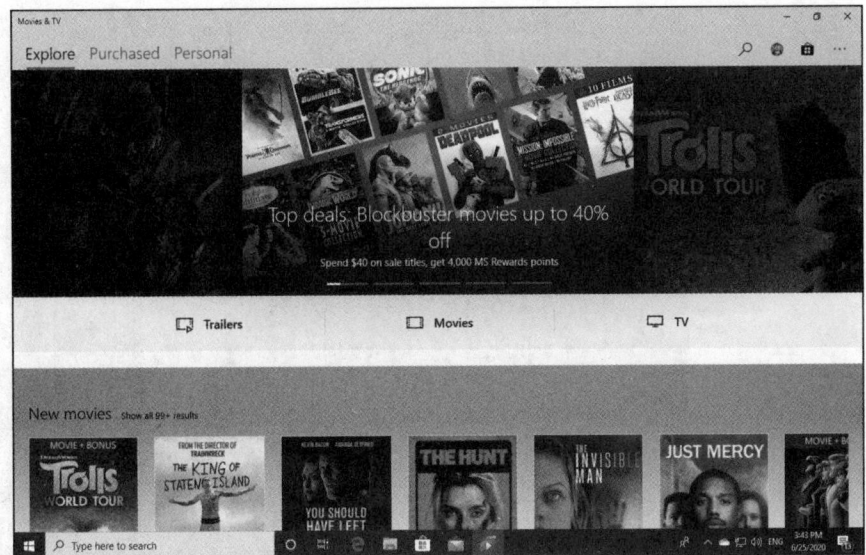

FIGURE 5-12:
Your own videos
appear under the
Personal menu,
up at the top.

Initially, Microsoft shows you only videos that they want you to buy or ones you've
already bought from them.

The menus at the top of Figure 5-12 cover movies (the ones you buy from Microsoft), TV shows (also the ones you buy from Microsoft), and videos (all other
kinds of video files). The Purchased heading lists all movies, shows, and videos
that you've bought from Microsoft. Personal includes anything on your computer.
Explorer should be called "we really want you to buy this stuff."

If you have videos in your OneDrive Videos folder, you have to knock Movies &
TV upside the head: As is the case with the Groove Music app, you can click the
ellipsis in the upper-right corner, choose Settings, choose where to look for videos
and add additional folders for the Movies & TV app to scan.

Double-click a video, and it plays in a letterboxed window, as shown in Figure 5-13.
You have all the basic controls for play, pause, go back 10 seconds, move forward
30 seconds, change the volume, and display subtitles.

The Movies & TV app isn't much more than a shell at this point, but as the samolians start rolling in, you can expect Microsoft to catch up with the competition.
Or maybe not.

FIGURE 5-13:
Playing your own
videos is easy.

5

Connecting with the Windows 10 Apps

Contents at a Glance

Chapter **1**

Introducing Edge

E's been abandoned by all except those who are forced to use it and those who don't know better. Even Microsoft has given up on IE, moving its efforts to the new Microsoft Edge. To make things even more confusing, Microsoft has decided to ditch the initial version of Microsoft Edge built into Windows 10 in favor of a new one based on the same rendering engine as Google Chrome. The old version of Edge worked only on Windows 10. The new Edge version can be downloaded from www.microsoft.com/en-us/edge and works on Windows, Mac, iPhone, iPad, and Android smartphones and tablets.

Microsoft is conceding that they lost the fight of web browsers as well as that of virtual assistants and music streaming services. When this book was written, the Windows 10 May 2020 update (version 2004) was current and the old version of Microsoft Edge was the default in Windows 10. Since then, Microsoft began pushing, through Windows Update, the new Chrome-based Edge. By the time you read this, the Chrome-based Edge will likely be the default browser in Windows 10 for all or most users. So in this chapter, I cover this latest version of Chrome-based Edge instead of the initial one shipped with Windows 10.

Born in a hellish crucible of Internet Explorer excess, Microsoft's new browser is fast and light, and — most importantly —has shed the baggage that IE carried for so long. That said, its initial version still suffered from many of the security problems that dogged IE, with security patches for both IE and Edge frequently appearing. Chrome-based Edge was launched in January 2020 as a separate download, and since then has evolved just as fast as Google Chrome. Sounds incredible, doesn't it?

ASK
WOODY.COM

I touch on Internet Explorer, lightly, in Book 3, Chapter 4. I don't recommend that you use it. In fact, I've been actively campaigning against its use since the days of Windows XP.

Why? Microsoft took its dominance in the web browser market as an excuse to release all sorts of Microsoft-only products, tie them into the browser, and convince developers to sing the IE song: ActiveX and Silverlight, Helper Objects, and Explorer Bars are all part of a lexicon that should have never appeared — one that should be crushed as quickly as possible.

What does that mean for you? The web pages you go to that used to be built for Internet Explorer are fading away. Rapidly.

The web programmers who were so caught up in Microsoft-proprietary technology have had their comeuppance. They're learning to build websites that are hospitable to all browsers. If they don't learn, their sites are going to wither. With Edge, all the browsers stand on a more or less level playing field. And that's truly refreshing.

All in all, Edge has provided an appreciated rebuke to IE. With this new Chrome-based version, Edge's popularity has risen a bit, and its market share is similar to that of Firefox — which is to say, not much but encouraging. Also, another great feat is that all Google Chrome extensions work in the new Microsoft Edge. Users are finally getting a Microsoft browser that is as extensible and personalizable as its competition.

ASK
WOODY.COM

If you don't know whether you have the new Edge or the old one, look at Figure 1-1. On the left you see the logo of the initial version of Edge built into Windows 10, and on the right you see the logo of the new Chrome-based version. This logo can be seen on the desktop, the taskbar, or in the Start menu entry for Microsoft Edge. If you have the old version, I recommend that you download the new one from www.microsoft.com/en-us/edge.

The logo of the old Edge   The logo of the new Edge

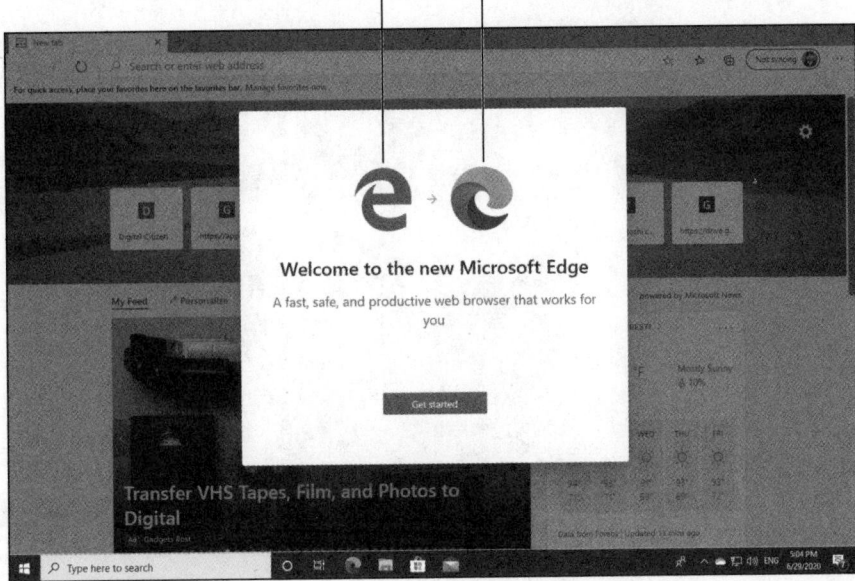

FIGURE 1-1:
See the logo of
the old Edge (on
the left) versus
the new Edge (on
the right).

A Walk through Microsoft Edge

Let's take a walk around the new kid on the block and kick a few tires. Try this:

1. **Click or tap the Microsoft Edge icon in the taskbar.**

Edge springs to attention, as in Figure 1-2.

2. **In the address bar, near the top, type the address of a website you like, and press Enter.**

I typed www.dummies.com/consumer-electronics.

3. **Click links on the web page. Try right-clicking. Convince yourself that Edge works just like any other browser you've ever used.**

For example, a right-click on the Browse Topics entry in Figure 1-3 displays the same basic navigation options you'd expect in any browser.

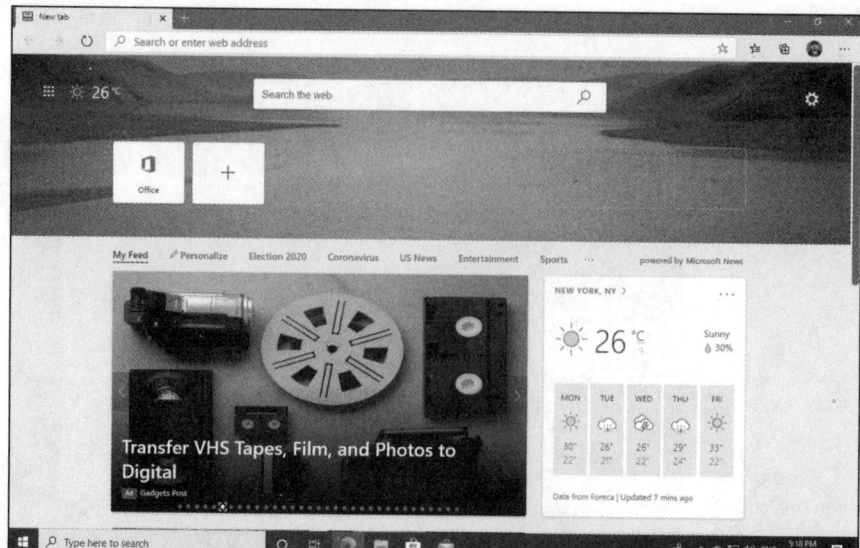

FIGURE 1-2:
Edge in all its
Spartan glory.

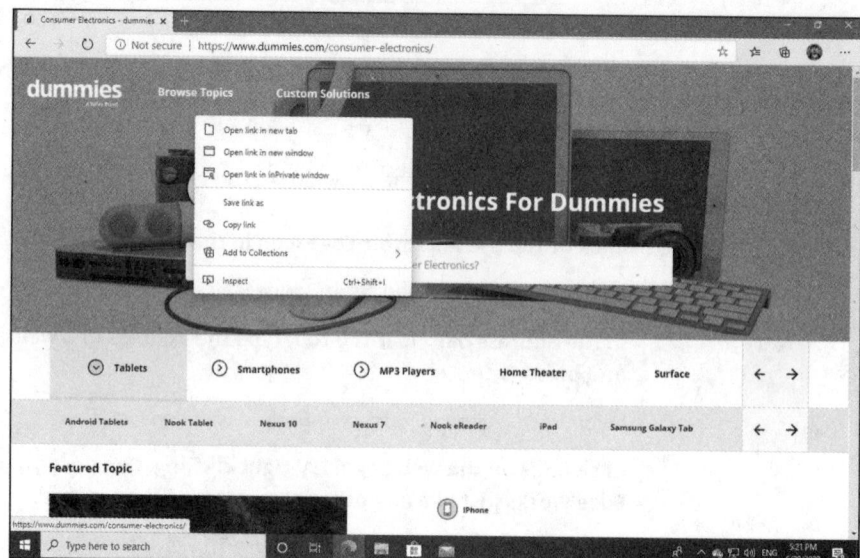

FIGURE 1-3:
Navigation is the
same in Edge
as in any other
browser.

4. **Type the address** www.digitalcitizen.life, **press Enter, and click one of the articles that interest you. Then click the ellipsis (. . .) in the top-right corner and choose Read Aloud.**

Read Aloud mode is activated (see Figure 1-4), and reads you the contents of the article you chose.

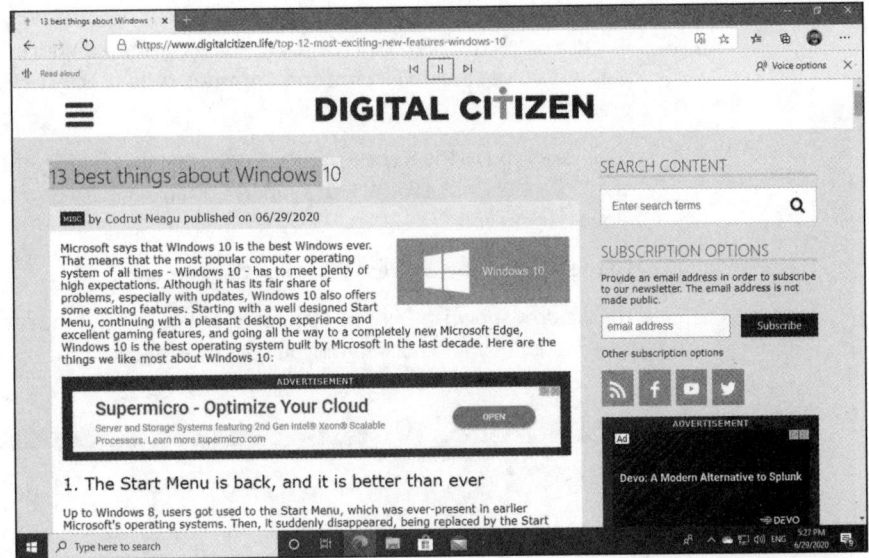

FIGURE 1-4:
Read Aloud reads back what you see on a web page.

5. **Click the X in the top-right to close Read Aloud mode. Then click the + at the top to start a new tab.**

Edge has an advertisement-laden new tab page, shown in Figure 1-5. The sites on the top are populated based on what you visited using Edge, while the My Feed tab is filled with ads custom-built for you by Microsoft.

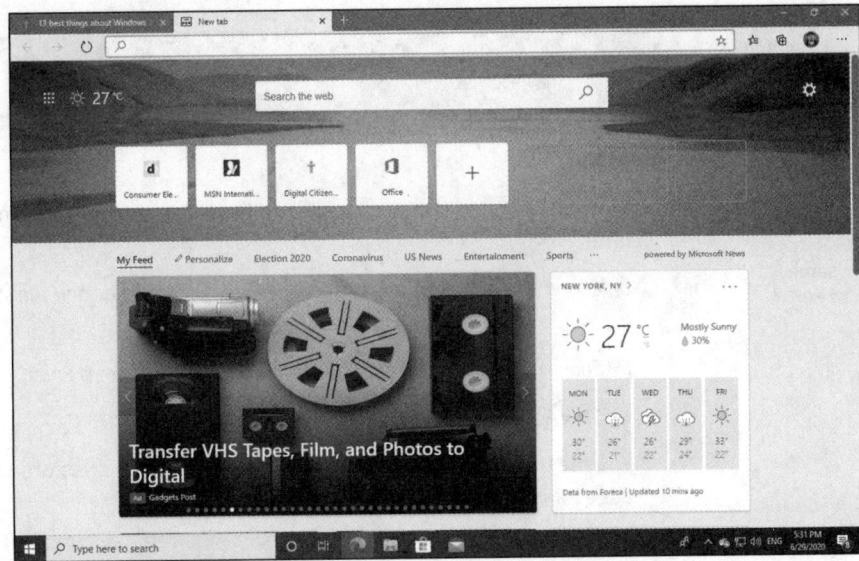

FIGURE 1-5:
The new tab page in Microsoft Edge.

6. **In the Search the Web box (which is just a combined address and search bar, as you've seen in Firefox and Chrome), type** cuoco-seattle.com **and press Enter.**

Edge takes you to the website for the Cuoco restaurant in Seattle, which is one of the few sites in the world that support Cortana, and it's easy to have it Read Aloud back to you.

7. **To the right of the address bar, click the star icon.**

The window shown in Figure 1-6 appears, so you can put the site (or some other folder of your choosing) in your Favorites folder.

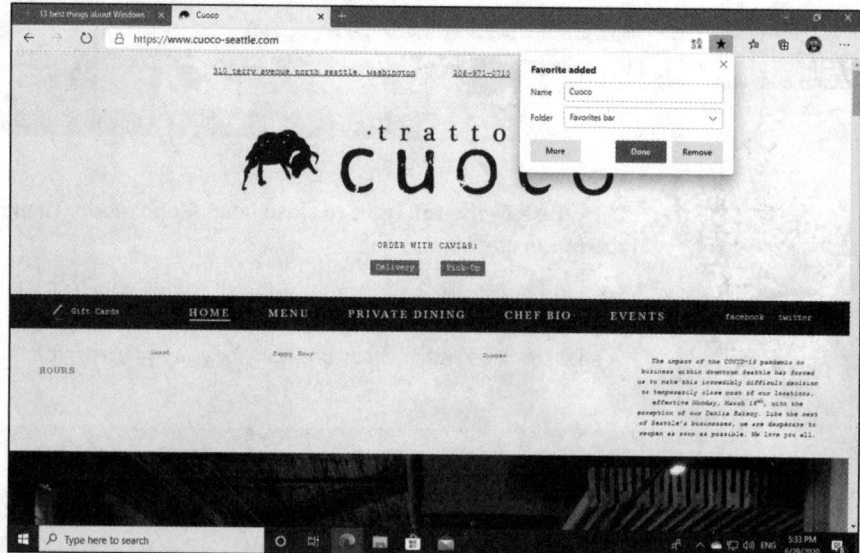

FIGURE 1-6:
The Cuoco restaurant in Seattle should be added to your Favorites.

8. **You can also add the site to your collections, so that you can revisit it later:**

a. Click the Collections icon to the right of the address bar and the star icon. Then click Start New Collection.

b. Enter a name for your collection of web pages, and press Enter.

c. Click Add Current Page.

You see the site shown in Figure 1-7. Collection items are stored by Microsoft Edge for later reading.

FIGURE 1-7:
Adding the site
to your reading
collections in
Microsoft Edge.

Working with the Immersive Reader

If you're tired of ads, you might want to use a useful Edge feature called Immersive Reader. When activated, it strips any article or web page of ads and all the junk that distracts your attention from the content. The only downside is that Immersive Reader doesn't work on all sites. Some were created so that you can never get rid of ads.

Here's how to use Immersive Reader in the new Edge:

1. **In Microsoft Edge, navigate to digitalcitizen.life and click any article that seems interesting to you.**

   I chose an article with the 13 best things about Windows 10. Ironic isn't it? :)

2. **At the right corner of the address bar, click the icon that looks like a book with a speaker on top of it. Alternatively, press the F9 key.**

   The web page turns into distraction-free page, where you see only the contents of the article, as shown in Figure 1-8.

3. **Read away and scroll down.**

4. **When you're finished reading, click the book with a speaker on top icon to return to normal mode.**

   Oh, behold the ads!

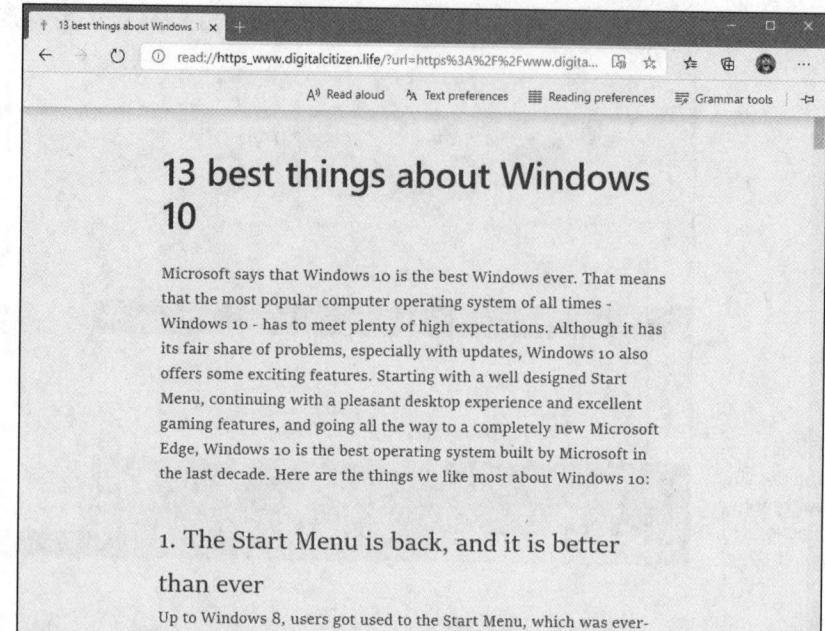

FIGURE 1-8:
Immersive Reader mode in Microsoft Edge.

A Sampler of Edge Settings

Microsoft Edge is young but growing rapidly. You can expect its settings to change as it gets some of the features you would expect from any browser. As the book went to press, this is what was on offer:

1. **Start Edge, and bring up an interesting page. Click the ellipsis icon in the upper-right corner.**

   You see the main settings pane shown in Figure 1-9. If you look at the settings on offer, they should strike you as being like most settings in any browser anywhere. The only novelties are Collections, which are links to articles you collect on the web, and Read Aloud, which I covered previously.

2. **At the bottom, click or tap Settings.**

   You see the next set of settings, organized in logical categories on the left, as shown in Figure 1-10. These settings do what you would expect.

FIGURE 1-9:
A simple set of
settings for Edge.

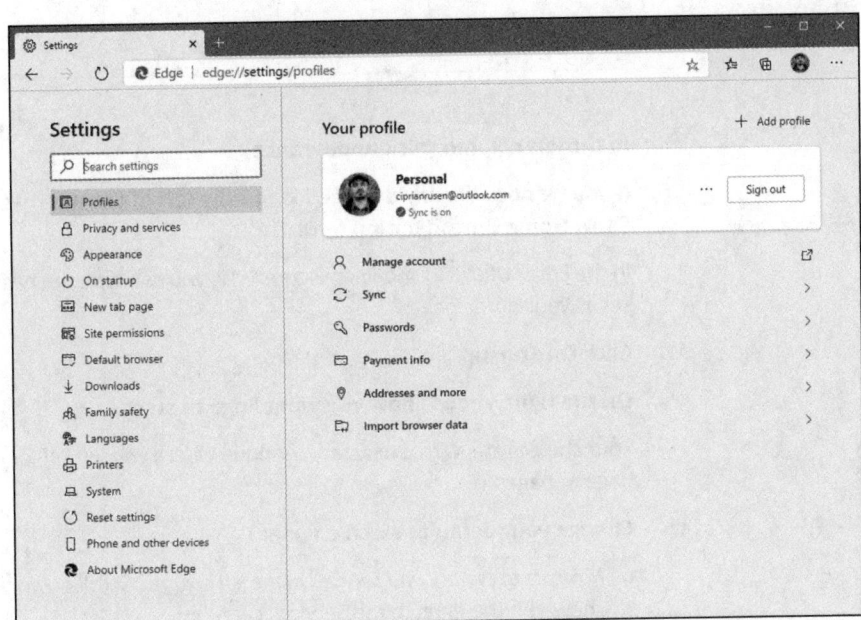

FIGURE 1-10:
A deeper dive
into Edge's
settings.

3. **On the left, click Privacy and Services. Then choose the kind of tracking prevention you want from Microsoft Edge.**

   The settings available are quite strict, as shown in Figure 1-11.

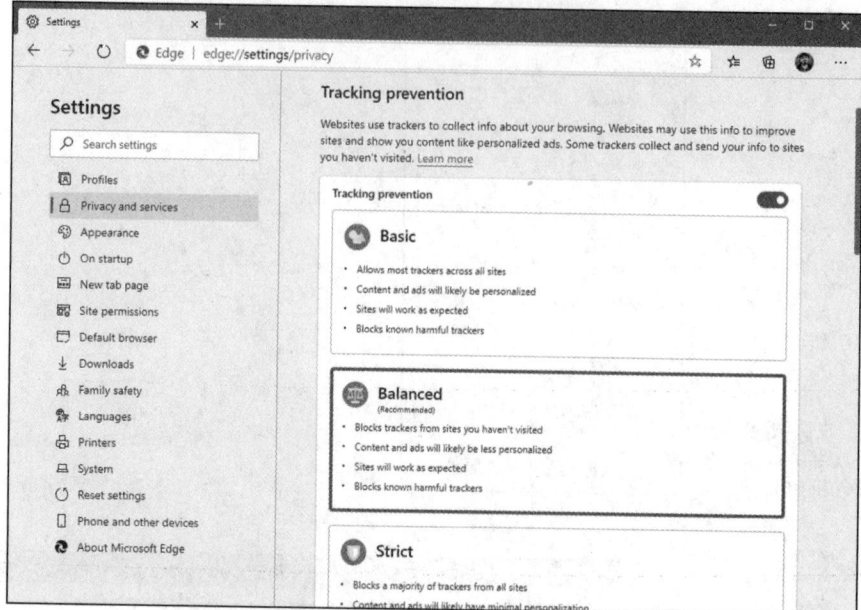

4. **In the left column, click Appearance.**

5. **If you want to display a Home icon to the left of the address bar, slide the Show Home Button option to On.**

   In the Enter URL field shown in Figure 1-12, you can type the page you want to set as your home page.

6. **Click On Startup.**

7. **On the right, choose how you want Edge to start.**

   Your choices are: with a new tab, continue where you left off, or open a specific page or pages.

8. **Change your default search engine:**

   a. *Navigate to Privacy and Services in the left column.* Scroll down the settings shown on the right, to the Services category.

   b. *Click the Address Bar entry.* Finally, you see the settings for managing search engines, as shown in Figure 1-13.

c. *Click the drop-down list for Search Engine Used in the Address Bar, and choose the search engine you want.*

Searches made from Edge will use the search engine you choose.

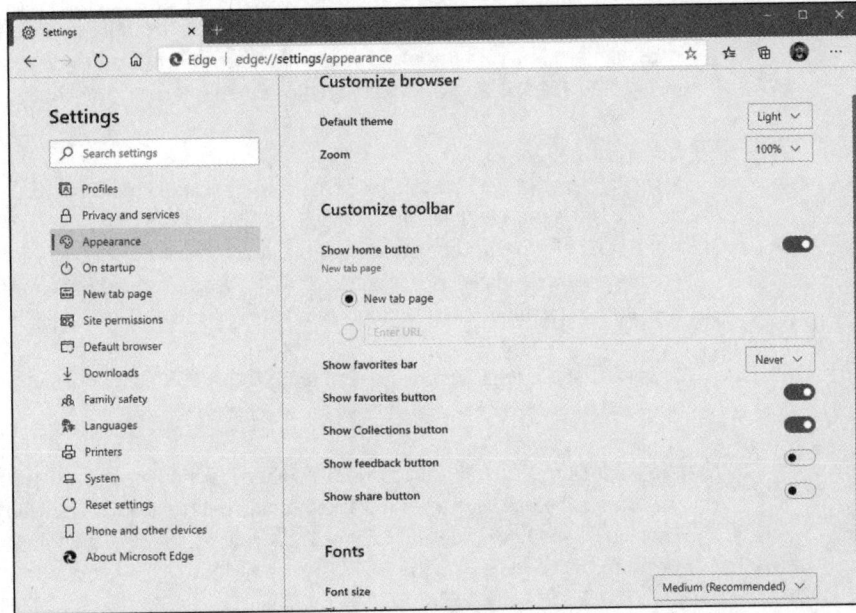

FIGURE 1-12:
Display the Home button and type the URL you want to be your home.

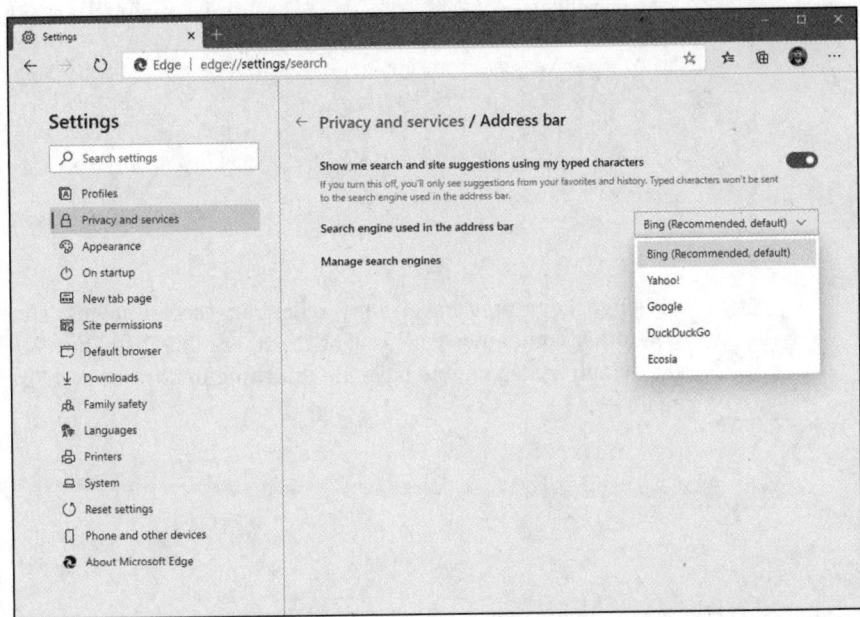

FIGURE 1-13:
Change the default search engine.

THE DO NOT TRACK SETTING

The privacy setting known as Do Not Track (DNT) has a long and torturous history. In 2009, a group of Internet privacy advocates created the Do Not Track specification as a way for you, the user, to tell the websites you're browsing that you do not want them to keep track of your visit — no cookies, don't store your IP address, and no monkey business with sending your information to advertisers. All six major Windows browsers — IE, Chrome, Firefox, Opera, Safari, and now Edge — can tell sites that you don't want to be tracked.

In mid-2012, as part of the IE 10 push, Microsoft decided that DNT would be enabled by default: Unless you took action to disable it, IE would send the DNT signal to every site you visit. The web standards world, particularly those representing advertisers, erupted. An online advertising advocacy group said DNT would "harm consumers, hurt competition, and undermine American innovation." Go figure.

Ends up that the original sorta-agreed-upon standard said that browsers were supposed to fly the DNT flag only if the user specifically chose it.

Microsoft didn't back off until early 2015, when it announced that IE would not send DNT unless the user had explicitly asked for it. That decision has carried forward to Edge; you won't have the Send Do Not Track requests setting On unless you specifically slide it to On.

It doesn't make much difference anyway. Conformance to the DNT spec was (and is) voluntary. About ten websites decided to obey the DNT. (Okay, I'm exaggerating, but aside from Twitter and Pinterest, there have been very few.) The others bowed to pressure from their advertisers and kept doing what they've always been doing.

Will DNT ever take hold and become a de facto Internet standard? Why, yes, I figure it'll happen just about the time advertisers stop advertising on the web. Give it, oh, a hundred years.

ASK WOODY.COM

Although Edge may have many redeeming social values, sharing the limelight with other companies isn't one of them. Frankly, little hurdles like this one with the default search engine have me returning to Chrome and Firefox, over and over again.

PROGRESSIVE WEB APPS

Universal Windows Platform (UWP) is almost dead, and a better alternative is on the horizon. Progressive web app (PWA) support is built into Microsoft Edge and Windows 10.

Progressive web apps aren't so much Google's much-better alternative to Win10-only Universal Windows programs (formerly known as Metro apps, Universal apps, Windows Store apps, or any of a half-dozen other monikers) as they are a genuine attempt to make browser-based applications look and feel more like regular old apps.

Chances are very good that you've never seen a PWA in action. But they're definitely here.

The theoretical benefits of PWA over UWP are enormous. Just for starters, UWP can run in only the stripped-down Windows 10 environment. A PWA, on the other hand, should be able to run on just about anything that supports a browser — particularly Chrome, or Chrome OS. Yeah, that includes Chromebooks.

The browser requirement has vanished in the past couple of years, banking on a concept called *service worker*. Horrible name, but web folks are good at horrible names.

It now looks to me as if there's going to be a headlong dash into developing PWAs — and that UWP's days are numbered. Time will tell.

Adding Edge Extensions

When Microsoft Edge first appeared in July 2015, we in the computer press expected a real browser. We didn't get one. The first version of Edge was barely functional. When Microsoft updated Edge in November 2015, we expected a real browser. We didn't get one. More than a year after its introduction, Edge picked up a few fundamental prerequisites for being a real browser. The version of Edge introduced in July 2016 could finally run extensions — the customizing programs that add immeasurably to a browser's capabilities. The new version from January 2020 is now capable of running *all* Google Chrome extensions – the largest repository for extending your web browser.

Think of extensions as the apps that you add to Edge. You can get them from Microsoft or Google. The choice is yours, and that's great.

REMEMBER

The new version of Edge has more extensions than the old version. However, the most important of all, by far, is the LastPass extension, which lets you use Last-Pass (see Book 9, Chapter 4) to store and retrieve all your passwords.

The method for installing and displaying extensions may well change by the time you read this, but the mechanism should look something like the following:

1. **Click the wave-like icon in the taskbar to display Microsoft Edge.**

2. **Click or tap on the ellipsis in the upper-right corner to display the Settings and More menu, and then choose Extensions.**

3. **Edge offers a button to Get Extensions for Microsoft Edge. Click it.**

   You see a web page like the one in Figure 1-14.

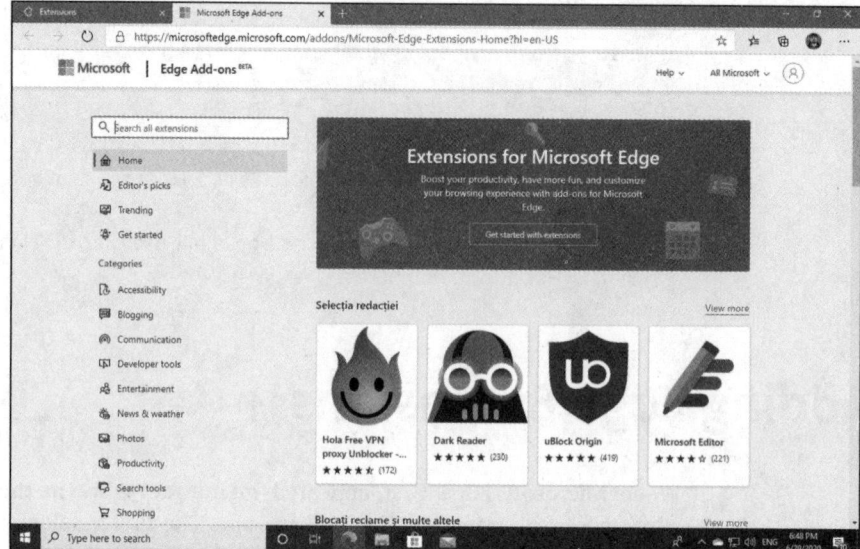

FIGURE 1-14:
Edge extensions
appear.

4. **Scroll down the list of extensions, and click the one that interests you.**

   I clicked the link for LastPass, and the screen shown in Figure 1-15 appeared.

5. **To download and install the extension, click or tap the Get button, and confirm your choice by clicking Add Extension.**

   You see the confirmation dialog shown in Figure 1-16. The extension installs itself and is activated automatically. In the case of LastPass, its icon appears to the right of the address bar.

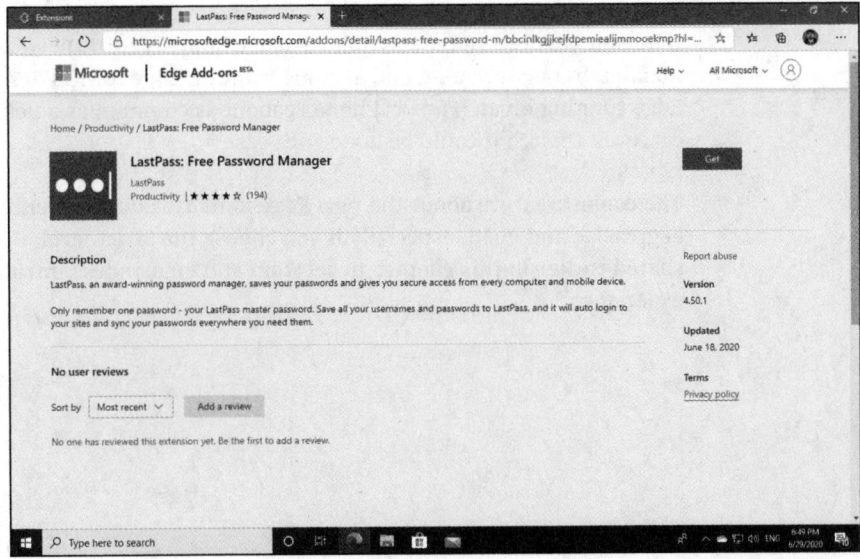

FIGURE 1-15:
The LastPass
extension.

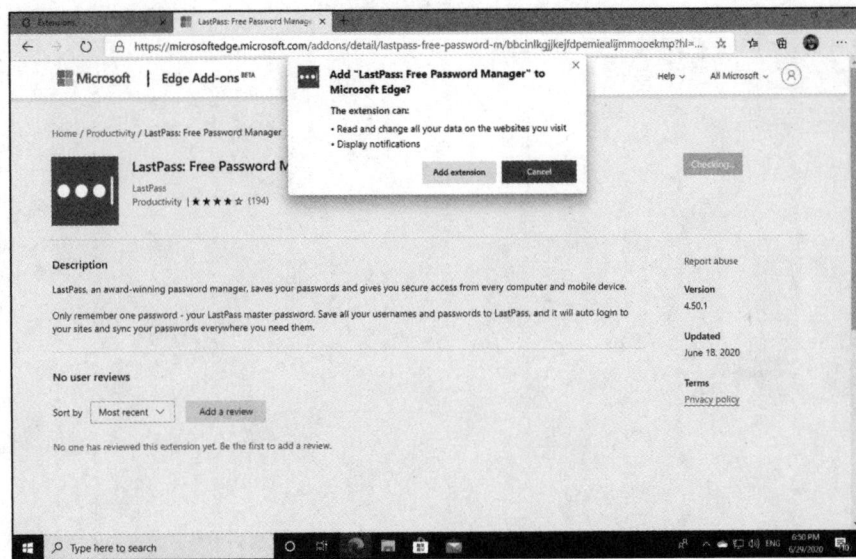

FIGURE 1-16:
Confirming that
you want to add
your extension to
Edge.

ASK
WOODY.COM

With this new Chrome–based version, Edge has been picking up steam. Microsoft has been releasing updates, improvements, and new features every month. And through the work they do on improving the rendering engine behind both browsers, Microsoft also contributes to Google Chrome. To be honest, the future looks exciting, and Edge might just become a worthy contender to Chrome's dominance.

In recent versions, Edge has gained added support for PDF viewing and synchronizing tabs between a PC and a phone running Edge for iOS or Android. Of course, Edge running on an iPad or iPhone is about as common as a tickhound riding on a cat, even though it could be done with ease.

The coolest feature about the new Edge is its tracking prevention, which is quite aggressive and good, especially if you choose the Strict level. Use the instructions shared earlier in this chapter to set it up and enjoy a less intrusive web browsing experience.

Chapter **2**

Using Skype in Windows 10

Everybody knows Skype, the instant-text-messaging, long-distance, telephone-killing video-chatting program. Not everybody knows that it started as something of a hacker's fantasy in 2003, in Estonia. Two of the key players in getting Skype to market, Janus Friis from Denmark and Niklas Zennstrom from Sweden, spent their earlier years getting Kazaa — the notorious file-sharing program — off the ground.

ASK
WOODY.COM

Microsoft bought Skype, lock, stock, and barrel camera, in 2011, for a paltry $8.5 billion — yes, that's *billion* with a *b*. The brass moved to Redmond, but most of the techies are still in Tallinn and Tartu, Estonia.

In spite of appearances, Skype is a Microsoft product. One hundred percent.

Skype was once known as a long-distance phone killer, but it's broadened enormously since then. In addition to voice, Skype also handles instant messaging (including SMSs to phones) and video calls, both one-on-one and conference call style. You can use Skype to call regular (landline) phones anywhere in the world, for an extra fee, which seems to change from year to year.

Microsoft's building Skype hooks into all sorts of products — your Windows 10 contacts come along for the ride, and Office is fully Skypeable. Skype works, and works well, with iPhones and iPads, Android smartphones, Android tablets, but historically it's had a difficult time with Windows.

WARNING

It took Microsoft a couple years to get around to building a Metro Skype app for Windows 8.1. The app was widely panned and shunned. The current Skype app for Windows 10 is a bit better, but many folks swear by the browser-based version (you can find it at www.web.skype.com) and feel it's better. However, there's also a Skype desktop app (or program) available for download online that offers all the features you can get from Skype. You find it at www.skype.com/en/get-skype/.

In this chapter, we look at the Windows 10 Skype app (which is what you get when you click Start⇨Skype), and Skype running on smartphones and tablets of various pedigrees.

Many hundreds of millions of people use Skype. The last official tally, from December 2010 — before Microsoft took over and stopped publishing statistics — put the number of registered users at 663 million. Much more reliable figures say that 300 million different users are on Skype every month, and they yak an average of 3 billion minutes *per day*.

ASK WOODY.COM

Microsoft stopped publishing statistics about Skype shortly after the takeover. Many people — including me — have concluded that the sudden lack of showmanship has a lot to do with Skype's declining popularity, at least relative to other messaging apps.

Skype's future seems to be inextricably linked to corporate accounts and enterprise versions of Office. For those of us with a choice, Skype isn't nearly as intriguing as it once was.

In July 2013, as part of the Snowden revelations, the *Guardian* newspaper reported that Microsoft had given the US National Security Agency access to Skype supernodes and helped the NSA crack Skype encryption. From the NSA the information was made available through Project Prism to the CIA and FBI. See www.theguardian.com/world/2013/jul/11/microsoft-nsa-collaboration-user-data.

Signing Up with Skype

Here's how to get started with Skype.

WARNING

If you're using a local account to sign in to Windows 10 — as opposed to a Microsoft account — and you crank up Skype, Skype prompts you immediately to use a Microsoft account. **If you want to use Skype, you must use it with a Microsoft account.**

1. **Click or tap Start ⇨ Skype.**

   There may even be a tile. Skype logs you in with your Microsoft account, or nudges you to type a Microsoft account.

2. **If you're starting Skype for the first time, click or tap Let's Go. Then, if you're not using a Microsoft account in Windows 10, click or tap Sign In or Create (see Figure 2-1)**

FIGURE 2-1:
It's time to sign
in to Skype.

3. **Do one of the following:**

   - If you have a Microsoft account, enter the details and click Next.
   - If you don't have a Microsoft account, create one now and then click Next.

   Skype may ask about updating your profile picture.

4. **To use the existing profile picture, click Continue. Otherwise, click Upload Photo, and choose another picture for Skype.**

Using Skype in
Windows 10

5. **When Skype asks you to test your audio:**

   a. *Choose the microphone that you want to use, and start speaking.* You should see the volume dots picking up your voice.

   b. *Choose the device that you want to use for Speakers, and click Continue.*

   c. *It's also a good idea to test the audio and make a free test call to see if your microphone and speaker settings work well.*

6. **When Skype asks you to test your video:**

   a. *Choose the webcam that you want to use and see if it works in the preview that's shown.*

   b. *When everything works, click Continue.*

7. **When Skype mentions finding contacts, click or tap OK.**

   Skype finally gets around to its main page, as shown in Figure 2-2.

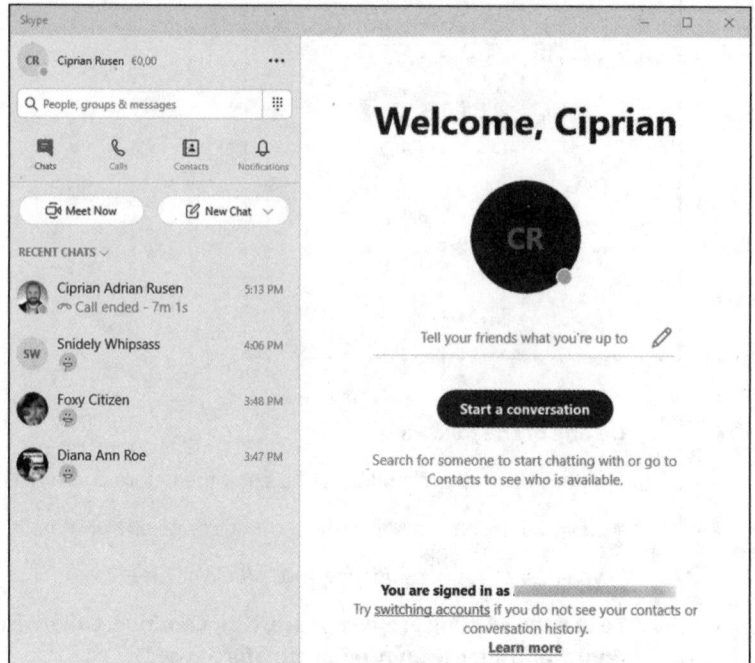

FIGURE 2-2:
Ready to start
Skypeing.

Following is a general bit of orientation. On the left in Figure 2-2, you have the following:

» In the upper left corner, click your picture or initials (if you don't have a picture for your Skype account) to display your profile.

» After you've had a few conversations, the people with whom you've had conversations appear in the left pane, under Recent Chats.

» The Contacts icon displays a list of your Skype contacts and bots (see Figure 2-3). Note that Skype contacts are not the same as your Windows 10 People app contacts. Nor are they the same as any other contact list you may have used. They're unique to Skype, although you can sync them in the Settings part of Skype.

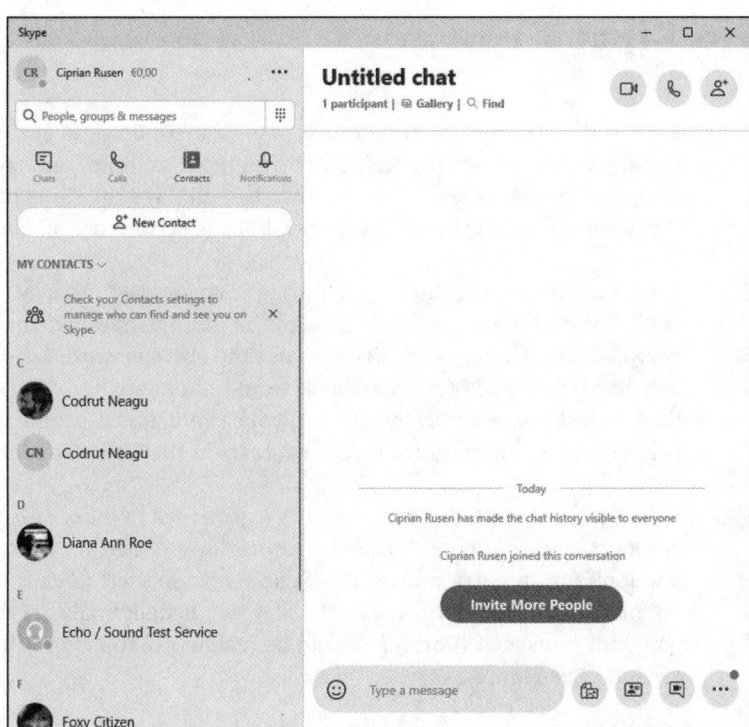

FIGURE 2-3:
Skype bots may provide worthwhile info. Some day.

» The bots, which look like regular contacts (refer to Figure 2-3), are automated responders that may or may not be of some help.

» The you icon (a picture of you in the upper-left corner or an icon with the initials of your Skype name) displays details about your account. It may be smart enough to pull your picture from your Microsoft account. (You can change the photo directly, though.)

» Click your picture, and then click the Settings link. The Settings section gives you some control over how Skype works. See the Settings section in this chapter.

On the right side of the screen you have room to keep track of your latest conversation with whichever contact you've chosen on the left.

Making First Contact

Each of the different versions of Skype — the Windows 10 app version, the desktop app version from the Skype website, iPad, Android, and so on — presents a slightly different way of working, but they all have the same basic core features. The locations on the screen may vary, but the actions are all similar.

TIP

A note about nomenclature: All through Skype (and other Microsoft products), you'll see the terms *contact* and *people* used interchangeably. There's no difference. Very confusing. Microsoft refuses to use the obvious word, *Friend*, because Microsoft has few friends (and Facebook would undoubtedly retaliate). Keep in mind that, unless you specifically allow it, your contacts, er, people in the Windows 10 People app are different from your contacts in the Skype app. I have no idea why.

TIP

Before you get started with Skype, it's a good idea to make sure your microphone, speakers, and (optionally) camera are working. To do so, click the Contacts icon and look for an entry called Echo/Sound Test Service. Click it and then click the phone icon, on the right; that's the way you usually make a call. Skype connects you with a test bot (Cortana should be jealous) that asks you to speak for a few seconds. See Figure 2-4.

If you can hear your voice in the playback, you're connected and ready to run. If you can't hear your voice, your microphone or your speakers aren't working or you chose the wrong device to act as your microphone or speakers.

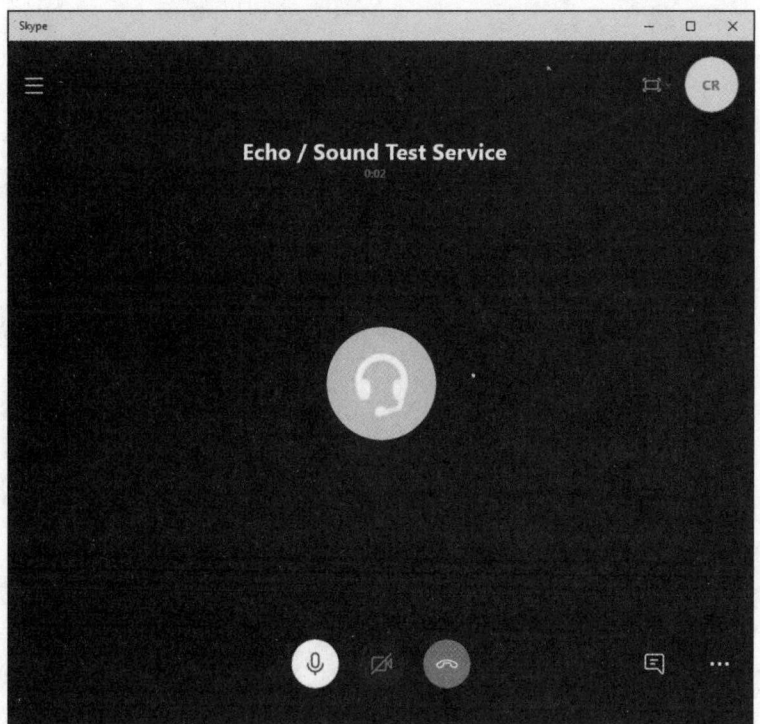

FIGURE 2-4:
Ding! Your sound
check service.

Adding a Contact

Before you can call someone, you have to make her an official contact — which means you have to ask for, and receive, permission to call.

REMEMBER

The methods for adding contacts vary depending on which version of Skype you're using. In most versions, you must go to the Contacts list before you can add a new contact. Here's how to add a contact in the Skype app from Windows 10:

1. **On the initial screen (refer to Figure 2-2), use the search box and find someone you want to add as a contact.**

Yes, trying to search by name can be a hassle. Skype lets you search by name (such as Woody Leonhard), Skype name (a name that predates the use of Microsoft accounts), or email (a Microsoft account).

When you've found someone you'd like to add, the screen should look like Figure 2-5.

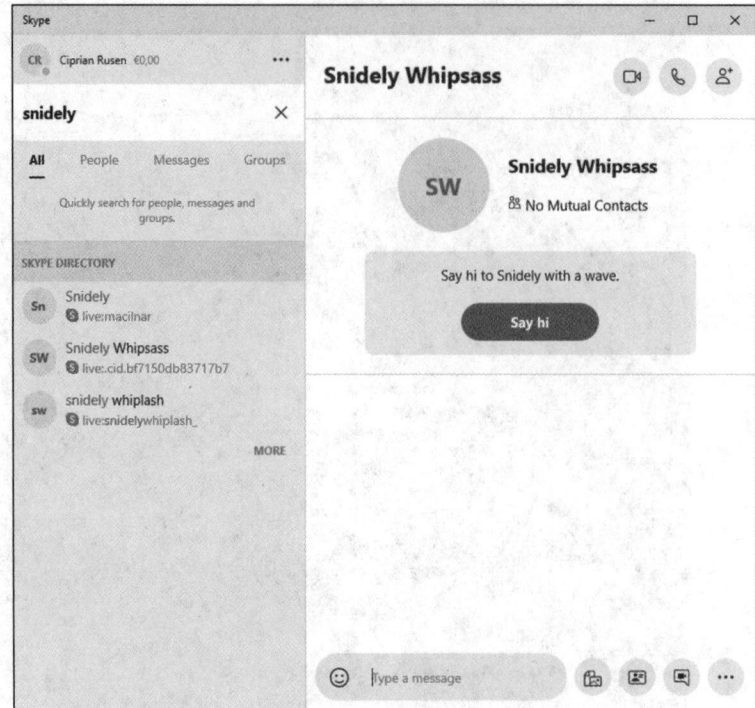

FIGURE 2-5:
A potential
new contact.

2. **Right-click the name and click View Profile.**

   Verify that the person you've selected is the one you want to converse with and, if so, click the link to Send Message.

   Skype formulates and offers to send a message to that person. If you click Say Hi, Skype sends the message and logs the fact on your call screen, as shown in Figure 2-6.

3. **Wait.**

   If your contact-to-be clicks or taps the invitation, and either responds to it (as on the Android) or clicks Accept (as in the Windows 10 version of Skype; see Figure 2-7), you'll suddenly find yourself able to communicate.

4. **Add a few more friends, er, Skype-enabled contacts, and you're ready to roll.**

   If you X out of Skype, the app continues to run — it can notify you of any incoming calls — but it shows you as Offline. There's an icon for Skype conveniently stuck in your taskbar, just to remind you that the program's alive and well.

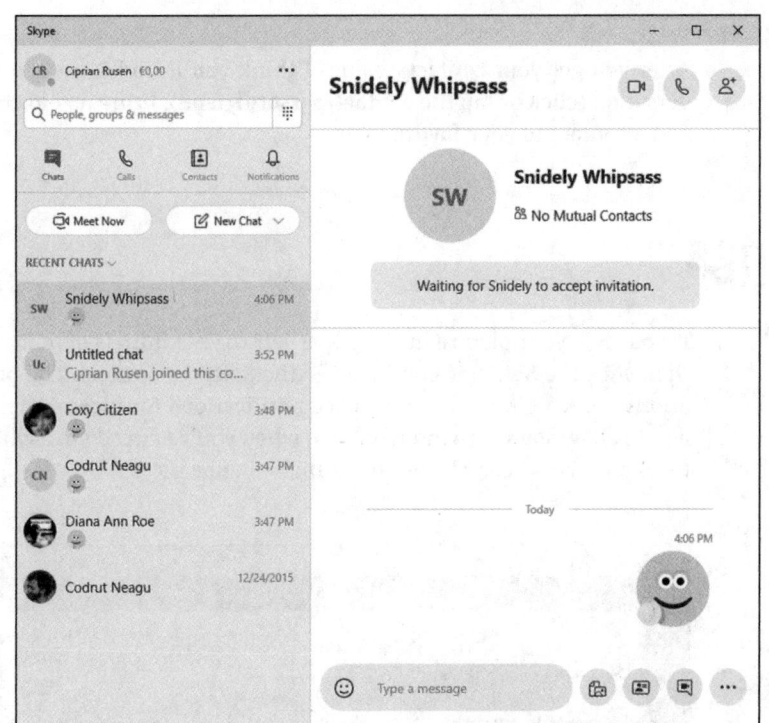

FIGURE 2-6:
Skype sends a message to the person you'd like to turn into a friend.

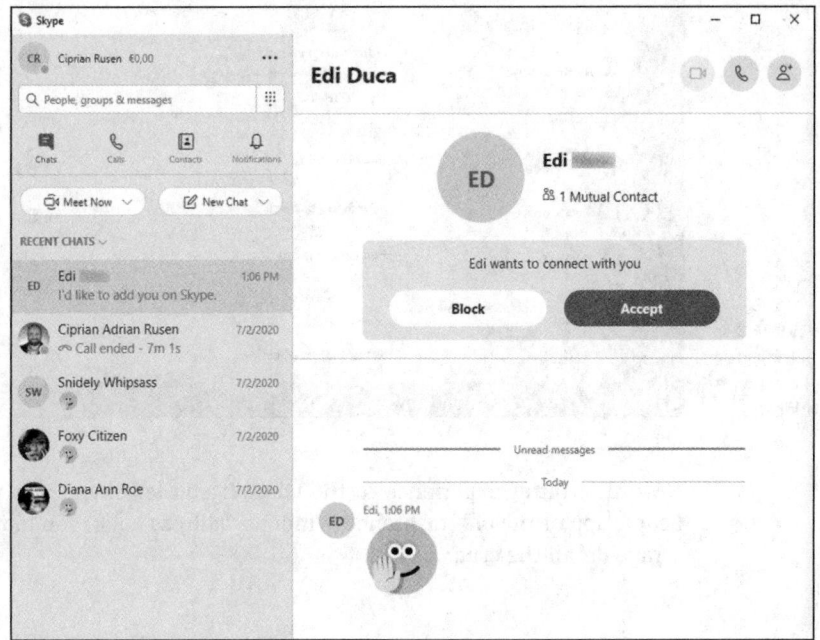

FIGURE 2-7:
Accept a contact request and the conversation can begin immediately.

After you get your contacts going, I think you'll find it easy to start a new conversation (click or tap the contact/person/friend), bring up old conversations, and add a contact to your favorites list.

Settings

If you click your picture in the upper left (or the initials of your Skype name) and then click the Settings link, you see the Settings options. The primary considerations are whether you want to see notifications for incoming message and calls and to allow sounds to interrupt you when you're already in a chat. Click Notifications and the screen shown in Figure 2-8 appears.

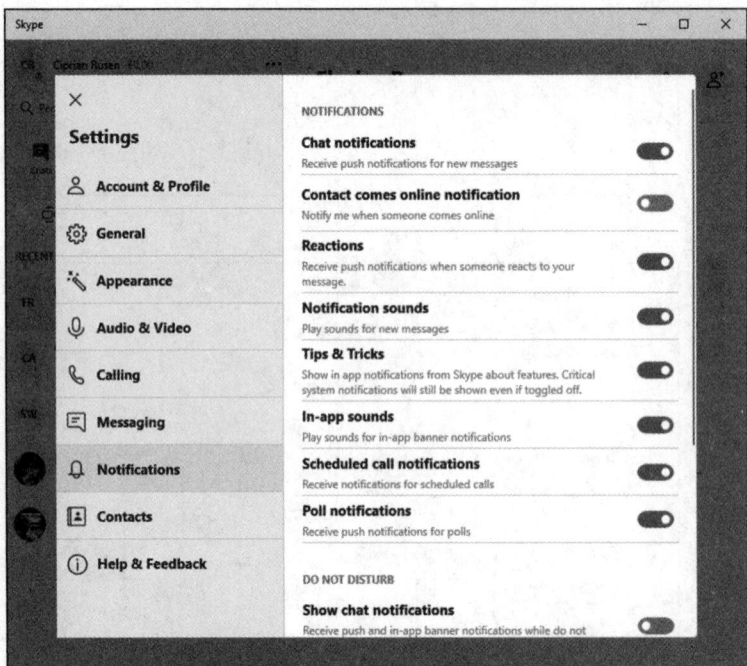

FIGURE 2-8:
Settings for controlling Skype notifications.

A word of bitter experience to the wise: If you let Skype into your Windows 10 People app or the old-fashioned Windows address book, you may never be able to scrape off all the crud.

Making Group Calls

The lockdown caused by the Covid-19 pandemic has forced many people to work from home and interact with their peers digitally. A useful feature of Skype is that it makes it easy to make group audio and video calls. You can start a group call in many ways. Here is the easiest method, from the Windows 10 Skype app:

1. **On the initial screen (refer to Figure 2-2), click or tap the Calls icon, near Chats.**

2. **Click the New Call button.**

   Skype shows a list with your contacts.

3. **Select the people you want to have a group call with by clicking the checkmark next to each name (refer to Figure 2-9), and then clicking the Call button in the upper-right corner.**

   If you select one person, you start a one-to-one call, instead of a group call. If you select two or more people, you start a group call. Your Skype voice call is initiated, and you can add a video feed to it even before the other participants answer by clicking the webcam icon at the bottom of the call screen.

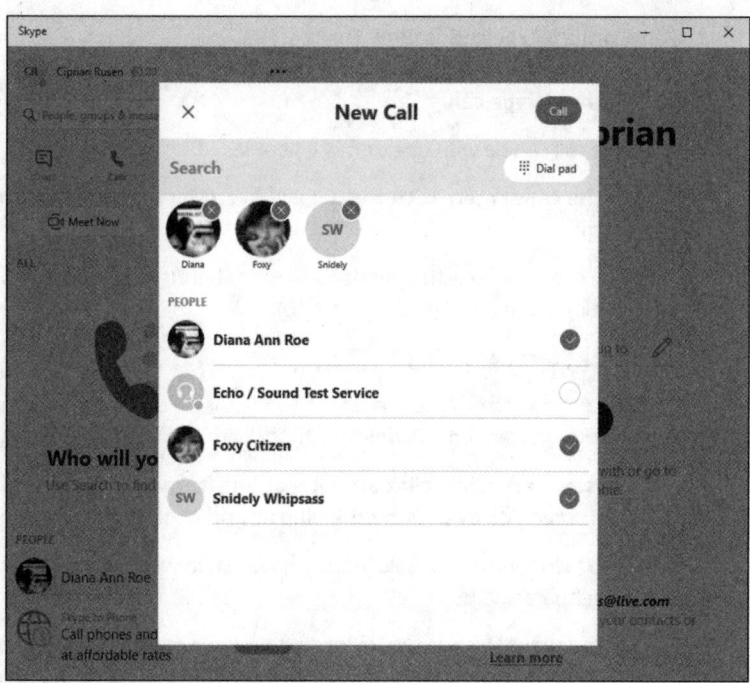

FIGURE 2-9: Starting a group call in Skype for Windows 10.

REMEMBER

The procedure for starting a group call is the same in all versions of Skype, including Skype for Android, iPhone, or iPad. The only difference is that the buttons are placed in different locations than in Skype for Windows 10.

WARNING

You can start group calls in Skype for Web (web.skype.com) too, but only if you load it in Google Chrome or Microsoft Edge. Microsoft doesn't provide this feature in other web browsers.

Recording Calls

If you have an important Skype call and want to make sure you don't forget anything that was said, it's a good idea to record your call. The same is true if you are a teacher or a trainer who delivers a lesson or presentation to others. You can record Skype calls with people using different platforms. Skype records everything during a call, including voice, everyone's combined video stream side-by-side, and screen sharing.

REMEMBER

Be sure to use the latest version of the Skype app for Windows 10, Mac, Android, or iPhone.

To record a Skype call, follow these steps:

1. **Start a Skype call.**

   The call can be with one or more people.

2. **In the call screen, click the ellipsis icon (More Options) in the lower-right corner.**

   You see a menu with options for screen sharing, subtitles, audio, and video settings, and more (see Figure 2-10).

3. **Click Start Recording.**

   Recording starts. A banner at the top of the screen displays a reminder, as well as the elapsed time. People in your call are also notified that you're recording.

4. **To stop recording, click Stop Recording from the banner at the top of the call screen or by clicking the ellipsis and choosing Stop Recording.**

   The recording is available in the conversation window to both you and the other participants.

5. **If you want to save the recording on your computer, right-click it in the chat window, choose Save As, and provide a name and location.**

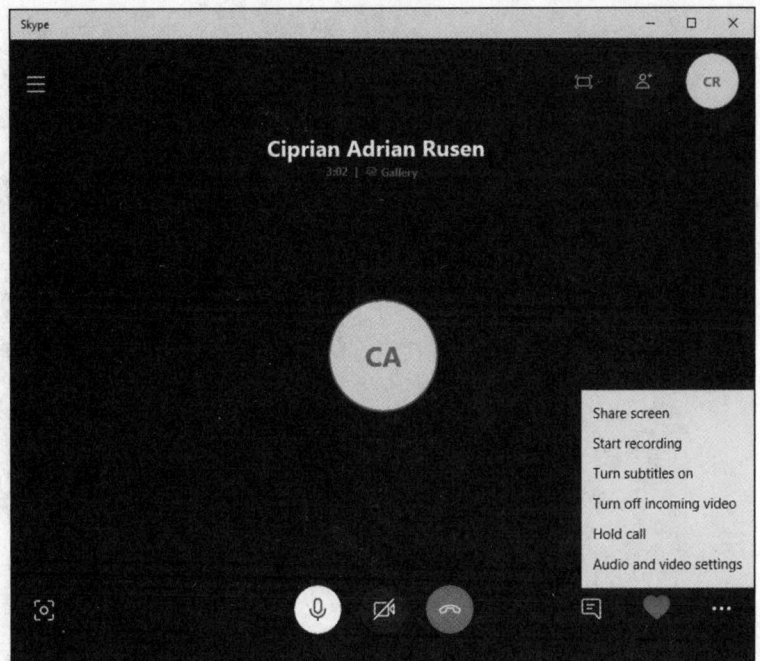

FIGURE 2-10:
Recording a
Skype call.

Using Skype in Windows 10

A Few Tips from Skype-ologists

By default, people can send a friend request to you if they have your old-fashioned Skype Name (they're being phased out), if they type your real name (Woody Leonhard) in the search box and can guess which result belongs to you, or if they have the email address for your Microsoft account. You don't have to respond to a contact request.

There's a nascent capability to include additional information in your Skype account. Someday, people may be able to search for you based on that information, but as I write this, the whole feature is garbled and doesn't work well.

To see what info Skype has about you, in Skype, click your picture on the top-left corner (or your initials). Then click Skype Profile, and you see something like Figure 2-11.

ASK
WOODY.COM

Microsoft sometimes pushes ads in the free versions of Skype. I have no idea how to turn them off.

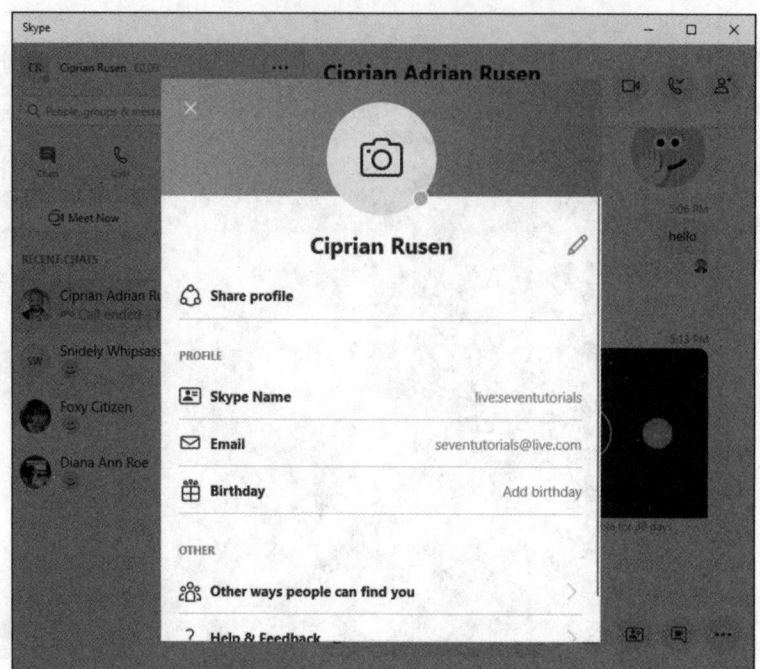

FIGURE 2-11:
Skype collects
minimal informa-
tion about you.

REMEMBER

When you install the old-fashioned desktop version of Skype, Windows 10 puts a folder labeled Skype in your Apps list. If you are using the Windows 10 Skype app, you'll see a Skype shortcut in the Apps list.

Skype-to-Skype calls are free. But Skype has many, many more options that aren't free. You will probably find that the Skype options are much, much cheaper than normal long-distance phone charges. There are promotions all the time and lots of ways to game the system. See www.skype.com just for starters.

WARNING

Finally, don't make the mistake of thinking that your Skype conversations are secure. The NSA can snoop on anything, any time. No matter what you think of Microsoft, Skype is not secure.

All in all, Skype is marginally useful — the desktop version works reasonably well — but the Windows 10 Skype app has a long way to go.

The folks at Skype are coming up with new capabilities all the time. For example, Skype Translator — which is slowly getting better — may offer real-time voice translations between English and other Romance languages . . . and possibly a few others, as well. We're still a long way from Arthur Dent's Babel Fish. Eurgh!

Exploring Skype Alternatives

Skype runs on just about anything: iPhone, iPad, Android smartphones and tablets, Windows 7 or later, macOS, Xbox One, BlackBerry — just about anything.

At the time the book went to press, Skype ran better on all those machines than the Windows 10 app. Don't have a Skype app on some random computer? No problem. Just go to www.skype.com and download it.

ASK WOODY.COM

I've played with Skype on dozens of machines, and I'm convinced that the best way to use voice-only Skype is on a smartphone, and the best way to use video Skype is on a tablet or a smartphone with a big screen.

If you have only a desktop PC connected to the Internet, you don't have many choices. If you're using a laptop with a built-in camera and mic, Skype will work and the picture may be great, but I bet you won't be impressed by the Skype sound quality. On the other hand, if you have an iPad, an Android tablet, or a big-screen smartphone, you're going to find that setting up and using Skype is a lead-pipe cinch. Download the app and install it, and everything just works.

WARNING

The only downside? Skype is tied to Microsoft accounts. Someday, you'll be able to connect using an everyday phone number. But we're still in username hell.

Alternatives to Skype? Jeeeeeeeeeeeez. Just about everybody does over-the-top voice and video calls these days. ("Over-the-top" means they run on the Internet directly, not through the phone company, and they're thus basically free.) A quick Google scan brings up the names of dozens of programs and program-less websites that can do the job. These are the big competitors:

» **Facebook Messenger,** www.facebook.com/video, works great with anyone who has Facebook. Text chat, video, emojis, stickers, the whole nine kilometers. Chatting competitor WhatsApp is now part of Facebook, swallowed up for a paltry $19 billion. If you and your friends are on Facebook, it's an excellent choice.

» **Viber,** www.viber.com (900 million users?), also lets you call regular (landline) phones all over the world. Your Viber ID is your phone number.

» **LINE,** www.line.me/en, may be the biggest chat app of all in terms of volume (400 million active users?), with a solid hold throughout Asia. It was built by the employees of NHN Japan, in response to the Tohoku earthquake in Japan in March 2011. With LINE, you can add contacts by scanning QR codes and phone numbers or by shaking phones simultaneously; it has Facebook-like posting capability, groups, locations, and just about any feature you can imagine. Easy to see why it's spread that fast.

Many others are available, including China-based Weibo (380 million?), WeChat (960 million?), Renren, ringID, Hike Messenger, and Tango. Google Hangouts (`https://plus.google.com/hangouts`) and Zoom seem to be everyone's favorites, with all sorts of problems real and imagined. Apple's FaceTime works tremendously well — but on only Mac, iPhone, and iPad.

ASK
WOODY.COM

What do I use? Glad you ask. I have many messaging programs set up on various machines, but most of the time I want to use my smartphone, not my desktop. Not long ago, most of the friends I wanted to call used LINE, so I almost always pulled up LINE. More recently, though, Facebook Messenger has taken the number-one spot for me, as shown in Figure 2-12.

FIGURE 2-12:
Facebook's Messenger is now my go-to messaging platform.

LINE is easy to navigate and reliable, works on (almost) any computer, tablet, or phone, and is drop-dead simple to set up and use — even for dummies. See the sidebar titled "The case for LINE." Facebook Messenger is a natural for folks who are using Facebook anyway. Tough choice.

THE CASE FOR LINE

Microsoft's support for Skype in "modern" Windows bobs up and down. My recommendation is that you use just about anything else until MS gets its act together. When I'm talking with people on an iPad, a Mac, or an iPhone, my first choice is FaceTime. If your friends are on Facebook, it's hard to beat Facebook's messaging app. But for a mixed environment, I swear by LINE. It works on just about everything.

LINE covers the gamut from plain old phone calls to text, images, video, and audio, and it's free. If you know people in Asia, chances are very good they already have it and depend on it — and they may be a bit surprised that you aren't using it, too.

LINE makes its money by selling zillions of sets of emojis and stickers. They're cheap and people love them. Right now, more than a billion stickers are sent every day. The company went public in Japan in July 2016 with a market capitalization of about $6 billion.

LINE has one significant limitation: When you create an account, it can be used on only one mobile device and one personal computer. If you want to run LINE on both your iPhone and your iPad, you need to use two different accounts: You can verify only one phone per mobile phone number or email address.

You can get the Windows (desktop) version of LINE here: `http://line.me/en/download`.

Chapter **3**

Navigating the Microsoft Store

I f you're familiar with buying programs in the Apple App Store or the Google Play Store, you already know about 90 percent of the procedures you'll find in the Microsoft Store.

ASK
WOODY.COM

That said, the selection, breadth, and quality of apps are considerably better in either the App Store or the Play Store. I hate to be the bearer of bad news, but developers these days go for iOS apps and Android apps long, long before they think about Windows 10. Whether that will change anytime soon remains to be seen. Microsoft's working on it, but they've been working on it for years.

The reason's simple: money. There are large fortunes to be made with cool apps in the App Store and the Play Store. There's also a reasonable amount of money in apps that are designed to run in Facebook and, increasingly, apps that run on the Internet (see the "Progressive Web Apps" sidebar). But the Microsoft Store is less than a backwater when most developers tally up the shekels.

Microsoft's Windows Store launched simultaneously with the release of Windows 8. As part of the release of the Windows 10 Fall Creators update — version 1709 — Microsoft, with great fanfare, changed the Windows Store to the Microsoft Store, peddling more and different items, hinting at a strong link to the brick-and-mortar Microsoft stores dotted around the world. The re-branding didn't accomplish much.

REMEMBER

The Microsoft Store is a big, extensible, very usable source of new programs for all of Windows 10.

Yes, you read that right. Although the Microsoft Store used to be the sole province of Metro style apps — what we call Universal Windows apps in this more enlightened age — now the Microsoft Store carries all kinds of apps, even ones that run exclusively on the Windows desktop. You can even buy Microsoft hardware in the Store. Games. Movies. Anything to turn a buck. Or a euro.

Apps make or break any computer these days, and Microsoft knows it. That's why you find some popular apps in the Microsoft Store — it's good for you and good for Microsoft, over and above the 30 percent commission Microsoft makes on every sale.

PROGRESSIVE WEB APPS

A revolution is going on – from web apps running in a browser, to web apps running outside the browser, to hosted web apps, which are pulled down dynamically on execution, to progressive web apps, which blur the distinction between web-based apps and native apps.

Progressive web apps (PWAs) aren't so much Google's much-better alternative to Win10-only Universal Windows programs (formerly known as Metro apps, Universal apps, Windows Store apps, or any of a half-dozen other monikers) as they are a genuine attempt to make browser-based applications look and feel more like regular old apps. Chances are good that you've never seen a PWA in action. But they're definitely coming. At some point.

The theoretical benefits of PWAs over UWPs are enormous. Just for starters, UWPs can run only in the stripped-down Windows 10 environment. PWAs, on the other hand, should be able to run on just about anything that supports a browser — particularly Chrome, or Chrome OS. Yeah, that includes Chromebooks.

The browser requirement has vanished in the past couple of years, banking on a concept called *service worker*. Horrible name, but web folks are good at horrible names. It looks to me like there will be a headlong dash into developing PWAs — and that UWP's days are numbered. Time will tell.

PWAs can be published in the Microsoft Store too. However, no cool names are available yet, and it remains to be seen how this type of app will work.

For many folks, the Microsoft Store continues to be a major disappointment. The big-name apps are appearing in the Microsoft Store at glacial speed — there wasn't even a legitimate Facebook app until more than a year after the original launch, and even now big-name players shun the Microsoft Store with glee. Slowly Microsoft's filling in some of the gaps — they're even paying developers with new ideas and cajoling old-timers as best they can — but don't be surprised if you hear about a cool Apple or Android app, and you can't find it in the Microsoft Store. Happens all the time. Increasingly, Microsoft itself is making cool apps for iOS and Android and neglecting its own Microsoft Store.

The only way you can get new Universal Windows apps for Windows 10 is to download and install the app from the Microsoft Store. Although large companies can put Universal apps on their Windows devices (using a technique known as *sideloading*), normal people like you and me have to go through the Microsoft Store: the alpha and omega of Universal Windows apps.

Checking out What a Universal Windows App Can Do

The longer Microsoft Store is available, the more apps you'll find there. The apps do all sorts of things, but each app also must meet a set of requirements before Microsoft will offer the app in the Microsoft Store.

Here's a short version of what you can expect from any app you buy (or download) via the Microsoft Store:

>> **You can get both Universal Windows apps (which are supposed to run on any version of Windows 10, including the extinct Windows 10 Mobile version on phones) and legacy-style apps (which run on the old-fashioned desktop) from the Microsoft Store.** If you want a new program for the desktop, you may be able to find it in the Microsoft Store, or you may be able to get it through all the old sources — shrink-wrapped boxes, monster download sites — to find and install what you want. But if you want a new Universal Windows app, you must get it through the Microsoft Store — unless you have a big company. (See the sidebar "Bypassing the Microsoft Store restrictions.")

>> **Universal Windows apps can be updated only through the Microsoft Store.** If your apps are set to update automatically — the default — when an update is available, the Store tile on the Start screen shows a number, indicating how many apps have updates available. See "Updating Your Microsoft Store Apps," later in this chapter.

>> **Apps that use any Internet-based services must request permission from the user before retrieving, or sending, personal data.**

>> **Each app must be licensed to run on up to five computers at a time.** For example, if you buy the latest high-tech version of Angry Birds, you can run that same version of Angry Birds on up to five Windows 10 devices — computers, tablets, laptops, Xbox One consoles, HoloLens augmented reality glasses, giant Surface Hubs — at no additional cost.

>> **Microsoft won't accept apps with a rating over ESRB** *Mature,* **which is to say adult content.**

>> **Apps can (thankfully) put only one tile on the Start menu.**

>> **Apps must start in five seconds or less and resume in two seconds or less.** Microsoft wants apps to be speedy, not sluggish; thus, it requires developers to make sure their apps meet this requirement.

In addition to the basic requirements for any app, you're also likely to find that the following is true of most apps:

TIP

>> **Microsoft's tools help developers create trial versions of their apps, so you can try before you buy.** The trial versions can be limited in many ways — for example, they work only on a certain number of pictures, messages, or files or only for a week or a month — before demanding payment. That's all part of the plan.

Where try-before-you-buy has a long and checkered history on the desktop, it's baked into many Microsoft Store apps. Microsoft is very strict about requiring the developer to explain precisely what has been limited and what happens if you fork over the filthy lucre.

ASK
WOODY.COM

>> **If an app breaks, you can complain to Microsoft, but the support responsibility lies 100 percent with the developer.** Although Microsoft acts as an agent in the distribution and sale of apps, Microsoft doesn't actually buy or sell or warrant anything at all. Even the license for using the tiled-style program goes between seller and buyer, with Microsoft out of the loop.

>> **Many apps attempt to get you to buy more — more levels, more features, more content.** Microsoft has that covered, just like Apple and Google: Orders generated by the app must go through the Microsoft Store. Only Microsoft can fulfill the orders. Ka-ching.

TIP

Don't confuse the Microsoft Store — which hooks directly into Windows 10 — with, uh, Microsoft stores, which existed in the real world for several years. Brick-and-mortar Microsoft stores were popping up all over the place until the Covid-19 pandemic. (Another bright idea borrowed from Apple, who in turn, got it from . . . Tandy?) The online version of a Microsoft store, www.microsoftstore. com, was as an online extension of the physical Microsoft stores. In the online Microsoft store, you can buy the new Microsoft Surface computers, applications that run on the desktop, as well as competitors' computers, Xboxes, headphones, mice, smartphones, Windows 10, Office — in short, everything you find at a meat-space Microsoft store.

BYPASSING THE MICROSOFT STORE RESTRICTIONS

Microsoft runs the Microsoft Store as a business — a tightly held business — and for that reason, it restricts what can be bought in the Microsoft Store. Microsoft can reject an application submitted to the Microsoft Store for a huge variety of reasons.

Here's the key point you need to understand about the Microsoft Store: With two exceptions, the Microsoft Store is the *only place* you can get Universal Windows apps or Windows 10 apps. See Book 1, Chapter 2 for a description of Windows 10 apps.

The exceptions:

- Big companies can bypass the restriction and put their own apps on Windows machines using a technique called *sideloading.* At least in theory, sideloading can be accomplished only on machines locked into a corporate network.

- If you *jailbreak* your PC, you may be able to put any tiled Windows 10 apps you like on your computer — Microsoft's censors no longer apply. On the other hand, jail-breaking your computer voids every warranty in existence and automatically disqualifies you from Microsoft support. Think: No security patches, lots of exposure. Because there are very few apps available for jailbroken Windows 10 machines, there's basically no incentive to jailbreak your computer.

Unlocking (which may or may not be accompanied by jailbreaking) allows you to switch carriers, if you bought your PC from a carrier who's locked in its services. Some carriers in the United States, for example, may offer a discounted price for your tablet in exchange for a multi-year Internet contract. If you unlock the computer (or tablet), you may (or may not) be able to hook it up to a different network. All sorts of penalties may apply. I don't recommend that you jailbreak your PC. But if you find an app that you really want and Microsoft won't let it into the Microsoft Store, jailbreaking may be your only option. Google is your friend.

However, the case I am describing is not valid for most PCs because they are not bought from carriers. Most desktop PCs are not locked, so they don't need jailbreaking, especially those built by users or bought from traditional PC vendors such as Dell, HP, and ASUS.

If you're familiar with the Apple view of life, the Microsoft Store is comparable to the App Store and the iTunes Store in general. Apple Stores, of-bricks-and mortar persuasion, are analogous to Microsoft stores.

Confused? Yeah. Such are the vagaries of Microsoft branding.

Browsing the Microsoft Store

When you're ready to venture into the Microsoft Store for Universal Windows apps, tap or click the Store tile, and you see something like Figure 3-1.

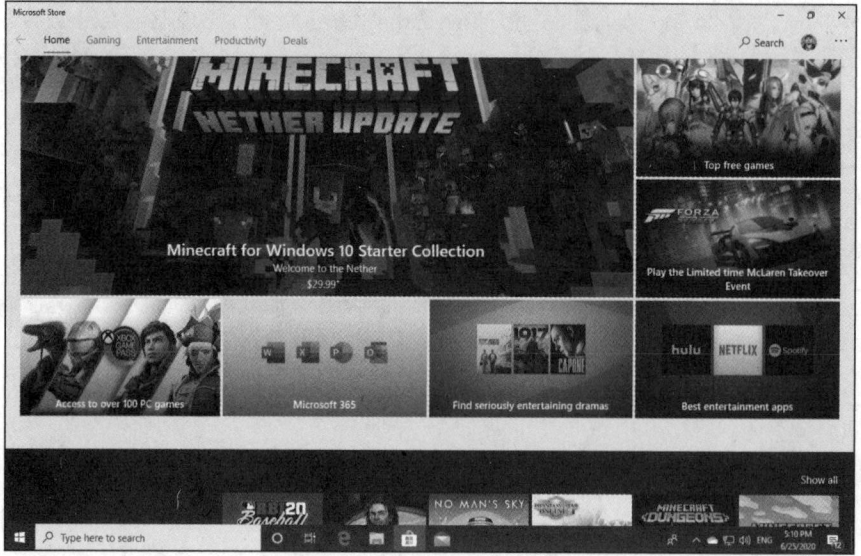

FIGURE 3-1:
Here's a peek
at the Microsoft
Store.

Moving around in the Microsoft Store is a little funky. The following tips can help you move around and find what you're looking for:

>> **You need a Microsoft account** to get anywhere beyond basic searching. You can't even download a free app unless you're logged in with a Microsoft account. (Microsoft needs it to keep track of what apps are on your machine.) If you logged in to Windows 10 with a local account, the Microsoft account requirement splats you right in the face, as in Figure 3-2. However, in newer versions of Windows 10, such as the May 2020 update, you can keep saying No and eventually download and install free apps from the Microsoft Store. Note that if you want to try paid apps and games, a Microsoft account is a must.

ASK
WOODY.COM

If you decide to use a local account but need to sign in with a Microsoft account to get updates or new apps from the Microsoft Store, set up a bogus Microsoft account (see Book 2, Chapter 5) and use the facility offered in Figure 3-2 to sign in to each app separately. That way, you'll be warned before you venture into another location that requires a Microsoft account.

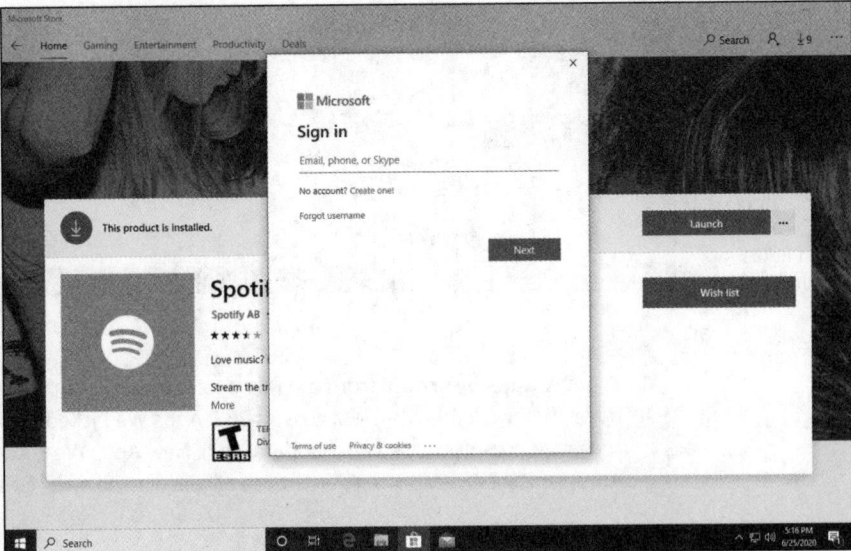

FIGURE 3-2:
You can only window-shop with a local account.

>> **To order an app,** tap or click the app's tile. The Store takes you directly to the ordering screen for the app. For example, if you tap or click the tile for the Adobe Photoshop app, you see the ordering page in Figure 3-3.

At the top, you see an overview of the app and its price. Scroll down, and you should see a list of hardware requirements, release history (except for Microsoft's own Universal Windows apps, which don't have histories), a list of permissions required, languages, and links to the manufacturer's site. Keep scrolling and you find the ratings and reviews.

ASK
WOODY.COM

The star rating shouldn't impress you — it's the accumulated wisdom of all the people who've bothered to rate the app. But the supported languages section, if there is one, may be of interest — and the permissions list is detailed and thorough. At the very least, you can vent your spleen on the Reviews page if the app doesn't live up to your expectations.

Navigating the
Microsoft Store

FIGURE 3-3:
The app-ordering page for the Adobe Photoshop Elements app.

» **To view apps by group,** from the Microsoft Store, click Productivity and scroll down, below all the ads for Microsoft 365. Apps We Picked for You (there's a reason Microsoft collects all that data, eh?), New Apps We Love, Essential Apps, Apps for Digestive Disorders — they're all just a scroll away (see Figure 3-4).

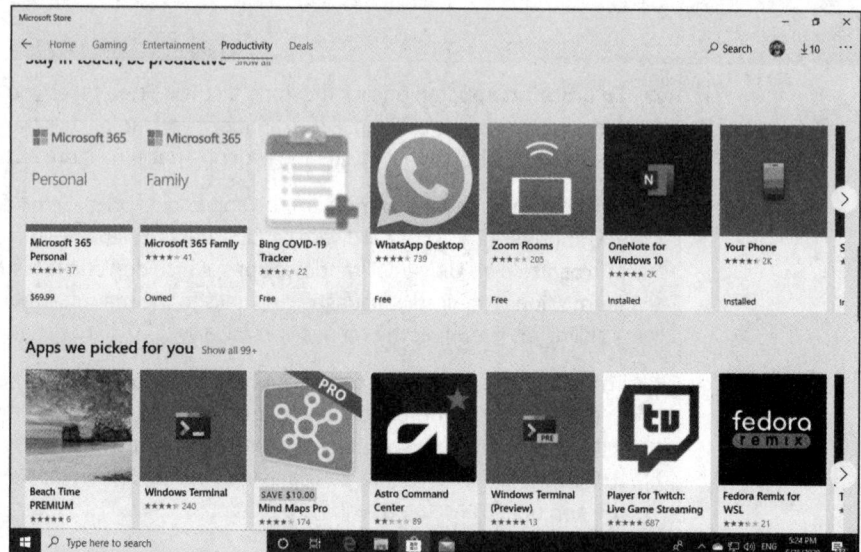

FIGURE 3-4:
Apps run quite a gamut, but they're a mile wide and an inch deep.

WARNING

Beware the marketing tricks. For example, the Future Managers app is a free shell whose sole purpose appears to be selling downloadable PDF "books." Whether that's its only feature is open to debate. The app permissions give this shell program the capability to access your Internet connection and your home or work networks, which is not comforting.

Microsoft has spent millions vetting the apps in the Microsoft Store, but you'll find crapware like the Future Managers everywhere. Ever wonder why first-tier developers don't want to put their stuff in the Microsoft Store?

Searching the Microsoft Store

You can search the Microsoft Store using the search box in the upper right, and/or by taking advantage of built-in categories. Here's how:

1. **Inside the Microsoft Store, type something in the search box in the upper right.**

I typed *news.*

2. **At the top, click Show All.**

Microsoft Store shows you a list of apps, grouped by categories, as in Figure 3-5.

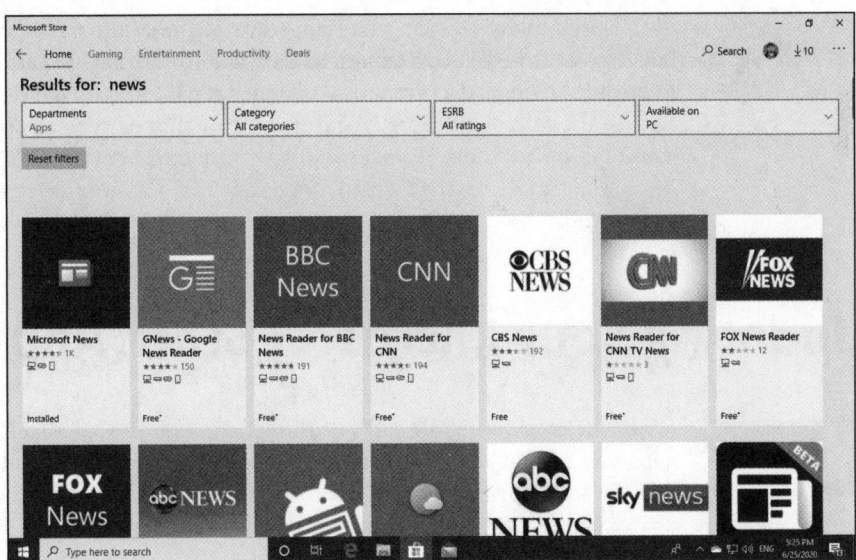

FIGURE 3-5:
Looking for news in the Microsoft Store.

3. **Choose a type and a category, should you feel so inclined.**

In Figure 3-6, I looked for News & Weather apps. Look at the quality of apps on offer. Aside from a small handful of readily identified major news organizations, there's an enormous collection of apps from organizations that, shall we say, aren't likely to be on your A list.

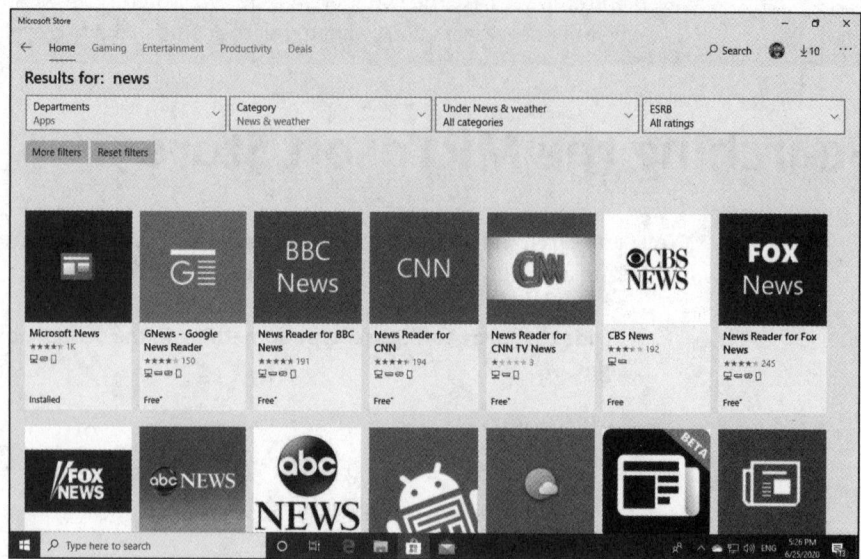

FIGURE 3-6:
Lots of, uh, big-name News apps on offer.

ASK
WOODY.COM

Ever wonder why so many bad apps are in the Microsoft Store? Preston Gralla has a great investigative report in *Computerworld* that explains it. Back in 2013, Microsoft "launched a promotion in which it paid $100 to developers for apps they sent to the Microsoft Store, regardless of quality or type of app. Each developer could get up to $200." www.computerworld.com/article/2600035/microsoft-windows/did-microsoft-help-seed-the-market-for-windows-store-scam-apps.html.

Updating Your Microsoft Store Apps

Microsoft is updating all sorts of things through the Microsoft Store — not just apps you bought or downloaded from the Microsoft Store, but also the built-in Windows 10 apps (UWP), and the list is likely to expand over time.

Sometimes, the Microsoft Store doesn't update itself (as it will in the normal course of events). You should check from time to time to make sure you have the latest updates for absolutely everything. Here's how:

1. **Start the Microsoft Store app.**

   It's probably on your taskbar.

2. **Up at the top of the window, next to your picture, click the ellipsis (the three dots) and choose Downloads and Updates.**

   In Figure 3-7, you can see my list of Windows 10 apps with updates available for installation.

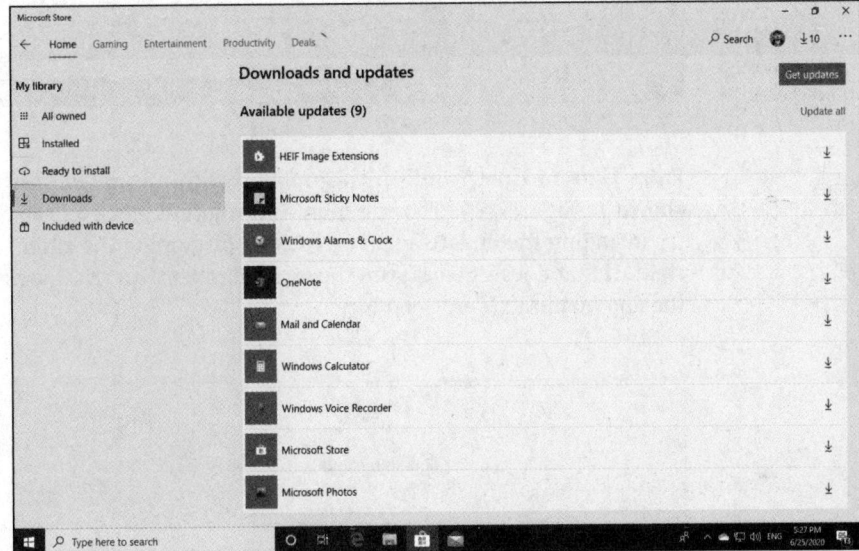

FIGURE 3-7:
Click Get Updates to make sure everything is up to speed.

3. **Click or tap Get Updates.**

   If you have any waiting updates, they start installing, as shown in Figure 3-8.

4. **If the Microsoft Store is updating your apps too slowly and seems to ignore your request to Get Updates, click Update All in the top-right corner.**

   This action forces the Microsoft Store to focus on updating all your apps, right now.

In the normal course of events, you'll want to update all your apps, but if you know of a bad update (and they happen), you can pick and choose which apps you want to bring up to date.

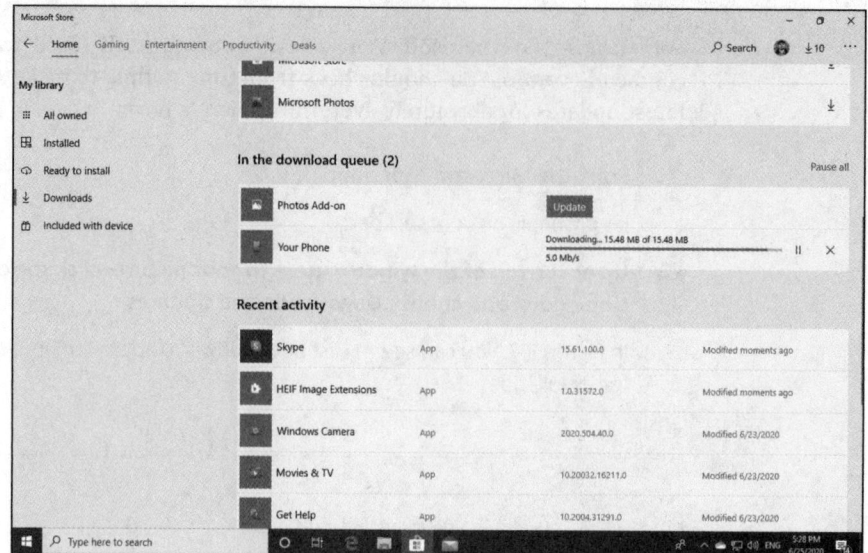

FIGURE 3-8:
The updates go through, although you can stop one of them if you press the X quickly enough.

From time to time, you'll hit a problem with an update. An error appears, as shown in Figure 3-9. Too see more information, click or tap See Details. Then, try installing the update again by clicking or tapping the circle-arrow icon on the right. If that doesn't work, run the error number through Google or try contacting the app manufacturer. Good luck.

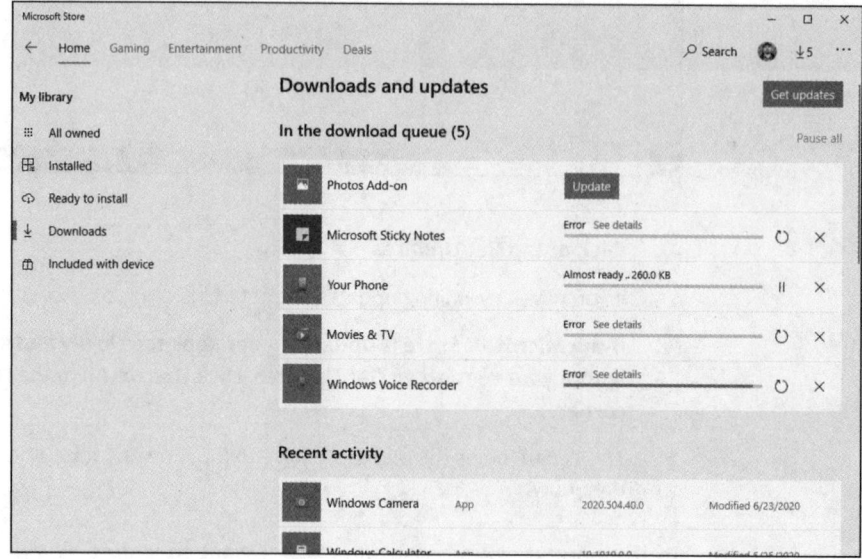

FIGURE 3-9:
From time to time, even app updates fail. Don't panic. Call Saul.

IN THIS CHAPTER

» Searching for games

» Playing with game mode

» Using the game bar

» Testing your connection to Xbox Live services

» Finding old games, reborn

Chapter **4**

Games, Games, and Games

The Windows 10 Start Menu is littered with tiles for commercial games — Candy Crush Soda Saga, March of Empires, even Minecraft (which is owned by Microsoft). Game makers have paid dearly for those spots. Microsoft Store offers tons of games, too. Many of them, including some free ones, are well worth trying.

If you're looking for old Windows standbys like Minesweeper and Solitaire, they're here too — but they're all gussied up, fabulously more playable, and touch friendly, unlike their elder counterparts. They're also freemium products. (See the "What is freemium?" sidebar.)

The free touch-savvy Minesweeper and Solitaire may be enough to convince you to buy a touch tablet. No joke.

Unfortunately, the old Windows 7 cheats don't work anymore, but the eye candy should more than compensate.

In this chapter, I also talk about a sampling of free games that you can download from the Microsoft Store and play directly on just about any Windows 10 computer. You don't need a monster graphics card, $600 joystick, or the reflexes of a trained fighter pilot to play.

The free games that come with Windows 10 run quite a gamut. Microsoft itself offers loads of free games, and some of them may be preinstalled on your computer. The poster child of the add-on bunch, Cut the Rope, runs on iPads and iPhones, but the game action on Windows 10 is faster — primarily because the whole game was rewritten (with Microsoft's help) in HTML5. You can read all about the technical dexterity on the UK Team blog for the Microsoft Developer Network, `https://blogs.technet.microsoft.com/jweston/2012/01/12/cut-the-rope-on-ie9/`.

If you're looking for Xbox games, you're in the wrong place. The Xbox ecosystem has some overlap with Windows 10, but by and large, Xbox gaming exists at a completely different level of complexity. If you're looking for an intro to that world, start at `support.xbox.com`.

I'm going to assume that you haven't coughed up the money to buy an Xbox One: If you have, you should approach Xbox gaming from the Xbox side, not the Microsoft Store Game app side.

Although the Xbox Console Companion app has some very cool capabilities — and more than a few top-ranked games — in my experience, they don't work that well if they live in an Xbox-free environment. That may change over time — Microsoft now supports attaching an Xbox console to your PC — but for now, the Xbox One provides a much, much better gaming experience than a tablet or PC connected to a TV set.

There's a new kid on the block: Steam. Serious gamers should consider subscribing to Steam, a digital game distribution center (PC, Mac, Linux, with limited help on iOS, Android, and PlayStation) combined with social networking, backups, tracking in-game achievements, micro-payments, and much more. You can even buy SteamOS machines, for Steam only. There's a reason why Steam accounts for almost 20 percent of worldwide PC game sales, and the number's increasing rapidly. See `http://store.steampowered.com/`.

Searching the Store for Games

Want to see what games will run on Windows 10? Head to the Microsoft Store. Here's how:

1. **Click or tap the Store shortcut, down on the taskbar.**

   The Microsoft Store appears.

2. **Tap the Gaming tab.**

   An enormous array of tiles for games appears, as shown in Figure 4-1. Choosing games is a black art, all by itself, but if you see a game that looks interesting, check it out.

3. **Scroll down to Top Free Games, and click or tap Show All.**

4. **Tap any game that interests you.**

   I chose Asphalt 9: Legends, as shown in Figure 4-2. The Microsoft Store displays a complete description of the game and presents you with an opportunity to install the app. The description may include a notice that you can buy stuff when you're inside the game (Offers In-App Purchases). Scroll down farther and the description of the game includes some indication of what's available and how much it costs.

5. **To install the app, tap Get (if the app is free) or the button with the price.**

Games, Games, and Games

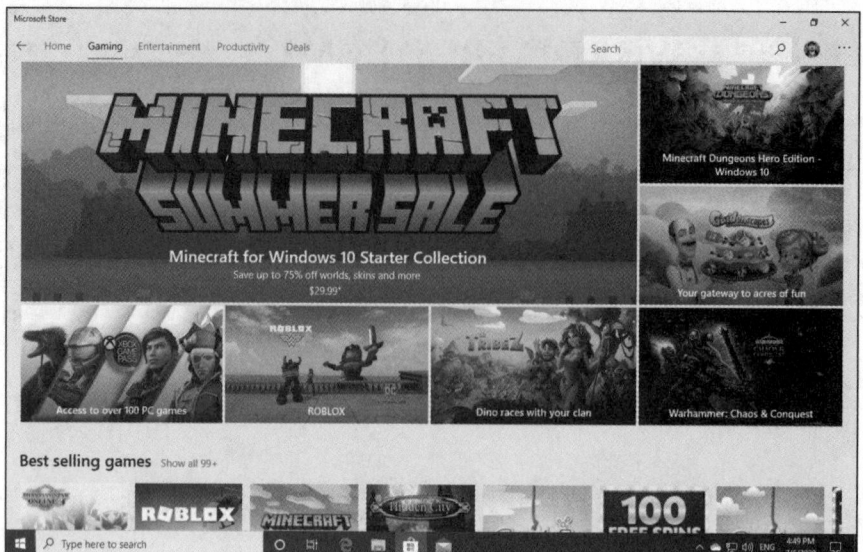

FIGURE 4-1:
Games
offered at the
Microsoft Store.

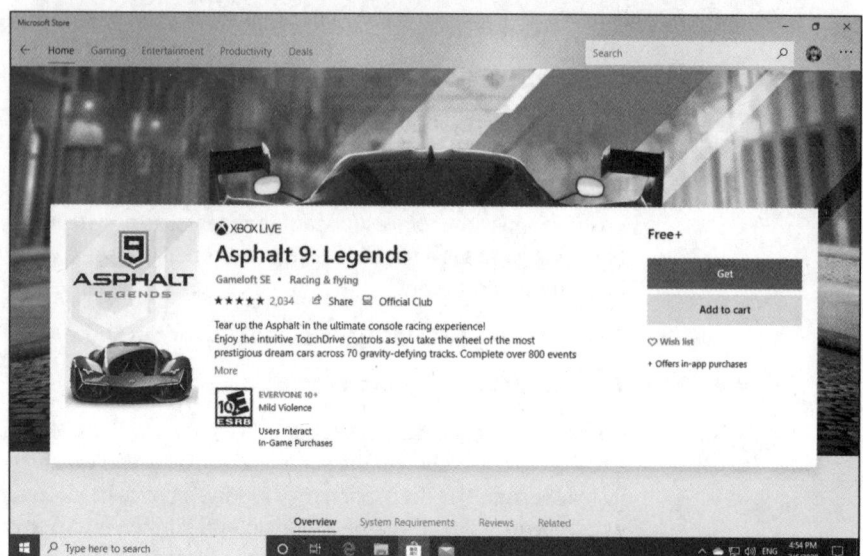

FIGURE 4-2:
If it tickles your
fancy, install it.

6. **If there's a charge, verify your billing details and provide a password.**

While it's downloading, you see a progress bar in the Microsoft Store. When
your app has finished downloading, it appears as an entry on your All Apps list,
just like any other freshly installed Windows 10 app.

Apps that are marked Xbox will, in general, play on plain old Windows 10 machines. For example, Despicable Me: Minion Rush works fine on Windows 10. It also works on Xbox.

7. **To run the game, click Start, look under Recently Added or through the list of apps and programs, and click its shortcut.**

   For example, the Asphalt 9: Legends! Game appears in the Recently Added list, just like any other Windows 10 app, as you can see in Figure 4-3. It's also in the full apps list, under *A* for *Asphalt.*

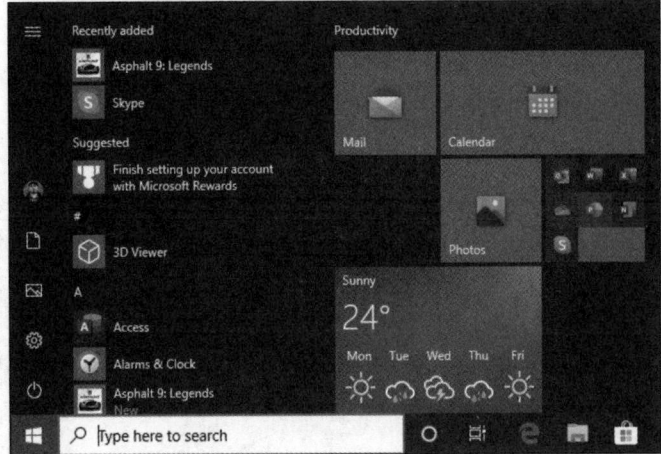

FIGURE 4-3: Games appear just like any other app in the Recently Added list.

Downloading and installing a game is one-click easy. Finding them and beating them are anything but.

Enabling Game Mode

In April 2017, with the Windows 10 Creators update (version 1709), Microsoft introduced a feature aimed at gamers. It's called — wait for it — game mode! Original, isn't it?

Game mode is a set of tools, options, and settings that make gaming more pleasant on Windows 10. According to Microsoft, game mode helps games render more frames on the screen while you play them by focusing your PC's processing power on the game, not background tasks.

The basic idea is that you enable game mode when you play a game to avoid extreme slowdowns, drops in frame rates, interruptions caused by notifications, and other annoyances. In theory, Windows 10 detects when you're playing a game, and enables game mode automatically. But that doesn't work every time, especially when you're playing an older title.

To check to see if game mode is enabled — and to enable it if necessary — follow these steps:

1. **Click or tap Start, and then click the Settings icon.**

   Windows 10 Settings opens.

2. **Click or tap Gaming. Then, on the left, choose Game Mode.**

3. **Set the Game Mode switch to On, as shown in Figure 4-4.**

4. **Close Settings and start the game.**

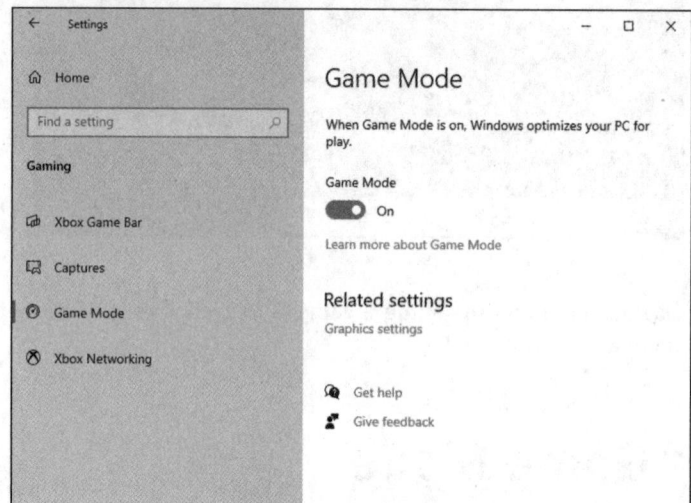

FIGURE 4-4:
Turning game mode on or off.

Using the Game Bar

Game mode in Windows 10 comes with a useful tool called the game bar. When you start a game, press Windows+G. The game bar appears over your game, with several widgets that offer useful settings as well as data, as shown in Figure 4-5.

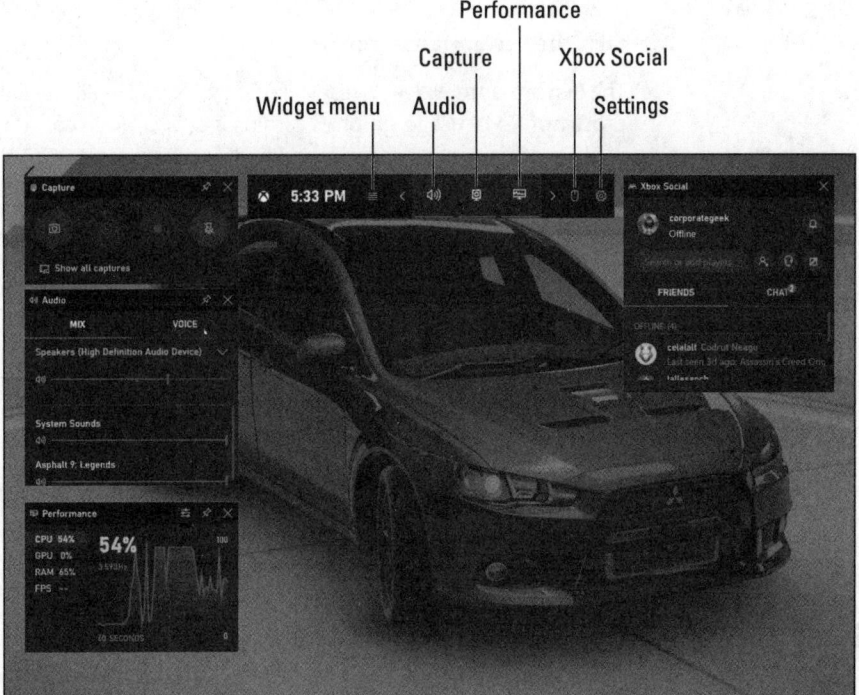

Performance

Capture   Xbox Social

Widget menu   Audio   Settings

FIGURE 4-5:
Use the game
bar while you
play games in
Windows 10.

To get familiarized with the game bar, follow these steps:

1. **Start a game that you want to play, and then press Windows+G.**

   The game bar appears (refer to Figure 4-5).

2. **On the game bar, click the Widget menu icon (labeled in Figure 4-5).**

   A menu with widgets that can be enabled and disabled is shown. The ones that are enabled have a start to the right of their name. Click their name to enable or disable them.

3. **Click the Audio icon to if see your audio devices are set for your game. Change the settings, if necessary.**

4. **Click the Capture icon.**

   The Capture widget is displayed. It has buttons for taking screenshots of your game, recording a video of your gameplay, and turning the microphone on and off.

5. **Click the Performance icon.**

   The Performance widget is displayed, showing you real-time data about the processor (CPU) usage, graphics card usage (GPU), RAM consumption, and the number of frames per second rendered on the screen (FPS). This data is useful to gamers who play demanding video games.

6. **Click the Xbox Social icon.**

   The Xbox Social widget gives you tools to chat with friends, see who is online, invite them to a party, and so on.

7. **Click the Settings icon (gear).**

   You get access to settings that you can use to personalize game mode and the game bar.

8. **To hide the game bar, click anywhere outside it or press Windows+G again.**

Testing Your Connection to Xbox Live Services

If you play online games and connect to Xbox Live services from Windows 10, you want your Internet connection to work well so that you don't encounter lag. Windows 10 has a tool hidden in the Settings apps that checks the following:

>> The status of your Internet connection.

>> Whether or not Xbox Live services are up and running.

>> The latency of your connection to Xbox Live services.

>> How many packets are lost when traveling between your Windows 10 gaming PC and Xbox Live services.

>> The NAT type, a network address translation service that lets you know exactly where your PC is on the Internet and delivers information to your PC while you play games and use Xbox Live services.

>> Server connectivity to Xbox Live servers. If you can't access the servers, you may not be able to play online games with your friends.

Here's how to test the quality of your Internet connection to Xbox Live services:

1. **Click or tap Start, and then click the Settings icon (gear).**

   Windows 10 Settings appear.

2. **Click or tap Gaming. Then on the left, choose Xbox Networking.**

   Windows 10 starts checking everything. After a while, you see Connection Status, Performance, and Xbox Live Multiplayer sections, as shown in Figure 4-6.

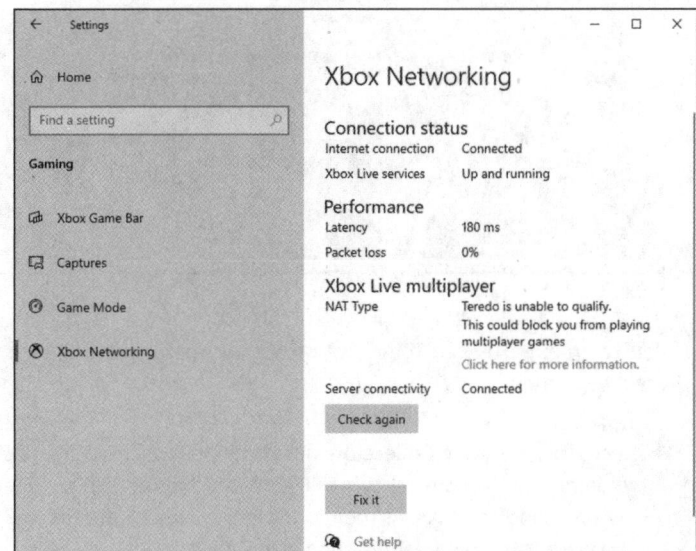

FIGURE 4-6: Using Xbox Network to test your connection to Xbox Live services.

Click the Fix It button in Xbox Networking when you encounter problems playing games online and when accessing Xbox Live services.

TIP

Bringing Back the Classics

Admit it. You want to play Solitaire on your new Windows 10 machine. And Minesweeper. Just like you did in Windows 3.1. (Windows 3.0, actually.) Well, you're in luck — and they're easy to find if you know where to look.

Just crank up the Microsoft Store, and in the search box in the upper right, type **"Microsoft Studios"** — including the quotation marks. Press Enter. In the Games section, click Show All. You get a list of all the apps published by Microsoft Studios, as shown in Figure 4-7.

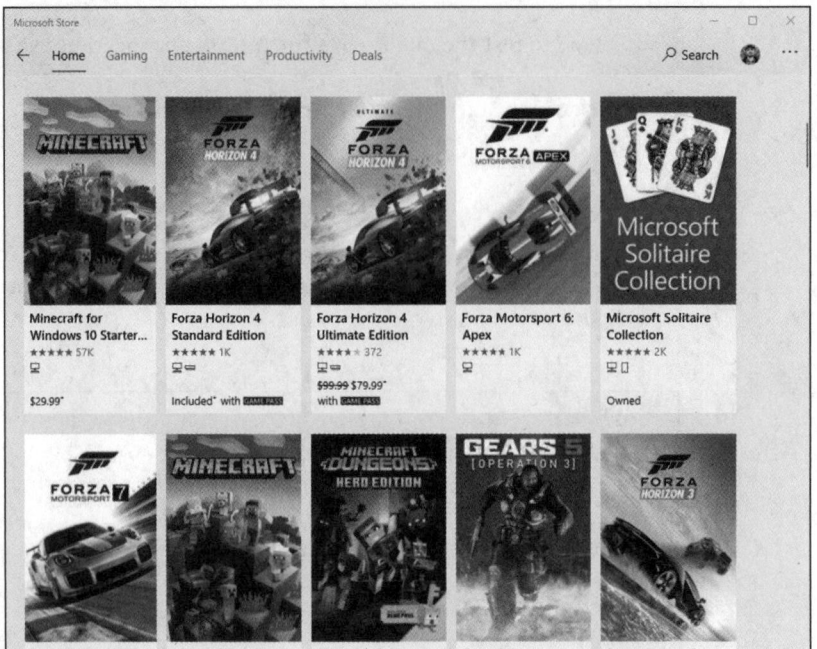

FIGURE 4-7:
These apps are
published by
Microsoft Studios.

If you're an experienced Windows user, you might want to pick up some or all of these free games:

>> **Microsoft Solitaire Collection** includes Klondike (the game you no doubt remember as Solitaire, shown in Figure 4-8), Spider Solitaire, FreeCell, Pyramid, and TriPeaks. As mentioned in the "What is freemium?" sidebar, if you want your Solitaire Collection without video ads, you have to pay for the privilege.

   None of the old cheats work in Solitaire — you can't switch how many cards you flip in the middle of a hand, or peek — but you can still play with hints, or choose between one-card and three-card draws.

>> **Microsoft Minesweeper,** the game that BillG loved to hate, works very much like it has for many years, in many versions of Windows. See Figure 4-9.

>> **Microsoft Mahjong** brings the classic click-clack to the screen.

>> **Microsoft Sudoku** is explained in the next section.

>> **Adera** is a story-driven adventure game that you can play with your kids.

There are many more, but those Microsoft Studio games should keep you going for hours. Or days.

FIGURE 4-8:
Klondike, the game you remember from when you were a kid.

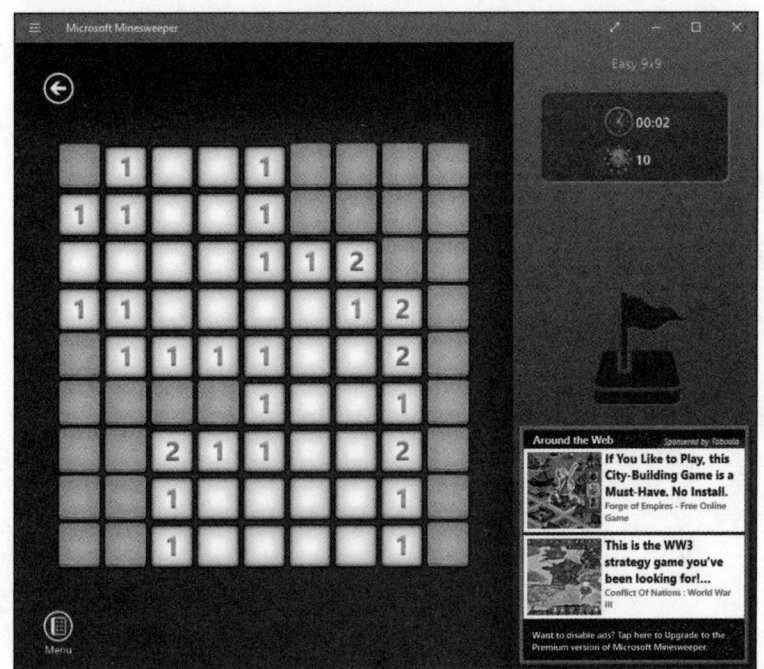

FIGURE 4-9:
Minesweeper works like the original but looks much better.

6

Socializing and Sharing from Windows 10

Contents at a Glance

Chapter **1**

Using OneDrive

I f you've used Windows for a while, you might recall the Microsoft online storage service known as SkyDrive. Those were the old days. Microsoft lost a trademark lawsuit in the UK with British Sky Broadcasting — the TV people — and instead of taking the lawsuit back for another appeal, the Redmond giant decided it was smarter to just stop using the term *Sky.* I'm astounded that a company can trademark the name *Sky,* but then again I'm still dealing with the idea that a company can trademark the name *Windows.*

Start with the basics: OneDrive is an online storage service, sold by Microsoft, which has some features woven into Windows, to make it easier to work with your files stored on Microsoft's servers in the cloud. (*Cloud* is another word for the web or the Internet.)

"In the cloud" is just a euphemism for "stored on somebody else's computer."

If you have a Microsoft account (such as an Outlook.com ID, or Hotmail ID, or any of a dozen other kinds of Microsoft accounts — see Book 2, Chapter 5), you already have "free" OneDrive storage space, ready for you to use.

The history of OneDrive is a tale of woe — starting many years ago as a rickety online file storage utility grafted onto Windows, with a heavy emphasis on photo storage, it morphed into a capable version in Windows 8.1, only to get its wings clipped in early versions of Windows 10. Now, with Windows 10 version 1709 (Fall

Creators update) and later, we finally have a full feature set and a reasonably solid cloud-storage capability.

OneDrive has many competitors — Dropbox (which I use, and did use for this book), Google Drive (see Book 10, Chapter 3), the Apple iCloud (which isn't quite the same, although you can get to it through a web browser), the Amazon Cloud Drive, Facebook storage, SugarSync, Box, SpiderOak, and cloud storage and sharing from many smaller companies. These competitors all have advantages and disadvantages — and the feature list changes from week to week. I talk about the tradeoffs in Book 8, Chapter 1.

In this chapter, I show you just about everything you need to know to make One-Drive work for you and in Windows 10.

What Is OneDrive?

OneDrive is an Internet-based storage platform with a significant chunk of space offered for free by Microsoft to anyone with a Microsoft account. Think of it as a hard drive in the cloud, which you can share, with a few extra benefits thrown in. One of the primary benefits: OneDrive hooks into Windows 10.

Microsoft, of course, wants you to buy more storage, but you're under no obligation to do so.

REMEMBER

As of this writing, OneDrive gives everyone with a Microsoft account 5GB of free storage (down from 15GB free in 2015), with 200GB for $2/month. Many Microsoft 365 subscription levels (formerly known as Office 365) have 1TB (1024GB) One-Drive storage, for as long as you're a subscriber. Back in 2015, the Office 365 subscriptions had unlimited storage, but Microsoft giveth and Microsoft taketh away.

Microsoft's offers change from time to time, but the general trend is that prices are going down, and it won't be too long before most online storage asymptotically approaches free.

The free storage is there whether you use your Microsoft account to log in to Windows, even if you never use OneDrive. In fact, if you have a Microsoft account, you're all signed up for OneDrive. See Figure 1-1.

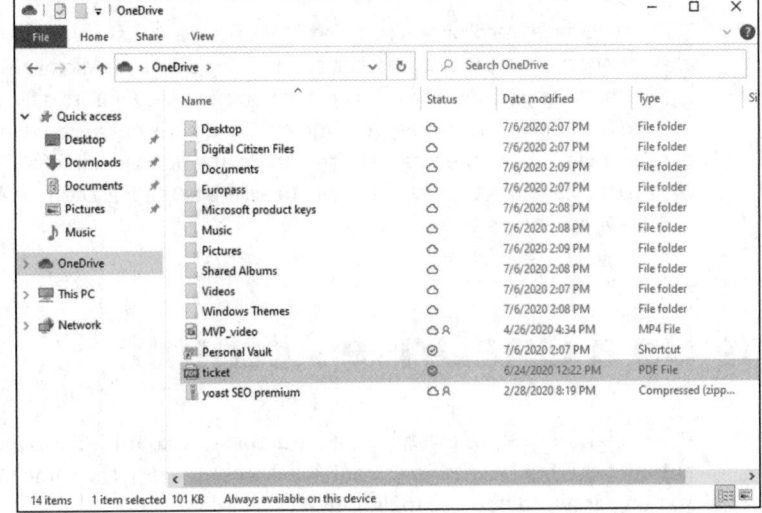

FIGURE 1-1:
OneDrive files
look and act a
lot like everyday
files, but they're
different.

Here's the full OneDrive shtick:

**ASK
WOODY.COM**

>> OneDrive does what all the other cloud storage services do — it gives you a place to put your files on the Internet. You need to log in to OneDrive with your Microsoft account (or, equivalently, log in to Windows 10 with your Microsoft account) to access your data.

>> OneDrive keeps a history of all changes you made to files over the past 30 days. That feature can be useful — and a lifesaver if you get hit by ransomware.

>> If you log in to a different device or computer (Windows, Mac, iPad, Android) using the same Microsoft account, you have access to all your OneDrive data.

>> You can share files or folders that are stored in OneDrive by sending or posting a link to the file or folder to whomever you want. So, for example, if you want Aunt Martha to be able to see the folder full of pictures of Little Billy, OneDrive creates a link for you that you can email to Aunt Martha. You can also specify that a file or folder is *Public,* so anyone can see it.

>> To work with the OneDrive platform on a mobile device, you can download and install one of the OneDrive apps — OneDrive for Mac, OneDrive for iPhone, iPad, or Android. The mobile apps have many of the same features that you find in File Explorer in Windows 10.

>> In Windows 10, you don't need to download or install a special program for OneDrive — it's already baked into the operating system.

>> If you have the program installed, OneDrive syncs data among computers, phones, and/or tablets that are set up using the same Microsoft account, as soon as you connect to a network. If you change a OneDrive file on your iPad, for example, when you save it, the modified file is put in your OneDrive storage area on the Internet. From there, the new version of the file is available to all other computers with access to the file. Ditto for Android devices.

Setting Up a OneDrive Account

If you sign in to Windows 10 with a Microsoft account, File Explorer gets primed automatically to tie into your OneDrive account, using the same Microsoft account ID and password you use to sign in.

But if you're using a local account (see Book 2, Chapter 5), life isn't so simple. You must either create a Microsoft account or sign in to an existing Microsoft account (and thus an existing OneDrive account) when you try to get into OneDrive. Here's the way to sign up for an account. You need to do it only once.

1. **On the taskbar, click the File Explorer icon.**

   You see File Explorer.

2. **On the left, click OneDrive.**

   You get a Set Up OneDrive splash screen, as in Figure 1-2.

3. **If you already have a OneDrive account on another computer, type the email address, click Sign In, enter the password, and click Sign In one more time.**

   OneDrive has you sign in with a Microsoft account. *Note:* You *must have a Microsoft account* to use OneDrive. It makes sense.

4. **If you don't have a Microsoft account or want to set up OneDrive with a new Microsoft account, leave the box blank and click Create Account.**

   Follow the advice in Book 2, Chapter 4 to get a Microsoft account set up.

   OneDrive gurgles and burps and makes changes to your File Explorer, adding some glue programs to both sides.

5. **When the OneDrive wizard says Your OneDrive Folder Is Here, click Change Location and select another folder on your computer (if you want that) or click Next to use the default path.**

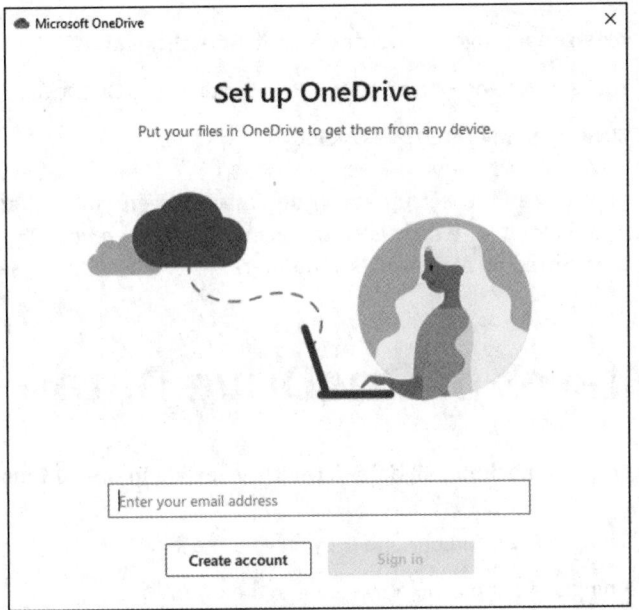

FIGURE 1-2:
If you're using a local Windows 10 account, hook it into OneDrive with a Microsoft account.

6. **Read the information about OneDrive on the next several screens, clicking Next on each one.**

Sooner or later, you see the most important screen in the OneDrive universe, Figure 1-3. That screen gives a hint of what lies ahead with Files On-Demand. See the next section for details.

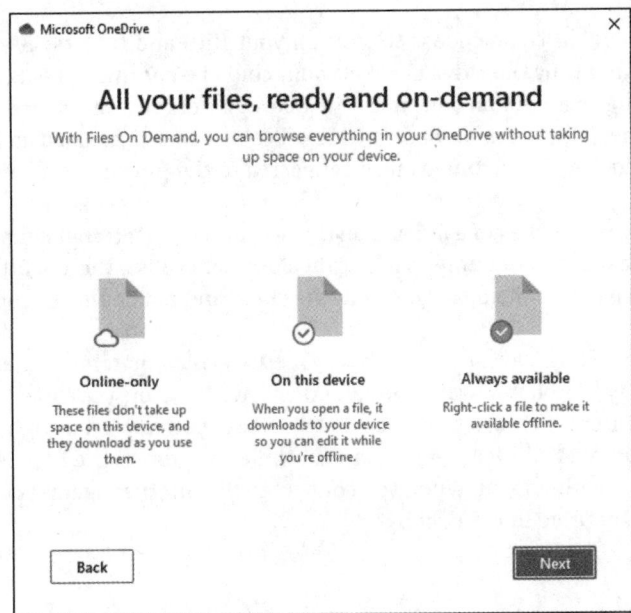

FIGURE 1-3:
A short explainer for Files On-Demand.

7. **When encouraged to Get the Mobile App, click Later.**

   You're presented with an opportunity to open your OneDrive folder.

8. **Click Open My OneDrive Folder.**

Now you're ready to set up synchronizing between your PC and your OneDrive files in the cloud — which is to say, syncing between your PC and the copies of your files stored on Microsoft's computers.

The Four States of OneDrive Data

On any given machine, all data in OneDrive exists in one of four states:

» Sitting in the cloud only; no copy on your machine.

» Sitting on your machine, synced with the cloud.

» Sitting on your machine, synced with the cloud, and you've told OneDrive that you **always** want a copy of it on your machine.

» In never-never land, in the process of syncing between your machine and Microsoft's computers.

That's the story behind the icons described in Figure 1-3. If you aren't confused, you obviously don't understand.

TIP

Why wouldn't you choose to sync all your files and folders? Because the amount of data in your OneDrive cloud account could be enormous (5GB, just for starters). Syncing that data on your machine takes up not only disk space but also time and Internet bandwidth because missing files will be downloaded and altered files will be uploaded, every time you're connected to the Internet.

If you have oodles of available disk space, and your Internet connection is reasonably fast (and not hampered by ridiculous data caps), there's little reason to keep OneDrive files sitting stranded in the cloud and not copied to your machine.

WARNING

An important caveat: If you have OneDrive files or folders that you use all the time, you probably want to make them available on your machine. That way, if your Internet connection goes down — say, you hop on a flight or a cruise with exorbitant Wi-Fi fees — you can continue to work on the files while the Internet goes on without you. When you connect to the Internet again, your files get synced with OneDrive in the cloud.

File Explorer tells you the status of every file and every folder in your OneDrive by using tiny icons in the Status column. You can see them in Figure 1-4:

>> **Blue cloud icon:** The file or folder is available when online and is stored in the cloud, not on your machine.

>> **White icon with a green checkmark:** The item is available locally on your machine, not just on Microsoft's servers.

>> **Green icon with a white checkmark:** The item is set to be always available on your device.

>> **Person icon:** The shared file or folders can be accessed by others. This icon is displayed alongside other status icons.

>> **Blue refresh icon:** The file is in the process of synchronizing with OneDrive.

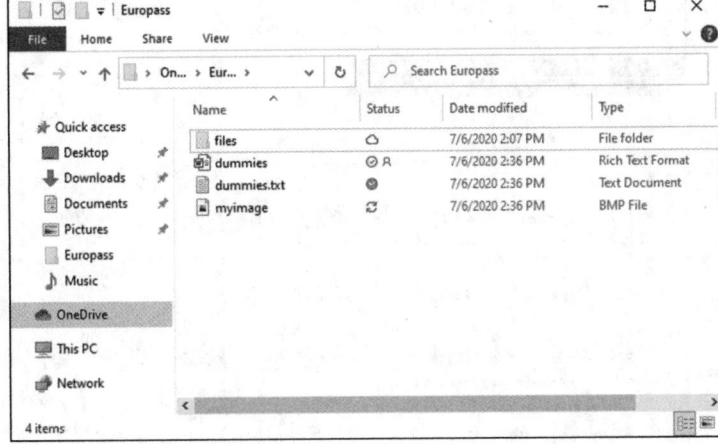

FIGURE 1-4:
The status icons for folders and files stored in OneDrive reassure you that OneDrive is installed and working.

To get OneDrive well and truly sorted, follow these instructions to step through the settings:

1. **Click or tap the cloud icon in the system tray, next to the time. Then, click the Help & Settings button.**

   You see the options shown in Figure 1-5.

2. **Choose Settings, and then click the Settings tab.**

   The Microsoft OneDrive Settings pane appears, as in Figure 1-6.

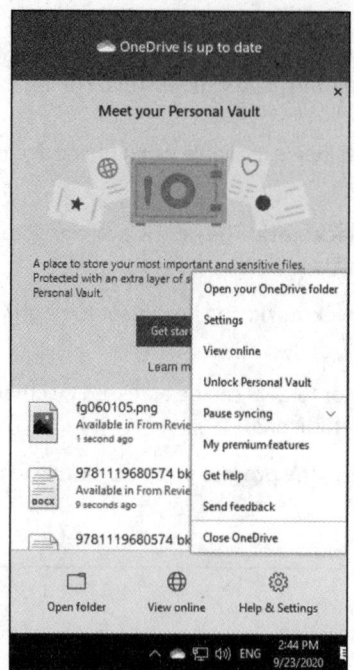

FIGURE 1-5:
Options for
controlling
OneDrive.

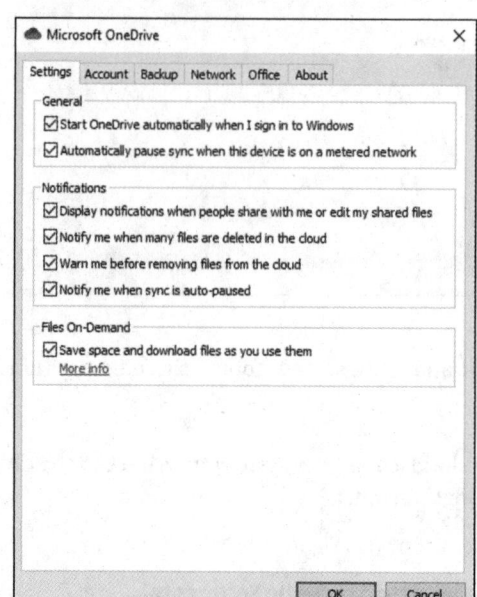

FIGURE 1-6:
Turn Files
On-Demand
on or off here.

3. **Make sure the box marked Save Space and Download Files as You Use Them is selected. Then click the Account tab.**

As long as OneDrive is working properly (not an absolute given), you're better off with that box selected.

4. **On the Account tab, click Choose Folders.**

That opens the Choose Folders pane, shown in Figure 1-7.

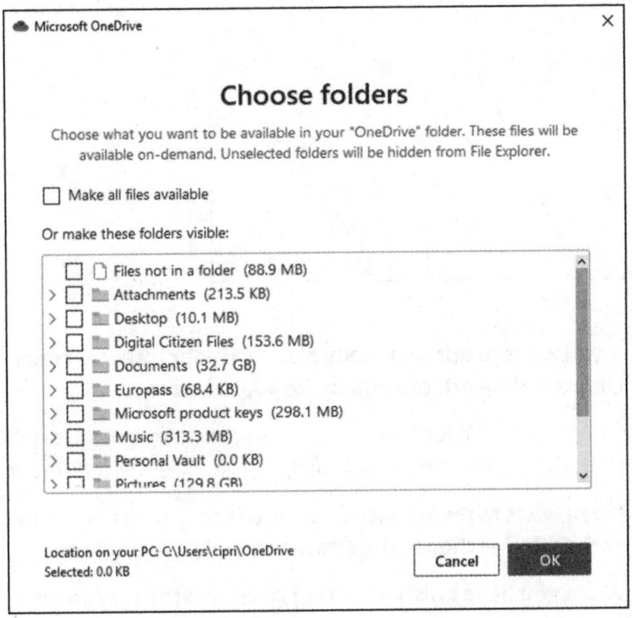

FIGURE 1-7:
Either sync all OneDrive data on this machine, or choose which folders get the Files On-Demand treatment.

5. **Do one of the following:**

- If you have a lot of room on your PC and have a reasonably good Internet connection, select the Make All Files Available box. Click OK.

- If you don't want to slavishly sync all OneDrive files onto this particular PC, deselect the Make All Files Available box, and select boxes next to the folders you want to sync. Click OK.

6. **Click the Backup tab and make your selections.**

The screen shown in Figure 1-8 appears. You can choose to back up your files in the Desktop, Documents, and Pictures folders to OneDrive, so that the files are protected even if your machine has problems. You also set whether you want to automatically save to OneDrive pictures and videos from the cameras, phones, and other devices you connect to your Windows 10 PC and screenshots.

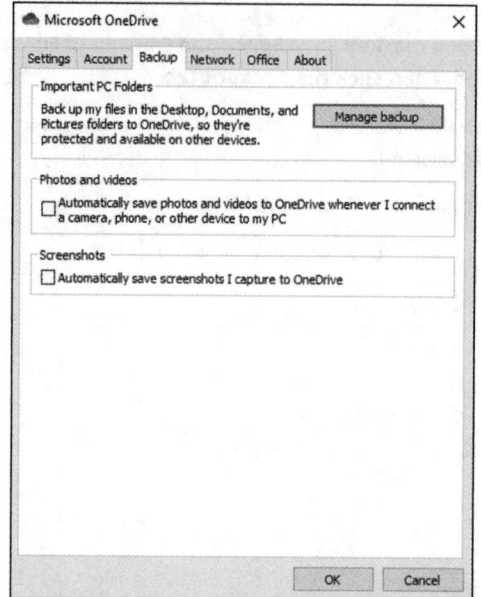

FIGURE 1-8:
Set what you
want OneDrive to
back up for you.

7. **If you want to automatically dial back your Internet connection speed, click the Network tab and make your choices.**

   You can set throttles for both uploading (sending your data to OneDrive in the cloud) and downloading (pulling data from OneDrive onto your machine).

8. **If you want to disconnect sync for Office files — it's set up automatically, with Office — choose the Office tab and go from there.**

9. **When you have finished setting OneDrive the way you want, click OK.**

   It may take a while for OneDrive to sync, but when it's finished, all folders you've chosen to sync will appear in File Explorer with the appropriate status icons. See Figure 1-9 for the result of making the choices in Figure 1-7.

From the screen shown in Figure 1-9, you can add a file to any of the folders. After you've added files, you can delete files or download any of them to your computer by simply dragging and dropping, the way you usually move files.

Anything you can do to files anywhere, you can do inside the OneDrive folder — as long as you use File Explorer or one of the (many) apps, such as the Microsoft Office apps (see Figure 1-10) that behave themselves with OneDrive.

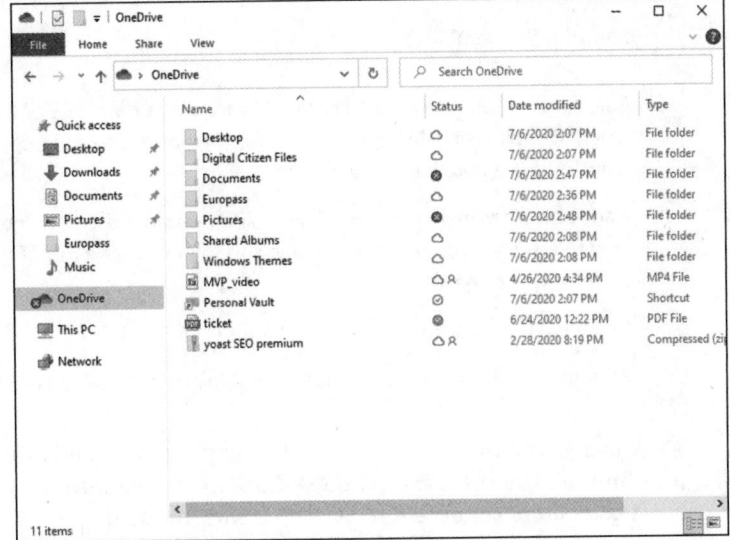

FIGURE 1-9:
The result of applying your OneDrive sync settings.

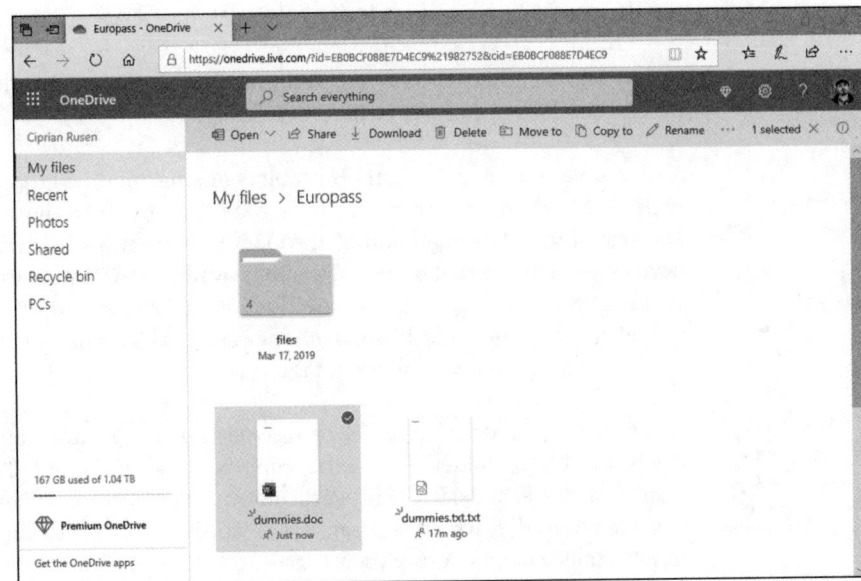

FIGURE 1-10:
If you open a DOC file from OneDrive, Online Word appears to handle it.

For example, you can

>> Edit, rename, and copy files, as well as move vast numbers of them. The OneDrive folder in File Explorer is by far the easiest way to put data into OneDrive and take it out.

>> Add subfolders inside the OneDrive folder, rename them, delete them, move files around, and drag and drop files and folders in and out of the OneDrive folder to your heart's content.

>> Change file properties (with a long tap or right-click).

>> Print files from OneDrive just as you would any other file in File Explorer.

TIP

What makes the OneDrive folder in File Explorer unique is that when you drag files into the OneDrive folder, those files are copied into the cloud. If you have other computers connected to OneDrive with the same Microsoft account, those other computers may or may not get copies of the files (depending on whether Make All Files Available in the PC's OneDrive is selected), but they can all access the files and folders through a web browser.

It may take a minute or two to upload the files. But plus or minus a bit-slinging delay, the files appear everywhere, magically.

So if you have other computers (or tablets or smartphones) that you want to sync with your computer, now would be a good time to go to those other computers and install whichever version of the OneDrive program is compatible with your devices. Remember that a OneDrive app is available for Windows (Windows 7, and Windows 8 only), macOS X and later, and iOS (for iPad and iPhone). There's also a OneDrive app for Android smartphones and tablets. That's the one I use on my Pixel Android phone, as shown in Figure 1-11.

WARNING

Know that if you delete a folder or file marked Cloud Only, you'll well and truly delete the file or folder —no extra copy is hanging around. That file or folder won't exist after a delete. However, OneDrive keeps deleted files for 30 days in a Recycle Bin folder, which you can access online. That's the only place where you have a chance of recovering them.

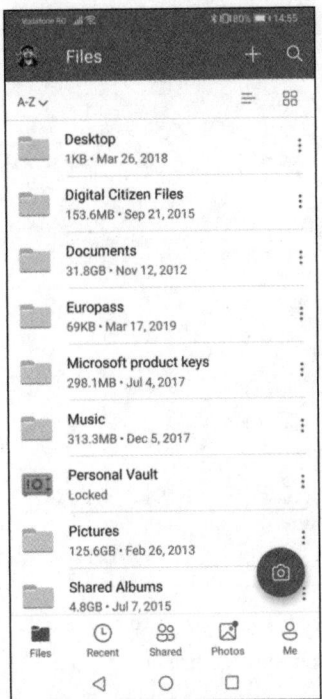

FIGURE 1-11:
The OneDrive app on an Android smartphone.

Changing the States of OneDrive Data

It's remarkably easy to change among the four states of OneDrive data:

» In the cloud only.

» On your machine, synced.

» On your machine, synced, and you've told OneDrive that you *always* want a copy of it on your machine.

» On your machine but not yet synced (a sync may be in progress).

In general, all you have to do is right-click a file or folder — even the entire OneDrive folder — and choose the correct option. For example, in Figure 1-12, I right-clicked the Math Problems folder, which is already synced.

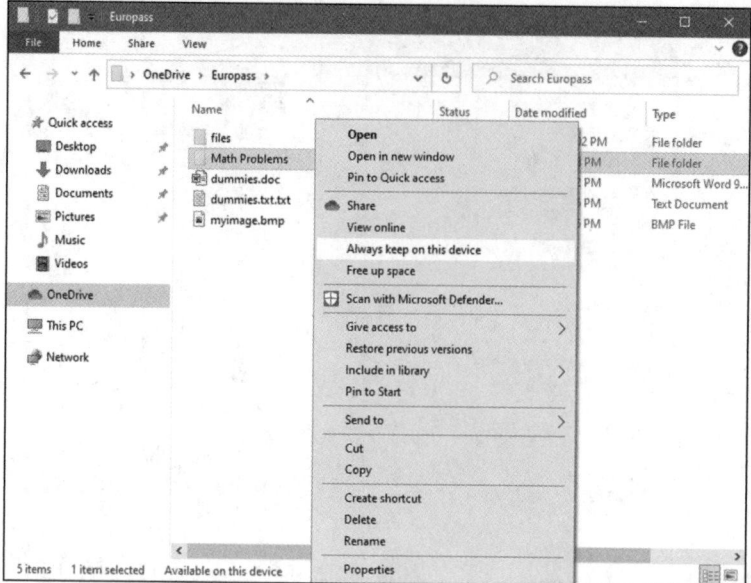

From this point, I have two state-changing choices:

>> If I select Always Keep on This Device, OneDrive will change the state to always
keep a copy on the PC.

>> If I select Free Up Space, OneDrive will delete the copy of the data currently on
the PC and set the state to cloud-only.

In general, after OneDrive has stopped syncing, you can change from one state to
another by right-clicking.

Sharing OneDrive Files and Folders

Sharing files and folders in OneDrive couldn't be simpler — although one little
advertising inanity exists.

To share a drive or folder, navigate in File Explorer to the file or folder that
you want to share and right-click it (or tap and hold down). Choose the Share
option with the blue OneDrive icon near it. The OneDrive Share dialog appears, as
shown in Figure 1-13. You can share the file or folder via email, copy a link to it

that you can paste into a chat app or some other place, set the sharing permissions, and so on. When you click Anyone with the Link Can Edit, you can also set an expiration date for the share. However, this option works only for premium (paid) versions of OneDrive.

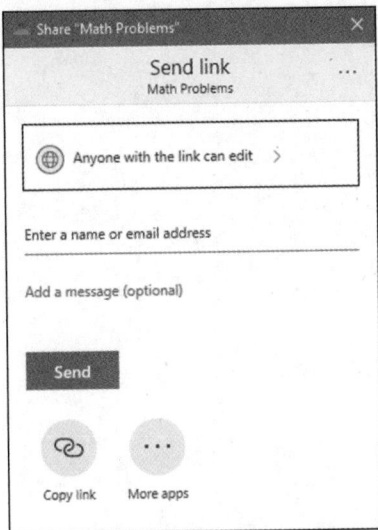

FIGURE 1-13:
The OneDrive
Share dialog.

OneDrive has many more capabilities. See the tutorial at `http://windows.` `microsoft.com/en-US/windows-10/getstarted-onedrive` for an overview.

Chapter **2**

Getting Started with Facebook

I f you don't yet have a Facebook account, more than 2 billion people are ahead of you.

**ASK
WOODY.COM**

I have friends who figure Facebook is some sort of fad that's going away soon. They'd rather be drawn and quartered than put anything on Facebook. "You lose your privacy," they say. "I don't see any need for it."

Of course, many of them said the same thing about mobile phones two decades ago. ATMs. Online banking. Two decades before that they lambasted the newfangled color television stuff — it'll never catch on, you know?

In the past decade, Facebook's become an important part of the daily routine of 1.4 billion people, and it claims more than 2 billion registered users who go online every month. 300 million photos are uploaded every day. It's been credited with starting revolutions. It's certainly a good source of news — almost as good as Twitter (see Chapter 3 in this minibook) — if you choose your sources carefully.

More than 40 percent of all American adults log in to Facebook *every day*.

REMEMBER

Facebook has fundamentally changed the way hundreds of millions of families interact, more so than any other invention since the telephone. It's altered the way people work. Businesses. Schools. Hospitals. Governments. Charities.

Facebook has even eaten into email, and instant messaging, for heaven's sake. Email usage has gone down the past couple of years because Facebook's one-to-many nature reduces the need for email messages, and its embedded mail and chat features are growing fast. To me, that's incredible. I grew up with email — sent my first email message in 1977 — and it boggles my mind that so many people prefer Facebook to email. But that's how it is.

I'm tempted to stand up and bellow a chorus from Bob Dylan's "The Times They Are A-Changin'."

You can ignore Facebook, if you want to, but someday your kids or grandkids or the young whippersnappers in the nursing home are going to ask why dad or grandpa or Uncle Fuddyduddy doesn't get off his duff and get with the system. It's the same argument people had with Luddites about typewritten letters and faxes a couple decades ago.

**ASK
WOODY.COM**

In this chapter, I only brush the surface of the capabilities available to Facebook users. You find a bit of depth about the Timeline because it's hard to find information about it. And I hit the privacy/security part hard because that's where you need to concentrate your efforts when you're just starting out.

TIP

As you get more adept at Facebook, you'll figure out about tagging photos, sharing things that have been posted to your home page or your Timeline, subscribing, setting up groups, chatting and video calling, setting up your own fan (or business, group, or charity) pages, posting events, searching, GPS location-based features, setting up your own lists — and much more. If Facebook intrigues you, I suggest you pick up a copy of *Facebook For Dummies*, by Carolyn Abram and Amy Karasavas. For a deeper look at the side of Facebook that's tailored for businesses, charities, and groups (including that knitting circle or bridge club), look at *Social Media Marketing All-in-One For Dummies* (published by John Wiley & Sons, Inc.), by Jan Zimmerman and Deborah Ng.

Facebook has apps that run on iPads, iPhones, Android tablets, and phones — and I use all of them. Its website, www.facebook.com, runs in every browser you can imagine — and many you probably can't.

Choosing a Facebook App

Facebook largely ignored Windows for a century or so in Internet years. Finally, on October 17, 2013, we saw the first Facebook Universal Windows app. It didn't work great, and it had fewer features than its iOS and Android counterparts. You can get a taste of how it looked in Figure 2-1. Windows 10 users did not appreciate it and left many negative reviews in the Microsoft Store, for good reason: The app offered an inferior experience to all its mobile alternatives.

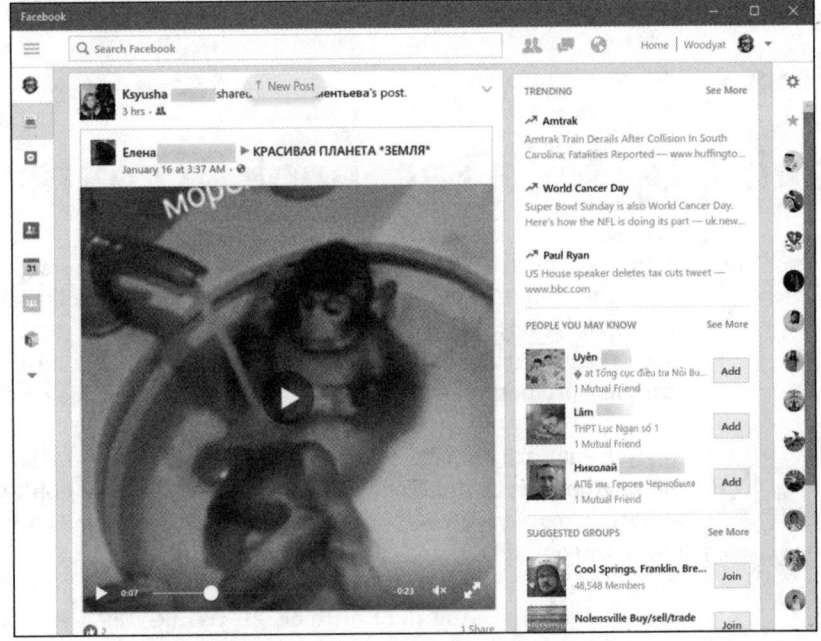

FIGURE 2-1:
The Universal Windows Facebook app left much to be desired.

On February 28, 2020, Facebook decided to stop faking that it supports Windows 10 and removed their Facebook app from the Microsoft Store. Today Windows 10 users can access www.facebook.com only in their web browser, where they get the full experience, which is a lot better than what the former Facebook app provided.

However, if you open the Microsoft Store, you'll find several Facebook apps available, as shown in Figure 2-2: Messenger (it works well for chatting with your Facebook friends), Facebook Watch, Messenger (Beta), and Instagram. The funny thing is that Instagram is listed as being published by Instagram, not Facebook Inc, like all the other apps. That's why it doesn't show up in the Microsoft Store when you search for apps published by Facebook. You have to search specifically for Instagram to find it.

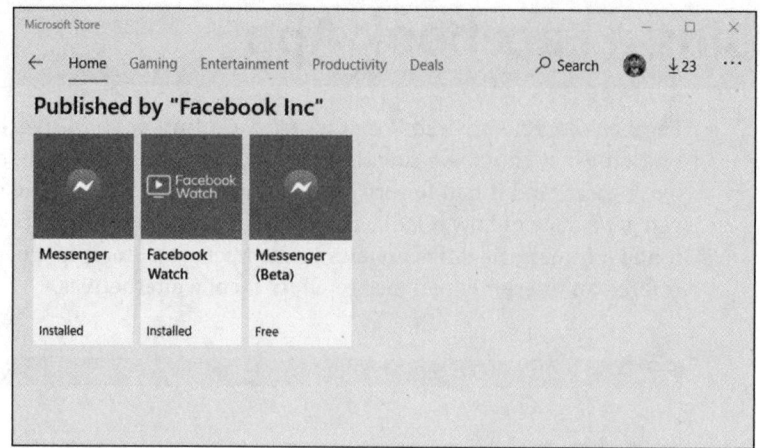

Signing Up for a Facebook Account

If you don't yet have a Facebook account, I suggest you sign up. Don't worry, nobody's going to steal your identity or mine your personal data. Yet. And Facebook's absolutely free — and will be free to use forever, we're assured, although some features may cost something someday, and a few business-oriented features like promoting posts or other kinds of advertising do cost real samolians.

WARNING

There's one cardinal rule about Facebook, which I call the *prime directive:* Don't put anything on or in Facebook — *anything* — that you don't want to appear in tomorrow morning's news. Or your ex-spouse's attorney's office. Or your boss's inbox. Or your kid's school class. Privacy begins at home, eh?

Now that you have the right attitude, all you need is a working email address, and as long as you state that you're at least 13 years old, you can have a Facebook account in minutes. Here's how:

1. **Use your favorite browser to go to** www.facebook.com.

   The Sign Up page appears, as shown in Figure 2-3.

2. **Fill in your name and email address (it must be a valid one that you can get to because a confirmation email goes to that address), give your new account a password, and make sure your birthday indicates that you're at least 13.**

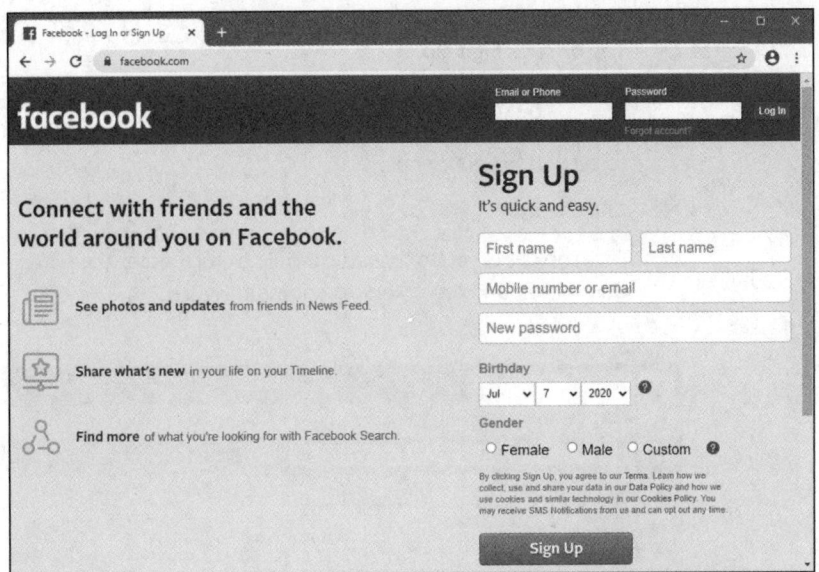

FIGURE 2-3:
The Facebook.
com Sign Up
page.

This is *not* the way to set up an account for a celebrity, band, business, charitable organization, or knitting group. In all those cases, you need to set up an individual account first — follow the instructions here — and then after your individual account is ready, you add a *page* (sometimes known as a *business page*) to your individual account. I know it's complicated, but Facebook works that way. Even Coca-Cola's page is attached to an individual — presumably either Mr. Coca or Ms. Cola signed up and then created a page for Coca-Cola afterward.

TIP

There's no reason to give personally identifiable information in this sign-up sheet. Facebook may balk if you try to sign up as Mark Zuckerberg, but it (probably) won't have any problem with Marcus Zuckerbergus (although, now that I've mentioned it, the name may be added to Facebook's blacklist). Some people have had trouble using their stage names, even when their stage names are, legally, their real names. Facebook has a policy that you have to use your real name, so if you feel so inclined, make sure whatever name you use looks real enough. (Apparently you can make up a silly middle name, though, and it's likely to be accepted.)

And if you figure your birthday is your business, the Internet Police aren't going to come knocking. The one item that has to be valid, though, is the email address — which can come from a free site, such as Hotmail/Outlook.com or Gmail.

3. **Tap or click Sign Up.**

   Facebook sends a confirmation email to the address you specified. After you click the confirmation button in the email, Facebook brings up a page that tries to get you to add friends.

4. **Click Next.**

   Facebook asks you to upload a profile picture, enter the name of friends, classmates, and coworkers, as shown in Figure 2-4.

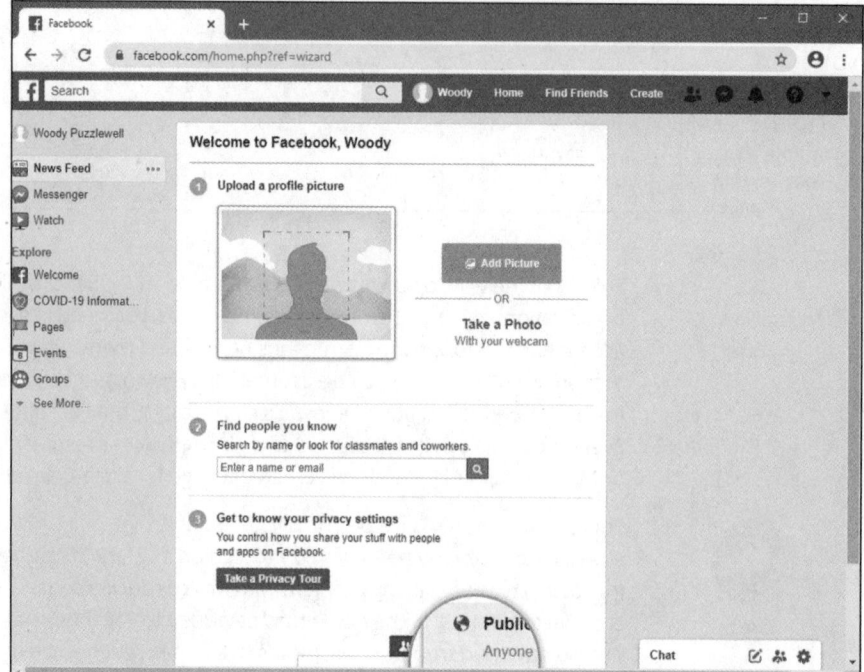

FIGURE 2-4: Adding your Facebook picture.

5. **Upload an appropriate picture, and then click Take a Privacy Tour, under point 3.**

6. **When you are finished with the privacy tour, click Finish.**

   Congratulations. You now have a Facebook account, and your first job is to lock it down. You should see a Welcome to Facebook page, like the one shown in Figure 2-5.

   Next, you set up some basic settings and get your security locked down.

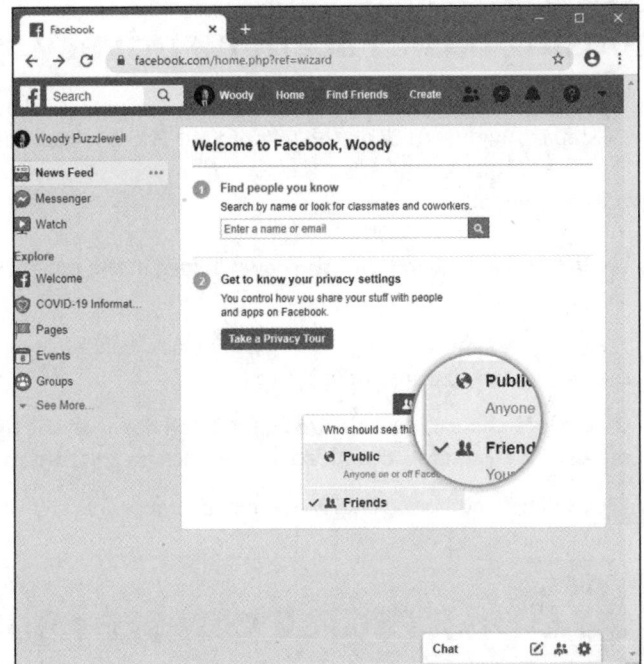

FIGURE 2-5:
A fresh Facebook
account.

WHAT, EXACTLY, IS A FRIEND?

Most people new to Facebook think that friends are, well, friends. Not so.

On Facebook, a *friend* is someone you're willing to interact with. If you're interested in interacting with somebody who has a Facebook account — let her see what you've posted (typed in the What's on Your Mind box), look at your *Timeline* (a historic bulletin board), or look at the pictures you've posted on Facebook — you send a *friend request.* The person who receives the friend request decides whether she wants to accept the request, decline it, or just sit on it.

Some of my Facebook friends are people I've never met and don't really know. They are, however, people I trust enough to allow them to look at my vacation pictures, say, and people who are interesting enough that I want to look at what they post on their profiles. If the concept of a friend is a bit overwhelming at this point, don't worry about it. Find two or three people you know who have Facebook accounts, send friend requests to them, and watch what happens when they respond.

Get your feet wet with the concept before you start friending everything with two legs. Or four. You can always add new friends (or delete them — *unfriend* them — for that matter), but it's easier to start out slowly while you're getting the hang of it. Too many friends at first can be overwhelming.

Choosing basic Facebook privacy settings

Before you try to figure out what you're doing — a process that will take several days — step through setting up the rest of your Facebook account.

Here's what you do:

1. **Log in to Facebook and click the down arrow, in the upper right, and then choose Settings.**

   The icon may appear dimmed, dark gray or black: Don't worry, it works. You see the General Account Settings page.

2. **On the left, choose Security and Login. Scroll down on the right until you see Use Two-Factor Authentication, and click its Edit button.**

   The two-factor authorization (2FA) signup offer appears.

WHAT OTHER PEOPLE CAN SEE ABOUT YOU

Ever since the FTC slapped Facebook's hands, repeatedly, for privacy problems — and Facebook submitted to a 20-year ongoing audit by the US Federal Trade Commission starting in November 2011 — Facebook has been quite forthcoming about its privacy policies.

Lots and lots of rumors circulate about what people can and can't see, so let me set the record straight.

If you look at someone's Timeline (or profile), the person you're looking up doesn't have any way to tell that you've looked. In fact, there's no way to tell how many times people have looked at a Timeline. There are lots of Facebook scams that offer to give you a list of who's visited your Timeline. They're just that — scams. It can't be done.

And if you confirm that only your friends can see your future posts, the amount of information that other people can see is very small.

Although the ubiquitous Facebook Like button sits on millions and millions of sites, Facebook doesn't give the people who run those sites any information at all about you. None. On the other hand, sites with the Like button allow Facebook to set third-party cookies on those sites. Facebook can trace your IP address as you go from site to site with the Like button. But the site itself doesn't get any information from Facebook.

2FA is an important adjunct to any account — and I find text message 2FA is the easiest option. When you use the text message 2FA, the first time you log on to Facebook from a new device, Facebook automatically sends a text message to your smartphone with a confirmation code. You must enter the confirmation code before you can proceed.

3. **Click Use Text Message (SMS). If you're asked to enter your Facebook password, do so.**

   The dialog box shown in Figure 2-6 appears.

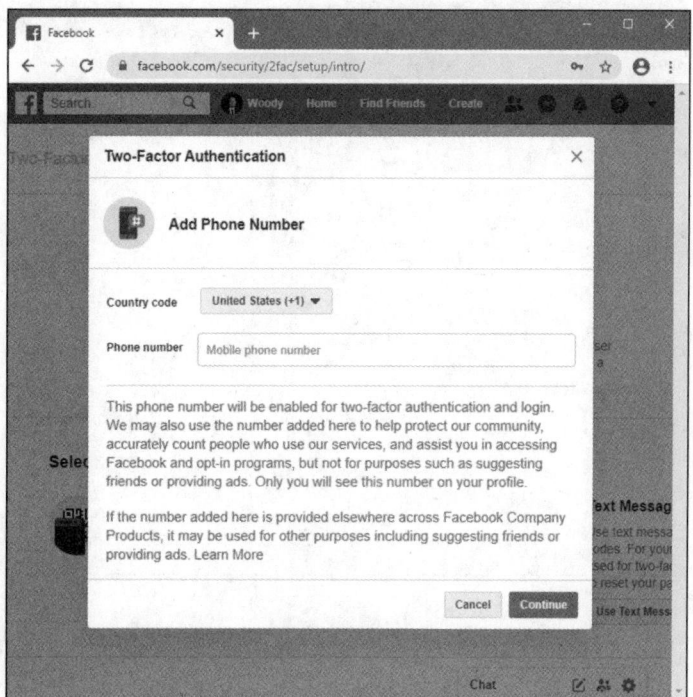

FIGURE 2-6:
The most
important
security setting
is two-factor
authentication.

4. **Type your phone number, and click Continue. When the confirmation message appears on your phone, type the code in the confirmation box, click Continue, and then click Done.**

   A pesky notification box asks you if it's okay to use your phone number to help you make connections and see ads that are more "relevant."

5. **Unless your friends are a whole lot friendlier than my friends, click Not Now.**

   You're informed that two-factor authentication is on.

6. Click the down arrow in the upper right, and then choose Settings again.

You go back to the General Account Settings page.

7. On the left, click or tap Privacy. Then, near the top, under Who Can See Your Future Posts? choose Edit.

The default sharing pane appears, as shown in Figure 2-7.

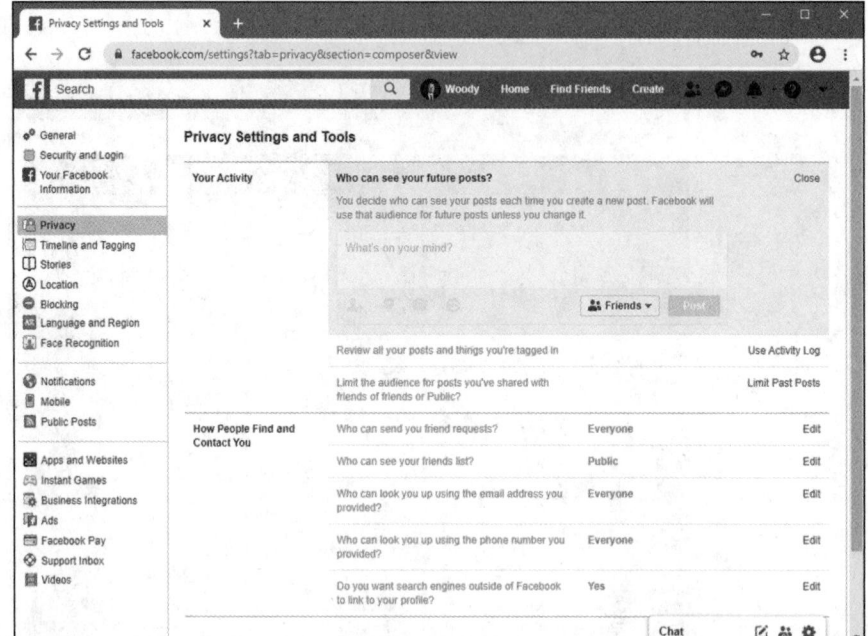

FIGURE 2-7:
The default is for
all future posts to
be visible only by
friends.

8. Make sure the drop-down box lists Friends, and then click Close.

You're back to the Privacy Settings and Tools page.

9. One final check: On the left, click or tap Apps and Websites.

A list of all apps that have permission to connect to your Facebook account appears, as shown in Figure 2-8. Don't be overly alarmed. Somehow, some-time, you gave those apps permission to hook into your Facebook account.

10. Delete any unwanted obtrusive app or website by selecting it and clicking Remove.

You can now click Home, up at the top, and go back to using Facebook.

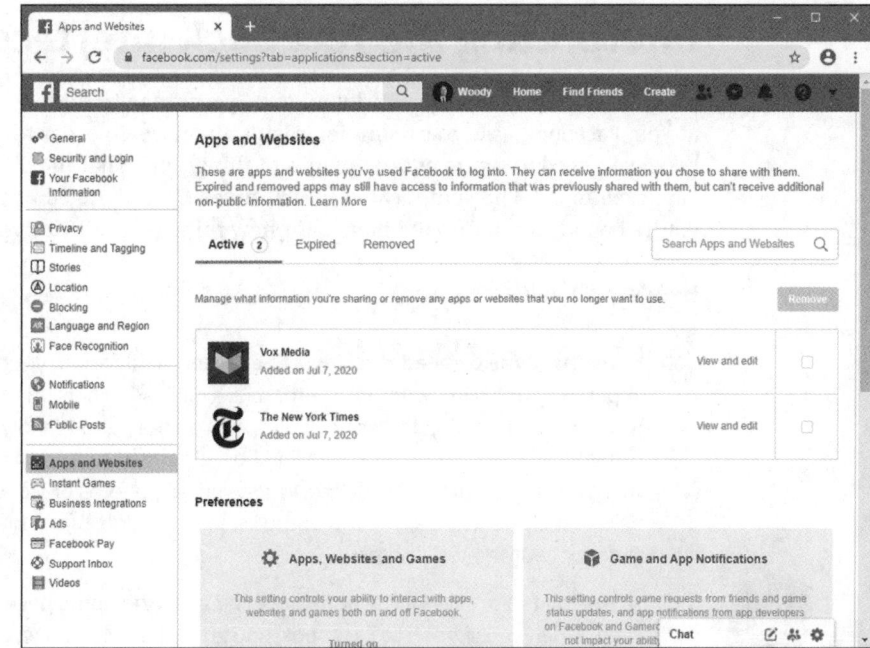

FIGURE 2-8:
Don't be too surprised if you see some bizarre apps and websites that have access to your Facebook data.

WHAT BUSINESSES CAN SEE ABOUT YOU

Many people starting out with Facebook are worried that businesses — particularly businesses that pay to advertise on Facebook — can see all their personal information.

Sorry. As much as I love a good conspiracy theory, it just isn't true.

Anybody who controls a business page can see the profiles of people who have visited the page and the people who have clicked the Like button on the page. So, for example, if you go to the Ford page (which is a very good one, by the way), Ford will know that one more female between 25 and 34 years old visited the page. Ford will also get one more visitor tallied by city, country, and major language. If you arrived at the page by clicking a Facebook ad, that fact is also counted. But that's it.

When a business pays for an ad, it chooses the demographics ("only show ads to males 18 to 24 living in Los Angeles") but there's no lingering information about who got served an ad, and no way to tie you, specifically, into a click on an ad. Facebook has that information. The advertiser does not.

Facebook guards your information jealously. It doesn't sell your info to businesses or give it away, unless you specifically permit an app to pull the data from Facebook. That's why the Windows 10 People app asks your permission before retrieving Facebook data — Facebook won't let Microsoft pull the data unless you specifically allow it.

Interpreting the Facebook interface lingo

Now that you've taken the whirlwind tour, permit me to throw some terminology at you. Facebook used to be simple; it isn't anymore. In order to work with Facebook, you need to figure out the names of things and what the different pieces are supposed to do. The complicated part? Names have changed over the years, and you're bound to run into old names for new things — and vice versa.

Here's my handy translator:

>> **Home page (also called the News Feed)** is primarily about your friends. The important stuff is in the middle — there are navigation aids on the left, and basically uninteresting things (including ads) on the right. When you type something in the What's on Your Mind box, it's added to the top of the list, as well as at the top of your Timeline. When you add photos or videos, thumbnails of the photos go at the top of the list in the middle of the home page. Ditto for your friends' photos.

REMEMBER

When you tap or click Home at the top, you go to the home page or the News Feed. When you sign on to Facebook, you go to the home page.

Facebook has a secret algorithm that it uses to figure out which items appear on your home page and in what sequence. If you're mystified why something's on the top of the page, but the really important stuff is down farther, well, I'm frequently mystified, too.

At this moment, your home page also has a drop-down box that lets you cycle between Facebook-generated Rooms, Top Stories, and Most Recent. *Rooms* are a new feature that allow up to 50 people to video chat at a time. There's no limit to how long you can talk.

>> **Timeline (replaces the old Wall and the old profile page)** is all about you. There's a big picture at the top, dubbed a cover, with your profile picture appearing to the left. Then there are all the settings you've made visible, followed by almost all the posts you've made over the years, in reverse-chronological order. I talk about the Timeline in the "Building a Great Timeline" section later in this chapter.

When you type something in the What's on Your Mind box, it's added to the top of the Timeline list, as well as at the top of your home page. Your friends can also post on your Timeline — in effect, leaving you a note.

REMEMBER

The Timeline appears when you tap or click your name at the top of the Facebook page. It also appears when someone clicks your profile picture in something you posted.

Building a Great Timeline

The Timeline — the place you go when you click your name — is where people usually go when they want to learn about you. If somebody clicks your picture in a post elsewhere in Facebook, he's sent to an abbreviated version of your Timeline.

When you bring up your own Timeline, you get to see a great deal more than what the world sees, as in Figure 2-9.

FIGURE 2-9:
Your Timeline is your resume in the Facebook world.

REMEMBER

Keep in mind the prime directive: Don't put anything on or in Facebook — *anything* — that you don't want to appear in tomorrow morning's news. Fill in the details sparingly.

Follow these steps to personalize your Timeline:

1. **Bring up your Timeline by tapping or clicking your name at the top of the Facebook screen.**

Depending on how much you've done to your Timeline, it might look like the one in Figure 2-9.

2. **Tap or click the Add/Update Cover Photo icon (camera).**

   Facebook takes you through the steps of either uploading a new photo or choosing from one that you've already uploaded.

3. **After you choose or upload a photo, tap or click it to drag the part you want to see into the fixed-size frame. Then tap or click Save Changes.**

WARNING

   If you don't have a suitable photo already, pre-fab Facebook cover photos are all over the Internet. Just be careful when you go out looking: Any website that has you click and log in to Facebook in order to deliver the photo may be gathering your Facebook login ID in the process. It's much safer to simply download the photo to your hard drive and then upload it yourself to Facebook.

   The Facebook cover photo is 850 pixels wide x 315 pixels tall. Facebook will actually accept any picture as long as it's at least 720 pixels wide. When you drag the uploaded picture to fit it into the fixed-sized frame, you're telling Facebook how to crop the picture to make it fit into the 315 x 850-pixel box. For best results, use a photo-manipulation program — or even Paint— to get the photo just right before you upload it.

4. **To change your *profile picture* — the little picture on the left that also appears on anything that you post, tap it or hover your mouse cursor and choose Update.**

   Remember that your profile picture gets squeezed down most of the time, so a highly detailed photo usually doesn't work very well.

5. **When you're finished editing your profile information, tap or click your name at the top of the screen to go back to the Timeline.**

   By now the layout of the Timeline is a little more comfortable, but now it's time to change the contents of the Timeline itself.

6. **Scroll down the Timeline and find an item that you don't want other people to see. Click or tap the ellipsis in the top-right corner of that item.**

   Facebook gives you the options shown in Figure 2-10.

7. **To remove that item from your Timeline, tap or click Hide from Timeline, and confirm your choice.**

   The item disappears immediately, replaced by a placeholder that only you can see.

FIGURE 2-10:
The options available for every item in your Timeline.

8. **On the left, where it says Add Profile Info, consider typing your school's name and clicking Save Changes. Or you can click Skip.**

   Either way, Facebook hits you with a barrage of questions that help flesh out your profile. Be careful to choose who can see all the information you add. By default, Facebook sets your School as public information, which is shared with everyone, not just your friends.

9. **Fill in the rest of the profile questions. Or don't. Your choice.**

REMEMBER

   As you get more adept at Facebook and figure out how to lock down your account, you may want to add more information to your profile. Cool, as long as you understand the consequences. For now, put in the minimum you feel comfortable about disclosing to the world at large. Remember, someday your boss or your son might read it.

   Each line you can enter — from your schools and marital status to your religious views — has a drop-down choice to limit access to that information.

   Access limitations are based on your lists. For example, if you identify Snidely Whiplash as a member of your family, Snidely can look at any items you've set to be visible to Family. Any friends who aren't on your Family list can't see the item.

ASK WOODY.COM

   For now, while you're still getting your feet wet, be very circumspect in what information you provide, *even if you limit access to the information to specific lists.* Give yourself awhile to get more friends. You can always update your profile.

If you've been using Facebook for a long time, your Timeline may go on and on and on. But I bet there's no chance you have your baby picture pinned.

10. **To add something to your Timeline that goes waaaaaay back (I'm talking years or decades, not centuries):**

a. Tap or click the Life Event link, just above the What's on Your Mind box. Facebook lets you identify the event, as shown in Figure 2-11.

b. Choose a category for your life event.

c. Follow the instructions to give a date, choose or upload a picture, and provide more details about the event.

d. Tap or click Share.

The item attaches itself to the appropriate place on your Timeline — even if it predates your joining Facebook.

It's your account. Take control over it.

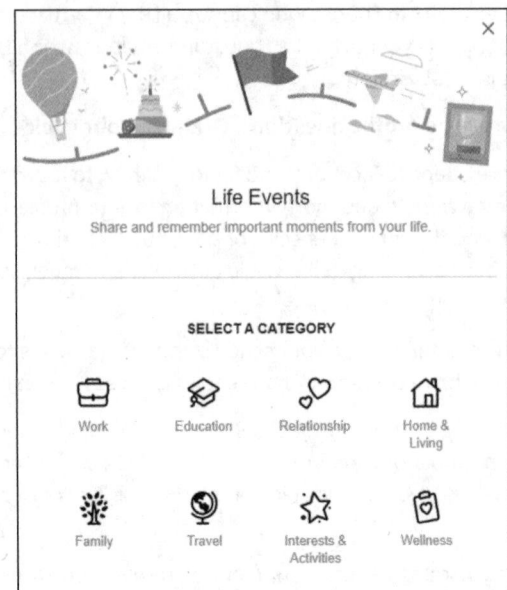

DOWNLOADING YOUR FACEBOOK DATA

Apps aren't allowed to download all your Facebook data. But you can.

Log in to Facebook. Tap or click the down arrow at the top all the way to the right, and choose Settings. On the left, click Your Facebook Information. Then, on the right, click the Download Your Information link. Choose the date range and the types of information you want, and then click Create File.

Then go have a latte. When you get back, check your email. You — eventually — receive a message from Facebook that says your information file is ready to download. Tap or click the indicated link to retrieve the download, and you go back to the Download Your Information page (getting vertigo yet?). Go to Available Copies, tap or click Download, enter your Facebook password, and choose a location; your browser downloads the zipped file. Finally.

Using the Facebook Apps for Windows 10

I find it much easier to set up a Facebook account — and particularly keep on top of the privacy settings — by using a web browser. For day-to-day use, though, most people rely on a mobile app. It's just simpler and faster to keep on top of Facebook comings and goings with your phone or tablet. However, for chatting directly with people, you may want to try the Messenger app for Windows 10, available in the Microsoft Store. It's simple to use, and it works well. See it in action in Figure 2-12.

To chat with someone, select the person from the left, or search for him or her. Type your message on the right, and press Enter or click Send.

Another app that you might want to try is Facebook Watch for Windows 10, shown in Figure 2-13. It is found in the Microsoft Store. Facebook Watch helps you discover videos from popular pages, friends, and other sources. Think of it as an inferior YouTube competitor that's a great time waster. An upside of Facebook Watch is that you can use it without logging in with your Facebook account.

To watch a video, just click it. To discover interesting videos, scroll through the available categories, or use the Search option and enter the subject, page, or person you're interested in.

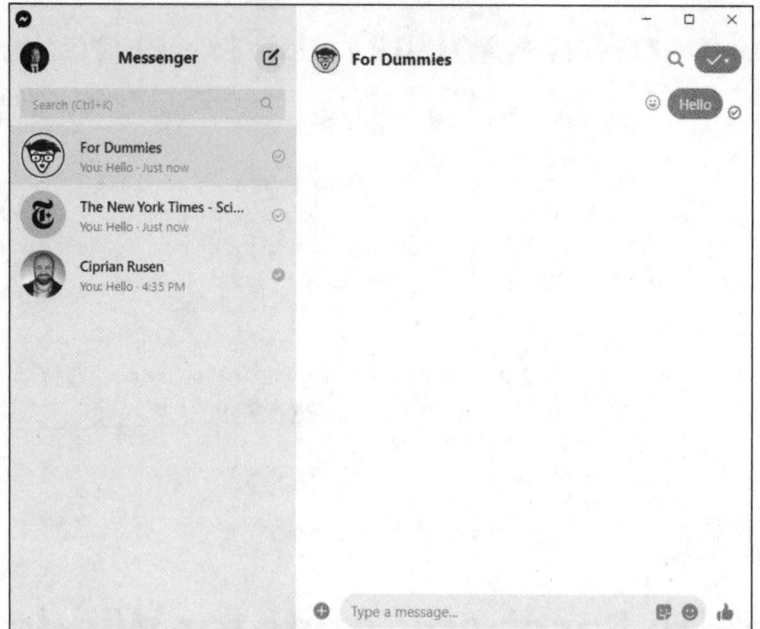

FIGURE 2-12:
The Messenger
app for
Windows 10.

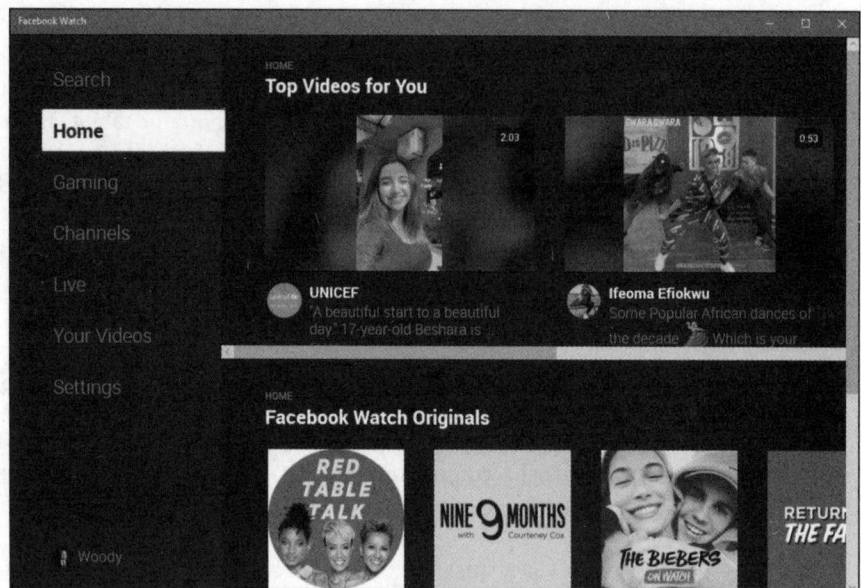

FIGURE 2-13:
The Facebook
Watch app for
Windows 10.

Chapter **3**

Getting Started with Twitter

The revolution will not be televised. It will be tweeted.

In March 2006, an amazing array of developers and entrepreneurs — originally intent on building a podcasting platform called Odeo — unleashed Twitter on an unsuspecting world. A decade later, Twitter has been credited with helping to overthrow totalitarian governments, spread fear and mayhem, aid and abet leaks of embarrassing government documents, shed light on official dirty dealings, establish a rallying point for the Occupy disenfranchised, and let everyone know what Lady Gaga had for breakfast.

That's quite an accomplishment. As of mid-2015, Twitter had more than 1.3 billion registered users, 310 million of them active every month, who send an average of 500 million or so tweets per day. By the end of 2017, Twitter had an estimated 330 million monthly active users. Those are all industry estimates because Twitter doesn't divulge much, even though it's listed on the New York Stock Exchange. Interestingly, 77 percent of the most active users hail from outside the US.

Twitter crashed on June 25, 2009, the day Michael Jackson died, after logging tweets at 100,000 per hour. The current record for tweet volume was set on August

3, 2013, when a screening of the movie *Castle in the Sky* in Japan generated 25,088 tweets *per second*.

I use Twitter all day, every day. I've used it to keep on top of important fast-breaking news, notify people around the world, quell tsunami fears, talk with other writers in the computer business, keep tabs on political organizations important to me, track down obscure pieces of Windows 10, and point people to my favorite funny videos.

Just about every tech writer you can name is on Twitter. Every major news outlet is on Twitter — and breaking news spills out over Twitter much sooner than even the newspaper wire services. The Royal Society. The Wellcome Trust. Lots of people who are on the ground, relaying news as it happens, use Twitter. And did I mention Justin Bieber?

In short, Twitter's a mixed bag — but an interesting one.

Twitter's fast, easy, and free. It works with every web browser. It works with almost every smartphone and tablet. There's a Windows 10 Universal Twitter app — an official one — that's not very inspiring, but it works.

Twitter's short, concise, sometimes vapid, but frequently illuminating and witty. And every single piece of it is limited to 280 characters.

Understanding Twitter

When I try to explain Twitter to people who've never used it, I usually start by talking about mobile phone messaging — texting. A message on Twitter — a *tweet* (see Figure 3-1) — is much like a text message.

FIGURE 3-1:
A typical tweet from an atypical source.

> **Dalai Lama** ✓
> @DalaiLama
>
> As a result of material development and modern education, people commonly seek happiness in external things, but neglect their minds. True, lasting happiness depends on our taming our unruly minds. This is not so much about intellectual development as cultivating a warm heart.
>
> 12:30 PM · Jun 12, 2020 · Twitter Web App

Twitter is a simple one-to-many form of communication, kind of like texting all the people who have agreed, in advance, that they want to receive your texts.

You usually send a text message to one person. If you have a business, you may send the same, identical text message to many people all at once. Now imagine a world in which these are true:

>> You have an ID, not unlike a phone number, and you can send any messages *(tweets)* that you like, any time you want. The messages are limited to 280 characters — short and sweet.

>> You get to choose whose texts you want to see on Twitter. In Twitter parlance, you can *follow* anybody. If you get tired of reading their tweets, it's easy to *unfollow* them as well.

You have some leeway in what counts toward the 280-character limit. For example, when you

>> Reply to a tweet, @names don't count toward the 280-character limit

>> Add attachments, such as photos, GIFs, videos, polls, or quote tweets, that media isn't counted as characters in your tweet

That's the whole shtick. Twitter has lots of bells and whistles — location tracking, if you turn it on, for example — but at its heart, Twitter is all about sending messages and wisely choosing whose messages you receive.

Spam texts and harassing phone calls may dog your days on the smartphone. On Twitter, while all is not happiness and light, in general the problems are much less frequent and less severe.

If you follow someone who posts a tweet, you see the tweet when you log on to Twitter. If you keep Twitter running on your PC, smartphone, or tablet, as I do, the tweet appears in your Twitter window. If you tweet, the people who follow you can see it.

WARNING

In fact, *anybody* can see *every* tweet — a fact that's proved highly embarrassing to an amazingly large number of people. (Twitter has a Protected Tweets feature that lets you manually approve every person who's permitted to receive your tweets. But, in general, when you let it all hang out on Twitter, it's all hung out, eh?)

In addition, when you send a tweet, you can identify keywords in the tweet by using the # character in front of the keyword, creating a *hashtag*. See Figure 3-2.

Hashtag   Hashtag

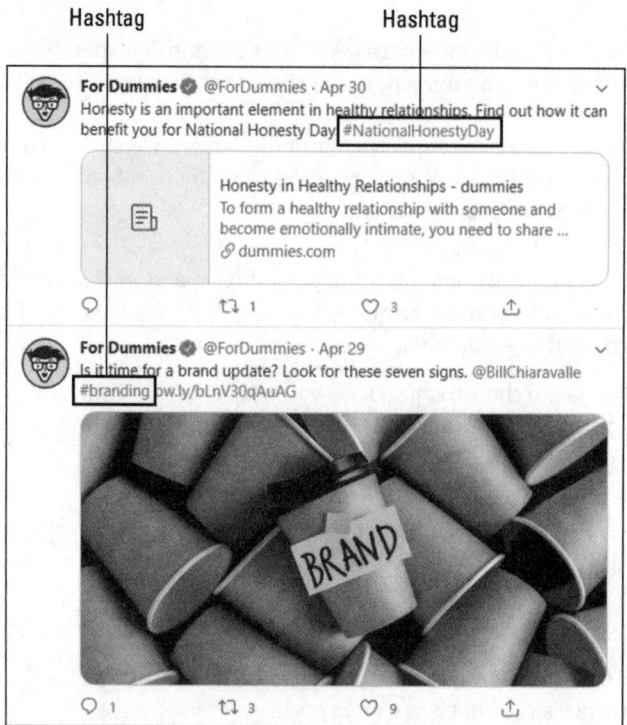

FIGURE 3-2:
Two sample
tweets with
hashtags.

ASK
WOODY.COM

You can tell Twitter that, in addition to the tweets from people you follow, you also want to see all tweets that contain specific hashtags. For example, if you ask to see all the tweets with the hashtag #ForDummies, Twitter delivers to your web page or Twitter reader every tweet where the author of the tweet specifically typed the characters #ForDummies.

Twitter (and other sites, such as `www.trendsmap.com`) keeps track of all the hashtags in all the tweets. It posts lists of the most popular hashtags, so you can watch what's really popular. Thus, hashtags not only make it easier for people to find your tweets but also publicize your cause — and many good causes have risen to the top of the hashtag heaps. Some odd ones, too, such as Lady Gaga kissing Marge Simpson, but I digress.

In fact, Twitter now keeps track of every phrase that's tweeted and compiles its trending lists from the raw tweets, with or without hashtags. You really don't need to use hashtags anymore. But you see them all the time in tweets, #knowwhatImean?

Google and Twitter have entered into a partnership whereby Google scans tweets so that they show up in Google searches.

TIP

The power of Twitter — outside of gossip and teenage angst — lies in choosing those you follow carefully. If they, in turn, receive information from reliable sources and then retweet the results, you'll have a steady stream of useful information, each in 280-character capsules.

For example, during the Egyptian political crisis in January 2011, which saw the downfall of President Hosni Mubarak, Twitter played a pivotal (if controversial) role in aiding communication among protestors. One of the government's first acts was to shut down access to Twitter and Facebook. The protestors found ways around the government's shutdown.

There's a fascinating re-creation of the tweeting and retweeting that followed the January 25 start of demonstrations in Cairo. Data about tweets with the hashtag #jan25 was assembled by the University of Turin, the ISI Foundation, and a research institute at Indiana University, to come up with the graph you see in Figure 3-3.

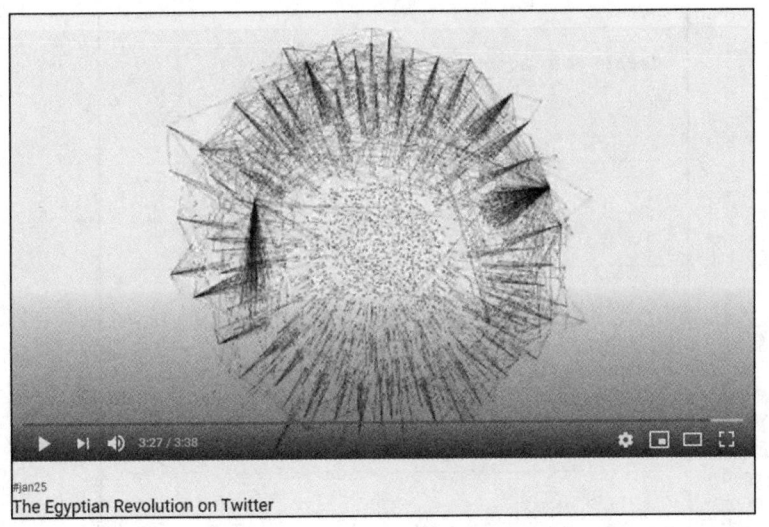

#jan25
The Egyptian Revolution on Twitter

FIGURE 3-3:
The interconnections among Twitter users during the Egyptian uprising.

Photo courtesy of http://youtu.be/2guKJfvq4uI

In the graph, the points represent individuals, and the lines are tweets that go from one individual to another. It's downright explosive.

That's how a one-to-many social network like Twitter works. If there's an important tweet (or even an unimportant, but popular one), it jumps from person to person.

My Twitter ID for computer-related news is @woodyleonhard, and you're welcome to follow me anytime you like.

Setting Up a Twitter Account

ASK WOODY.COM

Twitter has apps for all sorts of smartphones and tablets. I use it frequently on the iPhone, iPad, and Android phones and tablets. There's also a Twitter app for Windows 10 in the Microsoft Store. I mention the Twitter app for Windows 10 only occasionally in this chapter because it's woefully underpowered. If you want to get going with Twitter, it's much easier to start with a web browser, and that's the primary emphasis in this chapter.

Starting a new account at Twitter couldn't be easier. Here's what you do:

1. **Fire up your favorite web browser, go to** www.twitter.com, **and click Sign Up.**

    You see the Sign Up box, as shown in Figure 3-4.

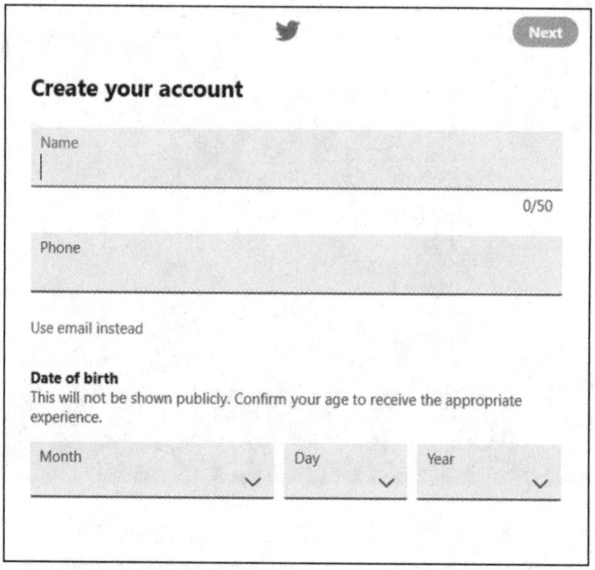

> Next
>
> **Create your account**
>
> Name
> _____
> 0/50
>
> Phone
> _____
>
> Use email instead
>
> **Date of birth**
> This will not be shown publicly. Confirm your age to receive the appropriate experience.
>
> Month ⌄   Day ⌄   Year ⌄

FIGURE 3-4:
All you need to sign up for Twitter is a valid phone number or email address.

2. **Enter the name you want to use, your phone (or email), and your date of birth. Then click Next.**

3. **Select how you want your Twitter experience to be (receive emails from Twitter, connect to others using your email address, or see personalized ads), and click Next.**

4. **Confirm that you want to create your account by clicking Sign Up.**

    Twitter sends a confirmation code to your phone or email address, depending on what you chose on Step 2.

5. **Type the confirmation code and then click Next.**

You're asked to set a password that has eight or more characters, as shown in Figure 3-5.

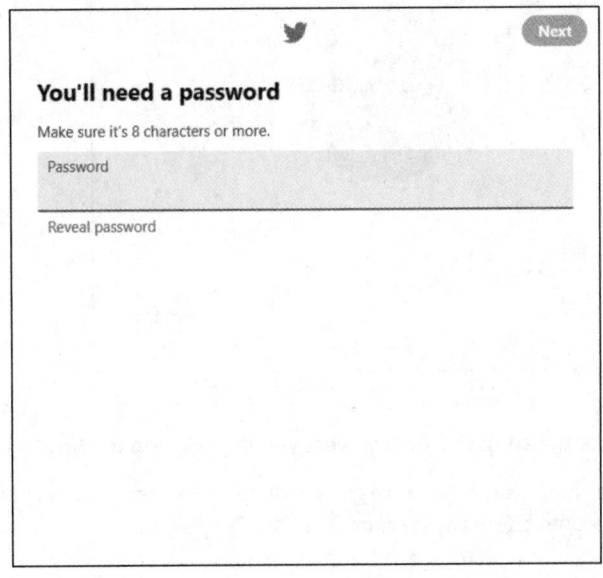
You'll need a password

Make sure it's 8 characters or more.

Password

Reveal password

FIGURE 3-5:
Use a safe password for your Twitter account.

6. **Type the password and click Next.**

You're asked to choose a profile picture for your Twitter account, as shown in Figure 3-6.

7. **Click the profile icon and choose the picture you want, or click Skip for Now.**

You're asked to describe yourself.

8. **Type a short bio (up to 160 characters) or click Skip for Now.**

You're asked to choose your interests from a long list that includes things such as sports, news, and music.

REMEMBER

Although you may be tempted to bypass typing your bio, give it some thought. If something about you is unique and you want the world to know — maybe you're an expert on 18th-century Tibetan bronzes — adding that to your bio may help someone else who shares the same interest find you. Your bio is accessible to anyone, so don't put anything in there that you don't want to be widely known.

FIGURE 3-6:
Choose a picture for your account.

9. **Click the subjects that interest you or click Skip for Now.**

   The things you choose are going to help Twitter recommend interesting accounts, as shown in Figure 3-7.

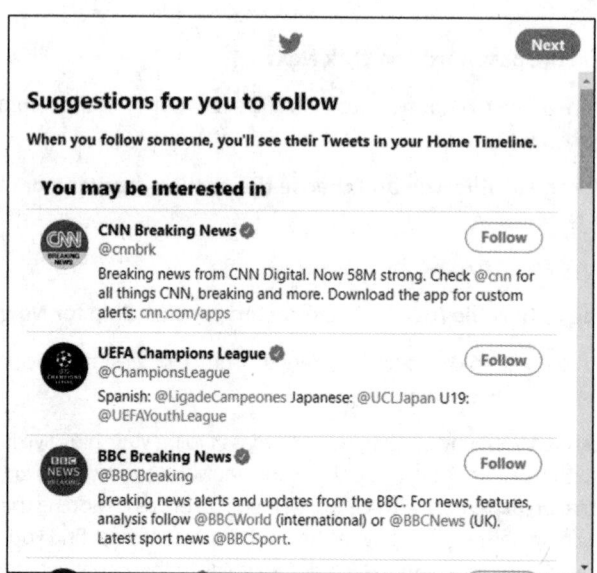

FIGURE 3-7:
Choose who you want to follow on Twitter.

10. **In the list of Suggestions for You to Follow, click Follow for the accounts that interest you.**

If you don't know who to follow, consider the following:

- @woodyleonhard, @ForDummies, @AndyRathbone, who wrote the original *Windows For Dummies,* @windowsblog to keep up on the Microsoft Party Line, and some of the major news services — @BBCWorld perhaps and @BreakingNews.

- Or try a few of the most-followed people on Twitter, @justinbieber at 112 million followers and @katyperry at 108 million and counting. You could even vote for fellow Nashvillian @taylorswift13 by adding to her trove of 87 million followers.

11. **When asked about turning on notifications, choose what you want.**

Keep in mind that Twitter notifications can become annoying, and it might be best to click Skip for Now.

You're finished and can finally use Twitter. :)

Celebrities and politicians don't have it so easy — many need to go through an independent confirmation step. But for normal dummies like you and me, the process is that easy.

At first, you probably just want to watch and see what others are tweeting to give you a sense of how tweeting is done. Create a practice tweet or two, and see how the whole thing hangs together.

TWO-FACTOR AUTHENTICATION (2FA)

Twitter supports two-factor authentication: Every time you start tweeting from a new device, it sends a confirmation text to your pre-established phone number, asking permission.

You can sign up for 2FA as part of the initial sign-up process. But if you have an account without 2FA, now's a good time to set it up.

To start using 2FA, log in to www.twitter.com, click the ellipsis on the left, and then choose Settings and Privacy. Go to Account, followed by Security. Then choose Two-Factor Authentication, and choose how you want to verify your logins: through text messages, an authenticator app, or a security key.

On the Twitter home page, type your first tweet in the What's Happening field, and click Tweet, as shown in Figure 3-8.

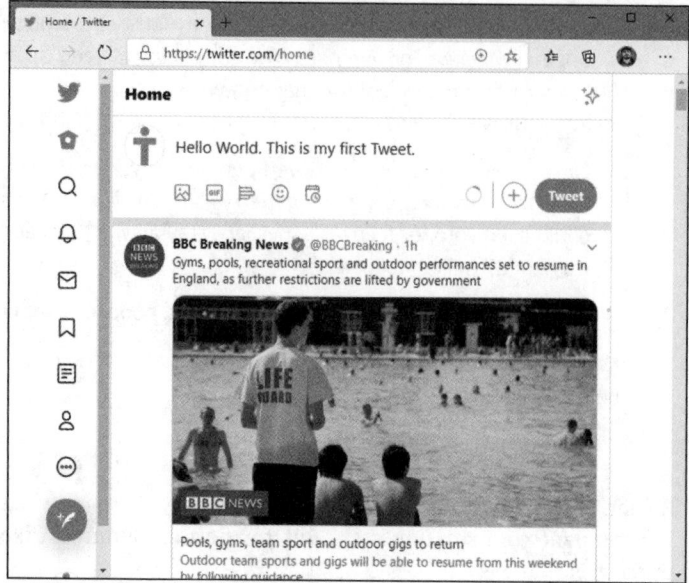

FIGURE 3-8:
Creating your
first tweet.

Tweeting for Beginners

On the surface, Twitter's easy and fun. Below the surface, Twitter's a remarkably adept application with lots of capabilities — and a few infuriating limitations.

Beware hacking

Before I dig in to the more interesting parts of Twitter, permit me to give you just one warning.

WARNING

There are unscrupulous people on Twitter, just as anywhere else. If you get a tweet from someone with a gorgeous picture who's trying to convince you to sign up for something or hand over your password, just ignore him, and he'll go away. If you get a tweet saying, "Somebody is writing bad things about you" or "Want to see a funny photo of you?" or "Find out who's been looking at your bio," ignore her.

Better yet, report her as a spammer. Tap or click the spammer's name. That takes you to the spammer's profile page. In the upper right, tap or click the ellipsis icon and choose Block or Report. To complete a report, choose the reason from the list, and then click Block.

If your Twitter account has been hacked — somebody talked you into clicking something that gets into your account or someone guessed your password — don't feel too bad about it. Fox News was hacked in July 2011. Mark Ruffalo (who plays The Hulk in *The Avengers*) got hacked in May 2012. Justin Bieber's account was hacked — back in the old days, when he had only 20 million followers. Ashton Kutcher. Taylor Swift. *The Huffington Post. USA Today.* Senator Chuck Grassley. Brett Favre. Miley Cyrus. *Reuters. Associated Press* (bombs at the White House). *Newsweek.* Queensland Police Department. Chipotle (it also faked a hack, as a publicity stunt). Burger King. US Central Command. Even Twitter's own Chief Financial Officer. And President Obama.

ASK WOODY.COM

It happens. If your account's been taken over, see the Twitter instructions at `https://help.twitter.com/en/safety-and-security/twitter-account-hacked`.

On the other hand, if you've posted some tweets you want to categorically disavow, you can always *claim* that your Twitter account was hacked. Sure to draw plenty of sympathy.

Using the @ sign and Reply

You see the @ sign everywhere on Twitter. In fact, I used it when listing the people you may want to follow. The @ sign is a universal indication that "what follows is an account name."

REMEMBER

In the not-so-great old days, sticking an @ and a username at the beginning of a tweet would limit the list of people who would automatically see the tweet. That's no longer the case. Anything you tweet goes out to all of the people who follow you. Easy peasy.

The Twitter viewer on the Internet has a Reply option to a tweet. On the Twitter website, hover your cursor over the speech bubble and you see a Reply button, as shown in Figure 3-9.

If you tap or click that Reply button, Twitter starts a new message with an @ sign followed by the sender's username. If you reply to the message in Figure 3-9, Twitter on the web creates a new tweet that starts: @ciprianrusen.

If you type a body to that message and click Tweet, the message goes to people who are following you and an extra copy is sent to @ciprianrusen.

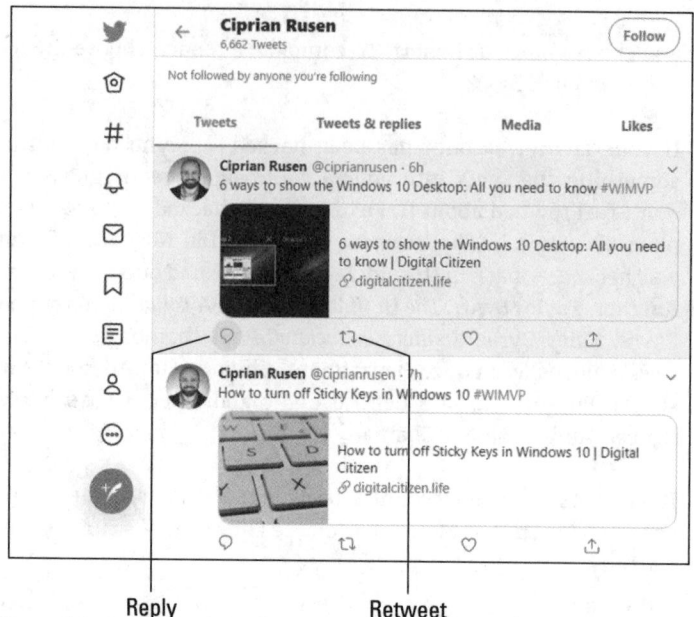

FIGURE 3-9:
Reply to a tweet.

Reply   Retweet

WARNING

A reply is *not* a private or hidden message. It's out in the open. Anybody who searches for your username or @ciprianrusen will see the message in its entirety.

Retweeting for fun and profit

If you receive a tweet and want to send it to all the people who follow you, the polite way to do so is with a *retweet* or *RT* for short. In order to give credit to the person who sent you the tweet, the retweet will include his username.

The circling arrows icon (refer to Figure 3-9) is for retweet. Tap or click the Retweet icon, and then choose a simple Retweet (Twitter builds a new tweet that copies the original tweet and adds the originator's username) or a Retweet with Comment. By retweeting a tweet precisely, you pass the information on to your followers yet preserve the attribution. If you want to add a comment to the original message, you can do so with ease, by choosing the appropriate option, as shown in Figure 3-10.

Direct Messaging

No discussion of the advanced part of Twitter would be complete without a mention of direct messaging (DM) — better known as *DM-ing*.

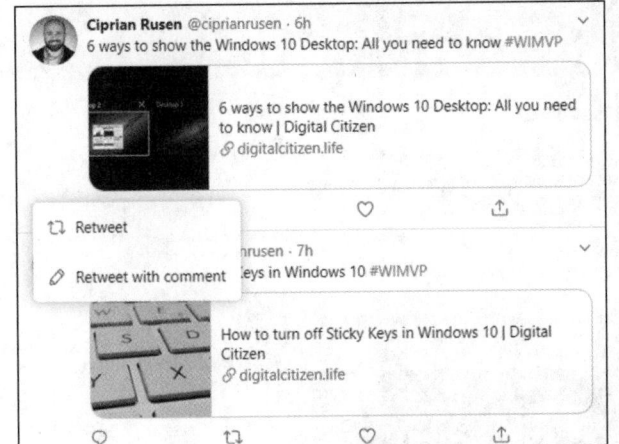

FIGURE 3-10:
Retweet a tweet
to all your
followers — with
or without your
own comment.

Unfortunately, people screw up DMs all the time, and the result can be embarrassing. I suggest you limit your use of DMs to situations where email may be a better approach, and that you studiously use the DM tools built into your Twitter account. You'll generally find DM hiding behind an icon that looks like a sealed envelope.

Hooking Twitter into Windows

I intentionally wrote this chapter to get you going on Twitter using the web directly. It's something of a lowest common denominator for Twitter access.

**ASK
WOODY.COM**

That said, when I'm using Windows 10, I always get at Twitter through a web browser.

Dozens of programs — many of them free or very cheap — run rings around the Twitter web interface. The names change every week, and the feature sets almost as quickly. There's a Twitter app available in the Microsoft Store, but (as of this writing) it's junk. If you're serious about using Twitter — particularly if you have more than one Twitter account or use both Twitter and Facebook — there are alternatives.

TIP

The Windows 10 app that I hear about most is Tweetium, which costs $3. It's a solid, usable Twitter alternative designed for heavy-duty use. See Figure 3-11.

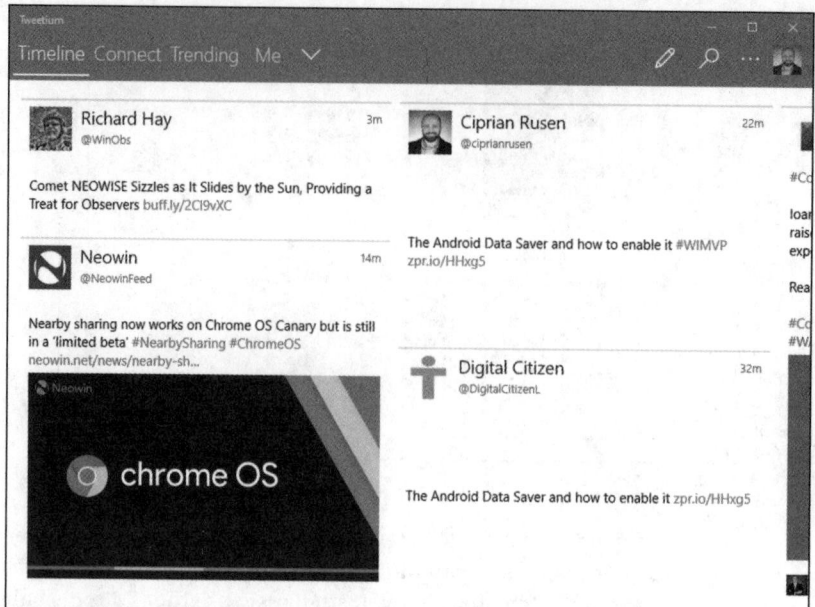

FIGURE 3-11:
Tweetium makes managing multiple Twitter accounts easy.

The Twitter web page and Tweetium support several key features:

» **Automatic URL shortening,** so something like `http://www.something` `oranother.com/this/and/that.php` ends up looking like `http://` `is.gd/12345`.

» **Multiple Twitter accounts** so people who keep their business and personal accounts separate can manage both simultaneously.

» **Picture attachments** with automatically generated links to picture sites. The best Twitter apps let you drag and drop pictures onto your tweets and take care of all the details.

» **Sophisticated search functions** so you can display not only your tweets and the tweets of those you follow, but also tweets on topics that interest you, such as #19thcenturydentistoffices.

Don't forget to download a Twitter app for your smartphone and tablet, too. I use both the Android and the iOS (iPhone/iPad) Twitter apps every day, and they're great.

Chapter **4**

Getting Started with LinkedIn

In some ways, LinkedIn resembles Facebook — keeping up with people and expanding connections are grist for the mill. But in other ways, LinkedIn is completely different; LinkedIn is focused on professional relationships, which LinkedIn calls *connections*.

You can use your LinkedIn connections to showcase products, look for a job, advertise a job, scout new business opportunities, find temporary help, stay up to date on companies that interest you — for any reason — or just replace your old Rolodex (does anybody still use a Rolodex?) or that tattered box of business cards on your desk.

With more than 690 million subscribers (169+ million are in the US) LinkedIn has more than reached critical mass. Many business people consider it a key part of their existence.

**ASK
WOODY.COM**

LinkedIn doesn't have a Windows 10 app. There's nothing in the Microsoft Store from LinkedIn. But that shouldn't stop you from using it on your Windows 10 computer. All it takes is a web browser.

In December 2016, Microsoft completed its purchase of LinkedIn for $26.2 billion — the largest software purchase, anytime, anywhere. LinkedIn founder and Silicon Valley heavyweight Reid Hoffman joined Microsoft's board. Some feared that LinkedIn would let other platforms — iOS, Android, Mac — wither, but that hasn't been the case. Right now, it appears that Microsoft's main drive for LinkedIn is to integrate it with Microsoft 365 (formerly known as Office 365) and use it as a benefit for business customers, particularly enterprises.

Signing Up for LinkedIn

Don't have a LinkedIn account? Got a few minutes?

Here's how to get started:

1. **Fire up your favorite browser, go to** www.linkedin.com, **and click Join Now.**

   You see a sign-up page like the one in Figure 4-1.

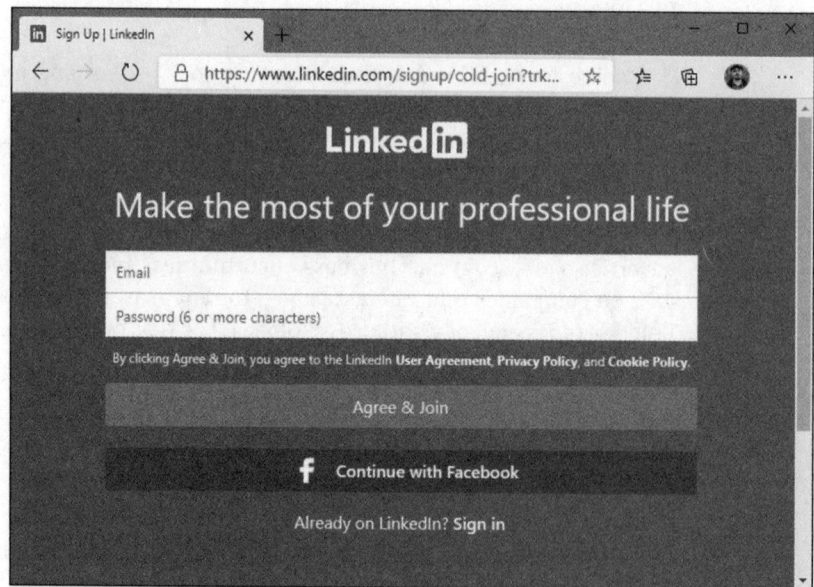

FIGURE 4-1:
Signing up for LinkedIn is easy.

2. **If you want to start a new account, fill in the blanks and click or tap Agree & Join.**

Make sure you use a real email address: LinkedIn uses it to verify your account. You're better off *not* using an email address that's associated with your current employer. Remember, even the walls have eyes — and you want to be able to get into your LinkedIn account even if you get fired!

3. **Next, enter your first name and last name, and click Continue.**

You're asked to do a quick security check to verify that you're a real person.

4. **Click Verify and follow the instructions on the screen to finish the security check.**

The first profile page appears, as shown in Figure 4-2, which asks you to enter your country and city/district.

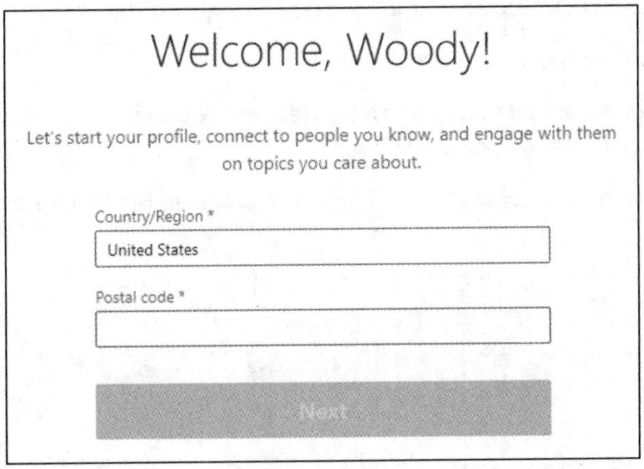

Welcome, Woody!

Let's start your profile, connect to people you know, and engage with them on topics you care about.

Country/Region *

United States

Postal code *

Next

FIGURE 4-2:
You must
complete a lot
of data for a
LinkedIn account.

5. **Enter your country, city, and postal code, and click Next.**

LinkedIn might ask for slightly different information based on your location. The screen shown in Figure 4-3 appears.

6. **Do one of the following and then click Continue:**

   - Enter the information requested from you.

   - If you're a student, click I'm a Student and then enter the required details.

   LinkedIn sends a verification code to your email address.

7. **Enter the verification code, and then click Agree & Confirm.**

8. **When asked if you are looking for a job, answer as you see fit.**

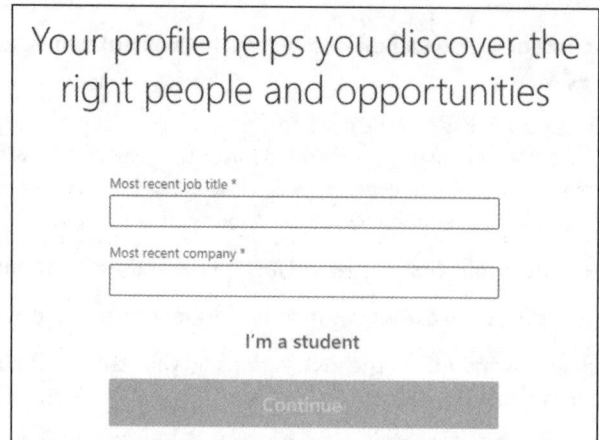

FIGURE 4-3:
Provide details
about whether
you're a student
or an employee.

9. **When LinkedIn asks to look inside your email account to find connections, click Skip and confirm that you want to Skip.**

   You can find your own contacts later.

10. **When asked if you want to connect with others, search for and select specific people, or click Skip.**

    You can add contacts later. The screen shown in Figure 4-4 appears.

FIGURE 4-4:
It is time to add
your LinkedIn
picture.

11. **Click Add Photo, choose a picture you like, and then save it.**

    LinkedIn says that you look great.

12. **Click Continue.**

13. **When asked if you want to connect with your teammates, click Skip.**

14. **When asked if you want to set up an alert for job opportunities, do so now, or click Skip.**

15. **When asked if you want the LinkedIn app for your mobile phone, enter your number and choose Text Me the Link or click Skip.**

16. **When LinkedIn recommends companies, people, or hashtags for you to follow, choose the companies or people you want, and then click Finish.**

    LinkedIn takes you to your profile page, which looks more or less like Figure 4-5. Remember that just about anybody can see anything you post.

The social part of LinkedIn involves establishing *connections* — links with people you know or know of. To start filling in your Connections list, tap or click the My Network button at the top of the home page, choose Connections, and add people based on their email addresses. You can also find people you know based on others' connections. Look for the little Connect buttons and links throughout the LinkedIn interface.

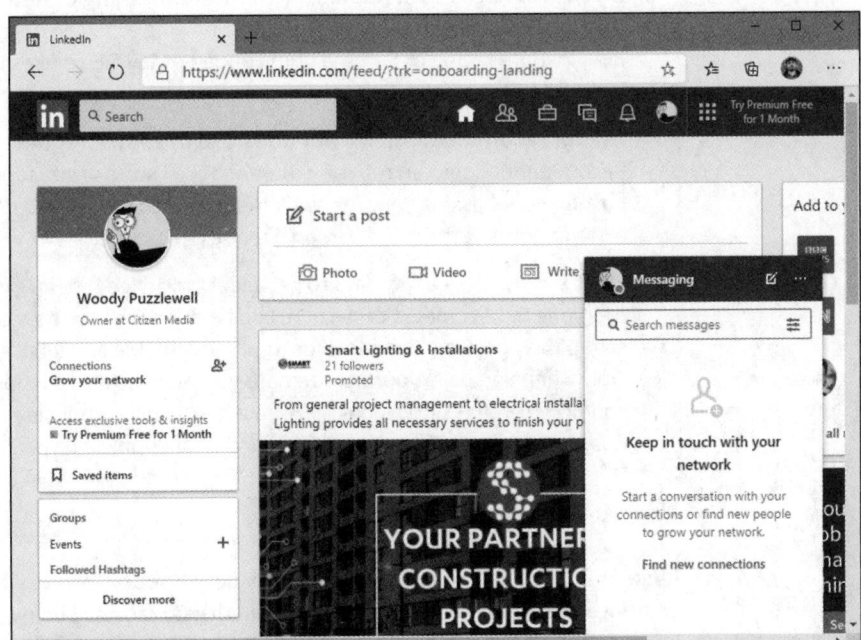

FIGURE 4-5:
LinkedIn takes you to the main page.

Using LinkedIn for Fun and Profit

Using LinkedIn with Windows is both an art and a science. Here are a few hints I've acquired over the years:

>> **Use your current job title to your advantage.** I'm not sure why, but LinkedIn seems to show your current job title almost everywhere. Anytime someone hovers his mouse cursor over your picture, for example, he sees your current job title and employer, and your location. Stock job titles (CEO, Analyst, Nice Guy — that's the one I use) don't have much sizzle. On the other hand, M2M Executive with Expertise in the Rapid Implementation of CRM Solutions (M.S., Ph.D., O.B.E.) certainly draws attention.

In some contexts, LinkedIn truncates your job title. Someone looking at your profile sees the entire title, but someone looking at search results, for example, sees only the first few words.

TIP

>> **Put a different, professional picture on your LinkedIn account.** Don't recycle your Facebook pic — you know, the one your friend took when you were completely plastered at the going-away party. Definitely a no-no in this arena. By all means, wear a suit and tie if you feel more comfortable that way, but casual is okay, too. Just remember that the people you want to impress will look at that mug and make decisions based on it.

>> **If you graduated with honors, or there's something of note about your degree, include it in the Degree field.** Showing a college degree, such as B.A. Phi Beta Kappa or Summa Cum Laude or M.S. E.E., makes a greater impression than just listing your degree. People will see it.

>> **Ask for recommendations, but don't use the stock request form.** Recommendations can make a difference in all sorts of situations, so don't be bashful about asking your friends to refer you. But when you do, take a few extra minutes, and write a personal request message.

>> **Start slowly.** Take a few days to get a feel for LinkedIn before you invite everyone to become a Connection. Look around and see how other people set up their profiles. Get a feel for what's acceptable and what's overly pushy. Only when you have your bearings are you ready to add all those old email contacts to your Connections list. And when you start building your Connections list, go slowly — just a handful of people a day.

Remember six degrees of separation?

LinkedIn allows you to follow industry leaders, groups, professional interests, and your local neighborhood firefly-collecting organization. The quality of the offered information varies widely, but it's king of the roost for business use.

After you have a few Connections put together, you can view them by clicking My Network. You may find that you have a lot of people connected only two or three steps away; see Figure 4-6.

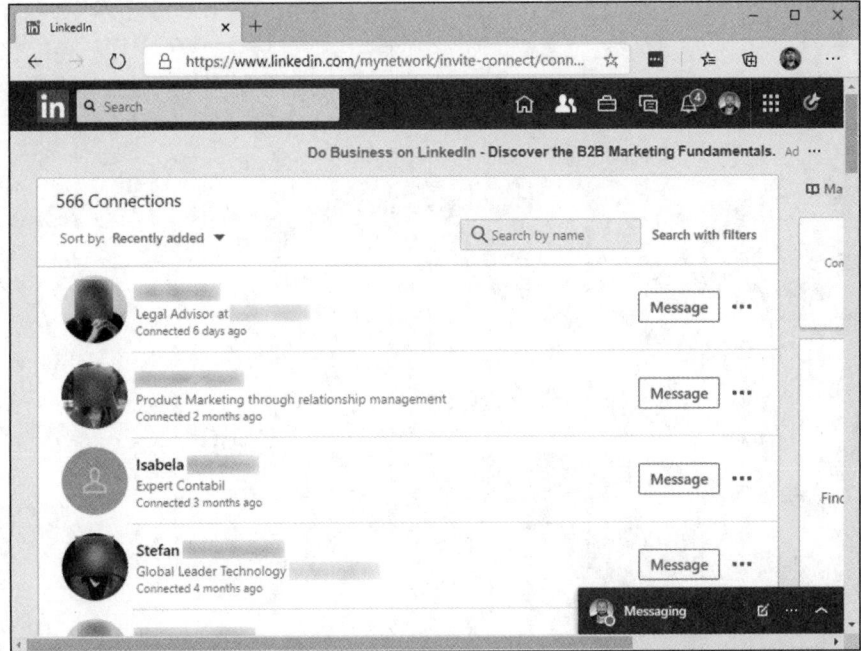

FIGURE 4-6: Even if you're only moderately well connected, you can have hundreds of people three hops away.

ASK
WOODY.COM

Social networking works. Even if you don't use LinkedIn very much, having it available just in case — just in case you're looking for a new job, or for an expert in a particular field — is well worth the effort.

7

Controlling Your System

Contents at a Glance

Chapter **1**

Settings, Settings, and More Settings

Windows 10 has settings. Boy howdy, does it have settings.

REMEMBER

The desktop's Control Panel (shown in Figure 1-1) — long the bastion of Windows settings, through many generations of Windows (see the nearby sidebar) — controls many of the aspects of how a Windows 10 PC works. The new Windows 10 Settings app (shown in Figure 1-2) controls several hundred settings. And — get this — there's overlap between the two, but some settings can be changed only in the Settings app, and other settings can be changed only on the old-fashioned Control Panel. However, with each new major update to Windows 10, more settings get migrated from the old Control Panel to the new Settings app.

This chapter straddles both sides of the fence, both the new Settings app and the old Control Panel. If you want to take control of your machine, unfortunately, you have to learn how to live in both worlds.

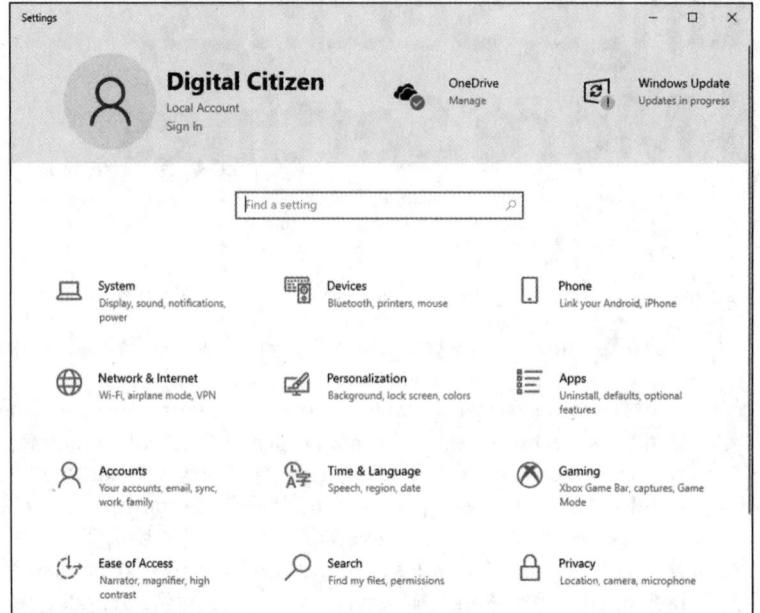

WHAT HAPPENED TO THE CONTROL PANEL?

Windows has had a Control Panel for as long as Windows has been Windows. It goes all the way back to Windows 1.0. You've probably probed it and sweated over it, as have I, for as long as you've been using Windows.

With Windows 10, Microsoft is clearing out some of the dead wood. I believe they started Windows 10 development hoping to move all those Control Panel settings and applets to a new Windows 10 app called Settings. If that was the intent, Microsoft missed the boat. There are gazillions of settings in the Control Panel. A googolplex of them. Look it up.

Whatever the intent, the final result is a bit schizoid. Some settings are in the Settings app, others are in the Control Panel, and some of them are kind of stretched between the two. Easy example: If you want to enable and control in detail how the File History backup feature works, you must go into the Control Panel. If you just want to enable File History, you can do that in the Settings app too.

The Control Panel is definitely headed into the bit bucket. But it remains to be seen if Microsoft will be able to fully eviscerate it this decade.

Introducing the Settings App

The Windows 10 Settings app (refer to Figure 1-2) is a remarkable collection of settings, arranged in a way that's infinitely more accessible — but sometimes less logical — than the old-fashioned desktop Control Panel. Click Start, Settings (the gear icon), and you see these options:

» **System:** This includes settings for changing the display and control notifications, analyzing your apps' usage, controlling Snap and multiple desktops, moving in and out of tablet mode, kicking in Battery Saver, controlling how long the screen stays active when not in use, analyzing how much storage space is being used, handling downloaded maps, assigning apps to specific filename extensions, and looking at your PC's name and ID. In Storage (shown in Figure 1-3), you can tell Windows 10 where to store certain kinds of files. There are also links to the Control Panel applets for admin tools, Power Options, Optimize Drives, BitLocker and Sysinfo.

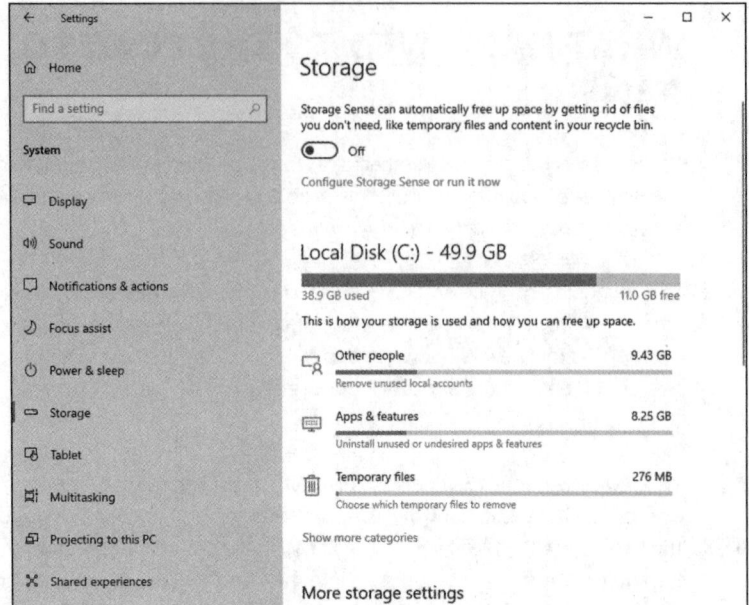

FIGURE 1-3:
System's Storage
pane lets you
free up space and
specify where to
put new files.

>> **Devices:** From here, you can control printers, scanners, and other connected devices; turn Bluetooth on and off; change mouse settings (with a link to the Control Panel app for mice); turn on and off autocorrect and text suggestions; manipulate the pen; and specify what AutoPlay program should kick in when you insert a drive or card.

>> **Phone:** Link your Android phone or iPhone to your Windows 10 account. This option is useful if you want to browse to a location using Microsoft Edge on your phone (not terribly likely), and continue using Edge on your PC.

>> **Network & Internet:** This lets you turn Wi-Fi off and on and change your connection, with lots of links to the Control Panel; set up the Windows Firewall (again through the Control Panel); go into airplane mode, thus turning off both Wi-Fi and Bluetooth connections; track how much data has been sent and received in the past month, by app; set up a VPN; work with a dial-up connection; and manually set a Proxy.

>> **Personalization:** This catchall category includes setting your wallpaper (background), themes, choosing accent colors, putting a picture on your lock screen, and controlling the Start menu.

>> **Apps:** Want to delete a program (app)? Here's where you do it. You can also set default apps (for opening graphic files, for example), and throttle some — but not all — apps that run when your computer starts. You also get to control the apps that start when Windows 10 starts and associate apps with websites.

>> **Accounts:** Disconnect a Microsoft account, set your account picture, and change information about your account with Microsoft's account database in the sky. Options enable you to add a new standard user, change your password, switch to a picture or PIN password, or switch between a Microsoft account and a local account. You can sync your settings among multiple computers that use your login (see Figure 1-4). There's also a section for connecting to a domain (typically a company or organization network) or Microsoft's Azure Active Directory in the cloud.

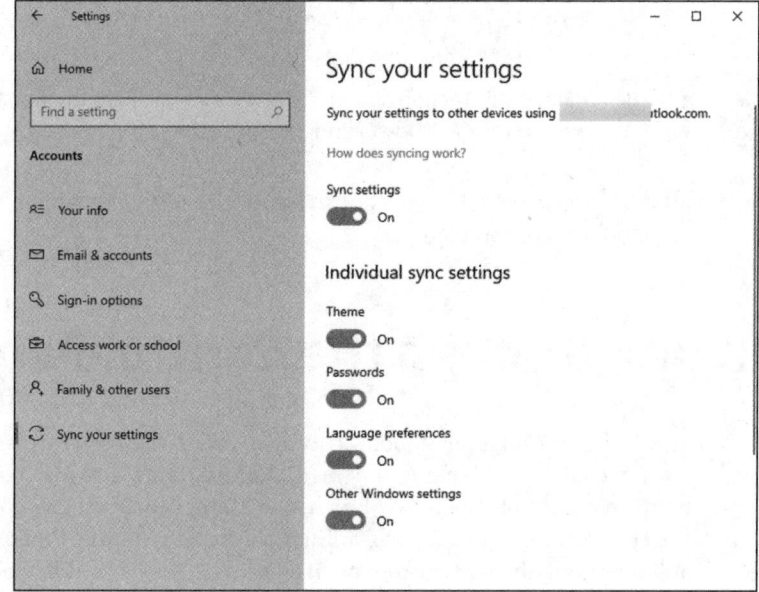

FIGURE 1-4:
Control exactly what gets synced among computers using your Microsoft account.

>> **Time & Language:** Set your time zone, manually change the date and time, set date and time formats, add keyboards in different languages, add new display languages, control how Windows 10 uses speech and spoken languages, and set up your microphone for speech recognition.

>> **Gaming:** Work with game mode, the Xbox game bar, and other gaming-related features. Your link to the Xbox-friendly part of your Windows 10 PC.

>> **Ease of Access:** Microsoft has long had commendable aids for people who need help seeing, hearing, or working with Windows. All the settings are here.

>> **Search:** Control how Search works in Windows 10, whether or not web results are filtered, and how indexing works, so that you get the search results you want.

>> **Privacy:** A grandstanding set of settings. You can turn off broadcasting of your advertising ID, which is a unique ID maintained by Microsoft to identify you, individually. You can block app access to your name and picture, turn on and off location tracking, and keep your webcam and microphone locked up. You can also control beacons and other sync proclivities, including giving Windows 10 permission to send your full health, performance, and diagnostics information to Microsoft.

>> **Update & Security:** Control how Windows 10 updates itself, when it installs updates, and so on. You can turn File History on and off from this location, under Backup, and use Go to Backup and Restore (Windows 7). The Windows Security (the former Windows Defender antivirus) settings and all recovery options live here too.

WARNING

Remarkably, this section also includes (be careful!) links to refresh or reinstall Windows on your PC. Don't accidentally choose one of these, okay?

All in all, it's a well-thought-out subset of the settings that you may want to use. But it's far from complete.

Spelunking through the Control Panel

The inner workings of Windows 10 also reveal themselves inside the mysterious (and somewhat haughtily named) Control Panel. You may be propelled to the (sniff) old Control Panel via a link in the aaaah new Settings app. But if you want to get in directly, of your own volition, click or tap in the Windows 10 search box and type **Con**. Choose Control Panel. Figure 1-5 shows the Control Panel window.

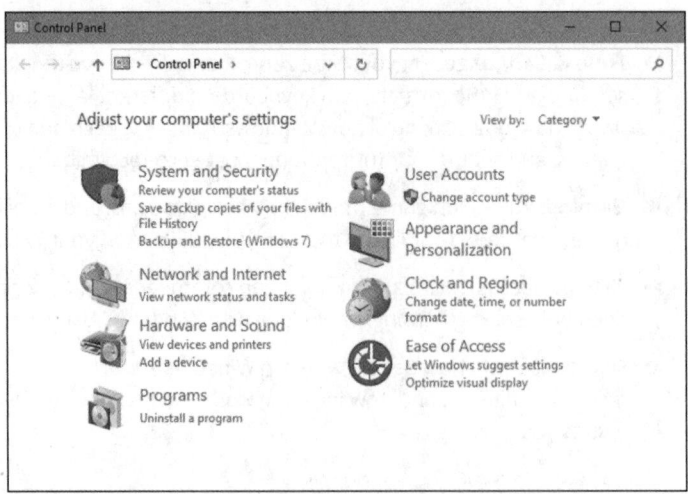

FIGURE 1-5:
The Windows 10 Control Panel is a sight to behold.

I cover various Control Panel components at several points in this book, but here is an overview. The main categories of the Control Panel span the breadth (and plumb the depths) of Windows 10-dom:

>> **System and Security:** Use an array of tools for troubleshooting and adjusting your PC and generally making your PC work when it doesn't want to. Check out the components of the Windows 10 Firewall. Change power options, retrieve files with File History, manage Storage Spaces, and rifle through miscellaneous administrative tools. Use this part of the Control Panel with discretion and respect.

>> **Network and Internet:** Configure network sharing settings. Set up Internet connections, particularly if you're sharing an Internet connection across a network or if you have a cable modem or digital subscriber line (DSL) service. There are even a few hooks for Internet Explorer, which you're not likely to need, because Microsoft Edge has edged it out.

>> **Hardware and Sound:** The "all other" category. Add or remove printers and connect to other printers on your network. Troubleshoot printers. Install, remove, and set the options for mice, game controllers, joysticks, keyboards, and pen devices. Power settings are here, too.

>> **Programs:** Add and remove specific features in some programs (most notably, Windows 10 and Office). Uninstall programs. Change the association between filename extensions and the programs that run them (so that you can, for example, have iTunes play WMA audio files). Most of the functionality here is available in the Settings app, but a few laggards are still in the old Control Panel.

>> **User Accounts:** This group is a very limited selection of actions that Microsoft hasn't yet moved to the Settings app. You must go here to remove an account or manage credentials associated with an account.

>> **Appearance and Personalization:** Font management is in this section, as well as a similar — but much prettier — section in the Settings app. You also get access to the Ease of Access Center and to File Explorer Options.

>> **Clock and Region (and Language):** Set the time and date — although double-clicking the clock on the Windows taskbar is much simpler — or tell Windows 10 to synchronize the clock automatically. You can also add support for complex languages (such as Thai) and right-to-left languages, and change how dates, times, currency, and numbers appear.

>> **Ease of Access:** Change settings to help you see the screen, use the keyboard or mouse, or have Windows 10 flash part of your screen when the speaker would play a sound. You also set up speech recognition here.

Many Control Panel settings duplicate options you see elsewhere in Windows 10, but some capabilities that seem like they should be Control Panel mainstays remain mysteriously absent.

ASK
WOODY.COM

If you want to change a Windows 10 setting, by all means try the Control Panel, but don't be discouraged if you can't find what you're looking for. The Settings app is growing into a better alternative each year.

Putting Shortcuts to Settings on Your Desktop

Want to see Windows 10's Update setting by simply clicking or tapping on the desktop? Enable or disable your microphone with two clicks? Turn off your webcam? Manage your Wi-Fi settings? It's easy.

ASK
WOODY.COM

I came up with a simple extension of a brilliant hack by Lucas López (@Whistler4Ever on Twitter), published by Sergey Tkachenko at Winaero, and unearthed by Steven Parker at Neowin. It's an easy way to put an icon on your Windows 10 desktop that opens to just about any Settings page, where you can change a setting in a nonce.

Here's how to make it work.

1. **Right-click (or tap and hold down) any blank place on the Windows 10 desktop. Choose New ⇨ Shortcut.**

    You see the New Shortcut Wizard shown in Figure 1-6.

2. **Choose one of the ms-settings apps listed in Table 1-1, and type it in the input box.**

    For example, as in Figure 1-6, to go to the Data Usage app, and type **ms-settings:windowsupdate** in the box marked Type the Location of the Item.

3. **Click Next, give the shortcut a name, and click Finish.**

    A new shortcut appears on your desktop. Double-click or tap it, and the Settings app appears, as in Figure 1-7.

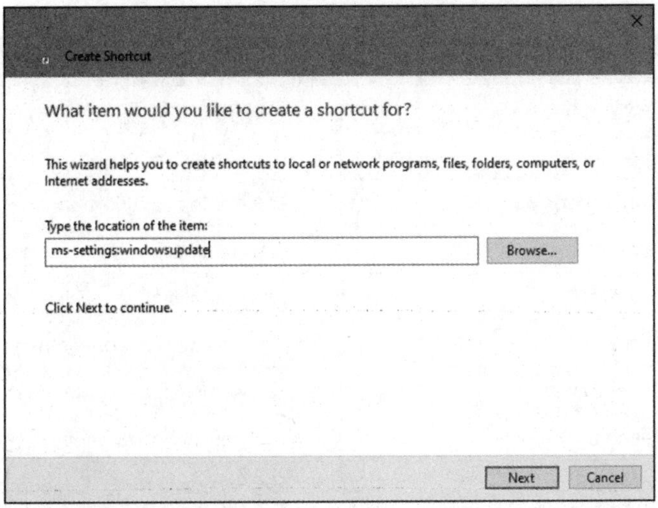

FIGURE 1-6:
Create a shortcut
to the Windows
Update pane in
the Settings app.

TABLE 1-1

Shortcuts to Settings App Panels

Settings App Page	Command
Battery Saver	ms-settings:batterysaver
Battery Use	ms-settings:batterysaver-usagedetails
Battery Saver Settings	ms-settings:batterysaver-settings
Bluetooth	ms-settings:bluetooth
Colors	ms-settings:colors
Data Usage	ms-settings:datausage
Date and Time	ms-settings:dateandtime
Closed Captioning	ms-settings:easeofaccess-closedcaptioning
High Contrast	ms-settings:easeofaccess-highcontrast
Magnifier	ms-settings:easeofaccess-magnifier
Narrator	ms-settings:easeofaccess-narrator
Keyboard	ms-settings:easeofaccess-keyboard
Mouse	ms-settings:easeofaccess-mouse
Other Options (Ease of Access)	ms-settings:easeofaccess-otheroptions
Lockscreen	ms-settings:lockscreen
Offline Maps	ms-settings:maps
Airplane Mode	ms-settings:network-airplanemode

(continued)

TABLE 1-1 *(continued)*

Settings App Page	Command
Proxy	ms-settings:network-proxy
VPN	ms-settings:network-vpn
Notifications & Actions	ms-settings:notifications
Account Info	ms-settings:privacy-accountinfo
Calendar	ms-settings:privacy-calendar
Contacts	ms-settings:privacy-contacts
Other Devices	ms-settings:privacy-customdevices
Feedback	ms-settings:privacy-feedback
Location	ms-settings:privacy-location
Messaging	ms-settings:privacy-messaging
Microphone	ms-settings:privacy-microphone
Motion	ms-settings:privacy-motion
Radios	ms-settings:privacy-radios
Speech, Inking & Typing	ms-settings:privacy-speechtyping
Camera	ms-settings:privacy-webcam
Region & Language	ms-settings:regionlanguage
Speech	ms-settings:speech
Windows Update	ms-settings:windowsupdate
Work Access	ms-settings:workplace
Connected Devices	ms-settings:connecteddevices
For Developers	ms-settings:developers
Display	ms-settings:display
Mouse & Touchpad	ms-settings:mousetouchpad
Cellular	ms-settings:network-cellular
Dial-up	ms-settings:network-dialup
DirectAccess	ms-settings:network-directaccess
Ethernet	ms-settings:network-ethernet
Mobile Hotspot	ms-settings:network-mobilehotspot
Wi-Fi	ms-settings:network-wifi

Settings App Page	Command
Manage Wi-Fi Settings	ms-settings:network-wifisettings
Optional Features	ms-settings:optionalfeatures
Family & Other Users	ms-settings:otherusers
Personalization	ms-settings:personalization
Backgrounds	ms-settings:personalization-background
Colors	ms-settings:personalization-colors
Start	ms-settings:personalization-start
Power & Sleep	ms-settings:powersleep
Proximity	ms-settings:proximity
Display	ms-settings:screenrotation
Sign-in Options	ms-settings:signinoptions
Storage Sense	ms-settings:storagesense
Themes	ms-settings:themes
Typing	ms-settings:typing
Tablet Mode	ms-settings://tabletmode/
Privacy	ms-settings:privacy

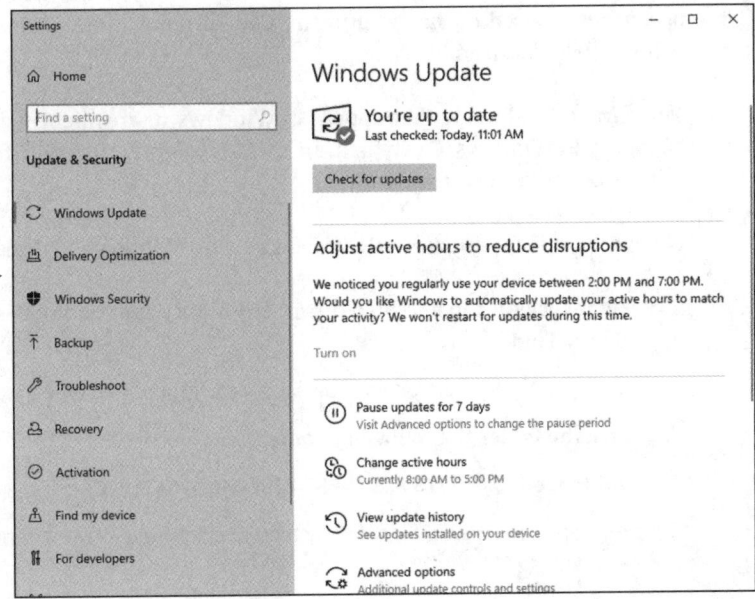

FIGURE 1-7:
The new shortcut takes you straight to the Windows Update pane.

God Mode

The Windows Vista–era parlor trick commonly called *God mode* is alive and well in Windows 10, as shown in Figure 1-8.

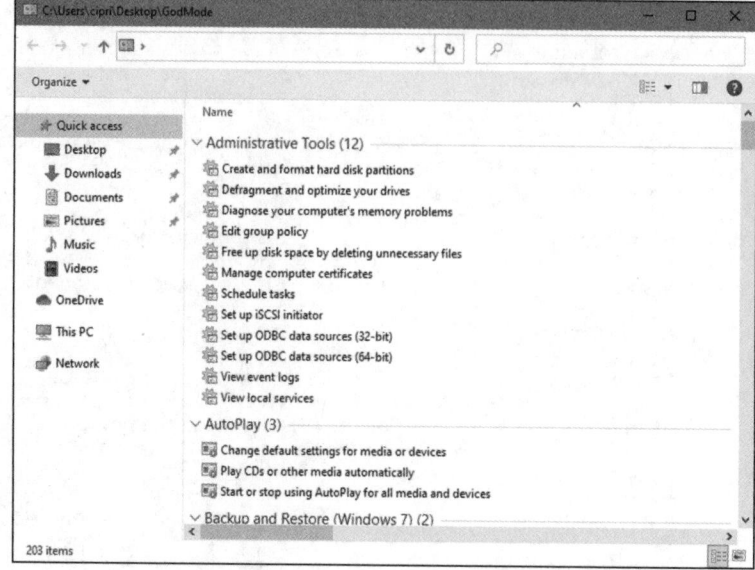

FIGURE 1-8:
God mode
is a massive
collection of
230 shortcuts
into all sorts
of Windows
settings, many of
which are quite
obscure.

I, for one, was quite surprised to see that it made the transition to Windows 10, because it's based on hooks into the Control Panel — and the Control Panel is being slowly disassembled.

The parts of God mode that appear in Windows 10 are slightly different from the elements in Windows 8.1 (which, in turn, is slightly different from Windows 7). But the overall effect is the same.

Follow these steps to access God mode on your Windows 10 desktop:

1. **Right-click (or tap and hold down) any empty spot on the desktop. Choose New ⇨ Folder.**

   A new folder appears on your desktop, ready for you to type a name.

2. **Give the folder the following name:**

   GodMode.{ED7BA470-8E54-465E-825C-99712043E01C}

   You can use any valid filename instead of GodMode — call it Parlor Trick if you like.

3. **Tap or click the folder to bring up the list you see in Figure 1-8.**

   It's a massive list of direct links into all sorts of settings. All of them seem to work.

Some of these may be useful. For example, the AutoPlay option, when accessed through God mode, brings up the old Windows 7/8 AutoPlay dialog box, which is considerably more advanced than the Windows 10 Settings version of AutoPlay (Start, Settings, Devices, AutoPlay).

Installing New Languages

Many Windows 10 users use more than one language. You may know English, but also Spanish, French, German, or Hindu. If you want to install a new display language, Windows 10 makes the process easy, especially in its latest versions. Here is how it works:

1. **Click Start and then Settings.**

   You see the Settings app.

2. **Go to Time & Language, and on the left, choose Language.**

3. **On the right, scroll down to Preferred Languages, and click or tap the +Add a Language button.**

   You are shown a surprisingly long list with all the languages available for Windows 10, as shown in Figure 1-9.

4. **Find and then click the language you want to install, and then click Next.**

   You can find the language by scrolling the list of languages or by using the search box at the top of the screen.

   You are shown several settings for the language you want, as shown in Figure 1-10. The language will be installed as both a display language and a keyboard language.

5. **(Optional) Select the Set as My Windows Display Language option.**

6. **Click Install.**

   Windows 10 shows a progress bar in the Language window, so that you see how long it has left until it finishes downloading the new language.

7. **When the new language is downloaded, you can set it as the default in the Windows Display Language drop-down list.**

To change between keyboard languages, press Windows+spacebar and then select the language you want from the list.

TIP

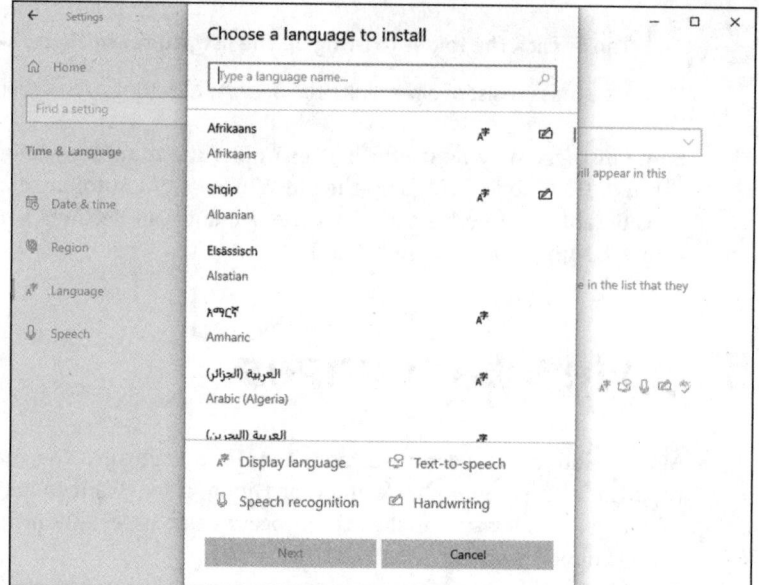

FIGURE 1-9:
Choose the language that you want to add to Windows 10.

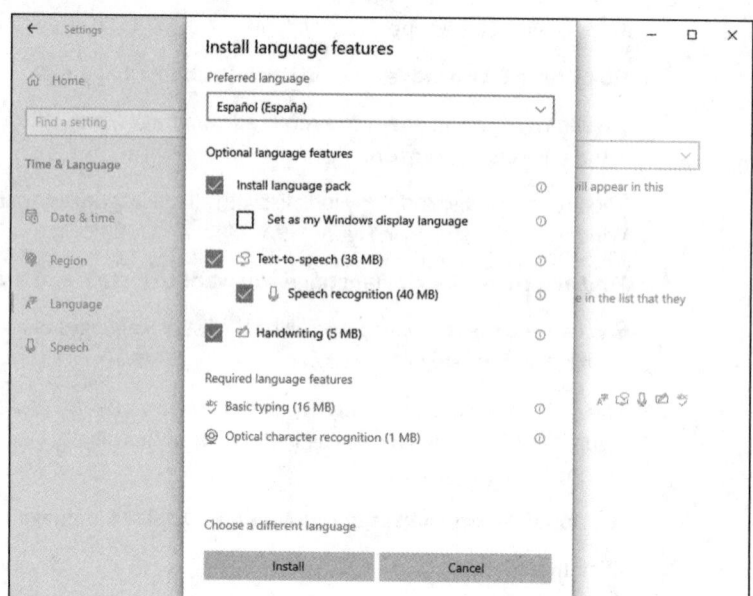

FIGURE 1-10:
The settings available when installing a new language.

Chapter **2**

Troubleshooting and Getting Help

Your PC ran into a problem that it couldn't handle, and now it needs to restart. You can search for the error online, but the *error message goes by so fast that you can't possibly read it.*

**ASK
WOODY.COM**

Wish I had a nickel for every time I've seen that "blue screen" message. People write to me all the time and ask what caused the message, or one like it, to appear on their computers. My answer? Could be anything. Hey, don't feel too bad: Windows couldn't figure it out either, and Microsoft spent hundreds of millions of dollars trying to avoid it.

Think of this chapter as help on Help. When you need help, start here.

Windows 10 arrives festooned with automated tools to help you pull yourself out of the sticky parts. The troubleshooters really do shoot trouble, frequently, if you find the right one. The error logs, event trackers, and stability graphs can keep you going for years — even the experts scratch their heads. Windows 10 abounds with acres and acres — and layers and layers — of help. Some of it works well. Some of it would work well, if you could figure out how to get to the right help at the right time.

This chapter tells you when and where to look for help. It also tells you when to give up and what to do after you give up. Yes, destroying your PC is an option. But you may have alternatives. No guarantees, of course.

ASK
WOODY.COM

This chapter includes detailed, simple, step-by-step instructions for inviting a friend to take over your computer, via the Internet, to see what is going on and lend you a hand while you watch. I believe that this Remote Assistance capability is the most powerful and useful feature ever built into any version of Windows.

Troubleshooting the Easy Way

If something goes bump in the night and you can't find a discussion of the problem and its solution in this book, your first stop should be the Troubleshooters. They don't call 'em Trouble fer nuthin'.

REMEMBER

Windows 10 ships with a handful of troubleshooters. *Troubleshooters,* as the name implies, take you by the hand and help you figure out what's causing problems — and, just maybe, solve them.

If you run into a problem and you're stumped, see whether Microsoft has released a pertinent troubleshooter by following these easy steps:

1. **Click in the search box, next to the Start button, and type (or say)** troubleshoot.

2. **Tap or click Troubleshoot Settings.**

   The Troubleshooting section from the Settings app is loaded. Windows 10 might recommend some troubleshooters.

3. **To see all the troubleshooters available, click or tap Additional Troubleshooters.**

   The list of troubleshooters is quite long, as shown in Figure 2-1.

4. **Click or tap the troubleshooter that can help with your problem, and then click the Run the Troubleshooter button.**

Frequently, troubleshooters just can't shoot the trouble, and they end up with an error message dialog box that says something like This Error Cannot Be Automatically Repaired. You can tap or click Next and end up with informative messages such as "The Error '5' Was Encountered." (I don't make this stuff up — that's an error message I received while running the connection troubleshooter.)

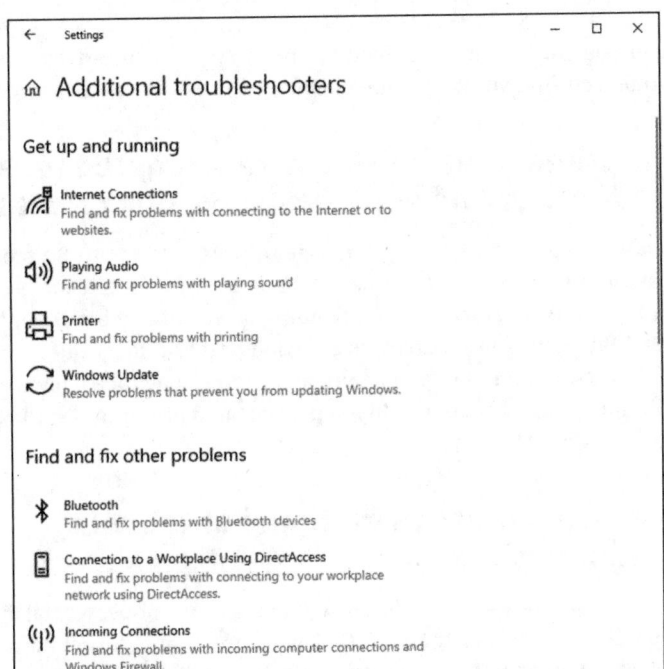

FIGURE 2-1:
Troubleshooting
wizards can cut
to the heart of a
problem, if you
can find one.

If you can't find a worthy troubleshooter, you may be able to unearth worthwhile content from your systems log using Event Viewer, a topic that I tackle in Book 8, Chapter 3.

Troubleshooting the Hard Way

No troubleshooter available to whisk you out of harm's way?

Yeah, that's a tough position to be in.

ASK
WOODY.COM

I've come up with two lists of solutions that you may find enlightening or at least helpful. One deals with installing Windows 10, and the other deals with updating Windows 10.

Tackling installation problems

This section is for folks who are using Windows 7 or Windows 8.1 and trying to upgrade to Windows 10 but can't, and for those who are trying to move from one version of Windows 10 to the next. I have categorized some installation problems, including initial setup problems, and offer a bit of advice and some pointers, should you find yourself trapped between the offal and the impeller.

The prime directive: If you're prompted for a product key and don't have one, don't sweat it

ASK
WOODY.COM

Assuming you're upgrading from a genuine Windows 7 or 8.1 machine, or if you're switching versions of Windows 10, if you're prompted for a product key, click Skip, Do This Later, or Next (depending on the dialog box). Don't bother trying to find a Windows 10 key. Chances are good that Windows will recognize the error of its ways and not bother you again, although it may take a few days for the activation routine to figure it out. If you get repeated prompts, see the upcoming section on activation problems.

The installer hangs for hours or reboots continuously

First, make sure that you've disconnected any nonessential hardware: Unplug any hard drives other than the C: drive. Yank that external hard drive, disconnect peripherals that aren't absolutely necessary, including extra monitors, Smart card readers, weird keyboards, whatever. If possible, consider turning off Wi-Fi and plugging into a router with a LAN cable (that worked for me).

TECHNICAL
STUFF

Second, make sure you have the right upgrade: 32-bit for 32-bit machines, 64-bit for most. If you started with Windows 7 Starter, Home Basic, Home Premium, or Windows 8.1 (standard, usually called Home), or you want to move to the next version of Windows 10 Home, you should install Windows 10 Home. If you started with Windows 7 Pro or Ultimate, Windows 8.1 Pro or Pro for Students, or Windows 10 Pro, you should install Windows 10 Pro. If you're working with any Enterprise version of Windows 7 or 8.1, the upgrade isn't free; it's dependent on your Software Assurance license terms.

Then try running the upgrade again.

If you continue to have the same problem, Microsoft's best advice is to use the Windows 10 media creation tool to create a USB drive (or DVD). See the Download Windows 10 page at www.microsoft.com/en-us/software-download/windows10 for details, but be aware of the fact that your genuine license depends on running the upgrade sequence correctly. Specifically, you must first upgrade the PC instead of performing a clean install, to make sure your old Windows 7 or 8.1 license is recognized as a valid license for the free Windows 10 upgrade. If you start with a valid Windows 10 machine and use the media creation tool to move to the next version, there should be no licensing problems. For full instructions on Installing

Windows 10 using the media creation tool, go to http://windows.microsoft.com/en-us/windows-10/media-creation-tool-install. Make sure that you follow the steps in order.

Error: "Something happened 0x80070005-0x90002"

The Windows 10 installer has such descriptive error codes, doesn't it? The 80070005 error code is a classic and generally means that the installer can't work with a file that it needs. Possible causes are many, but the general solution goes like this:

1. **Disable all antivirus and firewalls.**

2. **Reset Windows Update by going to KB 971058** (https://support.microsoft.com/en-us/kb/971058) **and following the instructions to reset.**

3. **Run the Windows 10 installer again (presumably through Windows Update).**

4. **If that doesn't work, turn your AV and firewall back on, and then follow the instructions at KB 947821** (https://support.microsoft.com/en-us/kb/947821), **which explains how to run DISM or the System Update Readiness Tool** (http://windows.microsoft.com/en-us/windows7/what-is-the-system-update-readiness-tool).

5. **Turn off your antivirus and firewall, and then try installing Win10 again.**

If that doesn't work, try any or all suggestions listed at http://answers.microsoft.com/en-us/insider/wiki/insider_wintp-insider_install/how-to-troubleshoot-common-setup-and-stop-errors/324d5a5f-d658-456c-bb82-b1201f735683.

Error: "The installation failed in the SAFE_OS phase"

This error comes in many variants: Errors 0xC1900101-0x20017, -0x30018, -0x20004 and others, "The installation failed in the SAFE_OS phase with an error during INSTALL_RECOVERY_ENVIRONMENT operation" or something similar.

This is another Windows installer error that dates back (at least) to the times of Windows 8. Many people report these errors occurring with freezes and crashes of varying intensity and length.

My advice is to wait, and if you have a spare weekend, you can try the comprehensive solutions presented by Gunter Born on his blog (http://borncity.com/win/2015/07/31/windows-10-upgrade-error-0xc1900101-0x20004/). But in general, this error is a mammoth, insurmountable time sink.

Trouble with video, sound, and other drivers

After basic installation problems — typically ending in hangs or reboots — the problem I hear about most involves lousy drivers. Sometimes the driver problem appears immediately after you install Windows 10. Sometime, the problems don't appear until you've rebooted the machine a few times and allowed Microsoft's forced updates to wipe out your stable drivers.

A lengthy post by Microsoft MVP and Answers Forum moderator Andre Da Costa steps through the finer points of installing drivers. (Go to http://answers.microsoft.com/en-us/insider/wiki/insider_wintp-insider_devices/how-to-install-and-update-drivers-in-windows-10/a97bbbd1-9973-4d66-9a5b-291300006293t.) He shows you how to install drivers the official way, through Windows Update, and the semi-official way, through Windows 10's Device Manager. Then he drops back a few yards and punts with instructions for using compatibility mode.

Da Costa doesn't cover the next phase of driver untangling, where you manually uninstall a driver and then prevent Windows 10 from automatically updating it, presumably to a bad (but newer!) driver. I talk about using the wushowhide program and KB 3073930 in my Computerworld post at www.computerworld.com/article/3143046/microsoft-windows/woodys-win10tip-apply-updates-carefully.html.

Unfortunately, wushowhide has to be handled in a specific way. It can only hide updates that are currently available, and you can't reboot between uninstalling the bad driver and running wushowhide. It's not a friendly solution.

If your problem lies with a faulty device driver being pushed by Microsoft, you may have to go straight to the manufacturer's website to get the right one and then install it manually.

Create a local account

Microsoft really, really wants you to use a Microsoft account. Over the years, they've made it increasingly difficult to create a local account — one that isn't hooked into Microsoft's stuff in the sky. It's an open point of debate as to whether using a local account also curtails Microsoft's snooping, given the ever-present Advertising ID, but that's another story. See www.computerworld.com/article/2956715/microsoft-windows/privacy-and-advertising-in-windows-10-both-sides-of-the-story.html.

To create a new local account, look in Book 2, Chapter 4.

Problems with installing updates

Windows 10's forced updates drive everybody nuts. If you're having problems, you aren't alone. Now that we've been using Windows 10 for several years, we've accumulated some coping experience.

Each new cumulative update is different, each situation unique, but a handful of tricks seem to work in specific situations — and a handful of tricks may jolt your system back into consciousness no matter how hard the cumulative update tries to knock it senseless.

ASK
WOODY.COM

Here are my recommendations for knocking an intransigent cumulative update upside the head. If you're having problems, run through these solutions and give them a try. If you can't get the Windows 10 beast to heel, follow the instructions at the end to find more personalized hope — or learn how to give up in disgust and live to fight another day.

This isn't an exhaustive list of problems and solutions. Quite the contrary. It's a short (and I hope understandable) list of the most common problems and most common solutions. Truth be told, it's a *massive* short list, but such is the nature of the beast. If you think Windows 10 updating is stable, you haven't been out very much.

Before you do anything else

Make sure your antivirus software is turned off. That's the number-one source of bad updates or no updates. If you're using Windows Defender (as I recommend many times in this book), you're fine. But if you got suckered into installing something different, turn it off.

Check for mundane hardware problems

Coincidences do happen. Just because your PC went to Hades in a multicolored hand basket right after you installed the latest cumulative update, it doesn't mean the update caused the problem.

It's the old *post hoc ergo propter hoc* fallacy.

Consider the possibility that your problem has nothing to do with the cumulative update. At the very least, someone with a cumulative update problem should right-click the Start button, choose Command Prompt, type the following in the box:

```
chkdsk /f
```

and press Enter. That will scan your main drive and fix any errors.

If you're having problems with a mouse or a keyboard, or a monitor or speaker, try plugging them into another computer to see if they're dead. Rudimentary, but it works in a surprisingly large number of cases.

Recover from a bricked PC

For most people, a bricked PC is the scariest situation. The cumulative update installs itself (possibly overnight, while you aren't looking), you come back to your machine, and nothing happens. It's dead, Jim.

REMEMBER

At least half the time, you can get back to a working machine by booting into safe mode, uninstalling the cumulative update, blocking it, then rebooting normally.

For the rundown on booting into safe mode, check out co-author Ciprian Rusen's article at *Digital Citizen* has the rundown on booting into safe mode in an article at www.digitalcitizen.life/4-ways-boot-safe-mode-windows-10. Unfortunately, booting into safe mode isn't as easy in Windows 10 as it was in Windows 8.1 (or 7, Vista, XP).

Once you're in safe mode, follow the instructions in the upcoming section "Make sure your problem is the patch" to uninstall the aberrant cumulative update. Then follow the instructions in the section "Break out of the endless update loop" to make sure you aren't tossed back into the fire. Reboot and you'll be back in your previous version of Windows 10.

Know when to give up

Some people, in some situations, report that going through the update process takes hours — many hours, with multiple restarts and all sorts of hangs. My best advice: Let the update run for three or four hours. If you come back to those spinning dots, it's time to pull the plug (literally turn off the electricity), reboot, and see if things worked or not.

You can always see what version you're running. In Windows 10's search box, type **about** and click or tap About Your PC. Scroll down to Windows Specifications, and compare the results (see Figure 2-2) to Microsoft's official Windows 10 update history list at https://support.microsoft.com/en-us/help/4555932.

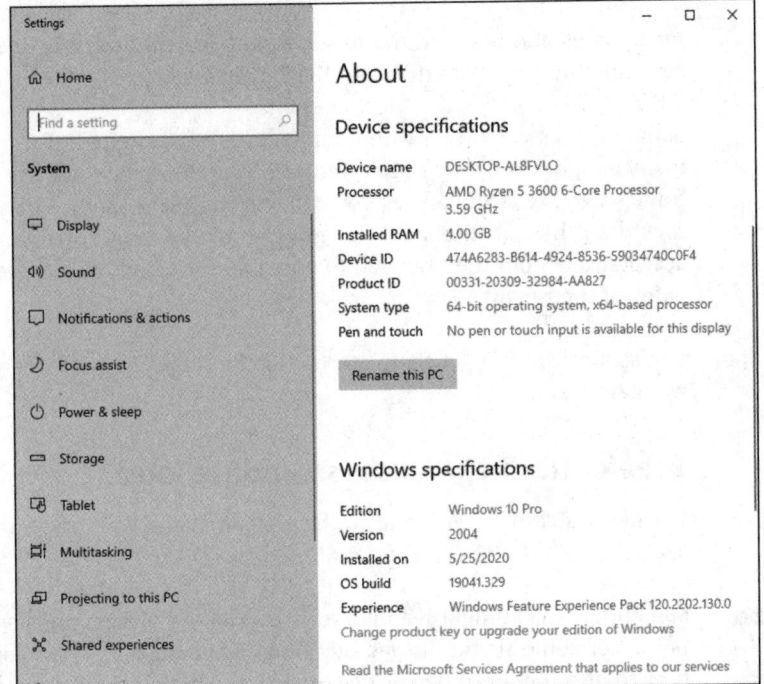

FIGURE 2-2:
Full version information is an "about" away.

See the later section "Just walk away and forget it" for commiseration.

Make sure your problem is the patch

First, restart your machine at least three times. I don't know why, but somehow rebooting numerous times sometimes shakes out the gremlins.

TIP

Second, try to uninstall the patch and see if the problem goes away. Click the Start button, the Settings icon, Update & Security, View Update History, Uninstall Updates. With a bit of luck, the aberrant update will appear at the top of the Microsoft Windows update list.

Double-click the update. When Windows asks "Are you sure you want to uninstall this update?" reply "No, I'm looking for my gefilte fish" or "Yes," whichever you feel appropriate. Windows 10 will take a while, maybe a long while, and then reboot. When it comes back, you should've retreated to the previous (presumably functional) version of Windows 10.

Immediately test to see if your problem went away. If it did, use wushowhide (instructions in the next section) to hide the bad patch. If your problem persists, chances are good the cumulative update didn't cause the problem. In that case, get onto the latest version. Reboot, and go to Start, Settings, Update & Security, Check for Updates and reinstall the patch. Your problem probably doesn't lie with this particular update. Note the operative term *probably*.

Some patches catch software manufacturers flat-footed. If a program you normally use goes belly-up right after installing the update, get over to the manufacturer's website as quickly as you can and complain *loudly*. Chances are good that they'll go through the stages of grief — denial, anger, bargaining, depression, acceptance — but then tell you to uninstall the Windows 10 patch or apply a new patch of their own.

The sooner you can get them started on the stages of grief, the sooner everybody will get a fix.

Break out of the endless update loop

It's like watching a PC bang its head against the wall, over and over and over again.

Sometimes the cumulative update fails. You see a message saying "Installation failed" or some such followed by "Undoing changes." When your system comes back to life an hour or two or five or six later, it goes right back to trying to install the same stupid cumulative update. You get the same error. Wash, rinse, repeat.

You might want to let your system go through the full self-mutilation cycle twice, just to see if you get lucky, but after that it's just too painful. You need to put Windows 10 out of its misery.

Fortunately, Microsoft has a tool that tells Windows Update to stop looking for the specific cumulative update that's causing problems. The tool wasn't built for stopping cumulative updates dead in their tracks, but it works nonetheless.

Here's how to use it:

1. **Go to KB 3073930** (https://support.microsoft.com/en-us/kb/3073930) **and download Microsoft's Wushowhide tool.**

   Click the link marked "Download the 'Show or hide updates' troubleshooter package now." Drag the downloaded file, wushowhide.diagcab, to any convenient location.

2. **Double-click wushowhide.diagcab to run it.**

3. **This part's important and easy to miss: Click the Advanced link. Deselect the Apply Repairs Automatically option (see Figure 2-3). Click Next.**

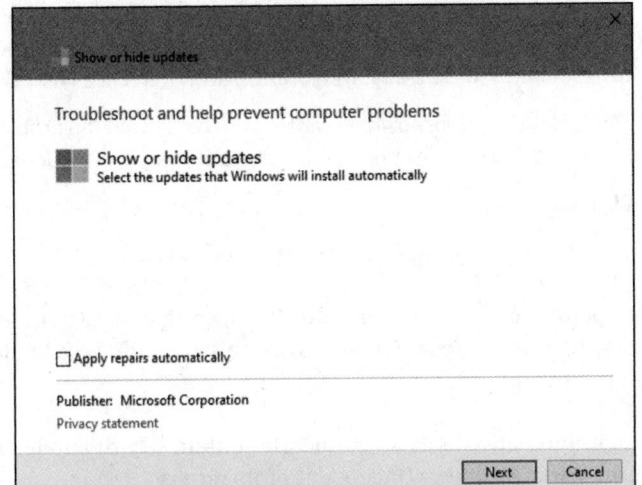

FIGURE 2-3:
To get
wushowhide
to hide updates,
go to the
Advanced
options and
turn off
Apply Repairs
Automatically.

4. **Wait for Wushowhide to look for all the pending updates on your system. When it comes up for air, click Hide Updates.**

There should be a box marked "Cumulative Update for Windows 10 Version 2004 for x64-based (or x-32 based) Systems (KB xxxxxxx)" or something similar. If you're curious whether you've found the wight wascally wabbit, look at Microsoft's Windows 10 update history log (`http://windows.microsoft.com/en-us/windows-10/update-history-windows-10`) and compare the KB numbers.

5. **Select the option for the latest Cumulative Update, click Next twice, and Close out of wushowhide.**

Windows 10 hides the update for you. The Windows Update program won't even see the update unless you specifically unhide it.

If you've found a solution to your problem (see the end of this section for some pointers) and want to reinstall the cumulative update, try this:

1. **Double-click wushowhide.diagcab to run it.**

2. **Deselect the Apply Repairs Automatically option (refer to Figure 2-3). Click Next.**

Wait for wushowhide to look for all pending updates on your system.

3. **When wushowhide comes up for air, click Show Hidden Updates.**

4. **Select the box marked "Cumulative Update for Windows 10 Version 2004 for x64-based (or x32-based) systems" and then click Next twice.**

This is weird, but wushowhide will tell you that it fixed the problems found. (See, I told you it wasn't built to hide cumulative updates, but nevermind.)

5. **Click Close.**

That should unhide the update you previously hid.

At this point, you can go back into Windows Update (Start, Settings, Update & Security, Check for Updates). Windows 10 will find the cumulative update and install it for you.

TIP

Although cumulative updates frequently contain security updates, sometimes you just have to put Windows Update out of its misery.

Fix error 0x80070020

Frequently, error number 0x80070020 accompanies a failed cumulative update installation and rollback. All too frequently, it's followed by another attempt to install the cumulative update, and another failure, with the same error code.

See the preceding section for advice on ending the loop. The steps there won't fix the error, but at least you can get your machine back. Usually.

Once you're back on your feet, you should try to figure out if any of your files are locked. (Error 0x80070020 generally means a file that the installer needed was locked.) Common culprits include corrupt Windows system files (see the next section), antivirus programs, and some video drivers.

Run SFC and DISM

Running SFC (System File Checker) and DISM (Deployment Image Servicing and Management) seems to be everyone's go-to suggestion for cumulative update installation problems. In my experience, it works only a small fraction of the time, but when it does, you come back from the brink of disaster with few scars to show for it.

System File Check, better known as sfc, is a Windows 10 program that scans system files, looking to see if any of them are corrupt. There are ways to run sfc — with switches — to tell sfc to replace bad versions of system files.

If sfc can't fix it, a second utility called Deployment Image Servicing and Management (DISM) digs even deeper. Microsoft recommends that you run both, in order, regardless of the dirt dug up (or missed) by sfc.

Be painfully aware that sfc has flagged files as broken, when in fact they weren't. You're looking for the automatic repair from sfc, not its diagnosis.

Here's how to run sfc:

1. **Right-click the Start button and choose Windows PowerShell (Admin).**

2. **Choose Yes when the UAC prompt is shown.**

3. **In PowerShell type** sfc /scannow **and press Enter.**

   Yes, there's a space between sfc and /scannow. It can take a couple of minutes or half an hour, depending on the speed of your storage drive. See Figure 2-4.

   If sfc reports "Windows Resource Protection did not find any integrity violations," you're out of luck. Whatever problem you have wasn't caused by scrambled Windows system files. If sfc reports "Windows Resource Protection found corrupt files and repaired them," you may be in luck. The problem may have been fixed. If sfc reports "Windows Resource Protection found corrupt files but was unable to fix some of them," you're back in the doghouse.

4. **Keep the same Windows PowerShell (Admin) app open, type** DISM /Online /Cleanup-Image /RestoreHealth **and press Enter.**

   Again, spaces before all the slashes, and note that's a hyphen between *Cleanup* and *Image*. Press Enter and let it run: half an hour, an hour, whatever. If DISM finds any corrupt system files, it fixes them.

5. **Reboot and see if your system was fixed.**

   It probably wasn't, but at least you've taken the first step.

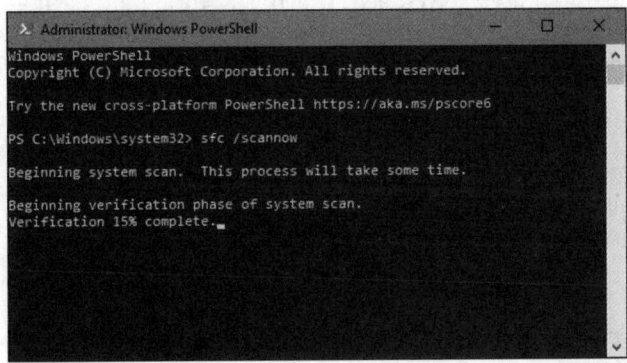

FIGURE 2-4:
The sfc /scannow command scans all your system files.

If you hit an odd error message or if one of the programs finds a bad file and can't fix it, refer to Microsoft's official documentation in KB 929833 (https://support.microsoft.com/en-us/kb/929833) for more information. (Don't feel too complacent: Microsoft has revised this particular KB many dozens of times.)

The result of the scans is placed in the C:\Windows\Logs\CBS\CBS.log file. (*CBS* stands for *Component Based Servicing*.) You may want to make a zip of that file, in case one of Microsoft's helpers needs to take a look.

Check the system event log

Everything, but everything (almost everything, anyway) gets posted to the system event log. The biggest problem with the log? People get freaked out when they see all the errors. That's why you rarely see a recommendation to check the log. It's hard to believe that an error in a system event log is a natural occurrence.

Fair warning: Telephone scammers frequently have customers look at their system event logs to convince them that their computer needs repair. Ain't so.

To bring up the system event log and interpret its results, look at Book 8, Chapter 3.

Refresh built-in Windows 10 programs

After the sfc /scannow run, this is the second-most-common general recommendation for fixing a bad Windows 10 cumulative update. It reaches into your computer, looks at each app installed in your user profile, and reinstalls a fresh, supposedly glitch-free copy.

Although it sounds like the process will fix only errant built-in Windows 10 apps, people have reported that it fixes all manner of problems with Windows 10, including icons that stop responding, Start menu and Cortana problems, balky apps, and halitosis.

The approach uses PowerShell, which is a world unto itself — a powerful command line adjunct to Windows 10. Here's how to refresh all sorts of apps, possibly knocking the Start Menu and Cortana back into shape, in the process:

1. **Right-click the Start button and choose Windows PowerShell (Admin).**

    That brings up PowerShell. You get a window that looks a lot like a command prompt window, except PS appears before the name of the current directory. (Refer to Figure 2-4.)

2. **Click Yes when UAC asks for your confirmation.**

3. **In the PowerShell window, type the following (all one line) and press Enter:**

    Get-AppXPackage -AllUsers | Foreach {Add-AppxPackage -Disable DevelopmentMode -Register "$($_.InstallLocation)\AppXManifest.xml"}

You see a bunch of red error messages. Don't panic! Ignore them. Yes, even the ones that say "Deployment failed with HRESULT: blah blah," "The package could not be installed because resources it modifies are currently in use," or "Unable to install because the following apps need to be closed."

When the Get-AppXPackage loop finishes — even with all those red warnings — you'll be returned to the PS PowerShell prompt.

4. **X out of PowerShell, reboot, and see if the demons have been driven away.**

Surprisingly, that approach does seem to clean up some Start, taskbar, and Cortana problems.

Even if an app refresh doesn't fix your machine, you've now undertaken the second standard approach (after sfc /scannow) that you'll find offered just about everywhere.

Check your Device Manager

Many problems can be traced back to non-Microsoft peripherals with drivers that don't work correctly. Many can be traced back to Microsoft peripherals that don't work correctly, too (`http://steamcommunity.com/app/292120/discussions/0/361787186425781965/`), but I digress.

First stop for bad devices is Device Manager, and it hasn't changed much since Windows XP:

1. **Right-click the Start button and choose Device Manager.**

2. **Look for yellow! icons.**

3. **If you find any, double-click the device that's causing problems, click the Driver tab, and see if you can find a newer driver, typically on the manufacturer's website.**

   Make sure that the new driver works better than the old one — Google is your friend — and that it's designed for Windows 10.

Failing that, usually Windows 8.1 and Windows 7 drivers work, but you never know for sure.

Just walk away and forget it

It's good to keep a bit of perspective. If the latest cumulative update won't install (or if it breaks something) and you can get your machine back to a normal state — using, perhaps, the uninstall/wushowhide sequence described at the beginning of this section — seriously consider doing nothing.

I know it's heresy, but the most recent cumulative update doesn't necessarily fix anything you need (or want!) to have fixed immediately.

Yes, security patches are tossed into the giant cumulative update maw, but Microsoft doesn't bother to split those out and let you install them separately. So you're stuck with an undifferentiated massive mess of fixes and security patches that may or may not be important for you.

There's no penalty for sitting out this particular cumulative update. The next one will come along, usually within a month, likely on Patch Tuesday (the anointed second Tuesday of the month) and it may well treat you and your machine better.

Or maybe not.

System Stability and Reliability Monitor

Reliability Monitor is a useful tool that can help pinpoint problems that you can only vaguely identify. Say your computer suddenly starts getting those blue screen messages saying "Your PC Ran into a Problem that It Couldn't Handle" and now "It Needs to Restart." You know for sure that your PC didn't have those problems last week. But something happened in the past few days, and now, suddenly, Windows 10 encounters more problems than Walter White hits in a season of *Breaking Bad*.

Windows 10 watches all, knows all, sees all — and keeps notes. Windows *events*, as they're called, get stored in a giant database, and you can look into that database with Event Viewer, which I describe in Book 8, Chapter 3.

One specific subset of the events gets collected into a report — a Reliability Monitor report — that you can understand at a glance.

WARNING

If you're looking at Reliability Monitor because somebody on the phone told you that he's trying to help you fix your computer, be very, very suspicious. Reliability Monitor will show that your computer has problems. Everybody's Reliability Monitor, sooner or later, shows problems. Scammers often use that fact to talk people into paying for services they don't need or allowing them to connect to your computer for nefarious reasons. Don't be conned! It's not unusual to have a string of problems showing in Reliability Monitor.

Here's the easy way to bring up Reliability Monitor:

1. **In the Windows 10 search box, next to the Start button, type** reliability. **At the top, tap or click View Reliability History.**

   The Reliability Monitor report appears, as shown in Figure 2-5.

2. **Tap or click any item in the list at the bottom of the report to bring up details.**

   You can also tap or click an event and merge reports by days or weeks by choosing the appropriate option at the top.

   Reliability Monitor calculates an aggregate score, based on how many problems appear in this graph, taken as a rolling (or in some cases, *roiling)* average. It's the Stability Score, shown as a number between 1 and 10, in the graph at the top.

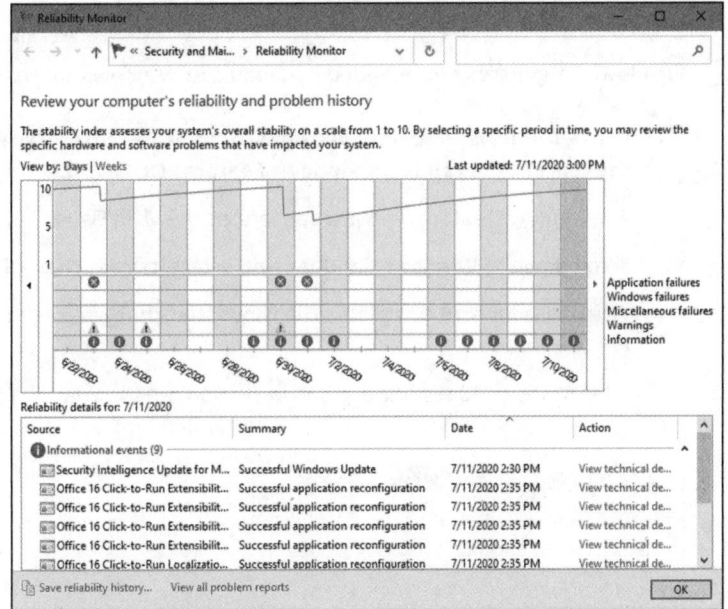

FIGURE 2-5:
The Reliability
Monitor report.

**ASK
WOODY.COM**

If you take the Stability score with a small grain of salt, you may be able to glean some useful information from the graph. For example, if you install a new driver and your system goes from ten to five that day, you can bet that the driver had something to do with the decline. Reliability Monitor shows you significant events for each day and leaves it to you to draw inferences.

Windows Sandbox

If you run Windows 10 Pro, Education, or Enterprise version 1903 (May 2019 update) or newer, you get access to Windows Sandbox. This useful app helps you run anything you want in an isolated environment, separate from your PC. The Sandbox is a virtual machine that simulates your Windows 10 PC without being directly connected to it.

While Sandbox is open, what you do in Sandbox remains there. Also, when you close it, everything you've done gets deleted. Suppose you receive a weird link via email, download a file from an untrusted source, or download an app with a weird name. Start Windows Sandbox, run the link or file there, and see what it does. If it's malware, it's gone the moment you close Windows Sandbox, and your machine is not affected. Isn't that better than having your PC locked down by ransomware or fighting off the Blue Screen of Death?

Windows Sandbox is not installed by default in Windows 10. Here's how to add it:

1. **In the Windows 10 search box, next to the Start button, type** features. **At the top, click or tap Turn Windows Features On Or Off.**

   The Windows Features window appears, as shown in Figure 2-6.

2. **Scroll down to Windows Sandbox and select its box. Then click OK.**

3. **When Windows 10 completes the requested changes, click or tap Restart Now.**

   After Windows 10 restarts and you log in, you can use Windows Sandbox.

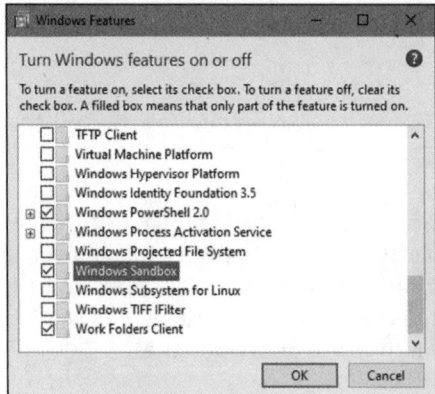

FIGURE 2-6: Adding Windows Sandbox to Windows 10.

To start Sandbox, click Start and then click Windows Sandbox. You see another copy of Windows 10 loaded as if it were an app.

Tricks to Using Windows Help

Slowly, Cortana is getting better at providing some Windows help — Microsoft's busy beefing up its database constantly. If you're very lucky, you can get Cortana to help by simply saying "Hey, Cortana" (or clicking down in the Cortana search bar, to the right of the Start icon) and trying to articulate your problem.

That's much easier said than done, of course.

To go straight to the source of Windows help, fire up your favorite browser and go to the https://support.microsoft.com website, as shown in Figure 2-7.

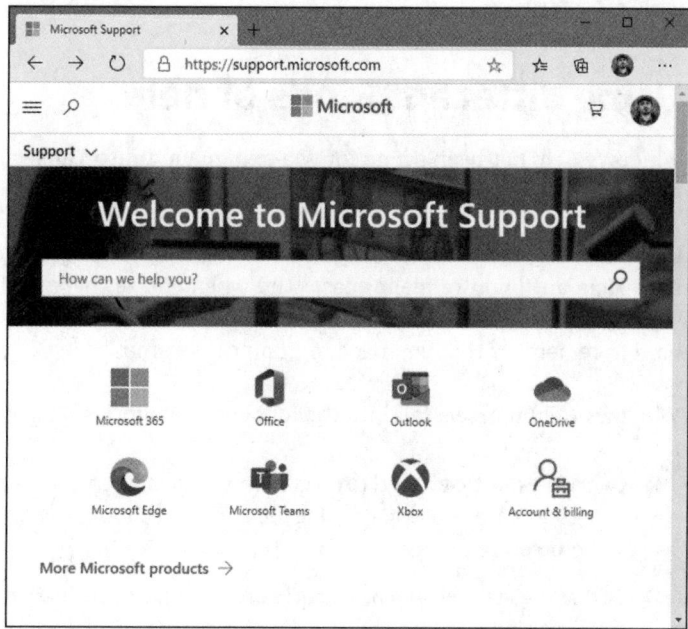

FIGURE 2-7:
Help from the
Internet is here!

Microsoft hopes to make finding what you need easier for you, even if you don't know the answer to your question in advance — a common problem in all versions of Windows Help.

The problem(s) with Windows Help

Windows Help offers only the Microsoft party line. If a big problem crops up with Windows 10 —think of problems with the Meltdown and Spectre vulnerabilities — you find only a milquetoast report in Windows Help. If a product from a different manufacturer offers a better way to solve a problem, you won't find that information in Windows Help. Want searing insight or unbiased evaluations? That's why you have this book, eh?

Windows Help exists primarily to reduce Microsoft support costs. Microsoft has tried hard to enable you to solve your own problems and to help you connect with other people who may be willing to volunteer. That's good. The new Answer Desk — where you get answers by chatting — is a great idea, but it's still too early to tell how well it will work. And all of it is under the cloud of the Microsoft Party Line. I spill the beans — and give you some much better alternatives — in the section "How to Really Get Help" later in this chapter.

Windows Help puts a happy face on an otherwise sobering (and bewildering!) topic.

Using different kinds of help

Windows Help has been set up for you to jump in, find an answer to your problem, resolve the problem, and get back to work.

Unfortunately, life is rarely so simple. So too with Help. You probably won't dive in to Help until you're feeling lost. And when you're there, well, it's like the old saying, "When you're up to your *<insert favorite expletive here>* in alligators, it's hard to remember that you need to drain the swamp."

Windows Help morsels fall into the following categories:

>> **Overviews, articles, and tutorials:** These explanatory pieces are aimed at giving you an idea of what's going on, as opposed to solving a specific problem.

>> **Tasks:** The step-by-step procedures are intended to solve a single problem or change a single setting.

>> **Walk-throughs and guided tours:** These marketing demos . . . uh, multimedia demonstrations of capabilities tend to be, uh, light on details and heavy on splash.

>> **Troubleshooters:** These walk you through a series of (frequently complex) steps to help you identify and resolve problems. I talk about troubleshooters earlier in this chapter.

The Windows Help index is quite thorough but, like any index, it relies heavily on the terminology being used in the Help articles themselves. That leads to frequent chicken-and-egg situations: You can find the answer to your question quite readily if you, uh, know the answer to the question — or if you know the terminology involved (which is nearly the same thing, eh?).

How to Really Get Help

You use Windows Help when you need help, right? Well, yes. Sort of.

In my experience, Windows Help works best in the following situations:

>> You want to understand what functions the big pieces of Windows 10 perform, and you aren't overly concerned about solving a specific problem (for example, *Windows Media Player).*

>> You have a problem that's easy to define (for example, *my printer doesn't print*).

>> You have a good idea of what you want to do, but you need a little prodding on the mechanics to get the job done (for example, *touch gestures*).

Help doesn't do much for you if you have only a vague idea of what's ailing your machine, if you want to understand enough details to think your way through a problem, if you're trying to decide which hardware or software to buy for your computer, or if you want to know where the Windows bodies are buried.

For example, if you type **how much memory do I need?**, the answers you see (Figure 2-8) talk about all sorts of things, but they don't tell you how much memory you need.

For all that, and much more, you need an independent source of information — this book, for example.

My website, AskWoody.com (www.askwoody.com), can come in handy, especially if you're trying to decide whether you should install the latest Microsoft security patch of a patch of a patch. The AskWoody Lounge, where you can post your own questions, has more than 70,000 recent searchable answers, absolutely free. Drop by from time to time to see what's happening.

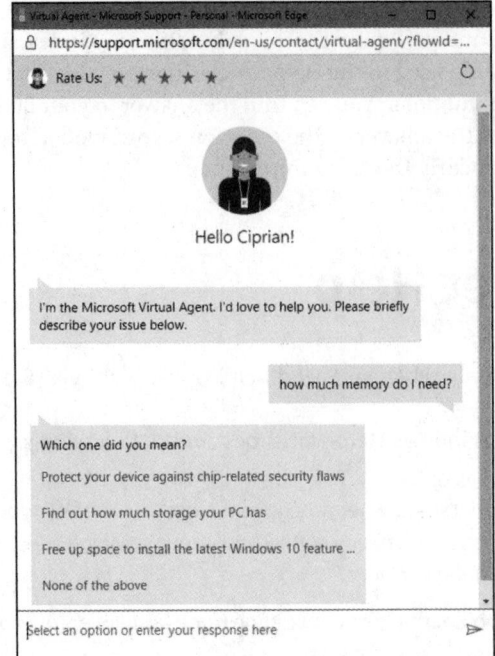

FIGURE 2-8:
The Windows
Help website's
Virtual Agent
doesn't always
answer the
question
you asked.

If you can't find the help you need in Windows Help and Support or at AskWoody. com, expand your search for enlightenment in this order:

1. **Use simple bribery, which is far and away the best way to get help.**

   Buttonhole a friend who knows about this stuff, and get her to lend you a virtual hand. Promise her a beer, a pizza, a night on the town — whatever it takes. If your friend knows her stuff, it's cheaper and faster than the alternatives — and you'll probably get better advice.

   If you can cajole your machine into connecting to the Internet — and get your friend to also connect to the Internet — Windows 10 makes it easy for a friend to take over your computer while you watch with the Remote Assistance feature, which I discuss a little later in this chapter.

2. **If your friend is off getting a tan at Patong Beach, you may be able to find help elsewhere on the Internet.**

   See the section "Getting Help Online," later in this chapter.

BEWARE OF "MICROSOFT" TECH SUPPORT SCAMS!

Somebody calls you, claims to be from Microsoft, and points you to a fancy website that says the caller's a Microsoft Registered Partner. The caller may even know your name or your phone number, or he may act like he knows what version of Windows or what computer you're using. The scammer offers to check whether your system is still under warranty. Invariably, it just went out of warranty, and oh golly, you have to pay $35 or $75 or $150 to get all your problems solved.

These folks are very clever. Many don't live in your home country, although they may sound like it. They may scrape your name from a tech support site and look up your phone number, or they may just make cold calls and figure there's likely to be a warm reception for anyone who says he's from Microsoft, and he wants to help.

The websites with Microsoft Registered Partner qualifications may look impressive, but anybody — even you — can become a Microsoft Partner; it takes maybe two minutes, and all you need is a free Hotmail or Outlook.com account or other Microsoft account. Drop by https://mspartner.microsoft.com/en/us/pages/membership/enroll.aspx, and sign up!

I have a general explanation of the scam in Book 9, Chapter 1, and a detailed report at http://windowssecrets.com/top-story/watch-out-for-microsoft-tech-support-scams.

3. **If you have a problem with a security patch — and can prove it — you may qualify for free support.**

   Microsoft used to have a website where you could request a free support ticket, but it has withdrawn the old site. Now, apparently you have to call (see the next step) and convince the person on the other end of the phone that you're having a problem with a security patch, and that your tech support call should be free.

   For the life of me, I can't find *any* email address — or pointer to an email address — for tech support at Microsoft.

REMEMBER

4. **As a last resort, you can try to contact Microsoft by telephone.**

   Heaven help ya.

REMEMBER

Microsoft offers support by phone — you know, an old-fashioned voice call — but some pundits (including yours truly) have observed that you'll probably have more luck with a psychic hotline. Be that as it may, the telephone number for tech support in the United States is 800-642-7676, and you may have to press 0 three or more times to get a live person. In Canada, it's 905-568-4494. Have your computer handy. Be prepared to pay.

Snapping and Recording Your Problems

Raise your hand if you've heard the following conversation:

Overworked Geek (answering the phone): "Hi, honey. How's it going?"

Geek's Clueless Husband: "Sorry to call you at work, but I'm having trouble with my computer."

OG: "What kind of trouble?"

GCH: "I clicked the picture, and it went into Microsoft, you know, and I tried to look at this report my boss sent me, but the computer said it couldn't."

OG: "Huh?"

GCH: "I'm sure you've seen this a hundred times. I clicked the picture, but the computer said it couldn't. How do I look at the report?"

OG: "Spfffft!"

GCH: "What's wrong? Why don't you say anything? You have time to help the other people in your office. Why can't you make time for me?"

OG wonders, for the tenth time that day, how she ever got into this crazy business.

At one time or another, you may have been on the sending or receiving end of a similar conversation — probably both, come to think of it. In the final analysis, one thing's clear: When you're trying to solve a computer problem, being able to look at the screen is worth 10,000 words. Or more.

Taking snaps that snap

TIP

Since the dawn of WinTime, you could take a snapshot of your desktop and put it on the Windows Clipboard by simply pressing the PrtScr or Print Screen key on your keyboard. Similarly, you can hold down the Alt key and press PrtScr, and Windows 10 puts a screenshot of the currently active window on the Windows Clipboard. From there, you can open Paint (or any of a hundred other picture-savvy programs, including Word), paste, and do what you will with the shot. That approach still works in Windows 10 — even in the Windows 10 apps — and in some circumstances, it's exactly the right tool for the job.

Windows Vista introduced the Snipping Tool, which is a more advanced tune on the same theme. With the Snipping Tool (see Figure 2-9), you tap or click New, and then drag and draw a rectangle around the area you want to capture, or use the Windows Snip mode to snag the window you want. You can also capture a free-form area anywhere on the screen or automatically capture the current window or the full screen.

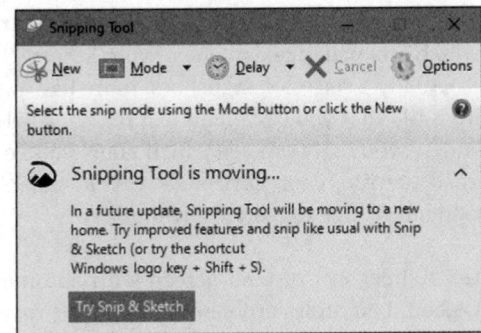

FIGURE 2-9:
The Snipping Tool can take screen-shots in a few steps.

The Snipping Tool has rudimentary tools for drawing on the captured screen, and the result can be copied to the Clipboard and/or saved as a PNG, GIF, JPG, or HTML file, or automatically attached to a newly generated email message.

To bring up the Snipping Tool, click or tap the Start button, then in the list of apps, click Windows Accessories, Snipping Tool. The Snipping Tool appears, and if you click New, it lets you click and drag around whatever you want to snip.

REMEMBER

Windows has a third screen capture option, and in many circumstances, it's much handier than its two older brethren. If you hold down the Windows key and press PrtScr or Print Screen on your keyboard, Windows takes a screenshot of the entire screen, converts it to a PNG file, and stores it in your Pictures\Screenshots folder. The file is given the name Screenshot (x).png, where the number x is increased by one with each shot.

Unlike the Snipping Tool, you can't select a part of the screen — you get the whole thing. Also unlike the Snipping Tool, you can't pick a format for the shot or a destination location. Still, for quick screens, it works well.

Starting with Windows 10 version 1809 (October 2018 update), another screenshot-taking tool is available: Snip & Sketch. It's a Windows 10 app, not a desktop app such as the Snipping tool, which it will replace someday. To start the new Snip & Sketch app and use it for taking screenshots, press Windows+Shift+S. Use the tools shown on the top of the screen to take the screenshots you need.

Recording live

If a screenshot's worth a thousand words, a video of the screen in action must be worth a thousand and one at least, right?

Windows 10 includes the magical *Problem Steps Recorder (PSR)*, recently renamed the Steps Recorder, which lets you take a movie of your screen. To a first approximation, anyway, it's actually a series of snapshots, more like an annotated slideshow. You end up with a file that you can email to a friend, a beleaguered spouse, or an innocent bystander, who can then see which steps you've taken and try to sort things out. To read the file, your guru must run Internet Explorer (unless Microsoft has finally updated Edge to read MHTML files).

Steps Recorder creates a slideshow of your screen with automatically generated detailed annotations, good, bad, ugly, problem-infested, or rosy-cheeked. If you have a rosy-cheeked background, anyway.

Steps Recorder is fast and easy, and it works like a champ.

Here's how to record your problems, er, screen:

1. **Make sure you remember which steps you have to take to make the problem (or rosy cheeks) appear.**

   Practice, if need be, until you figure out just how to move the whatsis to the flooberjoober and click the thingy to get to the sorry state that you want to show to your guru friend.

   Realize that anything appearing on the screen, even fleetingly, may be recorded, and your friend may be able to see it. So don't send your salary information, okay?

2. **In the search box to the right of the Start button, type** steps, **and tap or click Steps Recorder.**

   You can start the Steps Recorder from the Control Panel, but this method is a whole lot easier.

   The Steps Recorder, which resembles a full-screen camcorder, springs to life (see Figure 2-10). It isn't recording yet.

FIGURE 2-10:
The unassuming
Steps Recorder.

3. **Tap or click Start Record.**

   The recorder starts. You know it's going because the title flashes Steps Recorder — Recording Now.

   Note that the recorded slideshow will include the Steps Recorder window, so you may have to move it out of the way in order to show what you want to show.

4. **(Optional) If you want to type a description of what you're doing or why or anything else you want your guru friend to see while she's looking at your home movie:**

   a. *Tap or click the Add Comment button.* The recording pauses, and the screen grays out a bit. A Highlight Problem and Comment box appears at the bottom of the screen.

   b. *Tap or click the screen wherever your problem may be occurring and drag the mouse to highlight the problematic location.*

   c. *Type your edifying text in the box, and tap or click OK.* Recording continues.

5. **When you're finished with the demo, tap or click Stop Record.**

   Steps Recorder responds with the Recorded Steps dialog box, as shown in Figure 2-11. Take a good look at the file because what you see in the Save As box is precisely what gets saved — each of the screenshots, in a slideshow, precisely as presented. Remember, this isn't a video. It's an annotated slideshow.

6. **Tap or click Save, and then type a name for the file (it's a regular zip file).**

   The zip file contains an MHT file, which can be reliably read only by Internet Explorer — although you may have some luck reading the file in Firefox, if it's running the MAFF or UnMHT add-ons. (It's possible, by the time you read this, that Microsoft has built MHTML file reading capabilities into Edge, but don't hold your breath.)

7. **When you're finished, click the red X button to close the Steps Recorder.**

   Magical. Okay, Snagit (`http://techsmith.com/snagit.html`) does the screen recording shtick better, but still.

8. **Send the file to your guru friend.**

   Sneakernet — the old-fashioned way of sticking the file on a USB drive and hand-delivering it — works.

9. **Tell your friend to double-click the zip file when she receives it and then double-click the MHT file inside.**

   Internet Explorer or Microsoft Edge appears and shows the MHT file. You have several options; my favorite is to show the file as a slideshow (see Figure 2-12).

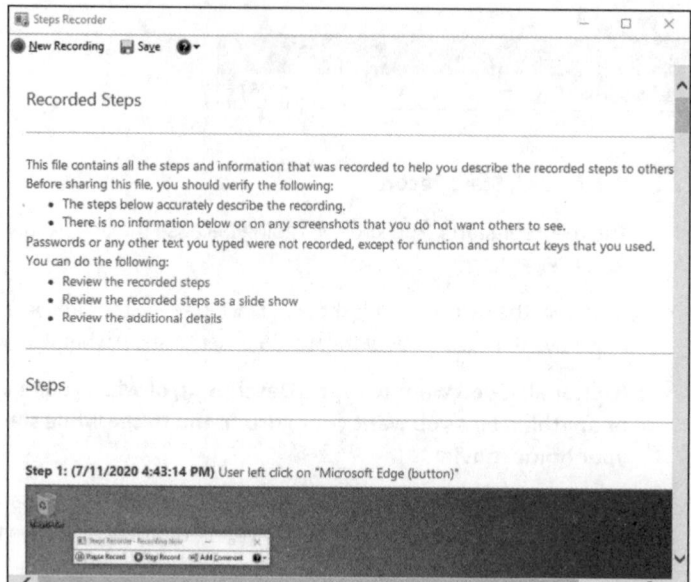

FIGURE 2-11:
Save the recording as soon as you finish it.

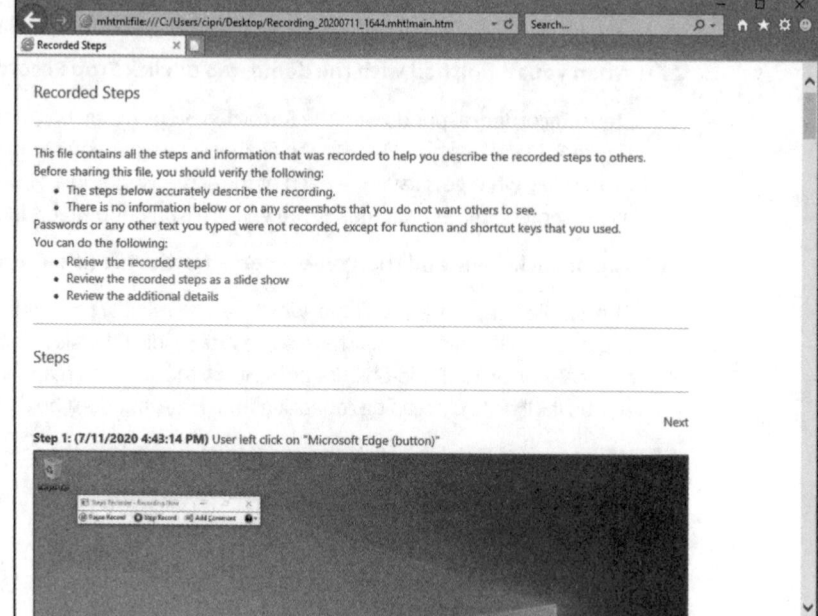

FIGURE 2-12:
The recording appears as a series of snapshots, with detailed accounts of what has been clicked and where.

Connecting to Remote Assistance

Windows has long boasted the Remote Assistance feature, which lets a person on one computer control a second computer, long distance, while both watch what's on the screen. It's a great puppet/puppet master capability that allows someone to solve your problems remotely while you watch. (Or, if you're the guru, Remote Assistance allows you to solve others' problems while they watch.)

WARNING

If you're looking at these instructions because someone you don't know wants to get into your computer, stop. Right now. *Seriously. Stop.* Ask yourself how much you know about the person who's trying to look at your PC. Do you trust her to take control of your PC — is it possible she'll pull a fast one on you, even drop an infected file? If you have any qualms at all, DON'T DO IT. Scammers love to talk people into using Remote Assistance because they get full control over the PC, and if they work fast enough (or talk fast enough to convince you that what they're doing is legitimate), they can easily plant anything they want on your computer.

Understanding the interaction

Windows 10 includes the Remote Assistance feature, which lets you call on a friend (or friendly guru) to take over your PC.

The basic interaction goes something like this:

1. You create an invitation file for your guru friend, asking him to look at your computer. Windows 10 creates a password for the invitation and shows it on your screen.

2. You send or give the file to the guru. Separately, you send your guru the password.

   The file can go any way you can imagine: Attach it to an email message, send it via an instant messaging program that allows you to transfer files, put it on a network shared drive, post it on your company's intranet, copy it to a shared folder on OneDrive, copy it to a USB key drive, burn it onto a CD, or strap it to a carrier pigeon. It's just a text file. Nothing fancy.

WARNING

   Similarly, you can send the password if you like, but it's smarter to call your guru and repeat it over the phone, just in case somebody's scraping your email.

3. Your guru friend receives the message or file and responds by clicking it and then typing the password.

4. Your PC displays a message saying that your guru friend wants to look at your computer.

5. If you give the go-ahead, your guru friend can see what you're doing — look, but not touch.

6. Your guru friend may ask whether he can take over your computer. If you give your permission, he takes complete control of your machine.

   He can start any program on your computer, bring stuff in from the Internet, go into Control Panel . . . the whole nine yards. You watch as your friend types and clicks, just as *you* would if you knew what the heck you were doing. Your friend solves the problem as you watch.

7. Either of you can break the connection at any time.

The thought of handing your machine over to somebody on an Internet connection probably gives you the willies. I'm not real keen on it either, but Microsoft has built some industrial-strength controls into Remote Assistance. Your guru friend must supply the password that you specify before he can connect to your computer. He can take control of your computer only if he requests it and you specifically allow it. And you can put a time limit on the invitation: If your friend doesn't respond within an hour, say, the invitation is canceled.

Making the connection

When you're ready to set up the connection for Remote Assistance, the following is what you need to do. (I'm writing this from the point of view of the Dummy requesting assistance from a guru. If you're the guru in the interaction, you have to kind of stand on your head and read backward, but, hey, you're the guru and no doubt you knew that already, huh?)

1. **Make sure that your guru friend is ready.**

   Call him or shoot him an email and make sure that he will have his PC on, connected to the Internet and running a reasonably new version of Windows. Also, make sure that he has his instant messenger program cranked up, will check email frequently, and/or will wait for you to hand him a file or make one available on your network.

   Make sure that you can contact your guru friend using your selected method: If you're using email, make sure that he's in your address book and send him a test message to make sure that you have his email address down pat; if you're going to send a floppy disk by carrier pigeon, make sure the pigeon knows the route and has had plenty of sleep.

2. **Start your machine (the PC that your Remote Assistance friend, the guru, will take over), and make sure it's connected to the Internet.**

   Make sure you aren't running any programs that you don't want the guru to see. Yes, that includes the Sudoku with the lousy score.

3. **In the Windows 10 search box next to the Start button, type** invite. **On the top, choose Invite Someone to Connect to Your PC and Help You or Offer to Help Someone Else.**

The Windows Remote Assistance dialog box appears, as shown in Figure 2-13.

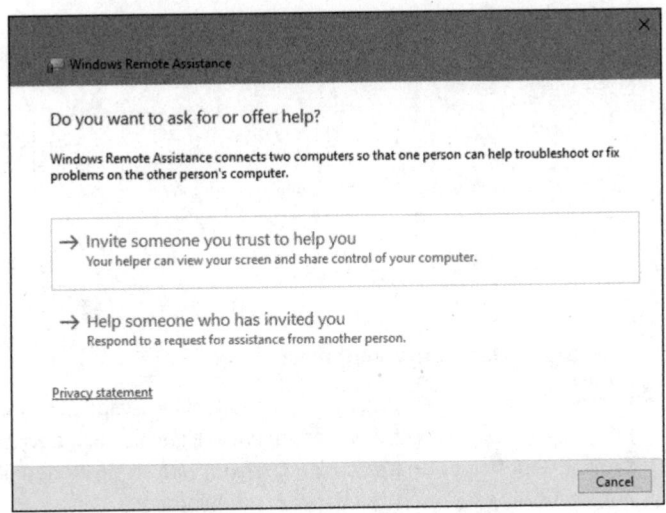

FIGURE 2-13:
Windows Remote
Assistance wants
to know whether
you're giving or
getting advice.

4. **Tap or click Invite Someone You Trust to Help You.**

You don't actually have to trust him but, well, you get the idea. Remote Assistance responds with the dialog box shown in Figure 2-14.

TECHNICAL STUFF

Easy Connect is an advanced version of Remote Assistance. It works for some people, if they're connecting with another person who's running Windows 7 or Windows 8, 8.1, or 10. Unfortunately, sometimes network routers get in the way. The big gain with Easy Connect is that you set it up once, and then you can reuse the same connection any time you like, without going through the invitation/password routine.

The method I describe in the following steps works whether your router likes it or not. If you want to try Easy Connect, choose that option in Figure 2-14 and see whether your guru can connect. If it works, it's, uh, easy.

5. **Choose Save This Invitation as a File.**

Even if you're going to email the file, it's easier to save the file first and then attach it to an email message.

Remote Assistance opens the Save As dialog box and prompts you to save the file Invitation.msrcIncident. You can change the name, if you like, but it's easier for your guru friend if you keep the filename extension *msrcIncident*.

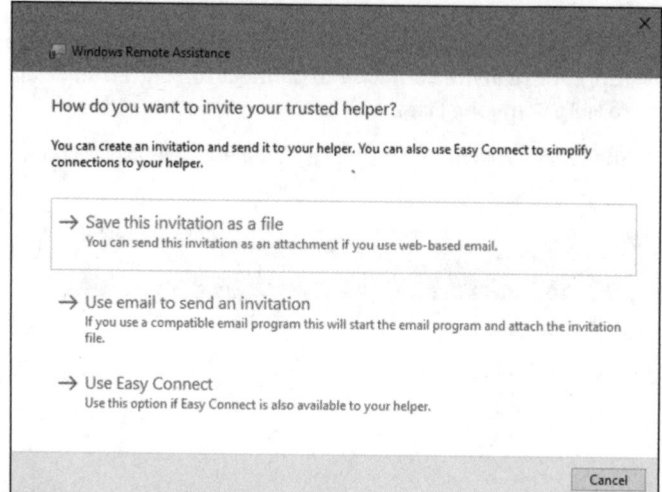

FIGURE 2-14:
The best choice
is to save the
invitation as a file.

6. **Save the file in a convenient place.**

Remote Assistance responds with an odd-looking dialog box, the Windows Remote Assistance control bar, as shown in Figure 2-15. It advises you to provide your helper (that's your guru friend) with the invitation file and the automatically generated 12-character password.

Windows 10 waits for your guru friend to contact you. You can continue to work, swear, play Minesweeper, or do whatever it takes to keep you sane until your friend can connect.

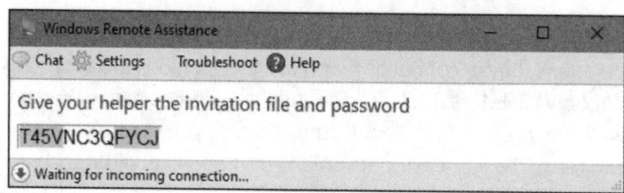

FIGURE 2-15:
Windows Remote
Assistance.

7. **Send the invitation file to your guru friend via email, in a shared OneDrive folder, or a USB slipped into his hamburger at lunch.**

8. **Tell your friend to double-click the invitation file to initiate the Remote Assistance session.**

Your friend's computer asks for the password that's in your Windows Remote Assistance control bar. He types it in the indicated box on his computer and clicks OK.

Windows Remote Assistance then asks whether it's okay to allow your guru friend to connect to your computer (see Figure 2-16).

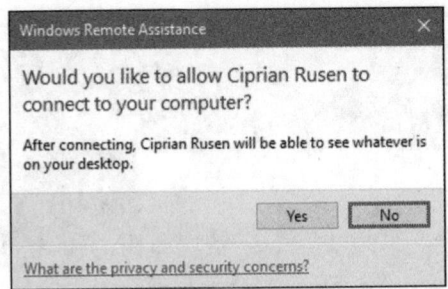

FIGURE 2-16:
Remote Assis-
tance requires
your explicit
permission.

9. **Tap or click the Yes button.**

Two things happen simultaneously:

- Your computer's Remote Assistance bar shows that you're connected, as shown in Figure 2-17.

- Your guru friend's computer sets up a window that shows him everything on your computer, as shown in Figure 2-18.

If your guru friend wants to take control of your PC, he needs to click the Request Control icon on his Remote Assistance bar. If he does that, your machine warns you that your guru friend is trying to take control, as shown in Figure 2-19.

10. **On your machine, tap or click Yes to allow your guru friend to take control of your PC.**

Your guru friend can now control your computer, move the mouse cursor, and type while you watch.

11. **Anytime either of you wants to sever the connection, tap or click the X button on the Remote Assistance bar.**

In addition, you — the person who requested the session — can cancel the session at any time by pressing Esc.

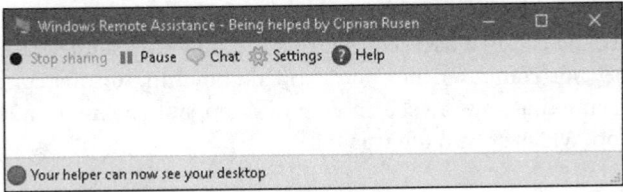

FIGURE 2-17:
Your computer
gets this Remote
Assistance bar.

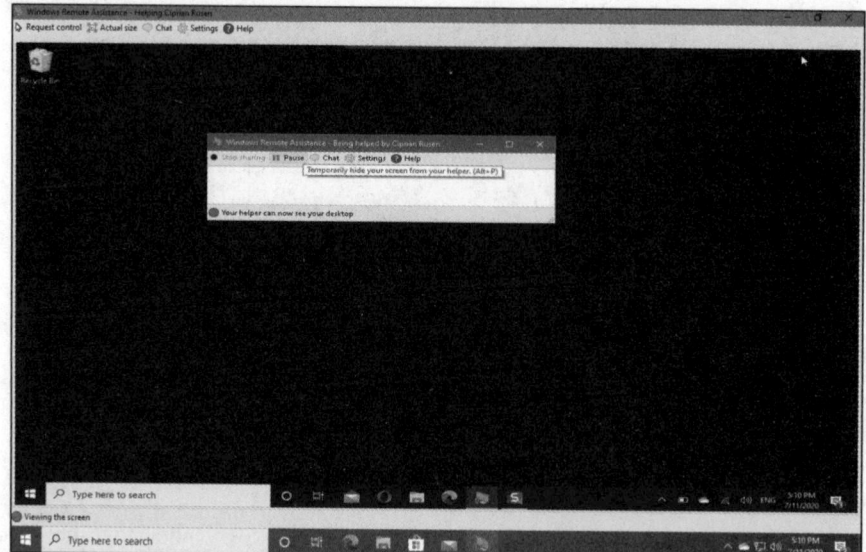

FIGURE 2-18:
Your guru friend
sees your entire
desktop in a
special Remote
Assistance
window.

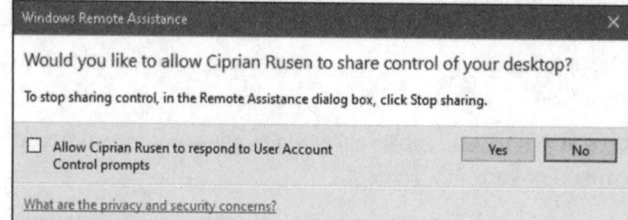

FIGURE 2-19:
Allow your guru
friend to take
over.

REMEMBER

After a Remote Assistance session is underway and you release control to your friend, your friend can do anything to your computer that you can do — anything at all, except change users. (If either logs off, the Remote Assistance connection is canceled.) Both of you have simultaneous control over the mouse pointer. If either or both of you type on the keyboard, the letters appear onscreen. You can stop your friend's control of your computer by pressing Esc.

TECHNICAL
STUFF

Your friend can rest assured that this is a one-way connection. He can take control of your computer, but you can't do anything on his computer. He can see everything that you can see on your desktop, but you aren't allowed to look at his desktop. Whoever said life was fair?

Limiting an invitation

Unless you change things, an invitation that you send requesting Remote Assistance expires after six hours. To change the expiration time, follow these steps:

1. **Bring up the Control Panel (right-click the lower-left corner of the screen and choose Control Panel); on the left, tap or click the System and Security link.**

2. **Under the System link, tap or click the Allow Remote Access link.**

3. **Make sure the Remote tab is displayed, and in the Remote Assistance box, tap or click the Advanced button.**

4. **In the Invitations box, choose the amount of time you want invitations to remain open.**

5. **Tap or click OK twice, and then tap or click the X to close the Control Panel.**

Troubleshooting Remote Assistance

Plenty of pitfalls lurk around the edges of Remote Assistance, but it mostly rates as an amazingly useful, powerful tool. The following are among the potential problems:

» You and your guru friend must be connected to the Internet or to the same local network. If you can't connect to the Internet — especially if that's the problem you're trying to solve — you're outta luck.

» Both of you must be running Windows 10, 8 or 8.1, 7, Vista, XP, Windows Server 2003, Windows Server 2008, Windows Server 2012, or another operating system that supports Remote Assistance. Sorry, your iPad doesn't qualify, but you can mix and match — you can be running Windows 10, while your friend is stuck with Windows 8.1. Go ahead and gloat.

» You must be able to give (or send) your guru friend a file so he can use the invitation to connect to your PC.

» If a firewall sits between either of you and the Internet, it may interfere with Remote Assistance. Windows Firewall (the firewall that's included in Windows 10, 8, 8.1, 7, and Windows Vista, as well as Windows XP Service Pack 2 and later) doesn't intentionally block Remote Assistance, but other firewalls may. If you can't get through, contact your system administrator or dig in to the firewall's documentation and unblock *Port 3389* — the communication channel that Remote Assistance uses.

You — the person with the PC that will be taken over — must initiate the Remote Assistance session. Your guru friend can't tap you on the shoulder, electronically, and say something like this (with apologies to Dire Straits): "You an' me, babe, how 'bout it?"

Getting Help Online

Microsoft is finally making it easier to chat with a real, live human being. But you may find the answers better (and less conformist to the Microsoft Party Line) if you hop on to the Microsoft Answers forum.

ASK
WOODY.COM

Lots of people join in on the forums to help (see the nearby sidebar). Many of the helpers are Microsoft MVPs (Most Valued Professionals) who work without pay, just for the joy of knowing that they're helping people. Microsoft gives the MVPs recognition and thanks, and some occasional benefits such as being able to talk with some people on the development teams. In exchange, the MVPs give generally good — sometimes excellent — support to anyone who asks.

REMEMBER

Realize that support techs aren't front-line programmers or testers. Mostly, they're quite familiar with the most common problems and have access to lots of support systems that can answer myriad questions that aren't so common. Some of the techs may even have copies of this book on their desks.

If you have a really, really tough question and the tech you talk to can't solve it, before you hit the reset button, ask to have your question escalated. Support, historically, has had three levels of escalation available, and in very rare cases, some problems are escalated to the fourth level — which is where the product devs (developers) live. Kind of like Dante's *Paradiso*. If your problem is replicable — meaning it isn't caused by bad hardware or cosmic rays — and the tech can't solve it, you should politely ask for escalation.

TIP

If you or someone you know is at the beginner stage, do both of you a favor and get Andy Rathbone's *Windows 10 For Dummies,* 4th Edition (Wiley). The book/DVD combination, in particular, will answer all your beginner's questions in terminology that you can understand.

ASK
WOODY.COM

I post constantly on www.AskWoody.com. If there's something new and important, I've probably written about it already, most likely for Computerworld. Check it out. If you're on Twitter, follow @woodyleonhard.

MICROSOFT ANSWERS FORUM

The Microsoft Answers forum is one of the great resources for Windows 10 customers. There are sections for just about every nook and cranny of every Microsoft product. You post questions, other people post answers, and it's free for everyone.

But it's important that you understand the limitations.

Most of the people on the Answers forum are not Microsoft employees — in fact, it's pretty rare to see Microsoft employees on the forum. (They're identified as Microsoft employees in their tag line.)

Although the typical forum denizen may be well intentioned, they aren't necessarily well informed. You must keep that in mind while wading through the questions and answers.

The Answers forum is a great place to go with immediate problems that may affect other people. It's one of the very few ways that you can register a gripe and expect that, if it's a valid gripe, somebody at Microsoft will actually read it — and maybe respond to it.

In particular, realize that both the moderators and the Microsoft *Most Valued Professionals, or MVPs* (also identified in their tag lines) are all volunteers. No, the Moderators are not Microsoft employees. No, the MVPs aren't paid by Microsoft either. They help on the forums out of the goodness of their hearts. Hard to believe that in this day and age, but it's true. So be kind!

Chapter **3**

Working with Libraries

ASK
WOODY.COM

Windows 10 brought several infuriating changes to earlier versions of Windows — I'd list the snooping "features" as the worst culprit. (See Book 2, Chapter 6.) In the same infuriated breath, I'd have to mention Microsoft's attempt to make it difficult to use libraries.

Libraries were a key selling point for Windows 7: They really do make it easier for you to organize and maintain your files. The feature continued untarnished in Windows 8. Unfortunately, Microsoft decided to stunt and bury them in Windows 8.1, and Windows 10 has nothing to make them easier to use. If I were a more cynical soul, I would guess that Microsoft is trying to get you to use OneDrive — and pay the piper for cloud storage.

It's silly, really, because libraries are the single best way to incorporate SD card storage and external hard drives into your everyday Windows life. When libraries are set up with the Public folders activated (as should've been the case in Windows 8.1 straight out of the box), they also give you a chance to share data with other people on your computer or on your home network, and you don't have to take a trip through Microsoft's cloud to do it.

In this chapter, I start with some concepts and then show you how to get libraries working on your Windows 10 machine. Then we can go into the advanced course.

Understanding Libraries

Lots of experienced Windows users get confused when they start thinking about libraries. That's because they have a long-imprinted misconception that data has to be located in one place. Your files are on your C: drive or on a DVD, or you download them from the Internet. You open a file, and if you don't find what you want, you look in another file in the same folder. If the folder doesn't have what you want, you go up one level and look again. All those concepts are locked into the idea that your data must be in just one place.

ASK
WOODY.COM

Although your files have to sit somewhere, Windows 7 introduced a concept that makes it easier to handle collections of files and folders. The concept lives on, half-buried, in Windows 10.

You know what a file is, right? (If not, I talk about it in Book 1, Chapter 1.) Files hold data. Typically, you have one photo or video in one file. You have one song in one file. You have one document, spreadsheet, or PowerPoint presentation in one file. Of course, there are lots of nuances, but at its heart, a *file* is just a collection of data that you stick in one place. Files can be empty. They can be huge too.

And you know about folders, yes? *Folders* are collections of files and other folders. Folders can also be empty. They can be huge too. They can have lots of little files or many big files, or any combination of little and big files and folders. You put a bunch of files and folders together in one place, and that place is a folder.

ASK
WOODY.COM

Note how I said *in one place.* The physical details may get a little hairy, but at least conceptually, all the data in the file is in one place. All the files in a folder are in one place. That's how libraries are different.

Libraries aren't all in one place. Libraries bring together folders that can be sitting just about anywhere: on your C: drive, on your D: drive, on a USB stick, an SD card, on an external drive, in the cloud (which is to say, on the Internet), even someplace else on your network, if you have one. A *library* is a collection of folders that's broken free of the "in one place" restriction. But libraries use pointers to make it *seem* like these files are all in one place.

Making Your Libraries Visible

When you bring up File Explorer in Windows 10, you're placed in a make-believe folder called Quick Access, which consists of folders that you have pinned, or that have been pinned for you (Desktop, Documents, Downloads, Music,

Pictures, Videos). File Explorer shows your most frequently used folders on top and recently used files on the bottom. See Figure 3-1.

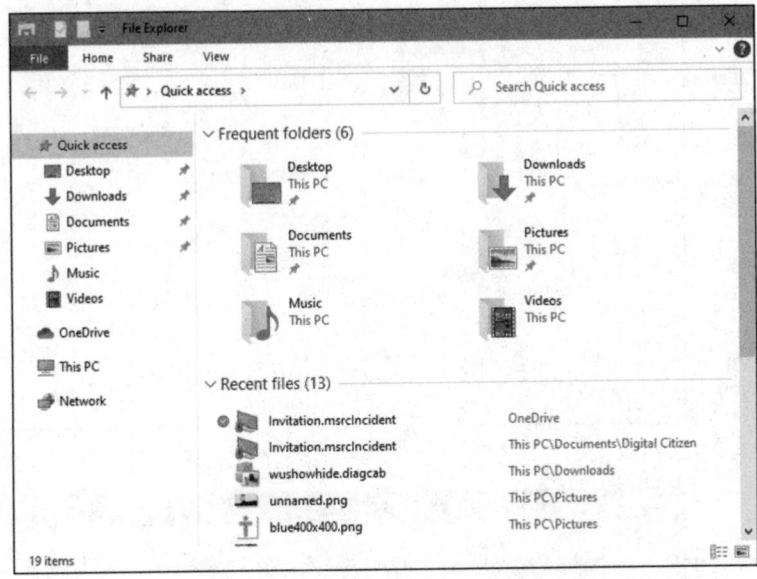

If you've used Windows 7 or Windows 8, you probably wondered what happened to your libraries — they used to appear on the left side of the screen as links to the Documents, Music, Pictures, and Videos libraries. Instead, you get the six folders (not libraries) listed at the top of Figure 3-1.

TIP

Here's how to bring back your libraries:

1. **Open File Explorer. Click the View tab.**

   You see the ribbon shown in Figure 3-2.

2. **Click or tap the large Navigation Pane icon on the left, and select Show Libraries.**

   Your four default libraries appear on the left, as in Figure 3-3.

Unfortunately, you aren't finished yet. One of the most important features of libraries in Windows 7 and Windows 8 was their capability to hook into the Public folders on your computer. The Public folders are a good place to put files that you want to share with other people on your computer or other people on your network.

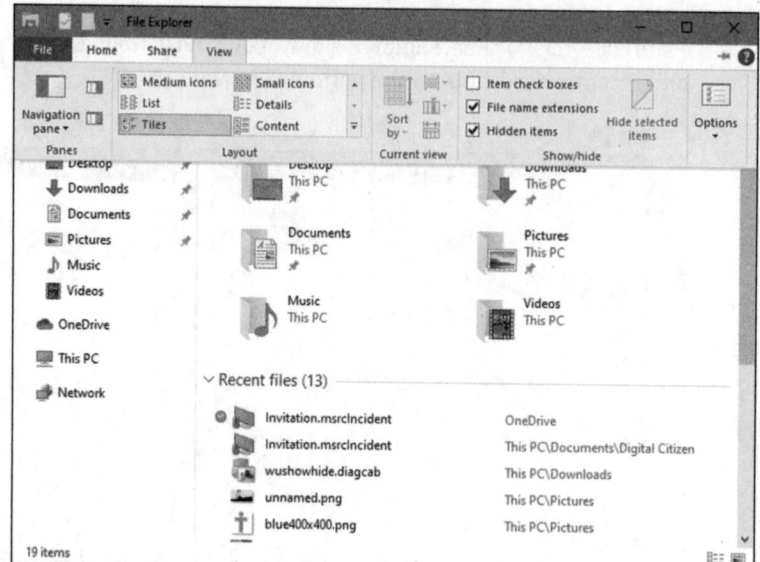

FIGURE 3-2:
Have File Explorer
show you
libraries.

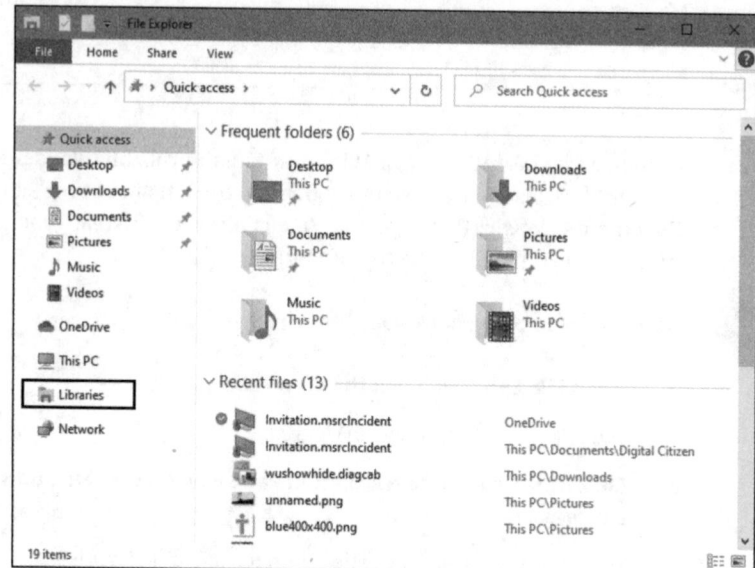

FIGURE 3-3:
Bringing back the
stunted version
of libraries.

ASK
WOODY.COM

In Windows 10, the default libraries aren't hooked up to the Public folders of the same type. You see later in this chapter why that's important. For now, just take my word for it, swear once or twice at Microsoft, and roll your Public folders into your libraries. Here's how:

1. **In File Explorer, navigate to your Public Documents folder.**

   To do so, double-click This PC, double-click Local Disk (C:), double-click Users, and then double-click Public. After all that double-clicking, you should come to a screen that looks like the one in Figure 3-4.

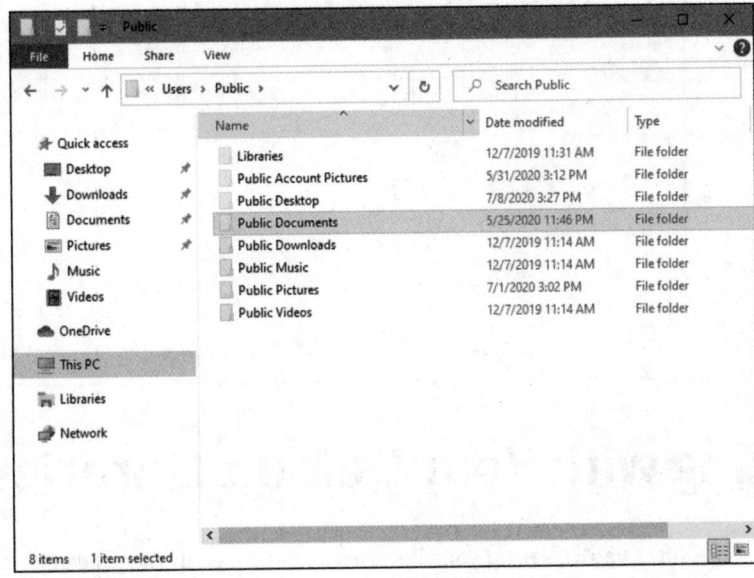

FIGURE 3-4: Adding Public folders to your libraries.

2. **Right-click the Public Documents folder, choose Include in Library, and then choose Documents.**

   Windows 10 reluctantly puts your Public Documents folder where it belongs.

3. **Repeat the steps for the Public Music folder (put it in the Music library), the Public Pictures folder (in Pictures), and the Public Videos folder (in Videos).**

4. **Close File Explorer (click the X in the upper-right corner), and restart it. Verify that all the Public folders now appear in their correct libraries, by clicking on each library one by one, as in Figure 3-5.**

   Give Microsoft a little epithet for that one.

Working with Libraries

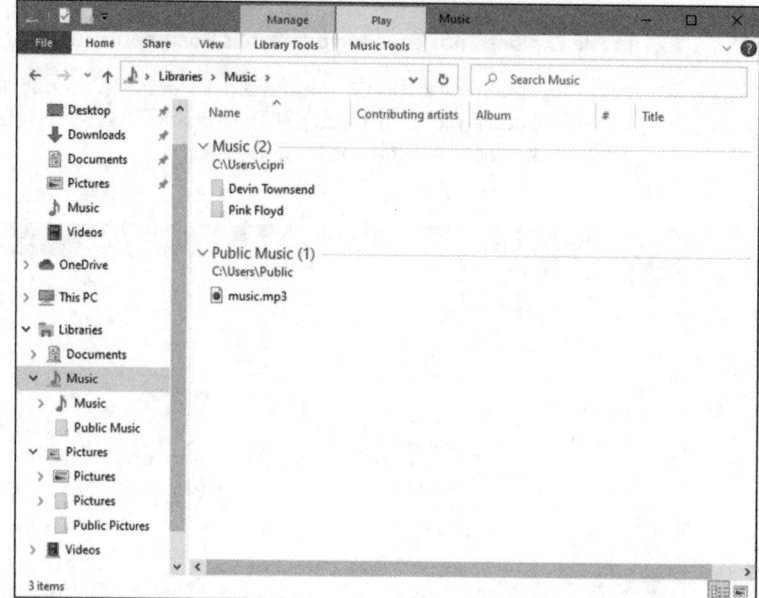

FIGURE 3-5:
Public folders
now appear
where they
should've been in
the first place.

Working with Your Default Libraries

After you've set up your libraries as described in the preceding section, when you start File Explorer and click Libraries on the left, icons for the libraries that you just built appear (see Figure 3-6).

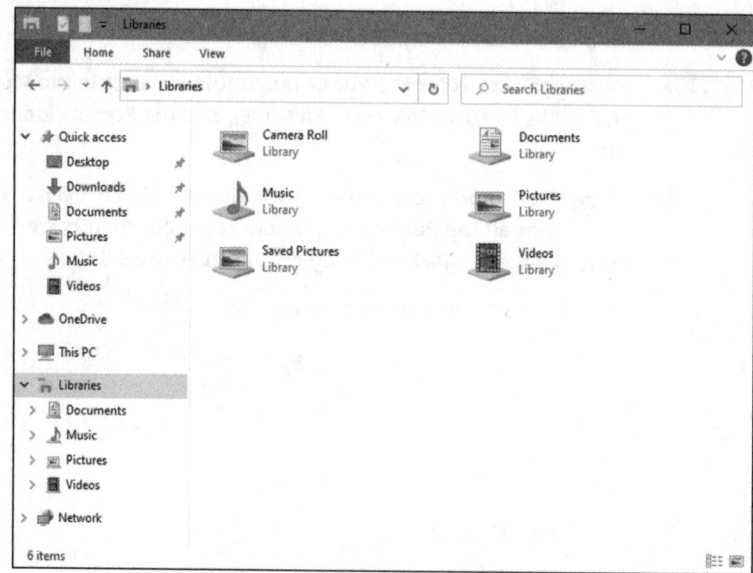

FIGURE 3-6:
The default
libraries in
Windows 10.

You may be tempted to think that Windows 10 magically identifies the kinds of files you're working with and shows them in the appropriate library — all your pictures appear in the Pictures library, for example. That isn't how libraries work.

The way we set up libraries in the preceding section makes them work this way:

>> Everything that appears in the Documents library comes from the Documents folder, mashed together with the Public\Documents folder, and the contents of Documents in OneDrive.

>> The same is true for the Music library.

>> Everything in the Pictures library comes from either the Pictures or Public Pictures folder, or from OneDrive.

>> And the same is true for the Videos library.

The converse is also true. Every file in the Music folder appears in the Music library, as does every file in the Public\Music folder. Windows 10 doesn't dig into the file and see whether it's a music file. The Music library doesn't consist of music files, necessarily. It's just a mash-up of all the files in those two folders.

LIBRARIES FOR OLD WINDOWS HANDS

If you've used any version of Windows Media Player (WMP), you already know about libraries. WMP starts with your Music folder and your PC's Public Music folder, and allows you to add other folders to its library. So, for example, you can add a folder full of music on an external hard drive to the WMP library or link to Music folders on other networked computers or even a Music folder on Windows Home Server.

When you add a folder to the WMP library, it doesn't copy the music anywhere. WMP merely provides easy access to all the files (the songs) in the library, keeps track of them, and lets you search and work with them as a group.

There are no limitations to the folders you can add to a WMP library: As long as your computer can get at the folders — the external drive is plugged in to the computer, say, or there are no security rules blocking access to another computer — WMP treats the music in those folders more or less the same way they'd be treated if they were sitting on your own PC.

Why would you want to bother with libraries? Ends up that they're pretty powerful after you get used to them. Probably the most valuable timesaver for most people is in the search that spans across multiple folders. Here are two examples:

>> If you want to search all your music for an album by Nickelback, go to the Music library and in the upper-right corner search for *Nickelback*.

>> If you want to search for documents and spreadsheets that contain the word *defenestrate*, bring up the Documents library, type **defenestrate** in the search box, and Windows returns all the documents in both \Documents and \Public\ Documents that contain the word.

REMEMBER

Imagine how that searching can make your life easier if you keep, say, all your music in a folder on one computer that's attached to your network. Set up your Music library to include that folder, and your searching just got a whole lot easier.

If you have a computer with an SD card, or an external hard drive, set up a \Documents folder on the SD card or external hard drive, and add it to your Documents library. That makes it easier to find documents on the SD card, store documents on the SD card, and generally keep your system running much, much easier: You don't have to think about where the data's stored because it's all in the library.

When an application running under Windows 10 looks for the Documents folder, the operating system hands it the entire Documents library. If you start a graphics program and choose File, Open, you don't go to your Pictures folder anymore. Instead, you open the Pictures library. Imagine. If you have a folder on another computer that contains documents you commonly use, and you add that folder to your Documents library, every time you crank up Word and choose File, Open, that folder is staring right at you. Unlike earlier versions, Windows Media Player doesn't need separate settings to handle libraries because Windows 10 takes care of everything.

Yes, Microsoft stacks the deck in more recent versions of Office and some other programs — a File, Open takes you to OneDrive. Blech. But few other programs work that way.

Think of libraries as Folders: The Next Generation.

Customizing Libraries

You can add more folders to a library above and beyond the folders that we added in the first section of this chapter. You can also change where a library saves data when you add items to it. Read on for the details.

Adding a folder to a library

The most common change I see people make to their libraries is to add a new folder to the Pictures or Music library. Typically, you have pictures or maybe music strewn in several locations, either on your computer or on your network. Here's how easy it is to add a folder from anywhere into your library:

1. **Using File Explorer, navigate to the folder you want to add.**

   It can be located just about anywhere.

2. **Tap and hold down or right-click the folder, select Include in Library, and choose the library.**

   In Figure 3-7, I added the Book Covers folder — located on the Desktop — to my Pictures library.

3. **Go back to the library, and make sure that the folder was added properly.**

   In Figure 3-8, you can see that the Book Covers folder is now in my Pictures library.

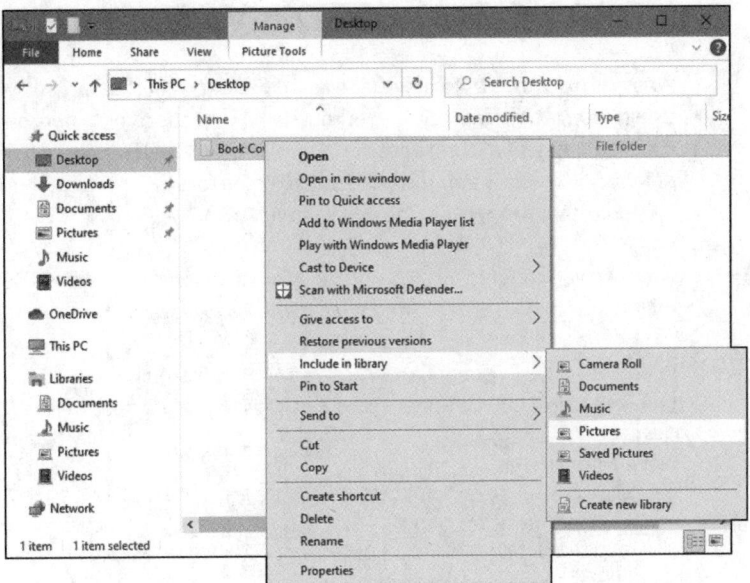

FIGURE 3-7: Adding a folder to a library is easy, if you start by going to the folder.

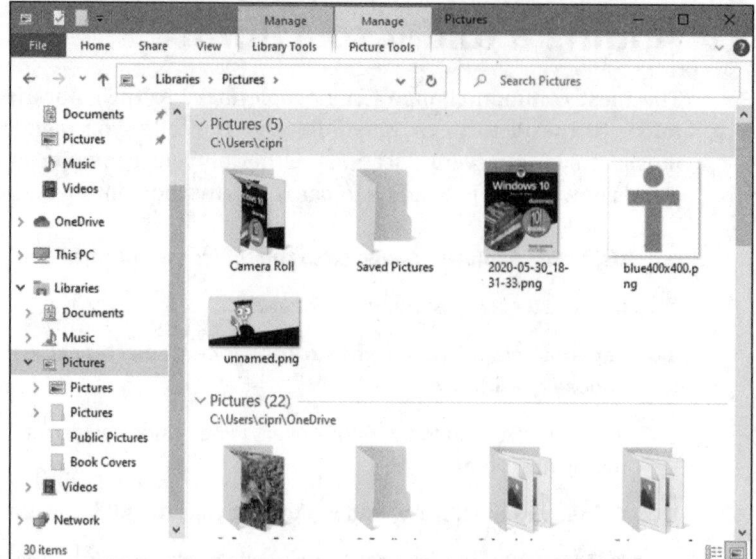

FIGURE 3-8:
Even though
the folder hasn't
moved, it's now
included in the
library.

LIBRARIES GO BETTER WITH TAGS

Whereas most music files have (at least rudimentary) tags associated with them, photos usually don't come with tags, other than the ones your camera puts on them — *EXIF data*, such as the time and date the picture was taken. Nor do videos. To keep massive amounts of media organized, you have to come to grips with tags, the index data (or *metadata)* that you can stick on every file you own.

Although you can't create a library based on tags, you can search on tags, and that makes it infinitely easier to keep large libraries organized.

Windows Media Player and the Windows 10 Photos app have good tools for handling tags. In general, you can assign your own tags to just about any file (except GIFs) as follows:

1. Locate the file in File Explorer, and make sure it's selected.

2. Open the Details pane (click or tap View, then in the Panes group choose Details), and edit the tags in the pane at the right.

Alternatively, you can right-click the file, choose Properties, and click the Details tab. Many free programs are available for editing tags on MP3 files, too.

At the risk of paraphrasing Beyoncé (and the Chipettes), if you like it, then you should put a tag on it. Whoa whoa whoa. If you want to find a file, put a tag on it!

It's important to realize that Windows *doesn't move anything.* The pictures are still in their old location — even over on a different computer. But the library has been expanded to include the folder in the remote location. If you search your Pictures library, in this case, Windows will look at the contents of not only the \ Pictures and \Public\Pictures folders but also the Book Covers folder — whether it's on your C: drive, an external hard drive, an SD card, someplace on your network . . . just about anywhere.

Libraries aren't exclusive. You can put one folder in multiple libraries. You can put a folder in one library and a subfolder of that folder in a different library. You can even put a OneDrive folder in your library.

If you ever want to remove a folder from a library, tap and hold down or right-click the library's name, and click Properties. In the Properties window, choose the folder that you want to remove from the library, click Remove, and then OK.

Changing a library's default save location

Want to challenge your brain a bit? Don't short-circuit on this one, but libraries *itself* is a library — a library that contains libraries.

When you drag, copy, or move a file (or folder) into a library, the file (or folder) must physically go somewhere — it must be placed in a real, physical folder. For example, if you save a new picture called Dummy.pic to the Pictures library, Windows 10 must put the file Dummy.pic someplace; it has to stick it in a real folder. Because the Pictures library isn't a real folder, Windows needs to figure out which folder inside the Pictures library should get the copy of Dummy.pic.

The folder is the *default save location* for the library. If you set up your libraries as described at the beginning of this chapter, the save location for the Documents library is your plain old everyday Documents folder. The save location for the Music library is the Music folder and so on.

It's easy to change the default save location for any of the libraries.

I change the save location of the Music library to the \Public\Music folder, so when I drag or save music into the Music library, it automatically ends up in a place where other people who use my PC, and other people on my network, can access that music easily.

Here's how to change the default save location:

1. **Start File Explorer, and click the Libraries link on the left.**

   The libraries appear (refer to Figure 3-6).

2. **On the left, tap or click a library. Then at the top, tap or click the Library Tools tab.**

   The Library Tools Manage tab opens and exposes the Manage Library ribbon, which looks like Figure 3-9.

3. **On the left, in the Navigation pane, tap or click whichever library you want to change.**

4. **At the top, tap or click Set Save Location and choose the folder that you want to set as the default save location.**

   Your change takes place immediately.

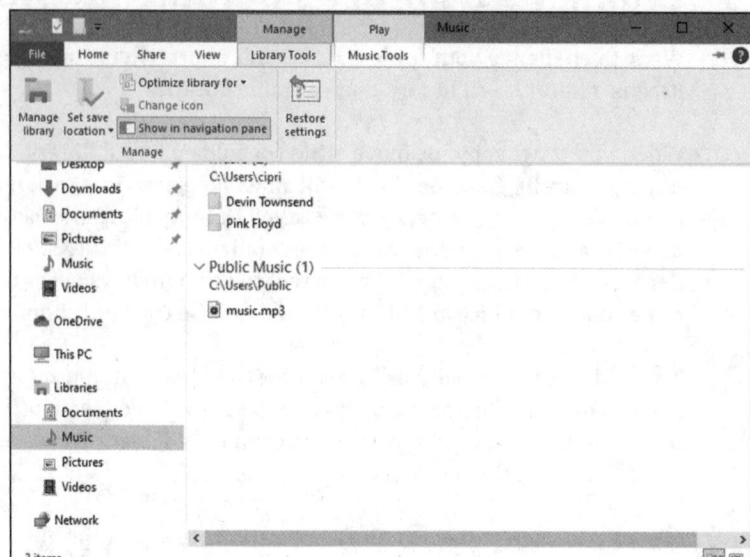

FIGURE 3-9:
Manage your libraries from this ribbon.

Creating Your Own Library

At the beginning of this chapter, I show you how to set up four libraries — the same four libraries that ship with Windows 7 and 8 — but you can add as many as you like.

ASK WOODY.COM

You may want to create your own library if, for example, you have a bunch of information about a house you want to sell. The info may include Word documents, an Excel spreadsheet, multiple photos, and maybe a video or two. You have the documents in a folder in Documents, the photos are in a separate folder in Pictures, and the video is in a separate folder in Videos. Here's how to make a library that ties them all together:

1. **Start File Explorer, and click the Libraries section on the left (refer to Figure 3-6).**

2. **Tap and hold down or right-click any blank location on the right, and choose New ➪ Library.**

   Windows 10 creates a new library, giving it the name New Library.

3. **Immediately type a name for the library, and press Enter (or tap the new icon).**

   In Figure 3-10, I typed the name *House for Sale* and pressed Enter, and File Explorer showed me my new empty library.

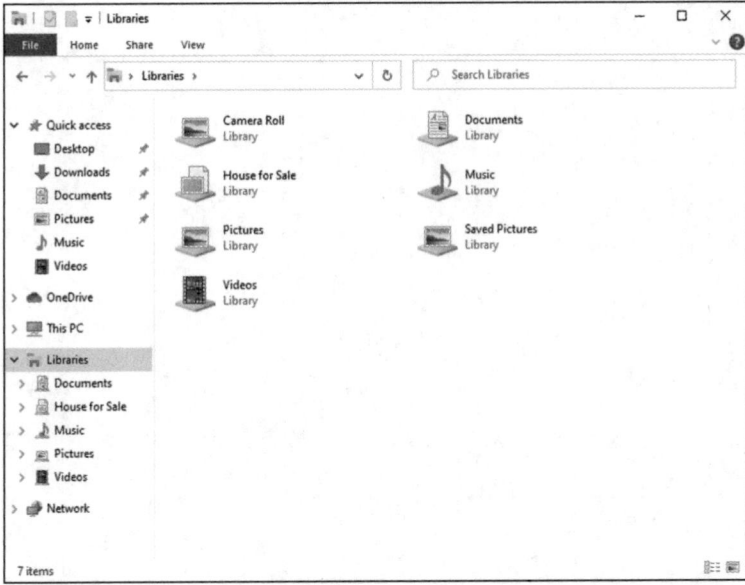

FIGURE 3-10: Start your own custom library.

4. **Double-click the new library, and then tap or click the Include a Folder button.**

Windows 10 lets you out to pick and choose your first folder.

5. **Navigate to the first folder you want to include, and tap or click Include Folder.**

The first folder becomes the default save folder.

6. **To add additional folders to the library, navigate to the folder, tap or right-click and hold down, choose Include in Library, and choose the name of the new library.**

The new library appears everywhere that the four default libraries appear, including the Navigation pane on the left of File Explorer and in the right-click menu for folders.

Chapter **4**

Storing in Storage Spaces

For people who want to make sure that they never suffer a data loss — in spite of dying hard drives or backup routines that don't run properly — the Storage Spaces feature may, in and of itself, justify buying, installing, and using Windows 10.

If you're using Drobo, ReadyNAS, or some other, expensive, network-attached storage device for file mirroring, you can toss your old hardware. Windows 10 handles it all as part of the operating system itself.

Some people prefer to back up to the cloud, but even if you do stick backups on the Internet, you'll feel a whole lot better knowing that the data you have here on earth is not going to disappear if a hard drive spins its last. On the other hand, if all your data is in the cloud, all the time, you don't need to worry about local drives failing, and you can give this chapter a pass.

In this chapter, I introduce you to the Windows 10 approach to drive virtualization and how it enables Storage Spaces to work. Then you walk through setting up Storage Spaces and the tips and tricks you need to know to make Storages Spaces work for you. Using Storage Spaces for backup is quick and easy, and it works.

Understanding the Virtualization of Storage

You're going to get sick of the term *virtualization* sooner or later. People who want to sell you stuff use the term all the time. But if you'll pinch your nose and wade through the offal, there's a solid core of real-world good stuff in this particular kind of virtualization technology.

Windows 10's Storage Spaces takes care of disk management behind the scenes, so you don't have to. You'll never even know (or care) which hard drive on your computer holds what folders or which files go where. Volumes and folders get extended as needed, and you don't have to lift a finger.

You don't have to worry about your D: drive running out of space because you don't *have* a D: drive. Or an E: drive. Windows 10 just grabs all the hard drive real estate you give it and hands out pieces of the hard drive as they're required.

REMEMBER

If you have two or more physical hard drives of sufficient capacity, any data you store in a Storage Spaces pool is automatically mirrored between two or more independent hard drives. If one of the hard drives dies, you can still work with the ones that are alive, and you never miss a beat — not one bit is out of place. Run out and buy a new drive, stick it in the computer, tell Windows 10 that it can accept the new drive into the Storage Spaces borg, wait an hour or two while Windows performs its magic, and all your data is back to normal. You never miss a beat. It's really that simple.

ASK WOODY.COM

When your computer starts running out of disk space, Windows tells you. Install another drive — internal, external, USB, eSATA, whatever — and, with your permission, it's absorbed into the pool. More space becomes available, and you don't need to care about any of the details — no new drive letters, no partitions, no massive copying or moving files from one drive to another, no homebrew backup hacks. For those accustomed to Windows' whining and whining, the Storage Spaces approach to disk management feels like a breath of fresh air.

When you add a new hard drive to the Storage Spaces pool, everything that was on that new hard drive gets obliterated. You don't have any choice. No data on the drive survives — it's all wiped out. That's the price the drive pays for being absorbed into the Storage Spaces borg.

Here's a high-level overview of how you set up Storage Spaces with data mirroring:

1. Tell Windows 10 that it can use two or more drives as a storage pool.

Your C: drive — the drive that contains Windows — cannot be part of the pool.

TIP

The best configuration for Storage Spaces: Get a fast solid-state drive for your system files and make that the C: drive. Then get two or more big, hunking drives for storing all your data. The big drives can be slow, but you'll hardly notice. You can use a mixture of spinning disks and solid-state disks if you like.

2. After you set up a pool of physical hard drives, you can create one or more Spaces.

In practice, most home and small business users will want only one Space. But you can create more if you like.

3. Establish a maximum size for each Space, and choose a mirroring technology, if you want the data mirrored.

The maximum size can be much bigger than the total amount of space available on all your hard drives. That's one of the advantages of virtualization: If you run out of physical hard drive space, instead of turning belly up and croaking, Windows 10 just asks you to feed it another drive.

TECHNICAL STUFF

For a discussion of the available mirroring technologies, see the sidebar "Mirroring technologies in Storage Spaces."

4. If a drive dies, you keep going and put in a new drive when you can. If you want to replace a drive with a bigger (or more reliable) one, you tell Windows to get rid of (or *dismount*) the old drive, wait an hour or so, turn off the PC, yank the drive, stick in a new one, and away you go.

It's that simple.

MIRRORING TECHNOLOGIES IN STORAGE SPACES

When it comes to mirroring — Microsoft calls it *resiliency* — you have four choices. You can

- Choose to *not mirror* at all. That way, you lose the automatic real-time backup, but you still get the benefits of pooled storage.

- Designate a space as a *two-way mirrored* space, thus telling Windows 10 that it should automatically keep backup copies of everything in the space on at least two separate hard drives and recover from dead hard drives automatically as well. It's important to realize that your programs don't even know the data's being mirrored. Storage Spaces takes care of all the details behind the scenes.

- Use *three-way mirroring,* which is only for the most fanatical people with acres of hard drive space to spare.

- Use another form of redundancy called *parity* that calculates check sums on your data and stores the sums in such a way that the data can be reconstructed from dead disks without having two full copies of the original file sitting around. This approach takes up less room than full mirroring, but there's higher overhead in processing input/output. Microsoft recommends that you use parity mirroring only on big files that are accessed sequentially — videos, for example — or on files that you don't update very often.

TIP

If you've ever heard of RAID (Redundant Array of Inexpensive Discs) technology, you may think that Storage Spaces sounds familiar. The concepts are similar in some respects, but Storage Spaces doesn't use RAID at all. Instead of relying on specialized hardware and fancy controllers — both hallmarks of a RAID installation — all of Storage Spaces is built in to Windows 10 itself, and Storage Spaces can use any kind of hard drive — internal, external, IDE, SATA, USB, eSATA, you name it — in any size, mix or match. No need for any special hardware or software.

Setting Up Storage Spaces

Even though you can set up Storage Spaces with just two hard drives — your C: system drive, plus one data drive — you don't get much benefit out of it until you move up to three drives. So, in this section, I assume that you have your C: drive, plus two more hard drives — internal, external, eternal, infernal, whatever — hooked up to your PC. I further assume that those two hard drives have absolutely nothing on them that you want to keep. Because they will get blasted. Guaranteed.

Ready to set up a Space? Here's how:

1. **Hook up your drives, log in to Windows 10 using an administrator account (see Book 2, Chapter 4), and then go into File Explorer and verify that Windows 10 has identified three drives.**

   In Figure 4-1, I have three drives. The C: drive has my Windows 10 operating system on it; C:'s the boot drive. The other two (E: and F:) have miscellaneous junk that I don't want to keep, and the D: drive is my old DVD disc reader.

2. **Bring up the Control Panel; tap or click System and Security, and then tap or click Storage Spaces.**

   Or go type **storage spaces** in the Windows 10 search box.

   If you choose either Storage Spaces or Manage Storage Spaces, you see the Storage Spaces dialog box, as shown in Figure 4-2.

3. **Tap or click the Create a New Pool and Storage Space link, and click Yes to the UAC prompt that shows up.**

   You have to create a storage pool first — that is, assign physical hard drives to the storage pool. Windows 10 offers to create a storage pool, as shown in Figure 4-3.

WARNING

4. **Select the check boxes next to the drives that you want to include in the storage pool. Note that if you accidentally select a drive that contains useful data, your data's going to disappear. Irretrievably.**

   And I do mean *irretrievably*. You can't use Recuva or some other disk scanning tool to bring back your data. After the drive's absorbed into the storage pool borg, it's gone.

Storing in Storage Spaces

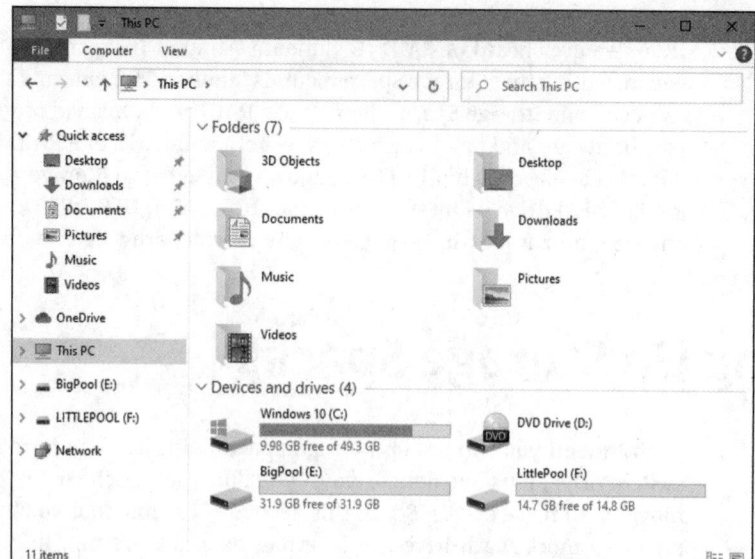

FIGURE 4-1:
Start with three drives, two for your storage pool.

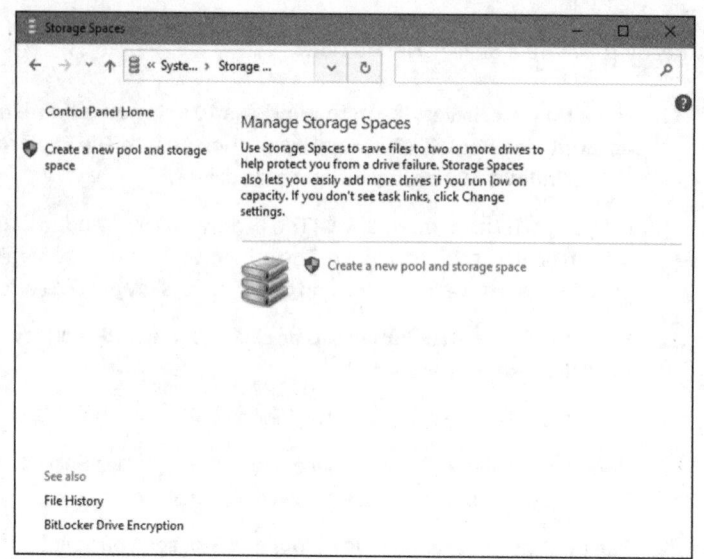

FIGURE 4-2:
Create a new storage pool.

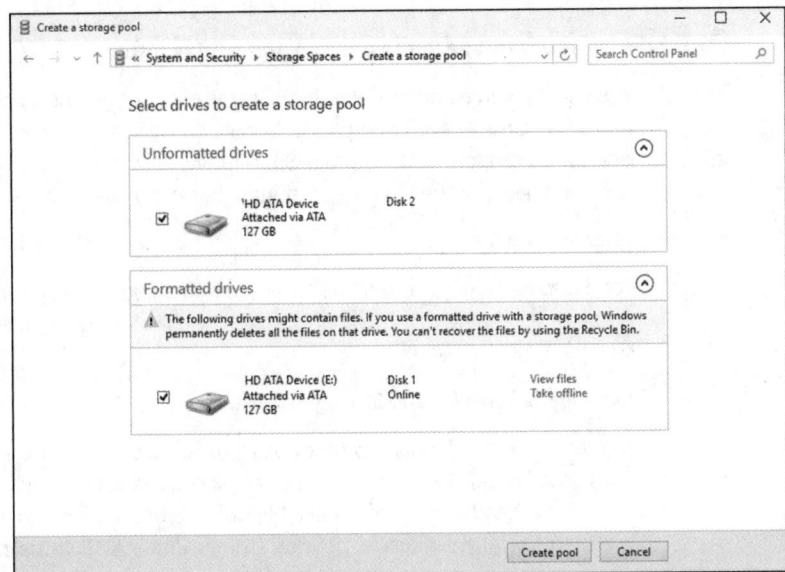

FIGURE 4-3:
Windows 10 allows you to pool any drives other than those that contain the boot and system partitions.

5. Tap or click Create Pool.

Windows 10 whizzes and wheezes and whirs for a while, and displays the Create a Storage Space dialog box, as shown in Figure 4-4.

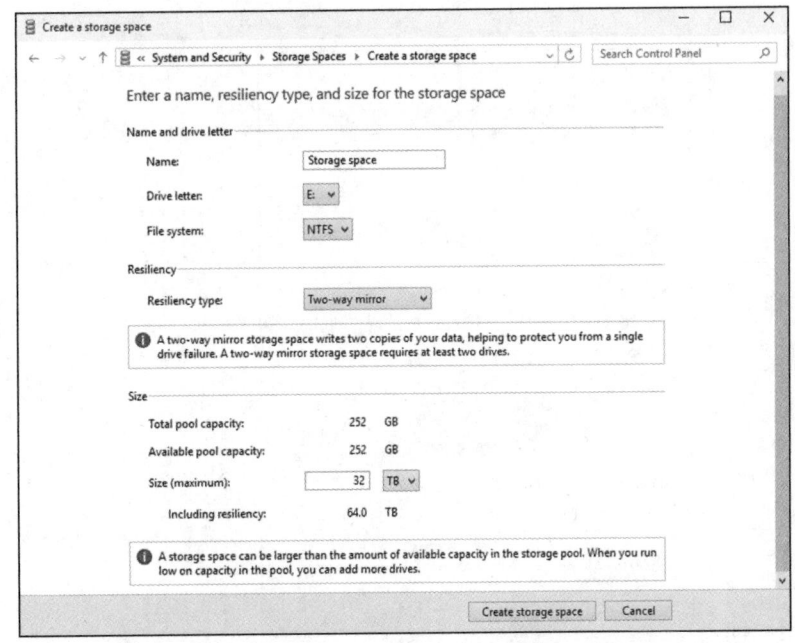

FIGURE 4-4:
Windows 10 wants you to give the new Storage Space a name and drive letter and choose the mirroring and the maximum size.

Storing in Storage Spaces

6. **Give your Storage Space a name and a drive letter.**

You use the name and the letter in the same way that you now use a drive letter and drive name — even though the Storage Space spans two or more hard drives. You can format the Storage Space drive, copy data to or from the drive, and even partition the drive, even though there's no real, physical drive involved.

7. **Choose a resiliency.**

For a discussion of your four choices — no mirroring, two-way, three-way, and parity — see the sidebar "Mirroring technologies in Storage Spaces" earlier in this chapter.

8. **Set a logical size for the Storage Space.**

As mentioned, the logical size of the Storage Space can greatly exceed the available hard drive space. There's no downside to having a very large logical size, other than a bit of overhead in some internal tables. Shoot for the moon. In this case, I turned less than 1 terabyte of actual, physical storage into a 32TB virtual monstrosity.

9. **Tap or click Create Storage Space.**

Windows 10 whirs and sets up a freshly formatted Storage Space.

10. **Go back out to File Explorer, and verify that you have a new drive, which is, in fact, an enormously humongous Storage Space.**

You see something like Figure 4-5.

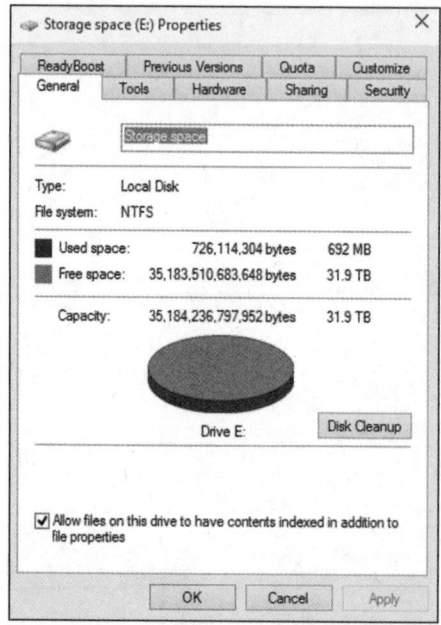

FIGURE 4-5: If it weren't for the fact that you just created it, you probably wouldn't be able to tell that the new Storage Space isn't a real drive.

Working with Storage Spaces

Have a new Storage Space? Good. Go kick some tires.

First, realize that to the outside world, your Storage Spaces looks just like any other hard drive. You can use the drive letter the same way you'd use any drive letter. The folders inside work like any other folders; you can add them to libraries or share them on your network. You can back it up. If you have a cranky old program that requires a simple drive letter, the Storage Spaces won't do anything to spoil the illusion.

That said, Storage Space drives can't be defragmented or run through the Check Disk utility.

Here's the grand tour of the inner workings of your Storage Spaces:

1. **Bring up the Control Panel; tap or click System and Security, and then tap or click Storage Spaces.**

Or go to the Windows 10 search box, and type **storage spaces**. If you choose either Storage Spaces or Manage Storage Spaces, the Storage Spaces dialog box appears, this time with a Storage Space.

2. **At the bottom, tap or click the down arrow next to Physical Drives.**

The full Storage Spaces status report appears (see Figure 4-6).

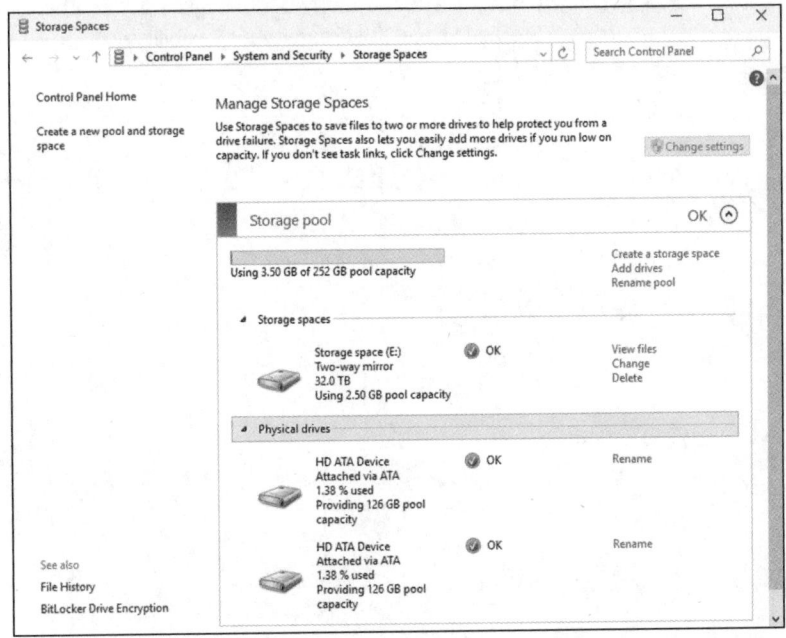

FIGURE 4-6:
Full details of your Storage Space and the storage pool it sits on.

Storing in Storage Spaces

The Storage Spaces report tells you how much real, physical hard drive space you're using; what the Storage Space looks like to your Windows 10 programs; and how your physical hard drives have been carved up to support all that glorious, unfettered space.

It's quite a testament to the Storage Space designers that all this works so well — and invisibly to the rest of Windows. This is the way storage should've been implemented years ago!

Storage Space Strategies

You can save yourself some headache by following a few simple tricks:

>> Use your fastest hard drive as your C: drive. (If you have a solid-state drive, use it for C:!) Don't tie it into a Storage Space.

>> If a hard drive starts acting up — you see an error report, in any of a dozen different places – pro-actively remove it from the Storage Space. See the Take Offline option in Figure 4-3. Replace it at your earliest convenience.

>> Remember, in a three-drive installation, where two drives are in the Storage Space, the two-way mirror option limits you to the amount of room available on the smallest Storage Space drive.

>> When you need to add more drives, don't take out the other drives. The more drives in Storage Space, the greater your flexibility.

Chapter **5**

Taking Control of Updates and Upgrades

E verybody complains about Windows 10's forced patches and upgrades. Rightfully so. From the days of the Windows 10 EternalBlue patching fiasco in February 2017, to the bricked computers brought to you by the January 2018 patches, Microsoft has proved over and over again that it can't be trusted to deliver reliable software fixes. Even the Windows 10 May 2020 update generated tons of problems for lots of users. It seems that no matter how much Microsoft fails and users complain, the company is not willing to improve its approach to Windows 10 updates.

That's where Automatic Update — the topic of this chapter — comes in. If you don't go into Windows 10 and change things, it automatically assumes that you want to install Microsoft's changes the moment they appear. In the past, that's led to all sorts of problems, and I doubt highly that it'll change in the future.

ASK WOODY.COM

I've been saying it, in print, for more than a decade: **Automatic Update is for chumps**.

Sure, your Sainted Aunt Martha, who's afraid to run anything other than Solitaire and Mahjong, should have her system set to get Windows 10 updates automatically. Folks who aren't interested in staying up to date on Windows's foibles should trust their machines to Microsoft's scheduled intrusions. Yes, you must update Windows 10 sooner or later. But there's absolutely no reason why you have to install security patches (or upgrade to a new version of Windows) at Microsoft's pace.

The Case Against Windows Automatic Update

Auto Update's an unnecessary risk for people who know how to use Windows and who keep current with Windows 10 developments. If you're knowledgeable enough to be reading this, you should seriously consider taking Windows patching into your own hands.

WARNING

The core problem: Microsoft still hasn't figured out how to deliver reliable Windows 10 patches. Patch Tuesdays have turned into massive beta-testing grounds where bugs crawl out of the woodwork and attack in unpredictable ways. With a few notable exceptions, I don't blame Microsoft for the mayhem — patching the mess we know as Windows, in all its varied glory, is an NP-complete problem (that is, it's "technically hard"). If everybody skipped Automatic Update, we'd be in an unholy mess. But folks who are willing and able to read the tea leaves don't need to expose themselves to the risks of marching in lock-step with the Auto Update cadence.

Few bad patches are particularly debilitating for most people, but they're a pain in the neck for some and positively agonizing for the unlucky. More to the point, the problems are avoidable if you just wait a couple of weeks for problem reports to die down and for Microsoft to get its patches patched.

Even if Microsoft isn't at fault — and if frequently isn't — the pointed finger comes as small consolation to folks who have their days disrupted by a weird conflict or their products clobbered.

ASK WOODY.COM

Patches are important, but you don't need Automatic Update to do them. Of course, you must get patched eventually; you just don't want to be in that initial unpaid beta-testing phase.

Certainly, the wait-and-watch approach has downsides. Foremost among them: If Microsoft patches a vulnerability in Windows 10 or Office and malware appears very quickly to take advantage of a previously unknown security hole, those who are deferring updates may be caught flat-footed.

That's happened in the past, but it has become uncommon. Sure, there are patches for *zero days* — Windows Update patches for security holes with known exploits — but this is a horse of a different color. Microsoft did a good job obfuscating its descriptions and preventing its patched code from fast reverse engineering. Could a massive reverse-engineered wave of malware roll out on some future Wednesday? Yes, and if it does, Automatic Update will save the day.

WARNING

As with everything associated with patching Windows, there are pros and cons. You have to weigh the possibility of a giant, quickly reverse-engineered attack against the certainty of buggy patches. History shows that the risk of blind patching on day one greatly exceeds the risk of delaying for a couple of weeks.

If you aren't particularly good at Windows 10 or you don't want to take the time to keep your machine fed (or both), use Automatic Update. That part's easy because you don't have to do anything. Windows 10, all by itself, will feed you patches as Microsoft releases them.

Terminology 101

Microsoft's terminology doesn't help. The official naming has changed several times — sometimes to devastating effect — but right now, as of this writing, these are the patching terms you need to know:

» **Cumulative updates:** Microsoft calls them **quality updates.** These updates include a combination of security patches, bug fixes, and little niggly things such as time zone changes.

   In theory, cumulative updates arrive once a month, on the second Tuesday of the month — so-called Patch Tuesday. Reality can be much, much messier. In January 2018, for example, Microsoft released, withdrew, and re-released cumulative updates more than a dozen times.

» **Version changes:** Microsoft calls them **feature updates,** but they're really upgrades. These move you from one version of Windows 10 to the next. As I explain in Book 1 Chapter 1, new versions of Windows 10 are supposed to appear every six months.

» **Changes to the updating software itself:** So-called **Servicing Stack updates** can appear out of the blue just about any time. They're intended to make it easier to install patches.

>> **Hotfixes:** Rarely identified as *hotfixes,* these are patches that fix bugs in earlier patches. Typically, they won't be pushed out of the Automatic Update chute. Instead, if you have problems with a patch, you must figure out how to download and install the hotfix.

>> **Microsoft Defender definition updates and the Microsoft Malicious Software Removal Tool (MSRT):** They happen all the time, and they're generally harmless. Don't worry about installing them.

>> **Drivers:** Microsoft's bad pushed drivers are legendary for creating mayhem where none was needed. Using criteria yet unknown, Microsoft occasionally pushes out driver updates for certain big-name hardware manufacturers.

WARNING

You should avoid installing a driver update from Microsoft. Instead, if you're having problems, go to your hardware manufacturer's website, and get it from them. If you aren't having problems with a piece of hardware, ABDF (ain't broke, don't fix).

>> **For Microsoft Surface computers, firmware changes:** One of the joys of owning a Surface device is that Microsoft pushes system software updates — both firmware changes and driver changes — via Windows Update. However, Microsoft's "tradition" of delivering faulty updates once every few months may be a reason to avoid Surface machines.

>> **Lots and lots of miscellaneous:** Microsoft occasionally pushes out patches specifically for Internet Explorer and Edge, the mini operating system known as .NET, and all sorts of additional pieces of Windows 10 flotsam and jetsam. In general, these patches are supposed to be rolled into cumulative updates. In practice, they squirt out in the middle.

TIP

Confusingly, some miscellaneous patches get released as quality updates — and are thus subject to Automatic Update rules. Others float around until they're picked up by the next month's cumulative update. The definition of *quality* is tenuous at best.

When it comes to version of Windows 10, the terminology's even worse. The simple fact is that Microsoft releases new versions of Windows 10, knowing full well that they aren't as stable as they should be. That's what unpaid beta testing is all about, eh?

The terms for version changes have changed three times since Microsoft first released Windows 10, but at this moment, here's what we have:

>> **Semi-Annual Channel (Targeted)** means that this version of Windows 10 is good enough to be sent out the door, but it isn't yet stable enough for business users — you know, real paying customers — to install.

>> **Semi-Annual Channel** refers to a version of Windows 10 that's been tested long enough for Microsoft to feel confident recommending it to business users. In the past, Microsoft has waited about four months for unpaid beta testing to pick up outstanding bugs.

I strongly recommend that you wait for any new version of Windows 10 to hit Semi-Annual Channel before you install it.

>> **Long-Term Servicing Channel** (LTSC) is a rare bird that applies to only Enterprise versions of Windows 10. In theory, it's a more stable version of Win10 suitable for environments that need to avoid update bugs. In practice, it's almost impossible to get real work done on an LTSC machine. Microsoft admonishes that LTSC is not intended for machines that run, say, Microsoft Office.

ASK
WOODY.COM

That's the general framework of the Windows 10 patching phenomenon. Your job, should you choose to accept it, is to make the rules work for you — not blind-side you.

Many so-called security experts will tell you that you should leave Windows Automatic Update turned on. I say "bah!" without a bit of hesitation, having witnessed the carnage with malignant Windows 10 patches.

REMEMBER

You must patch sooner or later. However, there's no pressing reason to patch immediately after Microsoft releases its updates.

The Great Divide: Home versus Pro

Windows 10 computers attached to a managed network get their updates through the network. If your Windows 10 PC is on a network (typically a business network) that runs WSUS, SCCM, or another update server, the network admin gets to decide which updates are applied and when. You don't have any choice.

For those of you who aren't hooked into a corporate server, the behavior of Windows Automatic Update is more or less in your hands.

When Windows 10 was first released back in 2015, Win 10 Home computers, by design, didn't have easy access to update- and upgrade-deferring settings. That omission was intentional because Microsoft uses the Windows 10 Home install base to test new versions before they're deemed ready for businesses. This situation has changed starting with the Windows 10 May 2019 update (version 1903): Microsoft decided to give Windows 10 Home users the possibility to postpone updates up to 35 days.

On the other hand, Windows 10 Pro users had the option to postpone updates up to 365 days, which was great. Unfortunately, Microsoft decided to remove this option in the Windows 10 May 2020 update (version 2004). Today, both Windows 10 Pro and Home users are unpaid beta testers who can postpone updates for only 35 days.

To see whether you have Win10 Home or Pro (or Education or Enterprise), type **about** down in the Windows 10 search box, near the Start button, and click or tap About Your PC. Scroll down the About window to Windows Specifications, and you should see something like Figure 5-1.

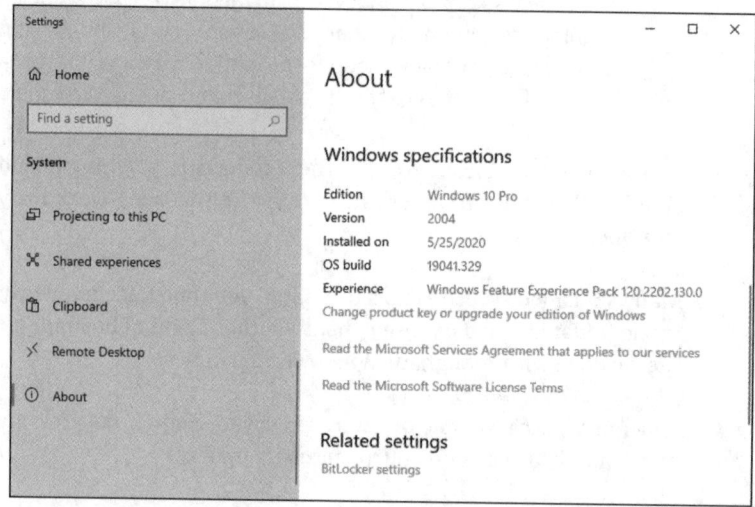

FIGURE 5-1:
Get version
details on the
About pane.

To be sure, there are ways to block Windows 10 updates for more than 35 days, but they're a bit sneaky and not well documented. As far as I'm concerned, the ability to readily control the rate of updates was the number-one reason to pay for Windows 10 Pro. Now, however, the infamous May 2020 update has removed most of the useful Windows Update control options people had available.

Keeping Your Windows 10 Machine Protected From Updates

The trick to blocking updates on Windows 10 Home or Pro machines lies in a little-known setting called metered connection. As originally conceived, Microsoft put the metered connection setting in Windows 10 to let you tell Windows that you're paying for your Internet access by the bit; that is, you don't want Windows 10 to download anything unless it absolutely has to. From that fortunate beginning arises the best option for Windows 10 users to block updates.

REMEMBER

What goes through a metered connection? Hard to say, specifically, and Microsoft has made no commitments. But experience has taught us that the metered connection setting guards against just about any patches, except Microsoft Defender updates — exactly what you would hope. No guarantees, of course, but metered connection looks like a decent, if kludgy, approach to blocking updates.

If you're using Windows 10 Home or Windows 10 Pro, version 1803 or later, and you want to (temporarily!) block updates and version upgrades, follow these steps:

1. **Figure out if you're using a Wi-Fi connection or a wired Internet (Ethernet) connection.**

The easiest way to do that is to click or Tap Start➪Settings, choose Network & Internet, and then on the left choose Ethernet. If you're on a wired Internet connection, you see something like Figure 5-2.

In some unusual situations, you may have access to the Internet through both an Ethernet cable and over Wi-Fi. If that's your situation, not to worry. These steps will get you covered.

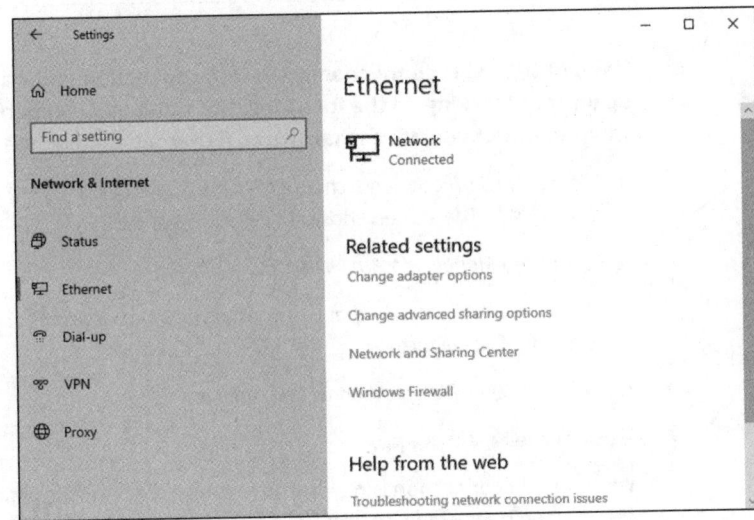

FIGURE 5-2:
A wired (Ethernet) connection looks like this.

2. **If you have a wired (Ethernet) connection:**

a. *Click the name of the connection.* In Figure 5-2, you would click Network (Connected). The Network Profile pane appears.

b. *Move the Set as Metered Connection slider to On.* See Figure 5-3.

c. *Click the left arrow at the top of the pane to return to the Network & Internet settings.*

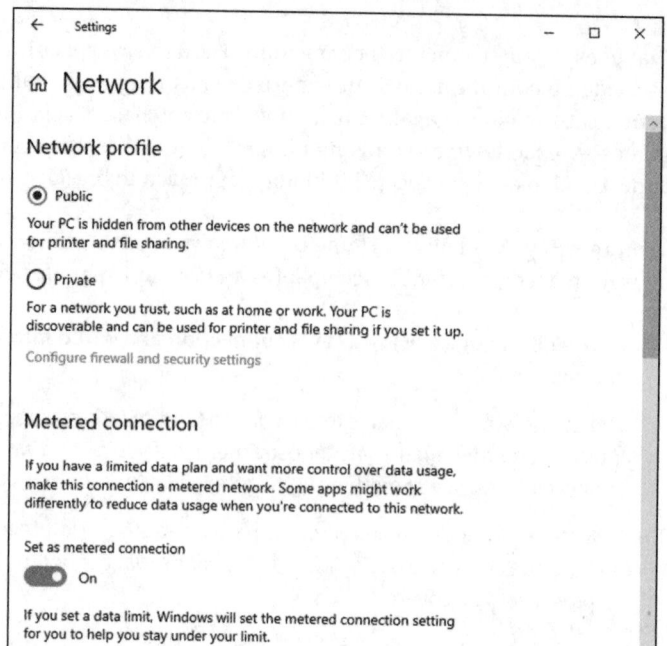

FIGURE 5-3:
Set your Ethernet
connection to
metered.

3. **If you have a Wi-Fi connection or a wired connection and want to double up on your blocking do the following. Note that you can perform this step whether or not you have a hardwired (Ethernet) connection:**

   a. *Click or Tap Start⇨Settings, choose Network & Internet, and then on the left choose Wi-Fi. The screen shown in Figure 5-4 appears.*

   b. *Click or tap Manage Known Networks.*

   c. *Click the Wi-Fi connection that you normally use and choose Properties.* You see something like Figure 5-5.

   d. *Move the Set as Metered Connection slider to On.*

4. **X out of the Settings app.**

   Your Internet connection is now set as metered and — unless Microsoft changes the rules — you're protected from both cumulative updates and from version upgrades.

WARNING

You must install updates sooner or later. See the last section of this chapter for advice on choosing the right time to drop the big one.

If you're using an earlier version of Windows 10 — versions 1507, 1511, 1607, 1703 or 1709 — you can follow these steps to set a Wi-Fi connection to metered. Unfortunately, in those earlier versions, you can't set a wired (Ethernet) connection to metered.

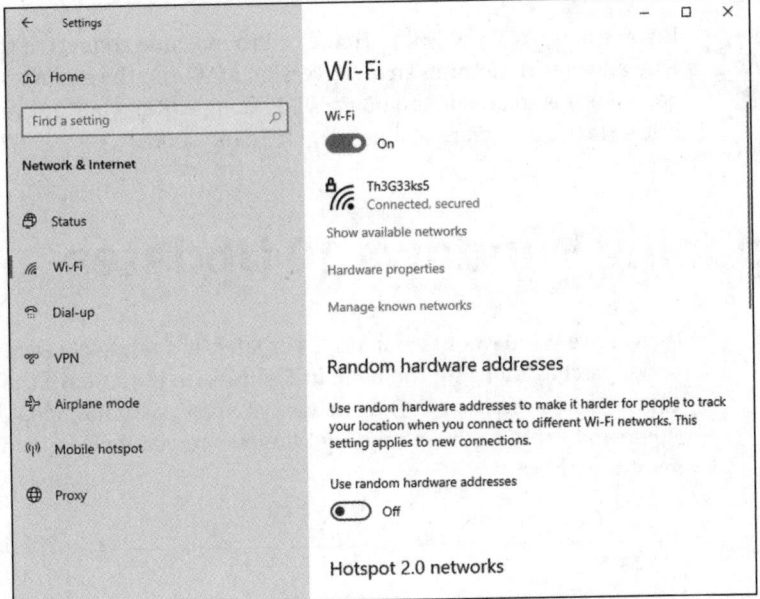

FIGURE 5-4:
If you have a
Wi-Fi connection,
it'll look like this.

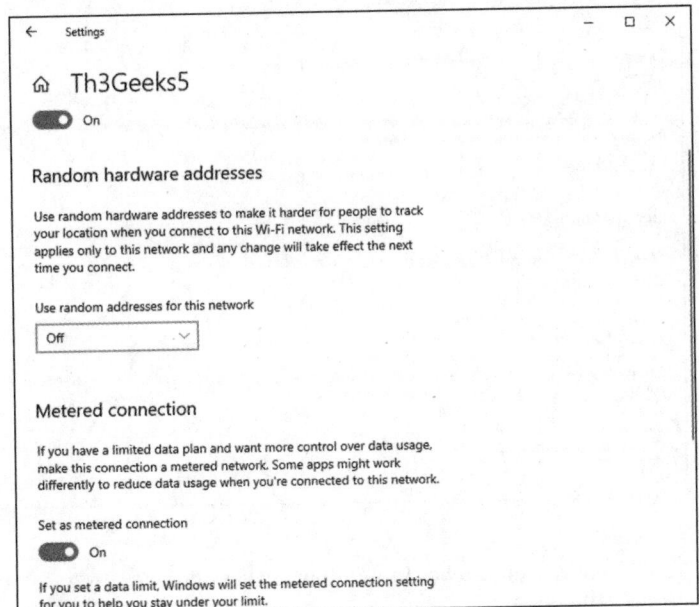

FIGURE 5-5:
Slide Set as
Metered
Connection
to On.

If you find your Windows 10 Home or Pro machine connected to the Internet by a wire and you're running an older version of Win10, the easiest solution (far from a good one!) is to go out and buy a Wi-Fi dongle. Run your machine on Wi-Fi, even if it's slower and more of a hassle. Life's too short.

Postponing Windows 10 Updates

If you have Windows 10 version 1703 or later, it's relatively easy to postpone those pesky patches by using the pane in Figure 5-6. If you have the Windows 10 May 2019 update (version 1903) or newer, you can postpone updates on Windows 10 Home and Pro. For those running Windows 10 version 1611 and earlier, you have my sympathies.

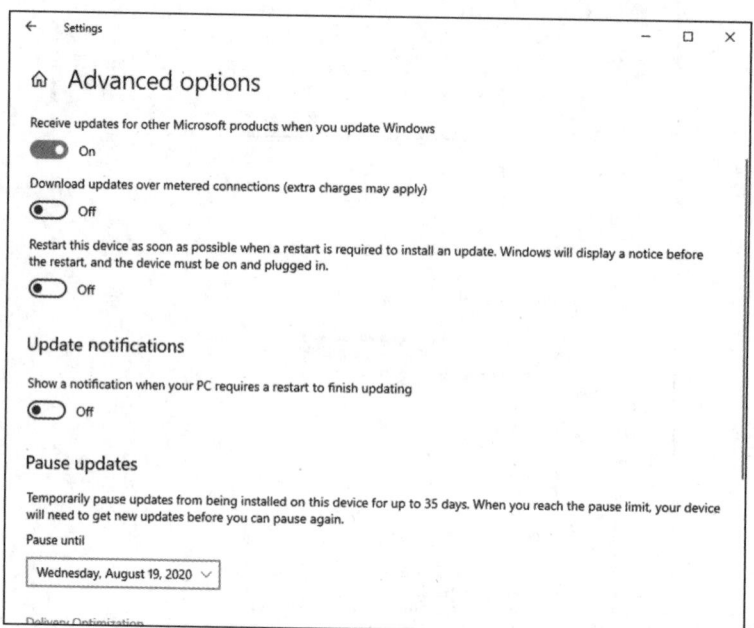

FIGURE 5-6:
Control Windows
Update from
here.

To take control of patches in Windows 10 version 2004 (May 2020 update) or later, do the following:

1. **Click or tap Start⇨ Settings (the gear icon), and then click Update & Security. On the right, click the Advanced Options button.**

   The Advanced Options pane appears (refer to Figure 5-6).

2. **In the first section, enable the switch for Receive Updates for Other Microsoft Products When You Update Windows.**

This step ensures that you'll get updates for Office at the same time you get updates for Windows 10.

3. **Under Update Notifications, turn on the switch for Show a Notification When Your PC Requires a Restart to Finish Updating.**

This step ensures that you're informed when Windows 10 needs a restart to finish updating; it doesn't perform the reboot without your knowledge.

4. **In the Pause Updates section, click the drop-down list, and select the last available date that is 35 days from today.**

In other words, when Microsoft releases a cumulative update for the current version of Windows 10, Windows Update must wait 35 days before installing it.

Because cumulative updates normally hit every Patch Tuesday (the second Tuesday of the month), setting this box to the last available date ensures that the update will be installed more or less right before the next cumulative update is available.

WARNING

Pause updates takes precedence over all the other settings. If you have Pause turned on, Windows 10 stops all updates but Microsoft Defender updates. The problem isn't in blocking all updates. The problem arises when you exceed the 35-day limit or when you turn off Pause Updates.

Microsoft is careful to mention that you can't reset the Pause Update setting. If you try to turn it off and turn it back on again, "this device will need to get new updates before you can pause again."

All sorts of things they don't teach you in Windows 10 school, eh?

Microsoft's documentation for all these settings is poor, aided by several different sets of terminology and changing policies. Many folks inside Microsoft think it's a mistake to give (sniff) users (sniff) the ability to block updates and upgrades — thinking that Mother Microsoft knows better.

Bah. Humbug. The tools are there. Use them to your advantage.

Keep Up on the Problems

If you block updates, you're on the hook to unblock them, sooner or later. As a general rule— and I've been fighting Windows patches since 3.1 days — it takes several days for the worst cumulative update bugs to shake out, and weeks for the more subtle problems to appear, get diagnosed and, in some cases, fixed.

Version upgrades, on the other hand, seem to follow Microsoft's original schedule. It really *does* take three or four months with the cannon fodder subjected to a new version to make sure it's stable enough to entrust your machine to the new order.

The deadline we're all fighting is the one imposed by the bad guys. If you delay patching long enough, something bad is going to hit, and it may hit quickly. We had a good example in February 2017, when Microsoft's patches, about six weeks after they were released, became crucial to block the WannaCry and NotPetya vulnerabilities. If you waited too long to patch in early 2017, your machine was wide open to some truly awful stuff.

Conversely, if you install all the patches soon after they're released, you expose yourself to the kind of problems we saw in January 2018. Back then, Microsoft released a bevy of Meltdown and Spectre patches that bricked large groups of PCs. It took Microsoft five days to identify the problem and pull the patch.

Microsoft spent weeks patching, pulling patches, re-patching, and re-re-patching — and it had to counter bugs in Intel's patches at the same time. The result was an abominable mess that left many Windows 10 users bewildered and more than a few staring at useless blue screens on bricked machines.

The upshot? Every month is different. If you block updates and upgrades, you must stay on top of the latest developments and judge for yourself when it's safe to patch.

Staying on top of patches has been my mission at AskWoody.com for more than a decade. It's also my prime directive at Computerworld. If you block patches, take a look around at the major technical websites and keep up on the latest shenanigans. While you're at it, be sure to drop by AskWoody.com occasionally for a full dose of no-bull advice.

Stopping Windows 10 Updates from Rebooting Your PC

Another annoying part of Windows 10 updates is that some require a reboot to get applied. Because of that, you may end up with your PC restarting while you work, deliver a presentation at an event, or play a game. To stop that from happening, Windows Update has a feature called Active Hours. You can set the interval during which you tend to use your PC, named Active Hours, and Windows 10 won't restart for updates during that time. Here's how to set it up:

1. **Click or tap Start and then Settings (the gear icon).**

2. **Click Update & Security. On the right, click the Change Active Hours button.**

   The Change Active Hours pane appears, as shown in Figure 5-7.

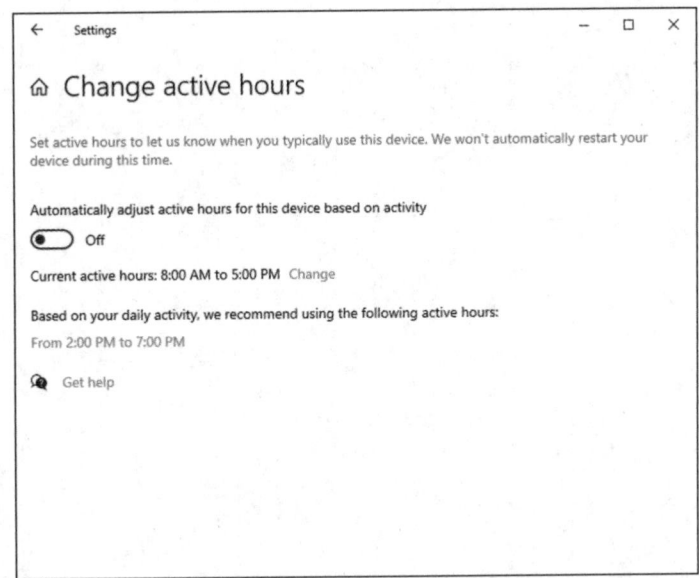

FIGURE 5-7:
Setting the hours
when Windows
10 doesn't restart
for updates.

3. **Do one of the following:**

   - *Click the Change link near Current Active Hours, set the Start Time and the End Time for your Active Hours, and click Save.* Windows 10 uses the interval you've set as your Active Hours during which it won't restart your PC for updates.

   - *Enable the switch for Automatically Adjust Active Hours for This Device Based on Activity.* Windows 10 will monitor how you use your PC and set Active Hours accordingly. I don't like this approach.

Chapter **6**

Running the Built-In Applications

N ew Windows 10 apps are just starting to appear in quantity and quality good enough to drive your everyday computing. We're still a long way from a Windows 10 PC where you use only apps instead of desktop apps, but the trend is definitely in that direction.

The big question, at this point, is whether Microsoft can get enough good Windows 10 apps in its store to fend off the rising tide of advanced Web Apps, which can be used with any browser, on any machine. The jury's still out on that one.

ASK
WOODY.COM

In this chapter, I introduce you to a handful of useful programs that you've already paid for. They aren't the greatest, but they're more than adequate in many situations — and when better, free alternatives exist, I tell you about them, too.

Even if they do come from a Microsoft competitor.

Keep your eyes open for new Microsoft Store–based apps that can match some of the functions in these free built-in Windows programs. As time goes by, the Windows 10 apps will get better — at least, that's the plan — although it's going to be difficult to beat the price on these guys.

Setting Alarms & Clock

The Windows 10 Alarms & Clock app works almost as well as the alarm and clock apps you'll find, free, for iPhones, iPads, and Android smartphones and tablets. Some paid apps add a few bells and whistles, but for most folks, the built-in free app works well enough.

To set an alarm or change an alarm you've already set, go into the Alarms & Clock app from Windows 10.

Permit me to take you on a guided tour through the app:

1. **Click or tap the Start button. Near the top of the app list in the middle, tap or click Alarms & Clock.**

   The basic Alarm app shows up, as in Figure 6-1. It's not particularly inspiring, but give it a chance and you may be surprised.

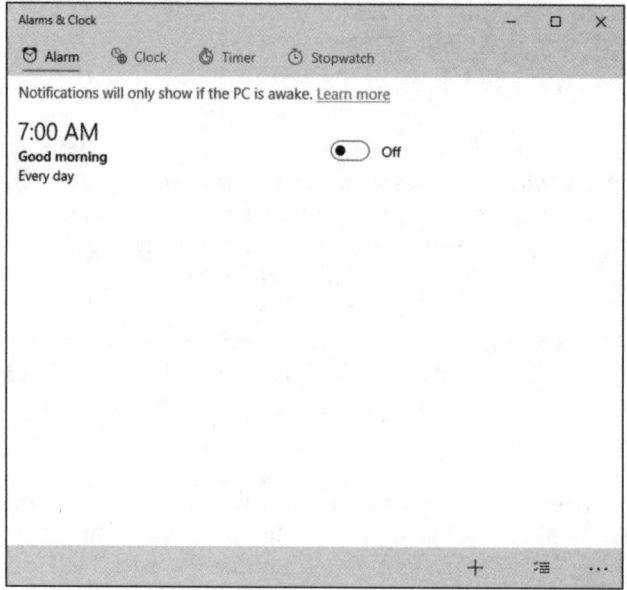

FIGURE 6-1:
The alarm clock starts with a 7:00 am weekday wakeup call, but it's turned off.

2. **Click or tap the + sign at the bottom of the alarm list, to add a new alarm.**

   Alarms & Clock shows you the standard alarm prompt, shown in Figure 6-2.

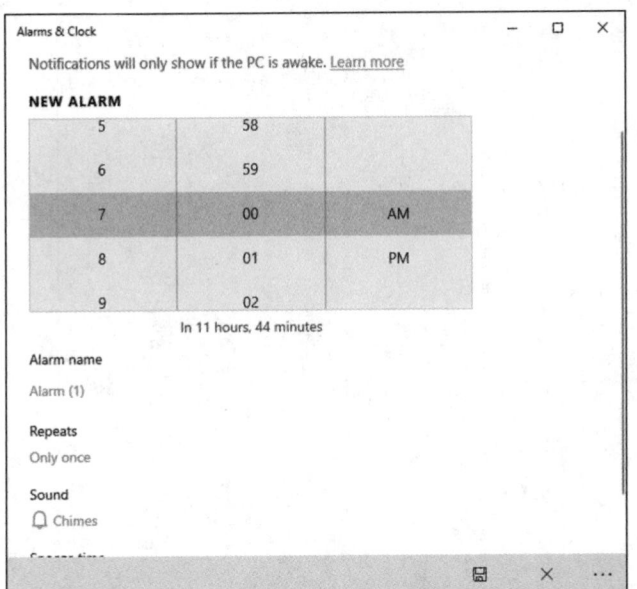

FIGURE 6-2:
Enter a new
alarm here.

3. **Fill in an alarm — say, for a few minutes from now. Then at the bottom, click the Save icon, which looks like a very snappy 1980s-style 3.5-inch diskette.**

Even if you've never seen a 3.5-inch diskette and can't remember what a joy it was to get one that was stuck out of a diskette drive, the alarm is added to the list in Figure 6-1.

How do you get rid of an alarm? Excellent question. Glad you asked. If you right-click or tap and hold down an alarm, nothing happens. But . . .

4. **To delete an alarm, click the icon at the bottom that looks like a double-decker hamburger with check marks, and then click the check box next to the alarm you want to delete. Finally, click the trash can.**

Yes, that's how you delete an alarm in Windows 10 app land.

5. **Click or tap the Clock icon.**

The world clock shows you the current time in your current location and makes it easy to add additional locations. Just click or tap the + sign, as in Figure 6-3.

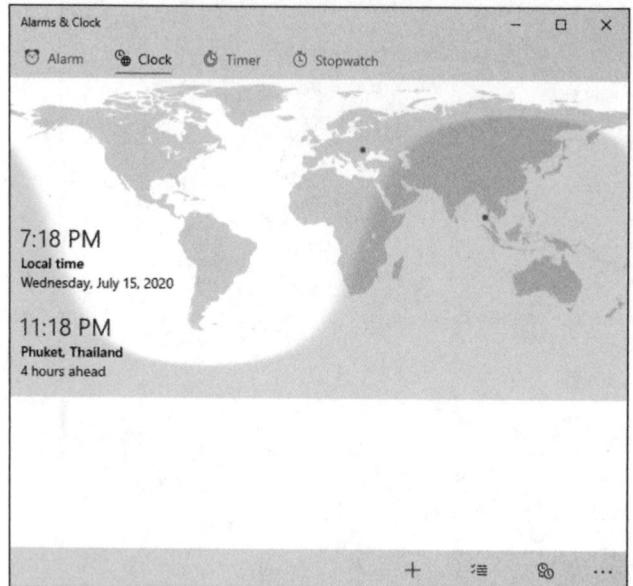

FIGURE 6-3:
It's easy to add
a location to the
world clock.

The Compare icon at the bottom — the one that looks like two analog clock faces — lets you compare a date and time in one location with another. So, for example, you can input a date and time for Phuket, Thailand, and have the app tell you what the date and time will be in your area.

The double-deck hamburger icon with check marks at the bottom of the Clock tab, like the previous one, lets you delete locations from the list.

6. **Click the Timer tab.**

You see the rather mundane countdown timer shown in Figure 6-4. Click the Start icon in the middle, and when it's finished counting down, a toaster notification appears on the Windows 10 desktop, and from there travels to the action/notification center described in Book 2, Chapter 3.

7. **Click the Stopwatch tab.**

It looks and works much like the Timer tab, except in reverse.

TIP

A little word to the wise: Both the countdown timer and the stopwatch keep working, even if you minimize the app or switch tabs. However, they don't work in the background if you close the Alarms & Clock app.

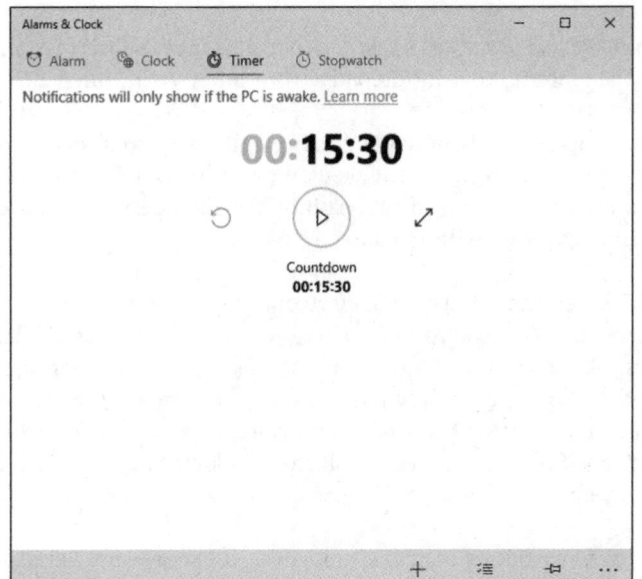

FIGURE 6-4:
The Timer is a straightforward countdown timer.

Getting Free Word Processing

With Office Online (www.microsoft.com/en-us/microsoft-365/free-office-online-for-the-web) and any browser, you have a free, useful word processor at hand anytime you're connected to the Internet. Word Online doesn't have all the bells and whistles of either the full-blown desktop version of Word, or of the tablet-based mobile version of Word, but it's good enough in almost every situation.

Two other free word-processing programs that ship with Windows 10:

» **Notepad:** For just plain text, use Notepad or its beefed-up (free) brother, Notepad++. I talk about Notepad in this chapter and Notepad++ in Book 10, Chapter 5.

» **WordPad:** If you need just a little bit of formatting, use WordPad. I talk about WordPad in this chapter.

Someday, one or the other may save your tail.

Running Notepad

Reaching back into the primordial WinOoze, Notepad was conceived, designed, and developed by programmers, for programmers — and it shows. Although Notepad has been vastly improved over the years, many of the old limitations remain. Still, if you want a fast, no-nonsense text editor (certainly nobody would have the temerity to call Notepad a word processor), Notepad's a decent choice.

Notepad understands only plain, simple, unformatted text — basically the stuff you see on your keyboard. It wouldn't understand formatting, such as bold, or an embedded picture if you shook it by the shoulders, and heaven help you if you want it to come up with links to web pages.

REMEMBER

On the other hand, Notepad's shortcomings are, in many ways, its saving graces. You can trust Notepad to show you exactly what's in a file — characters are characters, old chap, and there's none of this froufrou formatting stuff to mess up things. Notepad saves only plain, simple, unformatted text; if you need a plain, simple, unformatted text document, Notepad's your tool of choice. To top it off, Notepad is fast and reliable. Of all the Windows programs I ever met, Notepad is the only one I can think of that has never crashed on me.

The following tidbits of advice are all you'll likely ever need to successfully get in and around Notepad:

>> **To start Notepad,** click or tap the Start button, scroll way down to Windows Accessories, and choose Notepad. You can also double-click any text (.txt) file in File Explorer. You see something like the file shown in Figure 6-5.

>> **Notepad can handle files up to about 48MB in size.** (That's not quite the size of the *Encyclopedia Britannica,* but it's close.) If you try to open a file that's larger, a dialog box suggests that you open the file with a different editor.

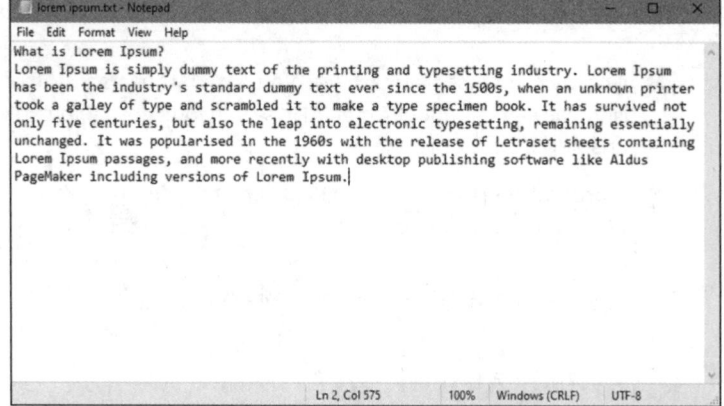

FIGURE 6-5:
Notepad rocks in a geriatric sort of way.

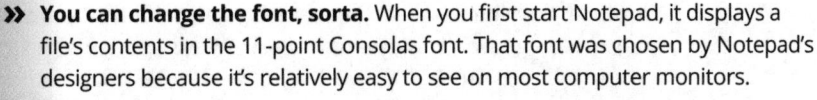
>> **You can change the font, sorta.** When you first start Notepad, it displays a file's contents in the 11-point Consolas font. That font was chosen by Notepad's designers because it's relatively easy to see on most computer monitors.

REMEMBER

Just because the text you see in Notepad is in a specific font, don't assume for a moment that the characters in the file itself are formatted. They aren't. The font you see on the screen is just the one Notepad uses to show the data. The stuff inside the file is plain-Jane, unformatted everyday text.

>> **To change the font that's displayed onscreen,** choose Format ➪ Font and pick from the offered list. You don't need to select any text before you choose the font because the font you choose is applied to all text onscreen, and it doesn't affect the contents of the file. The default Notepad font is *monospaced* — all the characters are the same width. If you change the font, text files that are designed for a fixed-width world can look very odd.

>> **You can wrap text, too.** Usually text extends way off the right side of the screen. That's intentional. Notepad, ever true to the file it's attached to, skips to a new line only when it encounters a line break — usually that means a *carriage return* (or when someone presses Enter), which typically occurs at the end of every paragraph.

Notepad allows you to wrap text onscreen, if you insist, so that you don't have to scroll all the way to the right to read every single paragraph. To have Notepad automatically break lines so that they appear onscreen, choose Format ➪ Word Wrap.

TECHNICAL
STUFF

>> **Notepad has one little geeky timestamp trick** that you may find amusing — and possibly worthwhile. If you type **.LOG** as the first line in a file, Notepad sticks a time and date stamp at the end of the file each time it's opened.

TIP

Many, many alternatives to Notepad exist: Programmers need text editors, and many of them take up the mantle to build their own. Over the years, I've used lots of them. Right now, I use Notepad++ — and yes, I do type text quite a bit. Native HTML. But that's another story.

Check out Notepad++ at `www.notepad-plus-plus.org`. It's free and works very well.

If you aren't quite so geeky, try another good alternative: Notepad Next. Check it out in the Windows Store. Free, of course.

Writing with WordPad

If you really want and need formatting — and you can't get connected to Office Online (see preceding section) for whatever reason —WordPad will do.

WordPad plays nice (at least, reasonably so), with DOCX format documents — the kind that is generated automatically in Word version 2007 and later. But if you have to edit a Word DOC or DOCX file with WordPad, whether it's from Word 97, 2000, 2002, 2003, 2007, or 2010, follow these steps:

1. **Make a copy of the Word document, and open the copy in WordPad.**

   Do not edit original Word doc files with WordPad. You'll break them as soon as you save them. Do not open Word docs in WordPad, thinking that you'll use the Save As command and save with a different name. You'll forget.

2. **When you get Word back, open the original document. On the Review ribbon, choose Compare, Combine, pick the WordPad version of the document, and click the Merge button.**

   The resulting merged document probably looks like a mess, but it's a start.

3. **Use the Review tab to march through your original document and apply the changes you made with WordPad.**

   This is the only reliable way to ensure that WordPad doesn't accidentally swallow any of your formatting.

WordPad works much the same as any other word processor, only less so. That said, WordPad isn't encumbered with many of the confusing doodads that make Word so difficult for the first-time e-typist, and it may be a decent way to start figuring out how simple word processors work.

To get WordPad going, click or tap the Start button, scroll down to Windows Accessories, and choose WordPad (see Figure 6-6).

Some people like the ribbon interface across the top of the WordPad window. I find it familiar (like Word 2007) but annoying (like, uh, Word 2007).

WordPad lets you save documents in any of the following formats:

» **Rich Text Format (RTF)** is an ancient, circa-1987 format developed by Microsoft and the legendary Charles Simonyi (yes, the space tourist) to make it easier to preserve some formatting when you change word processors. RTF documents can have some simple formatting but nothing nearly as complex as Word 97, for example. Many word-processing programs from many manufacturers can read and write RTF files, so RTF is a good choice if you need to create a file that can be moved to many places.

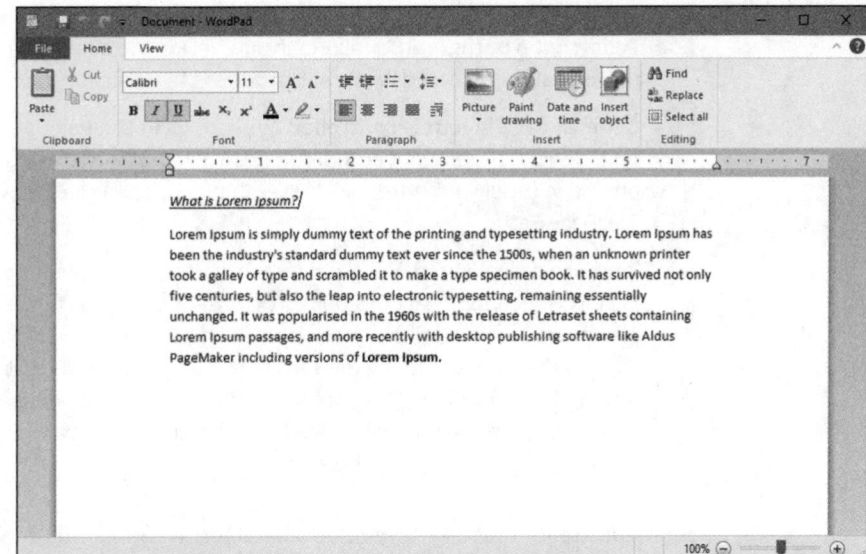

FIGURE 6-6:
WordPad includes
rudimentary
formatting and
the capability to
embed images
for free.

WARNING

>> **OOXML Text Document (DOCX)** is the new Microsoft document standard file format, introduced in Word 2007. If you're going to use the document in Word, this is the format to choose.

Note that WordPad can read and write DOCX files. Unfortunately, WordPad takes some, uh, liberties with the finer formatting features in Word: If you open a Word-generated DOCX file in WordPad, don't expect to see all the formatting. If you subsequently save that DOCX file from WordPad, expect it to clobber much of the original Word formatting.

>> **ODF Text Document (ODT),** the OpenDocument format, is the native format for LibreOffice and OpenOffice.

>> **Text Document (TXT)** strips out all pictures and formatting and saves the document in a Notepad-style, regular old text format. The two alternatives — MS-DOS format and Unicode — control the way WordPad handles non-Roman characters in the document.

If you're just starting out with word processing, keep these facts in mind:

>> **To format text,** select the text you want to format; then choose the formatting you want from the Font part of the Home tab, on the ribbon. For example, to change the font, click the down arrow next to the font name and choose the font you like.

>> **To format a paragraph,** simply click once inside the paragraph and choose the formatting from the Paragraph group in the Home tab, on the ribbon.

>> **General page layout is controlled by settings in the Page Setup dialog box.** General page layout includes things like margins and whether the page is printed vertically or horizontally, for example. To open the dialog box, choose File, Page Setup.

>> **Tabs are complicated.** Every paragraph starts with tab stops set every half inch. You set additional tab stops by clicking in the middle of the ruler. (You can also set them by clicking the tiny side arrow to the right of the word *Paragraph* and then clicking the Tabs button.) The tab stops that you set up work only in individual paragraphs: Select one paragraph and set a tab stop, and it works only in the selected paragraph; select three paragraphs and set the stop, and it works in all three.

ASK WOODY.COM

WordPad treats tabs like any other character: A tab can be copied, moved, and deleted, sometimes with unexpected results. Keep your eyes peeled when using tabs and tab stops. If something goes wrong, click the Undo icon (to the right of the diskette-like Save icon) or press Ctrl+Z immediately and try again.

WordPad has a few features worthy of the term *feature:* bullets and numbered lists; paragraph justification; line spacing; superscript and subscript; and indent. WordPad lacks many of the features that you may have come to expect from other word processors: You can't even insert a page break, much less a table. If you spend any time at all writing anything but the most straightforward documents, you'll outgrow WordPad quickly.

You may find Google Docs much more capable than WordPad, and it's free for personal use. See Book 10, Chapter 3 for details.

Taming the Character Map

Windows 10 includes the Character Map utility, which may prove a lifesaver if you need to find characters that go beyond the standard keyboard. Using the Character Map, you can ferret odd characters out of any font, copy them, and then paste them into whatever word processor you may be using (including WordPad).

Windows 10 ships with many *fonts* — collections of characters — and several of those fonts include many interesting characters that you may want to use. To open the Character Map, click or tap the Start icon, scroll down to Windows Accessories, and choose Character Map. You see the screen shown in Figure 6-7.

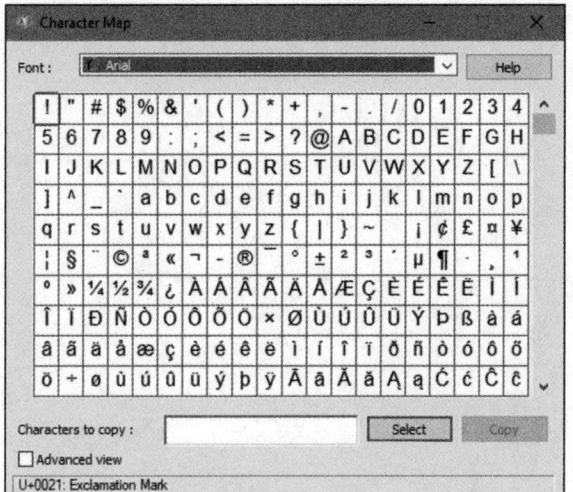

FIGURE 6-7:
Need a character
from a different
language? Use
the desktop
Character Map.
Klingon, anyone?

You can use many characters as pictures — arrows, check marks, boxes, and so on — in the various Wingdings and Webdings fonts. Copy them into your documents and increase the font size as you like.

Calculating — Free

Windows 10 includes a very capable Calculator app. Actually, Windows contains five capable calculators with several options in each one, plus a built-in units converter so you can translate furlongs per fortnight into inches per year. Before you run out and spend 20 bucks on a scientific calculator, check out the three you already own!

To run the calculator, click or tap the Start button then choose Calculator. You probably see the standard calculator, as shown in Figure 6-8.

To use the calculator, just type whatever you like on your keyboard or tap or click the keys, and press Enter when you want to carry out the calculation. For example, to calculate 123 times 456, you type or tap **123 * 456** and press Enter.

The calculator comes in five modes: standard, scientific (which adds *sin* and *tan*, and *x to the y,* and the like), graphing, programmer (hex, octal, Mod, Xor, Qword, Lsh), and data calculation. You can flip among those modes by clicking the hamburger icon in the upper-left corner (and shown in the margin).

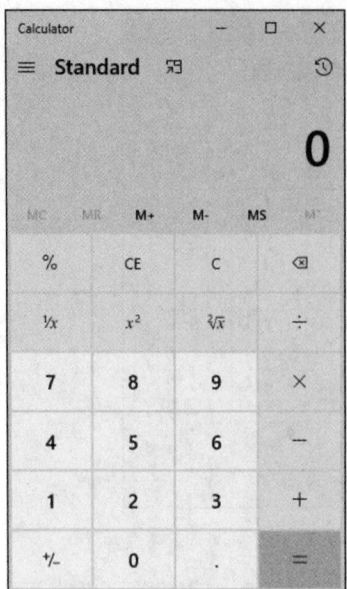

The calculator also has very extensive unit conversion capabilities. Choose Converter from the hamburger icon, and then choose one of the units converters —Currency, Volume, Length, Weight and Mass, Temperature, Energy, Area, Speed, Time, Power, Data, Pressure, or Angle. For example, if you choose Volume, you get something like Figure 6-9.

The fun part of the converters: They have little mind-jogging tips. For example, in Figure 6-9, you can see that 10 millimeters is about 2 teaspoons, but you can also see that it's about 0.68 tablespoons and 0.04 coffee cups. Play with it a bit, and you can see volumes in cubic yards and bathtubs, lengths in nautical miles, km and jumbo jet-lengths, weight in elephants, and much more.

I use Google for all the options. You can type **32 C in F** in Google and get the answer back immediately. (Google can calculate *1.2 euro per liter in dollars per gallon*, in one step — way beyond the Windows 10 Calculator.) Do a Google search for *mileage*, *lease payment*, or *amortization*, and you can find hundreds of sites with far more capable calculators.

A couple calculator tricks:

>> Nope, an X on the keyboard doesn't translate into the times sign. I don't know why, but computer people have had a hang-up about this for decades. If you want times, you must tap the asterisk on the calculator or press the asterisk key (*) or Shift+8.

>> You can use the number pad, if your keyboard has one, but to make it work, you have to get Num Lock going. Try typing a few numbers on your number pad. If the calculator sits there and doesn't realize that you're trying to type into it, press the Num Lock key. The calculator should take the hint.

Painting

If you've ever used the old Windows Paint program, be prepared to unlearn everything you knew. Or thought you knew. Microsoft's totally new Paint 3D program isn't anything like the venerable (since Windows 1.0!) Paint. Whereas Paint was a simple drawing stickman-worthy program, Paint3D lets you manipulate pictures — in 2D if you like, or 3D if you prefer — and work with other people to create worthwhile figures.

To crank up Paint 3D, click the Start button, scroll down to Paint 3D, and click its icon. You see something like the spread in Figure 6-10.

You can run through the tutorials (Start and New in Paint 3D) or just click New to dive in. What you see (Figure 6-11) is a clunky, old-fashioned 2D plane called the canvas.

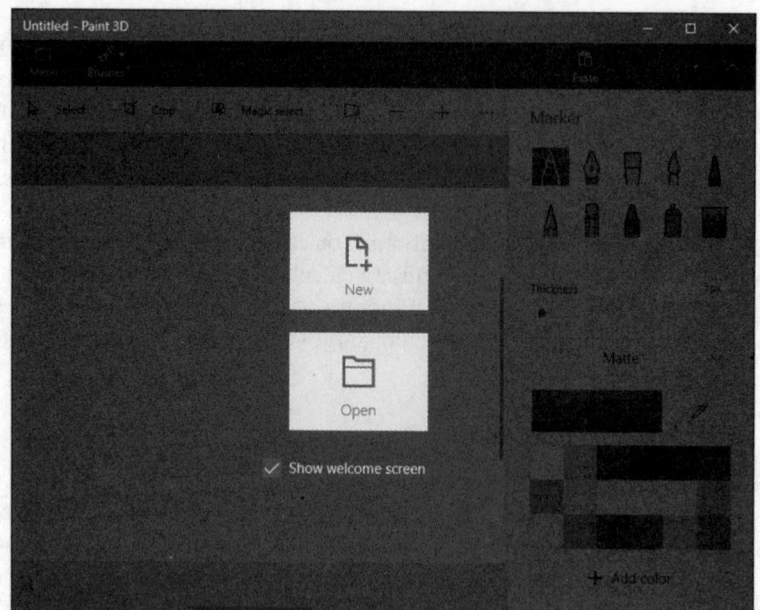

FIGURE 6-10:
The start of
something 3D.

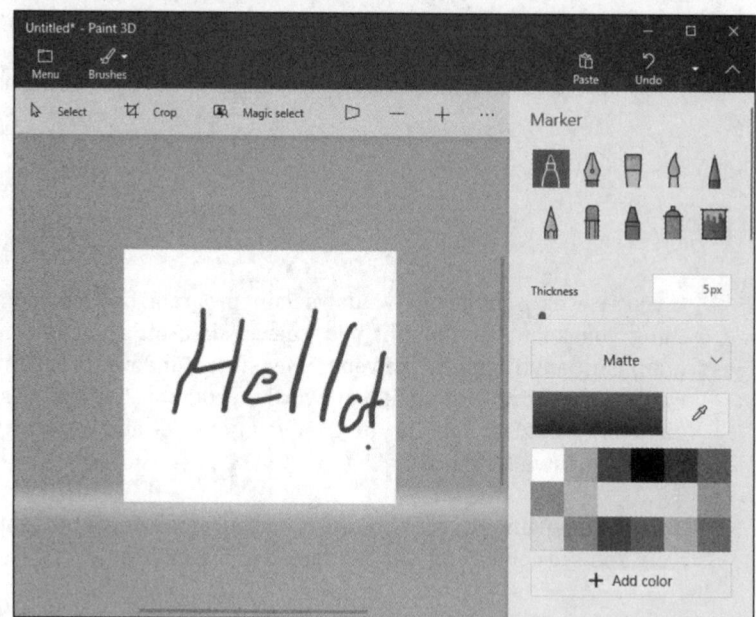

FIGURE 6-11:
The canvas.

You can draw on the canvas, as I did, just as you would in the old version of Paint, but that's a pretty boring start. So 1980s.

Click the cube (3D shapes) icon at the top to bring up the palette of 3D shapes, and then click one of the 3D shapes on the right. In Figure 6-12, I chose the donut shape, and then drew on the canvas.

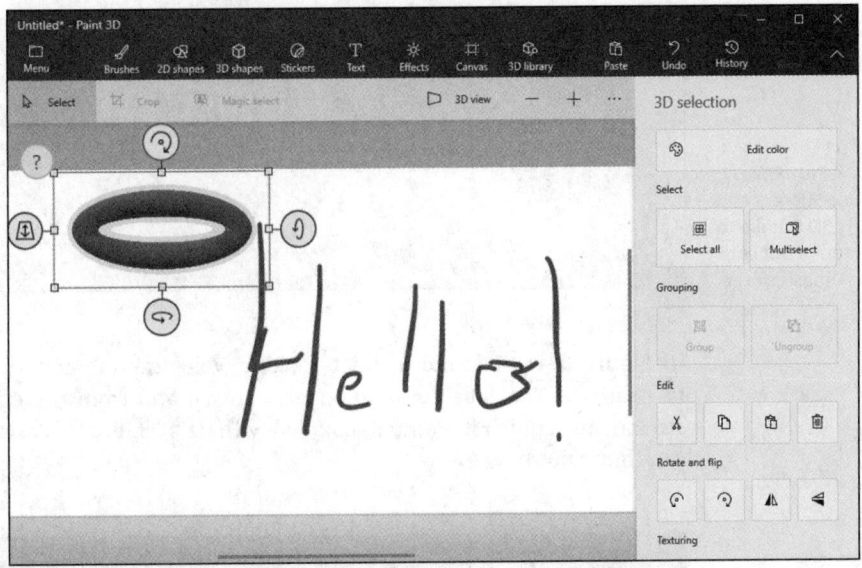

FIGURE 6-12: The 3D item grows dragging handles — in three dimensions.

ASK WOODY.COM

I'm lucky to get text lined up in two dimensions in the old Paint program, so I was relieved to see that Ctrl+Z undoes my steps (mistakes!) incrementally.

Two more features worth trying on a lazy summer afternoon:

>> **Stickers** (click the icon to the left of the T at the top) let you put stickers on the canvas or on one of your 3D objects. In Figure 6-13, I put a pair of eyeglasses on my work of art.

>> **3D Doodle** lets you draw free-form 2D objects, and Paint 3D extrudes them into 3D figures. In Figure 6-13, I drew a fish and Paint 3D made it fully rotatable.

If you aren't adept at 3D drawing — or if, like me, you can hardly draw a 2D blob — have no fear. Paint 3D comes with a huge amount of clipart. To open the floodgates, click the 3D Library icon that looks like the upper-left corner of a jigsaw puzzle. The Paint 3D library springs to life, and you can search for anything you can think of.

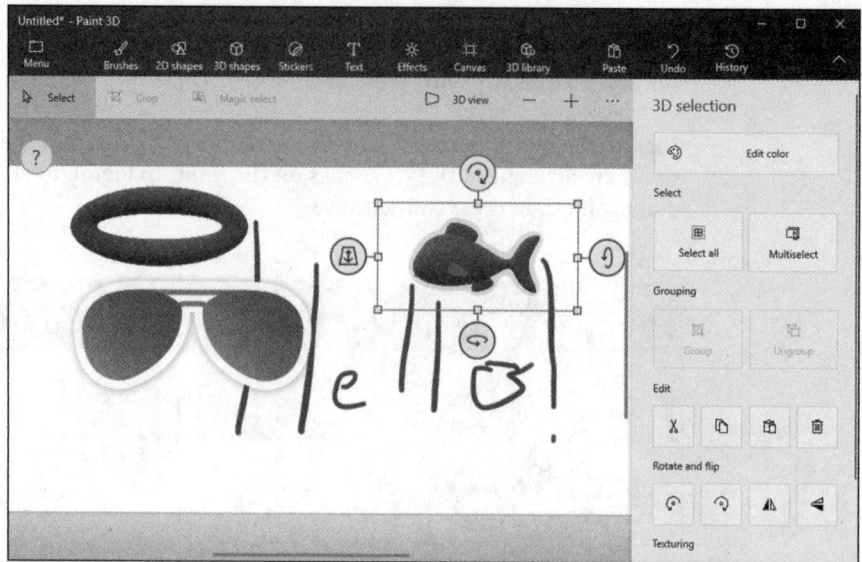

FIGURE 6-13:
Use Stickers and
3D Doodle to
round out your
masterpiece.

In Figure 6-14, I found a great-looking Velociraptor, and clicked it to add it to Paint 3D. That put the dino on my canvas, and from there I can twist, turn, stretch, and squish it. Paint 3D coupled with its 3D Library is astonishing — if you can figure out how to use it.

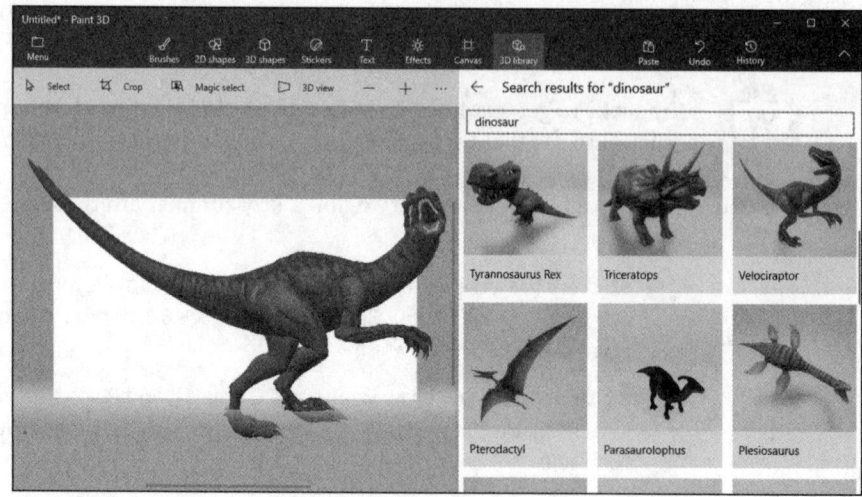

FIGURE 6-14:
The 3D library
contains an
enormous
collection of
high-quality
clipart.

Chapter **7**

Working with Printers

Ah, the paperless office. What a wonderful concept! No more file cabinets bulging with misfiled flotsam. No more hernias from hauling cartons of copy paper, dumping the sheets 500 at a time into a thankless plastic maw. No more trees dying in agony, relinquishing their last gasps to provide pulp as a substrate for heat-fused carbon toner. No more coffee-stained reports. No more paper cuts.

No more . . . oh, who the heck am I trying to kid? No way.

ASK
WOODY.COM

Industry prognosticators have been telling people for more than two decades that the paperless office is right around the corner. Yeah, sure. Maybe around *your* corner. Around *my* corner, I predict that PC printers will disappear about the same time as the last *Star Trek* sequel. We're talking geologic time here, folks. We're slowly getting rid of them, but like fax machines, they're not going to vanish with the next version of Windows.

The biggest problem? Finding a printer that doesn't cost two arms and three legs to, uh, print. Toner cartridges cost a fortune. Ink costs two fortunes. That bargain-basement printer you can get for $65 will probably print, oh, about ten pages before it starts begging for a refill. And four or five refills can easily cost as much as the printer.

Gillette may have originated the razor-and-blades business model, but it took the likes of HP, Brother, Canon, and Samsung to perfect it. Thank heavens Gillette hasn't figured out a way to put a microchip in the blades to guarantee their obsolescence.

TIP

There has been one important — even exciting — development in the laser/inkjet printer arena during the past ten years. Network connected printers — ones that attach to a network router, either through a wire or a Wi-Fi connection, bypassing PCs entirely — are finally affordable. Relatively. In my experience, anyway, printers attached to and used by one PC work best. Failing a one-to-one correspondence, network-attached printers have far fewer problems than the ones that are tethered to a specific machine on a network.

And 3D printers? Whoa, Nelly! They're here — and from what I've seen, they hook up just as easily as laser printers. Paying for them and running them is another story, of course.

And because you're here to learn about printers, you should know that Windows 10 has excellent printer support. It's easy after you grasp a few basic skills.

Installing a Printer

You have three ways to make a printer available to your computer:

>> Attach it directly to the computer.

>> Connect your computer to a network and attach the printer to another computer on the same network.

>> If the printer can attach directly to a network, connect your computer to a network and attach the printer directly to the network's hub, either with a network cable or via a Wi-Fi connection.

Having used all three attachment methods for many years, I can tell you without reservation that, if you have a home network, it's worth an extra $20 or $40 or more to get a Wi-Fi-connected printer.

REMEMBER

Connecting a computer directly to a network hub isn't difficult, if you have the right hardware. Each printer controller is different, though, so you have to follow the manufacturer's instructions.

Although choosing a new printer is beyond the scope of this book, you can find free tips — inkjet or laser, basic or multifunction? — at www.dummies.com.

Attaching a local printer

So you have a new printer and you want to use it. Attaching it *locally* — which is to say, plugging it directly into your PC — is the simplest way to install a printer, and it's the only option if you don't have a network.

All modern printers that connect to a PC have a USB connector that plugs in to your computer. (Network-attached printers work differently; see the next section.) In theory, you plug the connector into your PC's USB port and turn on the printer, and then Windows 10 recognizes it and installs the appropriate drivers. You're done.

If you're watching the desktop while Windows 10 is doing its thing, you see an icon flashing. If you're curious, click the flashing icon, and you see something like Figure 7-1.

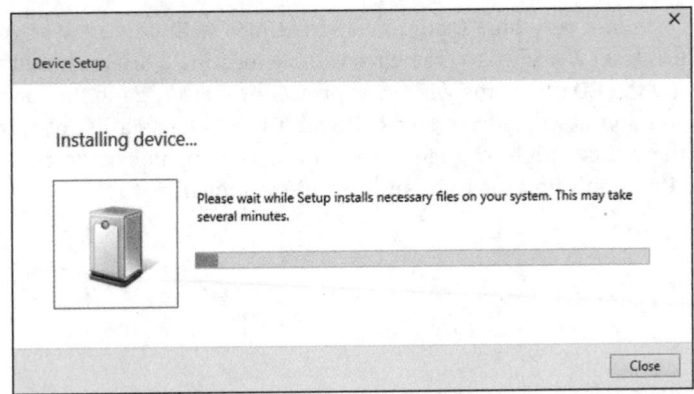

FIGURE 7-1: Let Windows 10 do all the work.

Working with Printers

I *don't* recommend that you install the manufacturer's software right off the bat, no matter what the instructions in the box with the printer may say. Most printers come with a CD loaded with . . . junk. Far better is to use the standard Windows drivers — in other words, just plug the thing in and print away — and resort to the manufacturer's CD only if it absolutely, definitely has something you need.

When the printer is installed properly, you can see the printer in your Devices list. To see your devices, click or tap the Start button and then the Settings icon, and then click Devices. You see a list similar to the one in Figure 7-2.

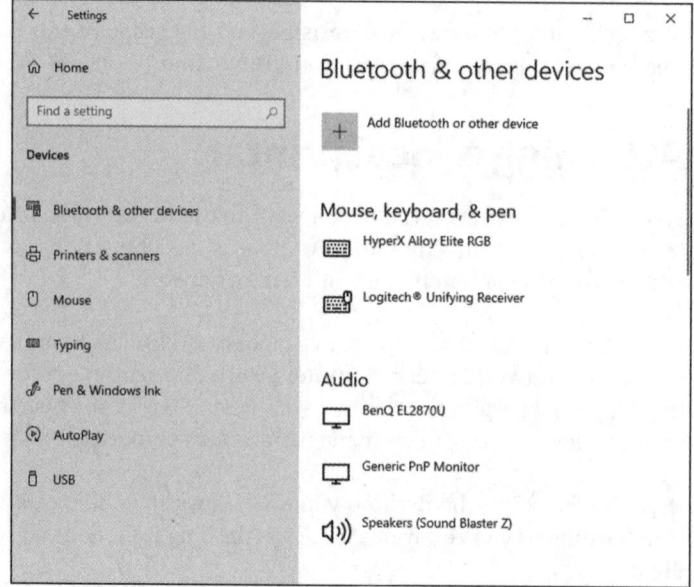

FIGURE 7-2:
Even my mechanical keyboard, HyperX Alloy Elite RGB is recognized as a device.

Once in a very blue moon, and sometimes with very new or very old models of printers, Windows 10 may have trouble locating a driver. If that happens, you can use the CD that came with your printer or, better, go to the manufacturer's website and download the latest driver. Table 7-1 has a list of websites. (Note that these links might change in the future as companies revamp or reorganize their sites. Can't find the right site? Google is your friend.)

TABLE 7-1 **Driver Sites for Major Printer Manufacturers**

Manufacturer	Find Drivers at This URL
Brother	www.brother-usa.com/brother-support
Canon	http://usa.canon.com/cusa/consumer/standard_display/support
Dell	www.dell.com/support/home/en-us//products
Epson	http://epson.com/cgi-bin/Store/support/SupportIndex.jsp
HP	http://support.hp.com/us-en/drivers/
Samsung	www.samsung.com/us/support/downloads

Connecting a network printer

Windows networks work wonders. When they work. Say that ten times real fast.

If you have a network, you can attach a printer to (almost) any computer on the network and have it accessible to all users on (almost) all computers in the network. You can also attach different printers to different computers and let network users pick and choose the printer they want to use as the need arises.

If you have printers attached to your network — for example, you may have a printer on a Windows 7 or 8/8.1 or Windows 10 machine that isn't set up to share devices — you can add it to your collection of shared printers. Here's how:

1. **Click the Start button and then the Settings icon. Choose Devices.**

2. **On the left, choose Printers & Scanners.**

   The Printer list appears, as shown in Figure 7-3.

3. **At the top, tap or click the Add a Printer or Scanner button.**

   Windows 10 looks all through your network to see whether any printers are available, and displays any printers that are turned on.

4. **Tap or click the name of your printer and then tap or click Add Device.**

   Windows 10 looks to see whether it has a driver handy for that particular printer. It whirs and clanks for a while and then tells you that you've successfully added the printer.

5. **To set the newly added printer as your default, scroll down and uncheck Let Windows Manage My Default Printer.**

6. **Click the name of your printer, and then click Manage.**

   Windows 10 displays several links and buttons for configuring your printer, as shown in Figure 7-4.

7. **Click the Set As Default button, and then close Settings.**

Your new printer appears in the Printers & Scanners list, and is set as the default.

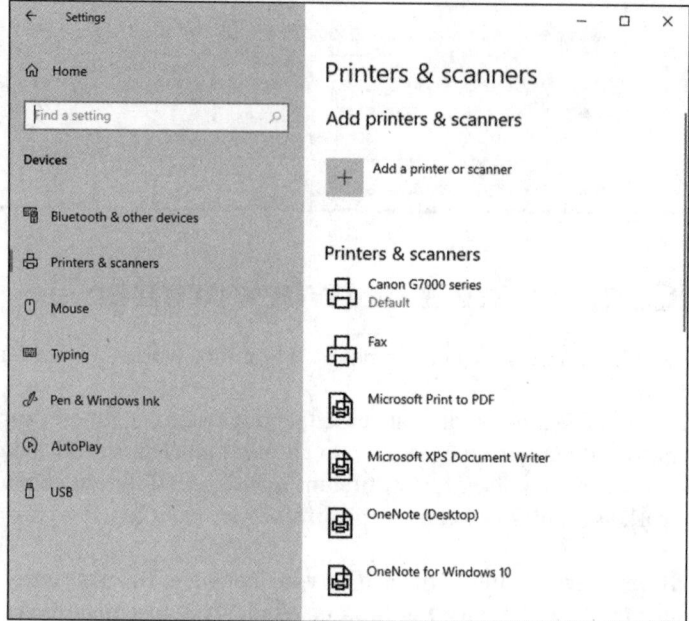

FIGURE 7-3:
All the printers accessible to this machine — most of which aren't printers but can work like printers.

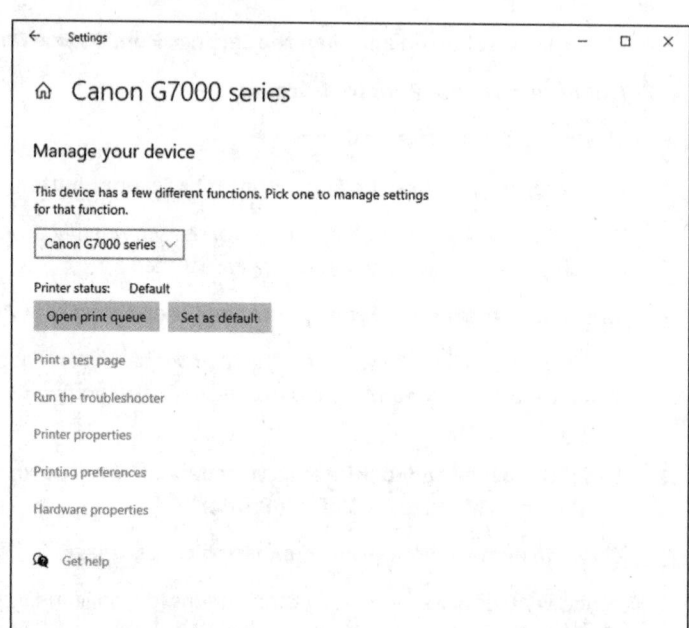

FIGURE 7-4:
Setting the new printer as the default.

Using the Print Queue

You may have noticed that when you print a document from an application, the application reports that it's finished before the printer finishes printing. If the document is long enough, you can print several more documents from one or more applications while the printer works on the first one. This is possible because Windows 10 saves printed documents in a *print queue* until it can print them.

If more than one printer is installed on your computer or network, each one has its own print queue. The queue is maintained on the *host PC* — that is, the PC to which the printer is attached.

If you have a network-attached printer, the printer itself maintains a print queue.

ASK WOODY.COM

Windows 10 uses print queues automatically, so you don't even have to know that they exist. If you know the tricks though, you can control them in several useful ways.

Displaying a print queue

You can display information about any documents that you currently have in a printer's queue by following these steps:

1. **Bring up the Control Panel by typing** Control **in the Windows 10 search box, near the Start button, and choosing Control Panel.**

2. **Under the Hardware and Sound category, click View Devices and Printers.**

   You see the list of devices like the one shown in Figure 7-5. Looks better than the modern-looking list in Figure 7-4, doesn't it?

3. **Click or tap the printer's name and then click or tap See What's Printing, in the bar on the top.**

   The print queue appears, as shown in the lower right of Figure 7-6. If you have documents waiting for more than one printer, you get more than one print queue report.

4. **To cancel a document, tap and hold down or right-click the document you want to cancel; choose Cancel.**

   In many cases, Windows 10 must notify the printer that it's canceling the document, so you may have to wait awhile for a response.

   The Owner column tells you which user put the document in the print queue. The jobs in the print queue are listed from the oldest at the top to the newest at the bottom. The Status column shows which job is printing.

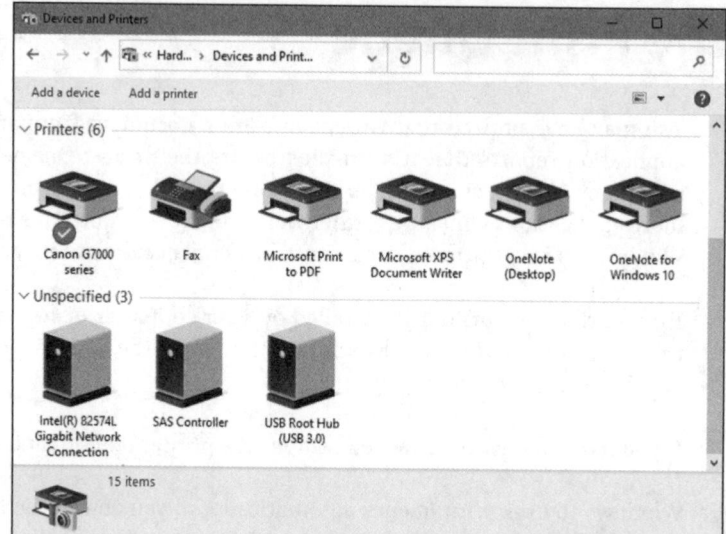

FIGURE 7-5:
A typical Devices
and Printers
listing.

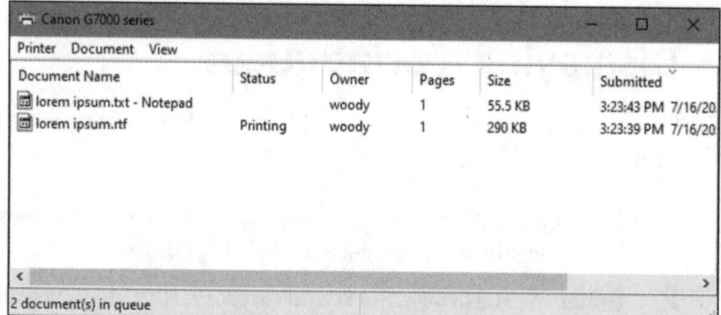

FIGURE 7-6:
All the documents
you have waiting
to print display in
the queue.

5. **Keep the print queue window open for later use, or minimize the print queue window and keep it on the taskbar.**

Keeping it open can be quite handy if you're running a particularly long or complex print job; Word mail merges are particularly notorious for requiring close supervision.

TIP

Pausing and resuming a print queue

When you *pause* a print queue, Windows 10 stops printing documents from it. If a document is printing when you pause the queue, Windows tries to finish printing the document and then stops. When you *resume* a print queue, Windows starts printing documents from the queue again. Follow these guidelines to pause and resume a print queue:

>> **To pause a print queue,** when you're looking at the print queue window (refer to Figure 7-6), choose Printer, Pause Printing.

>> **To resume the print queue,** choose the same command again. The check mark in front of the Pause Printing line disappears, and the printer resumes.

TIP

Why would you want to pause the print queue? Say you want to print a page for later reference, but you don't want to bother turning on your printer to print just one page. Pause the printer's queue, and then print the page. The next time you turn on the printer, resume the queue, and the page prints.

Sometimes, Windows 10 has a hard time finishing the document — for example, you may be dealing with print buffer overruns (see the "Troubleshooting Printing" section, later in this chapter) — and every time you clear the printer, it may try to reprint the overrun pages. If that happens to you, pause the print queue and turn off the printer. As soon as the printer comes back online, Windows is smart enough to pick up where it left off.

Also, depending on how your network is set up, you may or may not be able to pause and resume a print queue on a printer attached to another user's computer or to a network.

Pausing, restarting, and resuming a document

If you've followed along so far, here are some other reasons you may want to pause a document. Consider the following:

>> Suppose you're printing a web page that documents an online order you just placed, and the printer jams. You've already finished entering the order, and you have no way to display the page again to reprint it. Pause the document, clear the printer, and restart the document.

>> Here's another common situation where pausing comes in handy. You're printing a long document, and the phone rings. To make the printer be quiet while you talk, pause the document. When you're finished talking, resume printing the document.

Here's how pausing, restarting, and resuming work:

>> **Pause a document:** When you pause a document, Windows 10 is prevented from printing that document. It skips the document and prints later documents in the queue. If you pause a document while Windows is printing it, Windows halts in the middle of the document and prints nothing on that printer until you take further action.

>> **Restart a document:** When you restart a document, Windows 10 is again allowed to print it. If the document is at the top of the queue, Windows prints it as soon as it finishes the document that it's now printing. If the document was being printed when it was paused, Windows stops printing it and starts again at the beginning.

>> **Resume a document:** Resuming a document is meaningful only if you paused it while Windows was printing it. When you resume a document, Windows resumes printing it where it paused.

REMEMBER

To pause a document, right-click the document in the print queue, or tap and hold down, and choose Pause. The window shows the document's status as Paused. To resume or restart the paused document, right-click or tap and hold down that document, and choose Resume.

Canceling a document

When you *cancel* a document, Windows removes it from the print queue without printing it. You may have heard computer jocks use the term *purged* or *zapped* or something totally unprintable.

TIP

Here's a common situation when document canceling comes in handy. You start printing a long document, and as soon as the first page comes out, you realize that you forgot to set the heading. What to do? Cancel the document, change the heading, and print the document again.

To cancel a document, select that document. In the print queue window, choose Document, Cancel. Or tap and hold down, or right-click the document in the print queue window and choose Cancel. You can also select the document and press Delete.

REMEMBER

When a document is gone, it's gone. No Recycle Bin exists for the print queue.

Conversely, most printers have built-in memory that stores pages while they're being printed. Network attached printers can have sizable buffers. You may go to the print queue to look for a document, only to discover that it isn't there. If the document has already been shuffled off to the printer's internal memory, the only way to cancel it is to turn off the printer.

Troubleshooting Printing

The following list describes some typical problems with printers and the solutions to those sticky spots:

» **I'm trying to install a printer. I connected it to my computer, and Windows doesn't detect its presence.** Be sure that the printer is turned on and that the cable from the printer to your computer is properly connected at both ends. Check the printer's manual; you may have to follow a procedure (such as push a button) to make the printer ready for use.

» **I'm trying to install a printer that's connected to another computer on my network, and Windows doesn't detect its presence.** I know that the printer is okay; it's already installed and working as a local printer on that system! If the printer is attached to a Windows 7 or 8/8.1 PC, the PC may be set to treat the network as a public network — in which case, it doesn't share anything. To rectify the problem, right-click the printer and choose Sharing. (For details, see *Windows 7 All-in-One For Dummies, Windows 8 All-in-One For Dummies,* or *Windows 8.1 All-in-One For Dummies,* all by yours truly and published by John Wiley & Sons.)

» **I can't use a shared printer that I've used successfully in the past.** Windows 10 says that it isn't available when I try to use it, or Windows doesn't even show it as an installed printer anymore. This situation can happen if something interferes with your connection to the network or the connection to the printer's host computer. It can also happen if something interferes with the availability of the printer — for example, if the host computer's user has turned off sharing.

If you can't find a problem or if you find and correct a problem (such as file and printer sharing being turned off), but you still can't use the printer, try restarting Windows on your own system. If that doesn't help, remove the printer from your system and reinstall it.

To remove the printer from your system, click the Start button and then the Settings icon. Choose the Devices icon, and on the left, choose Printers & Scanners. On the right, click or tap the name of the device. A Remove Device button appears. Click it. Windows 10 asks whether you're sure you want to remove this printer. Tap or click the Yes button.

To reinstall the printer on your system, use the same procedure you used to install it originally. (See the "Connecting a network printer" section, earlier in this chapter.)

» I printed a document, but it never came out of the printer. Check the printer's print queue on the host PC (the one directly attached to the printer), if it's attached to just one PC, or the print queue on any attached PC if it's a network printer. Is the document there? If not, investigate several possible reasons:

WARNING

- *The printer isn't turned on, or it's out of paper.* Hey, don't laugh. I've done it. In some cases, Windows 10 can't distinguish a printer that's connected but not turned on from a printer that's ready, and it sends documents to a printer that isn't operating.

- *You accidentally sent the document to some other printer.* Hey, don't laugh — you've heard that one.

- *Someone else unintentionally picked up your document and walked off with it.*

ASK WOODY.COM

- *The printer is turned on but not ready to print, and the printer (as opposed to the host PC) is holding your whole document in its internal memory until it can start printing.* A printer can hold as much as several hundreds — even thousands — of pages of output internally, depending on the size of its internal memory and the complexity of the pages. Network attached printers frequently have 16MB or more of dedicated buffer memory, which is enough for a hundred or more pages of lightly formatted text.

If your document is in the print queue but isn't printing, check for these problems:

- *The printer may not be ready to print.* See whether it's plugged in, turned on, and properly connected to your computer or its host computer.

- *Your document may be paused.*

- *The print queue itself may be paused.*

- *The printer may be printing another document that's paused.*

- *The printer may be thinking.* If it's a laser printer or another type of printer that composes an entire page in internal memory *before* it starts to print, it appears to do nothing while it processes photographs or other complex graphics. Processing may take as long as several minutes.

  Look at the printer, and study its manual. The printer may have a blinking light or a status display that tells you it's doing something. As you become familiar with the printer, you develop a feel for how long various types of jobs should take.

- *The printer is offline, out of paper, jammed, or unready to print for some other reason.*

Catching a Runaway Printer

This topic must be the most common, most frustrating problem in printer-dumb.

REMEMBER

You print a document, and as it starts to come out the printer, you realize that you're printing a zillion pages you don't want. How do you stop the printer and reset it so that it doesn't try to print the same bad stuff, all over again?

Here's what you do:

1. **Turn off the printer. Pull the paper out of the printer's paper feeder.**

   Be careful with this step to avoid roller damage or paper tears that cause problems later.

   This step stops the immediate problem, uh, immediately.

2. **On the desktop, in the lower-right corner, look among the notification icons for one that looks like a printer; tap and hold down on it or double-click it.**

   The print queue appears (refer to Figure 7-6).

3. **Right-click (or tap and hold down) the runaway print job, and choose Cancel.**

   If this step deletes the bad print job, good for you.

4. **If it doesn't delete the bad print job, wait a minute and then turn off the printer and unplug it from the wall. (Really.) Reboot Windows 10. When Windows comes back, wait another minute, plug the printer back in, and turn the printer back on.**

   Your bad job is banished forever.

Maintaining Windows 10

Contents at a Glance

Chapter **1**

File History, Backup, Data Restore, and Sync

I f you're accustomed to using earlier versions of Windows to back up or restore data, to ghost a whole drive, or to set restore points, you're probably in this chapter looking for something that no longer exists.

REMEMBER

Although you can set manual restore points — much the same process as it was in Windows ME, many moons ago — the way to do so is buried deep inside Windows 10, and frankly, your need for them is highly debatable.

Microsoft has, in one stroke, made backup and restore much simpler and much less controllable. Or perhaps I should say *micro-manageable.*

In this chapter, I talk about how to back up your data: running backups, restoring them, being smart about where to store them, and accessing them if something goes wrong. (In Chapter 2 of this minibook, I talk about Refresh and Reset, two ways of bringing Windows 10 back to life. Refresh keeps all your data. Reset wipes out everything and returns your PC to its out-of-the-box state.)

ASK WOODY.COM

In this chapter and Chapter 2 of this minibook, I don't talk much about old Windows topics that just don't apply anymore. These include system repair discs, restore points, image backups, recovery mode, and safe mode. You can find vestiges of those features in Windows 10 if you look hard enough. But they aren't recommended anymore — and they're rarely supported.

What Happened to the Windows 7 Backup?

If you're an experienced Windows 7 user, you may be looking for specific features that have been renamed, morphed, or axed in the current version of Windows. Here's a little pocket dictionary to help you figure out the landmarks:

TECHNICAL STUFF

» **Shadow Copies (or Previous Versions) of files are now called File History.** It's functionally similar to the Apple Time Machine — just not as cool, visually.

» **Image Backup (or System Image or Ghosting) is buried deep.** If you really want to use Windows 10 to create a full disk image, tap or click the Start button, the Settings icon, and Update & Security. On the left, choose Backup, then on the right, click the link to Go to Backup and Restore (Windows 7). On the left, click Create a System Image, and go from there.

» **Windows Backup and the Backup and Restore Center are there, but they're hard to find.** They were in Windows 8, got tossed out of Windows 8.1, and now they're back in Windows 10. Click the Start button, the Settings icon, Update & Security, and Backup (on the left). Click the link to Go to Backup and Restore (Windows 7). Most of the time it's much smarter to use File History anyway, but if you're nostalgic — or you don't want to learn new tricks — the old way still works.

» **You can boot into safe mode if you really want to, but Microsoft makes it very difficult to get there.** Follow the instructions in Book 8, Chapter 2 to get into the Windows Recovery Environment.

ASK WOODY.COM

Microsoft is *deprecating* (killing, zapping) all the old backup, restore, system restore, and safe mode options, in favor of completely new (and much easier-to-use) backup and restore options.

All the while, the subtle push is there to store everything in OneDrive, so Microsoft can take care of backing up and restoring.

The Future of Reliable Storage Is in the Cloud

Microsoft wants you to put your data in the cloud.

It's more than a question of letting you shoot yourself in the foot. Microsoft has turned into a big-time fan of cloud storage. Cloud everything, for that matter. New features in Windows 10 are designed to make it easier for you to put your data in the cloud — preferably Microsoft's cloud, OneDrive, of course. Yes, part of the motivation is to get you to pay for cloud storage, or at least lock you into Microsoft's cloud offerings. But a big part of the reason for steering you to cloud storage is that it's better. That, in turn, translates into fewer support headaches.

ASK
WOODY.COM

Yes, you read that right. I'm telling you that cloud storage is better than local storage, for most people in most situations. One of the big reasons why: backup. You don't have to sweat backup when your data is in the cloud. I admit that there are rare examples of people who have lost data saved on one of the cloud storage systems — OneDrive, Google Drive, iCloud, Dropbox, Box, and so on.

REPLICATING WINDOWS HOME SERVER BACKUP

Some people used Windows Home Server (WHS). I loved it. Even wrote a book about it, *Windows Home Server For Dummies*. But Microsoft has given up on WHS, and it isn't coming back. (The guy who led the WHS effort — Charlie Kindel — has since gone to Amazon, where he led the Alexa team.)

With Windows 10 though, I'm not going to miss WHS too much. The absolutely best feature in WHS was its capability to back up data on connected PCs and keep redundant copies of the data. That way, any drive on the server or on my PC could fail, and all it took was a new drive and an hour or two to restore all my data as if nothing had happened.

And you can do all that and more by combining Storage Spaces (see Book 7, Chapter 4) with File History (see the nearby section "Backing Up and Restoring Files with File History").

WHS has one more significant feature that isn't replicable in Windows 10: It backs up not only your data but also the entire contents of all your drives. When I was using WHS, if my C: drive decided to crack into a million pieces, restoring it was quite simple. Without WHS, but using Windows 10 and File History, I can restore everything except my desktop applications.

As for Alexa versus Cortana — now that's a horse of a completely different color. Go, Charlie!

Contrast that with local storage. If you've been using computers for any length of time at all, chances are good that you've lost some data. If you know ten people who store data on their own PCs, I'd guess that ten of them have lost data.

So scoff at cloud storage if you like. Worry about the privacy problems. (Microsoft says it doesn't look into your files, although it does scan your photos.) Fret over maintaining an Internet connection. But contrast that with the possibility — no, the likelihood — that you'll lose data by managing it yourself. No contest, from my point of view.

I cover cloud storage later in this book, in the section "Storing to and through the Cloud."

Backing Up and Restoring Files with File History

ASK
WOODY.COM

Windows 10's File History backs up not only your data files but also many versions of your data files and makes it easy to retrieve the latest version and multiple earlier versions.

By default, File History takes snapshots of all the files in your libraries (see Book 7, Chapter 3), your desktop, your Contacts data, and your browser (Microsoft Edge, Internet Explorer) favorites. It can also take snapshots of OneDrive, if you set your user folders (Documents, Pictures, and so on) to use OneDrive as their default location. The snapshots get taken once an hour and are kept until your backup drive runs out of space.

You can change those defaults. I explain how later in this section.

Setting up File History

To use File History, Windows 10 demands that you have an external hard drive, a second hard drive, or a network connection that leads to a hard drive. In this example, I connect to an external USB drive that is connected to my PC. You can also use a cheap external hard drive, which you can pick up at any computer store, or use a hard drive on another computer on your network.

REMEMBER

If you have lots of photos in your Photos library or a zillion songs in the Music library, the first File History backup takes hours and hours (or longer!). If you have lots of data and this is your first time, don't even try to set up things until you're ready to leave the machine for a long, long time.

If you haven't yet set up your libraries, made them visible in File Explorer, and put the Public folders inside your libraries, mosey over to Book 7, Chapter 3 and bring back the library stuff Microsoft knocked out.

To get the desktop version of File History going, follow these steps:

1. **Click the Start button and then the Settings icon, and then go to Update & Security. On the left, choose Backup.**

   The File History appears in the Settings app, as shown in Figure 1-1. If you don't have a drive set up for File History, you can't turn on this feature.

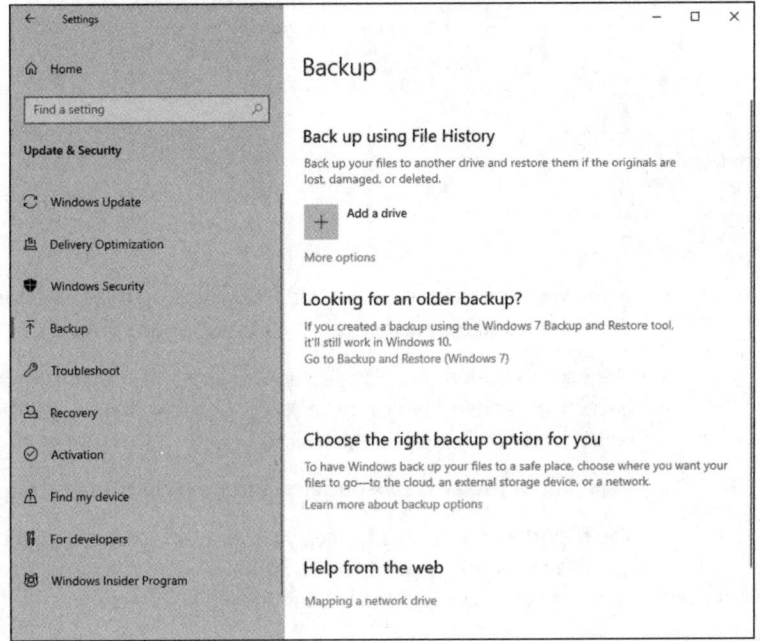

FIGURE 1-1:
File History is in
the Settings app.

2. **Attach your external drive and, after Windows 10 detects it, click or tap the +Add a Drive button.**

3. **Select the drive you want to use for File History.**

   File History is turned on, and the Automatically Back Up My Files switch is turned on, as shown in Figure 1-2. File History goes out to lunch for a long time. Possibly a very long time. It gathers everything in your user folders and libraries (see Book 7, Chapter 3) and on your desktop, all your Contacts, and your Internet Explorer and Microsoft Edge favorites.

   You can go back to work, or grab a latte or three. Go home. Take a nap. If you have lots of pictures in your library, you may want to consider rereading *War and Peace*.

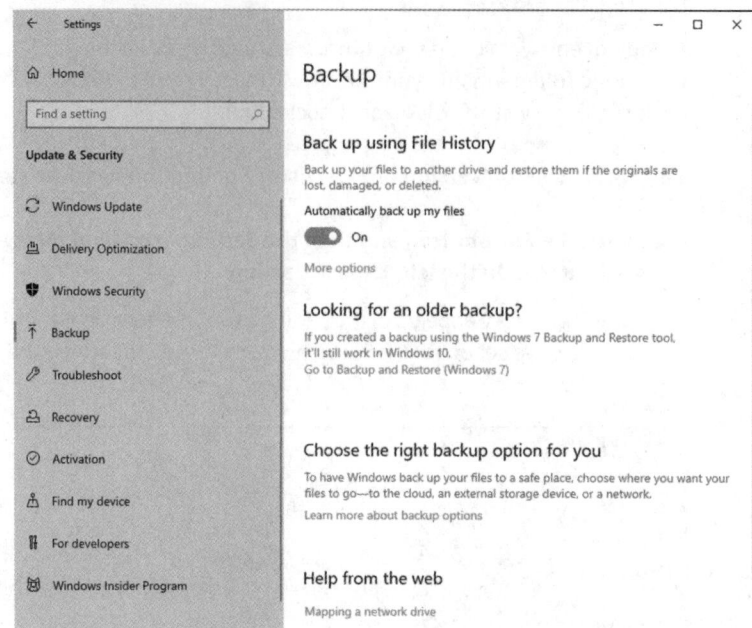

FIGURE 1-2:
With an external
drive connected,
it's time to turn
on File History.

4. **If you want to customize the folders backed up by File History and how they're backed up, click or tap the More Options link.**

The Backup Options are displayed (see Figure 1-3), including the size of the backup, how often File History backs up your files, how long it keeps them, and which folders it includes in the backup.

5. **Personalize the available options, and then close the Settings app.**

You may want File History to keep your backups for a month, three, or until space is needed, instead of forever. Otherwise, your backup drive may get full very fast because it also stores your deleted files, forever.

TIP

In Backup Options (refer to Figure 1-3), you can also manually start a File History backup by clicking the Back Up Now button.

Instead of relying on the File History program to tell you that the backup occurred, take matters into your own hands, and look for the backup with File Explorer. To find the backup files with File Explorer, follow these steps:

1. **On the desktop, tap or click the File Explorer icon on the taskbar.**

File Explorer opens.

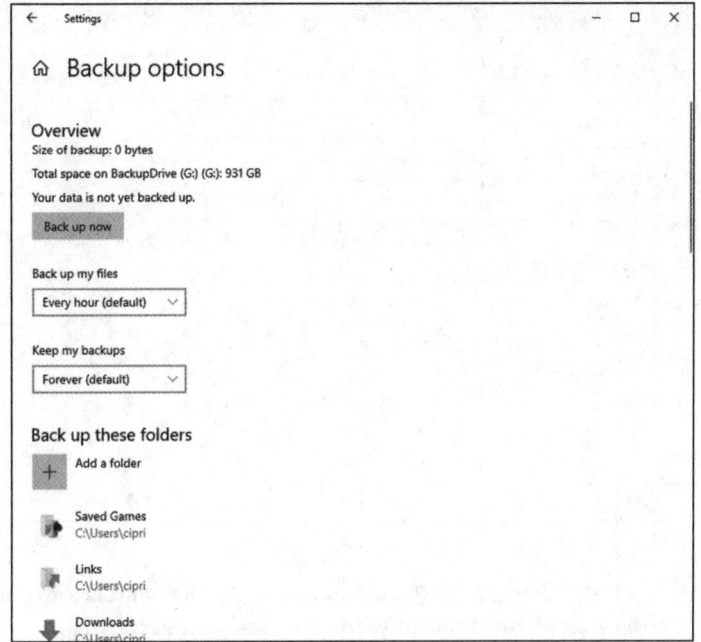

FIGURE 1-3:
Configure how
File History works
using the Backup
Options.

2. **Navigate to the drive that you just used in the preceding steps for a backup.**

   This may be an external or a networked drive; it may even be a second drive on your PC, although I don't recommend that.

3. **Tap or double-click your way through the folder hierarchy:**

   - File History

   - Your username

   - Your PC name

   - Data

   - The main drive you backed up (probably C :)

   - Users

   - Your username (again)

   - Desktop (assuming you had any files on your desktop that you backed up), or Pictures, or some other folder of interest

   A File Explorer screen like the one in Figure 1-4 appears.

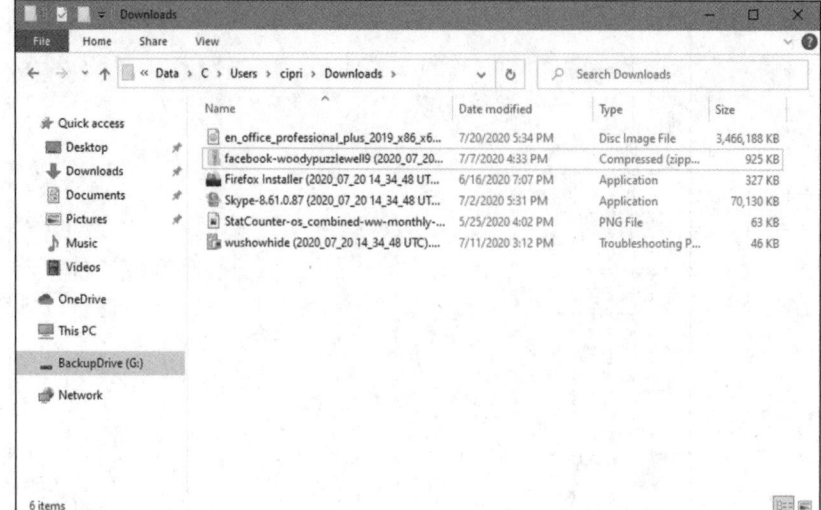

4. **Check whether the filenames match the files that are on your desktop, or in your Pictures folder, with dates and times attached.**

5. **Do one of the following:**

   - *If the files match, you can close File Explorer and close the File History dialog box.*

TIP

   Although you can restore data from this location via File Explorer, it's easier to use the File History retrieval tools. (See the next section for details.)

   - *If you don't see a list of filenames that mimics the files on your desktop, go back to Step 1 of the preceding Steps list and make sure you get File History set up right!*

REMEMBER

File History doesn't run if the backup drive gets disconnected or the network connection to the backup drive drops — but Windows 10 produces File History files anyway. As soon as the drive is reconnected or the network starts behaving, File History dumps all its data to the correct location.

Restoring data from File History

REMEMBER

File History stores snapshots of your files, taken every hour, unless you change the frequency. If you've been working on a spreadsheet for the past six hours and discovered that you blew it, you can retrieve a copy of the spreadsheet that's less than an hour old. If you've been working on your résumé over the past three months and decide that you really don't like the way your design changed five weeks ago, File History can help you there too.

TIP

If you're accustomed to the Windows 7 way of bringing back Shadow Copies, you need to unlearn everything you think you know about bringing back old files. Windows 10 works differently.

Here's how to bring back your files from cold storage:

1. **Bring up the Control Panel by typing** Control **in the Windows 10 search box, near the Start button, and choosing Control Panel.**

2. **Click or tap System and Security; then under File History, click or tap Restore Your Files with File History.**

   The File History Restore Home page, as shown in Figure 1-5, appears.

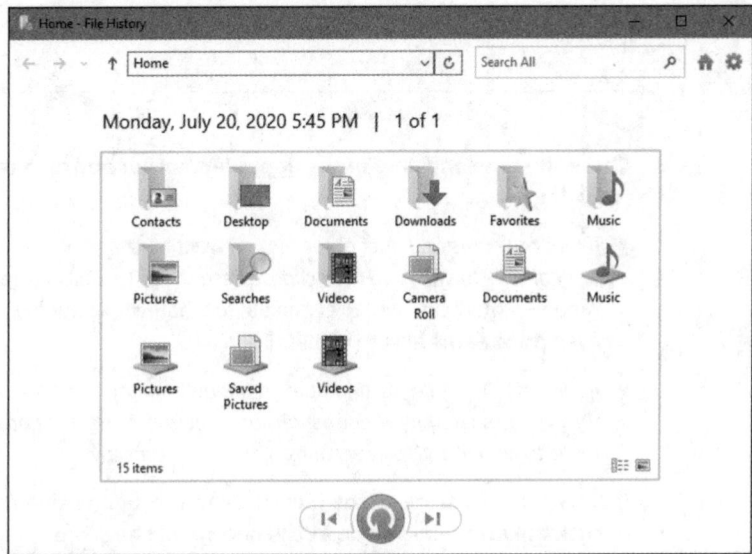

FIGURE 1-5:
You need to find the file you want to restore, starting at the top.

3. **Navigate to the location of the file you want to restore.**

   In Figure 1-6, I went to the Pictures library, where the file I want to resuscitate is stored.

TIP

You can use several familiar File Explorer navigation methods inside the File History program, including the up arrow to move up one level, the forward and back arrows, and the search box in the upper-right corner.

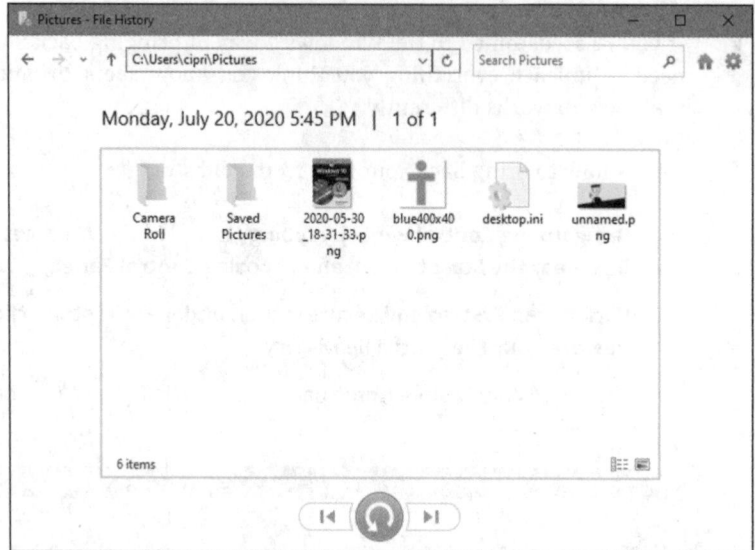

FIGURE 1-6:
First, find the
location. Then
find the correct
version.

4. **Check the time and date in the upper-left corner and do one of the following:**

   - *If that's the time and date of the file you want to bring back, tap and hold down or right-click the file, and then choose Restore.* Or (usually easier, if you have a mouse) simply click and drag the file to whatever location you like. You can even preview the file by double-clicking it.

   - *If this isn't the right time and date, at the bottom, tap or click the left arrow to take you back to the previous snapshot. Tap or click the left and right arrows to move to earlier and later versions of the files, respectively.*

5. **If you want to restore all the files you can see, at any given moment, tap or click the arrow-in-a-circle at the bottom of the screen.**

   I always restore by clicking and dragging. It's much easier to see exactly what's happening and avoid mistakes before they happen.

ASK
WOODY.COM

6. **Replace the files (which deletes the latest version of each file) or select which files you want to replace, as shown in Figure 1-7.**

If you accidentally replace a good file, be of good cheer. There was a snapshot of that file taken less than an hour ago. You just have to find it. Kinda cool how that works, eh? And that old copy stays around for a long time — years, if you have enough disk space, and your backup drive doesn't die.

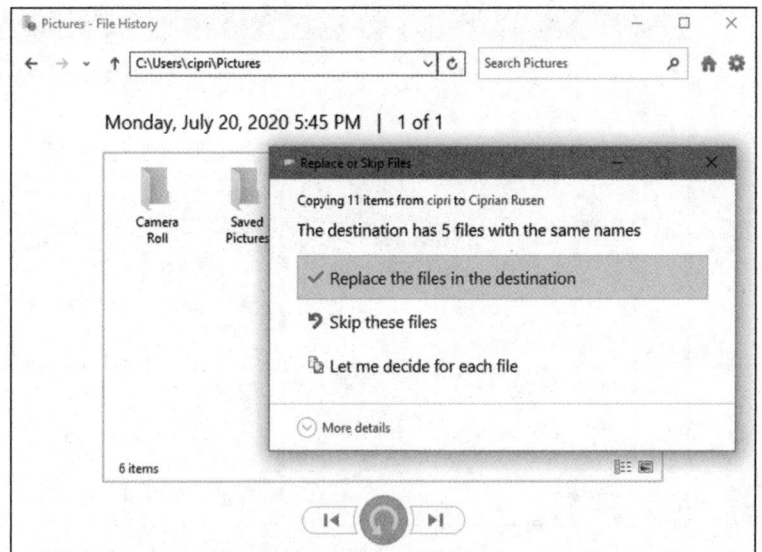

FIGURE 1-7:
You can restore an entire folder full of files all at once.

Changing File History settings

File History has several settings you may find valuable. You can find them in the Control Panel and the Settings app. The ones from the Settings app were covered earlier in this book. Those in the Control Panel are useful too, and it is time to get familiar with them.

REMEMBER

File History backs up *every file in every library* on your computer. If you have a folder that you want to have backed up, just put it in a library. Any library. Invent a new library if you want. You don't have to *use* the library; just put the folder in a library. File History takes care of all the details.

Microsoft makes it hard to find File History, just as they've made it hard to find libraries. Kind of goes hand in hand with trying to get you to use OneDrive.

TIP

I have an extensive discussion of libraries in Book 7, Chapter 3, but if you only want to stick an existing folder in a newly minted library, it's easy: Tap and hold down or right-click the folder, choose Include in Library, Create New Library, and give your new library a name. You're finished.

Here's how to change some other key settings:

1. **Bring up the Control Panel by typing** Control **in the Windows 10 search box, near the Start button, and choosing Control Panel. Click or tap System and Security, and then tap or click File History.**

The File History main page appears, as shown in Figure 1-8.

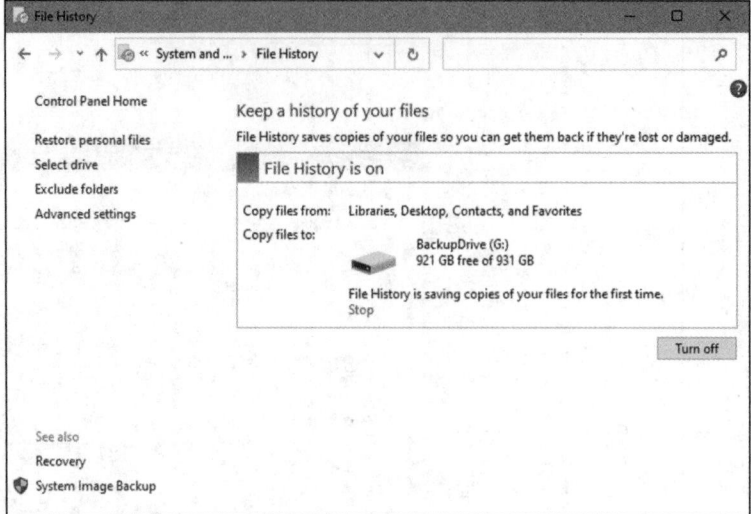

FIGURE 1-8:
File History
accessed from
the Control Panel.

2. **If you want to exclude some folders in your libraries so they don't get backed up:**

   a. *Choose Exclude Folders (on the left). File History opens a simple dialog box..*

   b. *Click the Add button and select a folder to put it in the exclude list.* For example, in Figure 1-9, I excluded a folder in the Documents library.

   c. *Repeat Step 2b until you have selected all the folders you want to exclude.*

   d. *Save your changes, and tap or click the back arrow to get back to the File History applet.*

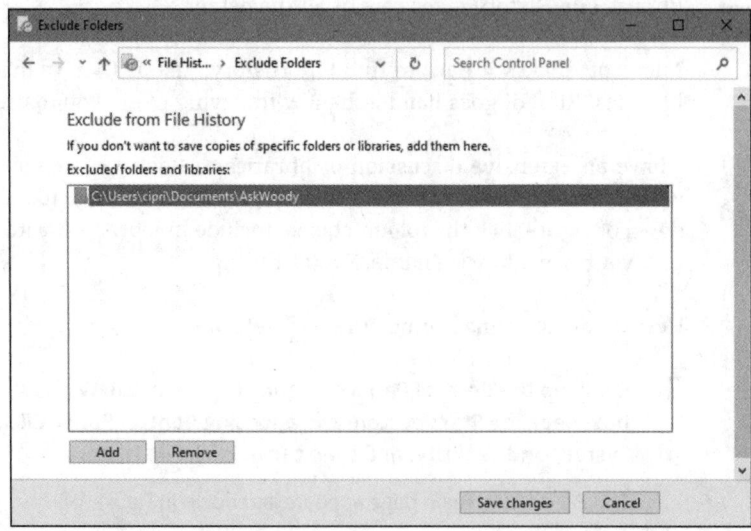

FIGURE 1-9:
Exclude individual
folders from File
History.

3. **To change how backups are made:**

   a. *Tap or click the Advanced Settings link (on the left).* The Advanced Settings dialog box in Figure 1-10 appears.

   b. *Change the frequency of backups and how long versions should be kept.* See my recommendations for these settings in Table 1-1.

4. **Tap or click Save Changes.**

   Your next File History backup follows the new rules.

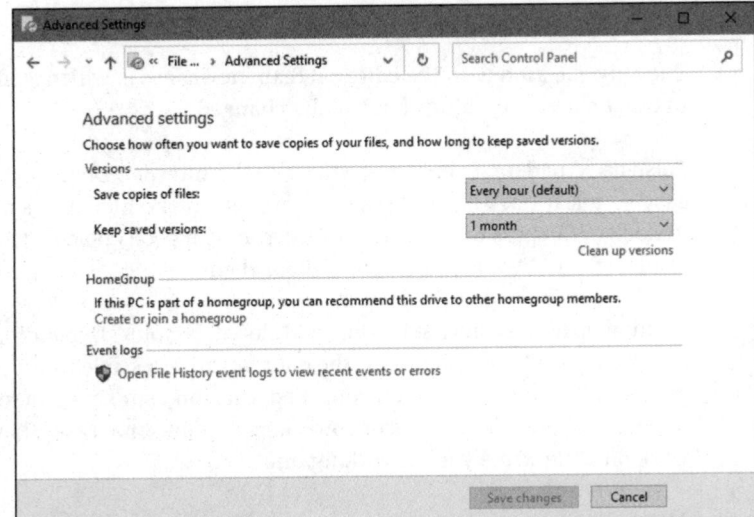

FIGURE 1-10:
Take control of
your backups
here.

TABLE 1-1 **File History Advanced Settings**

Setting	Recommendation	Why
Save Copies of Files	Every 30 Minutes	This is mostly a tradeoff between space (more frequent backups take a tiny bit of extra space) and time — your time. If you have lots of backups, you increase the likelihood of getting back a usable version of a file, but you have to wade through many more versions. I find 30 minutes strikes the right balance, but you may want to back up more frequently.
Keep Saved Versions	Forever (default)	If you choose Until Space Is Needed, File History won't raise a holy stink if you run out of room on your backup drive. By leaving it at Forever, File History sends notifications when the hard drive gets close to full capacity, so you can run out and buy another backup drive.

Storing to and through the Cloud

File History's a great product. I use it religiously. But it isn't the be-all and end-all of backup storage. What happens if my office burns down? What if I really, really need to get at a file when I'm away from the office?

REMEMBER

The best solution I've found is to have File History do its thing, but I also keep my most important files — the ones I'm using right now — in the cloud. That's how I wrote this book, with the text files and the screenshots both in Dropbox. I also handed off the files to my editors, and received edited versions back, through Dropbox.

The only question with the editorial team nowadays is which cloud storage vendor to use. Believe me, things have really changed.

I also back up data, from time to time, to the Internet. Doing so is fast, cheap, and easy — but it does have problems. I talk about the mechanics of using OneDrive in Book 6, Chapter 1, and OneDrive's certainly a good choice. But other choices are available, and I want you to know about them.

TIP

Backing up to the Internet has one additional, big plus: Depending on which package you use and how you use it, the data can be accessible to you, no matter where you need it — on the road, on your iPad, even on your smartphone. You can set up folders to share with friends or coworkers, and in some cases, have them help you work on a file while you're working on it, too.

Many years ago, only one big player — Dropbox — was in the online storage and sharing business. Now there's Dropbox, Microsoft OneDrive, Google Drive, the Apple iCloud (which is a bit different), and Amazon Cloud Drive — all from huge companies — and SugarSync, Box (formerly Box.net), SpiderOak, and many smaller companies.

What happened? People have discovered just how handy cloud storage can be. And the price of cloud storage has plummeted to nearly nothing.

REMEMBER

The cloud storage I'm talking about is specifically designed to allow you to store data on the Internet and retrieve it from just about anywhere, on just about any kind of device — including a smartphone or tablet. They also have varying degrees of interoperability and sharing, so, for example, you can upload a file and have a dozen people look at it simultaneously. Some cloud storage services (notably OneDrive and Google Drive) have associated programs (such as Microsoft Office and Google's G Suite) that let two or more people edit the same file simultaneously.

Considering cloud storage privacy concerns

I don't know how many times people have told me that they just don't trust putting their data on some company's website. But although many people are rightfully concerned about privacy issues and the specter of Big Brother, the fact is that the demand for storage in the cloud is growing by leaps and bounds.

The concerns I hear go something like this:

>> **I have to have a working Internet connection in order to get data to or from the online storage.** Absolutely true, and there's no way around it. If you use cloud storage for only offsite backup, it's sufficient to be connected whenever you want to back up your data or restore it. Some of the cloud storage services have ways to cache data on your computer when, say, you're going to be on an airplane. But in general, yep, you have to be online.

>> **The data can be taken or copied by law enforcement and local governments.** True. The big cloud storage companies get several court orders a day. The storage company's legal staff takes a look, and if it's a valid order, your data gets sent to the cops. Or the feds. Or the tax people.

Unless, of course, you're talking about the US National Security Agency and programs like PRISM, which basically allows the NSA to take any data it likes and prohibits the storage company from even talking about it. With a little luck, that's going to change, but it's hard to say how, or when, or even if.

REMEMBER

Moral of the story: If you're going to store data that you don't want to appear in the next issue of a certain British tabloid, it would be smart to encrypt the file before you store it. Word and Excel 2007 and later use very effective encryption techniques, and 7Zip (see Book 10, Chapter 5) also makes nearly unbreakable zips. Couple that with a strong password, and your data isn't going anywhere soon. Unless, of course, you're required by the court to give up the password, or the NSA sets one of its teraflops password crackers to the job.

>> **Programs at the cloud storage firm can scan my data.** True, once again, for most (but not all) cloud storage firms. With a few notable exceptions — Mega, Spider Oak, and others — cloud storage company programs can see your data. There's been a big push in the past few years to hold cloud storage companies responsible for storing copyrighted material: If you upload a pirate copy of *Men in Black 4,* the people who hold the copyright are going to get very upset.

Different cloud storage companies handle the task differently, but with the takedown of Megaupload in January 2012, everybody's concerned about incurring the wrath of the *MPAA* and *RIAA,* the companies that defend movie and music copyrights, respectively. Mega packed up, moved to New Zealand, and lives again, but the legal problems continue. The net result is that most cloud storage companies will be performing routine scans — either now or in the not-too-distant future — to see whether you're trying to upload something that's copyright-protected.

» **Employees at the cloud storage firm can look at my data.** True again. Certain cloud storage company employees *can* see your data — at least in the larger companies (Mega, SpiderOak, and a few others excepted). They must be able to see your data, in order to comply with court orders.

Does that mean Billy the intern can look at your financial data or your family photos? Well, no. It's more complicated than that. Every cloud storage company has very strict, logged, and monitored rules for who can authorize and who can view customer data. Am I absolutely sure that every company obeys all its rules? No, not at all. But I don't think my information is interesting enough to draw much attention from Billy, unless he's trying to swipe the manuscript of my next book.

» **Somebody can break in to the cloud storage site and steal my data.** Well, yes, that's true, but it probably isn't much of a concern. Each of the cloud storage services scrambles its data, and it'd be very, very difficult for anyone to break in, steal, and then decrypt the stolen data. Can it happen? Sure. Will I lose sleep over it? Nah. That said, you should enable two-factor authentication when it's offered (so the backup service sends an entry code to your email address, or sends you a message on your phone, requiring the entry code in order to get into your data). And if you want to be triple-sure, you can encrypt the data before you store it — 7Zip, among many others, makes it easy to encrypt files when you zip them.

Reaping the benefits of backup and storage in the cloud

So much for the negatives. Time to look at the positives. On the plus side, a good cloud storage setup gives you:

» **Offsite backups** that won't get destroyed if your house or business burns down.

» **Access to your data from anywhere,** using just about any imaginable kind of computer, including smartphones and tablets.

» **Controlled sharing** so you can password-protect specific files or folders. Hand the password to a friend, and he can look at the file or folder.

» **Broadcast sharing** from a Public folder that anyone can see.

» **Direct access from application programs that run in the cloud.** Both Google Apps and the many forms of Microsoft Office are good examples. That includes iWork (er, the Apple Productivity apps), if you're using Apple's iCloud. Office apps now have direct access to Dropbox data, too.

» **Free packages, up to a certain size limit,** offered by most of the cloud storage services.

Choosing an online backup and sharing service

So which cloud storage service is best? Tough question. I use four of them — three for PCs and Android, and iCloud for my Mac, iPad, and iPhone stuff — different services for different purposes.

Dropbox, Microsoft OneDrive, and Google Drive have programs that you run on your PC or Mac to set up folders that are shared. Drag a file into the shared folder, and it appears on all the computers you have connected (with a password) to the shared folder. Go on the web and log in to the site, and your data's available there too. Install an app on your iPhone or Android smartphone or tablet, and the data's there as well. Here's a rundown of what each cloud storage service offers:

» **Dropbox,** as shown in Figure 1-11, offers 2GB of free storage, with 2TB for $10 per month. It's very easy to use, reliable, and fast. I use it for synchronizing project files — including the files for this book. Dropbox also connects to Facebook to retrieve or post pictures (www.dropbox.com).

» **OneDrive** has 5GB of free storage, with 100GB for $2 per month. The amounts on offer change from time to time. Also note that many Microsoft 365 (formerly known as Office 365) subscription levels have 1TB of OneDrive storage, free, for as long as you're a subscriber. I talk about OneDrive in Book 6, Chapter 1 (www.OneDrive.com).

» **Google Drive (also known as Google One),** as shown Figure 1-12, has 15GB of free storage, with 100GB for $2 per month and 2TB for $10 per month. Google Drive isn't as slick as the other two, and there's no Facebook connection, but it works well enough. There's an optical character recognition facility and the capability to launch web apps directly. Most of all, it's fall-down simple to use Google Drive with Google's G Script and its apps, which include Gmail and several capable but not very Office-compatible writing and spreadsheet apps. See Book 10, Chapter 3 (www.drive.google.com). All the apps, to a greater or lesser extent, work while you aren't connected to the Internet.

FIGURE 1-11:
Dropbox popularized cloud storage.

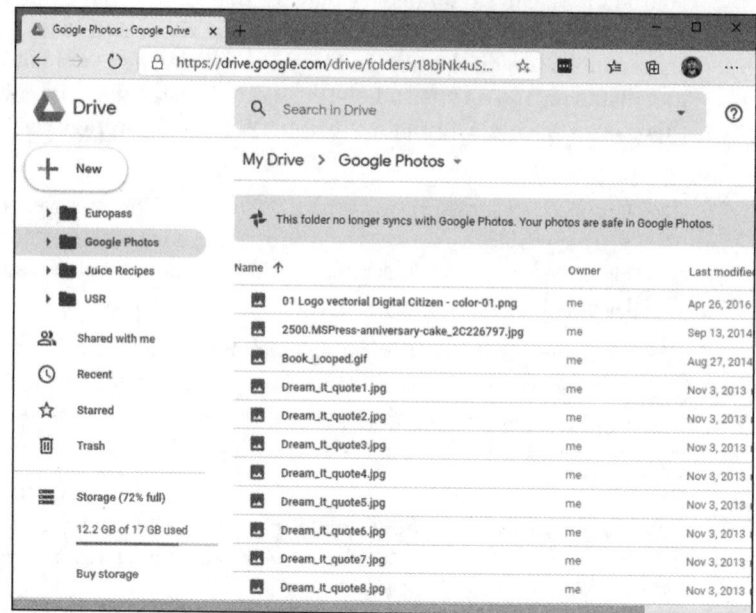

FIGURE 1-12:
Google Drive works very well with Google Apps.

» **Apple iCloud,** as shown in Figure 1-13, is really intended to be an Apple-centric service. The first 5GB is free, and then it's $0.99 per month for an additional 50GB. It works great with iPads and iPhones and even my new Mac, with extraordinarily simple backup of photos. In fact, photo and video backup and sharing take place automatically, and I don't have to do a thing. Music goes in easy as can be, and anything you buy from the iTunes store is in your storage, free, forever. But it's not really set up for open data sharing (www.icloud.com). Apple, too, is trying to bring its cloudy offerings down to the desktop. Stay up on the latest, if you're thinking about going with Apple.

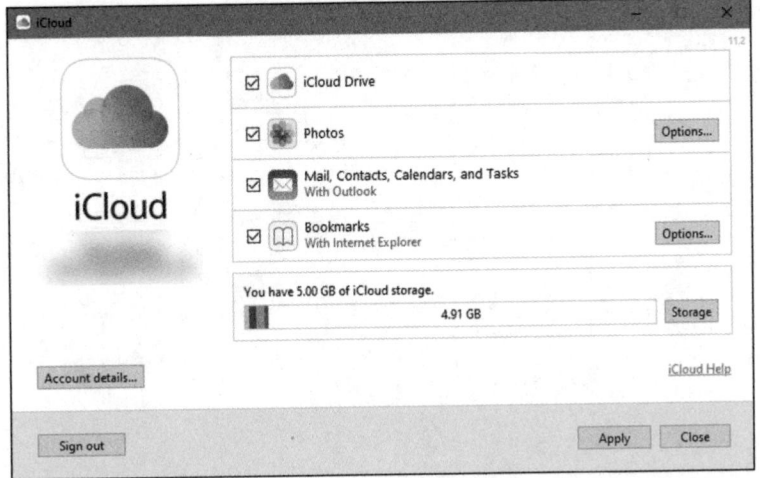

FIGURE 1-13: iCloud works with Apple products but makes it difficult to share files among PCs.

The other services have specific strong points:

» **Amazon Cloud Drive** ties in with Amazon purchases and the Kindle but not much else (www.amazon.com/clouddrive). If you pay for Amazon Prime ($100/year), you also get unlimited free photo storage.

» **SugarSync** lets you synchronize arbitrary folders on your PC. That's a big deal if you don't want to drag your sync folders into one location (www.sugarsync.com).

» **Box** is designed for large companies. It gives companies tools to control employee sharing (www.box.com).

» **SpiderOak** is the most secure of the bunch: It doesn't keep the keys to your files, and unlike the other services in this chapter, it's impossible for SpiderOak to see your files (www.spideroak.com).

ASK
WOODY.COM

Like so many other things in the PC business, cloud storage is changing very, very rapidly. If you're interested in backing up to the cloud — and sharing files on the Internet, too, by the way — stay on top of the latest at my site, AskWoody.com (www.askwoody.com).

Chapter **2**

A Fresh Start: Restore and Reset

I n this chapter, I look at how you can bring back to life a computer that's been possessed. (This chapter doesn't talk about bringing files back from the dead. That's the purview of Book 8, Chapter 1.) I also talk about a new way to clean all the crapware off your PC — even the junk installed by your PC's manufacturer.

ASK WOODY.COM

If you've worked with Windows for any length of time at all, you know that from time to time Windows PCs simply go out to lunch . . . and stay there. The problem could stem from a bad drive, a scrambled registry entry, a driver that's suddenly taken on a mind of its own, a revolutionary new program that's throwing its own revolution, or that dicey tuna sandwich you had for lunch.

Windows is a computer program, not a Cracker Jack toy, and it will have problems. The trick lies in making sure that *you* don't have problems too. This chapter walks you through the important tools you have at hand to make Windows 10 do what you need to do, to solve problems as they (inevitably!) occur.

REMEMBER

If you're the family's resident voodoo doctor — or the Windows go-to-gal in the office — this chapter can save your hide.

Microsoft has gone through a great deal of effort to make restoring a recalcitrant PC much simpler than ever before. The goal is to keep *you* out of the details and let Windows 10 handle it: Computer, heal thyself, as it were. To a large degree, Microsoft has succeeded.

The Three R's — and an SF and a GB

When resuscitating a machine with Windows 10 gone bad, consider the three R's — remove the latest update, reset but keep your programs and data, and reset with the data going bye-bye — as well as Start Fresh, a new method that first appeared in July 2016. Three of them are readily available, but they make major changes to your machine. One's not nearly so destructive, but it's harder to understand and use.

A NOTE ABOUT TERMINOLOGY

I *hate* the terminology Microsoft uses for its Windows-resuscitation technology.

If you and I get confused about Remove, Refresh, Restore/rollback, Recovery Environment, and Recombobulate (okay, I made up that last one), just imagine how confused normal, everyday users are going to get when they're confronted with choices that could, quite literally, obliterate all their data.

Further confusing the issue, Restore also applies to bringing back files. Refresh applies to network settings, in a different way. Recovery, in the Windows world, is a console that helps step you through the process.

It's important that you watch carefully when you apply any of these R's or the truly rejuvenating (there's another *R*) Start Fresh. The implications of your actions are spelled out reasonably well on all the screens that Windows uses. But you can still very easily get confused.

And for heaven's sake, don't tell your mom to reset (or remove) her PC when you meant to tell her to refresh it. You may not get invited over for dinner next Thanksgiving.

Here are the SF, GB, and three R's that every Windows 10 medic needs to know:

>> **Reset** has two variants that are as different as night and burning day.

- **Reset with Keep My Files** keeps some Windows settings (accounts, passwords, the desktop, Microsoft Edge and IE favorites, wireless network settings, drive letter assignments, and BitLocker settings) and all personal data (in the User folder). It wipes out all programs and then restores the apps available in the Windows Store (primarily the tiled apps). This one's pretty drastic, but at least it keeps the data stored in the most common locations — Documents folder, the desktop, Downloads, and the like. And as a bonus, the Reset with Keep My Files routine keeps a list of the apps it zapped and puts that list on your desktop, so you can look at it when your machine's back to its chirpy self.

- **Reset with Remove Everything** removes everything on your PC and reinstalls Windows. Your programs, data, and settings all get wiped out — they're irretrievably lost. This is the most drastic thing you can do with your computer, short of shooting it. (Did you see that viral video of the guy shooting his daughter's laptop? I digress.) Most hardware manufacturers have the command jury-rigged to put their crapware back on your PC. If you run Reset with Remove Everything on those systems, you don't get a clean copy of Win10; you get the factory settings version. Yes, that means you get the original manufacturer's drivers (see the "Why would you want factory drivers?" sidebar). But it also means you get the manufacturer's garbageware.

  If you like, you can tell Reset with Remove Everything to do a *thorough* reformatting of the hard drive, in which case, random patterns of data are written to the hard drive to make it almost impossible to retrieve anything you used to store on the disk. But in the end, you get the same crapware that came with a new computer.

If you've tried in previous years to bring back an older Windows machine from purgatory you may have encountered System Restore. In fact, System Restore still exists, but Microsoft doesn't want you to use it. Refresh is a combination of System Restore, safe mode, recovery console, and all sorts of minor earlier-system recovery techniques, wrapped into one neat one-click bundle — with none of the hassles, but none of the old controls.

» **Fresh Start** was new in the Windows 10 Anniversary update, version 1607. Fresh Start, like Reset, has two options: Keep My Files, and Remove Everything. There are two big differences between Start Fresh and the built-in Reset. First, Start Fresh isn't built into Windows 10; it runs from a web page that Microsoft can update frequently (go to http://go.microsoft.com/fwlink/?LinkId=808750). Second, Start Fresh is based on a blissfully clean, Microsoft-endorsed copy of Windows 10. No crapware in sight. Unfortunately, as of the Windows 10 May 2020 update (version 2004), Fresh Start is no longer available and its functionality has been moved to Reset This PC. Select Keep My Files, choose a cloud download or a local reinstall, change your settings, and set Restore Preinstalled Apps to No.

» **Go Back** tells Windows to unapply the last cumulative update. Use it when one of Microsoft's mighty forced cumulative updates (see Book 7, Chapter 5) crumbles your machine. I've had mixed results with Go Back, but when it works, it's a fast and easy solution to a congenital problem — Microsoft forcing Windows 10 updates down your throat.

TECHNICAL STUFF

» **Restore** (I prefer calling it *Rollback*) is hard to find — Microsoft doesn't want everyday users to find it — but it rolls Windows 10 back to an earlier *restore point,* which I describe in the "Restoring to an Earlier Point" section, later in the chapter. Restore doesn't touch your data or programs; it simply rolls back the registry to an earlier point in time. If your problems stem from a bad driver or a problematic program change you made recently, Restore may do all you need. If you're familiar with Windows 7 or earlier versions, Windows 10 Restore is almost identical to Restore in the earlier version; you just access it a little differently.

ASK WOODY.COM

Why does Microsoft make it hard to find Restore? As far as I know, the logic goes something like this: If you don't use Restore right, you can shoot your machine; in which case, you'll bother the folks at Microsoft mercilessly and accuse them of all sorts of mean things. Even if you *do* use Restore right, it fixes only a small percentage of all Windows-breaking problems, so if you try Restore and it doesn't work, you'll also bother the folks at Microsoft mercilessly — a classic lose-lose situation for the company. Importantly, there's nothing analogous to Restore with any competing operating system, tablet, or smartphone. The iPad doesn't have anything that resembles Restore; Android tablets and smartphones aren't in any shape to Restore; macOS wouldn't know a Restore from a hole in the ground; and my Linux friends start tittering obnoxiously anytime I say "Restore." In short, only Windows has a registry, and Restore works almost exclusively on the registry, so only Windows needs a Restore. There's not much competitive benefit — and lots of downside — to offering Restore to the average Windows consumer.

WHY WOULD YOU WANT FACTORY DRIVERS?

There's an obvious downside to running a Reset with Remove Everything: You get all the junk that manufacturers ship with their new computers. (Some of the apps they ship are supposed to make managing your PC better. Others are simply advertising, for which the advertiser pays. You really don't want either.)

The one upside to running a Reset with Remove Everything? You get back the original drivers. I can't say that I've ever seen a situation where a clean install of Windows 10 (see the "Starting Fresh" section) resulted in useless drivers — your monitor won't work or your keyboard or mouse turns belly up. In the very unusual situation where you get suboptimal drivers, one trip through Windows Update should get you up to speed.

But if you're worried that a truly clean install will leave you begging, then Reset with Remove Everything is for you.

REMEMBER

All three resuscitation methods play out in the *Windows Recovery Environment (WRE)*, a special proto-Windows system. If you run Reset or Refresh, you won't even know that WRE is at work behind the scenes, but it's there.

When there's trouble and Windows 10 can't boot normally, the operating system instead boots into WRE, not into Windows itself. WRE has the special task of giving you advanced tools and options for fixing things that have gone bump in the night.

Remove, Refresh, and Restore — and several more (Recycle, Reuse, Reduce?) are available in WRE.

I talk about WRE — and your advanced boot options — toward the end of this chapter.

Resetting Your PC

You don't really know or care about restore points, and you don't want to dig into Windows 10 to make it work right. Mostly, you just want a one-tap (or click) solution that reams out the old, replaces it with known good stuff, and might or might not destroy your files in the process — at your option. You get the manufacturer's drivers, but you also get the manufacturer's crapware. That's what Microsoft has tried to offer with Reset.

Reset runs in two different ways:

>> The **Keep My Files** option tries to work its magic without disturbing any of your personal data files.

>> The **Remove Everything** option blasts everything away, including your data. It's the scorched-earth approach, to be used when nothing else works, but you still want the hand-holding implicit in a factory refresh.

That's the view from 30,000 feet. Here are the details that you really need to know.

REMEMBER

Running a Keep My Files reset *keeps* all these:

>> **Many of your Windows 10 settings:** These include accounts and passwords, backgrounds, wireless network connections and their settings, BitLocker settings and passwords, drive letter assignments, and your Windows 10 installation key.

>> **Files in the User folder:** That includes files in every user's Documents folder, the desktop, Downloads, and so on. Refresh also keeps folders manually added to the root of the C: drive, such as C:\MyData. Reset with Keep My Files keeps File History versions, and it keeps folders stored on drives and in partitions that don't contain Windows (typically, that means Refresh doesn't touch anything outside of the C: drive).

TIP

Files that *aren't* kept can be retrieved for several weeks from the C:\Windows. old folder. Yes, Microsoft keeps a secret stash of the files that it really wants to delete — and it's up to you to find them, if something disappears unexpectedly.

>> **Windows 10 Apps from the Microsoft Store:** Their settings are saved too. So if you're up to the 927th level of Cut the Rope before you run a Reset with Keep My Files, afterward, you're still at the 927th level. Confusingly, if you bought a desktop app in the Microsoft Store, its settings get obliterated. Only your Windows 10/Universal/Metro apps come through unscathed.

Running Reset with the Keep My Files option *destroys* all of these:

>> **Many of your Windows 10 settings:** Display settings, firewall customizations, and file type associations are wiped out. Windows 10 must zap most of your Windows settings because they could be causing problems.

>> **Files — including data files — not in the User folder:** If you have files tucked away in some unusual location, don't expect them to survive the Reset.

>> **Desktop apps/programs:** Their settings disappear too, including the keys you need to install them, passwords in such programs as Outlook — everything. You need to reinstall them all.

The Reset routine, helpfully, makes a list of the programs that it identifies on the kill list and puts it on your desktop.

Here's how to run a Reset with the Keep My Files option:

1. **Make very, very sure you understand what will come through and what won't.**

   See the preceding bullet lists.

2. **Click or tap the Start button, the Settings icon, and Update & Security.**

3. **On the left, choose Recovery.**

   You see the recovery options shown in Figure 2-1.

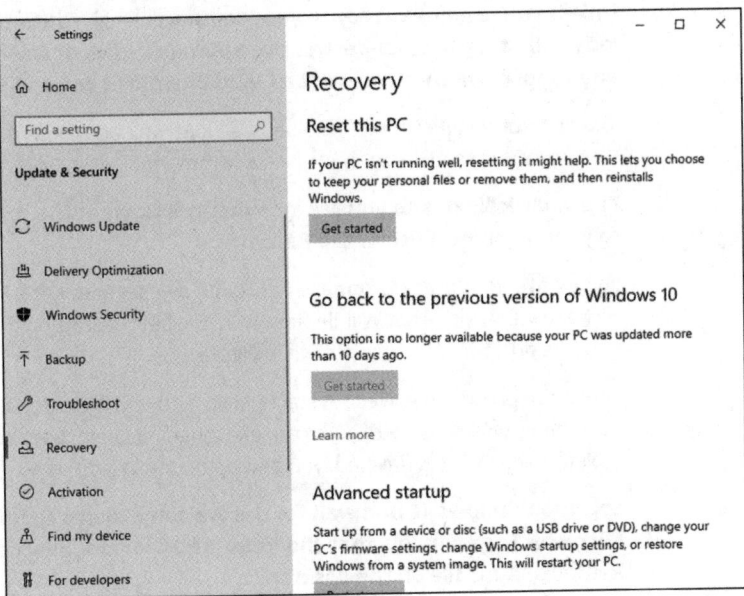

FIGURE 2-1:
Run Reset from the Settings app.

4. **Under the heading Reset This PC, tap or click Get Started.**

Windows 10 asks if you want to keep your files or obliterate everything. See Figure 2-2.

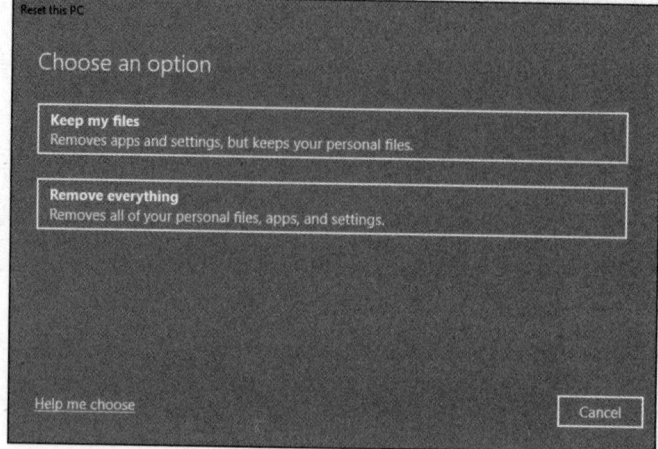

Reset this PC

Choose an option

Keep my files
Removes apps and settings, but keeps your personal files.

Remove everything
Removes all of your personal files, apps, and settings.

Help me choose

Cancel

FIGURE 2-2:
Well, whaddya say, punk? Keep 'em or blast 'em away?

5. **Unless you're going to recycle your computer — give it to charity or the kids — first try the less-destructive approach. Click or tap Keep My Files and choose whether you want a Cloud Download or Local Reinstall.**

The first option takes more time because it downloads the Windows 10 Setup files from Microsoft servers. You see a summary of your current settings.

6. **Tap or click Next, and you are informed about what the resetting will do to your machine. Click or tap Reset.**

If you have apps that won't make it through a Reset with Keep My Files option, click the View Apps That Will Be Removed link before starting the Reset. A list appears on your screen, as shown in Figure 2-3.

The entire process involves several restarts and takes about ten minutes on a reasonably well-seasoned PC. It can take longer, though, particularly on a slow tablet. When Reset is finished, you end up on the Windows 10 login screen.

7. **Log into Windows 10 and wait for the Welcome dialog to finish. Close the Microsoft Edge window that shows up, and then double-click the new Removed Apps file on the desktop.**

Microsoft Edge appears and shows you a list of all the programs it identified that didn't make it through the Reset.

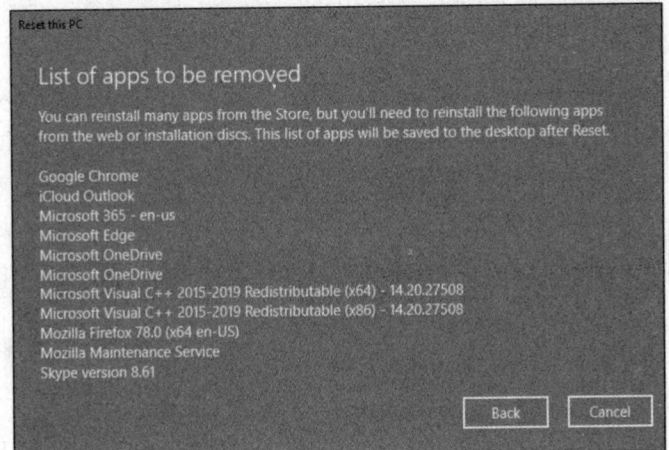

FIGURE 2-3:
These apps —
yes, even apps
from Microsoft —
won't survive a
reset.

If Windows 10 can't boot normally, you're tossed into the Windows Recovery Environment. See the last section in this chapter for a description of how to start Reset from the Windows Recovery Environment.

Resetting Your PC to Factory Settings

Reset with the Remove Everything option is very similar to running Reset with Save My Files except . . .

WARNING

Warning! Warning! Danger, Will Robinson! Resetting with Remove Everything on your PC wipes out everything and forces you to start all over from scratch. You even have to enter new account names and passwords, and reinstall everything, including Windows 10 apps. Your Microsoft account settings remain intact, as does any data you've stored in the cloud (for example, in OneDrive or in Dropbox). But the rest gets hurled down the drain.

In addition, when you're done, you'll have a factory fresh copy of Windows 10. If you're running one of the (many) Windows 10 PCs that ship with crapware pre-installed, all of it will suddenly reappear. (If you bought your PC from Microsoft's store, it won't have any crapware — that's the promise behind Signature Edition PCs. Microsoft won't sell anything that's sullied.)

If you're selling your PC, giving it away, or even sending it off to a recycling service, Reset with Remove Everything is a good idea. If you're keeping your PC, only attempt Reset with Remove Everything when you've run two or more Resets with Keep My Files, and they haven't solved the problem. Reset with Remove Everything is very much like a clean install. You're nuking everything on your PC, although you get your factory drivers back, along with all the factory-installed crud.

With that as a preamble, here's how to nuke, er, Reset your PC with Remove Everything:

1. **Make very, very sure you understand that your PC will turn out like a brand-new PC, fresh off the store shelves.**

   Absolutely nothing survives the wipeout.

2. **Click or tap the Start button, the Settings icon, and Update & Security.**

3. **On the left, choose Recovery.**

   You see the Reset options shown in Figure 2-1.

4. **Under the heading Reset This PC, tap or click Get Started.**

   Windows 10 asks if you want to keep your files or obliterate everything. Refer to Figure 2-2.

5. **Tap or click Remove Everything and choose whether you want a Cloud Download of Windows 10 (from Microsoft's servers) or a Local Reinstall.**

   The second option is the fastest. A summary of your current settings is displayed, as shown in Figure 2-4.

6. **Tap or click Next.**

   You are informed about what the resetting will do to your machine.

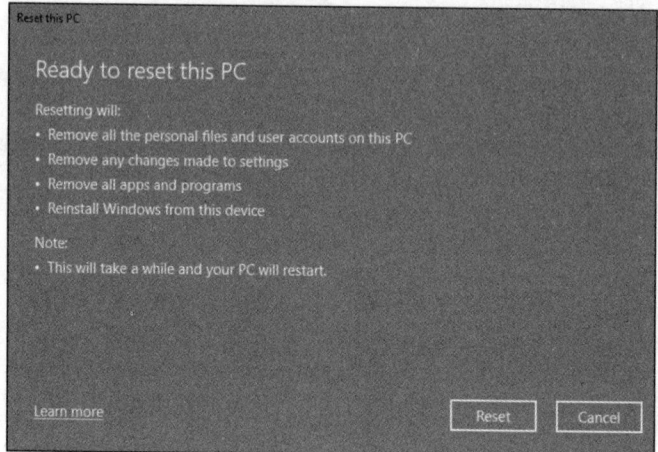

FIGURE 2-4:
What Reset your PC with Remove Everything does.

7. **Click Reset.**

   The entire process involves several restarts and takes about ten minutes on a reasonably well-seasoned PC. It can take longer, though, particularly on a slow tablet. When Reset is finished, you end up at the Welcome wizard, which allows you to personalize Windows 10 as if it were newly installed.

8. **Set up Windows 10 from scratch.**

   Now *that's* a complete, scorched-earth install.

Starting Fresh

If the thought of losing factory-installed drivers doesn't faze you (typically, a quick run through Windows Update will get them all reinstalled), and the thought of getting rid of all the manufacturer-installed junk thrills you, the Start Fresh option is for you. However, this option is available only for users who have a Windows 10 version between the Anniversary update (version 1607) and the May 2020 update (version 2004).

Here's how to get a really fresh copy of Windows 10:

1. **Click or tap the Start button, the Settings icon, and Update & Security. On the left, choose Recovery.**

   You see the Recovery options shown earlier in Figure 2-1.

2. **At the bottom, click the link under More Recovery Options to go to the Start Fresh website (which may change from time to time).**

3. **Download the Start Fresh tool and run it by double-clicking it.**

   Microsoft shows you an End User License Agreement.

4. **Read all 3,141 pages of the EULA, call your lawyer, and click Agree.**

   Start Fresh displays the dialog box shown in Figure 2-5.

5. **Do one of the following:**

   - *If you want to keep your files, select Keep Personal Files Only and then click Install.* The files will be kept in the User folder, and in the root of the C: drive; see the description about Keep My Files in the "Resetting Your PC" section. Start Fresh proceeds much like a Reset with Keep My Files, except it uses a clean copy of Win10.

   - *If you don't want to keep anything, select Nothing and then click Install.* Again, Start Fresh proceeds as in a Reset with Remove Everything, except it's done with a clean copy of Windows 10. You are not given the option of reformatting your hard drive.

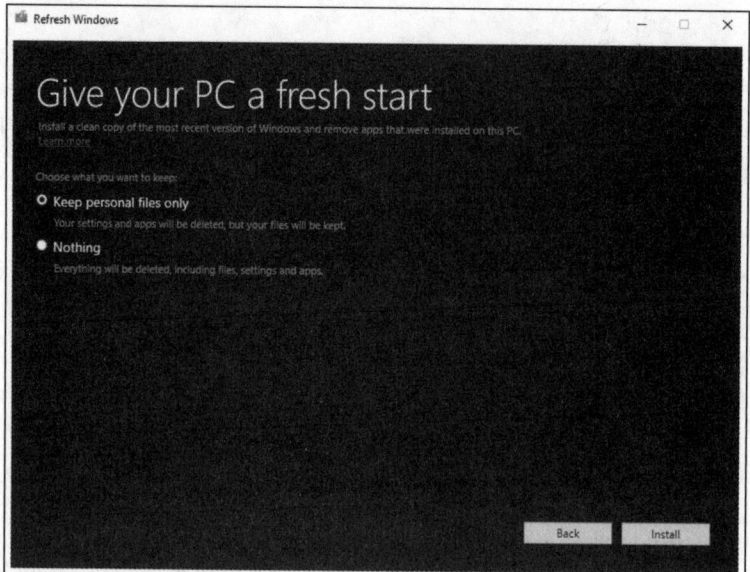

FIGURE 2-5:
The cleanup options in Start Fresh are like those in Refresh.

ASK
WOODY.COM

In either case, when you click Install, the installer does exactly that. You don't have a chance to say "oops." If you're caught in the middle of a reset and don't want it — and you see this admonition in time — you can X out of the installation dialog and choose to halt.

I hope Microsoft brings back the Start Fresh utility for Windows 10 in a future update. Some users, myself included, consider it useful.

Restoring to an Earlier Point

If you've used Windows 7 or earlier, you may have stumbled upon the System Restore feature. Windows 10 has full support for System Restore and restore points; it just hides all the pieces from you.

Why? Because Microsoft spends a fortune every year answering phone calls and email messages from people who bork System Restore. Instead of trying to handle all the picayune questions — and there are hundreds of thousands of them — Microsoft said, "That's enough!" and invented Reset, with and without Keep My Files.

**ASK
WOODY.COM**

With a few exceptions (see the next section on system image), Reset takes you all the way back in time to when you first set up your PC; it adds the Windows 10 apps that ship with the operating system, and it's careful not to step on your data. Aside from a few Windows 10 settings, that's about it. Reset is a sledgehammer, when sometimes the tap of a fingernail may be all that you need.

Smashing with a sledgehammer is easy. Tapping your fingernail requires a great deal more finesse. And that brings me to System Restore in Windows 10.

If you enable System Protection, Windows 10 takes snapshots of its settings, or *restore points*, before you make any major changes to your computer — install a new hardware driver, perhaps, or a new program. You can roll back your system settings to any of the restore points. (See the "System protection and restore points" sidebar.)

A restore point contains registry entries and copies of certain critical programs including, notably, drivers and key system files — a *snapshot* of crucial system settings and programs. When you roll back (or *restore)* to a restore point, you replace the current settings and programs with the older versions.

TIP

When Windows can tell that you're going to try to do something complicated, such as install a new network card, it sets a restore point — as long as you have System Protection turned on. Unfortunately, Windows can't always tell when you're going to do something drastic — perhaps you have a new CD player and the instructions tell you to turn off your PC and install the player before you run the setup program. So, it doesn't hurt one little bit to run System Restore — er, System Protection — from time to time, and set a restore point, all by yourself.

SYSTEM PROTECTION AND RESTORE POINTS

Windows 7 created restore points for your system drive (usually C:) by default. Windows 10 doesn't. Restore points take up space on your hard drive, and Microsoft would rather that you just trust in its cloud-based recovery options. But if you want to take your system into your own hands, properly maintained and used restore points can change a gut-wrenching Refresh or Reset into a simple rollback to an earlier restore point.

See the "Creating a restore point" section to see how to knock Windows upside the head and get it to start System Protection.

Enabling System Protection

System Protection is disabled by default in Windows 10, and you must turn it on manually. Here's how:

1. **Down in the Windows 10 search bar, type** restore.

The first result in the Search is Create a Restore Point.

2. **Tap or click the Create a Restore Point link.**

Windows 10 displays the System Properties window open to the System Protection tab, as shown in Figure 2-6.

3. **With the C: drive selected, click or tap the Configure button.**

The System Protection window appears.

4. **Choose Turn On System Protection, select the Disk Space Usage, and click OK.**

5. **Click or tap OK one more time to enable System Protection.**

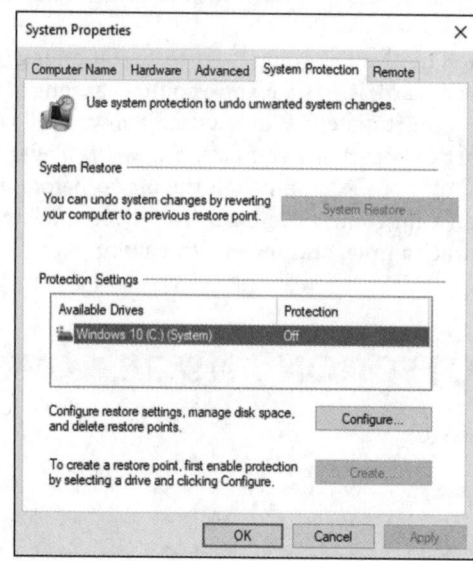

FIGURE 2-6:
The hard-to-find
System Restore
option.

Creating a restore point

Here's how to create a restore point:

1. **Wait until your PC is running smoothly.**

No sense in having a restore point that propels you out of the frying pan and into the fire, eh?

2. **Down in the Windows 10 search bar, type** restore point.

The first result is Create a Restore Point.

3. **Tap or click the Create a Restore Point link.**

Windows 10 brings up the System Properties window open to the System Protection tab (refer to Figure 2-6).

4. **At the bottom, next to the To Create a Restore Point . . . option, tap or click the Create button.**

The Create a Restore Point dialog box appears (see Figure 2-7).

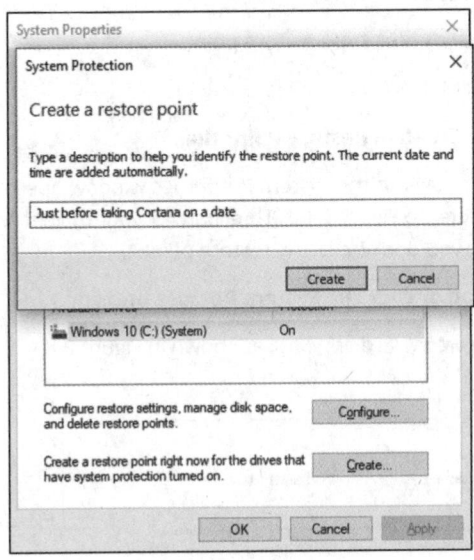

FIGURE 2-7:
Give your restore point a name.

5. **Type a good description, and tap or click Create.**

Windows 10 says that it's creating a restore point. When it's finished, it tells you that the restore point was created successfully.

6. **Tap or click Close, and then tap or click the X button to close the System Properties dialog box.**

Your new restore point is ready for action.

Rolling back to a restore point

If you don't mind getting your hands a little dirty, the next time you think about running Refresh, see whether you can roll your PC back to a previous restore point, manually, and get things working right. Here's how:

1. **Save your work, and close all running programs.**

   System Restore doesn't muck with any data files, documents, pictures, or anything like that. It works only on system files, such as drivers, and the registry. Your data is safe. But System Restore can mess up settings, so if you recently installed a new program, for example, you may have to install it again after System Restore is finished.

2. **Down in the search box, type** restore point.

   The first search result is Create a Restore Point.

3. **Tap or click the Create a Restore Point tile.**

   Windows flips you over to the System Properties window open to the System Protection tab (refer to Figure 2-6). Protection for your main drive should be On. (If it isn't On, you don't have any restore points.)

4. **Near the top, tap or click the System Restore button.**

   The System Restore Wizard appears, as shown in Figure 2-8.

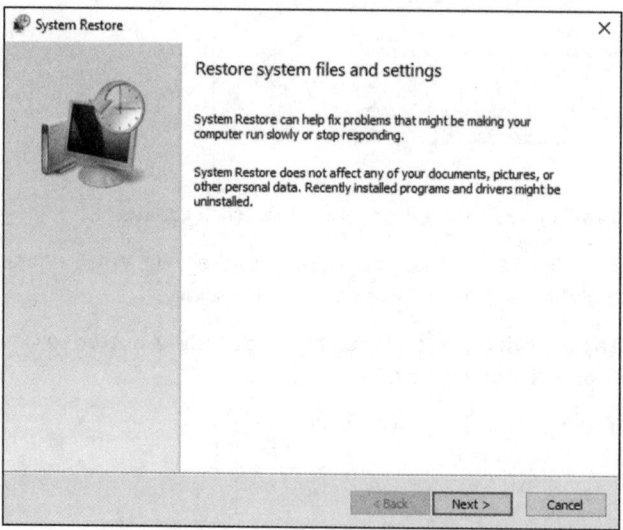

FIGURE 2-8:
See? Wizards are
in Windows 10.

5. **Tap or click Next.**

   A list of recent restore points appears, as shown in Figure 2-9.

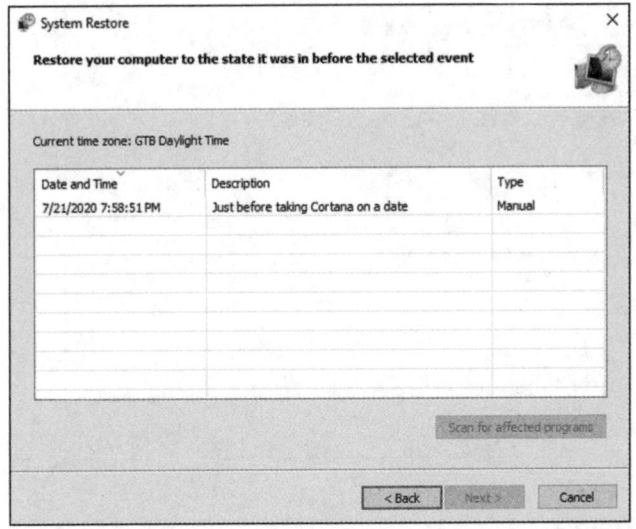

Current time zone: GTB Daylight Time

Date and Time	Description	Type
7/21/2020 7:58:51 PM	Just before taking Cortana on a date	Manual

FIGURE 2-9:
The latest restore
point isn't always
the best restore
point.

6. **Before you roll your PC back to a restore point, tap or click to select the restore point you're considering and then tap or click the Scan for Affected Programs button.**

   System Restore tells you which programs and drivers have system entries (typically in the registry) that will be altered and which programs will be deleted if you select that specific restore point. See Figure 2-10.

7. **If you don't see any major problems with the restore point — it doesn't wipe out something you need — tap or click Close, followed by Next.**

   (If you do see a potential problem, go back and choose a different restore point, or consider using Refresh, as I describe earlier in this chapter.)

   You're warned that rolling back to a restore point requires a restart of the computer and that you should close all open programs before continuing.

8. **Follow the instructions to save any open files, close all programs, tap or click Finish, and then confirm by tapping or clicking Yes.**

   True to its word, System Restore reverts to the selected restore point and restarts your computer.

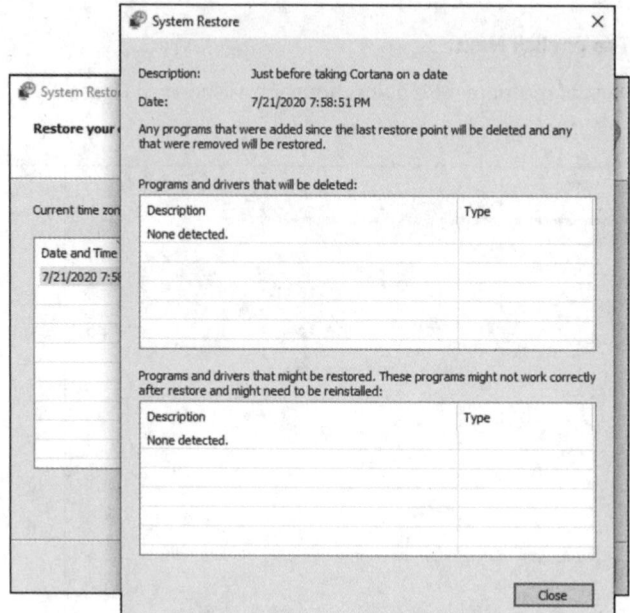

FIGURE 2-10:
Windows can scan the restore point to see what programs will be affected by rolling back to it.

ASK WOODY.COM

System Restore is a nifty feature that works very well. The folks at Microsoft figure it's too complicated for the general computer- and tablet-buying consumer public. They may be right but, hey, all it takes is a little help and a touch of moxie, and you can save yourself a Refresh — as long as System Protection is turned on.

Entering the Windows Recovery Environment

In Windows 10, the Windows Recovery Environment has become a very sophisticated, almost eerily intelligent fix-everything program that works very well.

Except, of course, when it doesn't.

The Windows 10 Recovery Environment appears when your machine fails to boot two times in a row. You know you're in the Windows Recovery Environment if you see a blue Choose an Option screen or a blue Troubleshoot screen like the one in Figure 2-11. (If you find yourself facing a blue Choose an Option screen, choose Troubleshoot!)

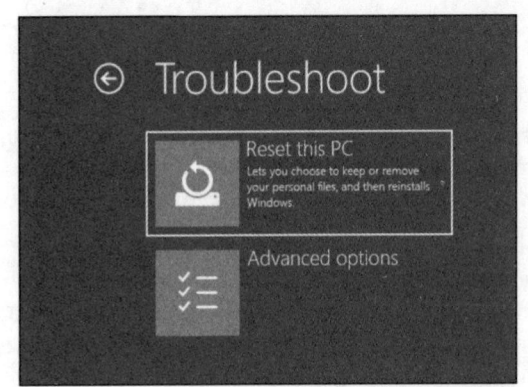

FIGURE 2-11:
The hallmark
of the Windows
Recovery
Environment.

From the Troubleshoot screen, you can run Refresh or Reset directly: They behave precisely as I describe earlier in this chapter. You can also choose Advanced Options, which brings you to several interesting — if little-used — options, as shown in Figure 2-12.

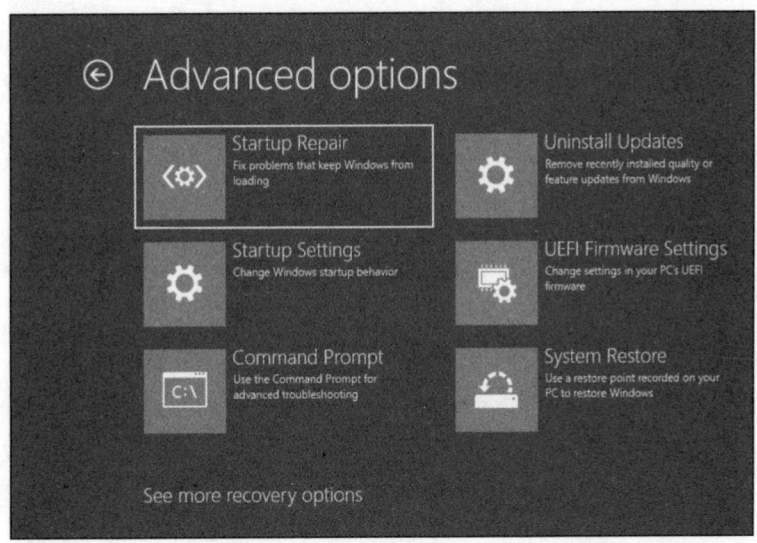

FIGURE 2-12:
Advanced boot
options.

REMEMBER

You can also get to this screen by choosing Advanced Startup from the Recovery list (refer to Figure 2-11). After you choose Advanced Startup, choose Troubleshoot and then choose Advanced Options.

Here's what the Advanced Options can do:

>> **System Restore** puts your system back to a chosen restore point, following the same steps in the section "Rolling back to a restore point," earlier in this chapter. It won't work, though, unless you've turned on system protection/restore points for one or more drives on your computer.

>> **System Image Recovery** via recimg.exe is no longer available, but if you're interested in a geeky solution, look at Push-button reset, as explained in the Microsoft Hardware Dev Center at https://docs.microsoft.com/en-us/windows-hardware/manufacture/desktop/push-button-reset-overview.

>> **Startup Repair** reboots into a specific Windows Recovery Environment program known as Start Repair and runs a diagnosis and repair routine that seeks to make your PC bootable again. I've seen this program run spontaneously when I'm having hardware problems. A Start Repair log file is generated at \Windows\System32\Logfiles\Srt\SrtTrail.txt. If you find yourself running Automatic Repair, you can't do anything: Just hold on and see whether it works.

>> **UEFI Firmware Settings** displays the BIOS for your computer, where you can set how its motherboard, processor, and other components work.

>> **Command Prompt** brings up an old-fashioned DOS command prompt, just like you get if you go into safe mode. Only for the geek at heart. You can hurt yourself in there.

>> **Startup Settings** reboots Windows and lets you go into safe mode, change video resolution, start debugging mode or boot logging, run in safe mode, disable driver signature checks, disable early launch antimalware scans, and disable automatic restart on system failure. Not for the faint of heart.

If you ever wondered how to do an old-fashioned F8 boot into safe mode, now you know.

Chapter **3**

Monitoring Windows

W indows 10 ships with a small array of tools designed to help you look at your system and warn you if something's wrong. In this chapter, I talk about two of them: Event Viewer and Reliability Monitor.

TIP

One long-time monitoring tool is gone. RIP. Windows Vista included a rag-tag, system performance benchmarking routine known as the Windows Experience Index (WEI). It continued through Windows 7 and Windows 8. I used WEI all the time as a quick way to check PCs in shops to see which ones were great and which were merely mediocre. It wasn't the best test, but it was good enough for quick comparisons, and every Windows PC had a copy.

Microsoft dropped the WEI in Windows 8.1. One of its biggest motivations: The original Surface Pro — Microsoft's flagship Windows machine at the time — scored a meager 5.9 on a scale from 1 to 9.9, which put it below just about any laptop or desktop you could mention. For years, I used a slapped-together Pavilion with a WEI of 4.8. It cost less than $300.

Instead of improving their devices — or changing the benchmark — Microsoft simply dropped the Windows Experience Index.

In Microsoft's zeal to make Windows less intimidating to new users, some of those tools are tucked away in rather obscure corners. But if you know what you're doing, you can find them and use them to help make your machine hum.

Or at least burble.

One of the tools, Event *Viewer*, is a favorite foil of scammers and charlatans, who use it to convince you that your PC needs fixing (for a fee, of course) when it's just fine. I talk about that in this chapter, too.

Viewing Events

Every Windows user needs to know about Event Viewer, if only to protect themselves from scammers and con artists who make big bucks preying on peoples' fears.

As I explain in Book 9, Chapter 1, scammers are calling people in North America, Europe, Australia, and other locations all around the world, trying to talk Windows users into allowing these con artists to take over victims' systems via Remote Assistance. The scammers typically claim to be from Microsoft or associated with Microsoft. They may get your phone number by looking up names of people posting to help forums.

Some of them just cold call: Any random phone call to a household in North America or Europe stands a good chance of striking a resonating chord when the topic turns to Windows problems. If you randomly called ten people in your town and said you were calling on behalf of Microsoft to help with a Windows problem, and you sounded as if you knew what you were talking about, I bet at least one or two of your neighbors would take you up on the offer. In my neighborhood, it'd probably be closer to nine. Maybe eleven or twelve.

The scam hinges around the Windows Event Viewer feature. It's an interesting, useful tool — but only if *you* take the initiative to use it, and don't let some fast talker use it to bilk you out of hundreds of bucks.

Using Event Viewer

Windows has had an Event Viewer for more than a decade. Few people know about it. At its heart, Event Viewer looks at a small handful of logs that Windows maintains on your PC. The logs are simple text files, written in XML format. Although you may think of Windows as having one event log file, in fact, there are many — Administrative, Operational, Analytic, and Debug, plus application log files.

Every program that starts on your PC posts a notification in an event log, and every well-behaved program posts a notification before it stops. Every system access, security change, operating system twitch, hardware failure, and driver hiccup all end up in one or another event log. Event Viewer scans those text log files, aggregates them, and puts a pretty interface on a deathly dull, voluminous set of machine-generated data. Think of Event Viewer as a database reporting program, where the underlying database is just a handful of simple flat text files.

REMEMBER

In theory, the event logs track significant events on your PC. In practice, the term *significant* is in the eyes of the beholder. Or programmer. In the normal course of, uh, events, few people ever need to look at any of the event logs. But if your PC starts to turn sour, Event Viewer may give you important insight to the source of the problem.

Here's how to use Event Viewer:

1. Right-click or tap and hold down the Start button. Choose Event Viewer.

Event Viewer appears.

2. On the left, choose Event Viewer, Custom Views, Administrative Events.

It may take a while, but eventually you see a list of notable events like the one in Figure 3-1.

3. Don't freak out.

Even the best-kept system (well, my production system anyway) boasts reams of scary-looking error messages — hundreds, if not thousands of them. That's normal. See Table 3-1 for a breakdown.

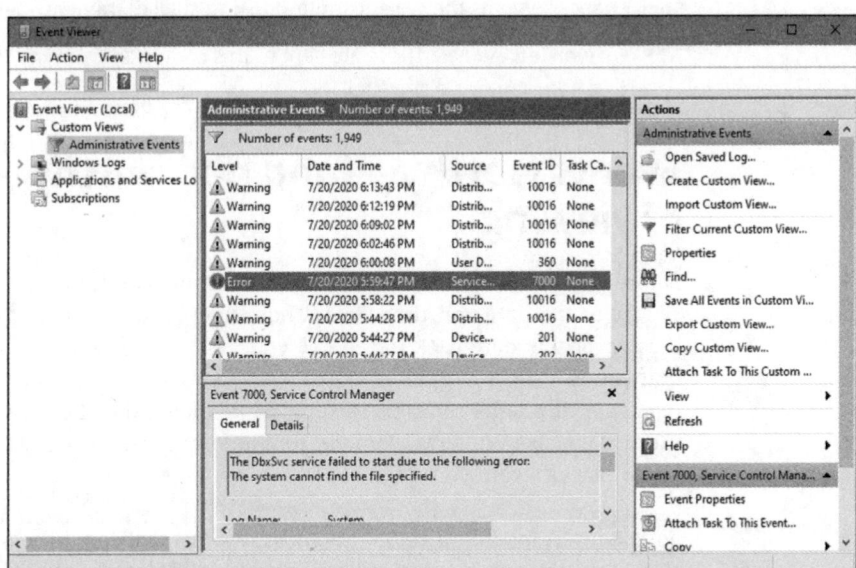

FIGURE 3-1:
Events are logged by various parts of Windows.

TABLE 3-1   **Events and What They Mean**

Event	What Caused the Event
Error	Significant problem, possibly including loss of data
Warning	Not necessarily significant, but might indicate that there's a problem brewing
Information	Just a program calling home to say it's okay

The Administrative Events log isn't the only one you can see; it's a distillation of the other event logs, with an emphasis on the kinds of things a mere human might want to see.

Other logs include the following:

>> **Application events:** Programs report on their problems.

>> **Security events:** They're called *audits* and show the results of a security action. Results can be either successful or failed depending on the event, such as when a user tries to log in.

>> **Setup events:** This primarily refers to domain controllers, which is something you don't need to worry about.

>> **System events:** Most of the errors and warnings you see in the Administrative Events log come from system events. They're reports from Windows system files about problems they've encountered. Almost all of them are self-healing.

>> **Forwarded events:** These are sent to this computer from other computers.

Events worthy — and not worthy — of viewing

Before you get all hot and bothered about the thousands of errors on your PC, look closely at the date and time field. There may be thousands of events listed, but those probably date back to the day you first installed the PC. Chances are good that you can see a handful of items every day — and most of the events are just repeats of the same error or warning. Most likely, they have little or no effect on the way you use Windows. An *error* to Windows should usually trigger a yawn and "Who cares?" from you.

For example, looking through my most recent event log, I see a bunch of error id 10010 generated by a source called DistributedCOM, telling me that the server didn't register with DCOM within the required timeout. Really and truly, no biggie. Fugeddaboutit.

That's exactly my advice. If you aren't experiencing problems, don't sweat what's in Event Viewer. Even if you *are* experiencing problems, Event Viewer may or may not be able to help you.

TIP

How can Event Viewer help? See the Event ID column? Make note of the ID number and look it up at www.eventid.net. They may be able to point you in the right direction or at least translate the event ID into something resembling plain English. Figure 3-2 shows the results when I went looking for event ID 10010 — the DCOM problem.

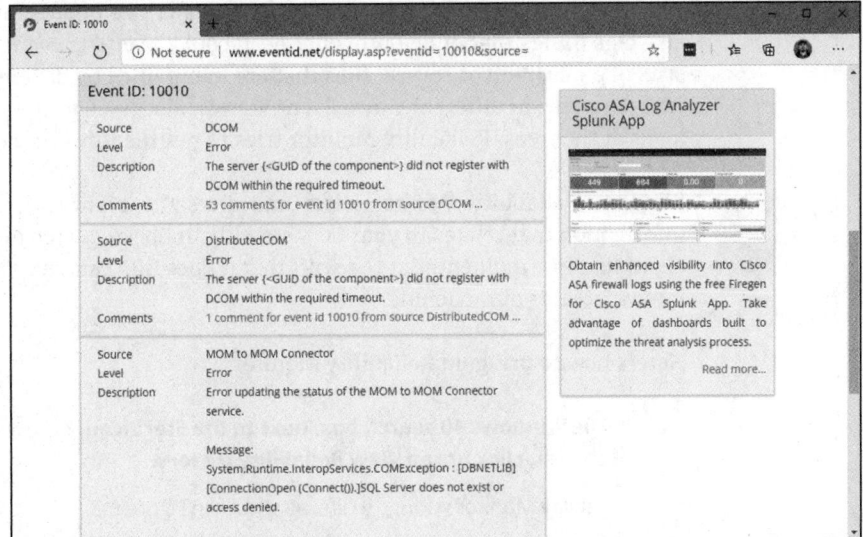

FIGURE 3-2:
The result of an eventid.net lookup for the error 10010.

You can click at the bottom of each description to see comments left by someone else who's tackled the problem.

If you're trying to track down a specific problem, and you see an event that may relate to the problem, use Google to see whether you can find somebody else who's had the same problem. Event Viewer can also help you nail down network access problems because the Windows programs that control network communication spill a large amount of details into the event logs. Unfortunately, translating the logs into English can be a daunting task, but at least you may be able to tell where the problem occurs — even if you haven't a clue how to solve it.

Gauging System Reliability

Every Tom, Dick, 'n Hairy Windows routine leaves traces of itself in the Windows event log. Start a program, and the ignoble event gets logged. Stop it, and the log gets updated. Install a program or a patch, and the log knows all, sees all. Every security-related event you can imagine goes in the Log. Windows Services leave their traces, as do errors of many stripes. Things that should've happened but didn't get logged, as well as things that shouldn't have happened but did. Soup to nuts.

The event log contains items that mere humans can understand. Sometimes. It also logs things that only a propeller head could love. The event log actually consists of a mash-up of several files that are maintained by different Windows system programs in different ways. Event Viewer, discussed in the preceding section, looks at the trees. Reliability Monitor tries to put the forest in perspective.

TIP

Windows Reliability Monitor slices and dices the event log, pulling out much information that relates to your PC's stability. It doesn't catch everything — more about that in a moment. But the stuff that it does find can give you instant insight into what ails your machine.

Here's how to bring up Reliability Monitor:

1. **In the Windows 10 search box, next to the Start icon, type** reli. **At the top of the list, click or tap View Reliability History.**

   Reliability Monitor springs to life, as shown in Figure 3-3.

2. **In the View By line, above the reliability graph, flip between Days and Weeks.**

   Reliability Monitor goes back and forth between a detailed view and an overview.

ASK WOODY.COM

Again, please don't freak out. There's a reason why Microsoft makes it hard to get to this report. It figures if you're sophisticated enough to find it, you can bear to see the cold, hard facts.

The top line in the monitor is supposed to give you a rating, from one to ten, of your system's stability. In fact, it doesn't do anything of the sort, but if you see the line drop like a wood barrel over Victoria Falls (as it has in Figure 3-3), something undoubtedly has gone bump in the night.

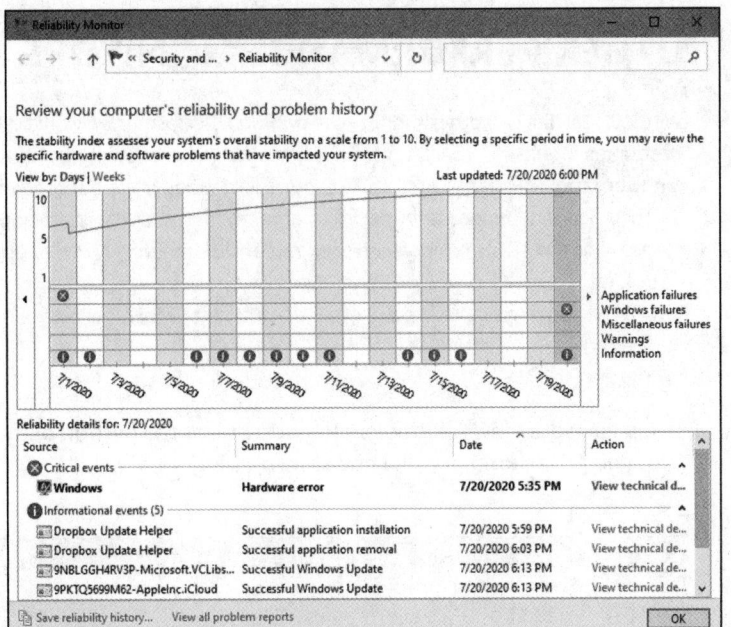

FIGURE 3-3:
When something
goes out to lunch,
it leaves a trace
in Reliability
Monitor.

REMEMBER

Your rating more or less reflects the number and severity of problematic event log events in four categories: Application, Windows failures, Miscellaneous failures, and Warnings. The Information icons (circled i's) generally represent updates to programs and drivers; if you installed a new printer driver, for example, there should be an Information icon on the day it was installed. Microsoft has a detailed list of the types of data being reported in its TechNet documentation at https://docs.microsoft.com/en-us/previous-versions/windows/it-pro/windows-vista/cc749583(v=ws.10). Here's what they say:

> Since you can see all of the activity on a single date in one report, you can make informed decisions about how to troubleshoot. For example, if frequent application failures are reported beginning on the same date that memory failures appear in the Hardware section, you can replace the faulty memory as a first step. If the application failures stop, they may have been a result of problems accessing the memory. If application failures continue, repairing their installations would be the next step.

If you tap or click a day (or a week), the box at the bottom shows you the corresponding entries in your Windows event log. Many events at the bottom have a more detailed explanation, which you can see by tapping/clicking the View Technical Details link.

If you click the View All Problem Reports link at the bottom of the Reliability Monitor, you get a summary like one shown in Figure 3-4.

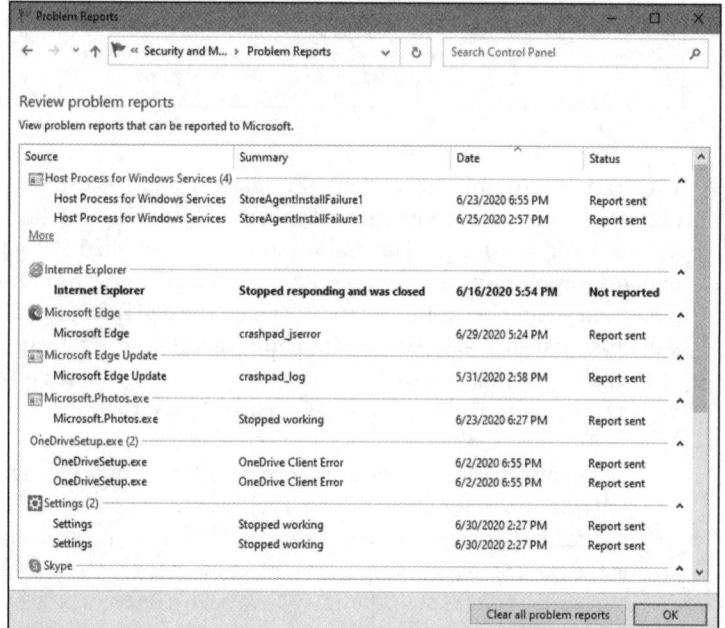

FIGURE 3-4: Here, you can find the key deleterious events and what they mean.

ASK
WOODY.COM

Reliability Monitor isn't meant to provide a comprehensive list of all the bad things that have happened to your PC, and in that respect, it certainly meets its design goals. It isn't much of a stability tracker, either. The one-to-ten rating uses a trailing average of daily scores where more recent scores have greater weight than old ones, but in my experience, the line doesn't track reality: My system can bounce like a Willy's in four-wheel drive, and it doesn't affect the rating; conversely, my system can be purring like a cat while my rating score goes to the dogs.

The real value of Reliability Monitor lies in showing you a time sequence of key events — connecting the temporal dots so you may be able to discern a cause and effect. For example, if you suddenly start seeing blue screens repeatedly, check Reliability Monitor to see whether something untoward has happened to your system. Installing a new driver, say, can make your system unstable, and Reliability Monitor can readily show you when it was installed. If you see your rating tumble on the same day that a driver update got installed, something's fishy, and you may be able to readily identify the scaly culprit.

The proverbial bottom line: Reliability Monitor doesn't keep track of everything, and some of it is a bit deceptive, but it can provide some worthwhile information when Windows 10 starts kicking. Reliability Monitor is well worth adding to your Win bag of tricks.

Chapter 4

Using System Tools

Windows 10 abounds with tools that can help you do everything from taking out the dog to making the perfect espresso — at least if your computer runs hot. In this chapter, I step you through three specific tools that can come in very handy:

REMEMBER

» The new-in-Windows-8, better-in-Windows-10 greatly improved and expanded *Task Manager* has turned into the Swiss Army knife of Windows applications. In Windows 7, you had to bring up, navigate to, and download and/or install a half dozen different tools to even come close to what the Windows 10 Task Manager does right out of the box. In earlier versions of Windows, many of the tools existed only in vestigial form.

» Windows includes all the tools you need to install a new hard drive, and the steps are easier than you think. All it takes is a trip to the *Disk Management* application. In this chapter, I show you how.

ASK WOODY.COM

» Finally, I have a bonus section on the virtual machine generator *(Hyper-V)* that ships with Windows 10 Pro only. (Sorry, if you have Windows 10 Home, you don't get it.) A *virtual machine (VM)* is a make-believe, fully contained PC that runs inside your regular PC. You can use it to run Windows XP programs, for example, without setting up a dual boot on your Windows 10 system. You can also use a VM to check out new tricks or try some different Windows settings without gumming up your working machine. Hyper-V works like a treat if you know how to treat it.

Tasking Task Manager

Windows 10 has a secret command post that you can get to if you know the right handshake, uh, key combination. Whatever. The key combination (or tap sequence) works all the time, unless Windows is seriously out to lunch.

Task Manager can handle any of these jobs:

REMEMBER

>> **Kill an app or a program.** That comes in very handy if, say, Spartan freezes and you can't get it to do anything. Doesn't matter if you're trying to kill a Windows 10 app (formerly Universal, formerly Metro) or an old-fashioned desktop app. Either way, one trip to Task Manager and *zap!*

Windows tries to shut down the application without destroying any data. If it's successful, the application disappears from the list. If it isn't successful, it presents you with the option of summarily zapping the application (called End Now to the less imaginative) or simply ignoring it and allowing it to go its merry way.

>> **Switch to any program.** This is convenient if you find yourself stuck somewhere — in a game, say, that doesn't let go — and you want to jump over to a different application. You can easily go to a Windows 10 app or desktop program.

>> **See which processes are hogging your processor.** There's a bouncing list of program pieces — called *processes* — and an up-to-the-second ranking of how much computer time each one's taking. That list is invaluable if your PC is working like a slug, and you can't figure out which program is hogging the processor.

>> **See which processes take up most of your memory, use your disk, or gab over the network.** Sometimes, it's hard to figure out which program's at fault. Task Manager knows all, sees all, and tells all.

>> **Get running graphs of CPU, memory, disk, GPU, or network usage.** They're cool and informative, and may even help you decide whether you need to buy more memory.

>> **See which tiled Microsoft Store apps use the most resources over a specified period of time.** Did the Camera take up the most time on your PC in the past month? Pinball?

ASK WOODY.COM

>> **Turn off auto-starting programs.** This used to be a huge headache, but now it's surprisingly easy. The simple fact is that almost everybody has automatically starting programs that take up boot time, add to your system overhead, cause aggravation, and may even be dangerous. Task Manager shows you major programs that start automatically and gives you the option to disable those programs.

>> **Send a message to the other users on your PC.** The message shows up on the lock screen when you log off.

>> **Force Task Manager to stay on top of all other windows.** This includes immersive, full-screen apps and games.

Who da man?

Here's how to bring out the full glory of the Swiss Army knife version of Task Manager:

1. **Do one of the following:**

   - *If you have a keyboard:* Press Ctrl+Alt+Delete; tap or click the Task Manager link in the screen that appears. Or right-click in the lower-left corner of the screen and choose Task Manager. Alternatively, press Ctrl+Alt+Escape to see Task Manager.

   - *If you don't have a keyboard:* In the search box, type **task**, and at the top of the list of search results, tap or click Task Manager.

   In either case, Task Manager appears in its compact view, showing a list with running apps (see Figure 4-1). To get the full list, click or tap More Details at the bottom. The full Task Manager is displayed, with the Processes tab open. Notably, the list includes all the running tiled apps, as well as all the running desktop programs.

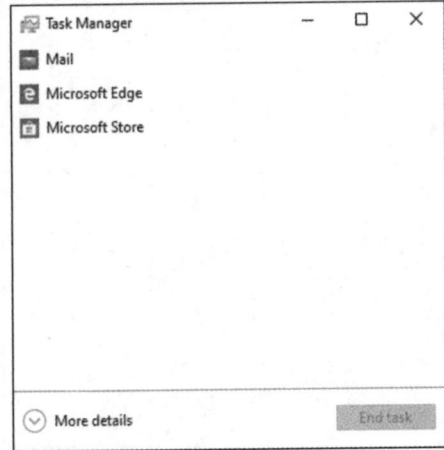

FIGURE 4-1:
Task Manager lets you control running programs.

2. **To kill one of your running programs, tap or click it and then tap or click End Task.**

   The program may continue for a few seconds — some programs hold on tenaciously — but in the end, almost every program succumbs to the preemptive force.

Task Manager Processes

If you see the compact view of Task Manager (refer to Figure 4-1), click or tap More Details. Then, in the Processes tab (shown in Figure 4-2), Task Manager groups running programs depending on the type of program:

» **Apps** are regular, everyday programs. They're ones you started or ones that are set up to start automatically. (You may think that *apps* mean just desktop apps or touch-friendly Windows 10 apps — but no. These are just programs, of any stripe — whatever happens to be running.)

» **Background processes** keep the pieces of your programs and drivers working.

» **Windows processes** are similar to background processes, except they're parts of Windows itself.

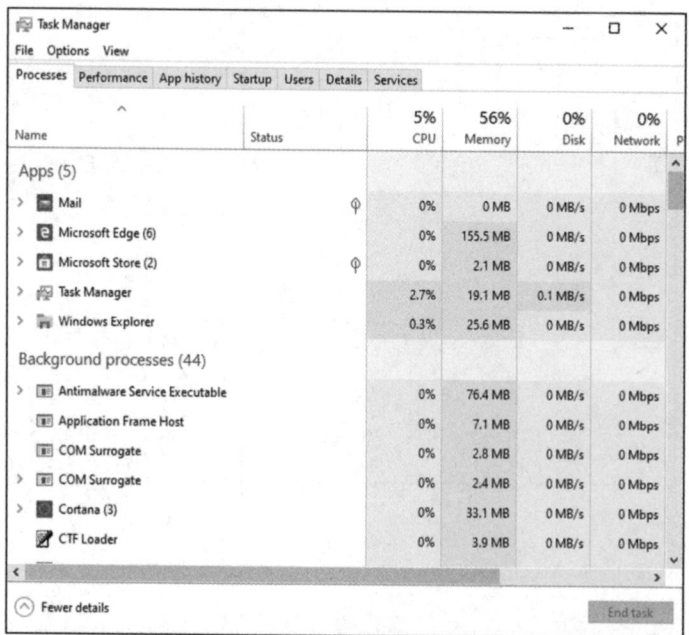

FIGURE 4-2:
Keep tabs on all the processes that run inside Windows 10.

TIP

You can tap or click a column heading (such as CPU, Memory, Disk, or Network), and Task Manager sorts on that particular value. To update the report, choose View, Refresh Now.

As you start new programs, they appear on the Apps list, and any background programs that they bring along appear on the Background processes list. Universal apps, in particular, go to sleep when they aren't being used, so they drop off the Task Manager list. One glance at the Processes tab should give you a good idea if any programs are hogging your machine — for CPU processor cycles, memory, disk access, or tying up the network.

Task Manager Performance

The Performance tab (see Figure 4-3) gives you running graphs of CPU usage, allocated memory, disk activity, video card GPU activity (if you have one or more separate GPUs), and the volume of data running into and out of your machine, on an Ethernet or a Wi-Fi connection or both. If you have Bluetooth turned on, it'll also show you activity over your Bluetooth connection.

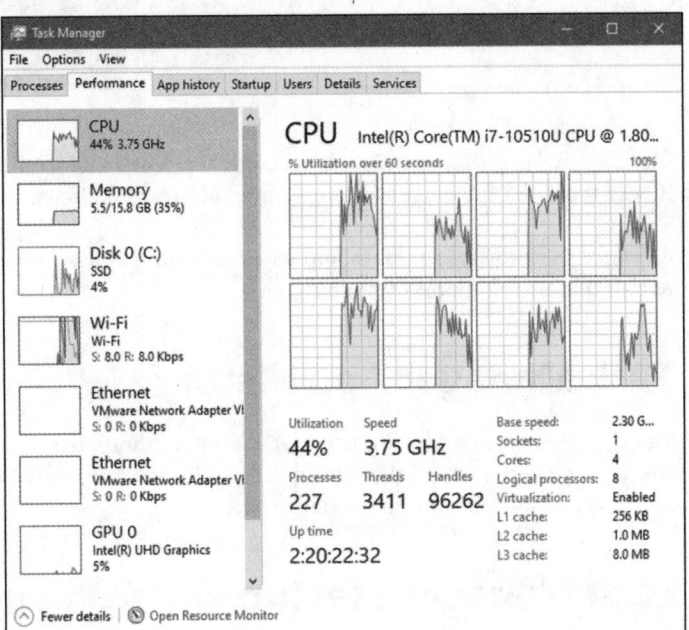

FIGURE 4-3:
Keep tabs on the key components of your PC's performance.

TECHNICAL STUFF

If you want to see much more detailed information — including utilization of each of the cores of a multi-core CPU — tap or click the Open Resource Monitor link at the bottom. See Figure 4-4.

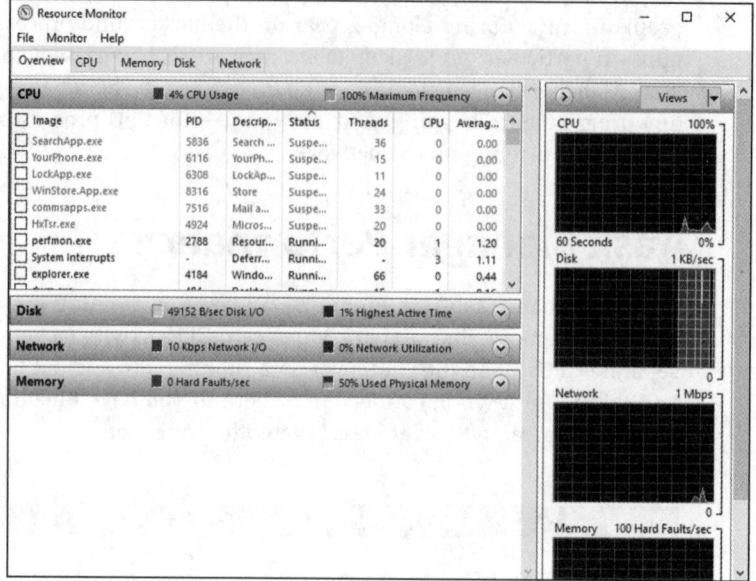

FIGURE 4-4:
Resource Monitor tells you at a glance what's going wrong with your machine.

I frequently keep Resource Monitor scrunched down and running on my desktop. It tells me about my current sorry state of affairs at a glance.

Resource Monitor is my go-to app when anything starts acting wonky. Which is not all that uncommon in Windows, eh?

Task Manager App History

The App History tab (see Figure 4-5) keeps a cumulative count of all the time you've spent on each of the various tiled Windows 10 apps from the Microsoft Store. Tap or click a column header to sort.

Task Manager Startup and Autoruns

No doubt you know that Windows 10 automatically runs certain programs every time you start it and that those programs can prove, uh, cantankerous at times. The Startup tab, as shown in Figure 4-6, represents a giant step forward for usability. It shows you all the programs that are started automatically each time you log in to Windows 10.

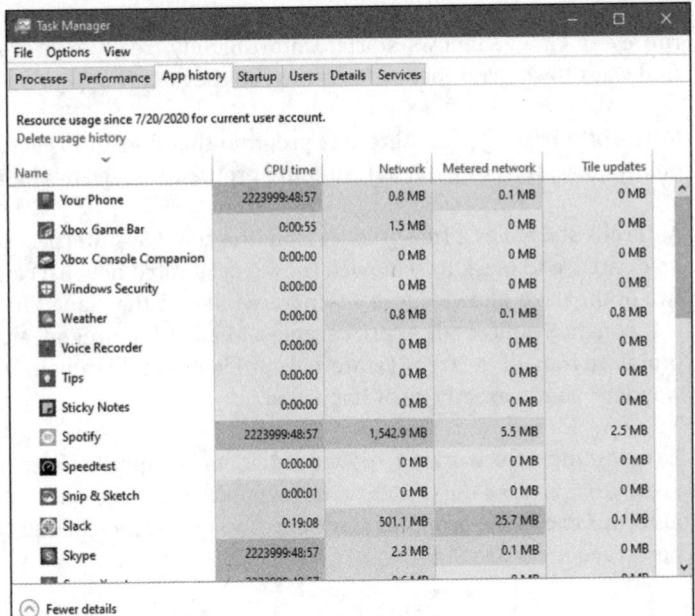

FIGURE 4-5:
A comprehensive
list of all the time
you've spent
using each of
the Windows 10
apps.

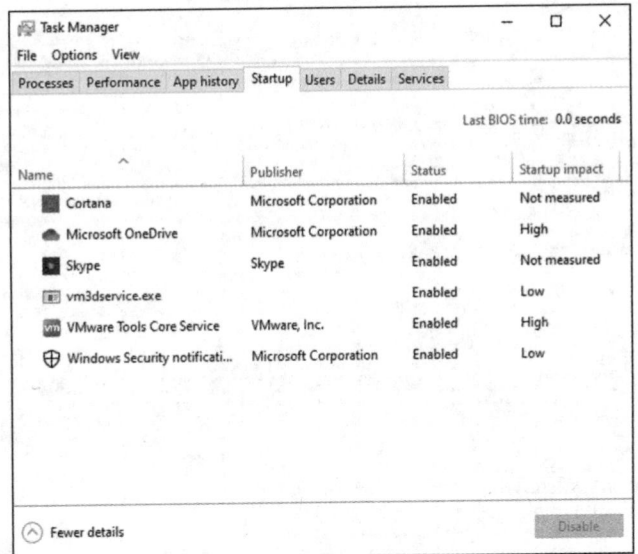

FIGURE 4-6:
A subset of those
cycle-stealing
auto-startup
programs.

If you want to disable an autorunning program, tap or click the program and then Disable, and reboot Windows 10.

The Task Manager Startup tab shows you the application programs, their helper programs, and sometimes problematic programs that use well-known tricks to run every time Windows starts. Unfortunately, really bad programs frequently find ways to squirrel themselves away, so they don't appear on this list.

REMEMBER

Microsoft distributes an Autoruns program that digs in to every nook and cranny of Windows, ferreting out autorunning programs — even Windows programs.

Autoruns started as a free product from the small Sysinternals company and owes its existence to Mark Russinovich (now a celebrated novelist) and Bryce Cogswell, two of the most knowledgeable Windows folks on the planet. In July 2006, Microsoft bought Sysinternals. Mark became a Microsoft Demigod, er, Fellow. Microsoft promised that all the free Sysinternals products would remain free. And wonder of wonders, that's exactly what happened.

To get Autoruns working, download it as a zip file from `http://technet.microsoft.com/en-us/sysinternals/bb963902.aspx` and extract the zip file. Autoruns.exe is the program you want. Tap or double-click to run it; no installation is required. See Figure 4-7.

FIGURE 4-7:
Autoruns finds many more sneaky autorunning programs than Task Manager. Compare this list to the one in Figure 4-6.

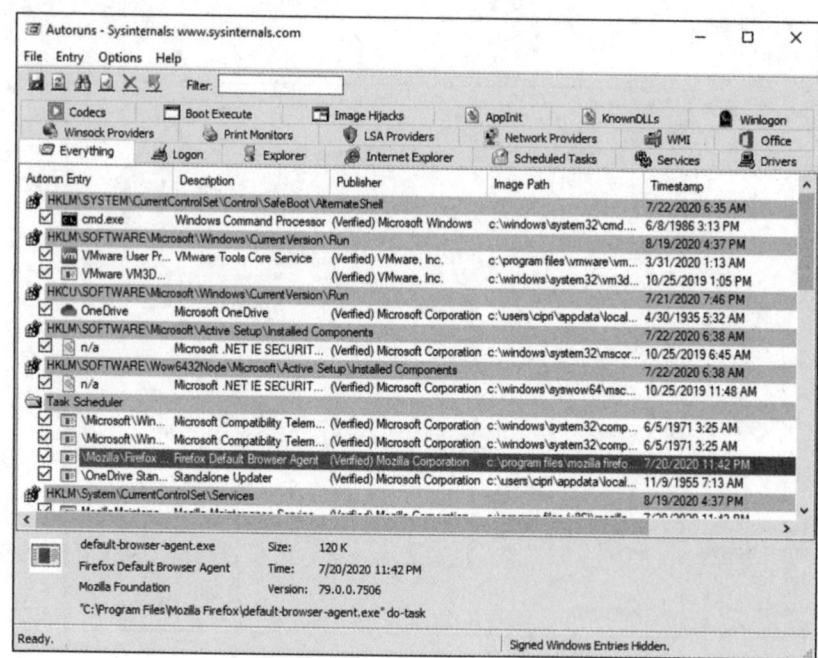

After Autoruns is working on your computer, the following tips can help you start using the program:

>> **Autoruns lists an enormous number of auto-starting programs.** Some appear in the most obscure corners of Windows. The Everything list shows all auto-starting programs in the order they're run.

>> **Autoruns has many options.** You can get a good overview on its product page at http://technet.microsoft.com/en-us/sysinternals/bb963902.aspx. The one I use most is the capability to hide all the auto-starting Microsoft programs. It's easy. Choose Options, and then select the Hide Microsoft Entries box. The result is a clean list of all the foreign stuff being launched automatically by Windows.

>> **Autoruns can suspend an auto-starting program.** To do so, deselect the box to the left of the program and reboot Windows. If you zap an auto-starting program and your computer doesn't work right, run Autoruns again and select the box. Easy.

Of course, you shouldn't disable an auto-starting program just because it looks superfluous, or even because you figure it contributes to global warming or slow startups, whichever comes first. As a general rule, if you don't know *exactly* what an auto-starting program does, don't touch it. It's not nice to fool with the support for those tiled Windows 10 apps.

On the other hand, if you concentrate on auto-starting programs that don't come from Microsoft, you may find a few things that you don't want or need — items that deserve to get consigned to the bit bucket.

Which programs deserve to die? Any that provide services you don't want. They go by various names, which change from time to time. Look for the Apple update checker, any utilities you no longer need or want, and perhaps the sync routines for cloud data services you no longer use. I've seen leftovers of antivirus programs that had been terminated with extreme prejudice long ago, game program helpers, communication tools for messaging systems long forgotten, and much more.

Task Manager Details and Services

If you used Task Manager in Windows 7 or earlier, you've seen this version, shown in Figure 4-8. The Details tab shows all the running processes, regardless of which user is attached to the process.

The Services tab, similarly, shows all the Windows 10 services that have been started. Once in a blue moon, you may find a Windows error message that some service or another (say, the printer service, or some sort of networking service) isn't running. This tab is where you can tell whether the service is really running.

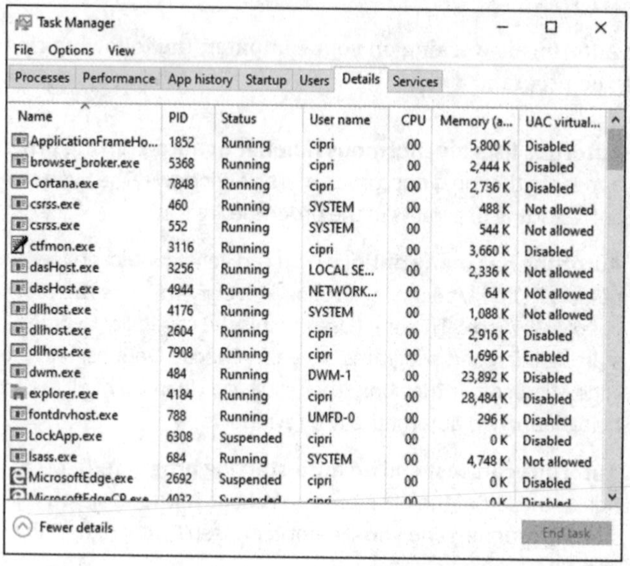

FIGURE 4-8:
All details about
every process
appear here.

Managing Startup Apps from Settings

Another easy way to manage startup apps is from Settings. One of the advantages of doing it from there is that it works well with tablets and other Windows 10 devices that have a touchscreen. Just tap your finger in a few places, and you're done. Here's how it works:

1. **Click or tap Start and then the Settings icon.**

   You see the home page of the Settings app.

2. **Go to Apps, and then to Startup (in the left column).**

   All apps that can be configured to start when you log into Windows 10 are displayed, as shown in Figure 4-9. For each app (or program), you see its name, the company who made it, its effect on the Windows 10 startup, and whether it's turned on or off.

3. **For each app that you want to stop from automatically starting with Windows, set its switch to Off.**

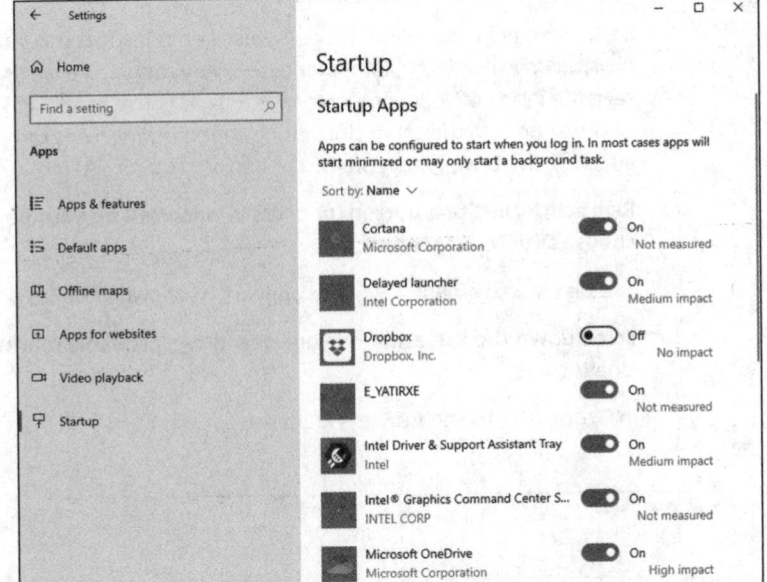

FIGURE 4-9:
Managing
Windows 10
startup apps
from Settings.

Installing a Second Hard Drive

You probably know how hard it is to install an external hard drive in a Windows 10 PC. Basically, you turn off the computer, plug the USB or eSATA cable into your computer, turn it on . . . and you're finished.

TIP

Yes, external hard drive manufacturers have fancy software. No, you don't want it. Windows 10 knows all the tricks. If you install one additional hard drive, internal or external, you can set up File History (see Chapter 1 of this minibook). Install two additional drives, internal or external, and you can turn on Storage Spaces (see Book 7, Chapter 4). None of the Windows programs need or want whatever programs the hard drive manufacturer offers.

Installing a second *internal* hard drive into a PC that's made to take two or more hard drives is only a little bit more complex than plugging an external drive into your USB port. Almost all desktop PCs can handle more than one internal hard drive. Some laptops can, too.

Here's how to do it:

1. Turn off your PC. Crack open the case, put in the new hard drive, attach the cables, and secure the drive, probably with screws. Close the case. Turn on the power and log in to Windows 10.

If you need help, the manufacturer's website has instructions. Adding the physical drive inside the computer case is very simple — even if you've never seen the inside of your computer — as long as you're careful to get a drive that will hook up with the connectors inside your machine. For example, you can attach an IDE drive to only an IDE connector; ditto for SATA.

2. **Right-click the Start button (or press Windows+X on your keyboard), and choose Disk Management.**

   The Disk Management dialog box appears, as shown in Figure 4-10.

3. **Scroll down the list, and find your new drive, probably marked Unallocated.**

   In Figure 4-10, the new drive is identified as Disk 1.

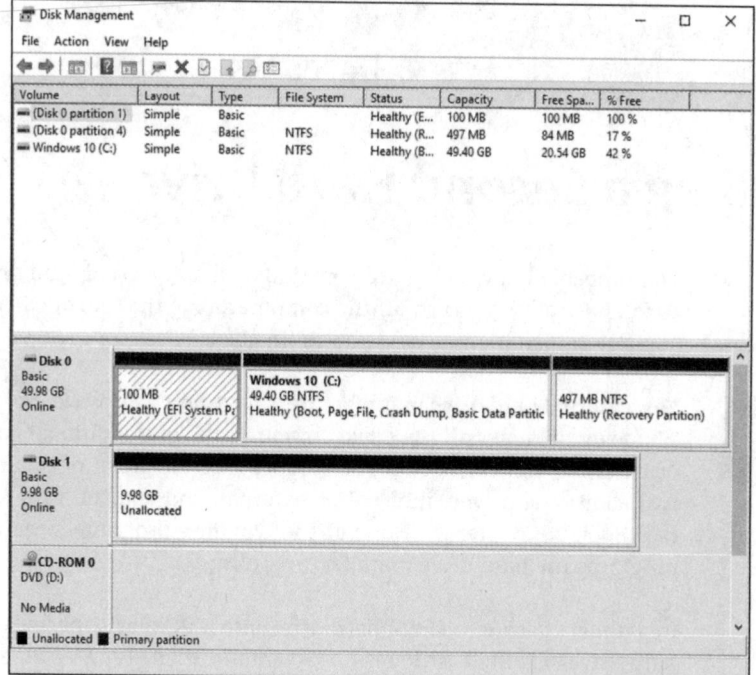

FIGURE 4-10:
Add the new drive here.

4. **On the right, in the Unallocated area, tap and hold down or right-click, and choose New Simple Volume.**

   The New Simple Volume Wizard appears, as shown in Figure 4-11.

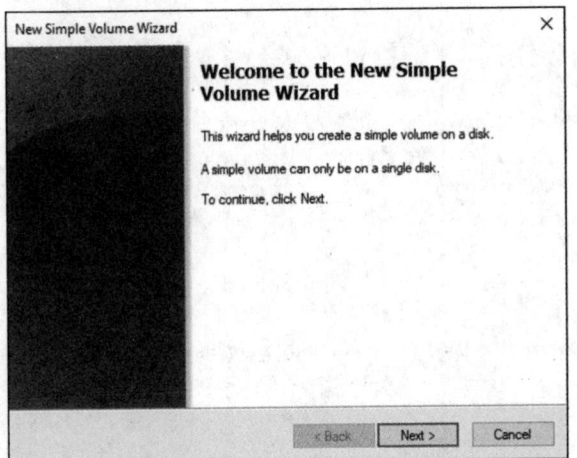

FIGURE 4-11:
The wizard takes
you through all
the steps.

5. **Tap or click Next.**

   You're asked to specify a volume size.

6. **Leave the numbers just as they are — you want to use the whole drive — and tap or click Next.**

   The wizard asks you to specify a drive letter. D: is most common, unless you already have a D: drive.

7. **If you really, really want to give the drive a different letter, go ahead and do so (most people should leave it at D:). Tap or click Next.**

   The wizard wants to know whether you want to use something other than the NTFS file system, or to set a different allocation unit. You don't.

8. **Tap or click Next; then tap or click Finish.**

   Windows 10 whirs and clunks, and when it's finished, you have a spanking new drive, ready to be used.

ASK
WOODY.COM

If you have three or more drives in or attached to your PC, consider setting up Storage Spaces. It's a remarkable piece of technology that'll keep redundant copies of all your data and protect you from catastrophic failure of any of your data drives. See Book 7, Chapter 4 for details.

Using System Tools

CHANGING YOUR C: DRIVE

Whoa nelly! If you've never seen a Windows 10 PC running an SSD (solid-state drive) as the system drive, you better nail down the door and shore up the, uh, windows. Changing your C: drive from a run-of-the-mill rotating platter to a fast, shiny new solid-state drive can make everything work so much faster. Really.

Unfortunately, getting from an HDD (hard disk drive) C: to an SSD C: isn't exactly 1-2-3.

Part of the problem is the mechanics of transferring your Windows 10 system from an HDD to an SSD: You need to create a copy (not exactly a clone) that'll boot Windows. Part of the problem is moving all the extra junk off the C: drive, so the SSD isn't swamped with all the flotsam and jetsam you've come to know and love in Windows.

Most of the drive cloning/backup/restore techniques developed over the past decade work when you want to move from a smaller drive to a bigger one. However, replacing your HDD C: drive with an SSD C: drive almost always involves going from a larger drive to a smaller one.

The Lifehacker website has an excellent rundown of the steps you need to take to get your old hard drive removed and have everything copied over to your new SSD. It's not a simple process. Check out www.lifehacker.com/5837543/how-to-migrate-to-a-solid+state-drive-without-reinstalling-windows.

Running a Virtual Machine

At its heart, a virtual machine (or VM) is a sleight of hand. A parlor trick. You set up a machine inside Windows that isn't really a machine; it's a program. Then you stick other programs inside the virtual machine. The programs think they're working inside a real machine, when they aren't — they're working inside another program.

Windows 10 Pro (and Enterprise) includes Hyper-V and all the ancillary software (drivers and such) you need to run a virtual machine inside Windows. If you have only Windows 10 Home, you need to look elsewhere. (Hint: Use Google, and find a copy of VirtualBox.) If you have Windows 10 in S mode, the *S* stands for *simply outta luck.*

In addition, to get the Hyper-V program going, you must be running the 64-bit version of Windows 10 Pro, with at least 2GB of memory. The hardware itself must be fairly up to date because it must support the *Second Level Address Translation*

(SLAT) capability. You can find a good overview of testing for SLAT on the How-To Geek site, `www.howtogeek.com/73318/how-to-check-if-your-cpu-supports-second-level-address-translation-slat`.

Why would you want to use a VM? Many reasons:

REMEMBER

>> Suppose you have an old program that runs only under Windows XP or Windows 95 (or even DOS, for that matter). You set up a VM, install XP or 95 (or DOS), and then stick the old program inside the VM. The old program doesn't know any better — it's fat, dumb, and happy working inside of XP. But you're watching from the outside. You can interact with the old program, type inside it, click inside it, give it disk space to play with, or attach it to a network interface card. A fake (virtual) one, of course, that works just like the real thing.

>> You want to try a different operating system. Maybe you want to play with Linux for a while or take Windows Server 2012 for a ride. Or you get nostalgic for the days of Windows Me. Or Microsoft Bob. Set up a virtual machine for each of the operating systems, and install the operating system in the VM. Then close each VM and save it. When you want to play with one of the OSs, just crank up the right VM, and you're on your way.

>> You need to isolate your real system while you try something that's tricky or experimental or potentially dangerous. If you have a VM that gets infected with a virus, the virus doesn't necessarily spread to your main machine. If you try a weird program inside a VM and it crashes, restarting the VM is much easier than restarting your PC, and if there are any bizarre side effects — say, weird Registry changes — they won't affect your main machine.

ASK WOODY.COM

>> I use VMs when I'm experimenting with hooking computers together. It's easy to set up several VMs, one running XP, say, another with Windows 7, another with Windows 8.1, and one more with Windows 10. Each of them thinks that it's connected to the other three. That way, I can test settings and figure out how to get them to communicate with each other.

Hyper-V is a complex product, worthy of a book unto itself. In this chapter, I just get you started and then point you to some sources of information that'll help you take full advantage of the product. Keep in mind that running multiple virtual machines can put a heavy drain on your system, especially on memory. Don't try running more than one machine on a 4GB machine; 8GB will keep you from nodding off.

Here's how to turn on Hyper-V:

1. **In the Search box on the Windows 10 taskbar, type** Hyper-V.

2. **Tap or click the Turn Windows Features On or Off link.**

The Windows Features dialog box appears, as shown in Figure 4-12.

FIGURE 4-12:
Hyper-V must be
turned on before
you can use it.

3. **Select the Hyper-V box and the two boxes below it, and then click OK.**

   Windows 10 installs two programs: Hyper-V Manager and Hyper-V Virtual
   Machine Connection.

4. **Reboot after the installation finishes.**

When Windows 10 comes back, you're ready to set up your first virtual machine.
Here's how:

1. **Tap or click the Start button, Windows Administrative Tools, and then
   Hyper-V Manager.**

   Hyper-V brings up the rather intimidating screen shown in Figure 4-13.

2. **On the right, in the Actions pane, tap or click Connect to Server. Choose
   the Local Computer button and then click OK.**

   Hyper-V brings up the even-more-intimidating dialog box shown in Figure 4-14.

3. **On the right, tap or click Virtual Switch Manager.**

   The Virtual Switch Manager for your PC appears, as shown in Figure 4-15.

   I assume you want your new VM to be able to communicate with the outside
   world — for an Internet connection, if nothing else — and it's easiest to set up
   that connection before you create the VM. The connection is done through a
   virtual switch, which ties a connection inside the virtual machine to a physical
   device on the outside, in the real world.

4. **On the right, in the list with the types of virtual switches, choose
   External. Then click Create Virtual Switch.**

   You're asked to set up properties for the new virtual switch, as shown in
   Figure 4-16.

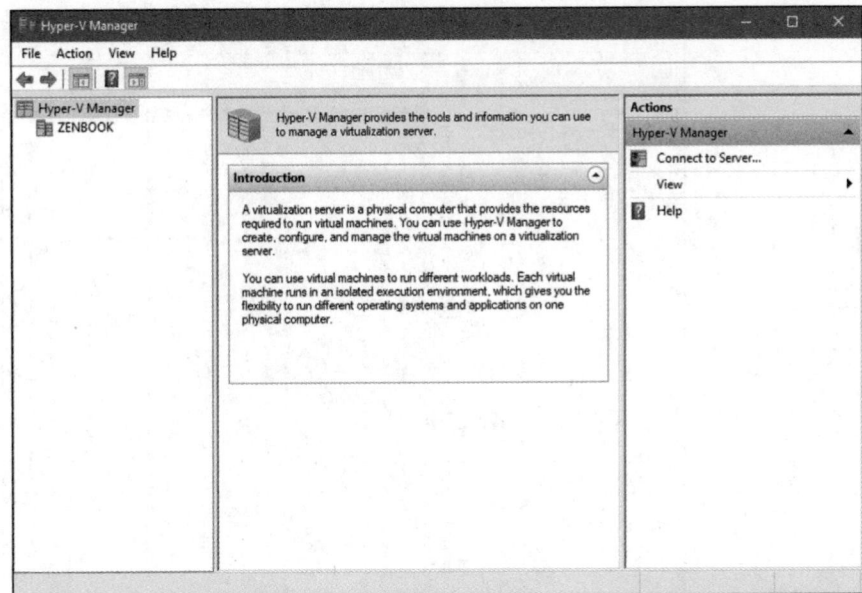

FIGURE 4-13:
Create new
virtual machines
here.

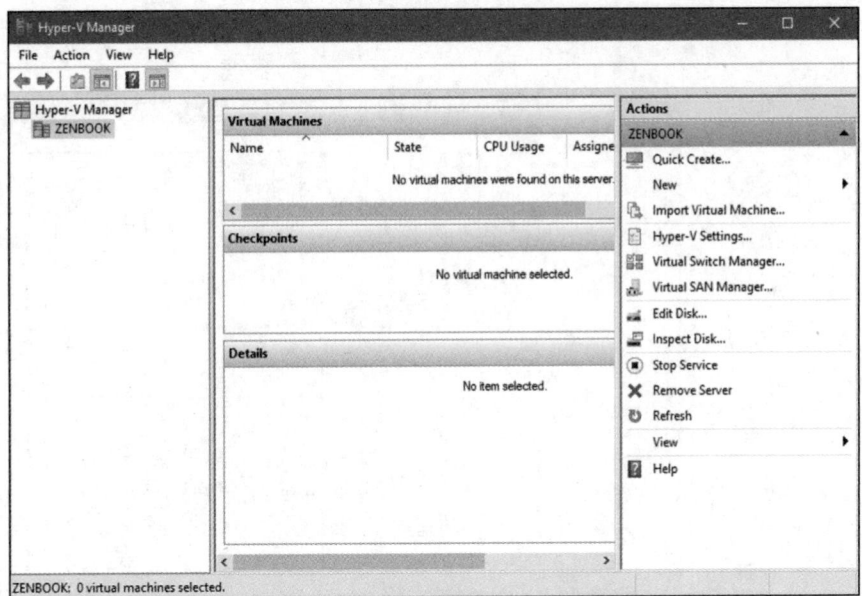

FIGURE 4-14:
Hyper-V shows
you its main
options.

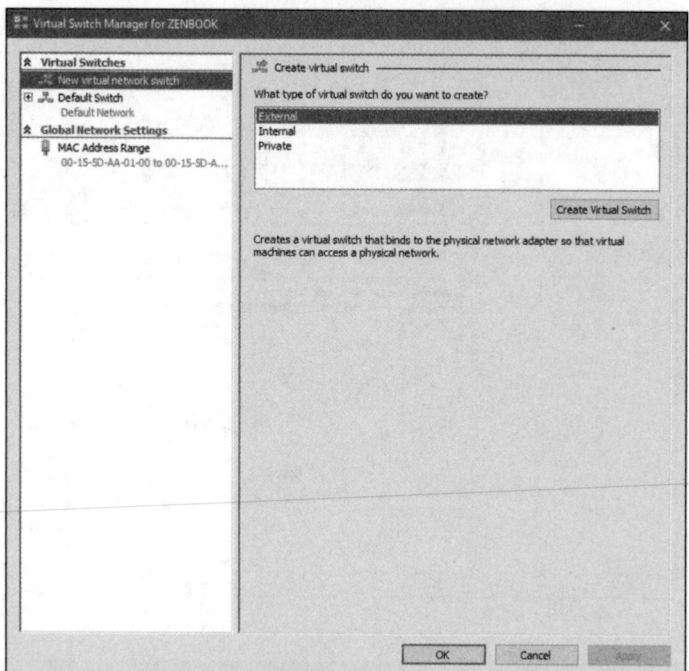

FIGURE 4-15:
Set up a virtual switch now, while it's easy.

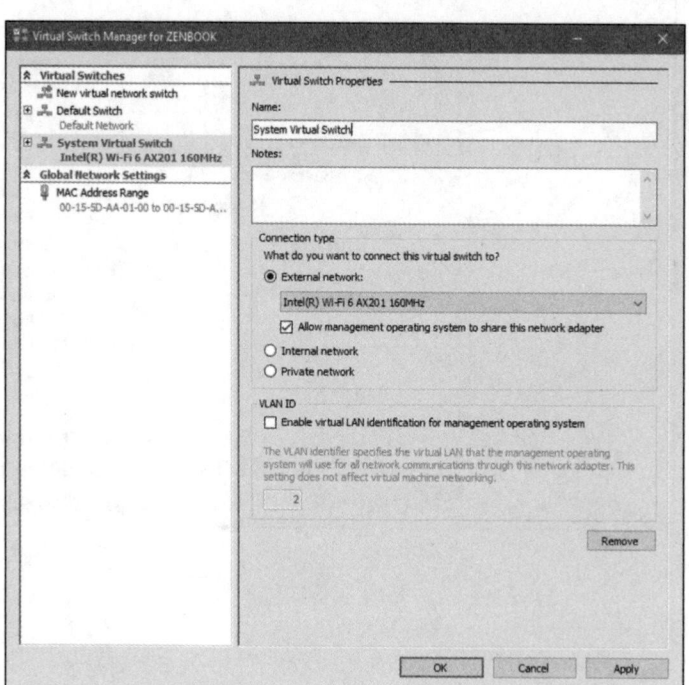

FIGURE 4-16:
Flesh out the virtual switch here.

5. **At the top, give the new virtual switch a name, and tap or click OK. If you receive a prompt asking you to confirm your pending changes, click Yes.**

Chances are good that you want your VM to connect to a physical network adapter in the outside world, so leave the default selections the way they are.

Hyper-V goes back to the Hyper-V Manager dialog box (refer to Figure 4-14).

6. **On the right, choose New, and then Virtual Machine.**

The New Virtual Machine Wizard starts.

7. **Tap or click Next.**

You're asked to specify a name and location for the VM, as shown in Figure 4-17.

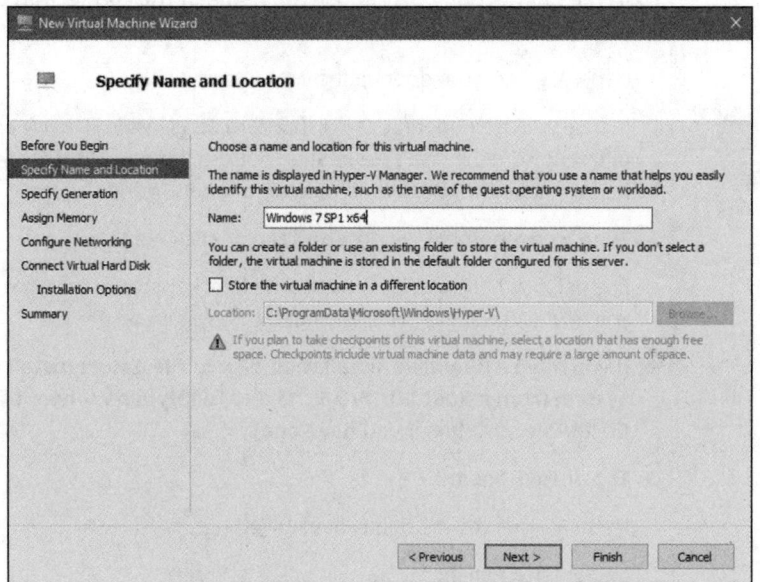

FIGURE 4-17:
Start by giving the
VM a name you
will recognize.

8. **Type a name that will immediately tell you what you're running on this VM; if you need to move the location of the VM (remember the VM is a program, and it needs to store its files somewhere), change the location.**

9. **Tap or click Next. Choose Generation 1, and then click Next again.**

VMs take up lots of room, and each time you take a snapshot, you store away the entire status of the VM — including any data on the disks, copies of installed programs, and all settings.

The wizard asks how much memory you want to assign for startup.

Using System Tools

10. **If you're going to run Windows 7, Windows 8.1, or Windows 10 and have at least 4GB of memory, set startup memory at 2048MB and select the Use Dynamic Memory for This Virtual Machine box.**

If you have many programs running, the memory constraints can slow things to a crawl.

Linux fans can get by with 512MB and no Dynamic Memory.

11. **Tap or click Next.**

You want enough memory so the VM doesn't start thrashing, but you don't want to specify too much in case you try to start many VMs at the same time.

Hyper-V wants to know whether you want to connect the VM to a network adapter. You set up the virtual connection already, so it's easy.

12. **In the Connection box, choose the name of the connection that you created in Step 5. Tap or click Next.**

Hyper-V wants you to set up the virtual hard disk.

REMEMBER

In case you're wondering, the virtual hard disks inside Hyper-V are quite different from the disk virtualization done in Windows Storage Spaces. Don't be confused. They work in completely different worlds.

13. **Type a new name if you like and tap or click Next.**

The defaults here are fine. You see the final key step in the wizard, which asks you how you want to install the operating system on the VM.

14. **If you have a Windows installation disk or file, select Install an Operating System from a Boot CD/DVD-ROM and tell Hyper-V where to find the Boot CD/DVD (or ISO file, if you have one).**

15. **Tap or click Next.**

Hyper-V gives you a last look at your settings.

16. **Tap or click Finish.**

Your new VM appears in the list of virtual machines shown on the main window, Figure 4-18.

To start the VM, tap or double-click it and, if necessary, click or tap Start. You see something like the VM in Figure 4-19, which runs Windows 7 in a VM inside Windows 10.

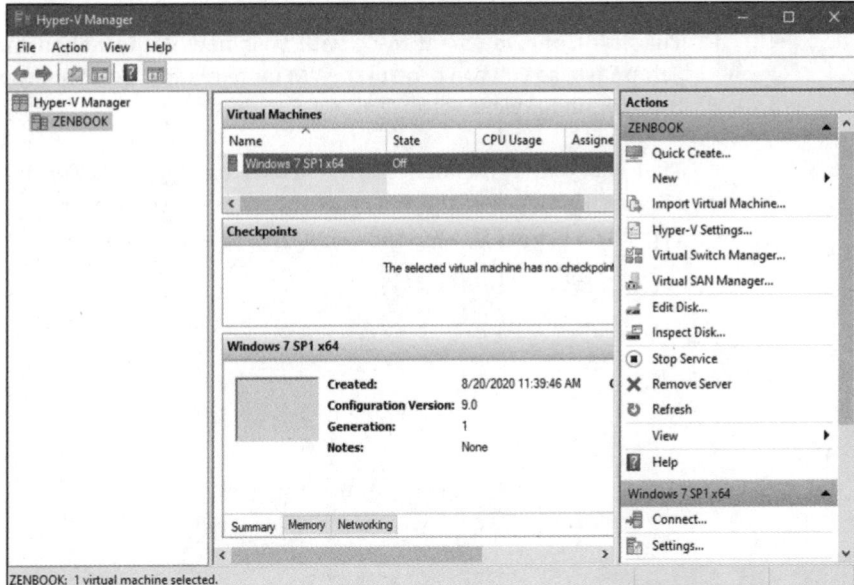

FIGURE 4-18:
The VM you created is now available.

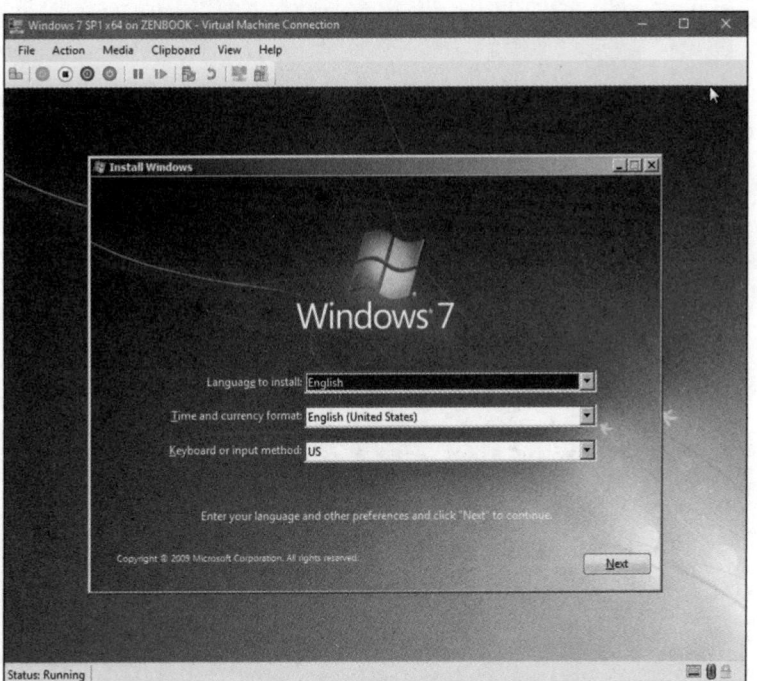

FIGURE 4-19:
A Windows 7 virtual machine running inside Windows 10.

REMEMBER

The first thing you want to do with your new VM is add an Integration Services Setup Disk, so you can control the VM more readily. To do so, choose Action, Insert Integration Services Setup Disk.

That just barely scratches the surface of Hyper-V. For more info, start at Microsoft's Hyper-V support center at https://technet.microsoft.com/en-us/library/mt169373(v=ws.11).aspx.

9

Securing Windows 10

Contents at a Glance

IN THIS CHAPTER

» Determining which hazards
and hoaxes to look out for

» Keeping up to date with reliable
sources

» Figuring out whether your system
is infected

» Protecting yourself

Chapter **1**

Spies, Spams, and Scams Are Out to Get You

Windows XP had more security holes than a prairie-dog field. Vista was built on top of Windows XP, and the holes were hidden better. Windows 7 included truly innovative security capabilities; it represented the first really significant break from XP's lethargic approach to security. Windows 8 included marginal security improvements to Windows itself, as well as better safety nets to keep you from shooting yourself in the foot and a fully functional, very capable antivirus program.

Windows 10's biggest contribution to your security? Microsoft finally, finally got rid of Internet Explorer. "Got rid of" is a bit of an overstatement; IE is still around, sitting in a formaldehyde jar, ready to be used if you really need it for compatibility. Alas, Microsoft Edge has inherited some IE deficiencies and now receives security fixes by the bushel. Windows 10 also has new built-in security capabilities that few Dummies will ever notice, but they work well in providing a more secure experience.

ASK
WOODY.COM

The single best security recommendation I can give you: Don't run Internet Explorer. Ever. Funny, that's the same advice I've been handing out since *Windows XP All-in-One For Dummies*. It's hard to recommend the initial version of Microsoft Edge, too, given its history of voluminous patches. However, the new one that rolled out with the Windows 10 May 2020 update, which is based on the same rendering engine as Google Chrome, is a better idea, including from a security standpoint.

Second-best security recommendation: Disable Flash and Java (see the "Disabling Java and Flash" section, later in this chapter) if you still have them around. Third: Use anything but Adobe Reader to open PDFs. By default, Windows 10 uses Microsoft Edge to open PDFs, and that's a good choice.

REMEMBER

Those four simple no's — no IE (or Edge), no Flash, no Java, no Adobe Reader — combined with periodic security updates, some restraint in clicking OK in every Tom Dick 'n Hairy dialog box, and casting a jaundiced eye when installing software of any type will protect you from at least 90 percent of all common infections.

Targeted infections, though, are another story. There's a lot of money to be made — and wealthy governments to please — with very narrowly defined information-gathering techniques. Unless you work for a defense contractor or a Tibetan relief organization or are trying to keep a big company such as Target away from the bad guys, you probably don't have much to worry about. But it doesn't hurt to keep your guard up.

In this chapter, I explain the source of real threats. (More details follow in the upcoming chapters in this minibook.) I bet it'll surprise you to find out that Adobe (Flash, Reader) and Oracle (Java) let more bad guys into Windows boxes than Microsoft. I also take you outside the box, to show you the kinds of problems people face with their computer systems and to look at a few key solutions. And I look even farther outside the box, to mass password leaks — think Verizon, Equifax, Uber, Target, Home Depot, MasterCard, Visa, Yahoo!, Facebook, LinkedIn, and the billion-plus compromised accounts (many with deciphered passwords, credit card numbers, and personal info) that are being sold every day like electronic trading cards. Then I concentrate on the problem *du jour* — ransomware, an increasingly distressing threat to every computer.

Most of all, I want you to understand that (1) you shouldn't take a loaded gun, point it at your foot, aim carefully, and pull the trigger, (2) any information online is vulnerable, and (3) if you're smart and can control your clicking finger, you don't need to spend a penny on malware protection.

Understanding the Hazards — and the Hoaxes

Many of the best-known Internet-borne scares in the past two decades — the WannaCrys, NotPetyas, Rustocks, Waledacs, Esthosts, Confickers, Mebroots, Netskys, Melissas, ILOVEYOUs, and their ilk — work by using the programmability built in to the computer application itself or by taking advantage of Windows holes to inject themselves into unprotected machines (see Figure 1-1).

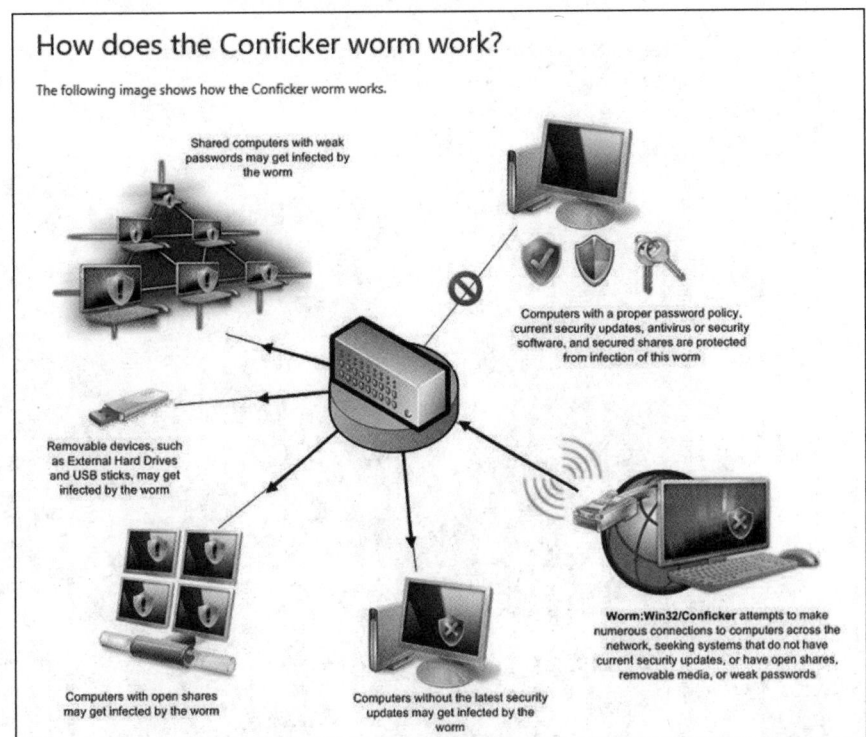

FIGURE 1-1:
The Conficker
worm employed
programmability
built into
Windows or
security holes
that had been
patched months
earlier.

Source: Microsoft

Fast-forward a dozen years or more, and the concepts have changed. The old threats are still there, but they've taken on a new twist: The scent of money, and sometimes political motivation, has made *cracking* (or breaking in to PCs for nefarious ends) far more sophisticated. What started as a bunch of miscreants playing programmer one-upmanship at your expense has turned into a profitable — sometimes highly profitable — business enterprise.

WARNING

Where's the money? At least at this moment — and for the foreseeable future — the greatest profits are made by using botnets and phishing attacks, to scramble data and demand a ransom. That's where you should expect the most sophisticated, most damnably difficult attacks. Unless you're running a nuclear reactor or an antigovernment campaign, of course. You get to choose the government.

The primary infection vectors

How do computers *really* get infected?

According to Microsoft's Security Intelligence Report 11, the single greatest security gap is the one between your ears. See Figure 1-2. (Pro tip: You can find the latest MS Security Intelligence Report at www.microsoft.com/sir. They're always intriguing.)

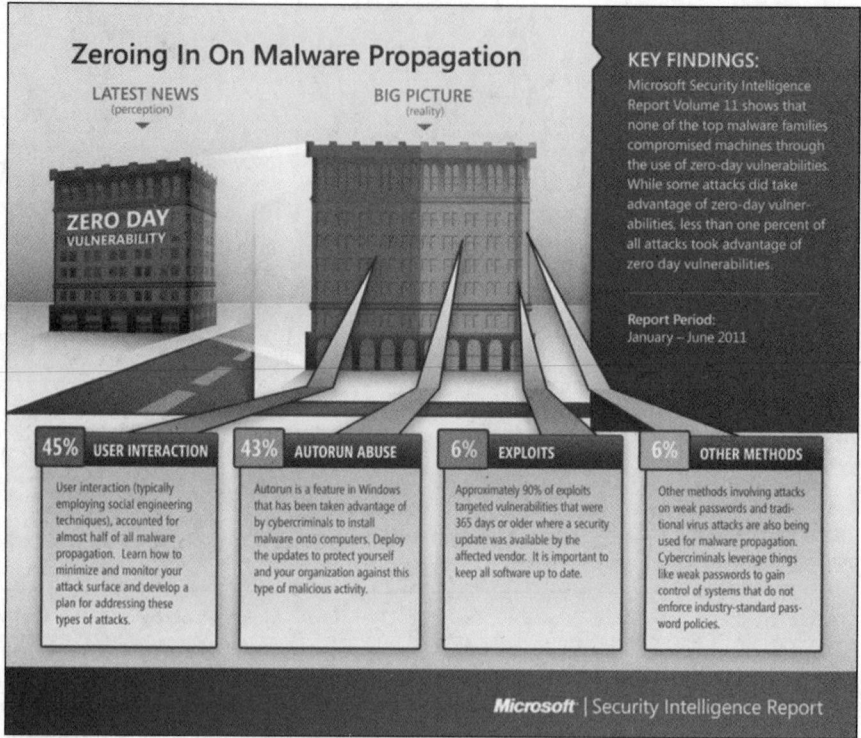

Zeroing In On Malware Propagation

LATEST NEWS
(perception)

BIG PICTURE
(reality)

ZERO DAY VULNERABILITY

KEY FINDINGS:

Microsoft Security Intelligence Report Volume 11 shows that none of the top malware families compromised machines through the use of zero-day vulnerabilities. While some attacks did take advantage of zero-day vulnerabilities, less than one percent of all attacks took advantage of zero day vulnerabilities.

Report Period:
January – June 2011

45% USER INTERACTION

User interaction (typically employing social engineering techniques), accounted for almost half of all malware propagation. Learn how to minimize and monitor your attack surface and develop a plan for addressing these types of attacks.

43% AUTORUN ABUSE

Autorun is a feature in Windows that has been taken advantage of by cybercriminals to install malware onto computers. Deploy the updates to protect yourself and your organization against this type of malicious activity.

6% EXPLOITS

Approximately 90% of exploits targeted vulnerabilities that were 365 days or older where a security update was available by the affected vendor. It is important to keep all software up to date.

6% OTHER METHODS

Other methods involving attacks on weak passwords and traditional virus attacks are also being used for malware propagation. Cybercriminals leverage things like weak passwords to gain control of systems that do not enforce industry-standard password policies.

Microsoft | Security Intelligence Report

FIGURE 1-2:
Most infections happen when people don't think about what they're doing.

Source: Microsoft

Many years ago, the biggest PC threat came from newly discovered security holes: The bad guys use the holes before you get your machine patched, and you're toast. They traded 'em like baseball cards. Those holes still get lots of attention, especially in the press, but they aren't the leading cause of widespread infection. Not even close. A large majority of infections happen when people get tricked into clicking something they shouldn't.

Narrow, targeted infections, though, tend to rely on previously unknown security holes. It's hard for the big boys to protect against that kind of attack. Little folks like you and me don't really stand a chance.

Then there are the EternalBlue-style exploits that borrow sneaky techniques developed by government think tanks. (Or are they stink tanks?) These security holes, discovered by well-funded governmental organizations, find their way from top-secret incubators to $50 black market Script Kiddie bundles.

Zombies and botnets

Every month, Microsoft posts a new Malicious Software Removal Tool that scans PCs for malware and, in many cases, removes it. In a recent study, Microsoft reported that 62 percent of all PC systems that were found to have malicious software also had backdoors. That's a sobering figure.

A *backdoor* program breaks through the usual Windows security measures and allows a scumbag to take control of your computer over the Internet, effectively turning your machine into a zombie. The most sophisticated backdoors allow creeps to adapt (upgrade, if you will) the malicious software running on a subverted machine. And they do it by remote control.

Backdoors frequently arrive on your PC when you install a program you want, not realizing that the backdoor came along for the ride.

Less commonly, PCs acquire backdoors when they come down with some sort of infection: The Conficker, Mebroot, Mydoom, Sobig, TDL4/Alureon, Rustock, Waledac, and ZeuS worms installed backdoors. Many of the infections occur on PCs that haven't kept Java, Flash, or the Adobe Acrobat Reader up to date. The most common mechanism for infection is a *buffer overflow* (see the nearby "What's a buffer overflow?" sidebar).

An evildoer who controls one machine by way of a backdoor can't claim much street "cred." But someone who puts together a *botnet* — a collection of hundreds or thousands of PCs — can take his zombies to the bank:

>> A botnet running a *keylogger* (a program that watches what you type and sporadically sends the data to the botnet's controller) can gather all sorts of valuable information. The single biggest problem facing those who gather and disseminate keylogger information? Bulk — the sheer volume of stolen information. How do you scan millions of characters of logged data and retrieve a bank account number or a password?

>> Unscrupulous businesses hire botnet controllers to disseminate spam, "harvest" email addresses, and even direct coordinated distributed denial-of-service (DDoS) attacks against rivals' websites. (A *DDoS attack* guides thousands of PCs to go to a particular website simultaneously, blocking legitimate use.)

There's a fortune to be made in botnets. The Rustock botnet alone was responsible for somewhere between 10 and 30 *billion* pieces of spam per day. Spammers paid the Rustock handlers, either directly or on commission, based on the number of referrals.

The most successful botnets run as *rootkits*, programs (or collections of programs) that operate deep inside Windows, concealing files and making it extremely difficult to detect their presence.

TECHNICAL STUFF

You probably first heard about rootkits in late 2005, when a couple of security researchers discovered that certain CDs from Sony BMG surreptitiously installed rootkits on computers: If you merely played the CD on your computer, the rootkit took hold. Several lawsuits later, Sony finally saw the error of its ways and vowed to stop distributing rootkits with its CDs. Nice guys. The researchers who discovered the problem, Mark Russinovich and Bryce Cogswell, were later hired by Microsoft.

Preinstalled software — crapware, installed by the folks who sell you computers — has opened significant security holes on machines made by Acer, Asus, Dell, HP, and Lenovo (see `https://duo.com/assets/pdf/out-of-box-exploitation_oem-updaters.pdf`). All the more reason to run the Windows 10 Start Fresh routine (see Book 8, Chapter 2) on any new machine as soon as you get it — or buy a Signature machine from the Microsoft Store.

ASK
WOODY.COM

Microsoft deserves lots of credit for taking down botnets in innovative, lawyer-laden ways. In October 2010, 116 people were arrested worldwide for running fraudulent banking transactions, thanks to Microsoft's tracking abilities. When the folks of Microsoft went after the ZeuS botnet, they convinced a handful of companies whose logos were being used to propagate the botnet to go to court. The assembled group used the RICO laws — the racketeering laws in the United States — to get a takedown order. On March 23, 2012, US Marshals took out two command centers — one in Illinois, the other in Pennsylvania — and effectively shut down ZeuS. Microsoft also led the efforts to take down the Waledac, Rustock, and Kelihos botnets.

That said, Microsoft has been roundly criticized by members of the security community for "hampering and even compromising a number of large international investigations in the United States, Europe, and Asia" while trying to dispense swift justice (www.krebsonsecurity.com/2012/04/microsoft-responds-to-critics-over-botnet-bruhaha).

WHAT ABOUT STUXNET?

Few computer topics have sucked in the mainstream press as thoroughly as the Stuxnet worm — the Windows-borne piece of malware that apparently took out several centrifuges in Iran's uranium enrichment facility.

Here's what I know for sure about Stuxnet: It's carried by Windows but doesn't do anything dastardly until it finds that it's connected to a specific kind of Siemens computer that's used for industrial automation. When it finds that it's connected to that specific kind of Siemens computer, it plants a rootkit on the computer that disrupts operation of whatever the computer's controlling. And that specific Siemens computer controlled the centrifuges at Iran's enrichment plant.

The people who wrote Stuxnet are very, very adept at both Windows infection methods and Siemens computer programming. David Sanger, chief Washington correspondent for *The New York Times,* claims convincingly in his book *Confront and Conceal* (published by Crown) that Stuxnet originated as a collaboration between the US National Security Agency and a secret Israeli military unit, and subsequent revelations have confirmed that's almost undoubtedly the case.

Phishing

Do you think that message from Wells Fargo (or eBay, the IRS, PayPal, Citibank, a smaller regional bank, Visa, MasterCard, or whatever) asking to verify your account password (Social Security number, account number, address, telephone number, mother's maiden name, or whatever) looks official? Think again.

Did you get a message from someone on eBay saying that you had better pay for the computer you bought or else he'll report you? Gotcha. Perhaps a notification that you have received an online greeting card from a family member — and when you try to retrieve it, you have to join the greeting card site and enter a credit card number? Gotcha again.

Phishing — sending email that attempts to extract personal information from you, usually by using a bogus website — has in many cases reached levels of sophistication that exceed the standards of the financial institutions themselves. Some phishing messages, such as the bogus message in Figure 1-3, warn you about the evils of phishing, in an attempt to persuade you to send your account number and password to a scammer in Kazbukistan (or New York).

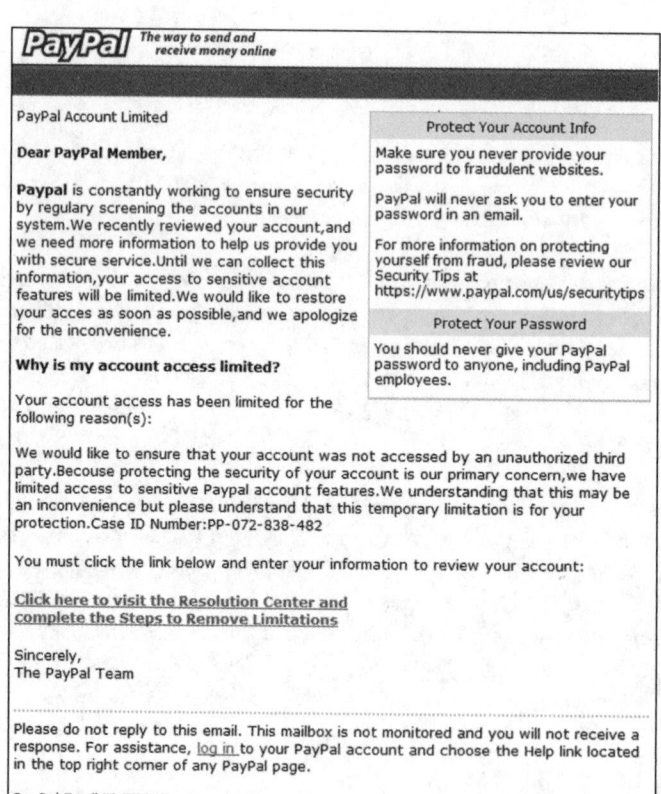

FIGURE 1-3:
If you click the link, you open a page that looks much like the PayPal page, and any information you enter is sent to a scammer.

Here's how phishing works:

1. A scammer, often using a fake name and a stolen credit card, sets up a website.

Usually it's quite a professional-looking site — in some cases, indistinguishable from the authentic site it tries to clone.

2. The website asks for personal information — most commonly, your account number and password or the PIN for your ATM card. See Figure 1-4 for an example.

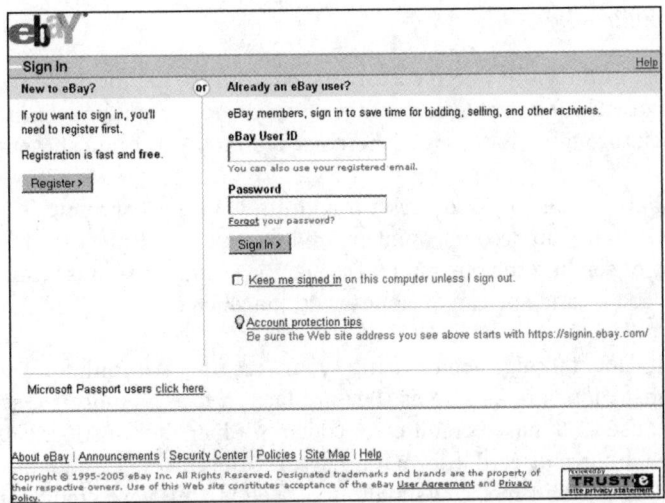

FIGURE 1-4:
This is a fake eBay sign-on site. Can you tell the difference from the original?

3. The scammer turns spammer and sends hundreds of thousands of bogus messages.

The messages include a clickable link to the fake website and a plausible story about how you must go to the website, log in, and do something to avoid dire consequences. The From address on the messages is spoofed so that the message appears to come from the company in question.

The message usually includes official logos — many even include links to the real website, even though they encourage you to click through to the fake site.

4. A small percentage of the recipients of the spam email open it and click through to the fake site.

5. If they enter their information, it's sent directly to the scammer.

Spies, Spams, and Scams
Are Out to Get You

6. The scammer watches incoming traffic from the fake website, gathers the information typed by gullible people, and uses it quickly — typically, by logging on to the bank's website and attempting a transfer or by burning a fake ATM card and using the PIN.

7. Within a day or two — or sometimes just hours — the website is shut down, and everything disappears into thin ether.

Phishing has become hugely popular because of the sheer numbers involved. Say a scammer sends 1 million email messages advising Wells Fargo customers to log in to their accounts. Only a small fraction of all the people who receive the phishing message will be Wells Fargo customers, but if the hit rate is just 1 percent, that's 10,000 customers.

Most of the Wells Fargo customers who receive the message are smart enough to ignore it. But a sizable percentage — maybe 10 percent, maybe just 1 percent — will click through. That's somewhere between 100 and 1,000 suckers, er, customers.

If half the people who click through provide their account details, the scammer gets 50 to 500 account numbers and passwords. If most of those arrive within a day of sending the phishing message, the scammer stands to make a pretty penny indeed — and she can disappear with hardly a trace.

ASK
WOODY.COM

I'm not talking about using your credit card online. Online credit card transactions are as safe as they are face to face — more so, actually, because if you use a US-based credit card, you aren't liable for any loss caused by somebody snatching your card information or any other form of fraud. I use my credit cards online all the time. You should, too. (See "Using your credit card safely online," later in this chapter, for more information.)

Here's how to fight against phishing:

>> **Use the latest versions of Microsoft Edge, Firefox, Opera, or Google Chrome.** All browsers contain sophisticated — although not perfect — antiphishing features that warn you before you venture to a phishy site. See the warning in Figure 1-5.

>> **If you encounter a website that looks like it may be a phishing site, report it.** Use the tools in Microsoft Edge, Firefox, Opera, or Chrome. Use all four if you have a chance! Chrome and Firefox use the same malicious site database. To report a site, go to www.google.com/safebrowsing/report_phish/.

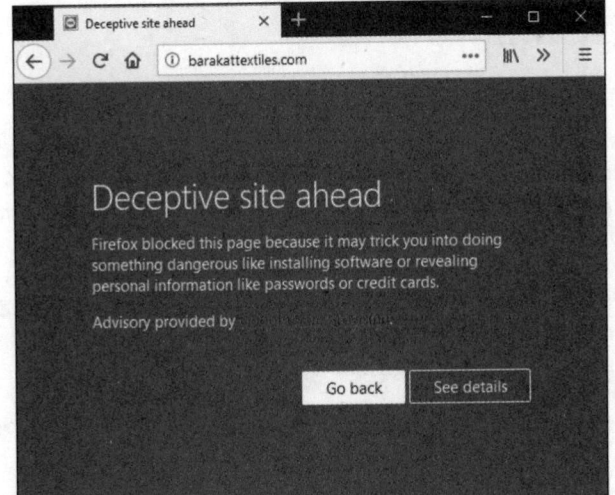

FIGURE 1-5:
If enough people report a site as being dangerous, you see a warning like this one from Firefox.

WARNING

>> **If you receive an email message that contains any links to the web, don't click them.** Nowadays, almost all messages with links to commercial sites are phishing come-ons. Financial institutions, in particular, don't send messages with links any more — and few other companies would dare. If you feel motivated to check out a dire message — for example, if it looks like somebody on eBay is planning to sue you for something you didn't do — open your favorite browser and type the address of the company by hand.

You can see which site a link *really* points to by hovering your cursor over the link. There's no tap equivalent just yet.

>> **Never include personal information in an email message and send it.** Personal information includes your address, Social Security or government ID number, passport number, phone number, or bank account information. Don't give out any of your personal information unless you manually log in to the company's website. Remember that unless you encrypt your email messages, they travel over the Internet in plain-text form. Anybody (or any government) that's "sniffing" the mail can see everything you've written. It's roughly analogous to sending a postcard, with the NSA as the addressee, and Google and Microsoft on the cc list.

>> **If you receive a phishing message that may be new or different, check the linked site by using the PhishTank database at** www.phishtank.com. If you don't see your phishing site listed, submit a copy to PhishTank (which is a service of OpenDNS).

419 scams

Greetings,

I am writing this letter to you in good faith and I hope my contact with you will transpire into a mutual relationship now and forever. I am Mrs. Omigod Mugambi, wife of the late General Rufus Mugambi, former Director of Mines for the Dufus Diamond Dust Co Ltd of Central Eastern Lower Leone . . .

I'm sure you're smart enough to pass over email like that. At least, I hope so. It's an obvious setup for the classic 419 ("four one nine") scam — a scam so common that it has a widely accepted name, which derives from Nigerian Criminal Code Chapter 38, Article 419.

Much more sophisticated versions of the 419 scam are making the rounds today. The basic approach is to convince you to send money to someone, usually via Western Union. If you send the money, you'll never see it again, no matter how hard the sell or dire the threatened consequences.

REMEMBER

There's a reason why everybody gets so much 419 scam email. It's a huge business. Some people reckon it's the third to fifth largest revenue-generating business in Nigeria. I have no way of verifying independently whether that's true, but certainly these folks are raking in an enormous amount of money. And they don't all work out of Nigeria: 419 scams are a significant source of foreign exchange in Benin, Sierra Leone, Ghana, Togo, Senegal, and Burkina Faso, plus just about anywhere else you can mention. Some even originate in the United States although, as you see shortly, there are big advantages to working out of small countries.

Here's one of the new variations of the old 419. It all starts when you place an ad that appears online. It doesn't really matter what you're selling, as long as it's physically large and valuable. It doesn't matter where you advertise — I've seen reports of this scam being played on Craigslist advertisers and major online sites, tiny nickel ad publishers, local newspapers, and anywhere else ads are placed.

**ASK
WOODY.COM**

The scammer sends you an email from a Gmail address. I got one recently that said, "I will like to know if this item is still available for sale?" I wrote back and said, yes, it is, and he'd be most welcome to come and look at it. He wrote back:

> "Let me know the price in USD? I am OK with the item it looks like new in the photos I am from Liverpool U.K., i am sorry i will not be able to come for the viewing, i will arrange for the pickup after payment has been made, all documentation will be done by the shipper, so you don't have to worry about that. Thanks"

Three key points: the scammer

> » Is using a Gmail address, which can't be traced with anything short of a court order.

> » Claims to be out of the country, which makes pursuing him very difficult.

> » Claims that he has a shipper who will pick up the item. The plot thickens.

Also, his grammar falls somewhere between atrocious and unintelligible. Unfortunately, that isn't a sure sign, but it's not bound to inspire confidence.

I wrote back and gave him a price, but I expressed concern about the shipper. He wrote that he would send the shipper from the UK for pickup and said, "I will be paying the PayPal charges from my account and i will be paying directly into your PayPal account without any delay, and i hope you have a PayPal account."

I gave him a dormant PayPal account, listing my address as that of the local police station. He wrote back quite quickly:

"I have just completed the Payment and i am sure you have received the confirmation from PayPal regarding the Payment. You can check your paypal email for confirmation of payment.a total of 25,982usd was sent, 24,728usd for the item and the extra 1,200usd for my shipper's charges, which you will be sending to the address below via western union" and then he gave me the name of someone in Devon, U.K. "You should send the money soon so that the Pick Up would be scheduled and you would know when the Pick Up would commence, make sure you're home. I advise you to check both your inbox and junk/spam folder for the payment confirmation message."

I then received a message that claimed to be from Service-Intl.PayPal.Com:

"The Transaction will appear as soon as the western union information is received from you, we have to follow this procedure due to some security reason . . . the Money was sent through the Service Option Secure Payment so that the transaction can be protected with adequate security measures for you to be able to receive your money. The Shipping Company only accept payment through Western Union You have nothing to doubt about, You are safe and secured doing this transaction and your account will be credited immediately the western union receipt of *1,200USD* is received from you."

There's the hook. Of course, the message didn't come from PayPal, much less from www.paypal.com. I strung the scammer along for several days. Ultimately he threatened me with legal action, invoking PayPal and the FBI as antagonists; see Figure 1-6.

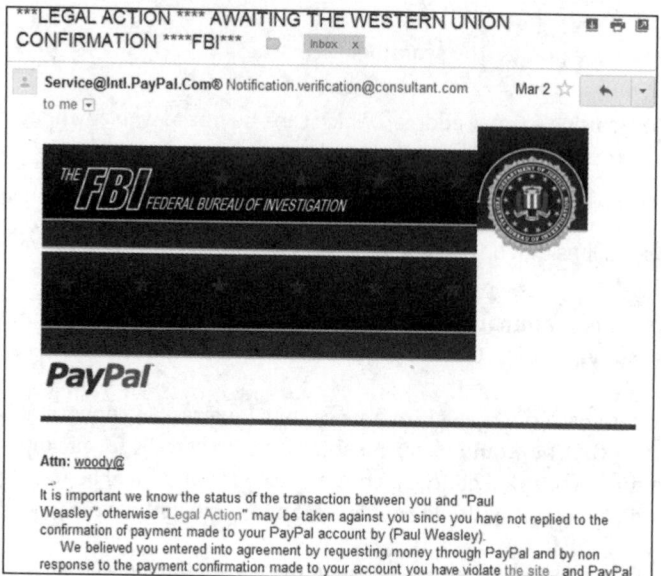

FIGURE 1-6:
Oh me, oh
my, he's going to
send the FBI.

In the end, the scammer and his cohorts were quite sloppy. Most of the time when scammers send email from "PayPal," they use a virtual private network (VPN; see Chapter 4 in this minibook) to make it look like the mail came from the United States. But on three separate occasions, the scammer I was conversing with forgot to turn on the VPN. Using a very simple technique, I traced all three messages back to one specific Internet service provider in Lagos, Nigeria.

So I had three scamming messages with identified IP addresses, the name of a large Internet service provider in Nigeria, and a compelling case for PayPal (to defend its name) and Western Union (which was being used as a drop) to follow up.

REMEMBER

I sent copies of the messages to Western Union and PayPal. I got back form letters — it's unlikely that a human even read them. I wrote to the ISP, MTN Nigeria. They responded, but the upshot is disheartening:

> "All our 3G network subscribers now sit behind a small number of IP addresses. This is done via a technology called network address translation (NAT). In essence, it means that 1 million subscribers may appear to the outside world as one subscriber because they are all using the same IP address."

So now you know why Nigerians love to conduct their scams over the Nigerian 3G network. No doubt MTN Nigeria could sift through its NAT logs and find out who was connected at precisely the right time, but tracing a specific email back to an individual would be difficult, if not impossible — and it would certainly require a court order.

If you know anybody who posts ads online, you may want to warn him.

I'm from Microsoft, and I'm here to help

This kind of scam really hurts me because I've made a career out of helping people with Microsoft problems.

WARNING

Someone calls and says she's been referred by Microsoft to help with your Windows problem. She's very convincing. She says that she heard about your problem from a post you made online, or from your Internet service provider, or from a computer user group. She even gives a website as reference, a very convincing site that has the Microsoft Registered Partner logo.

You explain the problem to her. Then one of two things happens. Either she requests your 25-character Windows activation key or she asks for permission to connect to your computer, typically using Remote Assistance (see Book 7, Chapter 2).

ASK
WOODY.COM

If you let her onto your machine, heaven knows what she'll do. (Believe me, these guys are fast and convincing: It's like playing three-card Monte with a tech support guru.)

If you give her your activation key — or she looks up your validation key while she's controlling your PC — she'll pretend to refer to the "Microsoft registration database" (or something similar) and give you the bad news that your machine is all screwed up, and it's out of warranty, but she can fix it for a mere $189.

REMEMBER

As proof positive that your machine's on its last legs, she'll probably show you Event Viewer. As I mention in Book 8, Chapter 4, Event Viewer on a *normal* machine shows all sorts of scary warnings. And that Microsoft Registered Partner stuff? Anybody can become a Microsoft Registered Partner — it takes maybe two minutes, and all you need is a Microsoft account — a Hotmail, Live, or Outlook.com ID. Don't believe it? Go to `https://epe.mspartner.microsoft.com/EPE/portal/en-US` and fill out the forms.

The overwhelming con give-away — the big red flag — in all this: *Microsoft doesn't work that way.* Think about it. Microsoft isn't going to call you to solve your problems, unless you've received a very specific commitment from a specific individual in the organization — a commitment that invariably comes only after repeated phone calls on your part, generally accompanied by elevation to lofty levels of the support organization on multiple continents, frequently in conjunction with high-decibel histrionics. Microsoft doesn't respond to random online requests for help by calling a customer. Sorry. Doesn't happen.

TIP

If you aren't sure whether you're being conned, ask the person on the other end of the line for your Microsoft Support Case *tracking number* — every MS tech support interaction has a tracking number or Support ID. Then ask for a phone number and offer to call him back. Con artists won't leave trails.

If the con is being run from overseas — much more common in these days of nearly free VoIP cold calling — your chances of nailing the perpetrator runs from extremely slim to none. So be overly suspicious of any "Microsoft Expert" who doesn't seem to be calling from your country.

Microsoft knows all about the tech support scams — one of their blog posts claimed that "three million [Microsoft] customers this year alone" had been hit by scummy scammers. In the first legal action of its kind, in late December 2014, Microsoft sued Omnitech Support, a division of Customer Focus Services, "and related entities," claiming unfair and deceptive business practices and trademark infringement.

Microsoft's filing says the scammers "have utilized the Microsoft trademarks and service marks to enhance their credentials and confuse customers about their affiliation with Microsoft. Defendants then use their enhanced credibility to convince consumers that their personal computers are infected with malware in order to sell them unnecessary technical support and security services to clean their computers."

The Customer Focus Services website says it's "A pioneer in India-based offshoring with over a decade of experience in call center outsourcing . . . [with] Multi-location delivery (offshore and onshore) centers in India (Bangalore)." Wonder how long it'll take for them to fold up their company in the US, and continue overseas?

If you've already been conned — given out personal information or a credit card number — start by contacting your bank or the credit card issuing company and follow its procedures for reporting identity theft.

0day exploits

What do you do when you discover a brand-new security hole in Windows or Office or another Microsoft product? Why, you sell it, of course.

When a person writes a malicious program that takes advantage of a newly discovered security hole — a hole that even the manufacturer doesn't know about — that malicious program is a *0day exploit*. (Fuddy-duddies call it "zero-day exploit." The hopelessly hip say "zero day" or "sploit.")

0days are valuable. In some cases, very valuable. The Trend Micro antivirus company has a subsidiary — *TippingPoint* — that buys 0day exploits. TippingPoint works with the software manufacturer to come up with a fix for the exploit, but at the same time, it sells corporate customers immediate protection against the exploit. "TippingPoint's goal for the Zero Day Initiative is to provide our customers with the world's best intrusion prevention systems and secure converged networking infrastructure." TippingPoint offers up to $10,000 for a solid security hole.

Rumor has it that several less-than-scrupulous sites arrange for the buying and selling of new security holes. Apparently, the Russian hacker group that discovered a vulnerability in the way Windows handles WMF graphics files sold its new hole for $4,000, not realizing that it could've made much more. In 2012, *Forbes Magazine* estimated the value of 0days as ranging from $5,000 to $250,000. You can check it out at the following URL: `www.forbes.com/sites/ andygreenberg/2012/03/23/shopping-for-zero-days-an-price-list- for-hackers-secret-software-exploits/`.

Bounties keep getting bigger. Google's Pwnium competition offers up to $2.7 million for hacks against its Chrome OS, and significant bonuses for other cracks. The Zero Day Initiative (from TippingPoint) now offers more than $500,000 in prize money for the best cracks in the Pwn2Own contest — and an additional $400,000 for the separate Mobile Pwn2Own.

According to *Forbes*, some government agencies are in the market. Governments certainly buy 0day exploits from a notorious 0day brokering firm. The problem (some would say "opportunity") is getting worse, not better. Governments are now widely rumored to have thousands — some of them, tens of thousands — of stockpiled 0day exploits at hand.

ASK
WOODY.COM

How do you protect yourself from 0day exploits? In some ways, you can't: By definition, nobody sees a 0day coming, although most antivirus products employ some sort of heuristic detection that tries to clamp down on exploits based solely on the behavior of the offensive program. Mostly, you have to rely on the common-sense protection that I describe in the section "Getting Protected," later in this chapter. You must also stay informed, which I talk about in the next section.

Staying Informed

When you rely on the evening news to keep yourself informed about the latest threats to your computer's well-being, you quickly discover that the mainstream press frequently doesn't get the details right. Hey, if you were a newswriter with a deadline ten minutes away and you had to figure out how the new Bandersnatch 0day exploit shreds through a Windows TCP/IP stack buffer — and you had to explain your discoveries to a TV audience, at a presumed sixth-grade intelligence level — what would you do?

The following sections offer tips on getting the facts.

Relying on reliable sources

Fortunately, some reliable sources of information exist on the Internet. It would behoove you to check them out from time to time, particularly when you hear about a new computer security hole, real or imagined:

REMEMBER

>> **The Microsoft Security Response Center (MSRC) blog** presents thoroughly researched analyses of outstanding threats, from a Microsoft perspective (`https://msrc-blog.microsoft.com/category/msrc/`).

The information you see on the MSRC blog is 100-percent Microsoft Party Line — so there's a tendency to add more than a little "spin control" to the announcements. Nevertheless, Microsoft has the most extensive and best resources to analyze and solve Windows problems, and the MSRC blog frequently has inside information that you can't find anywhere else.

>> **SANS Internet Storm Center (ISC)** pools observations and analyses from thousands of active security researchers. You can generally get the news first — and accurately — from the ISC (`http://isc.sans.org`).

TIP

Take a moment right now to look up those sites and add them to your Firefox or Chrome Bookmarks or Microsoft Edge Favorites. Unlike the antimalware software manufacturers' websites, these sites have no particular ax to grind or product to sell. (Well, okay, Microsoft wants to sell you something, but you already bought it, yes?)

From time to time, Microsoft also releases security advisories, which generally warn about newly discovered 0day threats in Microsoft products. You can find those, too, at the MSRC blog.

ASK WOODY.COM

It's hard to keep all the patches straight without a scorecard. I maintain an exhaustive list of patches and their known problems and also the Microsoft patches of the patches (of the patches) on www.AskWoody.com. I also write about them frequently in *Computerworld*, and tweet about them all the time @woodyleonhard.

Ditching the hoaxes

Tell me whether you've heard any of these:

>> "Amazing Speech by Obama!" "CNN News Alert!" "UPS Delivery Failure," "Hundreds killed in *[insert a disaster of your choice]*," "Budweiser Frogs Screensaver!" "Microsoft Security Patch Attached."

>> A virus hits your computer if you read any message that includes the phrase "Good Times" in the subject line. (That one was a biggie in late 1994.) Ditto for any of the following messages: "It Takes Guts to Say 'Jesus'," "Win a Holiday," "Help a poor dog win a holiday," "Join the Crew," "pool party," "A Moment of Silence," "an Internet flower for you," "a virtual card for you," or "Valentine's Greetings."

>> A deadly virus is on the Microsoft *[or insert your favorite company name here]* home page. Don't go there or else your system will die.

>> If you have a file named *[insert filename here]* on your PC, it contains a virus. Delete it immediately!

They're all hoaxes — not a breath of truth in any of them. Fake news that's really and truly fake.

WARNING

Some hoaxes serve as fronts for real viruses: The message itself is a hoax, a red herring, designed to convince you to do something stupid and infect your system. The message asks (or commands!) you to download a file or run a video that acts suspiciously like an .exe file.

I'm not talking about YouTube videos, or Vimeo, or links to any of the other established video sites. Steer clear of attachments that appear to be videos, but in fact turn out to be something else. If you tell Windows to show you filename extensions (see Book 3, Chapter 1), you have most of the bases covered.

Other hoaxes are just rumors that circulate among well-intentioned people who haven't a clue. Those hoaxes hurt, too. Sometimes, when real worms hit, so much email traffic is generated from warning people to avoid the worm that the well-intentioned watchdogs do more damage than the worm itself! Strange but true.

ASK WOODY.COM

Do yourself (and me) a favor: If somebody sends you a message that sounds like the following examples, just delete it, eh?

>> A horrible virus is on the loose that's going to bring down the Internet. (Sheesh. I get enough of that garbage on the nightly news.)

>> Send a copy of this message to ten of your best friends, and for every copy that's forwarded, Bill Gates will give *[pick your favorite charity]* $10.

>> Forward a copy of this message to ten of your friends and put your name at the bottom of the list. In *[pick a random amount of time],* you will receive $10,000 in the mail, or your luck will change for the better. Your eyelids will fall off if you don't forward this message.

>> Microsoft (Intel, McAfee, Norton, Compaq — whatever) says that you need to double-click the attached file, download something, don't download something, go to a specific place, avoid a specific place, and on and on.

If you think you've stumbled on the world's most important virus alert, by way of your uncle's sister-in-law's roommate's hairdresser's soon-to-be-ex-boyfriend (remember that he's the one who's a really smart computer guy, but kind of smelly?), count to ten twice and keep these four important points in mind:

REMEMBER

>> No reputable software company (including Microsoft) distributes patches by email. You should never, ever, open or run an attachment to an email message until you contact the person who sent it to you and confirm that she intended to send it to you.

>> Chances are very good (I'd say, oh, 99.9999 percent or more) that you're looking at a half-baked hoax that's documented on the web, most likely on the Snopes urban myths site (www.snopes.com).

>> If the virus or worm is real, Brian Krebs has already written about it. Go to www.krebsonsecurity.com.

>> If the Internet world is about to collapse, clogged with gazillions of email worms, the worst possible way to notify friends and family is by email. D'oh! Pick up the phone, walk over to the water cooler, or send a carrier pigeon, and give your intended recipients a reliable web address to check for updates. Betcha they've already heard about it anyway.

Try hard to be part of the solution, not part of the problem, okay? And if a friend forwards you a virus warning in an email, do everyone a big favor: Shoot him a copy of the preceding bullet points, ask him to tape it to the side of his computer, and beg him to refer to it the next time he gets the forwarding urge.

Is My Computer Infected?

So how do you know whether your computer is infected?

The short answer is this: Many times, you don't. If you think that your PC is infected, chances are very good that it isn't. Why? Because malware these days doesn't usually cause the kinds of problems people normally associate with infections.

Whatever you do, don't fall for the scamware that tells you it removed 39 infections from your computer but you need to pay in order to remove the other 179 (see "Shunning scareware," a little later in this chapter).

Evaluating telltale signs

Here are a few telltale signs that may — *may* — mean that your PC is infected, or that one of your online accounts has been hacked:

>> **Someone tells you that you sent him an email message with an attachment — and you didn't send it.** In fact, most email malware these days is smart enough to spoof the From address, so any infected message that appears to come from you probably didn't. Still, some dumb old viruses that aren't capable of hiding your email address are still around. And, if you receive an infected attachment from a friend, chances are good that both your email address and his email address are on an infected computer somewhere. Six degrees of separation and all that.

>> **You suddenly see files with two filename extensions scattered around your computer.** Filenames such as kournikova.jpg.vbs (a VBScript file masquerading as a JPG image file) or somedoc.txt.exe (a Windows program that wants to appear to be a text file) should send you running for your antivirus software.

Always, always, always have Windows 10 show you filename extensions (see Book 3, Chapter 1).

>> **Your antivirus software suddenly stops working.** If the icon for your antivirus product disappears from the notification area (near the clock), something killed it — and chances are very good that the culprit was a virus.

>> **You can't reach websites that are associated with antimalware manufacturers.** For example, Firefox or Edge or Chrome works fine with most websites, but you can't get through to www.microsoft.com, www.symantec.com, or www.mcafee.com. This problem is a key giveaway for several infections.

Where did that message come from?

In my discussion of 419 scams, I mention that I can trace several scammer messages back to Nigeria. If you've never traced a message before, you'll probably find it intriguing — and frustrating.

You know that return addresses lie. Just like an antagonist in the TV series *House.* You can't trust a return address because "spoofing" one is absolutely trivial. So what can you do?

If you receive a message and want to know where it came from, the first step is to find the header. In the normal course of events, you never see message headers. They look like the gibberish in Figure 1-7.

FIGURE 1-7:
The header for
the 419 message
in Figure 1-6.

Here's how to find a message's header:

ASK
WOODY.COM

» **Outlook 2010 or newer** will show you the header, but only if you know the secret handshake. Open the message. In the message window, click File and then Properties. The header is listed in the box marked Internet Headers.

» **In Gmail,** click the down arrow next to the message subject and choose Show Original. That shows you the entire message, including the header.

» **In Hotmail or Outlook.com,** click the down arrow next to Reply, which is near the sender and subject.

Other email programs work differently. You may have to jump on to Google to figure out how to see a message's header.

After you have the header, copy it, and head over to the ipTracker site, www. iptrackeronline.com/header.php. Paste the message's header into the top box, and tap or click Submit Header for Analysis. A report like the one in Figure 1-8 appears.

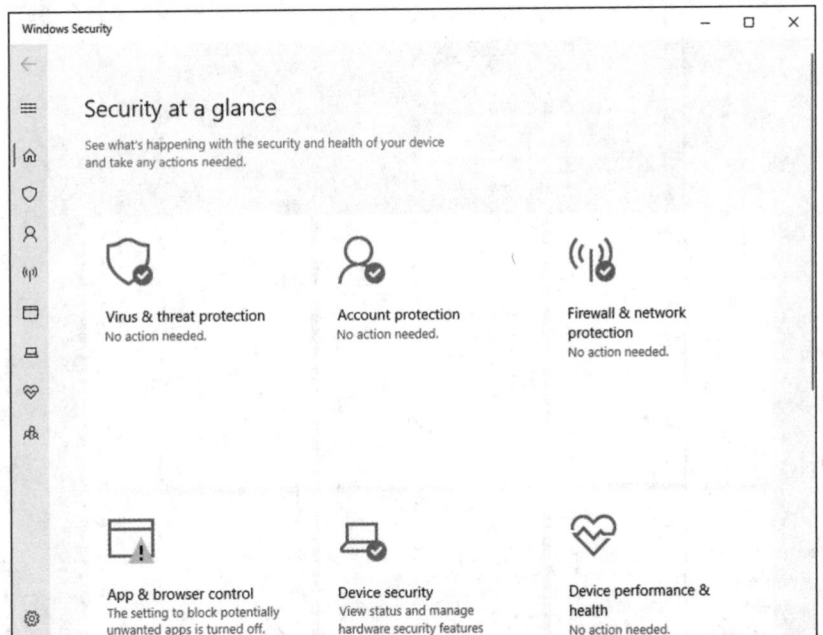

FIGURE 1-8:
Confirmation that
a message came
from Nigeria.

WARNING

Realize that the header can be faked, too. Really clever scammers can disguise the origin of a message by faking the header. It's difficult, though, and scammers tend not to be, uh, the brightest bulbs on the tree.

What to do next

If you think that your computer is infected, follow these steps:

1. **Don't panic.**

   Chances are very good that you're not infected.

2. **DO NOT REBOOT YOUR COMPUTER.**

   You may trigger a virus update when you reboot. Stay cool.

3. **In the search box to the right of the Start button, type** windows security, **and then choose Windows Security.**

   If you aren't using Windows Security, get your antivirus package to run a full scan.

   The Windows Security main interface appears (see Figure 1-9). See Chapter 3 in this minibook for details about Windows Defender.

Spies, Spams, and Scams
Are Out to Get You

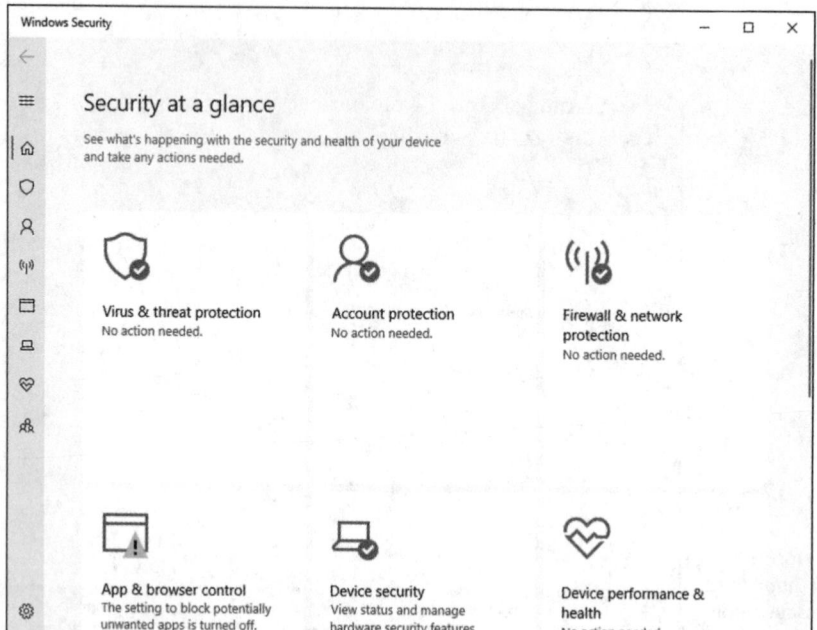

FIGURE 1-9:
Windows
Security, ready
for action.

4. **On the right, tap or click Virus & Threat Protection and then Scan Options.**

5. **Choose Full Scan and click or tap Scan Now.**

   A full scan can take a long time. Go have a latte or two.

6. **If Step 5 still doesn't solve the problem, go to the Malwarebytes Removal forum at** `http://forums.malwarebytes.org/index.php?showforum=7` **and post your problem on the Malware Removal forum.**

   Make sure that you follow the instructions precisely. The good folks at Malwarebytes are all volunteers. You can save them — and yourself — lots of headaches by following their instructions to the letter.

7. **Do not — I repeat — do not send messages to all your friends advising them of the new virus.**

   Messages about a new virus can outnumber infected messages generated by the virus itself — in some cases causing more havoc than the virus itself. Try not to become part of the problem. Besides, you may be wrong.

ASK
WOODY.COM

In recent years, I've come to view the mainstream press accounts of virus and malware outbreaks with increasing skepticism. The antivirus companies are usually slower to post news than the mainstream press, but the information they post tends to be much more reliable. Not infallible, mind you, but better. I also cover security problems at `www.AskWoody.com`.

Shunning scareware

A friend of mine brought me her computer the other day and showed me a giant warning about all the viruses residing on it (see Figure 1-10). She knew that she needed XP Antivirus, but she didn't know how to install it. Thank heaven.

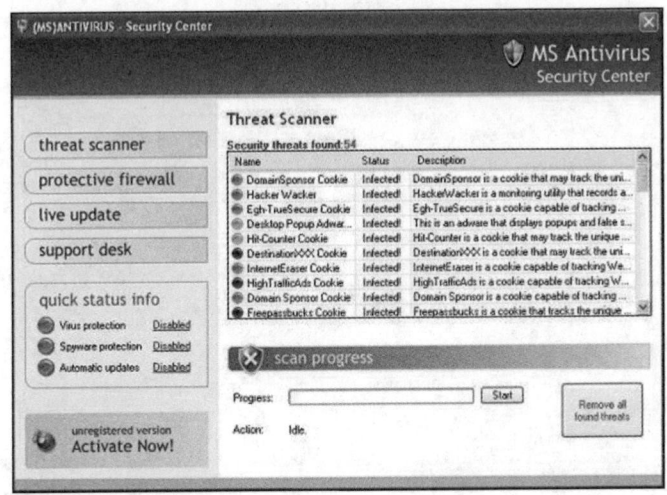

FIGURE 1-10:
Rogue antimalware gives you reason to pay.

Another friend brought me a computer that always booted to a Blue Screen of Death that said

```
Error 0x00000050 PAGE_FAULT_IN_NON_PAGED_AREA
```

It took a whole day to unwind all the junkware on that computer, but when I got to the bottom dreck, I found Vista Antivirus 2009.

ASK
WOODY.COM

I've received messages from all over the world from people who want to know about this fabulous new program, Antivirus Pro 2017 or WioniAntiVirus Pro (or XP Antivirus or MS Antivirus Security Center or Total Win7 Security or similar wording). Here's what you need to know: It's malware, plain and simple, and if you install it, you're handing over your computer to some sophisticated folks who will install keyloggers, bot software, and the scummiest, dirtiest stuff you've ever seen on any PC.

Here's the crazy part: Most people install this kind of scareware voluntarily. One particular family of rogue antivirus products, named Win32/FakeSecSen, has infected more than a million computers; see Figure 1-11.

Spies, Spams, and Scams
Are Out to Get You

FIGURE 1-11:
Win32/Fake
SecSen scares
you into thinking
you must pay
to clean your
computer.

The exact method of infection can vary, as will the payloads. Almost always, people install rogue antivirus programs when they think they're installing the latest, greatest virus chaser — and they're hastened to get it working because they just *know* there are 179 more viruses on their computers that have to be cleaned.

If you have it, how do you remove it? For starters, don't even bother with Windows Add or Remove Programs. Any company clever enough to call a piece of scum Antivirus 2021 won't make it easy for you to zap it. I rely on www.malwarebytes. com — but removing some of these critters is very difficult (see Chapter 4 in this minibook).

TECHNICAL STUFF

One of my favorite antimalware industry pundits, Rob Rosenberger, has an insightful analysis of this type of scareware in the article "Two decades of virus hysteria contributes to the success of fake-AV scams," at www.vmyths.com/2009/03/22/ rogue-av.

Getting Protected

The Internet is wild and woolly and wonderful — and, by and large, it's unregulated, in a Wild West sort of way. Some would say it cannot be regulated, and I agree. Although some central bodies control basic Internet coordination questions — how the computers talk to each other, who doles out domain names such as Dummies.com, and what a web browser should do when it encounters a particular piece of HyperText Markup Language (HTML) — no central authority or Web Fashion Police exists.

In spite of its Wild West lineage and complete lack of couth, the Internet doesn't need to be a scary place. If you follow a handful of simple, common-sense rules, you'll go a long way toward making your Internet travels more like Happy Trails and less like *Grand Theft Auto V*.

Protecting against malware

"Everybody" knows that the Internet breeds viruses. "Everybody" knows that really bad viruses can drain your bank account, break your hard drive, and give you terminal halitosis — just by looking at an email message with *Good Times* in the Subject line. Right.

TIP

In fact, botnets and keyloggers can hurt you, but hoaxes and lousy advice abound. Every Windows user should follow these tips:

>> **Don't install weird programs, cute icons, automatic email signers, or products that promise to keep your computer oh-so-wonderfully safe.** Unless the software comes from a reputable manufacturer whom you trust and you know precisely *why* you need it, you don't want it. Don't be fooled by products that claim to clean your Registry or clobber imaginary infections.

You may think that you absolutely must synchronize the Windows clock (which Windows does amazingly well, no extra program needed), tune up your computer (gimme a break), use those cute little smiley icons (gimme a bigger break), install a pop-up blocker (Edge, Firefox, Opera, and Chrome do that well), or install an automatic email signer (your email program already can sign your messages — read the manual, pilgrim!). What you end up with is an unending barrage of hassles and hustles.

The Microsoft Store goes a long way in culling the junk, but even there you can find awful programs.

» **Never, ever, open a file attached to an email message until you contact the person who sent you the file and verify that she did, in fact, send you the file intentionally.** You should also apply a bit of discretion and ask yourself whether the sender is smart enough to avoid sending you an infected file. After you contact the person who sent you the file, don't open the file directly. Save it to your hard drive and run Windows Security on it before you open it.

» **Follow the instructions in Book 3, Chapter 1 to force Windows 10 to show you the full name of all the files on your computer.** That way, if you see a file named something.cpl or iloveyou.vbs, you stand a fighting chance of understanding that it may be an infectious program waiting for your itchy finger.

» **Don't trust email.** Every single part of an email message can be faked, easily. The return address can be spoofed. Even the header information, which you don't normally see, can be pure fiction. Links inside email messages may not point where you think they point. Anything you put in a message can be viewed by anybody with even a nodding interest — to use the old analogy, sending unencrypted email is much like sending a postcard. Those of you who live in the United States or send mail to or from the United States now know that Uncle Sam himself has been looking at all your mail — the NSA has been sharing the information with the DEA and IRS and lying about it (see the *Forbes* magazine series by Jennifer Granick).

» **Check your accounts.** Look at your credit card and bank statements, and if you see a charge you don't understand, question it. Log in to all your financial websites frequently, and if somebody changed your password, scream bloody murder.

Disabling Java and Flash

ASK
WOODY.COM

As I'm fond of saying, "It's time to run Java out of town." More precisely, I think developers should stop developing programs that require the Java Runtime Environment, or JRE, to run on your computer.

I also salute the rapid change from Flash, for automating websites, to HTML5, which does a better job in a faster and more secure way. Few sites still use Flash, but they're rapidly dying. If you know of a site that requires Flash, send those at the site a nasty message telling them that they're showing enormous disrespect for their customers. And you can quote me.

REMEMBER

If you use Firefox, get the free NoScript Firefox extension (www.noscript.net), which automatically blocks both Java and Flash in Firefox. You can allow Java and Flash to run, on a case-by-case basis, but for general surfing, NoScript and Firefox are the safest ways to go.

Google's browser Chrome has some serious malware-blocking capabilities, combined with custom-built Java and Flash engines that make surfing with Chrome (debatably) the safest choice of the Big Three.

TECHNICAL STUFF

When I talk about Java, I'm not talking about JavaScript. Although the two names are very similar, they're as different as chalk and cheese. JavaScript is a language that automates actions on web pages. Java (in our case, the JRE) is a set of programs inside your computer that web pages can call. JavaScript is relatively benign (although it has been exploited). Java has led to millions of infections.

Flash is now blocked by default in all major web browsers, but not in older ones. Here's a quick checklist for disabling Flash in older browsers such as Internet Explorer and the initial versions of Microsoft Edge:

1. **If you're still running Internet Explorer — ill advised, but sometimes necessary — look under Windows Accessories in the Start menu:**

   a. *Click the Tools (gear) icon.*

   b. *Choose Manage Add-Ons.*

   c. *On the left, click Toolbars and Extensions.*

   d. *If you see an entry for Shockwave Flash, click it and choose Disable.*

   Generally, IE won't have a Flash entry, but it may if you upgraded from an older version.

2. **In older versions of Microsoft Edge, those that are not based on the Chrome rendering engine, Flash may not be disabled by default. Turn it off like this:**

   a. *Click the ellipsis in the upper-right corner.*

   b. *Click Settings.*

   c. *On the left, click Advanced.*

   d. *Slide Use Adobe Flash Player to Off.*

   If you absolutely must use Flash, it's probably best to use it in Edge by flicking this setting On. Mutter under your breath the whole time about ancient websites and be sure to turn it Off when you're done.

Using your credit card safely online

Many people who use the web refuse to order anything online because they're afraid that their credit card numbers will be stolen and they'll be liable for enormous bills. Or they think the products will never arrive and they won't get their money back.

If your credit card was issued in the United States and you're ordering from a US company, that's simply not the case. Here's why:

>> **The Fair Credit Billing Act protects you from being charged by a company for an item you don't receive.** It's the same law that governs orders placed over the telephone or by mail. A vendor generally has 30 days to send the merchandise, or it has to give you a formal written chance to cancel your order. For details, go to the Federal Trade Commission (FTC) website (www.consumer.ftc.gov).

>> **Your maximum liability for charges fraudulently made on the card is $50 per card.** The minute you notify the credit card company that somebody else is using your card, you have no further liability. If you have any questions, the Federal Trade Commission can help (www.consumer.ftc.gov/articles/0213-lost-or-stolen-credit-atm-and-debit-cards).

The rules are different if you're not dealing with a US company and using a US credit card. For example, if you buy something in an online auction from an individual, you don't have the same level of protection. Make sure that you understand the rules before you hand out credit card information. Unfortunately, there's no central repository (at least none I could find) of information about overseas purchase protection for US credit card holders: Each credit card seems to handle cases individually. If you buy things overseas using a US credit card, your relationship with your credit card company generally provides your only protection.

REMEMBER

Some online vendors, such as Amazon, absolutely guarantee that your shopping will be safe. The Fair Credit Billing Act protects any charges fraudulently made in excess of $50, but Amazon says that it reimburses any fraudulent charges under $50 that occurred as a result of using its website. Many credit card companies now offer similar assurances.

Regardless, take a few simple precautions to make sure that you aren't giving away your credit card information:

>> **When you place an order online, make sure that you're dealing with a company you know.** In particular, don't click a link in an email message and expect to go to the company's website. Type the company's address into Edge or Chrome or Firefox, or use a link that you stored in your Edge Favorites or the Chrome or Firefox Bookmarks list.

>> **Type your credit card number only when you're sure that you've arrived at the company's site and when the site is using a secure web page.** The easy way to tell whether a web page is secure is to look for a picture of a lock (see Figure 1-12). Secure websites scramble data so that anything you type on

the web page is encrypted before it's sent to the vendor's computer. In addition, Firefox tells you a site's registration and pedigree by clicking the icon to the left of the web address.

WARNING

Be aware that crafty web programmers can fake the lock icon and show an https:// (secure) address to try to lull you into thinking that you're on a secure web page. To be safe, confirm the site's address and click the icon to the left of the address at the top to show the full security certificate.

>> **Don't send your credit card number in an ordinary email message.** Email is just too easy to intercept. And for heaven's sake, don't give out any personal information when you're chatting online.

>> **Don't send sensitive information back by way of email.** If you receive an email message requesting credit card information that seems to be from your bank, credit card company, Internet service provider, or even sainted Aunt Martha, do not send the information in an email. Insist on using a secure website, and type the company's address into your browser.

FIGURE 1-12:
The lock indicates
a secure site.

TIP

Identity theft continues to be a problem all over the world. Widespread availability of personal information online only adds fuel to the flame. If you think someone may be posing as you — to run up debts in your name, for example — see the US government's main website on the topic at `www.consumer.ftc.gov/features/feature-0014-identity-theft`

Defending your privacy

"You have zero privacy anyway. Get over it."

That's what Scott McNealy, former CEO of Sun Microsystems, said to a group of reporters on January 25, 1999. He was exaggerating — Scott has been known to make provocative statements for dramatic effect — but the exaggeration comes awfully close to reality. (Actually, if Scott told me the sky was blue, I'd run outside and check. But I digress.)

Spies, Spams, and Scams
Are Out to Get You

I continue to be amazed at Windows users' odd attitudes toward privacy. People who wouldn't dream of giving a stranger their telephone numbers fill out their mailing addresses for online service profiles. People who are scared to death at the thought of using their credit cards online to place an order with a major retailer (a very safe procedure, by the way) dutifully type their Social Security numbers on web-based forms.

I suggest that you follow these few important privacy points:

>> **Use work systems only for work.** Why use your company email ID for personal messages? C'mon. Sign up for a free web-based email account, such as Gmail (www.gmail.com), Yahoo! Mail (www.mail.yahoo.com), or Hotmail/Outlook.com (www.hotmail.com and www.outlook.com).

In the United States, with few exceptions, anything you do on a company PC at work can be monitored and examined by your employer. Email, website history files, and even stored documents and settings are all fair game. At work, you have zero privacy anyway. Get over it.

>> **Don't give it away.** Why use your real name when you sign up for a free email account? Why tell a random survey that your annual income is between $20,000 and $30,000? (Or is it between $150,000 and $200,000?)

All sorts of websites — particularly Microsoft — ask questions about topics that, simply put, are none of their business. Don't put your personal details out where they can be harvested.

>> **Follow the privacy suggestions in this book.** You know that Google keeps track of what you type in the Google search engine, and Microsoft keeps track of what you say to Cortana or type in Bing. You know that both Google and Microsoft scan your email — and that Google, at least, admits to using the contents of emails (on free accounts) in order to direct ads at you. You know that files stored in the cloud can be opened by all sorts of people, in response to court orders, anyway.

>> **Know your rights.** Although cyberspace doesn't provide the same level of personal protection you have come to expect in *meatspace* (real life), you still have rights and recourses. Check out www.privacyrights.org for some thought-provoking notices.

Keep your head low and your powder dry!

THE DOUBLECLICK SHTICK

A website plants a cookie on your computer. Only that website can retrieve the cookie. The information is shielded from other websites. ZDNet.com can figure out that I have been reading reviews of digital cameras. Dealtime.com knows that I buy shoes. But a cookie from ZDNet can't be read by DealTime and vice versa. So what's the big deal?

Enter Doubleclick.net, which is now a division of Google. For the better part of a decade, both ZDNet.com and Dealtime.com have included ads from a company named Doubleclick.net. Unless ZDNet or DealTime has changed advertisers, you see Doubleclick.net featured prominently in each site's privacy report.

Here's the trick: You surf to a ZDNet web page that contains a Doubleclick.net ad. DoubleClick kicks in and plants a cookie on your PC that says you were looking at a specific page on ZDNet. Two hours (or days or weeks) later, you surf to a DealTime page that also contains a Doubleclick.net ad — a different ad, no doubt — but one distributed by DoubleClick. DoubleClick kicks in again and discovers that you were looking at that specific ZDNet page two hours (or days or weeks) earlier.

Now consider the consequences if a hundred sites that you visit in an average week all have DoubleClick ads. They can be tiny ads — 1 pixel high or so small that you can't see them. All the information about all your surfing to those sites can be accumulated by DoubleClick and used to target you for advertising, recommendations, or whatever. It's scary.

Want to look at who's watching you? Install the Ghostery browser add-in (www.ghostery.com). It shows you exactly which cookies are tracking you on every page you visit.

Reducing spam

Everybody hates spam, but nobody has any idea how to stop it. Not the government. Not Bill Gates. Not your sainted aunt's podiatrist's second cousin.

You think legislation can reduce the amount of spam? Since the US CAN-SPAM Act (www.fcc.gov/cgb/consumerfacts/canspam.html) became law on January 7, 2003, has the volume of spam you've received increased or decreased? Heck, I've had more spam from politicians lately than from almost any other group. The very people who are supposed to be enforcing the antispam laws seem to be spewing out spam overtime.

By and large, Windows is only tangentially involved in the spam game — it's the messenger, as it were. But every Windows user I know receives email. And every email user I know gets spam. Lots of it.

Why is it so hard to identify spam? Consider. There are 600,426,974,379,824,381,952 different ways to spell *Viagra*. No, really. If you use all the tricks that spammers use — from simple swaps such as using the letter *l* rather than *i* or inserting e x t r a s p a c e s in the word, to tricky ones like substituting accented characters — you have more than 600 quintillion different ways to spell Viagra. It makes the national debt look positively tiny.

Hard to believe? See www.cockeyed.com/lessons/viagra/viagra.html for an eye-opening analysis.

Spam scanners look at email messages and try to determine whether the contents of the potentially offensive message match certain criteria. Details vary depending on the type of spam scanner you use (or your Internet service provider uses), but in general, the scanner has to match the contents of the message with certain words and phrases stored in its database. If you've seen lots of messages with odd spellings come through your spam scanner, you know how hard it is to see through all those sextillions, er, septillion variations.

Spam is an intractable problem, but you can do certain things to minimize your exposure:

>> **Don't encourage 'em.** Don't buy anything that's offered by way of spam (or any other email that you didn't specifically request). Don't click through to the website. Simply delete the message. If you see something that may be interesting, use Google or another web browser to look for other companies that sell the same item.

>> **Opt out of mailings only if you know and trust the company that's sending you messages.** If you're on the Costco mailing list and you're not interested in its email anymore, click the Opt Out button at the bottom of the page. But don't opt out with a company you don't trust: It may just be trying to verify your email address.

>> **Never post your email address on a website or in a newsgroup.** Spammers have spiders that devour web pages by the gazillion, crawling around the web, gathering email addresses and other information automatically. If you post something in a newsgroup and want to let people respond, use a name that's hard for spiders to swallow: *woody (at) ask woody (dot) com*, for example.

>> **Never open an attachment to an email message or view pictures in a message.** Spammers use both methods to verify that they've reached a real, live address. And, you wouldn't open an attachment anyway — unless you know the person who sent it to you, you verified with her that she intended to send you the attachment, and you trust the sender to be savvy enough to avoid sending infected attachments.

>> **Never trust a website that you arrive at by clicking through a hot link in an email message.** Be cautious about websites you reach from other websites. If you don't type the address in the Microsoft Edge address bar, you may not be in Kansas anymore.

>> Most important of all, if spam really bugs you, stop using your current email program and **switch to Gmail or Hotmail/Outlook.com.** Both of them have superb spam filters that are updated every nanosecond. You'll be very pleasantly surprised, I guarantee.

Ultimately, the only long-lasting solution to spam is to change your email address and give out your address to only close friends and business associates. Use a fake phone number or email address or both whenever you can. Even that strategy doesn't solve the problem, but it should reduce the level of spam significantly. Heckuva note.

Dealing with Data Breaches

In recent years we've seen a breathtaking rise in the number of data breaches — where scumbags have broken into company computers and stolen data for millions of customers. Verizon, 14 million. Equifax, 143 million. Home Depot, 56 million. JP Morgan Chase, 76 million. Target, 70 million. eBay, 145 million. Adobe, 36 million. Evernote, 50 million. Activision, 14 million. Sony, 77 million (and almost every key internal document). T.J. Maxx, 94 million. AOL, 92 million, then 20 million more. Kmart, 7-Eleven, JC Penney, Dow Jones, Snapchat, Staples, Facebook, Twitter, and on and on.

Usually the thieves get away with email addresses and some personal information. If you're one of the unfortunate victims and your password was stolen, you can hope that the password was stored in a very secure way. Sometimes you're lucky. Sometimes you're not.

Researchers recently found a database with 1.2 *billion* stolen IDs.

ASK
WOODY.COM

Lots of people want to know what they can do to keep from being the next statistic. The short answer is, mostly you need to constantly monitor your credit card statements, bank statements, and other financial accounts, to catch problems as quickly as you can. That's a fatalistic analysis of the situation: You can't do much to stop it, so you have to watch to see if the cows have run out of the barn.

RANSOMWARE

So you're staring at a screen that says all of your files have been scrambled, and you need to pay a Bitcoin or two to get them back. Don't panic. It happens.

First, realize that data you've stored in the cloud (in Dropbox or OneDrive or Google Drives) may be safe — not all ransomware is smart enough to reach out to online services. Second, if the ransomware just locks up your browser or your computer, chances are good you can bypass it easily. (For browser locks, press Ctrl+Shift+Esc; then in Task Manager, on the Processes tab, click the browser and choose End Task. For screen locks, just unplug your computer.)

But if you have a data-scrambling version — you can't open your Office docs or look at your photos — you should go through all the normal channels to look for a solution. Take a picture of your screen (the police may want it), disconnect your computer from any and all networks, unplug any external drives, and if you can get a web browser going, head to the No More Ransom! Crypto Sheriff site, www.nomoreransom.org/crypto-sheriff.php. Follow the instructions there to see what kind of ransomware you have and whether there's a known antidote.

If there are no effective rescue programs, you have a tough decision to make. Most authorities, including Microsoft, recommend against paying the ransom — it only encourages other cretins to take up a new profession. In addition, a substantial percentage of ransomware hijackers don't send a key even after you've paid them (indeed, some scares are set up so there's no feasible way for the perp to send a key). Wish I had a definitive solution for you, but I don't.

That said, there are a few things you should be doing to keep the bad guys guessing as much as possible:

1. **Don't use the same password twice.**

   Yeah, I know. *Everybody* reuses passwords. I do, too. But I try to reuse passwords on sites that aren't important — and leave the unique passwords for financial sites.

2. **Use a password-remembering program such as LastPass, 1Password, Dashlane, RoboForm, or IronKey.**

   The only chance you have at remembering passwords *just on your financial sites* is to rely on some computer assistance. There are plenty of pros and cons to the products and methods — do you want to trust the cloud, can you remember to take a USB drive everywhere, where and how securely should the master password be manipulated — but the bottom line is that you need some sort of automated password helper.

I use LastPass. See Book 10, Chapter 5.

3. **Assume that the bad guys have your email address and some additional identifying information.**

   They may even have the passwords to your not-sensitive websites. Act accordingly.

4. **If you receive notification that your account has been compromised, don't worry so much about changing the password on the hacked account — look to your other accounts, to see if any of those need changing.**

   After the deed's done, there isn't much you can do — kind of like putting the toothpaste back in the tube. But you can, and should, take a hard look at what might've been taken, and move to mitigate the disclosure.

A lot of companies offer free credit monitoring after they've had a data breach and some banks may offer the service free. Credit monitoring sites scan the credit reporting sites for unusual activity and report to you if they detect any suspicious activity.

Chapter **2**

Fighting Viri and Scum

Windows 8 was the first version of Windows to ship with a complete antivirus/antispyware/antimalware package baked right into the product. Windows 10 brings along all those goodies and expands on them surprisingly well. They're more than enough for just about anybody except spies and organizations that have secrets to keep from the North Koreans.

ASK
WOODY.COM

You don't need to buy an antivirus, firewall, or anti-everything product. Windows 10 has all you need. It's already installed and working, and it doesn't cost a penny.

On the other hand, you need to hold up your end of the bargain by not doing anything, uh, questionable. I wanted to say stupid, but some of the tricks the scummeisters use these days can get you even if you *aren't* stupid. Chapter 1 in this minibook helps you understand the tactics online creeps use and keep your guard up.

I start this chapter with a very simple list of do's and don'ts for protecting your computer and your identity. They're important. Even if you don't read the rest of this chapter, make sure you read — and understand, and follow — the rules in each list.

Basic Windows Security Do's and Don'ts

TIP

Here are the ten most important things you need to do, to keep your computer secure:

>> **Check daily to make sure Windows Security is running.** If something's amiss, a red X appears on the Windows Security shield, down in the desktop's notification area, near the time. To check its status, double-click the shield in the notification area. If Windows Security is running and all's well, green check marks appear, as shown in Figure 2-1.

Windows 10 should tell you if Windows Security stops, either via a toaster notification from the right side or a red X on the flag in the lower-right corner of the desktop. But if you want to be absolutely sure, there's no better way than to check it yourself. Only takes a second.

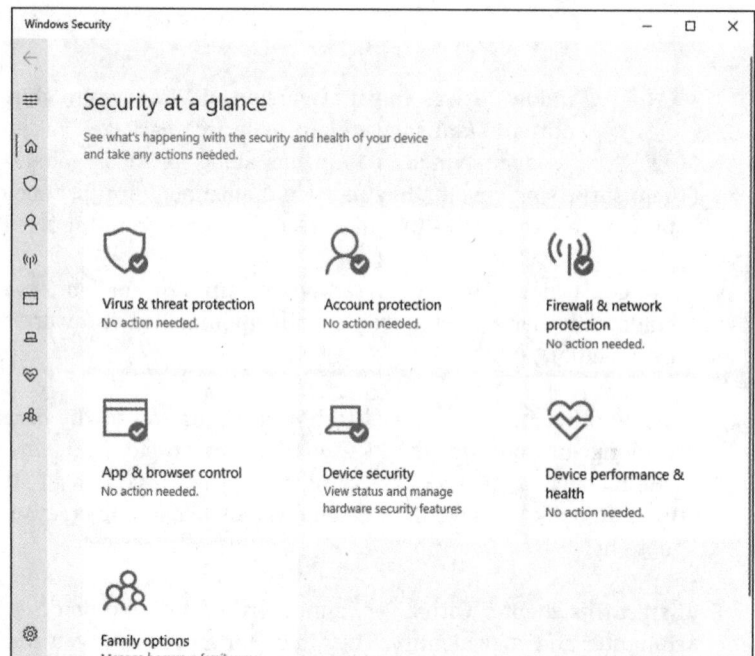

FIGURE 2-1:
Windows
Security is up
and running.

>> **Don't use just any old browser.** My go-to browser these days is Firefox. It's as secure as Google Chrome or Microsoft Edge, or Opera. It doesn't snoop as much as Chrome, has lots of custom extensions, and performs well — despite what Microsoft may splash on your screen about Edge.

Most Windows infections come in the door through Java, Flash, or Adobe Reader (see Chapter 1 in this minibook), and they usually get in through Internet Explorer.

>> **Use anything other than Adobe Reader to look at PDF files.** All the major browsers have their own PDF readers, just because Adobe Reader has caused so many infections. For a stand-alone reader, download and install an alternative to Adobe Reader.

>> **Every month or so, run Microsoft Defender Offline (MDO).** MDO scans for rootkits. You can find it in the scan options offered by Windows Security, in its Virus & Threat Protection section.

>> **Every month or so, run Malwarebytes.** The Malwarebytes program gives you a second opinion, possibly pointing out questionable programs that Windows Security doesn't flag.

>> **Delete chain mail.**

TIP

I'm sure that you'll be bringing down the wrath of several lesser deities for the rest of your days, but do everyone a favor and don't forward junk. Please.

If something you receive in an email sounds really, really cool, it's probably fake — an urban legend or a come-on of some sort. Look it up at www. snopes.com.

>> **Keep up to date with Windows 10 patches and (especially) patches to other programs running on your computer.** Windows 10 should be keeping itself updated, although you can take control of Windows Update if you're reasonably vigilant.

>> **Check your credit cards and bank balances regularly.**

ASK WOODY.COM

I check my charges and balances every couple of days and suggest you do the same.

Credit monitoring services keep a constant eye on your credit report, watching for any unexpected behavior. Most companies that get hacked will offer free credit monitoring to potentially ripped-off customers. Many big banks offer the service free, too.

>> **If you don't need a program any more, get rid of it.** Use the uninstall feature in Windows 10. If it doesn't blast away easily, use Revo Uninstaller in Book 10, Chapter 5.

REMEMBER

>> **Change your passwords regularly.** Yeah, another one of those things everybody recommends, but nobody does. Except you really should. See the admonitions in Book 2, Chapter 4 about choosing good passwords, but especially look at LastPass and RoboForm, which I describe in Chapter 4 in this minibook.

Here are the ten most important things you *shouldn't* do, to keep your computer secure:

>> **Don't trust any PC unless you, personally, have been taking close care of it.** Even then, be skeptical. Treat every PC you may encounter as if it's infected. Don't stick a USB drive into a public computer, for example, unless you're prepared to disinfect the USB drive immediately when you get back to a safe computer. Don't use the business center computer in a hotel or FedEx if you have to type anything sensitive. Assume that everything you type in a public PC is being logged and sent to a pimply-face genius who wants to be a millionaire.

>> Don't install a new program unless you know precisely what it does, and you've checked to make sure you have a legitimate copy.

Yes, even if an online scanner told you that you have 139 viruses on your computer, and you need to pay just $49.99 to get rid of them.

If you install apps from the Microsoft Store, you're generally safe — although the Store has its share of crappy programs. But any programs you install from other sources should be vetted ten ways from Tuesday, downloaded from a reputable source (such as www.cnet.com, www.softpedia.com, www.majorgeeks.com, www.tucows.com, www.snapfiles.com), and *even then* you need to ask yourself whether you really need the program, and *even then* you have to be careful that the installer doesn't bring in some crappy extras like browser toolbars.

Similarly, Firefox and Chrome add-ons are generally safe, as long as you stick to the well-known ones.

>> **Don't use the same password for two or more sites.** Okay, if you reuse your passwords, make sure you don't reuse the passwords on any of your email or financial accounts.

True confession time. Yes, I reuse passwords. Everybody does. LastPass (see Chapter 4 in this minibook) makes it easier to create a different password for every website, but I'm lazy sometimes.

Email accounts are different. If you reuse the passwords on any of your email accounts and somebody gets the password, he may be able to break into everything, steal your money, and besmirch your reputation. See the nearby "Don't reuse your email password" sidebar.

>> **Don't use Wi-Fi in a public place unless you're running exclusively on HTTPS-encrypted sites or through a virtual private network (VPN).**

If you don't know what HTTPS is and have never set up a VPN, that's okay. Just realize that anybody else who can connect to the same Wi-Fi station you're using can see *every single thing* that goes into or comes out of your computer. See Chapter 4 in this minibook.

>> **Don't fall for Nigerian 419 scams, "I've been mugged and I need $500 scams," or anything else where you have to send money.** There are lots of scams — and if you see the words *Western Union* or *Postal Money Order,* run for the exit. See Chapter 1 in this minibook.

>> **Don't tap or click a link in an email message or document and expect it to take you to a financial site.** Take the time to type the address into your browser. You've heard it a thousand times, but it's true.

>> **Don't open an attachment to any email message until you've contacted the person who sent it to you and verified that she intentionally sent you the file.** Even if she did send it, you need to use your judgment as to whether the sender is savvy enough to refrain from sending you something infectious.

No, UPS didn't send you a non-delivery notice in a zip file, Microsoft didn't send you an update to Windows attached to a message, and your winning lottery notification won't come as an attachment.

>> **Don't forget to change your passwords.** Yeah, another one of those things everybody recommends, but nobody does. Except you really should.

WARNING

>> **Don't trust anybody who calls you and offers to fix your computer.** The "I'm from Microsoft and I'm here to help" scam has gone too far. Stay skeptical, and don't let anybody else into your computer, unless you know who he is. See Chapter 1 in this minibook.

>> **Don't forget that the biggest security gap is between your ears.** Use your head, not your tapping or clicking finger.

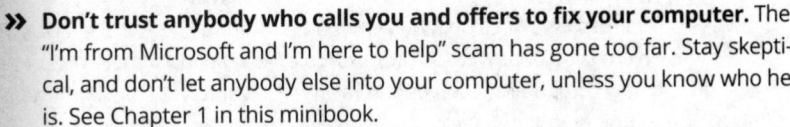

DON'T REUSE YOUR EMAIL PASSWORD

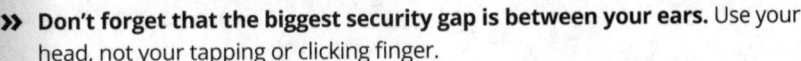

Say you have a Gmail account. You run over to an online classified advertising site and sign up for an account there. You're lazy, so you use the same password for both accounts.

A day, month, or year later, you place an ad on the classified advertising site. You have to provide your email address. Hey, no problem.

The next week, somebody breaks into the classified advertising site and steals the information from 10,000 accounts. Unbeknownst to you, the people who created and maintain the classified advertising site stored the passwords and email addresses in a way that can be cracked.

(continued)

(continued)

The person who broke into the site posts his booty on some underground file-sharing site, and within minutes of the break-in, two dozen people are trying every combination of your Gmail address and password, trying to break into banking sites, brokerage sites, PayPal, whatever.

If they hit on a financial site that requires only an email address in order to retrieve the account information, bingo, they use your Gmail address and ask for a new password. They log in to Gmail with your password and wait for the password reset instructions. Thirty seconds later, they're logged in to the financial site.

If they hit a site that will send you a password reset code and send it to your email address, there you go again.

The site doesn't have to be financial. Just about any site that stores your Social Security number, or includes sensitive information like a hospital site, might be similarly vulnerable.

Happens every day.

Your best bet: Turn on two-factor authentication (2FA) wherever it's offered. If you have an email account with 2FA, for example, you can insist that access to that account from a new computer must respond to an SMS sent to your smartphone. There are many variations, but two-factor authentication can save your tail.

Making Sense of Malware

Although most people are more familiar with the term *virus*, viruses are only part of the problem — a problem known as malware. *Malware* is made up of the elements described in this list:

>> **Viruses:** A computer virus is a program that replicates. That's all. Viruses generally replicate by attaching themselves to files — programs, documents, or spreadsheets — or replacing genuine operating system files with bogus ones. They usually make copies of themselves whenever they're run.

REMEMBER

You probably think that viruses delete files or make programs go belly-up or wreak havoc in other nefarious ways. Some of them do. Many of them don't. Viruses sound scary, but most of them aren't. Most viruses have such ridiculous bugs in them that they don't get far in the wild.

>> **Trojans:** Trojans (occasionally called Trojan horses) may or may not be able to reproduce, but they always require that the user do something to get them started. The most common Trojans these days appear as programs downloaded from the Internet, or email attachments, or programs that helpfully offer to install themselves from the Internet: You tap or double-click an attachment, expecting to open a picture or a document, and you get bit when a program comes in and clobbers your computer, frequently sending out a gazillion messages, all with infected attachments, without your knowledge or consent.

>> **Worms:** Worms move from one computer to another over a network. The worst ones replicate quickly by shooting copies of themselves over the Internet, taking advantage of holes in the operating systems (which all too frequently is Windows).

LIES, DAMN LIES, AND MALWARE STATISTICS

Computer crime has evolved into a money-making operation, with some espionage tacked on for good measure, but when you hear statistics about how many viruses are out and about and how much they cost everyone, take those statistics with a grain of salt.

As *The New York Times* puts it so accurately, "A few criminals do well, but cybercrime is a relentless, low-profit struggle for the majority . . ." (www.nytimes.com/2012/04/15/opinion/sunday/the-cybercrime-wave-that-wasnt.html).

Here's what you need to know about those cost estimates:

- There's no way to tell how much a virus outbreak costs. You should expect that any dollar estimates you see are designed to raise your eyebrows, nothing more.

- Although corporate cyberespionage certainly takes place all the time, it's very hard to identify — much less quantify. For that matter, how can you quantify the effects of plain-old, everyday industrial espionage?

- Instead of flinging meaningless numbers around, it's more important to consider the amount of hassle people and companies encounter when they have to clean up after a group of cybercretins. One hundred thousand filched credit card credentials may not lead to lots of lost money, but it'll certainly cause no end of mayhem for lots of people.

Although the major antivirus companies release virus-catching files that identify tens of millions of signatures, most infections in any given year come from a handful of viruses. The threat is real, but it's way overblown.

>> **Ransomware:** Ransomware takes control of your files and folders by encrypting them. Then it tries to force you to pay large amounts of money to get them back. And even if you pay, you can't be sure that you can get your data back.

Viruses, trojans, worms, and ransomware are getting much, much more sophisticated than they were just a few years ago. Lots of money can be made with advanced malware, especially for those who figure out how to break in without being detected.

Some malware can carry bad *payloads* (programs that wreak destruction on your system), but many of the worst offenders cause the most harm by clogging networks (nearly bringing down the Internet itself, at times) and by turning PCs into zombies, frequently called *bots*, which can be operated by remote control. (I talk about bots and botnets in Chapter 1 in this minibook.)

ASK
WOODY.COM

The most successful pieces of malware these days run as *rootkits* — programs that evade detection by stealthily hooking into Windows in tricky ways. Some nominally respectable companies (notably, Sony) have employed rootkit technology to hide programs for their own profit. Rootkits are extremely difficult to detect and even harder to clean. Microsoft Defender Offline, discussed later in this chapter, is a great choice for clobbering the beasts.

UNDERSTANDING HOW WINDOWS SECURITY WORKS

Windows Security in Windows 10 is a fully functional, very capable, fast, small anti-malware program that works admirably well. There's absolutely no reason to spend any money on any other antimalware/anti-whatever program. You have the best inside Windows 10, already working, and you don't have to lift a finger, or pay a penny.

Windows Security in Windows 10 is built on the foundation set by Windows Defender and Microsoft Security Essentials, which I've raved about for years. It incorporates all the MSE pieces (so there's no reason to install Microsoft Security Essentials on a Windows 10 machine), while adding new features, including the capability to work with the new UEFI boot system to validate secure boot operating systems.

I talk about Windows Security, UEFI, and secure boot in Chapter 3 in this minibook.

All these definitions are becoming more academic and less relevant, as the trend shifts to *blended-threat* malware. Blended threats incorporate elements of all three traditional kinds of malware — and more. Most of the most successful viruses and malware you read about in the press these days — WannaCry, Petya and NotPetya, Conficker, Rustock, Aleuron, and the like — are, in fact, blended-threat malware. They have come a long way from old-fashioned viruses, and are increasingly being built into $99 Script Kiddie kits.

Deciphering Browsers' Inscrutable Warnings

One last trick that may help you head off an unfortunate online incident: Each browser has subtle ways of telling you that you may be in trouble. I'm not talking about the giant Warning: Suspected Phishing Site or Reported Web Forgery signs. Those are supposed to hit you upside the head, and they do.

I'm talking about the gentle indications each browser has that tell you whether there's something strange about the site you're looking at. Historically, if you're on a secured page — where encryption is in force between you and the website — you see a padlock. That simple padlock indicator has grown up a bit, so you can understand more about your secure (or not-so-secure) connection with a glance.

Chrome

Chrome browsers have three different icons that can appear to the left of a site's URL, as shown in Figure 2-2.

Here's what they mean:

>> The **gray padlock** says that there's a secure connection in place, and it's working. As long as you're looking at the correct domain — you didn't mistype the domain name, for example — you're safe.

If the site has an Extended Validation certificate (see the nearby "What is extended validation?" sidebar), you also see green highlighting when you click the padlock.

>> The **gray Not Secure text with an i icon** says that Chrome has set up a secure connection, but there are parts of the page that can, conceivably, snoop on what you're typing. That's what the "not secure" warning means.

>> The **red question mark with the Not Secure text** tells you that there are problems with the site's certificate or that insecure content on the page is known to be high risk. When you hit a red question mark, you have to ask yourself whether the site's handlers just let the certificate lapse (I've seen that on sites and other sites that shouldn't go bad) or if there's something genuinely wrong with the site.

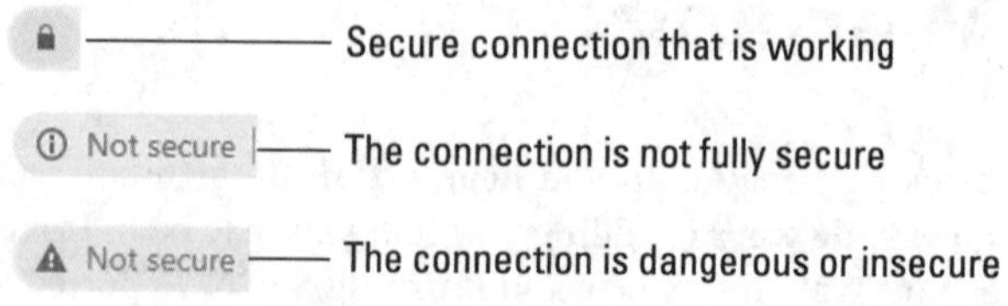

FIGURE 2-2:
The three different HTTPS padlocks in Chrome.

ⓘ Not secure ⊢—— The connection is not fully secure

⚠ Not secure ——— The connection is dangerous or insecure

WHAT IS EXTENDED VALIDATION?

Companies have to pay to get a secure certificate and use it correctly on their sites before the major browsers will display a padlock for those sites.

Unfortunately, in recent years, there have been many problems with faked, stolen, or otherwise dubious certificates. Part of the difficulty lies in the fact that just about anybody can get a website security certificate. Several years ago, a couple people applied for a security certificate for Microsoft.com. They sweet-talked their way into having a certificate issued.

Starting in April 2008, a second level of certification, an *Extended Validation certificate*, was put into effect. To buy an EV certificate, the organization or individual applying for the certificate has to jump through many hoops to establish its legal identity and physical location, and prove that the people applying for the certificate do, in fact, own the domain name that they're trying to certify.

EV certificates aren't infallible, but they're much more trustworthy than regular certificates.

Firefox

Firefox handles things a little differently. Firefox puts a box to the left of the URL — called a Site Identity Button or padlock (see Figure 2-3) — that's color-coded to give you an idea of what's in store. If you tap or click the button, you see detailed information about the security status of the site.

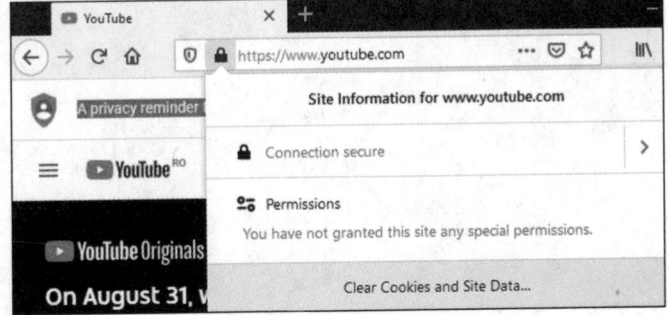

FIGURE 2-3: Firefox gives detailed, site security information.

The three colors indicate the following:

>> **Gray padlock:** Secure site, with full encryption that prevents eavesdropping. Click the gray padlock to find out if the website is using an EV certificate.

>> **Gray padlock with a yellow warning triangle:** The connection between you and the website is only partially encrypted and it doesn't prevent eavesdropping.

>> **Gray padlock with a red strike over it:** The site uses an insecure protocol or it is known to be high risk.

Chapter **3**

Running Built-In Security Programs

Windows 10, right out of the box, ships with a myriad of security programs, including a handful that you can control.

This chapter looks at the things you can do with the programs on offer: Windows Security, Controlled Folder Access for blocking ransomware, SmartScreen, UEFI (don't judge it by its name alone), User Account Control, and Windows Defender Firewall. What you find in this chapter is like a survey of the tip of an iceberg. Even if you don't change anything, you'll come away with a better understanding of what's available, and how the pieces fit together. With a little luck, you'll also have a better idea of what can go wrong, and how you can fix it.

Working with Windows Security

Fast, full-featured, and free, Microsoft Windows Security draws accolades (or at least lukewarm endorsements) from experts and catcalls from competitors.

If you've ever put up with a bloated and expensive security suite exhorting/ extorting you for more money, or you've struggled with free antivirus packages that want to install a little toolbar here and a funny monitoring program there — and *then* ask you for money — you're in for a refreshing change . . . from an unexpected source.

Windows Security takes over antivirus and antispyware duties and tosses in bot detection and anti-rootkit features for good measure. In independent tests, Microsoft has consistently received high detection and removal scores for Windows Security (and Microsoft Security Essentials, Windows Security's kissing cousin) for years.

ASK WOODY.COM

This tool has been rebranded many times over the years: from Microsoft Security Essentials to Windows Defender to Windows Defender Antivirus to Windows Defender Security to Windows Security, or from Windows Defender Offline to Microsoft Defender Offline. To make things even more confusing, Microsoft is not consistent about how it names this product in its Windows 10 notifications. Sometimes you see notifications from Windows Security but other times from Windows Defender Antivirus. If you search for Windows Defender in recent versions of Windows 10, you get shortcuts to Windows Defender Firewall, previously known as the Windows Firewall, and not to the antivirus product that you used to know. No matter what Microsoft calls it, Windows Security is just an improved version of the former Windows Defender and now encompasses more security tools in one easy-to-use app.

Windows Security conducts periodic scans and watches out for malware in real time. It vets email attachments, catches downloads, deletes or quarantines at your command, and in general, does everything you'd expect an antivirus, anti-malware, and/or anti-rootkit product to do.

Is Windows Security the best antivirus package on the market? No. It depends on how you define *best*, but Microsoft has no intention of coming out on top of the competitive antimalware tests. I think Lowell Heddings said it best, in his "How-To Geek" article (www.howtogeek.com/225385/what's-the-best-antivirus-for-windows-10-is-windows-defender-good-enough/) in January, 2020:

"Other antivirus programs may occasionally do a bit better in monthly tests, but they also come with a lot of bloat, like browser extensions that actually make you less safe, registry cleaners that are terrible and unnecessary, loads of unsafe junkware, and even the ability to track your browsing habits so they can make money. Furthermore, the way they hook themselves into your browser and operating system often causes more problems than it solves. Something that protects you against viruses but opens you up to other vectors of attack is not good security.

Just look at all the extra garbage Avast tries to install alongside its antivirus.

Windows Defender does not do any of these things — it does one thing well, for free, and without getting in your way. Plus, Windows 10 already includes the various other protections introduced in Windows 8, like the SmartScreen filter that should prevent you from downloading and running malware, whatever antivirus you use. Chrome and Firefox, similarly, include Google's Safe Browsing, which blocks many malware downloads."

I think *Windows 10 All-in-One For Dummies* readers tend to be experienced and involved and would agree wholeheartedly with Fred's assessment.

REMEMBER

The beauty of Windows Security is that it just works. You don't have to do anything — although you should check from time to time to make sure it hasn't been accidentally (or maliciously) turned off. To check whether Windows Security is running, go to the search box next to the Start button, type **sec**, and in the list of apps choose Windows Security. If you see green check marks (see Figure 3-1), you're doing fine.

TIP

Microsoft maintains an active online support forum for Windows Security at Microsoft Answers, `http://answers.microsoft.com/en-us/windows/forum/windows_10-security`.

When you use Windows Security, you should be aware of these caveats:

REMEMBER

» It's *never* a good idea to run two antivirus products simultaneously, and Windows Security is no exception: If you have a second antivirus product running on your machine, Windows Security has been disabled, and you shouldn't try to bring it back.

» If you don't like your antivirus product and don't particularly want to keep paying and paying and paying for it, use the Windows 10 tool to get rid of it. Click or tap the Start icon, the Settings icon, and then Apps. On the left choose Apps & Features. Wait for the list to fill out. Then pick the program you want to remove and choose Uninstall. Reboot your machine, and Windows Security returns.

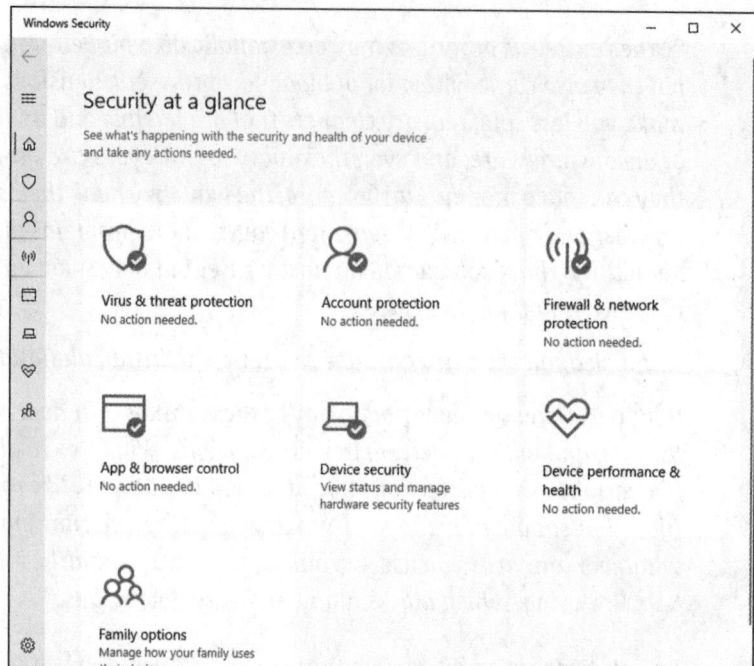

WARNING

» In summary, Windows Security works great, but if you get a second antivirus program that's designed to run continuously, do *not* run Windows Security and the usurper at the same time.

» You may see updates listed for Windows Security if you go into Windows Update and look. Just leave them alone. They'll install all by themselves.

» No matter how you slice it, real-time protection eats into your privacy. How? Say Windows Security (or any other antivirus product) encounters a suspicious-looking file that isn't on its zap list. In order to get the latest information about that suspicious-looking file, Windows Security has to phone back to Mother Microsoft, drop off telltale pieces of the file, and ask whether there's anything new. You can opt out of real-time protection, but if you do, you won't have the latest virus information — and some viruses travel very fast.

Adjusting Windows Security

Unlike many other antivirus products, Windows Security has a blissfully small number of things that you can or should tweak. Here's how to get to the settings:

1. **In the search bar to the right of the Start button, type** sec. **At the top, tap or click Windows Security.**

   The main Windows Security screen appears (refer to Figure 3-1).

2. **Tap or click Virus & Threat Protection. Click the Scan Options link.**

   You see the options for manually running a Full scan, Custom scan, or Microsoft Defender Offline scan. (See the "Microsoft Defender Offline" sidebar.) If you go back to Virus & Threat Protection, you can also change Windows Security's behavior. If you really want to turn off the main antivirus protections, you can do so here.

3. **If you have any reason to fear that your machine's been taken over by a rootkit, select the Microsoft Defender Offline scan, click or tap Scan Now, and confirm your choice.**

   You are signed out of Windows 10. Go have a cup of coffee, and by the time you come back, Microsoft Defender Offline will show you a list of any scummy stuff it caught.

MICROSOFT DEFENDER OFFLINE

Microsoft Defender Offline (MDO) sniffs out and removes *rootkits,* which are malicious programs that run underneath Windows. Rootkits can be devilishly difficult to identify. The "best" ones may not even have symptoms. They sit in the background, swipe your data, and send it out to listening posts.

MDO should occupy a key spot in your bag of tricks. It works like a champ on Windows 7, Windows 8, and Windows 10 systems and should be able to catch a wide variety of nasties that evade detection by more traditional methods.

It's important to understand that MDO is not a Windows application, even though Microsoft makes and distributes it. MDO is self-contained. When you choose to run an offline scan, Windows reboots for you, and MDO looks at your system without interference from the installed copy of Windows. MDO runs all by itself and, when it's done, brings Windows back.

To find rootkits, a rootkit detector has to do its job when Windows isn't running. If the detector were running on Windows, it would never be able to see underneath Windows to catch the rootkits. That's why it has to run offline — without Windows.

Running Windows Security manually

Windows Security works without you doing a thing, but you can tell it to run a scan if something on your computer is giving you the willies. Here's how:

1. **In the search bar next to the Start button, type** sec. **At the top, tap or click Windows Security.**

   The main Windows Security screen appears (refer to Figure 3-1).

2. **Tap or click Virus & Threat Protection. At the bottom of the Virus & Threat protection pane, click Check for Updates, in the Virus & Threat Protection Updates section.**

3. **On the resulting Protection Updates pane (see Figure 3-2), tap or click Check for Updates to get the latest antimalware definitions.**

   When you tap or click Check for Updates, Windows Security retrieves the latest signature files from Microsoft but doesn't run a scan. If you want to run a scan, you need to go back to the Virus & Threat Protection screen and run it.

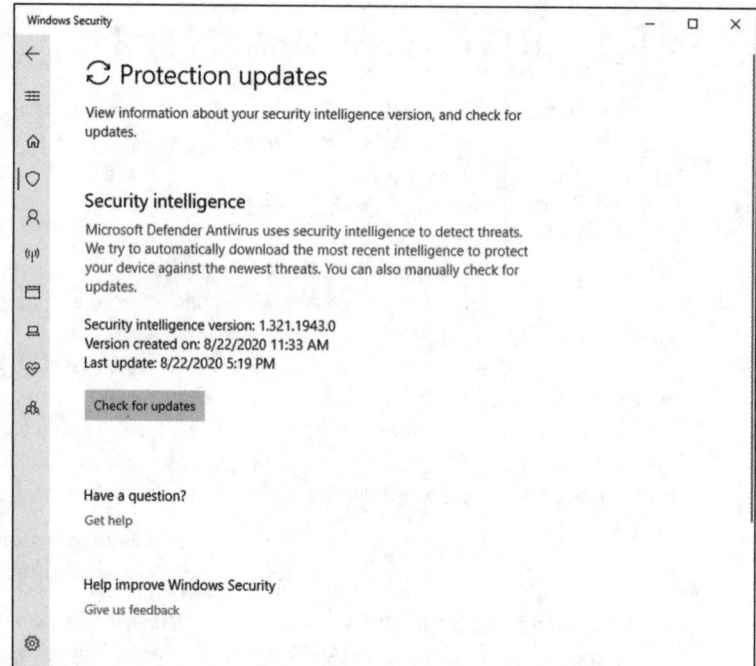

FIGURE 3-2: The current status of Windows Security signature file updates.

4. **Tap or click the Back arrow in the top-left corner. Then, to perform a manual scan, click Scan Options. Choose one of the following three options (see Figure 3-3):**

- To perform a *quick* scan, which looks in locations where viruses and other kinds of malware are likely to hide, select the Quick Scan option and then tap or click Scan Now.

- To run a *full* scan, which runs a bit-by-bit scan of every file and folder on the PC, select Full Scan and then click Scan Now.

- To run a *custom* scan, which is like a full scan but you get to choose which drives and folders get scanned, choose Custom Scan and then click Scan Now.

Windows Security   — □ ×

←

Scan options

Run a quick, full, custom, or Microsoft Defender Offline scan.

No current threats.
Last scan: 8/22/2020 5:30 PM (quick scan)
0 threats found.
Scan lasted 17 seconds
16829 files scanned.

Allowed threats

Protection history

◉ **Quick scan**

Checks folders in your system where threats are commonly found.

○ **Full scan**

Checks all files and running programs on your hard disk. This scan could take longer than one hour.

○ **Custom scan**

Choose which files and locations you want to check.

○ **Microsoft Defender Offline scan**

Some malicious software can be particularly difficult to remove from your device. Microsoft Defender Offline can help find and remove

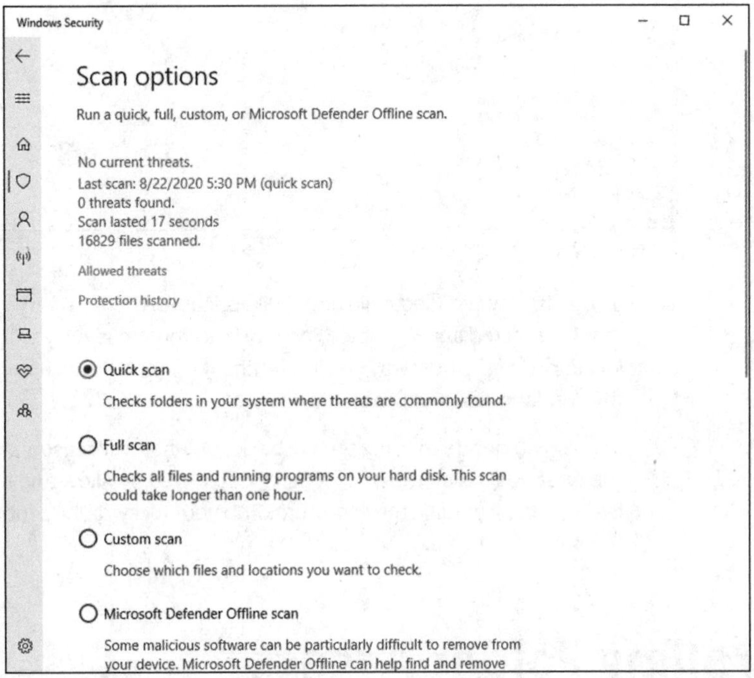

FIGURE 3-3:
Scan settings for Windows Security.

5. **To see what Windows Security has caught and zapped historically, tap or click the Protection History link in the Virus & Threat Protection pane.**

The screen shown in Figure 3-4 appears. Once upon a time, Windows Security would flag infected files and offer them up for you to decide what to do with the offensive file. It appears as if that behavior has been scaled back radically. As best I can tell, in almost all circumstances, when Windows Security hits a dicey file, it *quarantines* the file — sticks it in a place you won't accidentally find — and just keeps going. You're rarely notified (although a toaster notification may slide out from the right side of the screen), but the file just disappears from where it should've been.

CHAPTER 3 **Running Built-In Security Programs**　　**803**

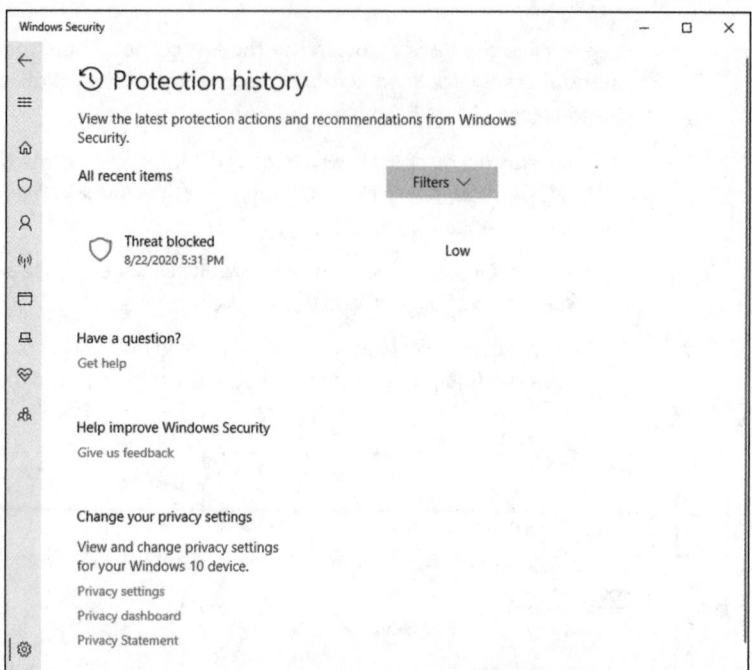

FIGURE 3-4:
A full history
of the actions
Defender's taken
appear here.

If you just downloaded a file and it disappeared, there's a very good chance that it's infected and Windows Security has whisked it away to a well-guarded location, and the only way you'll ever find it is in the Protection History tab of the Windows Security program.

Should you decide to bring the file back, for whatever reason, click the name of the threat, and then tap or click Actions followed by Allow and then Yes in the UAC (User Account Control) prompt. Rub your lucky rabbit's foot a couple of times while you're at it.

Controlling Folder Access

Ransomware — software that scrambles files and demands a payment before unscrambling — has become quite the rage. It's an easy way for Script Kiddies to monetize their malware. I talk about ransomware in Book 9, Chapter 1.

Microsoft has come up with a way to preemptively block many kinds of ransomware by simply restricting access to folders that contain files the ransomware may want to zap.

WARNING

There's just one problem. Restricting, or controlling, folder access is a pain in the neck — it blocks every program unless you specifically give a specific program access. So, for example, you can turn off access to your Documents folder but allow access to Word and Excel. That may work well until you want to run Notepad on a file in the Documents folder. Oh-oh.

That's the reason why Microsoft doesn't turn on Controlled Folder Access (CFA) by default. If you really want CFA, you must dig deep and find it. If you do make the effort, the monkey's on your back to (1) stick CFA on all the right folders and (2) whitelist any program that may need to use files in the CAFs folders.

To enable CFA, you need to jump through the following hoops:

1. In the search bar to the right of the Start button, type sec. At the top, tap or click Windows Security.

2. Tap or click the Virus & Threat Protection icon, scroll way down, and click or tap Manage Ransomware Protection.

The CFA settings screen appears, as shown in Figure 3-5.

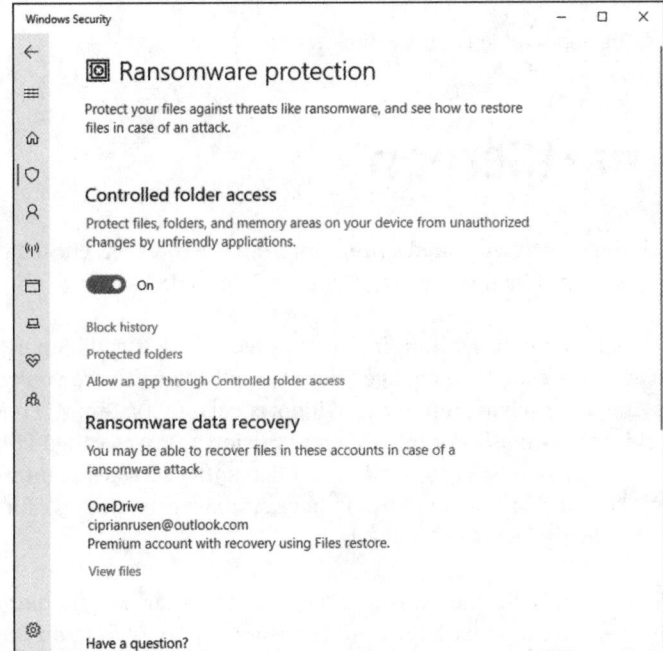

FIGURE 3-5: You have to set up controlled folder access manually — and doing so is problematic on many systems.

3. **Set Controlled Folder Access to On, and click the Protected Folders link. Click Yes when asked to confirm your choice.**

   You see a list of all folders protected by CFA — Documents, Pictures, Videos, Music, Desktop.

   Realize that ransomware frequently attacks files in other locations.

4. **If you want to add another folder to the blocked list, click the Add a Protected Folder button and navigate to and select the folder. Repeat as necessary.**

   Note that Windows has an automatically created (but not fully disclosed!) set of programs that it deems to be friendly.

5. **Click the Back arrow in the upper-left corner to return to the window shown in Figure 3-5.**

6. **If you have any programs that need access to those folders, and the apps aren't automatically identified as friendly, click the Allow an App through Controlled Folder Access link, and then Yes.**

7. **Click the Add an Allowed App button, click Browse All Apps, and then navigate to and select the app. Repeat as necessary.**

   The app is added to the whitelist.

Judging SmartScreen

Have you ever downloaded a program from the Internet, clicked to install it — and then, a second later, thought, "Why did I do that?"

Microsoft came up with an interesting technique it calls SmartScreen that gives you an extra chance to change your mind, if the software you're trying to install has drawn criticism from other Windows customers. SmartScreen was built in to an older version of Internet Explorer, version 7 (it was called Phishing Filter). It's now part of Windows 10 in both of Microsoft's browsers: Internet Explorer and Edge. Google Chrome and Firefox have similar technologies, but the SmartScreen settings apply only to IE and Edge.

SmartScreen is not the same as Smart Search. SmartScreen, discussed here, offers some benefits to most Edge (and IE) users. Smart Search, on the other hand, is a pernicious piece of snooping malware (did I put you off sufficiently?) that Microsoft sneaks into Windows 10. Follow along here to use SmartScreen.

One part of SmartScreen works with Windows Security. In fact, sometimes I've seen an infected file trigger a toaster notification from Windows Security, and later had the same infected file prompt the SmartScreen warning shown in Figure 3-6.

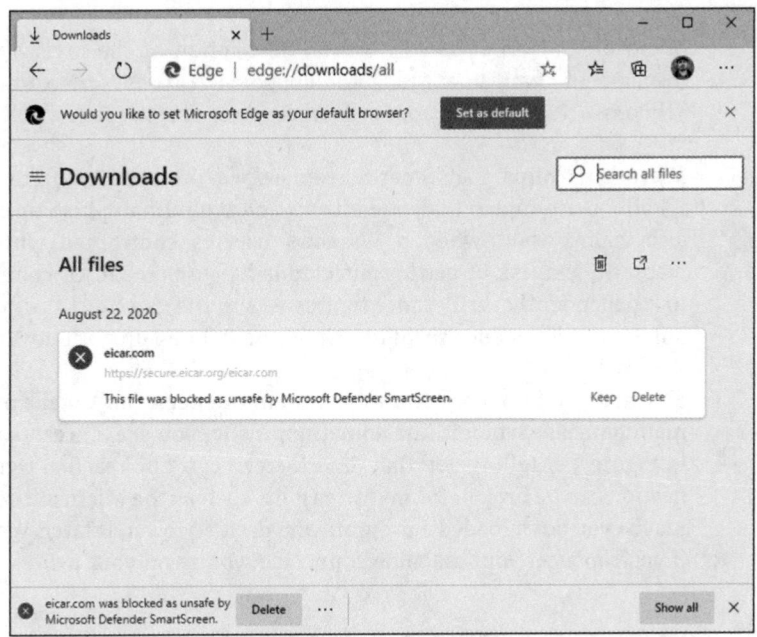

FIGURE 3-6:
SmartScreen may take the credit for the bust, but Windows Security did the work.

If you don't run the program, it gets stuffed into the same location that Windows Security puts its quarantined programs — out of the way where you can't find it, unless you go in through Windows Security's Protection History tab (refer to Figure 3-4).

There's a second part of SmartScreen that works completely differently. Something like this:

1. You download something — anything — from the Internet.

   Most browsers and many email programs and other online services (including instant messengers) put a brand on the file that indicates where the file came from.

2. When you try to launch the file, Windows 10 checks the name of the file and the URL of origin to see whether they're on a trusted whitelist.

3. If the file doesn't pass muster, you see the notification in Figure 3-6.

4. The more people who install the program from that site, the more trusted the program becomes.

Again, Microsoft is collecting information about your system — in this case, about your downloads — but it's for a good cause.

Microsoft has an excellent, official description of the precise way the tracking mechanism works at `https://support.microsoft.com/en-us/help/17443/windows-internet-explorer-smartscreen-filter-faq`.

REMEMBER

Microsoft claims that SmartScreen helped protect IE9 users from more than 1.5 billion attempted malware attacks and 150 million phishing attacks. Microsoft also claims that, when a Windows user is confronted with a confirmation message, the risk of getting infected is 25-70 percent. Of course it's impossible to independently verify those figures — and the gap from 25-70 percent gapes — but SmartScreen does seem to help in the fight against scumware.

So what can go wrong? Not much. If SmartScreen can't make a connection to its main database when it hits something fishy, you see a green screen like the one in Figure 3-7 telling you that SmartScreen can't be reached right now. The connection can be broken for many reasons, such as the Microsoft servers go down or maybe you downloaded a program and decided to run it later. When that happens, if you can't get your machine connected, you're on your own.

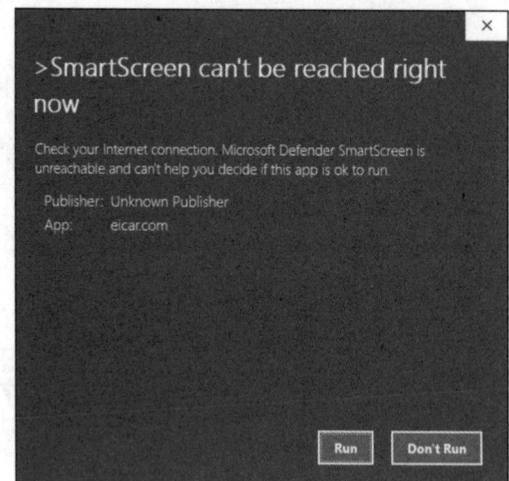

FIGURE 3-7:
If SmartScreen can't phone home, it leaves you on your own.

Turning off SmartScreen is an option when you install Windows 10. You can also turn it off manually. Normally, overriding a SmartScreen warning requires the okay of someone with an administrator account. You can change that, too. Here's how:

1. **In the search box to the right of the Start button, type** smartscreen. **At the top of the resulting list, click or tap Reputation-Based Protection.**

   The Windows Security Reputation-Based Protection pane appears (see Figure 3-8).

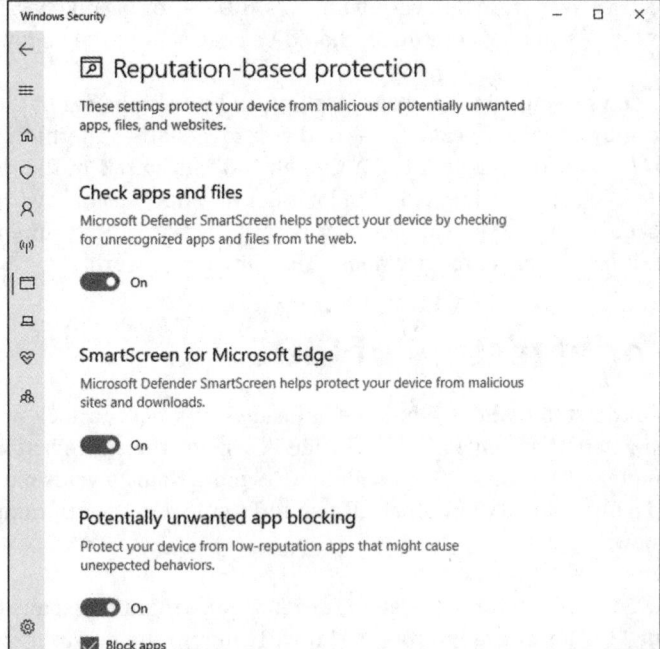

2. **(Optional) To change the default warning behavior for SmartScreen when you run it on downloaded files, change the first group of buttons.**

   You won't receive a warning when something bad is downloaded by Google Chrome, Firefox, or any non-Microsoft browser, but you will be warned if you try to open or run the file.

3. **(Optional) To change the default warning behavior for Edge (and Internet Explorer), adjust the second block of buttons.**

4. **(Optional) Turn off SmartScreen for potentially unwanted Apps as well as for Microsoft Store apps by setting the appropriate switches to Off. Click Yes when asked to confirm your choice.**

Don't forget to close Windows Security when done. Also, its icon in the notification area will be busy with warning messages.

Booting Securely with UEFI

If you've ever struggled with your PC's BIOS — or been kneecapped by a capable rootkit — you know that BIOS should've been sent to the dugout a decade ago.

Windows 10 pulled the industry kicking and screaming out of the BIOS generation and into a far more capable — and controversial — alternative, *Unified Extensible Firmware Interface* (UEFI). Although UEFI machines in the time of Windows 7 were unusual, starting with Windows 8, every new machine with a Runs Windows sticker is required to run UEFI; it's part of the licensing requirement. Windows 10 continues the same requirement. 'Tis a brave new world.

A brief history of BIOS

ASK
WOODY.COM

To understand where Windows is headed, it's best to look at where it's been. And where it's been with BIOS inside PCs spans the entire history of the personal computer. That makes PC-resident BIOS more than 30 years old. The first IBM PC had a BIOS, and it didn't look all that different from the inscrutable one you swear at now.

The Basic Input/Output System, or *BIOS*, is a program responsible for getting all your PC's hardware in order and then firing up the operating system (OS) — in this case, Windows — and finally handing control of the computer over to the OS. BIOS runs automatically when the PC is turned on.

Older operating systems, such as DOS, relied on the BIOS to perform input and output functions. More modern operating systems, including Windows, have their own device drivers that make BIOS control obsolete, after the OS is running.

Every BIOS has a user interface, which looks much like the one in Figure 3-9. You press a key while the BIOS is starting and, using obscure keyboard incantations, take some control over your PC's hardware, select boot devices (in other words, tell BIOS where the operating system is located), overclock the processor, disable or rearrange hard drives, and the like.

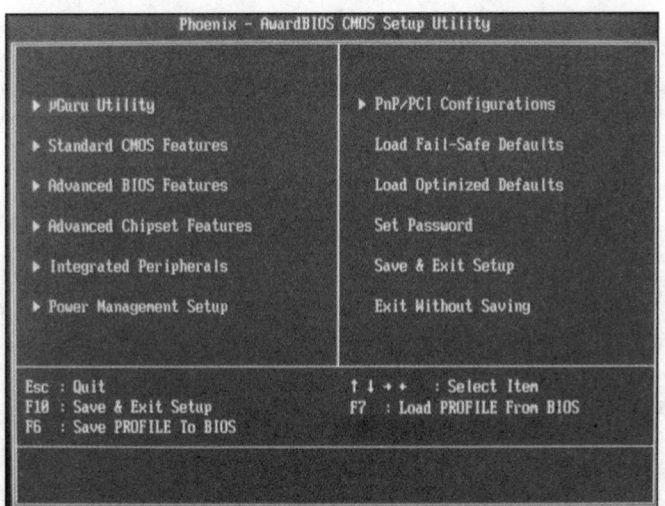

FIGURE 3-9:
The AwardBIOS
Setup Utility.

The PC you're using right now may or may not have UEFI, and even if it does have UEFI, you may not be able to get to it. Windows 10 runs just fine on BIOS systems, but it can protect you even better — especially from rootkits — if your PC supports UEFI.

How UEFI is different from/better than BIOS

BIOS has all sorts of problems, not the least of which is its susceptibility to malware. Rootkits like to hook themselves into the earliest part of the booting process — permitting them to run underneath Windows — and BIOS has a big Kick Me sign on its tail.

TECHNICAL
STUFF

UEFI and BIOS can coexist: UEFI can run on top of BIOS, hooking itself into the program locations where the operating system may call BIOS, basically usurping all the BIOS functions after UEFI gets going. UEFI can also run without BIOS, taking care of all the run-time functions. The only thing UEFI can't do is perform the *POST* power-on self-test or run the initial setup. PCs that have UEFI without BIOS need separate programs for POST and setup that run automatically when the PC is started.

Unlike BIOS, which sits inside a chip on your PC's motherboard, UEFI can exist on a disk, just like any other program, or in non-volatile memory on the motherboard or even on a network share.

UEFI is very much like an operating system that runs before your final operating system kicks in. UEFI has access to all the PC's hardware, including the mouse and network connections. It can take advantage of your fancy video card and monitor, as shown in Figure 3-10. It can even access the Internet. If you've ever played with BIOS, you know that this is in a whole new dimension.

FIGURE 3-10:
The UEFI
interface on an
ASUS PC.

Compare Figure 3-9 with Figure 3-10, and you'll have some idea where technology's been and where it's heading.

BIOS — the whole process surrounding BIOS, including POST — takes a long, long time. UEFI, by contrast, can go by quite quickly. The BIOS program itself is easy to reverse-engineer and has no internal security protection. In the malware maelstrom, it's a sitting duck. UEFI can run in any irascible, malware-dodging way its inventors contrive.

Dual boot in the old world involves a handoff to a clunky text program; in the new world, it can be much simpler, more visual, and controlled by mouse or touch.

More to the point, UEFI can police operating systems prior to loading them. That could make rootkit writers' lives considerably more difficult by, for example, refusing to run an OS unless it has a proper digital security signature. Windows Security can work with UEFI to validate OSs before they're loaded. And that's where the controversy begins.

How Windows 10 uses UEFI

TECHNICAL STUFF

A UEFI *Secure Boot* option validates programs before allowing them to run. If Secure Boot is turned on, operating system loaders have to be signed using a digital certificate. If you want to dual boot between Windows 10 and Linux, the Linux program must have a digital certificate — something Linux programs have never required before.

After UEFI validates the digital key, UEFI calls on Windows Security to verify the certificate for the OS loader. Windows Security (or another security program) can go out to the Internet and check to see whether UEFI is about to run an OS that has had its certificate yanked.

In essence, in a dual boot system, Windows Security decides whether an operating system gets loaded on your Secure Boot–enabled machine.

That curls the toes of many Linux fans. Why should their operating systems be subject to Microsoft's rules, if you want to dual boot between Windows 10 and Linux?

If you have a PC with UEFI and Secure Boot and you want to boot an operating system that doesn't have a Microsoft-approved digital signature, you have two options:

>> You can turn off Secure Boot.

>> You can manually add a key to the UEFI validation routine, specifically allowing that unsigned operating system to load.

ASK WOODY.COM

Some PCs won't let you turn off Secure Boot. So if you want to dual boot Windows 10 and some other operating system on a Windows 10-certified computer, you may have lots of hoops to jump through. Check with your hardware manufacturer.

Controlling User Account Control

User Account Control (*UAC*) is a pain in the neck, but then again, it's supposed to be a pain in the neck. If you try to install a program that's going to make system-level changes, you may see the obnoxious prompt in Figure 3-11.

UAC's a drama queen, too. The approval dialog box in Figure 3-11 appears front and center, but at the same time, your entire desktop dims, and you're forced to deal with the UAC prompt.

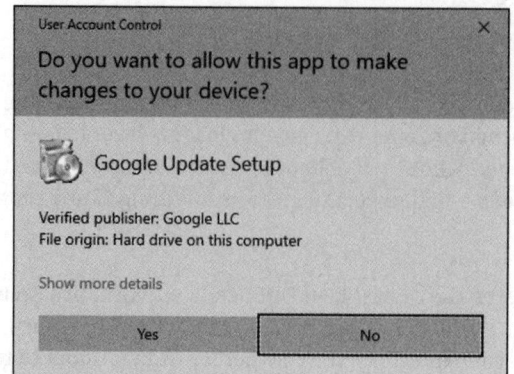

FIGURE 3-11:
User Account
Control tries to
keep you from
clobbering your
system.

REMEMBER

UAC grabs you by the eyeballs and shakes once or twice for a good reason: It's telling you that a program wants to make changes to your system — not piddling things like changing a document or opening a picture, but earth-shaking things like modifying the Registry or poking around inside system folders.

If you go into your system folders manually or if you fire up the Registry Editor and start making loose and fancy with Registry keys, UAC figures you know what you're doing and leaves you alone. But the minute a program tries to do those kinds of things, Windows whups you upside the head, warns you that a potentially danger-ous program is on the prowl, and gives you a chance to kill the program in its tracks.

Windows lets you adjust User Account Control so it isn't quite as dramatic — or you can get rid of it entirely.

To bring up the slider and adjust your computer's UAC level, follow these steps:

1. **In the search box next to the Start button, type** user account. **At the top of the ensuing list, choose Change User Account Control Settings.**

   The slider shown in Figure 3-12 appears.

2. **Adjust the slider according to Table 3-1, and then tap or click OK.**

   Perhaps surprisingly, as soon as you try to change your UAC level, Windows 10 hits you with a User Account Control prompt (refer to Figure 3-11). If you're using a standard account, you have to provide an administrator username and password to make the change. If you're using an administrator account, you have to confirm the change.

3. **Tap or click Yes.**

   Your changes take effect immediately.

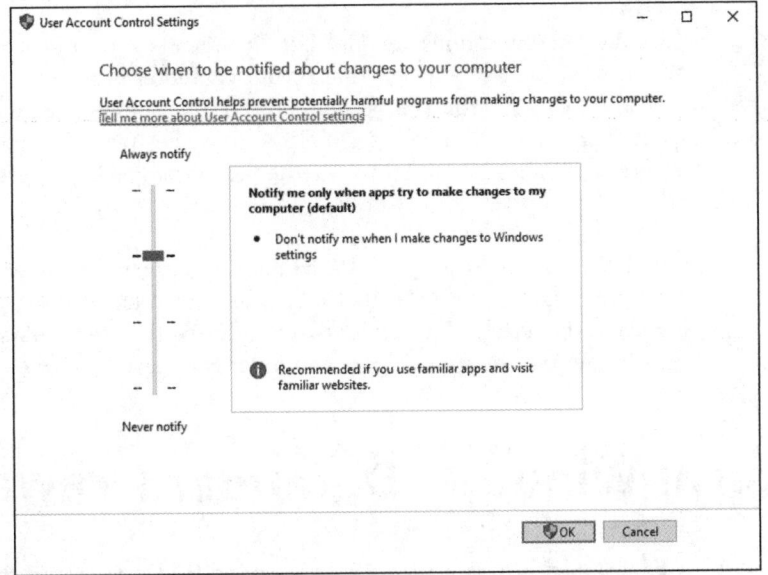

FIGURE 3-12:
Windows allows
you to change
the level of UAC
intrusiveness.

Running Built-In
Security Programs

TABLE 3-1   User Account Control Levels

Slider	What It Means	Recommendations
Level 1	Always brings up the full UAC notification whenever a program tries to install software or make changes to the computer that require an administrator account, or when you try to make changes to Windows settings that require an administrator account. You see these notifications even if you're using an administrator account. The screen blacks out, and you can't do anything until the UAC screen is answered.	This level offers the highest security but also the highest hassle factor.
Level 2	Brings up the UAC notification whenever a program tries to make changes to your computer, but generally doesn't bring up a UAC notification when you make changes directly.	The default — and probably the best choice.
Level 3	This level is the same as Level 2 except that the UAC notification doesn't lock and dim your desktop.	Potentially problematic. Dimming and locking the screen present a high hurdle for malware.
Level 4	UAC is disabled — programs can install other programs or make changes to Windows settings, and you can change anything you like, without triggering any UAC prompts. Note that this doesn't override other security settings. For example, if you're using a standard account, you still need to provide an administrator's ID and password before you can install a program that runs for all users.	Choosing Level 4 automatically turns off all UAC warnings — not recommended.

TECHNICAL STUFF

This description sounds simple, but the details are quite complex. Consider. Microsoft's Help system says that if your computer is at Level 2, the default setting in Windows, "You will be notified if a program outside of Windows tries to make changes to a Windows setting." So how does Windows tell when a program is outside Windows — and thus whether actions taken by the program are worthy of a UAC prompt at Levels 2 or 3?

UAC-level rules are interpreted according to a special Windows security certificate. Programs signed with that certificate are deemed to be part of Windows. Programs that aren't signed with that specific certificate are outside Windows and thus trigger UAC prompts if your computer is at Level 1, 2, or 3.

Poking at Windows Defender Firewall

A *firewall* is a program that sits between your computer and the Internet, protecting you from the big, mean, nasty gorillas riding around on the information superhighway. An *inbound firewall* acts like a traffic cop that, in the best of all possible worlds, allows only good stuff into your computer and keeps all the bad stuff out on the Internet, where it belongs. An *outbound firewall* prevents your computer from sending bad stuff to the Internet, such as when your computer becomes infected with a virus or has another security problem.

Windows includes a usable (if not fancy) inbound firewall. It also includes a snarly, hard-to-configure, rudimentary outbound firewall, which has all the social graces of a junkyard dog. Unless you know the magic incantations, you never even see the outbound firewall — it's completely muzzled unless you dig in to the Windows doghouse and teach it some tricks.

REMEMBER

Everybody needs an inbound firewall, without a doubt. You already have one, in Windows 10, and you don't need to do anything to it.

Outbound firewalls tend to bother you mercilessly with inscrutable warnings saying that obscure processes are trying to send data. If you simply click through and let the program phone home, you're defeating the purpose of the outbound firewall. On the other hand, if you take the time to track down every single outbound event warning, you may spend half your life chasing firewall snipes.

I have a few friends who insist on running an outbound firewall. They uniformly recommend Comodo Firewall, which is available in a free-for-personal-use version at http://personalfirewall.comodo.com.

HARDWARE FIREWALLS

Most modern routers and wireless access points include significant firewalling capability. It's part and parcel of the way they work when they share an Internet connection among many computers.

Routers and wireless access points add an extra step between your computer and the Internet. That extra jump — named network address translation — combined with innate intelligence on the router's part can provide an extra layer of protection that works independently from, but in conjunction with, the firewall running on your PC.

**ASK
WOODY.COM**

I think outbound firewalls are a complete waste of time. Although I'm sure some people have been alerted to Windows infections when their outbound firewall goes bananas, 99.99 percent of the time, the outbound warnings are just noise. Outbound firewalls don't catch the cleverest malware, anyway.

Understanding Firewall basic features

All versions of Windows 10 ship with a decent and capable, but not foolproof, *stateful* firewall named Windows Defender Firewall (WDF). (See the nearby sidebar, "What's a stateful firewall?")

The WDF inbound firewall is on by default. Unless you change something, Windows Defender Firewall is turned on for all connections on your PC. For example, if you have a LAN cable, a wireless networking card, and a 4G USB card on a specific PC, WDF is turned on for them all. The only way Windows Defender Firewall gets turned off is if you deliberately turn it off or if the network administrator on your Big Corporate Network decides to disable it by remote control or install Windows service packs with Windows Defender Firewall turned off.

WARNING

In extremely unusual circumstances, malware (viruses, Trojans, whatever) have been known to turn off Windows Defender Firewall. If your firewall kicks out, Windows lets you know loud and clear with balloon notifications near the system clock on the desktop, toaster notifications, and a crescendo from Ride of the Valkyries blaring on your speakers.

You can change WDF settings for inbound protection relatively easily. When you make changes, they apply to all connections on your PC. On the other hand, WDF settings for outbound protection make the rules of cricket look like child's play.

WDF kicks in before the computer is connected to the network. Back in the not-so-good old days, many PCs got infected between the time they were connected and when the firewall came up.

Speaking your firewall's lingo

At this point, I need to inundate you with a bunch of jargon so that you can take control of Windows Defender Firewall. Hold your nose and dive in. The concepts aren't that difficult, although the lousy terminology sounds like a first-year advertising student invented it. Refer to this section if you become bewildered when wading through the WDF dialog boxes.

As you no doubt realize, the amount of data that can be sent from one computer to another over a network can be tiny or huge. Computers talk with each other by breaking the data into *packets* (or small chunks of data with a wrapper that identifies where the data came from and where it's going).

On the Internet, packets can be sent in two ways:

>> **User Datagram Protocol (UDP):** UDP is fast and sloppy. The computer sending the packets doesn't keep track of which packets were sent, and the computer receiving the packets doesn't make any attempt to get the sender to resend packets that vanish mysteriously into the bowels of the Internet. UDP is the kind of *protocol* (transmission method) that can work with live broadcasts, where short gaps wouldn't be nearly as disruptive as long pauses, while the computers wait to resend a dropped packet.

>> **Transmission Control Protocol (TCP):** TCP is methodical and complete. The sending computer keeps track of which packets it has sent. If the receiving computer doesn't get a packet, it notifies the sending computer, which resends the packet. These days, almost all communication over the Internet goes by way of TCP.

TECHNICAL STUFF

Every computer on a network has an *IP address*, which is a collection of four sets of numbers, each between 0 and 255. For example, 192.168.0.2 is a common IP address for computers connected to a local network; the computer that handles the Dummies.com website is at 208.215.179.139. You can think of the IP address as analogous to a telephone number. See Book 2, Chapter 6 for details.

Peeking into your firewall

When you use a firewall — and you should — you change the way your computer communicates with other computers on the Internet. This section explains what Windows Defender Firewall does behind the scenes so that when it gets in the way, you understand how to tweak it. (You find the ins and outs of working around the firewall in the "Making inbound exceptions" section, later in this chapter.)

TECHNICAL STUFF

When two computers communicate, they need not only each other's IP address but also a specific entry point called a *port* — think of it as a telephone extension — to talk to each other. For example, most websites respond to requests sent to port 80. There's nothing magical about the number 80; it's just the port number that people have agreed to use when trying to get to a website's computer. If your web browser wants to look at the Dummies.com website, it sends a packet to 208.215.179.139, port 80.

Windows Defender Firewall works by handling all these duties simultaneously:

>> **It keeps track of outgoing packets and allows incoming packets to go through the firewall if they can be matched with an outgoing packet.** In other words, WDF works as a stateful inbound firewall.

>> **If your computer is attached to a private network, Windows Defender Firewall allows packets to come and go on ports 139 and 445, but only if they came from another computer on your local network and only if they're using TCP.** Windows Defender Firewall needs to open those ports for file and printer sharing. It also opens several ports for Windows Media Player if you've chosen to share your media files, for example.

>> **Similarly, if your computer is attached to a private network, Windows Defender Firewall automatically opens ports 137, 138, and 5355 for UDP, but only for packets that originate on your local network.**

>> **If you specifically told Windows Defender Firewall that you want it to allow packets to come in on a specific port and the Block All Incoming Connections check box isn't selected, WDF follows your orders.** You may need to open a port in this way for online gaming, for example.

>> **Windows Defender Firewall allows packets to come into your computer if they're sent to the Remote Assistance program, as long as you created a Remote Assistance request on this PC and told Windows to open your firewall (see Book 7, Chapter 2).** Remote Assistance allows other users to take control of your PC, but it has its own security settings and strong password protection. Still, it's a known security hole that's enabled when you create a request.

>> **You can tell Windows Defender Firewall to accept packets that are directed at specific programs.** Usually, any company that makes a program designed to listen for incoming Internet traffic (Skype is a prime example, as are any instant-messaging apps) adds its program to the list of designated exceptions when the program is installed.

>> **Unless an inbound packet meets one of the preceding criteria, it's simply ignored.** Windows Defender Firewall swallows it without a peep. Conversely, unless you've changed something, any and all outbound traffic goes through unobstructed.

Making inbound exceptions

Firewalls can be infuriating. You may have a program that has worked for a hundred years on all sorts of computers, but the minute you install it on a Windows 10 machine with Windows Defender Firewall in action, it just stops working, for absolutely no apparent reason.

You can get mad at Microsoft and scream at Windows Defender Firewall, but when you do, realize that at least part of the problem lies in the way the firewall has to work. (See the "Peeking into your firewall" section, earlier in this chapter, for an explanation of what your firewall does behind the scenes.) It has to block packets that are trying to get in, unless you explicitly tell the firewall to allow them to get in.

Perhaps most infuriatingly, WDF blocks those packets by simply swallowing them, not by notifying the computer that sent the packet. Windows Defender Firewall has to remain stealthy because if it sends back a packet that says, "Hey, I got your packet, but I can't let it through," the bad guys get an acknowledgment that your computer exists, they can probably figure out which firewall you're using, and they may be able to combine those two pieces of information to give you a headache. It's far better for Windows Defender Firewall to act like a black hole.

Some programs need to listen to incoming traffic from the Internet; they wait until they're contacted and then respond. Usually, you know whether you have this type of program because the installer tells you that you need to tell your firewall to back off.

TIP

If you have a program that doesn't (or can't) poke its own hole through Windows Defender Firewall, you can tell WDF to allow packets destined for that specific program — and only that program — in through the firewall. You may want to do that with a game that needs to accept incoming traffic, for example, or for an Outlook extender program that interacts with smartphones.

To poke a hole in the inbound Windows Defender Firewall for a specific program:

1. Make sure that the program you want to allow through Firewall is installed.

2. In the search box next to the Start button, type firewall. Choose Allow an App through Windows Firewall.

Windows Defender Firewall presents you with a lengthy list of apps that you may want to allow (see Figure 3-13): If a box is selected, Windows Defender Firewall allows unsolicited incoming packets of data directed to that program and that program alone, and the column tells you whether the connection is allowed for private or public connections.

REMEMBER

These settings don't apply to incoming packets of data that are received in response to a request from your computer; they apply only when a packet of data appears on your firewall's doorstep without an invitation.

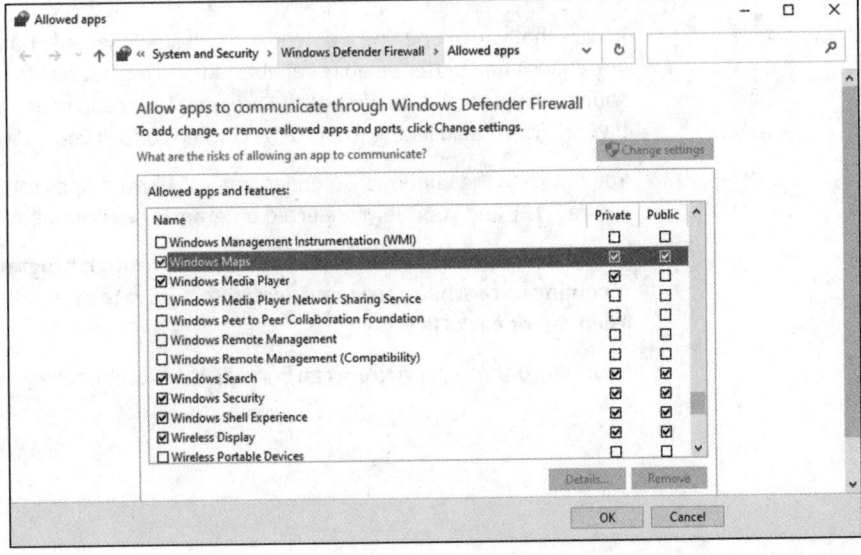

FIGURE 3-13:
Allow installed programs to poke through the firewall.

In Figure 3-13, the tiled Windows Maps app is allowed to receive inbound packets whether you're connected to a private or public network. Windows Media Player, on the other hand, may accept unsolicited inbound data from other computers only if you're connected to a private network: If you're attached to a public network, inbound packets headed for Windows Media Player are swallowed by the Windows Defender Firewall Black Hole (patent pending).

3. **Do one of the following:**

   - *If you can find the program that you want to poke through the firewall listed in the Allow Programs list,* select the check boxes that correspond to whether you want to allow the unsolicited incoming data when connected to a home or work network and whether you want to allow the incoming packets when connected to a public network. It's rare indeed that you'd allow access when connected to a public network but not to a home or work network.

   - *If you can't find the program that you want to poke through the firewall,* you need to go out and look for it. Tap or click the Change Settings button at the top, and then tap or click the Allow Another App button at the bottom. You must tap or click the Change Settings button first and then tap or click Allow Another Program. It's kind of a double-down protection feature that ensures you don't accidentally change things.

   Windows Defender Firewall goes out to all common program locations and finally presents you with the Whack a Mol . . . er, Add an App list like the one shown in Figure 3-14. It can take a while.

4. **Choose the program you want to add, or browse to its location, and then tap or click the Add button.**

WARNING

   Realize that you're opening a potential, albeit small, security hole. The program you choose had better be quite capable of handling packets from unknown sources. If you authorize a renegade program to accept incoming packets, the bad program could let the fox into the chicken coop. If you know what I mean.

   You return to the Windows Defender Firewall Allowed Apps list (refer to Figure 3-13), and your newly selected program is now available.

5. **Select the check boxes to allow your poked-through program to accept incoming data while you're connected to a private or a public network. Then tap or click OK.**

   Your poked-through program can immediately start handling inbound data.

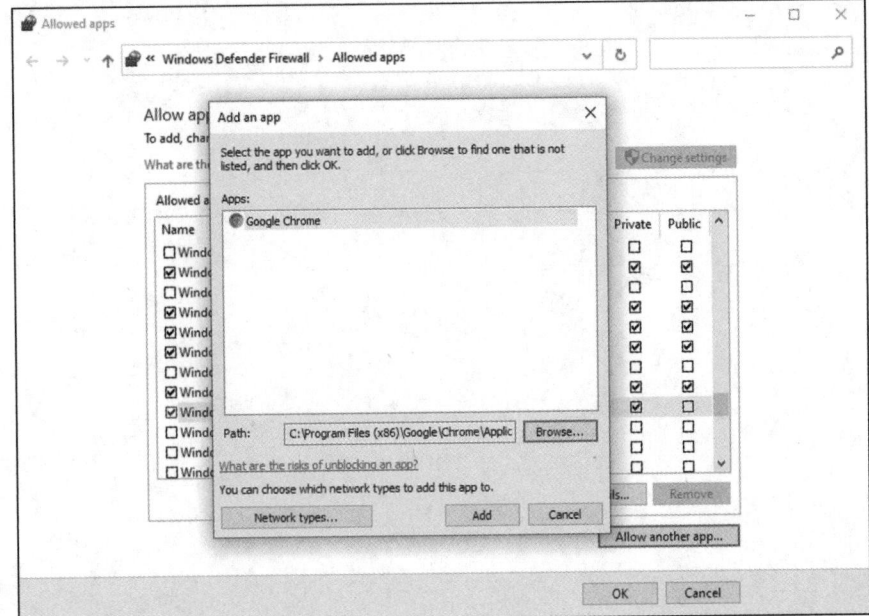

FIGURE 3-14:
Allow a program
(that you've
thoroughly
vetted!) to break
through the
firewall.

**TECHNICAL
STUFF**

In many cases, poking through Windows Defender Firewall doesn't solve the whole problem. You may have to poke through your modem or router as well — unsolicited packets that arrive at the router may get kicked back according to the router's rules, even if Windows 10 would allow them in. Unfortunately, each router and the method for poking holes in the router's inbound firewall differ. Check the site www.portforward.com/routers.htm for an enormous amount of information about poking through routers.

Chapter **4**

Top Security Helpers

I n Chapter 3 of this minibook, I talked about built-in Windows security programs that are available to every Windows 10 owner. In this chapter, I cast the web out a bit further to include one Microsoft encryption program you have to pay for — *BitLocker*, which is in Windows 10 Pro — and a handful of free-for-personal use programs that belong on every Windows 10 user's desktop.

Windows covers lots of security bases, but it doesn't touch them all.

ASK
WOODY.COM

Two very good programs store all your passwords and automatically fill in the username/password prompts at the websites you visit. One of them, *LastPass*, is based in the *cloud*, which means you can get at it even when you're on a dive boat in the Similans. The other, *RoboForm*, stores its data on your computer or on a USB drive. Both are slowly gaining features of the other — RoboForm in the cloud, LastPass in files on your computer — but their heritage stays true. I take you through the pros and cons of both approaches in this chapter.

Sometimes you — or one of your friends — will get an infection that even Windows Security (and Microsoft Defender Offline) can't handle. Usually it's because you (or they) installed a program they didn't research. If you (err, *they*) get hit bad, there's one place to turn. *Malwarebytes*, a combination of software and a very competent website, can crack just about any infection.

Ninite is free and does an amazing job of helping you keep all your software up to date.

If you want to connect to a website and make sure nobody can snoop on your connection — particularly important if you access financial sites from public Wi-Fi setups, like in a coffee shop or a bank — you should figure out how to use a VPN. I've been using *Private Internet Access, PIA*, for years, and although it isn't free, it works great.

Finally, I know of one specific Java and Flash blocker that works very well in the Firefox browser. *NoScript* can be customized in many ways. Although there are more-or-less similar choices for Google Chrome and Microsoft Edge, NoScript works the best of them all. It's the primary reason why I use Firefox as my main browser.

All these programs are free (or nearly so), well known, and tested — and they need to be part of your Windows system.

ASK
WOODY.COM

Deciding about BitLocker

REMEMBER

BitLocker encrypts an entire drive. (Actually, it encrypts a *volume* — typically a piece of a drive that's been lopped off to treat as if it were a drive all by itself.) Unlike Encrypting File System (see the nearby "Encrypting File System [EFS]" sidebar), you have to encrypt full drives (or, more accurately, volumes) or nothing at all. BitLocker runs *underneath* Windows: It starts before Windows starts. The Windows partition on a BitLocker-protected drive is completely encrypted. Even if a thief gets his hands on your laptop or hard drive, he can't view anything on it — not even your settings or system files.

ENCRYPTING FILE SYSTEM (EFS)

Microsoft *Encrypting File System* (EFS) works with or without BitLocker. EFS is a method for encrypting individual files or groups of files on a hard drive. EFS starts after Windows boots: It runs as a program under Windows, which means it can leave traces of itself and the data that's being encrypted in temporary Windows places that may be sniffed by exploit programs. The Windows directory isn't encrypted by EFS, so bad guys (and girls!) who can get access to the directory can hammer it with brute-force password attacks. Widely available tools can crack EFS if the cracker can reboot the, uh, crackee's computer. Thus, for example, EFS can't protect the hard drive on a stolen laptop or notebook. Windows has supported Encrypting File System since the halcyon days of Windows 2000.

BitLocker and EFS protect against two completely different kinds of attacks. Given a choice, you probably want BitLocker.

BitLocker To Go is quite similar to BitLocker, except it works on USB drives.

BitLocker is part of Windows 10 Pro. It is not part of the regular version of Windows 10. If you have Windows 10 and you want to get BitLocker, you have to upgrade to Windows 10 Pro. There's no other way to get it.

**ASK
WOODY.COM**

I talk about the various versions of Windows 10 in Book 1, Chapter 3. Suffice it to say that some people feel their information is sufficiently valuable that BitLocker, all by itself, justifies paying the extra bucks for Windows 10 Pro.

Here's how to encrypt your hard drive with BitLocker:

1. **In the search box, next to the Start button, type** bitlocker; **then click or tap Manage BitLocker.**

   The BitLocker Drive Encryption dialog box appears, as shown in Figure 4-1.

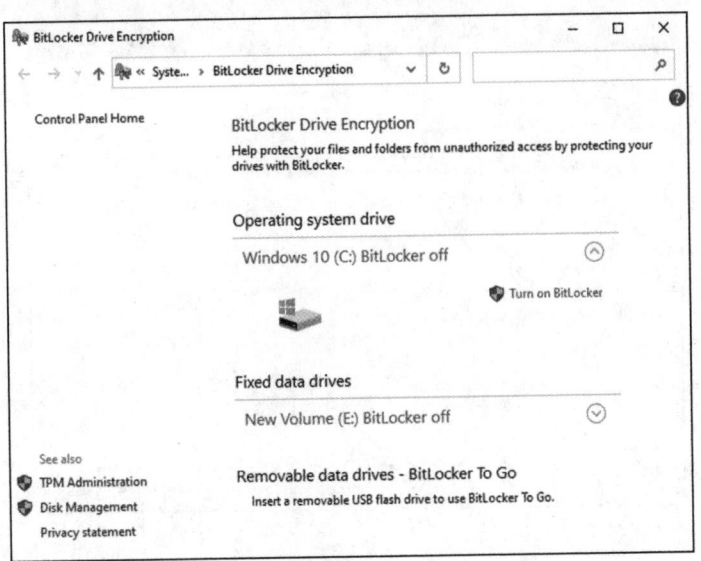

FIGURE 4-1:
Manage everything from the BitLocker Drive Encryption window.

2. **Next to the drive (volume) you want to encrypt, tap or click Turn On BitLocker.**

   The BitLocker Drive Encryption setup dialog box appears, and Windows 10 checks whether Windows 10 meets the requirements for running BitLocker.

TECHNICAL STUFF

If your PC doesn't have a built-in Trusted Platform Module system, you see a message that says *Your administrator must set the 'Allow BitLocker without a compatible TPM' option.* The only easy way to solve that problem is to run the Local Group Policy Editor program, gpedit.msc. If you need advice, check out the Tom's hardware article at `www.tomshardware.com/forum/id-3491730/turn-bitlocker-tpm.html`.

If everything is well, you are asked to back up your BitLocker recovery key.

3. **Choose how you want to save it, and tap or click Next.**

   After encrypting your PC with BitLocker, the recovery key is the only way to access your files when you have problems unlocking your PC. See Figure 4-2. You have options to save the key to your Microsoft account (on Microsoft's servers in the cloud), to save the key to a file on your computer (it is a good idea to back it up to a USB memory stick), or to print the recovery key.

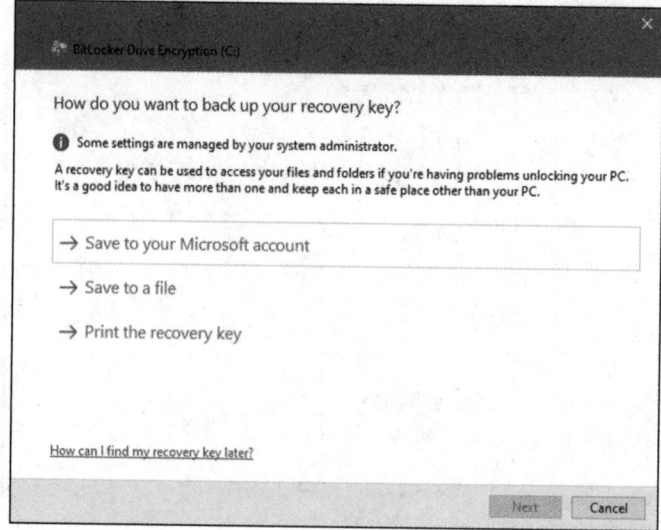

FIGURE 4-2: Choose how you want to save the BitLocker recovery key.

4. **When asked to choose between encrypting used disk space only or the entire drive, make your choice and tap or click Next.**

   If you want the encryption to finish faster, choose to encrypt only the used disk space. Encrypting the entire drive may take many hours.

5. **Choose which encryption mode to use, and then tap or click Next.**

   The new encryption mode, which I highly recommend, uses a more secure type of encryption and is available as of Windows 10 version 1511 (released in November 2015).

6. **Select the option to Run BitLocker System Check and tap or click Continue (instead of clicking Start Encryption).**

   The system check ensures that BitLocker can read the recovery and encryption keys correctly before encrypting the drive, which is a great idea.

7. **When asked to restart your computer, close all your open apps and files, and then click Restart Now.**

   When you log back into Windows 10, BitLocker encrypts your drive automatically, in the background. You can continue using your PC as usual. The BitLocker icon is shown in the system tray, on the right-side of the taskbar. If you click it, you see the progress of the encryption process.

In case you were wondering, yes, you can use BitLocker on Storage Spaces. BitLocker encrypts the entire Storage Space.

Managing Your Passwords

You can find no end of advice on creating strong passwords, using clever tricks, stats, mnemonics, and such. But all too frequently people (myself included in this rebuke) tend to reuse little passwords at what people think are inconsequential sites. It's a big mistake. If somebody hacks into that small-time site and steals your password — a process that's frighteningly common these days — any other place where you've used that same password is immediately vulnerable.

WARNING

There have been some spectacular examples of ultra-secure sites getting hacked in the past few years, where the hacker stole a username and password off a little inconsequential site and then discovered that the same username and password opened the doors to a trove of top-secret — even politically sensitive — corporate email or customer bank account information. The usernames and passwords were stolen from seasoned security professionals and admins at sensitive sites. You'd think they'd know better.

Using password managers

I don't know about you, but I have dozens of usernames and passwords that I use fairly regularly. There's just no way I can remember them all. And my monitor isn't big enough to handle all the yellow sticky notes they'd demand.

TIP

That's where a password manager comes in. A *password manager* keeps track of all your online passwords. It can generate truly random passwords with the click of a button. Most of all, it remembers the username and password necessary to log in to a specific website.

Every time I go to www.ebay.com, for example, my password manager fills in my username and password. Amazon, too. Facebook. Twitter. My bank. Stock brokerage house. I have to remember the one password for the password manager, but after that, everything else gets filled in automatically. It's a huge timesaver.

A password manager won't log into Windows for you, and it won't remember the passwords on documents or spreadsheets. But it does keep track of every online password and regurgitates the passwords you need with no hassle.

Which is better: Online or in-hand?

ASK
WOODY.COM

I have used two password-remembering programs for many years. I like — and trust — them both. The big difference between them? One was originally designed to run on a USB drive; the other has always been in the cloud, which is to say, on the Internet:

» **RoboForm Desktop,** which can store passwords on your hard drive or on a USB drive, works with all the major web browsers and has simple tools for synchronizing passwords between your hard drive and a USB drive.

» **LastPass,** which stores passwords on its website, uses an encryption technique that guarantees your passwords won't get stolen or cracked. I talk about the encryption method in the section "Liking LastPass," later in this chapter.

Although it started as a USB-toting application, RoboForm now offers **RoboForm Everywhere,** which synchronizes in the cloud, like LastPass.

Which one is better? It depends on how you use your computer.

If you always use the same computer or you can always remember to sync and take your RoboForm2Go USB drive with you, RoboForm works great.

Unfortunately, I don't meet either of those two criteria, so in recent years, I've been using LastPass. Of course, there's an additional security concern because your data's stored on the LastPass servers and not on the USB drive in your pocket. In addition, you need an Internet connection to get to LastPass — but then if you don't have an Internet connection, you probably don't need LastPass, either.

The new RoboForm Everywhere syncs to the cloud, too.

Opinions run all over the place, but I prefer the LastPass interface, as opposed to RoboForm Everywhere's. You should feel comfortable using either.

Several alternatives to LastPass have appeared recently, particularly since the LastPass company was sold to LogMeIn in October 2015. (LogMeIn recently got rid of its free version, and many worry that LastPass will suffer the same fate. Fortunately, the LastPass free option still survives.) If you're concerned, consider getting KeePass (www.keepass.com) or 1Password (www.1password.com), both of which work well.

Rockin' RoboForm

RoboForm Desktop (www.roboform.com) has all the features you need in a password manager. It manages your passwords, of course, with excellent recognition of websites, automatically filling in your login details, but it'll also generate random passwords for you, if you like, fill in forms on the web, and create backups either on a USB drive or on another computer on your network.

RoboForm stores all its data on a disk in AES-256 encrypted format. If somebody steals your RoboForm database, you needn't worry. Without the master key — which only you have — the whole database is gibberish.

RoboForm has versions for Windows, Mac, Linux, iPhone, iPad, Android smartphones and tablets. You need to buy a separate license for each computer, device, or USB drive.

WHAT IS AES-256?

The most effective encryption method that's commonly used on PCs conforms to the US National Institute of Standards and Technology's Advanced Encryption Standard 256-bit specification.

AES is the first widely available, open encryption technique (yes, you can look at the program) that's been approved by the US National Security Agency for Top Secret information. Of course, that fact has led to speculation that the NSA has cracked the algorithm, so it can decrypt AES-256 data, but there doesn't seem to be any corroboration. I guess the conspiracy theory makes for good beer-drinking banter but not much more.

It's been estimated that if you took all the computer horsepower currently on the face of the earth and set it to work on a single AES-256 encrypted file, cracking the encryption would take far longer than the age of the universe.

There's a trial version of RoboForm that expires after 30 days.

RoboForm Everywhere will store all your information on RoboForm's servers, so you can download it and use it anywhere — even on an unlimited number of computers. The trick is the price: RoboForm Everywhere costs $19.95 per year. (The first year's discounted to $9.95.) Prices may change, of course.

Liking LastPass

LastPass (`www.lastpass.com`) stores everything in the cloud on the LastPass servers. Like RoboForm, LastPass keeps track of your user IDs, passwords, automatic form-filling information (think name, address, phone, credit card number), and other settings and offers them to you with a click.

Using LastPass can't be simpler. Download and install it, and it'll appear with a red ellipsis in the upper-right corner of your browser (see Figure 4-3).

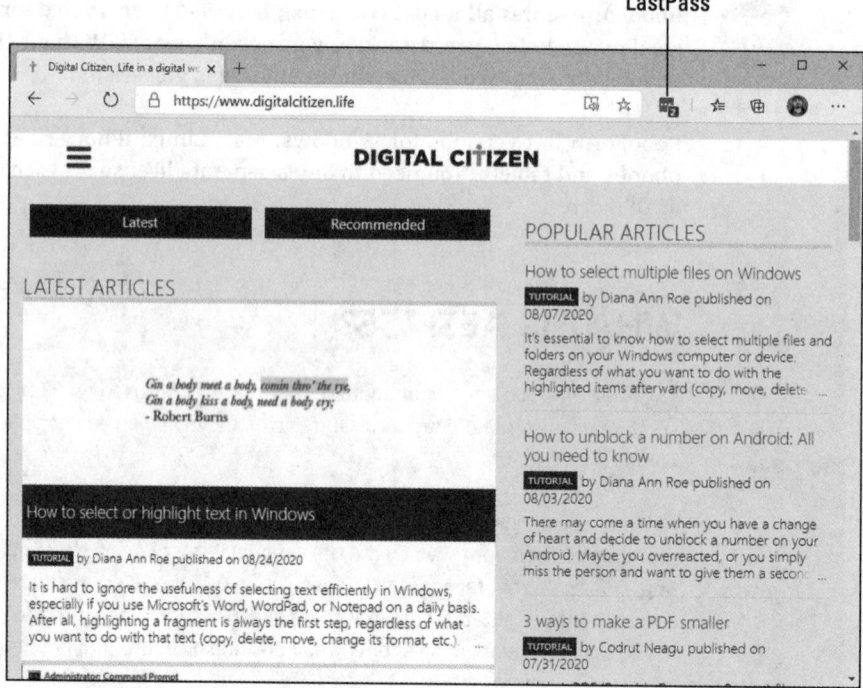

FIGURE 4-3:
LastPass is on the job if you can see a red asterisk in the upper-right corner.

You don't really need to do anything. LastPass will prompt you for the master password when you start using your browser. If LastPass is turned off, the star icon turns gray. Tap or click it, provide the master password, and the LastPass icon turns red again, ready to roll.

When you go to a site that requires a username and password, if LastPass recognizes the site, it fills them both in for you. If LastPass doesn't recognize the site, you fill in the blanks and click, and LastPass remembers the credentials for the next time you surf this way.

Form filling works similarly.

TIP

You can maintain two (or more) separate usernames and passwords for any specific site — say, you log in to a banking site with two different accounts. If LastPass has more than one set of credentials stored for a specific site, it takes its best guess as to which one you want but then gives you the option of using one of the others. In this screenshot, I have two separate credentials for the site — that's why a 2 is on the LastPass icon.

Any time you want to look at the usernames and passwords that LastPass has squirreled away, tap or click the red LastPass icon. You have a chance to look at your *Vault* — which is your password database — or look up recently used passwords and much more. You can even keep encrypted notes to yourself.

REMEMBER

The way LastPass handles your data is quite clever. All your passwords are encrypted using AES-256. They're encrypted and decrypted *on your PC*. Only you have the master password. So if the data is pilfered off the LastPass servers or somebody is sniffing your online communication, all the interlopers get is a bunch of useless bits.

You can also store secure notes, form-filling information such as your credit card information and address, and other data in LastPass.

LastPass is free for individual use, and it works on all major PC and mobile platforms. If you want some advanced features, such as sharing passwords with others or autofilling passwords for Windows 10 apps, you need the Premium edition, which costs $24 a year.

Keeping Your Other Programs Up to Date

You have Windows Update to keep Windows working and patched.

WARNING

But what about all the other programs on your PC? Considering that something like 80 percent of all new infections come from *third-party* programs (read: software written by some company other than Microsoft), keeping those other programs updated is a crucial task.

I used to recommend Ninite, which not only downloads and installs a wide variety of software but also creates a file that you can use to update all programs at once.

ASK WOODY.COM

Here's how to run Ninite:

1. **Go to the Ninite site at** `https://ninite.com/`.

2. **Select the boxes next to the programs you want to install.**

   Nearly a hundred programs are on offer. See Figure 4-4.

3. **At the bottom of the screen, click Get Your Ninite.**

   Ninite creates a custom program and downloads it to your machine.

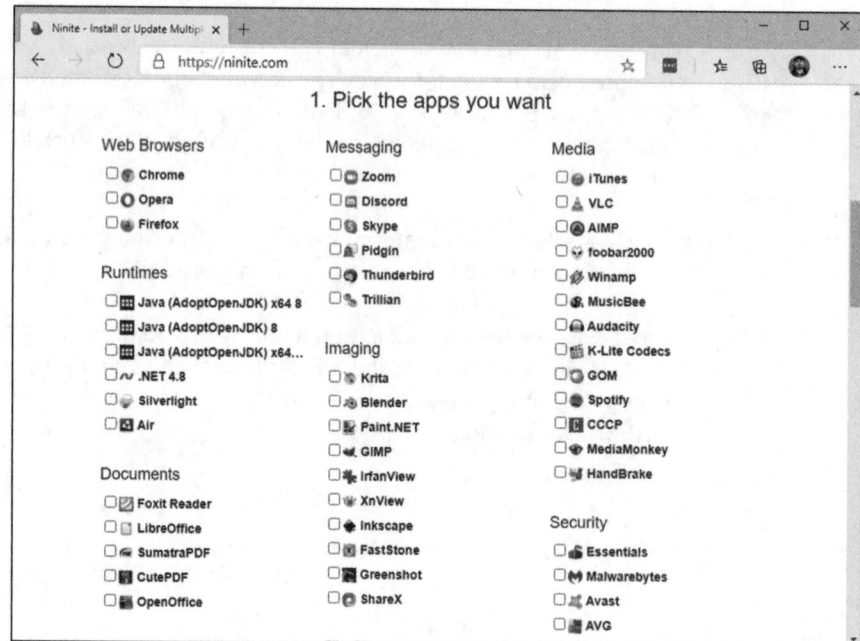

FIGURE 4-4:
An enormous number of programs are just a few clicks away.

4. **Run the downloaded EXE file, and click Yes when the UAC prompt appears.**

   The first run can take a long time, so be patient. When the custom installer is finished, all programs you selected are installed on your machine. No junk. No garbage. Just the programs. See Figure 4-5.

5. **Save the downloaded EXE file so you can run it again.**

   Every time you run it, all your applications are updated.

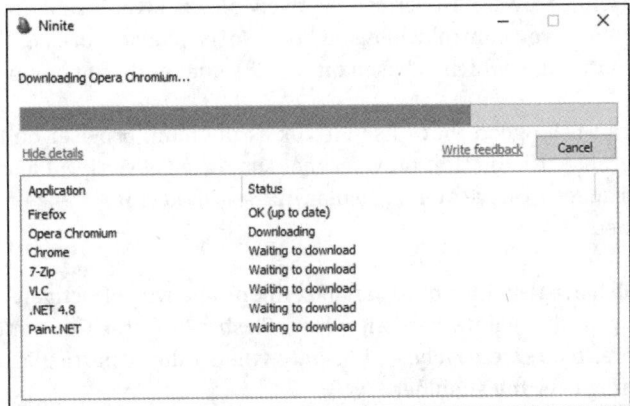

FIGURE 4-5:
Ninite's clean installer delivers programs without extraneous junk.

Top Security Helpers

Although you can manually run the free Ninite program anytime and the latest versions of all your apps get installed, Ninite Updater ($9.99/year, `https://ninite.com/updater/`) proactively watches your installed programs and warns you of any available updates. Ninite Updater even works with programs that you installed manually — as long as they're among the currently supported apps.

Blocking Java and Flash in Your Browser

Adobe's Flash, long a vector of widespread infections, is on its way out. As of this writing, Microsoft's Internet Explorer and Edge browsers are actively phasing out Flash and Java:

>> Edge asks for permission to run Flash on most sites the first time it visits. Since the end of 2019, IE has Flash turned off by default. By the end of 2020 — at least, according to the current plan — Flash won't run on IE or Edge.

>> Google Chrome already requires manual permission to run Flash on all but a handful of high-volume sites. As of July 2019, Flash is off by default for all sites.

>> Similarly, Firefox is gradually curtailing use of Flash, with a goal of ending Flash at the end of 2020.

Java's future is less clear, but it's still a favorite conduit for Windows-based malware.

Giorgio Maone has done the world a favor by bringing the NoScript add-on to the Firefox browser. NoScript selectively blocks Java, JavaScript, Flash, and other plug-ins — you control when and how. NoScript also works in Chrome or the new Microsoft Edge, which is based on the Chrome rendering engine.

ASK
WOODY.COM

NoScript is so good that I use Firefox as my main browser on the desktop, simply because it's the first browser that supported NoScript. I also like the fact that Firefox doesn't have any particular interest in keeping track of where I go on the Internet.

Google has a new improved sandbox in Chrome that effectively keeps Flash safely tied up in a separate cocoon, where Flash can't crash or control the PC. I use Chrome, too, extensively — but only when I don't particularly care if Google's watching over my shoulder.

Edge has a simple switch that lets you turn off Java (actually the Java Runtime Environment). See Book 5, Chapter 1 for details.

Installing and using NoScript is easy. Here's how:

1. **Start Firefox, and in the upper-right corner, tap or click the hamburger (three-line) icon and choose Add-Ons.**

   The standard Firefox add-ons page appears.

2. **In the search box, in the upper right, type** noscript **and press Enter or tap the magnifying glass icon.**

   Firefox comes up with a list of about a zillion add-ons, and the first is NoScript Security Suite.

3. **Click or tap NoScript Security Suite, and then Add to Firefox. When asked, confirm your choice to add NoScript.**

   Firefox downloads and installs NoScript. The NoScript S icon appears in the upper-right corner of Firefox.

4. **Click Okay, and then tap or click the NoScript S icon. Tap or click the Options button and choose Options.**

The NoScript Options dialog box appears, as shown in Figure 4-6.

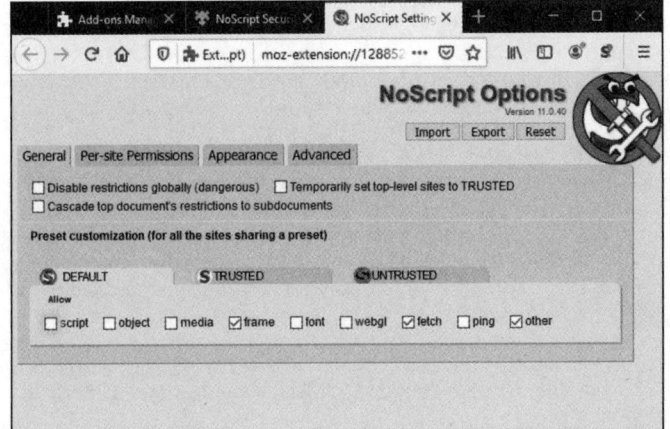

FIGURE 4-6: NoScript's default configuration really locks things down.

5. **Consult Table 4-1, and decide whether you want to change any of the settings. If you do, select or deselect the appropriate box(es) and tap or click OK.**

6. **Review the annotated directions at** www.noscript.net/screenshots.

You may have to click the S icon and select Temporarily Allow All on This Page for the video to run.

ASK WOODY.COM

Getting used to NoScript may take a while. You're going to find that some of the sites you visit all the time — including financial sites and most sites with ordering baskets — won't work unless you allow scripts on the site. You may even hate me for recommending it to you. Fair enough.

At the same time, you should feel much more secure, knowing that the largest source of Windows infections are being blocked before they even have a chance to get into your PC.

REMEMBER

NoScript is absolutely free. The effort's supported a little bit by those cloying Clean Your Registry and other ads, when they appear, but primarily by donations from people like you and me. If you use NoScript, take a minute to make a donation via the Donate button. You'll be helping to make the web a safer place for everybody. And, yes, PayPal is already on NoScript's allowed whitelist.

TABLE 4-1   **NoScript Restrictions**

Forbid	And You Block
Java	Both JavaScript and Java. In spite of the names, Java (which is a complex programming language that interacts with the Java Runtime Environment on your PC) and JavaScript (which is a much simpler language that runs on your PC all by itself) are very different. Historically, JavaScript was used by malicious websites to wreak havoc. More recently, Java — particularly aided by bugs in the Java Runtime Environment — has become a very fertile ground for attacks. Shopping sites, such as Amazon and eBay, use Java programs to keep track of your shopping cart and purchases. Email sites, such as Hotmail/Outlook.com and Gmail, also need Java, as do forums. You have to tell NoScript to back off on those sites.
Flash	Any Flash videos on a site won't play. If you think that means you can't watch videos on YouTube, you're wrong: YouTube has spent years converting the vast majority of its videos to other formats, including formats that work with NoScript. If you have NoScript set to block Flash and you go to a YouTube site, YouTube is smart enough to understand that it can't play Flash, and will switch to a different format if it's available. The web is finally getting rid of Flash. Slowly.
Silverlight	Microsoft's answer to Flash is so bad that Microsoft *itself* isn't allowing Silverlight into the tiled full-screen part of Windows 10. That should tell you something. Don't need it. Don't want it.
Other Plug-ins	A motley assortment of plug-ins get stopped in their tracks including, notably, any PDF rendering plug-ins. Select this box, and you can't read PDF files directly in your browser; you have to go through the extra step of downloading the PDF file and opening it in a viewer, preferably one other than Adobe Acrobat Reader, which has been plagued with security holes for years. Choosing this box also blocks QuickTime files.

Fighting Back at Tough Scumware

ASK
WOODY.COM

Windows Security works great. But sometimes you need a second opinion. Sometimes you get hit with an infection that's so nasty, absolutely nothing will clean it up.

That's when you want to check out Malwarebytes (www.malwarebytes.org).

REMEMBER

Malwarebytes is a last resort. If your system is running normally, there's no reason to bother with it. In fact, if your system is really messed up, you can probably fix things with a full scan in Windows Security (see Chapter 3 in this minibook) or Microsoft Defender Offline — or even a System Refresh (see Book 8, Chapter 2). If you've tried all that and still can't get your furshlinger machine to work properly, time to haul out the big guns.

Malwarebytes has long been my software (and site) of choice for going after absolutely intractable infections — viruses, Trojans, scumware, spyware, retroware, introware, sticky gooey messyware, you name it, Malwarebytes can probably get rid of it.

When you're ready to tear out your hair, you've run Windows Security and Microsoft Defender Offline, and performed Refresh, and you *still* can't get rid of the beast that's plaguing your system, do the following:

1. **Go to the Malwarebytes support forum at** `http://forums.malwarebytes.com`, **see whether anyone has the same problem, and if so, log in and talk to him.**

2. **If that doesn't work, go to the Malwarebytes Anti-Malware Free site at** `http://malwarebytes.org/products/malwarebytes_free`, **and install the free version of its antimalware package.**

   During the installation phase, Malwarebytes disables parts of Windows Security. Not to worry. You don't want to run two antivirus packages at the same time.

3. **Run Malwarebytes:**

   - *If it doesn't get rid of your problem, post your results on the support forum.* Start at `http://forums.malwarebytes.com/index.php?showtopic=9573`, and follow the instructions precisely.

   - *If Malwarebytes fixes your problem, pay for its Premium package.* It's only $40, and you're helping to keep the Malwarebytes effort solvent.

You should run Malwarebytes manually: Don't let it run all the time because you'll hit inevitable conflicts with Windows Security. When Malwarebytes is finished with a manual scan, it returns Windows Security to its full and upright position.

Securing Your Communication with PIA

If you're serious about protecting your surfing from prying eyes, and you ever use a public, unencrypted Wi-Fi connection, the onus is on you to lock your connections down. The best way I know to protect against surreptitious sniffing — and a dozen other problems — involves a technology known as Virtual Private Networking, or VPN.

Firesheep (see the sidebar) has raised the hackles — and the awareness — of Wi-Fi users all over the world.

Https isn't the only way to subvert Firesheep in particular and sidejacking in general. If you connect to a wireless access point that uses WPA2 encryption, you're protected. (At least at this point, nobody I know has figured out a way to sidejack over a WPA2 encrypted Wi-Fi connection.) But if you're using a public hotspot with no password required, you're definitely at risk.

FIRESHEEP AND SIDEJACKING

In October 2010, white hat hacker Eric Butler released a startling Firefox add-on called Firesheep. If you run Firesheep on your computer, and other people using the same network aren't careful, you can sniff other people logging into websites. Click a link inside Firefox, and you can take over the login credentials for the other person.

Eric Butler describes it this way: "When logging into a website, you usually start by submitting your username and password. The server then checks to see if an account matching this information exists and if so, replies back to you with a cookie, which is used by your browser for all subsequent requests. While most websites protect your username and password by forcing you to log on over a secure (https) connection, many websites immediately drop back into unsecure (http) communication. If the cookie comes back to you over an unsecured connection, anybody snooping on your conversation can make a copy of the cookie and use it to interact with the website in precisely the same way you do — a process known as sidejacking. Firesheep makes it point-and-click easy to monitor Wi-Fi signals, looking for cookies shouted out in the clear. It specifically sidejacks interactions with Amazon, CNET, Dropbox, Facebook, Flickr, Windows Live (including Hotmail), Twitter, WordPress, and Yahoo!, among many others."

Put simply, if you use an unencrypted Wi-Fi hotspot, you need to take the bull by the horns and protect your own transmissions. Fortunately, that's reasonably easy, using a technology called Virtual Private Networking, or VPN.

What's a VPN?

You may have heard of VPN, but figured it was just too difficult for regular Windows users to hook together. Big companies have VPN, but they also have experts to keep them running. Ends up that we little guys have good choices now, too.

VPN started as a way for big companies to securely connect PCs over the regular phone network. It used to take lots of specialty hardware, but if you worked for a bank and had to get into the bank's main computers from a laptop in Timbuktu, VPN was the only choice. Times have changed. Now you can get free or low-cost VPN connections that don't require any special hardware on your end, and they work surprisingly well.

When you set up a VPN connection with a server, you create a secure tunnel between your PC and the server. The tunnel encrypts all the data flowing between your PC and the server, provides integrity checks so no data gets scrambled, and continuously looks to make sure no other computer has taken over the connection.

VPNs prevent sidejacking because the connection between your PC and the wireless access point runs inside the tunnel: Firesheep or any other sniffer can see the data going by, but can't decipher what it means. VPNs do much more than simply foil Firesheep attacks: They provide complete end-to-end security, so nobody — not even your Internet service provider — can snoop on your communication, or look to see if you're using a service such as BitTorrent that may give them conniption fits. If you're traveling in a country subject to governmental eavesdropping, VPN is a must.

With a VPN, data goes into the tunnel from your PC, out of the tunnel at the VPN server, then to whatever location you're accessing, back into the VPN server, and out at your PC. There's a very effective cloaking device that hides your data everywhere in between. The people running the VPN server can match you up with your data stream, but nobody else can.

Setting up a VPN

I've used free VPNs from OpenVPN and its hidden VPN. They both work, but I've had problems with speed in both cases. I'm also getting to the point (Saints preserve me) where I would like to have VPN protection for my mobile phone connection. There are also times when I would like to connect to a VPN server in Europe, not in the States.

I've been using Private Internet Access from London Trust Media for several years (www.privateinternetaccess.com/). PIA runs on Windows, of course, but it also runs on macOS, Linux Ubuntu, iPhone, iPad, and Android phones. They have server clusters located in 46 different countries, all over the world.

It isn't free. The basic package costs $40 a year or $70 for two years. It has three additional VPN protocols. Those protocols can come in handy if you have an ISP or if you travel or live in a country that tries to block VPN: The VPN blockers snag the older PPTP protocol, but they don't catch the newer OpenVPN, L2TP/IPSec, or Chameleon protocols, the ones provided in PIA.

Here's how hard it is to get VPN running on your computer (or phone, for that matter). Go to the PIA order site, and sign up. You get an email message with a link. Click the link, and you go to your account's control panel. Click the link to Get Started. On the left, click the link for the protocol you want to install. Installing PPTP is easy — the instructions step you through a simple trip to PC Settings — but the other protocols take more work. That's it. Windows does all the heavy lifting.

Once PIA is installed, you have to enter your username and password once. Then whenever you want to connect to the VPN, you click the PIA icon. Wait a minute or two, right-click the icon in the tray near the time, and choose the location of the VPN server (see Figure 4-7), or click Connect Auto. From that point on, your communication is cloaked. Easy.

FIGURE 4-7:
Private Internet Access makes industrial-strength protected communication easy.

Connect Finland
Connect Switzerland
Connect France
Connect Germany
Connect Belgium
Connect Austria
Connect Ireland
Connect Italy
Connect Romania
Connect Turkey
Connect South Korea
Connect Hong Kong
Connect Singapore
Connect Japan
Connect Israel
Connect Mexico
Connect Brazil
Connect India

Settings
Send Slow Speed Complaint
Help

Exit

10

Enhancing Windows 10

Contents at a Glance

Chapter **1**

Working Remotely with Windows 10

The COVID-19 pandemic has changed the way we live and work. During the lockdown, millions of people had to work from home and required equipment that they may not have had in their homes: a webcam, a second display, a better keyboard, a computer desk, or even an office chair. They also had to familiarize themselves with apps and tools for remote work.

In this chapter, I walk you through some basics about remote work. I start with how to enable Remote Desktop and use it to connect remotely to another computer. This discussion ties in with Book 9, Chapter 4, where I discuss VPN. You may have to use VPN to connect to your company's network, and then use Remote Desktop to connect to a computer in your company's office. If that's the case, this chapter has you covered.

I also share how to connect a second display to your laptop or PC so that you can work productively on two monitors at once. In addition, I share some essential tips about what to look for in a webcam and how to set it up.

Finally, for readers who work with people from all over the world, I share some tips on how to keep track of time zones. You don't want to call someone on Skype or Teams at the wrong hour, do you?

Enabling Remote Desktop Connections

Remote Desktop connections allow Windows devices to connect to one another through the Internet or your local network. When you are connecting remotely to another Windows PC, you see that computer's desktop. You can also access its apps, files, and folders as if you were sitting in front of its screen. This is useful for IT professionals and business users who must work remotely, especially during a lockdown.

If you want to connect remotely to the Windows 10 PC you are on, from another PC, or you want to let others connect to it, you must enable Remote Desktop. Here's how:

1. **Click or tap the Start button, and then the Settings icon.**

   The Settings app opens.

2. **Open the System category of settings, and on the left, click or tap Remote Desktop.**

   On the right, you see the Remote Desktop settings, like in Figure 1-1.

3. **Click the switch to Enable Remote Desktop and confirm your choice.**

   You may also want to dwell and click Advanced Settings, to see how Remote Desktop is configured to work in Windows 10.

4. **Close Settings.**

WARNING

This procedure works only on Windows 10 Pro or Enterprise. If you run another edition, such as Windows 10 Home, you can't enable this feature. In Windows 10 Home, if you open the Remote Desktop section in the Settings app, you see a message stating that *Your Home edition of Windows 10 doesn't support Remote Desktop.*

REMEMBER

Don't forget that you turn on Remote Desktop to let other computers connect remotely to yours. You do not need to enable Remote Desktop if you want to connect from your computer to another. However, that computer to which you want to connect must have Remote Desktop enabled for the remote connection to work.

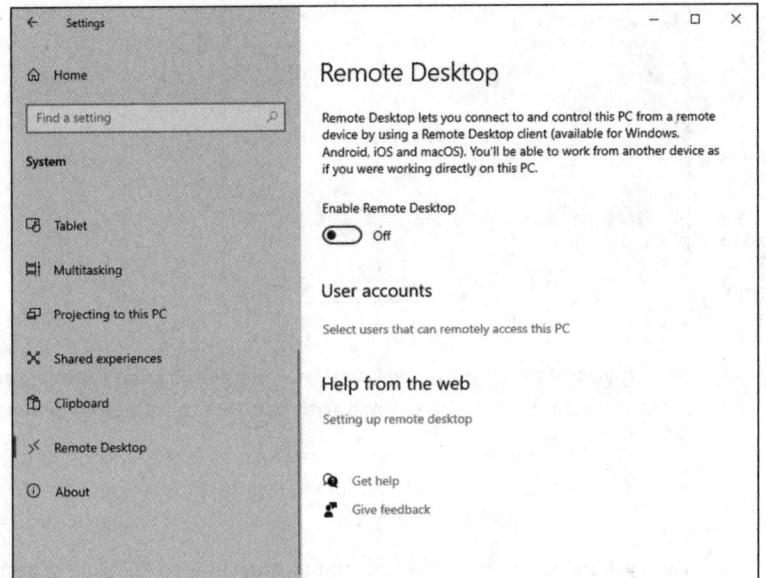

Connecting with Remote Desktop Connection

If Remote Desktop is enabled on the PC that you want to connect to and you know the IP address and details of a user account that exists on that computer, you can connect to it from your Windows 10 PC by using the built-in Remote Desktop Connection app. Here's how to establish a remote desktop connection from Windows 10:

1. **In the search box next to the Start button, type** remote, **and click or tap the Remote Desktop Connection result.**

The Remote Desktop Connection app opens, asking you to enter the address of the computer that you want to connect to, as shown in Figure 1-2.

2. **Enter the IP address of the computer you want to connect to, and click or tap Connect.**

Remote Desktop Connection may take some time to establish the connection, after which it asks for the username and password to use to connect to that PC.

3. **Enter the details of the user account to use to connect to the remote PC, and then tap or click OK.**

FIGURE 1-2:
The Remote
Desktop app
allows you to
connect to other
computers.

4. **If you see a warning message that problems exist with the security certificate of the PC you want to connect to, tap or click Yes to continue.**

   When the connection is established, you'll see the desktop of the remote PC as if it were your own. A toolbar at the top displays connection information, as shown in Figure 1-3.

5. **When you've finished working on the remote PC, click or tap the X button in the toolbar on the top of the screen.**

TIP

If you want to control how the Remote Desktop connection works, click or tap Show Options (refer to Figure 1-2), and configure the available settings. You can also set the username so that you don't have to enter it manually every time. Also, connect to only trusted computers.

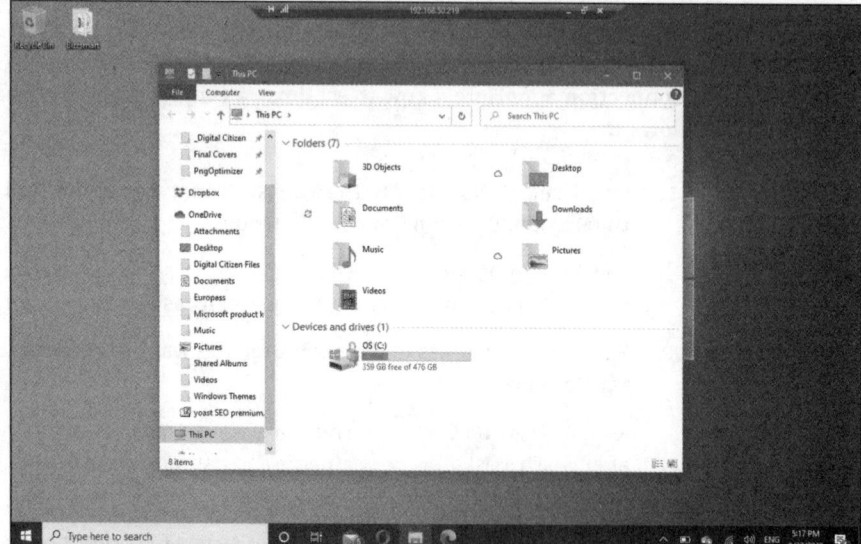

FIGURE 1-3:
You see the
desktop of the
remote computer
and you use it as
your own.

Connecting a Second Monitor

Working on two screens at the same time can increase productivity, especially in times of lockdown, when you have to work from home. To connect a second display to your Windows 10 laptop or PC, first check out the ports on the display and on your Windows device. Figure 1-4 shows you how all the video ports look.

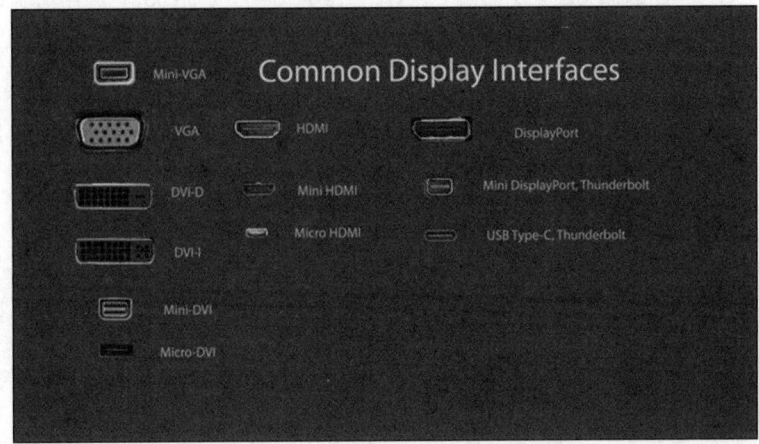

FIGURE 1-4:
All the ports used by monitors, new and old.

Wikipedia

There are two possible situations:

>> **Your monitor and your laptop or PC share the same video port.** Buy a cable that has the same video port on both ends (HDMI, DisplayPort, USB Type-C, and so on).

>> **Your monitor and your laptop or PC do not share a common video port.** Buy an adapter to convert the video signal from your laptop or PC to the external monitor. Depending on what video ports you have on your laptop or PC and monitor, you might need a DisplayPort-to-VGA, HDMI-to-DisplayPort, USB-C-to-HDMI, VGA-to-HDMI, DVI-to-HDMI, or Mini DisplayPort-to-DisplayPort adapter. You can find inexpensive adapters in electronics shops for almost any type of video connection.

After you have the necessary cable, do the following to connect the second monitor:

1. Using the appropriate cable, connect the monitor to your Windows 10 laptop or PC.

2. **Turn on the second monitor by plugging it into a power outlet and pressing its power button.**

   Windows 10 takes a few seconds to detect the external monitor. Note that the external monitor may not display anything after it's detected.

3. **Press Windows+P to display the Project options (see Figure 1-5).**

   You can view the desktop only on your PC screen (the main display) or only on your second screen, view the same desktop on both screens, or extend the desktop and have two different desktops side by side.

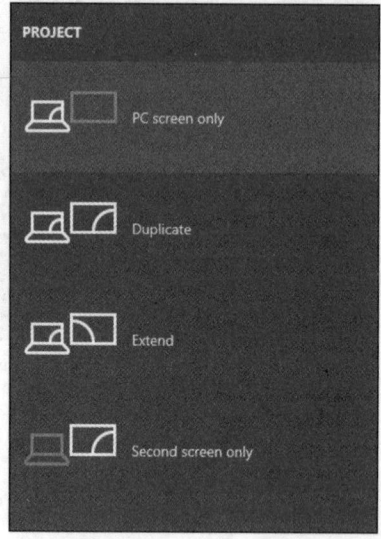

FIGURE 1-5:
The Project options in Windows 10.

4. **Press Windows+P to cycle through the Project options and view the results.**

   You can also click to select an option. The image changes with each selection.

TIP

If you want more help on this subject, check out a great Digital Citizen tutorial that covers all possible scenarios for connecting a second display, including establishing a wireless connection to a Smart TV or a Miracast-enabled display. Read it here: www.digitalcitizen.life/connect-external-monitor-laptop-windows-10.

Installing a Webcam

During the COVID-19 lockdown, webcams became a hot item. Millions of people began working from home and had to rely on webcams to join countless conference calls. If you are in the market for a webcam, realize that most people don't need a high-end model with 4K video recording. A simple webcam with 720p or Full-HD video recording should suffice.

Installing a webcam is as simple as plugging it into a USB port on your computer and waiting for Windows 10 to detect it and install its drivers. One of my favorite webcams is Microsoft LifeCam HD-3000 (shown in Figure 1-6). It covers the basics, is affordable, and is plug-and-play.

FIGURE 1-6: Microsoft LifeCam HD-3000.

Some webcams include software to activate features that may be useful to you. That's why it's a good idea to do an Internet search for the Support page of the webcam's manufacturer and download from there the latest software and drivers for your webcam model. Install the webcam's software, and you should have no problems using it for Skype, Zoom, Teams, and Google Meet video calls.

ASK
WOODY.COM

I prefer webcams from proven manufacturers, such as Microsoft, Logitech, and Razer. Their webcams have many options at diverse price points.

Adding Clocks to the Taskbar

If you work with a team from a multinational corporation, it's a good idea to set Windows 10 to display a clock from that corporation's time zone. That way, you can quickly check the time in the country of your team members. Here's how it works:

1. **Right-click the clock in the bottom-right corner of the screen.**

   A large menu appears, with many options for customizing the taskbar.

2. **Choose Adjust Date/Time.**

   The Settings app opens, displaying options about adjusting the date and time.

3. **Scroll down to Related Settings on the right, and click Add Clocks for Different Time Zones.**

   The Date and Time window appears, as shown in Figure 1-7.

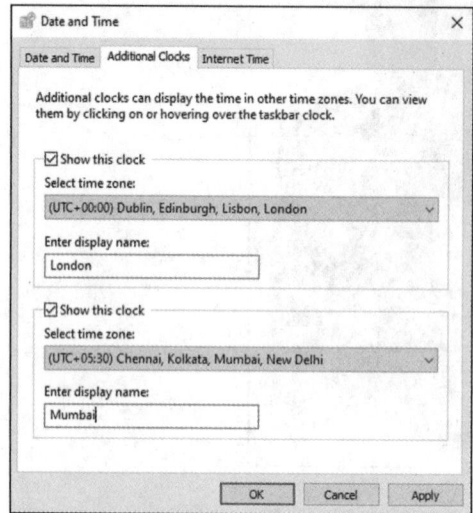

FIGURE 1-7: The Date and Time window where you add clocks to the taskbar.

4. **Select the first Show this Clock box, choose a time zone from the list, and enter the name of the city/country that interest you.**

5. **Select the second Show this Clock box, choose another time zone from the list, and enter the name of the city/country that interests you.**

6. **Click OK.**

To see the additional clocks, move your mouse cursor over the clock on the task-bar. You can also click the clock and see the additional clocks just above the calendar, as shown in Figure 1-8.

FIGURE 1-8:
The clocks you added appear just above the calendar.

Working Remotely with Windows 10

CHAPTER 1 Working Remotely with Windows 10   853

Chapter **2**

Using Android, iPhone, and Kindle with Windows 10

I love my iPad (actually, iPads). My wife loves her iPhone. I also love my Android-based Samsung Galaxy Note, and our various Android tablets, including the best reading tablet around, the Amazon Kindle. I also love to read e-books and save trees by keeping the many books that I own in digital format instead of on paper. And if you don't keep a large library of physical books, there's also more room in the house.

ASK WOODY.COM

How do I reconcile all that technological promiscuity with my decades-long Windows-centric background? That's easy. I don't.

I say pick the right tool for the job, and if you don't like what's happening now, wait a few months and see what crops up. It's never been truer than it is right now: There's more than one way to skin the computing cat. As long as you don't get bogged down in the "Windows first and best" mentality, or hide behind a fear of learning new things, there's a big, exciting world out there. Yes, even if you use Windows. Let me steer you through it.

REMEMBER

If you think that the iPhone rules the smartphone roost, you're wrong. Android phones (that is, smartphones that run the Android operating system) outsell iPhones by a very wide margin, in almost every country. That's true for tablets, too — although the numbers change from version to version, location to location.

iPhone outsells Android in the US, in dollar terms, and the Apple App Store certainly dwarfs Google's Play Store for both number of apps and profitability for developers — people just spend more on Apple apps than on Google apps. Android and iOS run neck-and-neck by most ways of reckoning, but iOS certainly commands more public attention and money.

The e-reader market helps: Few people realize it, but the Amazon Kindle is an Android tablet. Inside the understated exterior and behind the gorgeous eye-friendly display beats a heart of pure Android.

Android's market share is increasing, too. As I write this chapter, more than 2.5 billion Android devices are in use every month. In 2017, more than 38 percent of devices used to access the Internet were using Android. That's a whole lotta Android.

ASK WOODY.COM

This area is seeing lots of activity right now — in fact, with so many people doing so many things, Android may be the target of more change than any other platform in history, including the iPad and (emphatically) Windows. No matter how you look at it or whose statistics you believe, if you combine mobile (phone and tablet) and desktop operating systems, iOS and Android are soaring while Windows and macOS are crumbling. So my emphasis in this chapter is on showing you Android techniques that are likely to survive as long as Windows 10 remains on the market.

Which could be an eternity, in Internet time.

Android isn't Android isn't Android. The Android device you buy today may not be capable of running the new Android of tomorrow. That's true of Windows tablets (just ask a Surface 2 customer), iPad, and iPhone, too. But Android seems to be less upgradable than its competitors. Be careful.

What, Exactly, Is Android?

You know all about Windows — at least if you've managed to get this far in the book — and you probably have at least a nodding acquaintance with macOS (pronounced "mack oh ess"), the operating system that runs on Macs, and iOS ("eye oh ess"), the operating system that drives iPads and iPhones.

APPLE'S WALLED GARDEN VERSUS ANDROID'S OPEN SOURCE

When you deal with iPhones and iPads (and the iCloud, iMacs, Apple TVs, iPods, and all those other iThingies), you're living in a walled garden. Apple controls it from beginning to end. That's one of the reasons why all the different iDevices work together so well — the hardware and software come from the same company, they're designed to fit together, and Apple's designers are absolutely first-class. But you pay for the privilege.

On the other hand, Android devices come from a huge array of manufacturers, many with very different ideas of what's right and what's almost right. Although Google is in the driver's seat — Google bought Android, give or take a patent claim or two or ten, and has released Android to the world — hardware manufacturers, to a first approximation, are free to take Android in any direction they like.

Android is open source under the Apache License, which means that not only is the program free, the source code for the program is free and readily available as well. (It's a little more complicated than that; for details, see www.apache.org.)

TIP

Android is different. Just for starters, it's open source, at least to some degree, based on a modified version of Linux. That means individuals and companies have free access to the programs that make up Android; they can modify the code and release their own versions of Android, on devices of their own devising.

That's both a blessing and a curse: Upgrading some Android devices is easy; others are difficult, and for some it's impossible. Apple doesn't promise that older hardware will run newer software, in all cases, but the Android situation is fractured and confusing.

Android started in 2003, envisioned as an advanced operating system for digital cameras. By 2004, the core group — Andy Rubin, Rich Miner, Nick Sears, and Chris White, all experienced developers — had run out of money. One of Rubin's friends, Steve Perlman, loaned the group $10,000 out of his own pocket, wired them an undisclosed additional amount, simultaneously turning down a stake in the company.

Both Rubin and Perlman worked for Apple in the early 1990s. Perlman's a wealthy entrepreneur and inventor. He says he handed over the $10,000 cash "to help Andy."

Google's Larry Page learned about Rubin's project, and the two companies started a six-month-long mating dance that ended in July 2005, with the Android team moving over to Google, a rumored $50 million changing hands.

Getting clear on Android

Android isn't free-as-in-beer.

Microsoft claims to hold patents on certain parts of Linux, and claims (with varying degrees of justification) that those patents are violated in Google's implementations of Android. Microsoft thus demands ransom, er, royalty payments from large hardware vendors that use Android. In 2013, Samsung alone paid more than a billion dollars to Microsoft, just to avoid a patent court battle.

More than that, manufacturers running Android still have to pay the Google piper: Details are top secret, but apparently Google requires phone and tablet manufacturers to preinstall more than 20 Google apps on every Android device. A recent scan of a Samsung smartphone came up with these obvious Google apps: Android Auto, Chrome, Gmail, Google, Maps, Drive, YouTube, YouTube Music, Google Duo, Photos, Google Assistant, and Google Play Store. You can uninstall some individual apps, of course, but few people do.

Apparently, there are even requirements about where some of those apps must appear on the fresh-out-of-the-box smartphone and tablet screens. Ka-ching.

Making Windows talk to your Android phone or tablet

If you're trying to get your Android phone or tablet to interact with your Windows 10 PC, you need to know several tricks.

First, just plug it in. Every Android device I know about can connect to a USB port. Chances are good that Windows 10 will recognize the device and install a driver for it. On your smartphone, you may be asked to choose how you want to use the USB connection that was just established. Choose to use it to transfer files, as shown in Figure 2-1.

After it's installed, you can access all the files on the Android device through File Explorer. One of my older Android smartphones looks like Figure 2-2. The pictures are in the folder \Phone\DCIM\Camera.

The Android device shows up in the This PC section of File Explorer. Depending on what kind of device you attached, you may see one or two folders: The one marked Internal Storage is for the phone or tablet itself; the other, marked SD card, is for any additional storage you have on the phone or tablet. (There is no additional storage on an iOS device.)

From File Explorer, you can cut or copy files, moving them to your PC. You can edit or delete them. And you can print them.

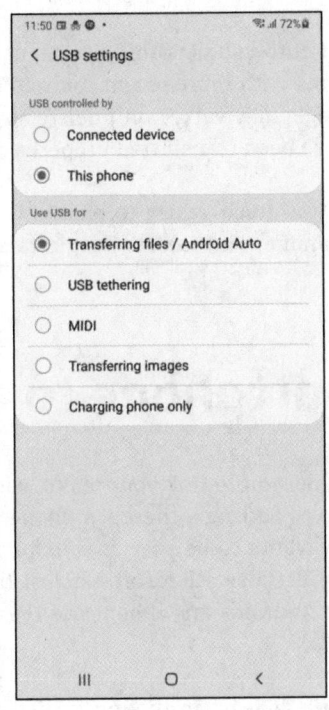

FIGURE 2-1:
On an Android smartphone, choose to use the USB connection for transferring files.

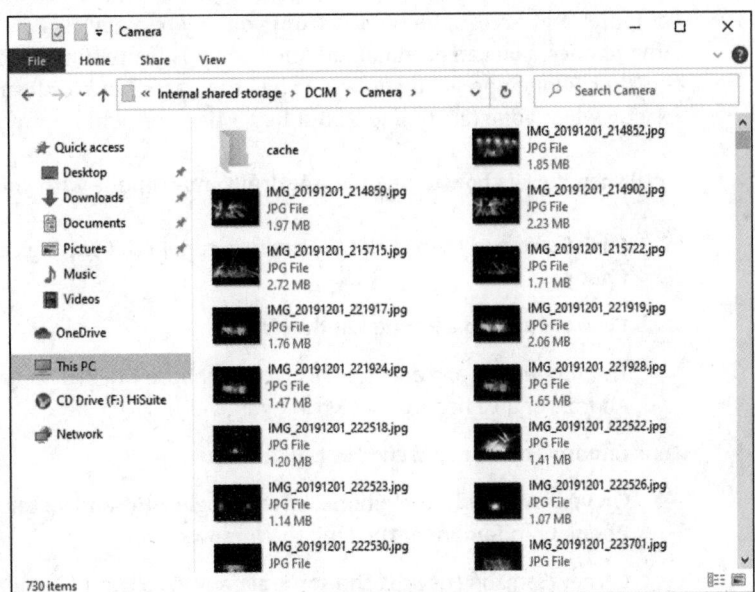

FIGURE 2-2:
If the device installs properly, you can get at files through File Explorer.

WARNING

Copying files from your computer onto your Android device, using File Explorer, usually works. I've had no problems with more recent Android smartphones and tablets, but your device, and mileage, may vary. Don't delete any precious photos until you know for sure that they've been transferred properly.

TIP

Take it from a pro: For photos, it's much easier to enable Google Photos (see Book 4, Chapter 3) on your PC, phone, and tablet. Google takes care of all the syncing details.

Linking an Android Smartphone to a PC

When you install Windows 10, it asks you to link your phone with your PC, using the Your Phone app. When you go to Settings, there's a Phone section that asks you to add your phone. Microsoft wants to be part of your phone too, no matter what you do with Windows 10. Because Microsoft has lost the mobile war, it decided to link Windows 10 to the Androids and iPhones of the world and annoy users in new ways.

The Your Phone app sounds useful, at least in theory: It displays live notifications from your Android device and allows you to respond to messages from your computer and access the photos from your mobile device. And with select Samsung phones, you can even launch Android apps from Windows 10. Unfortunately, the Your Phone app is buggy, and it has the nasty habit of losing the connection exactly when you start to like it. But hey, Microsoft will improve it over time.

Until then, here's how to link your Android smartphone with your Windows 10 PC:

1. **Click Start, and then click the Settings icon. In the Settings app, go to Phone.**

2. **Click or tap Add a Phone (on the right).**

   The Your Phone app opens, asking you to choose whether you want to link an Android or an iPhone, as shown in Figure 2-3.

3. **Choose Android, and click or tap Continue.**

4. **On your Android smartphone, open Google Play, and install the Your Phone Companion (or the Link to Windows) app.**

   On new Samsung devices, the app is already installed.

5. **On your Android smartphone, open the Your Phone Companion app. On your Windows 10 PC, select the Yes, I Finished Installing Your Phone Companion option, as shown in Figure 2-4.**

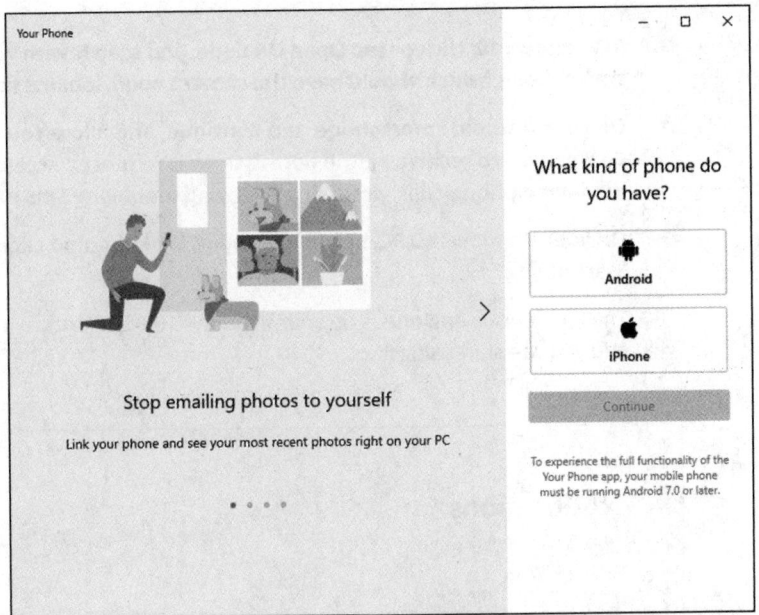

FIGURE 2-3:
Choosing which
smartphone you
have: Android or
iPhone.

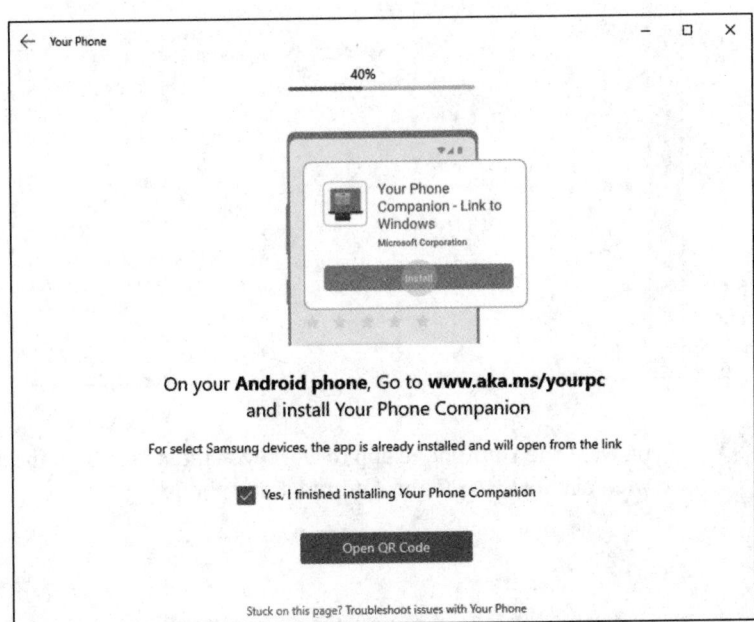

FIGURE 2-4:
Linking an
Android
smartphone to a
Windows 10 PC.

CHAPTER 2 **Using Android, iPhone, and Kindle with Windows 10**   861

6. **In Windows 10, click or tap Open QR Code, and scan it with your Android smartphone, which should have the camera open, looking for the QR code.**

7. **On your Android smartphone, tap Continue, and allow Your Phone Companion to receive all the permissions it requests: accessing contacts, managing phone calls, accessing files, and managing SMS messages.**

8. **On your Windows 10 PC, select Pin App to Taskbar, and click or tap Get Started.**

   The Your Phone app opens on your Windows 10 PC, as shown in Figure 2-5, and you can start using it.

If you want the Your Phone app to work, you must use the same Microsoft account on your Windows 10 PC and Android smartphone.

REMEMBER

Linking an iPhone to a PC

The Your Phone app works with iPhones too — at least in theory. The problem is that the app doesn't do much, even though the link process is similar to Android. You go through the same setup steps, but on your iPhone you install the mobile Microsoft Edge browser (see Figure 2-6) instead of the Your Phone Companion app.

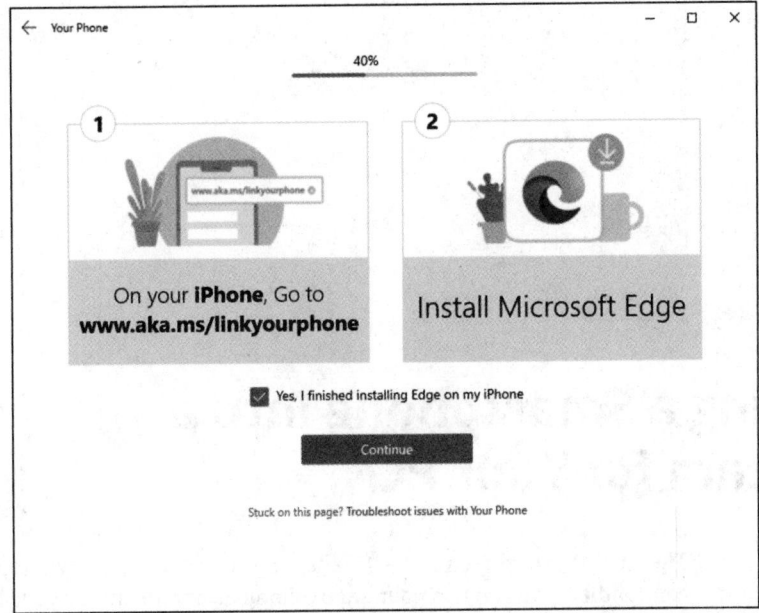

FIGURE 2-6: Linking your iPhone to your Windows 10 PC involves installing Microsoft Edge.

After the setup is finished, open the Your Phone app on Windows 10 and note how empty it is. At the time of this writing, it was literally lots of white space (as shown Figure 2-7). The only functionality that Microsoft supports is sending links to web pages from the mobile Microsoft Edge to the desktop Edge in Windows 10. That's too little to be worth the hassle of going through the link process, and I hope that Microsoft will wake up and provide a real service.

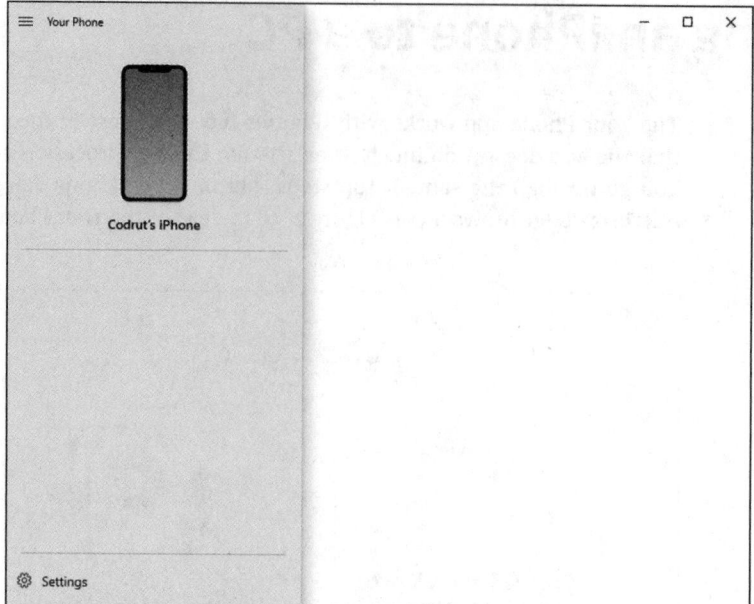

Turning a Smartphone into a Webcam for Your PC

The COVID-19 pandemic has made webcams an expensive and difficult to find commodity. You can use your smartphone as a webcam for your PC. Simply install a specialized app both on your Windows 10 PC and your Android smartphone or iPhone. Many solutions are available; the one I like best is DroidCam. Head over to www.dev47apps.com and download the app on both of your devices (PC and phone).

The DroidCam setup is easy and involves having both your smartphone and your Windows 10 PC in the same network. If you need help setting it up, the folks at Digital Citizen have a detailed tutorial that's updated regularly at www.digitalcitizen.life/turn-android-smartphone-webcam-windows.

DroidCam has both free and paid versions, and I have found that the free version is good enough for most people. See it in action in Figure 2-8.

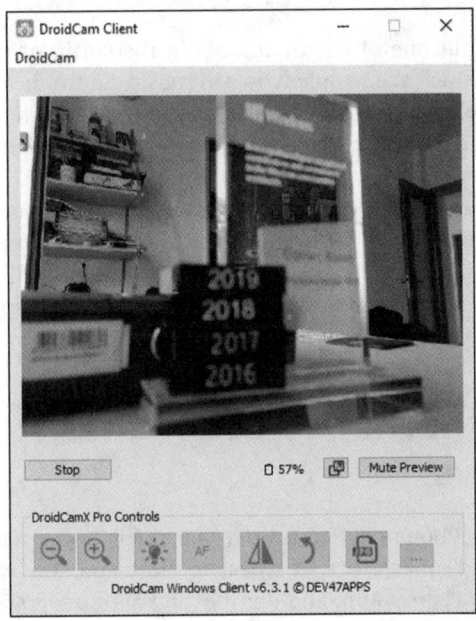

FIGURE 2-8:
DroidCam helps
you use your
smartphone as
a webcam for
your PC.

Running iTunes on Windows — or Maybe Not

iTunes is an Apple program that syncs your Windows PC or Mac with iPods and other mobile devices. You don't really *need* iTunes for your iPad or iPhone anymore because Apple has made them freestanding devices, ready to connect directly to iCloud. (iCloud is Apple's storage service that stores and syncs your iPhone or iPad data over the Internet.) But if you overlook the fact that iTunes is simply one of the worst Windows applications ever created, it has some good points, too.

**ASK
WOODY.COM**

Never mind me. I've been complaining about iTunes running on Windows for more than a decade now. (iTunes on the Mac is a completely different kettle of fish.) iTunes on Windows does have a sharing capability that allows one PC on your home network to play music available to iTunes on another. Still, as a Windows program, iTunes leaves much to be desired.

I'm most assuredly not dissing the iTunes Store, the online shop where you can buy music, video, apps, and more from Apple, all of which are formatted to work on Apple's devices. The iTunes Store has its own problems, but it's revolutionized the way I buy music. In 2009, in response to Amazon launching a DRM-free

MP3 store, iTunes put one of the final nails in the coffin of music *digital rights management* — in which the people who sell music control how it's played, even after you buy it. Apple made an incredible array of music relatively affordable and easy to access — and it's made a bundle of money out of the effort.

iCloud is fine, but just try to sync something outside the iCloud domain. Back up songs that originated outside iTunes and you should plan on paying the iTunes Match piper. Apple has built a walled garden. Truth be told, all three of the cloud consumer giants — Apple, Google, and Microsoft — have spent just as much effort building walls as building bridges.

Deciding whether to use iTunes for Windows

As long as all your iPhone or iPad music, videos, or books reside in (or can be retrieved from) the iTunes Store, it's best to start and stay with iCloud. Don't install the Windows iTunes app, and don't even try to understand it. Just follow the instructions to set up iCloud at www.apple.com/icloud/setup. You end up with an iCloud Settings app like the one in Figure 2-9.

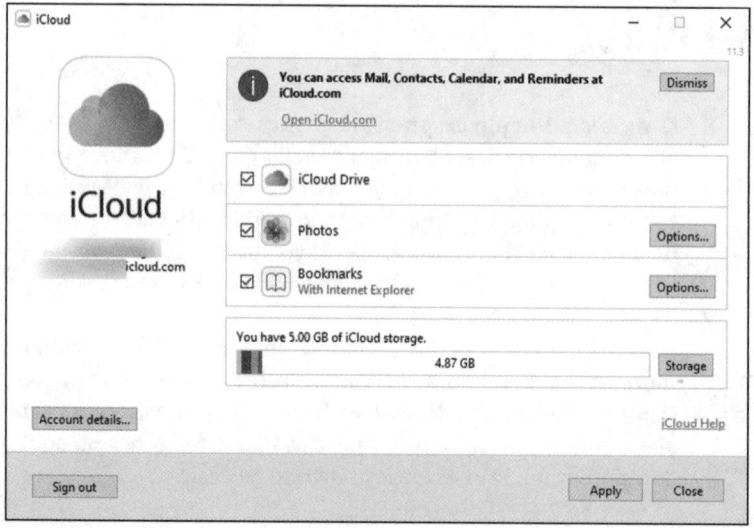

FIGURE 2-9:
Install the iCloud app and you can control it from this Settings pane.

TIP

Switching your iPad or iPhone over to using iCloud is simple: In the iPad or iPhone Settings app, on the left, tap your Apple ID, and then tap iCloud. Make sure you have the right account set up. (You don't want to hassle with mismatched accounts.). Scroll to the bottom and tap iCloud Backup. Slide the iCloud Backup switch to On. Then wait — my initial backup took two hours.

Here are two reasons why you may want iTunes:

>> **iTunes is the easiest way to sideload non-iTunes stuff from your PC onto your iPhone or iPad.** For example, if you've acquired books, movies, or TV shows from someplace other than the iTunes Store, it's easier to use the iTunes Windows app to put them on an iPad.

>> **If you've paid for iTunes Match, running iTunes on your PC is the only way to pull music from iCloud and use it on your PC.** If you have a sizable collection of music, see the nearby sidebar "Music on iCloud — iTunes Match."

MUSIC ON ICLOUD — ITUNES MATCH

If you are willing to pay $25 per year and have lots of (upload) time on your hands, iTunes Match lets you upload *all* your music — it doesn't matter where it came from. That music will then be available on all your iPhones and iPads as well as through iTunes for Windows on all your PCs (and through iTunes on the Mac as well). Yes, that's a good reason to install iTunes on your Windows PC. But it's also a good reason to hook your iPad directly into iCloud, so you can retrieve all your music, all the time.

Apple doesn't copy your music, *per se.* It uses sophisticated software to identify the music you have on your PC and match it with the 43 million songs Apple already has on file — millions of exceedingly high-quality recordings. If Apple can't match your music (live recordings of Juice Newton, anyone?), it stores the unidentified tracks on Apple's servers and makes them available to you directly. These unidentified tracks are counted against your free allowance of 5GB of iCloud storage. Ship too many oddball songs to iCloud, and you end up paying for storage. The songs that iTunes Match identifies are stored without eating into your free 5GB.

After you sign up for the service and let iTunes scan your music, you can download up to 25,000 matching tracks — all in 256KB (high-quality) MP3 files. You can either replace your current tracks or keep the old ones. If you stop paying $25 per year, the music's all yours; you just can't pull it down from iCloud anymore — so you can't stream to your iPhone, iPad, or iTunes.

iTunes Match is one of the great bargains on the Internet — and one of the few good reasons for installing iTunes for Windows. Google Play's Music app is righteous, too, as an iOS app. Your music gets stored on the Google side of the fence, of course, but it's free (for now) and reasonably easy to use. That said, subscription or streaming apps such as Apple Music, Spotify, and Pandora — where you pay per month and get any music you like — are eating up the music industry. YouTube's taking out a hunky chunk, too.

Installing iTunes

iTunes used to be one of the snarliest Windows programs I had ever used: It took over the computer and didn't let go until it was good and ready. It was slow to switch services — double-clicking anything resulted in odd behavior. All in all, it did not look or work like a Windows app. This nightmare ended in 2018 when Apple finally deployed a new iTunes app in the Microsoft Store that works like any other Windows 10 app. The new version looks better, is responsive, and is relatively easy to use. Plus, it works in Windows 10 with S mode.

Here's how to get your Windows PC iTuned:

1. **Open the Microsoft Store and search for iTunes or go to `apple.co/ms` in your web browser.**

   Apple redirects your browser to a different page, but that's okay. You end up in the right place, which looks like Figure 2-10.

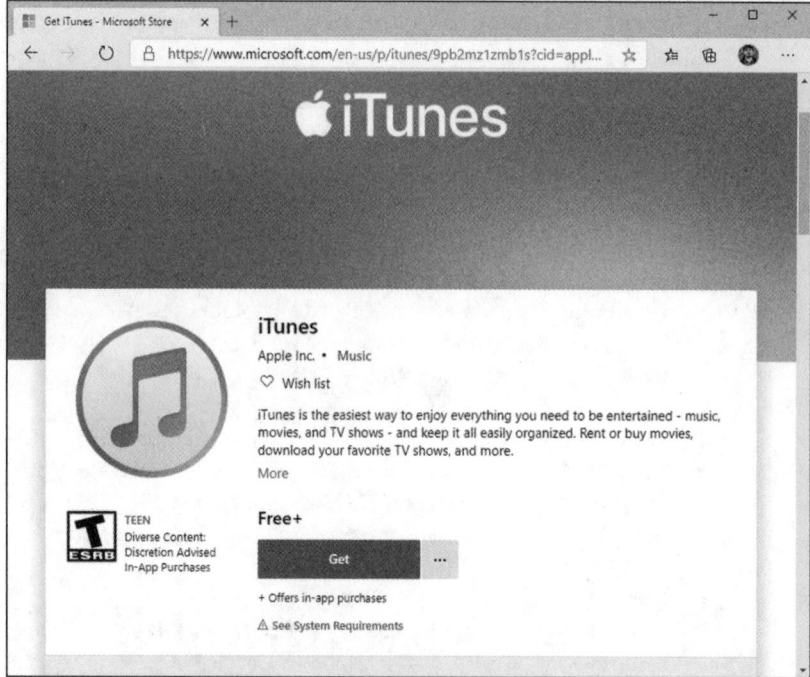

FIGURE 2-10:
The landing page
for installing
iTunes.

2. **Tap or click the Get button, and then tap or click Install.**

   A progress bar appears during the installation process. A message appears when the product is installed.

3. **Tap or click Launch, and then agree to the iTunes license terms.**

Your browser downloads the correct version — 32-bit or 64-bit. Depending on which browser you're using, you may have to tap or click something to save and run the downloaded file.

4. **Click or tap Agree to let iTunes access your Music folders.**

You can quit at this point, or you can continue to use iTunes for the first time, as described in the next section.

Setting up iTunes

Before you use iTunes for the first time, you have to run through the iTunes Setup Assistant program. Here's how to minimize your ongoing headaches:

1. **If you quit immediately after iTunes was installed (see the preceding section) or if iTunes was preinstalled on your PC, tap the iTunes shortcut in the Start menu's All Apps list.**

If you didn't quit iTunes, you automatically come to this step after iTunes has been successfully installed.

2. **Provide your Apple ID.**

To do that, in the Library tab, click or tap Sign in to the iTunes Store (as shown in Figure 2-11).

FIGURE 2-11:
Accessing the
iTunes Store with
your Apple ID.

3. **Enter your Apple ID and password, and then click or tap Sign-in.**

4. **Click or tap Go to the iTunes Store, or click the Store tab on the top.**

    You now have access to Apple's Music store, where you can buy anything you want.

Moving files from Windows 10 to an iPhone

Most videos you see on the Internet can be downloaded and stored permanently on your iPad or iPhone. This process can be useful if you will be someplace that doesn't have an Internet connection or want watch the same video over and over (hey, you have kids, yes?) and don't want to pay for repeatedly downloading the same clip.

Google, which owns YouTube, has a contrary opinion. It says, "You shall not download any Content unless you see a 'download' or similar link displayed by YouTube on the Service for that Content." Not surprisingly, few videos have download links. Clearly, YouTube can prevent you from using its content for commercial purposes. But it isn't so clear if YouTube can prevent you from recording something playing on your computer. That's a very small step removed from downloading. Fair use or piracy? You be the judge.

Many products will scrape videos off the Internet. KeepVid (www.keepvid.com) was one of the first, but it's fallen into disrepute lately, with Google reporting security problems.

ASK WOODY.COM

My preference is the Firefox Video DownloadHelper add-in (http://addons. mozilla.org/en-US/firefox/addon/video-downloadhelper). When you install Video DownloadHelper in Firefox, it watches to see whether a scrapable video is on the page you're viewing. If there is, an icon starts rotating. Tap or click the icon to download the video. Easy.

The trick with KeepVid, Video DownloadHelper, or any other video scraper you find is that you need to have it produce videos in MP4 format. Although MP4 isn't a format as much as a group of formats (details too boring to recount here), MP4 files most of the time play just fine on an iPad. Or anywhere else, for that matter.

REMEMBER

Of course, you shouldn't scrape copyrighted material or material on sites that expressly forbid it.

Another reason for moving some of your music from Windows 10 PC to the iPhone is to listen to it on the go, without consuming your data plan with music streaming, or when you'll be disconnected from the Internet for a few days.

Here's how to get music from Windows 10 into your iPhone. The procedure for moving music and videos from your PC to your iPhone or iPad is similar:

1. **Using the Lightning cable, connect your iPhone to your Windows 10 PC.**

2. **Start iTunes by clicking the drop-down list at the top left and choosing Music, or by pressing Ctrl + 1 on your keyboard.**

   iTunes appears on your iPhone in the list of devices on the left.

3. **Click your iPhone, and then click or tap Music in the column on the left.**

   You should see the music files on your iPhone. If no music is stored on your iPhone, the list may be empty at first.

4. **Under Settings, click Music.**

   Options for synchronizing the Music on your PC with your iPhone appear, as shown in Figure 2-12.

5. **Choose whether you want to sync the entire music library or only selected playlists, albums, and genres.**

6. **Select the Sync Music option and click or tap Apply.**

7. **After iTunes syncs your music, click Music under your iPhone's name to see all the synced files.**

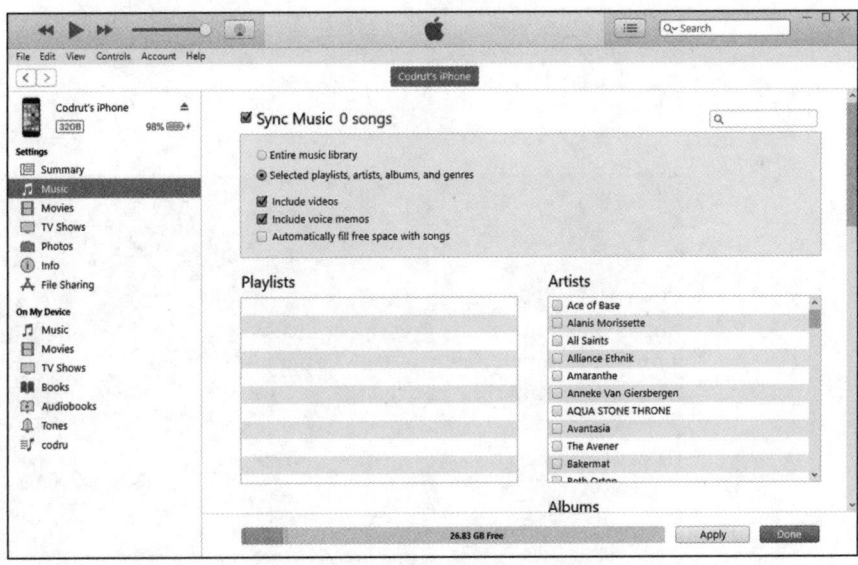

FIGURE 2-12: Options for using iTunes to sync music from your PC to your iPhone.

Controlling Windows 10 from an iPhone or iPad

More than a dozen PC remote control apps are available in the Apple App Store. Some of them work surprisingly well, including the following:

>> **LogMeIn for iOS:** A favorite among reviewers, LogMeIn must run on both the iPad and the Windows machine. If you go with LogMeIn Free on the Windows PC, you can't transfer files, print remotely, hear sounds from the PC, or share desktops. To do any of that, you have to spend an additional $350 per year (gulp!) for the Windows PC's software.

>> **GoToMyPC:** GoToMyPC also draws good reviews but becomes pricey quickly. Figure on spending $20 per month per computer after the initial, 30-day free trial.

>> **Splashtop:** A lesser-known product that works well on a Wi-Fi system, Splashtop connects PCs on the same network. Going outside the local network can be more difficult. I use Splashtop to play videos on my iPad that aren't in MP4 format.

TIP

>> **TeamViewer:** My favorite remote control program (free for non-commercial use), TeamViewer can run in one of two ways. Install the TeamViewer program on your Windows PC and let it control the interaction, or run the program on your PC manually when you want to access the Windows PC from your iPhone or iPad. Figure 2-13 shows Windows 10 on an iPhone. When you run the program manually, it generates a random user ID and password, which you use on the iPhone to initiate the session.

FIGURE 2-13: TeamViewer lets you control your PC from an iPhone or iPad — and it's free.

After TeamViewer is connected, you can use the iPhone or iPad keyboard, pinch to expand or reduce the size of the screen, tap with two fingers to emulate a right-click, use the buttons on the top of the screen for Alt and Ctrl and Esc, and much more. Even Flash animations come through remarkably quickly.

Wrangling E-Book Files

Most popular e-book readers, including, notably, the Kindle line, are based on a modified fork of Android. That's why I put this discussion in the same chapter with Android.

Someday, a single format will exist for all electronic books. In my utopian future, you will buy a book in one format, and that format will just work, no matter what device you want to use to read it.

Unfortunately, the world isn't at that point yet. In fact, it isn't even close. The single biggest headache you're likely to have with electronic books revolves around book formats, and how to get one device to show you books that were made for a competing device.

REMEMBER

If you can afford to stick with just one device and bookstore — only buy books from Amazon and read them on the Kindle, for example, or only buy books from the iTunes Store and read them on the iPad — I salute you. Your life will be considerably less complicated. Most people aren't so lucky.

If you're one of them, you can simplify e-book management by buying your books online through your PC's web browser, using a program called *calibre* to convert files into whatever format your reader requires, and then syncing your e-books with your e-reader on your PC. (You can also read any e-book on your Windows computer, but that may be beside the point, huh?)

Introducing popular e-book formats

Here are the most popular book file formats:

>> **EPUB** comes closest to being a universal format. The iPhone and iPad handle EPUB natively, there are many third-party Windows EPUB readers (more about that after this list), the almost-disappeared Nook reads EPUB natively, and many Android apps read EPUB. The only major holdout for the EPUB format is Kindle.

REMEMBER

Given a choice, unless you live in a Kindle-only world, get your books in EPUB format.

>> **MOBI and PRC** formats are the Kindle's bread and butter. Amazon has a format converter — *KindleGen* — that changes EPUB files into MOBI. It works surprisingly well.

>> **PDF** is the original format for publications that have to survive a transition from one kind of computer to another. Although every common device can read PDF, most readers just display the original document without trying to reflow pages or add any features, such as note-taking. Reading a PDF file in most readers is a frustrating and headache-inducing experience.

Reading e-book files on your PC

Whether or not you have e-books you bought with an e-reader, you can read anything on a Windows 10 PC. Sometimes, though, you have to get a little creative and bring in apps that can do the heavy lifting.

Microsoft's Edge browser can read EPUB files, as long as they aren't password protected.

TIP

PDF viewers are also a dime a dozen. The viewer that Microsoft built into Microsoft Edge, shown in Figure 2-14, works reasonably well, with new features added in each new version of Windows 10.

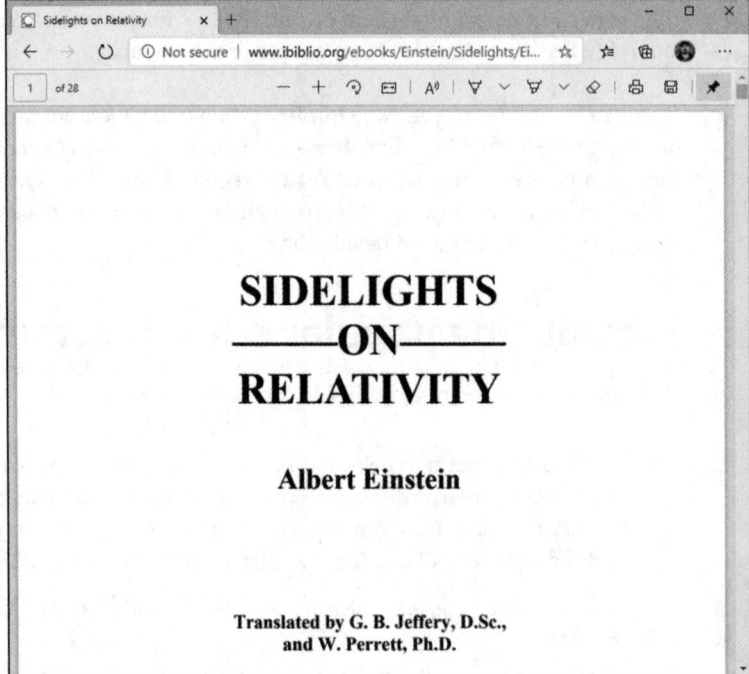

FIGURE 2-14:
Microsoft Edge has a built-in PDF viewer.

Organizing your e-book files with calibre

Before you lose any sleep over different book file formats, realize that one desktop app has been translating among the formats for years. In fact, calibre's more than a Babel fish; it's also a book manager — for free. See Figure 2-15.

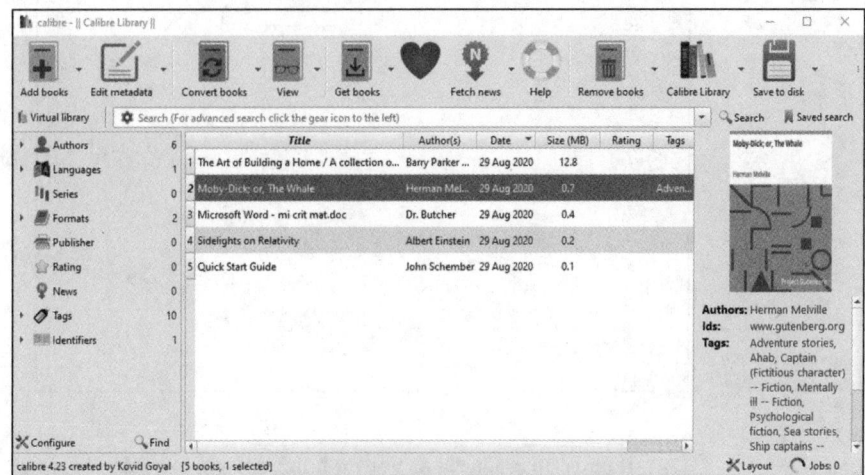

FIGURE 2-15:
The calibre app translates and organizes.

Much like Windows Media Player or iTunes, calibre keeps track of all your books, translates them into the correct format if need be, and offers the files up for easy transfer to the reader of your choice.

Here's a quick look at calibre's capabilities:

1. **Bring up your favorite browser, go to** `http://calibre-ebook.com`, **and download and install calibre.**

    The installer doesn't have any options.

2. **Tap or click the Calibre — E-Book Management shortcut on the desktop and run calibre for the first time.**

3. **Choose the language for Calibre, and the location where your books will be stored. Click or tap Next.**

    The default folder works fine.

4. **When asked, choose your e-book device, as shown in Figure 2-16.**

    Don't panic — calibre converts any format to any other. This step just makes it easier to choose your most common format.

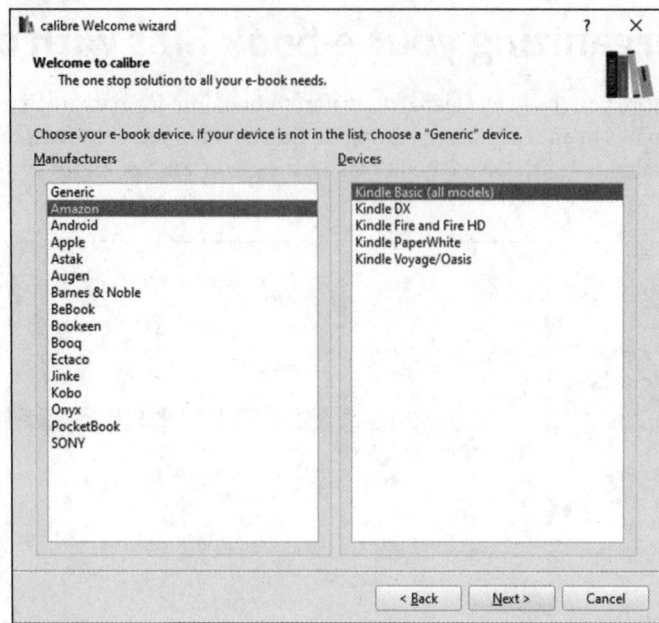

FIGURE 2-16:
Choose the
device you use
most commonly.

5. **If you chose Kindle as your e-book device, provide the email of your Amazon account. Click or tap Next and then Finish.**

    calibre scans your Documents library for books — just about any format you can imagine — and lists each book (refer to Figure 2-15).

TIP

Note that calibre lists books, not files. If you have a book in two different formats — say, a MOBI file and an EPUB file — it appears as only one book on this main screen.

6. **To see and edit the details about an individual book:**

    a. *Right-click the book and choose Edit Metadata, Edit Metadata Individually.* (Someday, calibre will have a touch option; for now, it's mouse only.) calibre shows you an enormous amount of information about the book, including the available formats, as shown in Figure 2-17.

    b. *If you want, edit the data and click or tap OK to save your changes.*

    c. *Click X to close the dialog box.* You return to the calibre library (refer to Figure 2-15).

7. **To convert a book to a different format:**

    a. *Right-click the book and choose Convert Books, Convert Individually.* A Convert dialog box appears, as shown in Figure 2-18.

    b. *In the upper right, choose the format you want to convert the book to; in the lower right, tap or click OK.* The calibre app converts the book to the format you choose and places the new file next to the old ones.

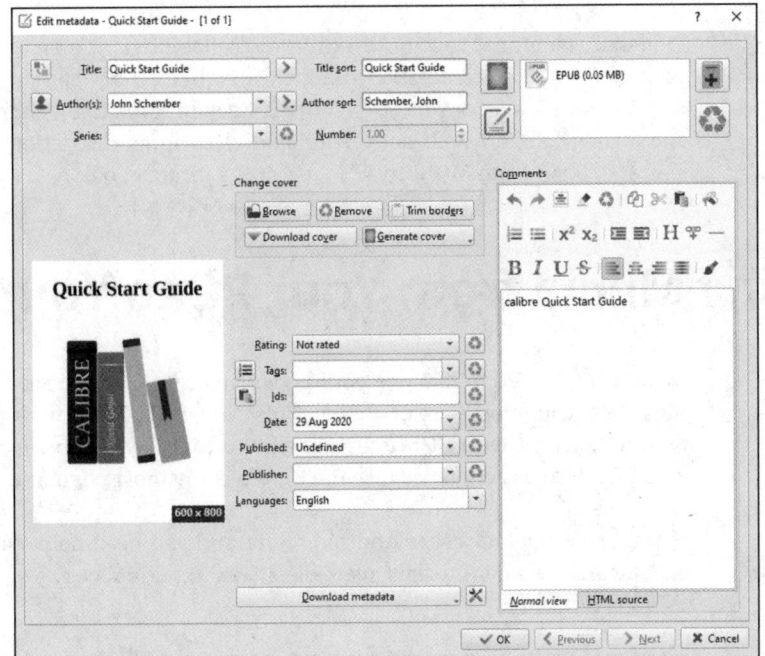

FIGURE 2-17:
The calibre app displays, and allows you to edit, a lot of data about each book.

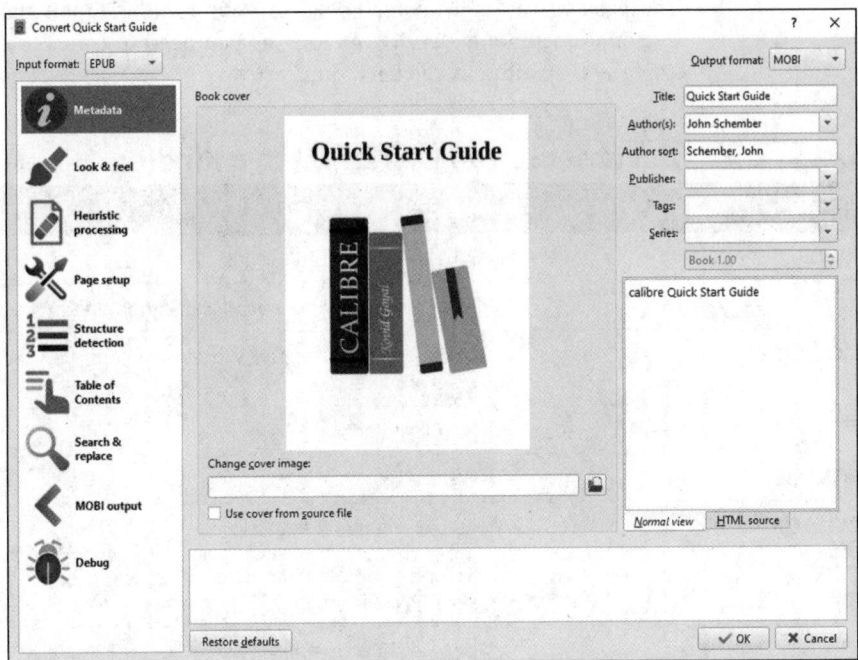

FIGURE 2-18:
Choose the new book format in the Output Format box.

This discussion just touches on calibre's capabilities; it's an amazingly versatile program. For a more detailed rundown of what calibre can do, start at `http://manual.calibre-ebook.com/gui.html`. Keep in mind that calibre translates from one format to another. It doesn't relax digital rights restrictions: If you translate a pirated book from MOBI to PDF, it's still a pirated book.

Getting Media from Your PC to Your Kindle

If you use your PC to manage your books and music, you need a way to get those files onto your e-reader or tablet. This section is here to help. Unfortunately, the methods for each device are specific to that device. So I focus on the Kindle e-reader in this section because the Kindle is the most popular e-reader out there.

If you use a Nook or other Android tablet and you need help syncing files, check out the articles and tutorials available at `www.dummies.com`.

Emailing books from your PC to your Kindle

The easiest way to transfer books to your Kindle? Email them via the Kindle Personal Documents Service. As long as you need to transfer a file type listed in Table 2-1, emailing is the best, quickest way.

TABLE 2-1   **Documents You Can Email to a Kindle**

File Type (Filename Extension)	What It Is
MOBI	Kindle native MOBI format
TXT	Plain text files (looks surprisingly good on the Kindle)
DOC, DOCX	Doesn't handle complex Word documents very well, but simple ones are fine
RTF	Rich Text Format
HTML	Web pages
ZIP, X-ZIP	Kindle unpacks the files
PDF	Second-generation Kindle devices (Kindle 2 or later, Fire, and so on) show PDF files directly
JPG, GIF, BMP, PNG	Images show up fine

Here's how to transfer a file:

1. **On your Kindle's home screen, tap the gear icon in the upper right and choose More on the right.**

    Kindle shows you several settings options, starting with Help & Feedback.

2. **Tap My Account.**

    Kindle shows you the registration information, including an email address, such as woody_217b64@kindle.com.

3. **Write down the email address.**

4. **In Windows (or on any computer for that matter), send a message to that email address, from the email address that you use to log in to Amazon, and attach to the message the file you want to transfer.**

    The file ends up in your Kindle's Documents folder.

**ASK
WOODY.COM**

Amazon has a Send to Kindle application that lets you right-click a file in the desktop File Explorer and choose Send To, Kindle. That sends the file to your Kindle, using the email method described earlier. You can also print from any desktop application and choose Send to Kindle. I don't use either because emailing is simple and clean, and I don't have to worry about the Amazon application gumming up things.

Receiving emailed books from a friend

If you want a friend to send books or documents to your Kindle, you have to give her permission by adding her email address to your allowed list. Here's how to let others email books and documents directly to your Kindle:

1. **Sign on to** www.amazon.com **with the same ID you use on your Kindle.**

    If you're already logged in, your personalized Amazon screen appears, as shown in Figure 2-19. (If you aren't logged in, click Sign In and get with the system.)

2. **Click, tap and hold down, or hover your mouse cursor over Your Account, and choose Your Content and Devices.**

    Amazon shows you a list of all the titles you've bought and placed on your Kindle.

3. **At the top, click the Preferences tab, scroll down to the section marked Personal Document Settings, and click it to extend it.**

    You see the options shown in Figure 2-20.

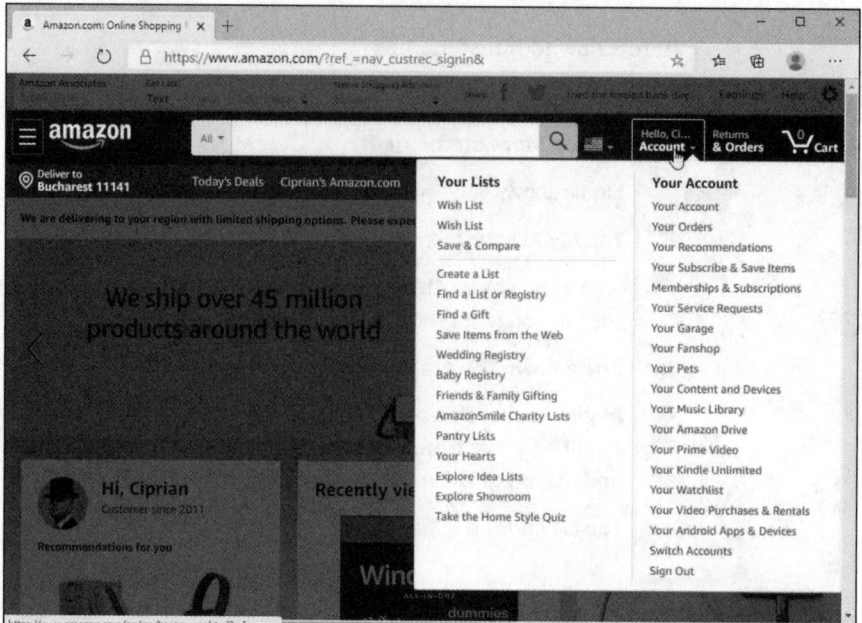

FIGURE 2-19:
Your Account settings are here.

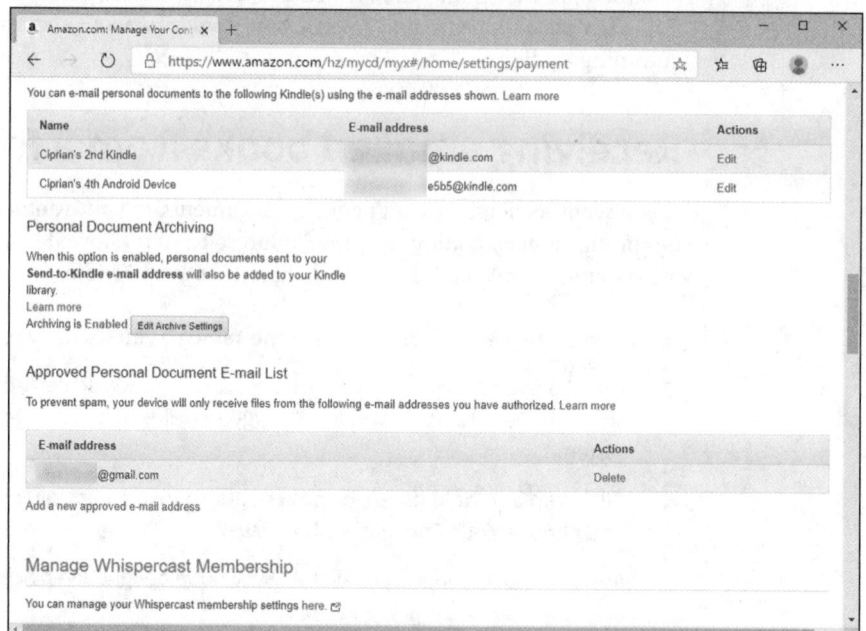

FIGURE 2-20:
Add your send-enabled friends to this list.

4. **At the bottom, tap or click the Add a New Approved Email Address link.**

   A box that lets you add email addresses appears.

5. **Type the address of anyone you want to allow to send stuff directly to your Kindle, and tap or click Add Address.**

   To add multiple addresses, simply repeat the preceding steps. The changes take effect immediately.

Adding music to your Kindle

To get music into the Kindle, you need to connect it to your PC and drag the files across. Follow these simple steps:

1. **Plug a standard mini-USB cable into your Kindle (one may have come with the device), and stick the other end in your PC.**

2. **Slide the Start Screen slider on your Kindle.**

   A screen appears telling you that You Can Now Transfer Files from Your Computer to Kindle. Windows hums and haws for a while and may ask (in a toaster notification on the right side) what you want to do with newly inserted hard drives.

3. **If a Windows 10 notification appears, ignore it.**

4. **Bring up File Explorer by tapping or clicking the Explorer icon on the taskbar.**

   It may take a minute, or two, or even three, but sooner or later, your Kindle appears on the left side of File Explorer, somewhere on the list of other hard drives on your computer.

5. **Find a favorite MP3 file, or a folder full of MP3 files, and drag it from your PC into the Kindle \Music folder.**

   All the music in the \Music and \Audible folders is available to the Kindle music player.

6. **When you've transferred all the music that's fit to play, tap Disconnect on the Kindle and unplug the USB cable.**

   Your music is loaded and ready to rock.

7. **On the Kindle's home page, tap Music.**

   A list of all the MP3 files appears in either the \Music or \Audible folder.

8. **(Optional) To create a playlist, tap the Playlists link and follow the instructions to build a playlist.**

9. **To simply play your music, tap the Shuffle and Play button, or simply tap a song.**

The song starts playing. Individual controls are available for volume, pause, fast-forward, rewind, shuffle, and cycle. After the music starts, you can go back to the Kindle's home page and read books. The music keeps going even after the screen has gone dark.

10. **To turn off the music, tap Music and, at the bottom of the screen, tap the Pause button.**

TIP

You can also copy your music to the Amazon Cloud Drive and play it on your Kindle from there: To play iTunes music, for example, download it to your computer and upload to your cloud drive. See the Amazon Music website, www.amazon.com/cloudplayer, for details.

If you own a Kindle, Amazon gives you free Amazon Cloud Drive storage for everything you've bought from Amazon, plus 5GB of free storage for things you've acquired elsewhere — even songs from iTunes. Very slick.

If you're an Amazon Prime member, you can stream Prime Music songs for free onto your Android phone or tablet, iPhone, iPad, PC, or Mac. See the Prime Music site at www.amazon.com/gp/help/customer/display.html?nodeId=201530920 for details.

Chapter **3**

Getting Started with Gmail, Google Apps, and Drive

n spite of the rivalry between Microsoft and Google, Google's so important to today's computer users that Microsoft builds hooks into Windows 10 that try to get you to add your Gmail account to their Mail app and add your Gmail contacts to the People app. Of course, Google is happy to return the favor, with easy ways to put your Hotmail/Outlook.com mail inside Gmail, and to import your Hotmail/Outlook.com contacts into Gmail.

There's a reason why Microsoft wants you to put your Google eggs in its basket. Google has very good competitors to the Microsoft online stables, including the following:

» Chrome OS, as explained in the nearby sidebar, obviates the need to run Windows for many people.

» *Microsoft Hotmail/Outlook.com,* the *Windows 10 Mail app,* the mail part of Microsoft's Outlook and Outlook 365, the *Outlook Web App,* and a zillion other Microsoft mail programs all compete with *Google Gmail,* in different ways.

>> The *Microsoft Windows 10 Calendar app* and the various *Office Outlook calendars* compete with *Google Calendar*.

>> The *Microsoft Windows 10 People* app and *Hotmail/Outlook.com contacts* compete with Google Gmail contacts.

Worth noting: Every app in *italics* in the preceding list is *free* if you're running Windows 10. Absolutely free. Microsoft and Google give away the apps to draw you in to their corners, with the hope of selling you something in the future.

ASK
WOODY.COM

You can use Gmail to send and receive mail using your own private domain, although you have to pay for a G Suite account to make it work. So, for example, I can use Gmail to handle all the mail coming into and going out of AskWoody.com without changing my email address and without anyone knowing that I'm using Gmail: All the mail going out says it's from Woody@AskWoody.com, and all the mail sent to Woody@AskWoody.com ends up in my Gmail Inbox. It's a feature in G Suite, costs $6/user/month, and except for one step, it's easy. See the last section in this chapter, "Moving Your Domain to Google," for details.

All this wrangling takes place against a backdrop of increased competition from Apple and new assaults from Facebook. All the companies really want to get you hooked on their ways of working.

REMEMBER

Don't forget that free services aren't free in the sense of being zero-sum. The companies offering the free service gather information about you, unabashedly, and show you targeted ads, in the hope of selling you something. As a poster named *blue_beetle* on the site MetaFilter (www.metafilter.com/95152/Userdriven-discontent) put it so succinctly, "If you're not paying for it, you're not the customer; you're the product being sold."

In the following section, I briefly look at Google alternatives to Microsoft products from the perspective of a Windows user.

Finding Alternatives to Windows with Google

Google has a handful of free online products and offerings that warrant your attention. Microsoft has two or three handfuls, but that's the subject of the rest of this book.

CHROME OS — THE WINDOWS KILLER

In the course of a few years, Chromebooks have jumped from scoffed-at toys to genuine Windows rivals. I talk about Chrome OS and the Chromebooks that run them in Book 1, Chapter 1. In general, if I know people who are looking for a computer and they don't need to do anything that's directly tied to Windows (which describes 80% of my friends or more), I usually recommend that they get a Chromebook. They're easier to use, less prone to infection, and all-in-all a whole lot less hassle for me to support.

Chromebooks run Chrome OS which is, to a first approximation, just the Chrome browser you've used before.

To a second approximation, Chrome OS can support overlapping resizable windows (each resembling a Chrome window on Windows or macOS), as well as apps built to the Google Package App Platform. That's what gives specific apps (such as Gmail, Sheets, Docs) the ability to run even when the OS is offline. Chrome OS also includes a built-in media player and a file manager.

My wife and I went shopping for a car. At one dealership we met a salesman who was carrying a Chromebook, using it to make the sale. I asked him how he liked it and was bowled over by the response. He not only liked it; he loved it. Bought one to use at home.

As he stepped me through the virtues of the Chromebook — it's like he was trying to sell me one — he showed me how the corporate IT guys had built a Chrome-based support system for salespeople at the dealership. "We used to have PCs, and they sucked." Alas, that's a refrain I hear dozens of times a day. You probably do, too. "The Chromebook works all the time. The PCs would go up and down, or get slower as they got older. I don't have to worry about updates. The printer's always there. No waiting when I start the machine, it takes like two seconds and it's ready to go. And if I can't find this computer, like I forgot it in the meeting room or in a car, I pick up another one, log in, and everything's just the way I left it."

I asked him if he missed Office. You could've heard the snort at the other end of the showroom. "You're joking, right?" He has one sales spreadsheet to fill out every week, and his boss prefers that he use Google's Sheets, which is one click away on a Chromebook. "I don't need to worry about messing up anything or emailing an attachment. Hell, I don't even need to save it." As for email? "We use Gmail, like I've been using at home for years."

(continued)

(continued)

Then the kicker. "One of the guys at corporate told me they saved enough on the maintenance contract to pay for the system." Probing a little deeper, I discovered that the car manufacturer had hardware maintenance agreements with a large national chain. When the salespeople had problems with their old PCs, they called "The Computer Guy," who promptly ran out and fixed it. Several times a week, on average. With the Chromebooks, The Computer Guy shows up only when the secretary's machine goes down or when the service department needs fixing. Now, if a Chromebook starts misbehaving, the salespeople just pick up a different one, log in, and they're off to the races.

If that anecdote reverberates with you in your environment, Chromebooks may be a good alternative to Windows laptops.

Here are the key Google products, other than Chrome OS, that serve as alternatives to Microsoft offerings:

>> **Gmail:** A free, online mail service, like Microsoft's Hotmail/Outlook.com. Features change constantly, but it's fair to say that if you find a feature you like in Hotmail/Outlook.com, it'll be in Gmail soon — and vice versa. Some people prefer one interface over the other; I'm ambivalent but for now I've settled on Gmail, primarily because I prefer the interface. If you use Google's Chrome web browser, you can even use Gmail when you aren't connected to the Internet.

>> **Google Drive:** A service from Google that gives you up to 15GB of free online storage, more than Microsoft OneDrive's 5GB free allotment, with occasional discounts for various promotions. I talk about the different online storage services in Book 8, Chapter 1. Google Drive's main advantage is its ability to work easily with Google Apps.

>> **Google Apps:** Contains online programs for creating and editing word processing documents (Docs), spreadsheets (Sheets), fill-in-the-blank forms (Forms), presentations (Slides), and drawings. The programs have been gaining new features rapidly, and they're designed to work collaboratively — two or more people can edit the same document at the same time, with no ill effect and no weird restrictions. And you can get at your docs from your PC, Mac, tablet, or phone. Slick, and you don't need to do a thing.

G Suite is the official name of the paid version of all those programs and services. All the pieces — Gmail, Google Drive, Docs, Sheets, Slides, and more — are free for personal use, but if you use them for business, you need to pay for G Suite. In addition to the free-for-personal use features, G Suite includes more Google Drive space and a collection of administrative tools. G Suite subscribers also get additional features in Docs and Sheets. For details, go to https://gsuite.google.com/pricing.html, which is shown in Figure 3-1.

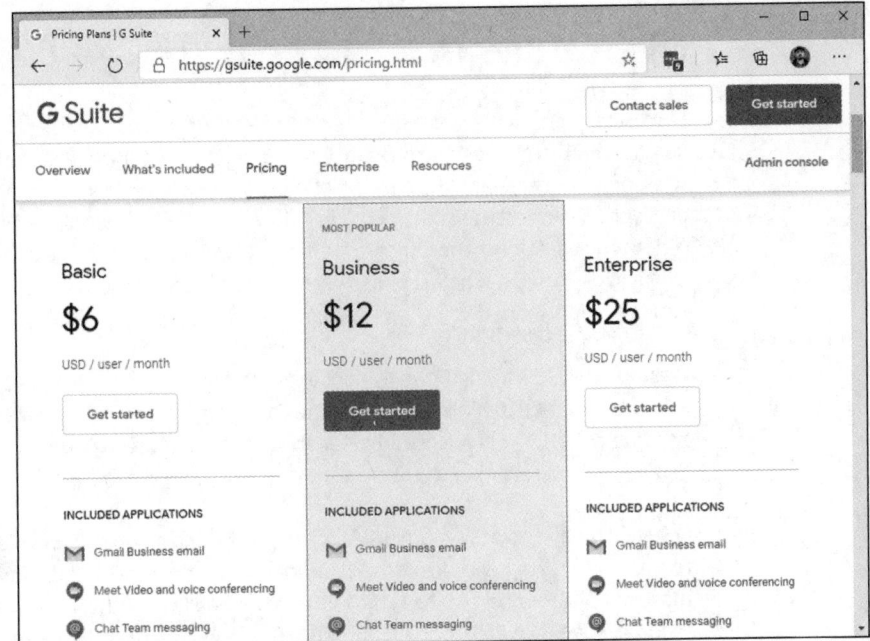

FIGURE 3-1:
G Suite runs from $5 to $25 per person per month, with discounts for yearly payments.

G Suite for Education and G Suite for Government are identical to G Suite but are available only to bona fide educational institutions or government institutions, respectively.

There's another big difference between the free Google apps and G Suite. If you don't pay for using the Google apps, Google takes a peek inside your emails and your stored files, scanning them to target ads in your direction. If you pay for G Suite or have an official, free G Suite for Education or G Suite for Government account, Google does *not* scan your email or your stored data to target ads.

Concerned about the privacy? There are some if's, and's, and but's. For one, *every* online email program (Outlook.com/Hotmail, Yahoo! Mail, AOL, whatever) scans your mail for viruses, spam and scams — some more thoroughly than others. The online storage providers also scan for malware than can clobber their systems. That's part of the ballgame. Scanning, in and of itself, isn't bad. The email provider is protecting both you and itself.

Second, Google's snooping is expressly for the purpose of directing ads. They aren't sniffing for your bank account numbers, and any organization that wants access to your data has to go through the usual channels — which usually involve a search warrant.

Third, if you use encryption to either protect the body of your email message or to lock up files stored in Google's cloud, Google won't go to the trouble of cracking the encryption. If you want something safe, lock it up yourself.

CHAPTER 3 **Getting Started with Gmail, Google Apps, and Drive** 887

OFFICE 365 IN A NUTSHELL

Microsoft's entry in the office application wars is *Office 365*. Over the years, Office 365 has grown from a stub of an offering into a multi-billion-dollar baby. Depending on the subscription level — and the amount of money you pay — Office 365 can include web-only apps, which aren't very exciting, full-blown installed versions of the Office apps, click-to-run versions of the Office apps that are updated continuously, and touch-centric Office apps, including the full-blown version of Office for Windows 10, Office for iPad, and Office for Android.

Preston Gralla has a lengthy analysis of the differences between G Suite and Office 365 (recently renamed to Microsoft 365) in the *ComputerWorld* review, www.computerworld. com/article/3170112/enterprise-applications/smackdown-office-365-vs-g-suite-productivity.html. The short version goes like this.

G Suite is small and easily administered; it covers the high points; and it doesn't try to reach into the more obscure corners. Office 365, on the other hand, offers the best (and most complex) support in the business. I'm continually amazed at how well Microsoft has built out Office 365, rolling feature upon feature into the mix, yet keeping the whole package remarkably stable, usable, and manageable.

If you need to create complex Office-standard documents, Office 365 has no equal. But if you have less-stringent requirements and a willingness to part with 100% absolute Office document compatibility, G Suite offers a good, inexpensive, and reliable alternative.

Pull out your calculator (or your Google Sheet), and do the math. On the Google side, personal use is free; in business settings, it costs $6 per month per person, plus the price of the necessary copies of Office (rent or buy). On the Microsoft side, if you want all the standard Office apps installed on your PC, plan on paying at least $69.99 per person per year. (Each copy can be installed on up to 5 machines.) For business, plan on $5 per person per month for up to 25 users.

I cover Office 365 (recently renamed Microsoft 365), Office for Windows 10, for iPad and for Android, as well as the desktop Office apps at www.askwoody.com.

Setting Up Gmail

If you don't yet have a Gmail account, get one. Doing so is free and easy. Besides, every new Gmail account gets 15GB of free cloud storage. Here's how to set up an account:

1. **With your favorite browser, go to** `www.gmail.com`.

   At the bottom or on the left is the Create an Account button.

2. **Tap or click the Create an Account button and choose whether the account is for yourself or to manage your business.**

   The sign-up form in Figure 3-2 appears.

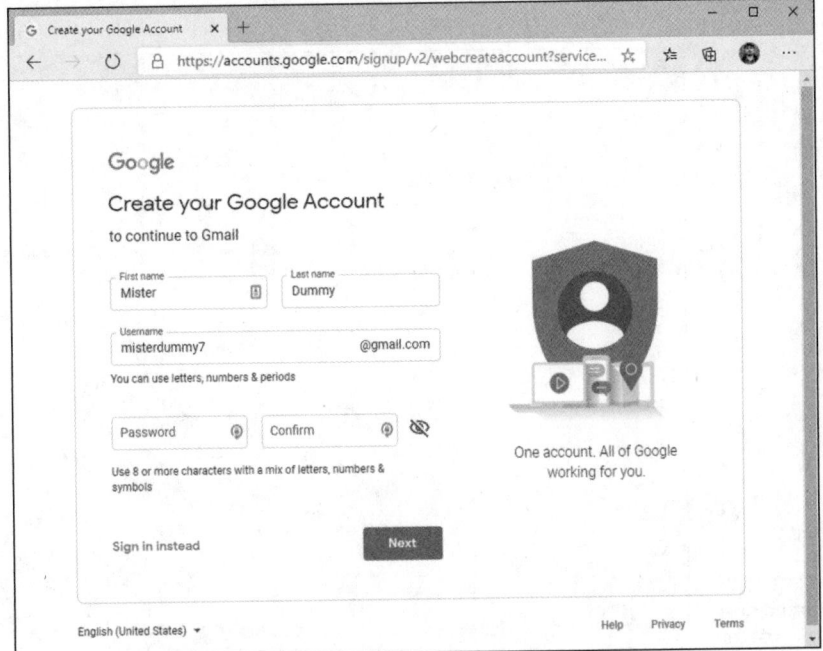

FIGURE 3-2:
Signing up for a Google account is free and easy.

3. **Fill in the form as creatively as you want, and then click or tap Next.**

   If you type a real phone number, Google can use it to help you get into your account if you're locked out, or for two-factor authorization. Similarly, your current email address may help you get back into your account if somebody hijacks it.

   REMEMBER

   In some countries, you're required to give a valid mobile number, and Google sends you an SMS to verify that phone number before you can sign in. Currently, the United States, most of the countries in Europe, and India require valid mobile numbers, but the requirement can change from day to day. If you're reticent to give Google your phone number, remember that it could save your tail one day, if you get locked out, or if you elect to have two-factor authorization added to your account (challenging you with an SMS message every time you log in from a new computer). Google says it "won't use this number for anything else besides account verification."

4. **On the next form, type the mandatory data requested from you (gender, birthday, and so on), and then click or tap Next.**

5. **Agree to the privacy and terms, click or tap Create Account, and confirm your choices (if asked).**

   You now have an official Google account and a new Gmail address. Google dangles the default Gmail screen in front of you (see Figure 3-3), and it's already populated with at least one email message.

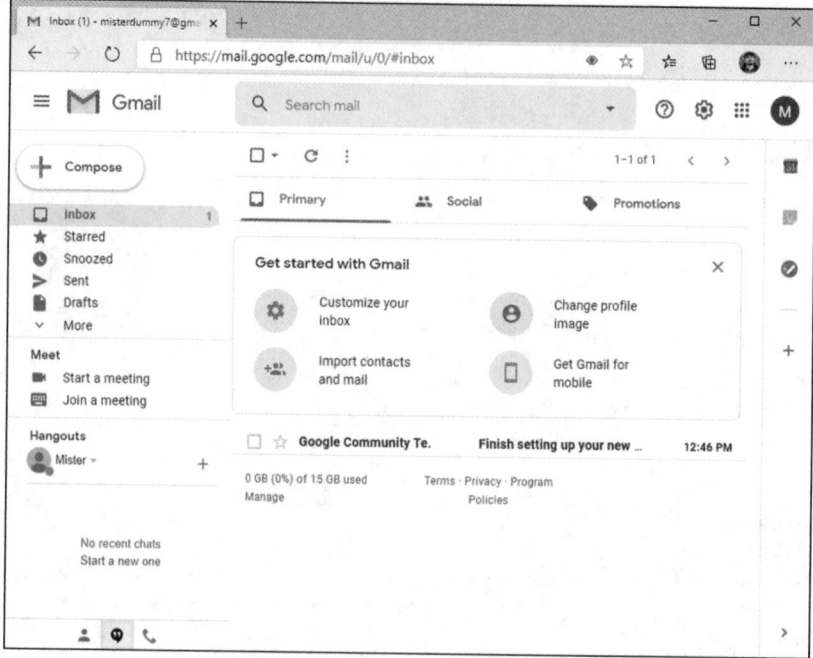

FIGURE 3-3:
Your brand-new Gmail account comes with an email message.

A good way to get started is to simply send an email to yourself. Follow these simple steps for an orientation:

1. **In the upper-left corner, tap or click Compose.**

   The mail composition pane shown in Figure 3-4 appears.

2. **In the To field, type your new Gmail address, add a subject, write a message, and try formatting parts of the message using the string of formatting icons at the top of the typing box.**

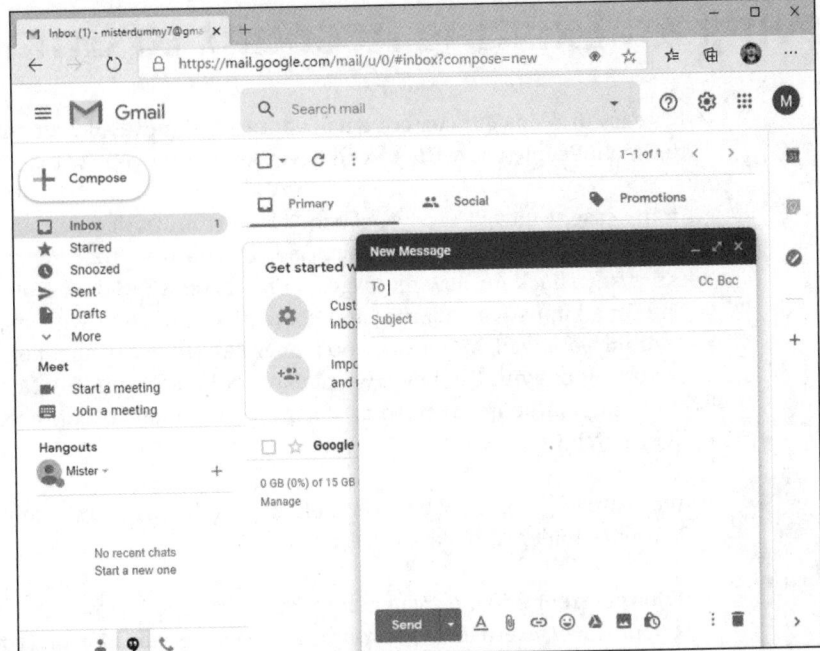

FIGURE 3-4:
Create a new
email message
here.

3. **When you tire of talking to yourself, in the lower-left corner, tap or click Send.**

   Wait a minute or two. If you get bored, click the round arrow at the top, to force your browser to look again.

4. **When the message arrives, play with it a bit.**

   Gmail is different from other mail programs. For starters, it groups messages by the subject. With one click, change to a conversation view that looks like the list seen in forum messaging. Its folders — called *labels* — work differently from other mail programs. Some people like the organization, some people hate it, but it's well worth taking some time to see whether this method feels better to you than the method you're using now.

TIP

After you have a few messages under your belt, hop over to the Gmail learning center at `http://support.google.com/mail` and figure out the options Gmail has to offer. They're extensive and impressive. It probably won't surprise you to know that Gmail has search down cold — you can find any message in seconds, if you know the tricks. But you may be surprised to see how Gmail can work offline — when you aren't connected to the Internet (but you have to use the Chrome browser) — and its support for huge (25MB!) messages.

Moving an Existing Account to Gmail

It's easy to keep your current email address but move all your email handling over to Gmail. People you write to will never know that you switched to Gmail.

In my case, I moved woody@askwoody.com from Outlook on my own email servers to Gmail — and nobody knew, not a soul. Now that my mail is on Gmail, I can easily check for new messages on my Pixel XL, Galaxy Note, iPhone, or iPad. The Gmail apps (calendaring and contacts came along with mail) for Apple and Android work well. Moreover, I no longer have to worry about backing up .pst files or putting up with Outlook's weird ways of handling IMAP. Nor do I have to fret over program hangs. Instead of storing every bloody bit of incoming mail in .pst files, I archive selectively.

REMEMBER

And searches? Oh my! Where Outlook might take about three minutes to search its Sent Files folder, Gmail takes seconds.

If your current email provider supports POP3 (and it probably does), all you need is your email username, password, and POP server address (your mail provider should have it). Here are the details for moving all your mail to Gmail:

1. **If you don't have a Gmail account, go to the Gmail site, click the big red Create an Account box (in the upper-right corner), and follow the instructions.**

2. **Sign in to Gmail. Click the gear icon (right side, above your messages) and choose See All Settings. Select the Accounts and Import tab. Next to Check Mail from Other Accounts, click Add a Mail Account.**

   A dialog box appears.

3. **Type the email address you want to use with Gmail and then click Next. Choose Import Emails from My Other Account (POP3) and click Next. Enter your username and password plus the details for your mail provider's server.**

   In the Add a Mail Account dialog (Figure 3-5), I typically select the Leave a Copy of Retrieved Message on the Server box; it gives me an emergency out, should something go bump in the night. I also select the Always Use a Secure Connection (SSL) when Retrieving Mail box.

   For Label Incoming Messages, pick an address from the drop-down list or create one. I don't automatically archive incoming messages.

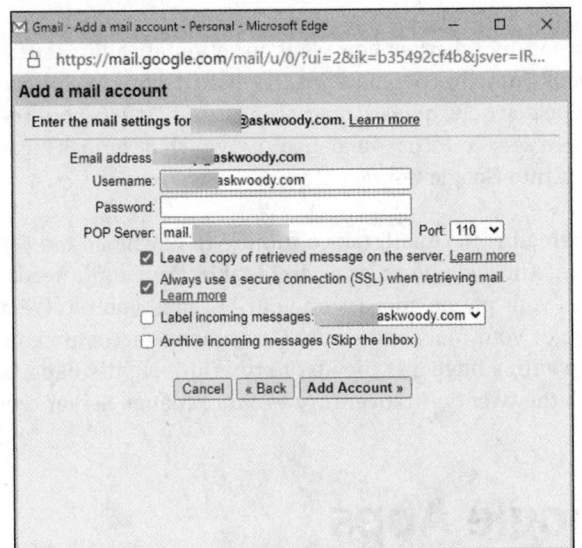

FIGURE 3-5:
Adding a POP3
account to Gmail
is easy, if you
know the server
name.

4. **When you're done, click Add Account.**

   Gmail starts sucking up all the mail it can find. If you've set up Outlook to leave copies of mail on the server, importing can take hours.

   While Gmail copies your mail over to its servers, you get a dialog box that asks whether you want to be able to send mail using your original email address (in my case, that's woody@askwoody.com).

5. **Click Yes, and then click Next Step.**

   You'll see another dialog box that confirms the details about your previous username.

6. **Click Next Step again.**

   Gmail asks whether you want to send outbound mail through Gmail or through your original email provider. Having been bitten by ISPs that block port 25, I always opt to send via Google; as long as I have an Internet connection, my mail always goes out.

7. **Click Send via Google, and then click Next Step.**

   Gmail asks you to verify the email address you'll be using from now on (either your old address or a new address).

8. **Click Send Verification and then check your old (Outlook) inbox for an email with a verification code. Type the code in the next Gmail dialog box and click Verify.**

   You're done!

Consider carefully whether you want to automatically export all your contacts from Outlook into the Google Contacts list. If you do, follow the steps in the How-To Geek article at www.howtogeek.com/201988/how-to-import-export-contacts-between-outlook-and-gmail/, which describes creating a CSV file and importing it into Google Contacts.

After your email is in Gmail, take a minute to download the Gmail apps for your iPhone, iPad, Android phone, or Android tablet. You don't need to do a thing: Mail you send on your phone appears on your PC; mail you receive on your iPad is on your Galaxy or your Mac; and so on. For someone accustomed to lugging around a big laptop with a huge .pst file just to run Outlook, it's like a breath of fresh air. Welcome to the twenty-first century — no Exchange Server required!

Using the Google Apps

After you get a free Google account (see the preceding section), take a few minutes to see what Google Drive can do for you. Remember that the Google apps for creating documents, spreadsheets, presentations, fill-in-the-blanks forms, and drawings may be referenced in some places as being part of Google Drive or G Suite (which is the official name of the paid service but colloquially includes all free-for-personal-use Google stuff) or both. If you see the name *Google Docs* or *Google Apps* while working with *Google Drive*, it's only because Google is slow in getting its names sorted out.

Everything's free, of course.

True confession time. I use Google Docs and Sheets for almost all daily work. I still use Word for writing books, and I have an old Excel spreadsheet to help with doing taxes. But other than that, I've made the change to Docs and Sheets, and I've never looked back.

Here's how to start with the Google apps:

1. **With your favorite browser, go to** www.drive.google.com.

2. **If you aren't logged in to Google, provide your Google account and password. Tap or click Get Started.**

    There's a short tutorial that you can click through. Then the Google Drive page appears, as shown in Figure 3-6.

    You can get Google Drive on your machine by clicking the Settings button, choosing Get Drive for Desktop, and then clicking the Download button next to For Individuals.

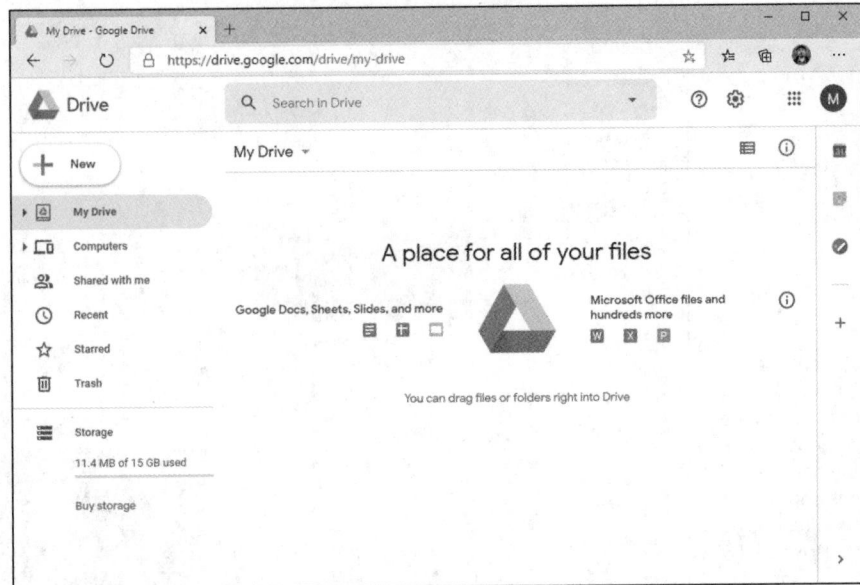

FIGURE 3-6:
Google Drive is familiar to anyone who's used a cloud drive.

3. **Install Google's Backup and Sync app using the instructions shown by the setup wizard, and provide your Google account details.**

4. **Open File Explorer and note that you have a new folder called Google Drive in the Quick Access section. Drag an assortment of files into the Google Drive folder.**

   Try grabbing a simple Word document, a spreadsheet, some graphics files, some PowerPoint slides, and maybe a PDF. Get a handful of them so you can experiment with the Google Drive apps.

5. **Go back to your browser, and again go to** `www.drive.google.com.`

   All the files you put in the Google Drive folder appear, as shown in Figure 3-7.

6. **Open one of the documents (a Word document, Excel spreadsheet, or PowerPoint slide, if you have one) that you copied into the Google Drive folder.**

   If you have the corresponding Office program installed and working on your computer, Google Drive opens the document inside the correct program.

   If you didn't spend the exorbitant amount of money for Office — there's no Office or Office-wannabe on your computer — and the document's fairly simple, as you can see in Figure 3-8, Google Drive does a reasonably good job of *rendering* it — showing it on the screen.

WARNING

More complex documents, though, can have all sorts of problems, from missing pieces to jumbled text. Although Google Drive does yeoman's work trying to display Office documents, it's far from 100-percent accurate — and it doesn't play well with complex templates, and doesn't work at all with macros.

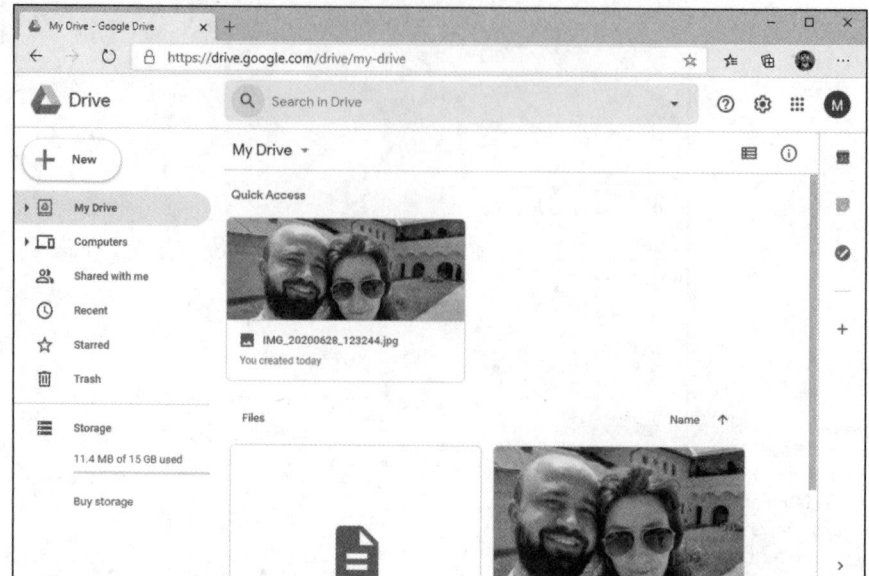

FIGURE 3-7:
Files you drag or copy into the Google Drive folder on your desktop appear inside Google Drive on the Internet.

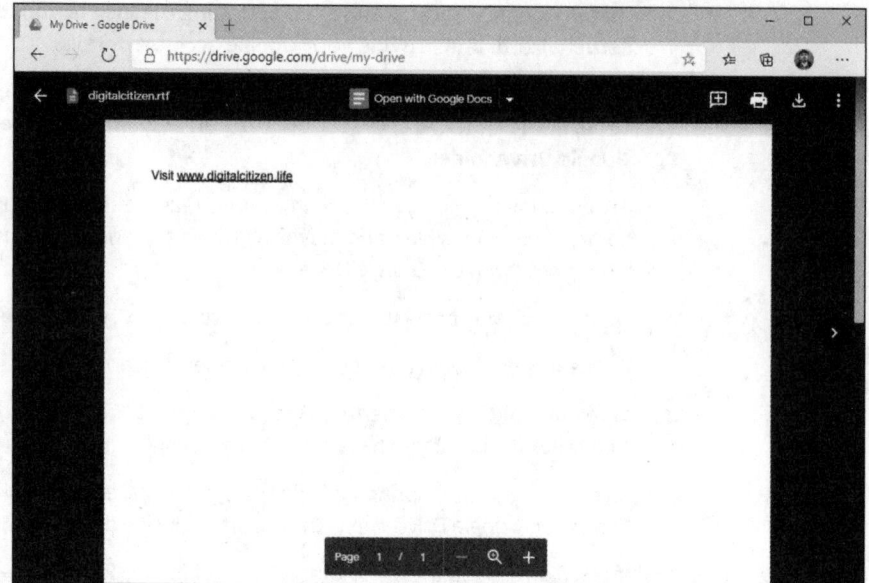

FIGURE 3-8:
Simple Microsoft Office documents render well, but more complex documents like this one may have problems.

7. To edit the document, double-click it.

A copy of the document is saved in Google format (Word documents become .gdoc; Excel files become .gsheet; and PowerPoint slides become .gslides, for example), which you can then edit.

At this point, converting from the Office format to the Google format is a one-way trip. At least as of this writing, you can't change a Google document back to an Office document, although Google has at times offered a File, Export as Word option. Although you can treat Google documents just like any other file — copy or email them, for example — they can be edited only by Google Drive applications.

WARNING

Don't be surprised if the Google applications fall over when converting documents from Microsoft format (or even PDF) to Google format. You'll see something like *An error has occurred and we cannot save your changes.* The conversion feature is very much a work in progress.

8. To create a new document, on the Google Drive home page, tap or click the New button and choose what kind of document you want.

You can create a new document, presentation, spreadsheet, fill-in-the-blanks form (which is stored as a spreadsheet), or drawing (which is stored as a .gdraw file). See Figure 3-9.

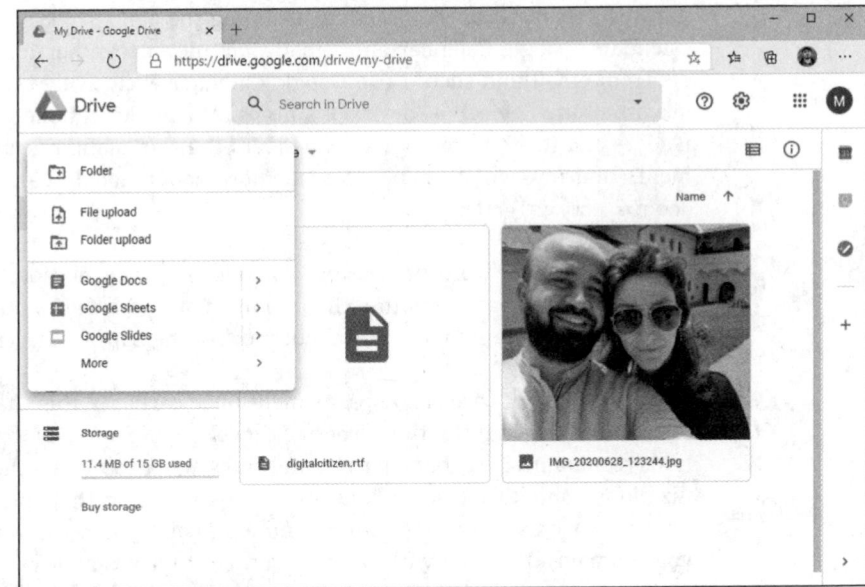

FIGURE 3-9:
It's safer to create new documents from inside Google Drive, rather than importing and switching from the Microsoft Office format.

9. Edit the file using the Google Drive apps' comparatively limited tools (although the spreadsheet app does support Pivot Tables).

In fact, more than one person can edit the file simultaneously.

10. When you're finished, close the browser tab.

Your files are saved automatically, and the latest versions appear almost immediately in the Google Drive folder on your desktop.

TIP

After you play with Google Drive a bit, take a few minutes to read the manual. You can find the Google Docs help system at `http://support.google.com/docs`.

Moving Your Domain to Google

The terminology's confusing. Permit me to review quickly.

Google has a bunch of apps — word processing, spreadsheet, presentation, drawing, and fill-in-the-blanks forms. The apps are tied together with a Dropbox-like, online file storage and synchronization app.

Google Calendar, which I didn't cover in this chapter, is a stand-alone calendar with lots of advanced features, including the ability to sync with many other calendars. Google Calendar and Gmail play nicely together for adding reminders from email and such. I use Google Calendar exclusively because it's easy to use on all my devices — desktop, laptops, MacBooks, iPhone, Samsung Galaxy, iPad — and it fits in nicely with calendars that are maintained by local governments and my son's school. To read more about Google Calendar, go to `www.google.com/calendar`.

REMEMBER

All those apps — word processing, spreadsheet, presentation, forms, drawing, and calendar — together with 15GB or more of online synced storage, are available free for anybody, anytime. I talk about most of the apps in this chapter.

ASK
WOODY.COM

The basic Google Drive and apps, as mentioned earlier in this chapter, are all free for personal use, all the time. The next level up, G Suite ties together organizations (companies, yes, but charities and clubs and all sorts of other kinds of organizations), and it's particularly useful for organizations that operate with a single domain, such as AskWoody.com or Dummies.com. When your organization (and your domain) hooks up with G Suite, you get to use Gmail for handling all your mail and you aren't tied to @gmail.com email addresses.

Why would an individual or small group want G Suite? Good question. The most persuasive arguments I know are these:

>> It's simple, effective, cheap (or free, if you're with a nonprofit or school), and easy, especially if you know and like Gmail.

>> If the Google apps do everything you need — straightforward documents, spreadsheets, presentations — you can save yourself and your organization a ton of money by not buying Microsoft Office.

   This, to me, is the crucial question: Do you need to spend the money to get all the frills in the Office apps, or do the Google apps give you enough of what you need? Tough question, and one only you can answer after you try it for a while.

>> If you set things up properly, you can share documents with everyone in your group, and it doesn't take any extra work. In fact, you can all collaborate on a document at the same time with basically zero effort.

>> Group members can work on the device they prefer; whether the device is a PC, a Mac, an iPad, a Pixel, a Chromebook, or an abacus (okay, I exaggerated a little bit), the Google apps have you covered. And you can switch from machine to machine, location to location, without any concerns about syncing or dropping files.

>> Google's reliability is second to none. It isn't up 100 percent of the time, but it's mighty close.

WARNING

Before you go screeching to your terminal to sign up for G Suite, understand that, although the day-to-day use of G Suite is as simple as using Gmail, setting it up has a couple of gotchas. Converting to the free versions of the Google apps isn't too difficult, but moving your group's domain (such as AskWoody.com) to use Gmail has a few tough spots. It'd be wise to make sure you understand the steps before you commit yourself.

Also ensure that you understand what will and won't happen with your email after you switch. For example, G Suite doesn't move your old messages over to Gmail: If you want your old messages to come across, you have to run its migration program. You can find a comprehensive discussion about moving to Gmail at `http://learn.googleapps.com/gmail`.

I assume that you already have a domain name for yourself or your organization. If not, you can register a domain name with thousands of different, web-hosting companies. I use `www.greengeeks.com`, but your friends may have better recommendations.

In general terms, here's how to get your domain grafted onto the free version of Google Apps (read all the steps before you get started):

1. **Go to** `gsuite.google.com/pricing.html`, **and under your chosen G Suite version, tap or click the Get Started button.**

    If you want to use your domain with Gmail, you have to sign up for G Suite.

2. **Enter your business name, country, and number of employees. Tap or click Next.**

    Google asks a few questions about you and your organization, and then asks if you already have a domain name (see Figure 3-10). if you would like to use a domain name you already have, or if you want to buy a new domain through Google (at $12/year, which is quite reasonable).

    If you already have a domain that you're very attached to, you'll have to jump through some hoops to prove that you own the domain. On the other hand, if you aren't very infatuated with your domain name and would like to start with a new one, buying it from Google is much simpler — Google will even set up your email automatically.

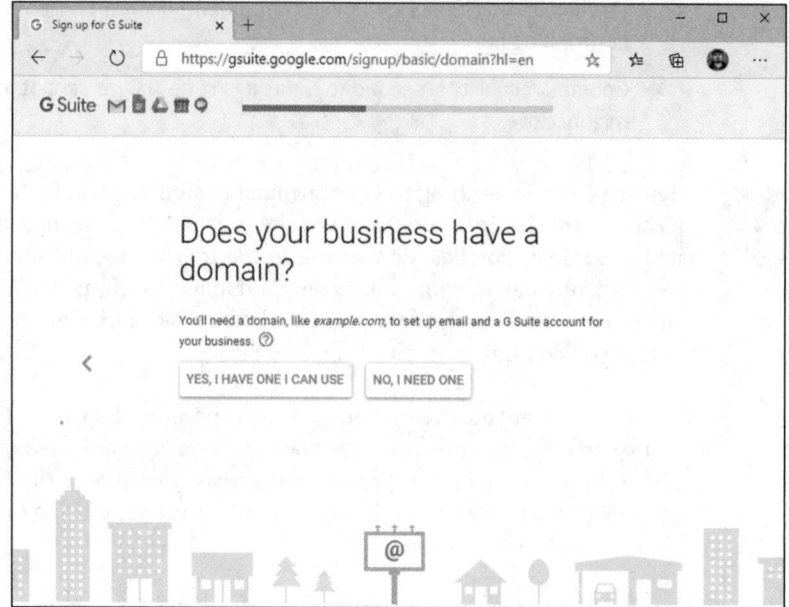

FIGURE 3-10:
Use your current
domain name, or
have Google set
one up for you.

3. **Choose your preference, enter the requested details, and click Next.**

4. **Type your chosen username and password for the G Suite for Work account, and then click Agree and Continue.**

5. **You will have three weeks to verify that you do, in fact, own the domain that you're moving over to G Suite.**

   You have to put a unique identifier inside a file that you upload to your website. That unique number tells Google that you do, indeed, own your domain.

   Look at the video at `http://support.google.com/a/bin/answer.py?hl=en&answer=60216` for details.

   Although you may be uncomfortable performing the upload yourself, if you have a person who helps you with your website, he may well find it to be a piece of cake. Google has detailed instructions for more than 50 different web hosts. Yes, it has step-by-step instructions for Go Daddy, in case you were wondering.

6. **After you verify that you own the domain and Google confirms that it's received the verification, change your site so it starts routing email to the Google Apps servers.**

   You do that by changing the so-called MX Records that are associated with the domain.

   This part's easier than Step 3, but it takes some concentration, especially if you're not accustomed to bumping around inside your domain's records. Details are at `http://support.google.com/a/bin/answer.py?hl=en&answer=140034`.

7. **Wait for the changes to take effect.**

   Usually that's less than an hour. In my case, it took only a few minutes. Mail starts flowing to your Gmail account, and you can use it immediately.

8. **If you want to move any mail over from your current program to Gmail, follow the instructions at** `http://learn.googleapps.com/gmail`.

All in all, setting up your own domain for Gmail is a bit of a pain — it's definitely non-trivial, takes time and jumping some difficult hurdles — but after you're over the hump, using Gmail for all your mail can be a liberating experience.

Chapter **4**

Using Web-Based Outlook.com (nee Hotmail)

T wo months before Microsoft shipped the original Windows 8, the folks in Redmond dropped a bomb on the online email world. Hotmail — one of the best-recognized brands on the planet — would be put out to pasture, replaced by something completely different. Yes, Microsoft tossed out a brand as well-known as Coca-Cola or the terms *taxi* or *Visa and* replaced it with . . . Outlook.com.

**ASK
WOODY.COM**

If you think that the name Outlook.com was chosen because Microsoft's new flagship online email service-formerly-known-as-Hotmail looked or acted like Outlook in Office, or Outlook Express, or the Outlook Web App, or Outlook on the iPad, or the Windows 10 Mail app (which, at one point, was also known as Outlook), or anything else that's ever been called *Outlook,* you'd be wrong, of course. Outlook.com started out as the old Hotmail, with a few internal changes and a new, tiled-style boxy interface. In the years that followed, Microsoft has gradually made Outlook.com look and behave more like the other Outlooks — and made the other Outlooks look and behave more like Outlook.com. The match-up is still not perfect.

How thorough is the change? Well, right now, if you point your web browser to www.hotmail.com, you end up at outlook.live.com/owa — the Outlook Web Access Live login location. Windows Live IDs are now called Microsoft accounts, and Windows Live itself was abandoned during the days of Windows 7, but whatever. It's been that way for years, despite Microsoft getting rid of the Live brand.

Doesn't make any sense, does it?

In this chapter, I step you through Outlook.com, with a nod and a wink to Hotmail, which is now officially dead.

Getting Started with Outlook.com

When working with Outlook.com, any Microsoft account will do. I talk about Microsoft accounts in Book 2, Chapter 5. If you already have an @hotmail.com, @live.com, @outlook.com, or an older @msn.com email address, it's already a Microsoft account. If you don't yet have a @hotmail.com or @outlook.com email address, getting one is easy. Follow these steps:

1. **On the old-fashioned desktop, with your favorite web browser, go to** www.outlook.com.

   The main screen allows you to Sign In, Get Premium (a highly dubious opportunity to spend more money), or Create Account.

2. **To get a new Microsoft account, tap or click Create Free Account.**

   The sign-up form appears, as shown in Figure 4-1.

3. **Choose a unique username, and choose whether you want @outlook.com or @hotmail.com. Click Next.**

4. **Enter the password you want to use, and deselect any options about receiving tips and offers about Microsoft products and services. Then click Next.**

   Depending on what Microsoft has dreamed up today and the phase of the moon, you may be required to enter a name, a country, a birthday (over 18 is a good idea), and maybe even a telephone number. Microsoft insists that it uses the phone number only to set up additional authorization. See Figure 4-2.

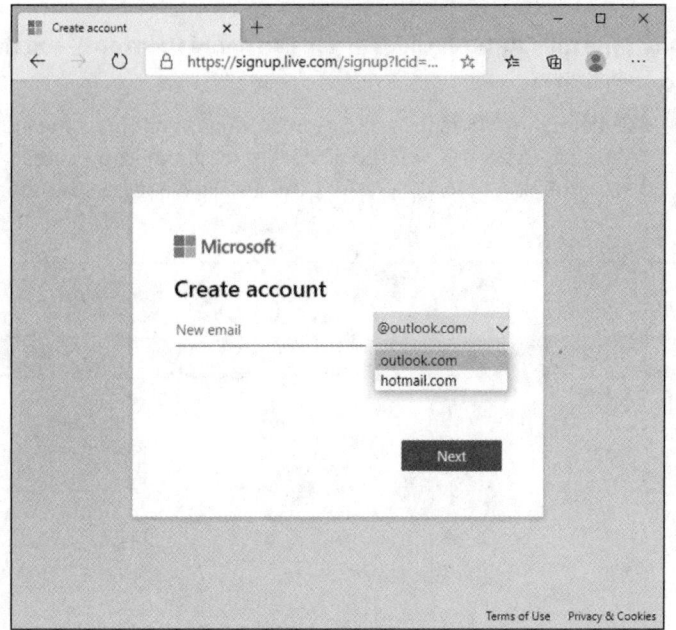

FIGURE 4-1:
Sign up for an @
hotmail.com or
@outlook.com
email address.

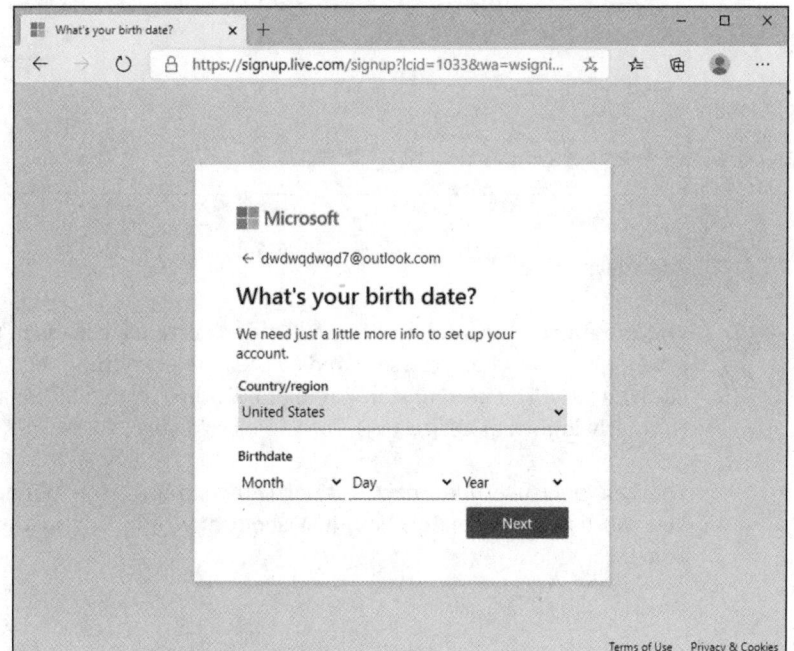

FIGURE 4-2:
Microsoft asks for
personal details,
such as your
country and
birth date.

5. **Type the CAPTCHA codes (if you can figure them out), and then tap or click Next.**

   Outlook.com whirrs for a minute or so, depositing you at the Inbox screen shown in Figure 4-3. You should have at least a welcome message from Microsoft and a few options for getting started and personalizing your account.

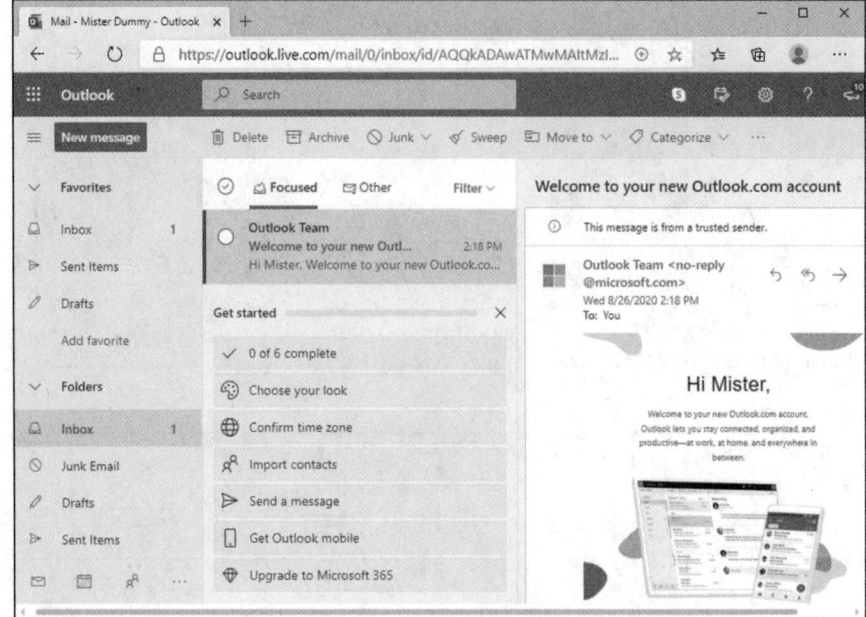

The ads are free.

TIP

Yes, the entire fourth column of your Outlook.com page is taken up by advertising if you maximize your browser window to show everything. Want to get rid of the ads? It's easy. You need to start paying for Outlook.com. You can get Microsoft 365 (formerly known as Office 365), for $70 or so a year.

You can now use your new Outlook.com account as a Windows 10 login ID. You can use it for email, Xbox, just about anything. It's just another Microsoft account.

TWO-FACTOR AUTHENTICATION (2FA)

Having a password is fine, but it isn't as secure as you think. Passwords have a nasty way of getting exposed with major website hacks — particularly if you reuse the same password — when folks sit down at your computer or when the inevitable yellow sticky note falls into the wrong hands.

Enter two-factor authentication, or 2FA. With 2FA, Microsoft can double-check and make sure that you're really you. In short, you're offered several ways to identify yourself. One is to provide a phone number where Microsoft can send an SMS (text) message. Another is to provide an email address where Microsoft can send a verification email. You can also use a specialized app that generates authentication codes for you, such as Microsoft Authenticator. Although the form doesn't make it clear, you don't need to give a phone number, as long as you give an alternate email address — and you don't have to give an email address as long as you give a phone number. Either or both can be fake.

Having a phone number or alternate email address or both on file with Microsoft makes it easier and more secure to reset your password if you lose it; Microsoft sends the reset key in an SMS to your phone or in an email. Only you can decide if the additional convenience (and greater security) of having a working SMS phone number or alternate email address on file is worth the dent in your privacy. Note that SMSs and email accounts can be hacked, too!

More details in Book 2, Chapter 5.

Take a quick spin around Outlook.com, starting from the welcome screen, which you see when you log in to Outlook.com (www.outlook.com) using your favorite @hotmail.com, @live.com, or @outlook.com email address (refer to Figure 4-3):

>> **Default folders on the left are Inbox, Junk E-Mail, Drafts, Sent Items, and Deleted Items.** You click each folder to open it. Make sure you understand what each one is supposed to contain:

- *Inbox* gets all your mail as it comes in. If you don't do anything with it, the message stays in your inbox.

- *Junk Email* holds mail that was sent to you but that Outlook.com has identified as being junk. Outlook.com and Gmail have effective junk identifiers, but occasionally a message will get tossed in here that really isn't junk. If that happens, tap or click the box next to the "good" junk message, and at the top, choose Move To ➪ Inbox.

  You can also drag and drop the message into whatever folder you like.

TIP

If you get a piece of junk mail in your Inbox, don't delete it. You can help the Hotmail filters and other Hotmail users by marking the message as Junk. Just select the box next to the message, and at the top, tap or click Junk.

- *Drafts* holds mail that you were working on but didn't send.

- *Sent Items* contains copies of everything that's gone out.

- *Deleted Items* is the place where messages go when you delete them.

   You can create new folders. Just tap or click the New Folders link at the bottom of the Folders list.

» **The Search box on the top is the most important location on the Outlook. com main page.** People go nuts trying to organize their mail. The Search function finds things amazingly quickly. But that's the topic for the next section.

» **The Sweep feature enables you to move all the messages sent from a specific address into a folder.** Select one message from the sender you want to move, choose Sweep, and then choose how you want to sweep.

- *Delete All from Inbox* takes the action on all mail sent from this sender that's currently residing in the Inbox.

- *Delete All from Inbox Folder and Any Future Messages*— the future messages end up in the Junk folder.

- *Always Keep the Latest One, Delete the Rest.*

- *Always Delete Email Older than 10 Days* applies to mail from only this particular sender. It does, however, apply to all current and future messages from the sender. Discretion advised.

TIP

You may find it amusing or instructive to see how Outlook.com compares to the Windows 10 Mail app. While they're dressed up to appear similar, they work in different ways. Flip back to Book 4, Chapter 1 and see how the Outlook.com interface differs from the Mail's app interface. In addition to cosmetic differences, there are functional differences:

» Outlook.com runs in your browser, doesn't store any information on your computer, and works only when you're connected to the Internet.

» The Windows 10 Mail app runs on your computer, stores a small subset of your email on your computer, and will continue to work (albeit on a subset of messages) whether or not you're connected to the Internet.

ASK WOODY.COM

I continue to have problems with both Windows 10 Mail and Outlook.com, running on various browsers. Don't be too surprised if something doesn't work right for you. (There's a reason why I use Gmail, eh?)

In the next section, I talk about organizing mail so you can use it effectively.

A BRIEF HISTORY OF HOTMAIL

Hotmail blazed new ground as the first, major, free, web-based email service when Sabeer Bhatia (a native of Bangalore and a graduate of Caltech and Stanford) spent $300,000 to launch it in 1996.

On December 29, 1997, Microsoft bought Hotmail for $450 million, cash, and the service has never been the same. Microsoft struggled with Hotmail for many years, adding new users like flies, but always suffering from severe performance problems and crashes heard round the world. Ultimately, Hotmail was shuffled under the Microsoft Network (MSN) wing of the corporate umbrella, its free services were clipped, and its user interface was subjected to more facelifts than an aging Hollywood actor, which is saying something.

As MSN lost its luster and competitors, such as Gmail and Yahoo! Mail, battered at the, uh, Gates, the Hotmail, subscription-based, income model died almost overnight, and the company's market share fell precipitously. Why pay for 20MB of Hotmail message storage when Google gave away 1GB for free? Hotmail became the number-one candidate for a Live makeover and the poster child for Microsoft's entire Live effort. Now that Live is dead, Hotmail has to stand on its own.

Microsoft has gone through a series of well-intentioned but horrendously implemented rebrandings and a few minor upgrades, passing through (get out your scorecard) MSN Hotmail, Windows Hotmail, Windows Live Hotmail, Microsoft Hotmail, and now Outlook. com. Hotmail's final facelift, pre-Outlook.com, came in early 2012. Few people cared, and among the ones who did, the reaction was not universally positive.

Although email as a whole isn't an endangered species, it isn't growing very quickly. Social networking sites pick up a substantial portion of traditional, one-to-one email traffic, and instant messages, SMSs (texts), and VoIP/Skype calls eat away at the numbers.

Bringing Some Sanity to Outlook.com Organization

TIP

Here's my number-one tip for Outlook.com users:

If you have an Archive folder, don't create any new folders.

That way lies madness.

Yes, you can create a folder hierarchy that mimics the filing cabinets in the Pentagon. You can fret for an hour over whether an email about your trip to the beach should go in the Trips folder or the Beaches folder — or both. You can slice and dice and organize 'til you're blue in the face, and all you'll have in the end is a jumbled mess.

Here's my number-two tip for Outlook.com users:

If you don't have an Archive folder, create one — and use it.

If you want to save that message about your trip to the beach or that gorgeous Pinterest mail, just click the message and then click the Archive tab at the top of the Outlook.com window. Or drag the message to the Archive folder. See Figure 4-4.

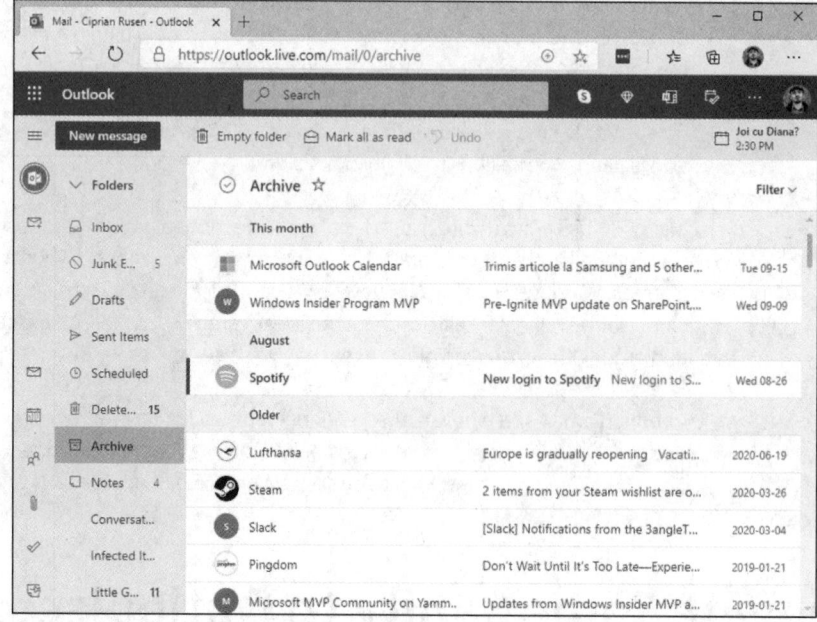

FIGURE 4-4:
The Archive folder (note the Archive heading at the top) can hold any mail you don't want to delete.

The first time you click the Archive button at the top of the Outlook.com window, Outlook.com offers to set up a new Archive folder for you. After that, anything you archive goes into that folder.

REMEMBER

If you want to find all the messages about Trips, use the search box. If you want to find all the messages about Beaches, use the search box. And if you want to find all the messages about Trips *and* Beaches . . . wait for it . . . use the search box!

People get caught up in categories or flags as a way to organize and sort mail. Outlook.com comes with built-in color-coded flags and categories.

If you work well that way, hey, knock yourself out. But note that there's only one kind of flag; you can't set up different flag colors as you can in many other email programs. My general approach is to blast through email as quickly as I can, responding to what needs responding and filing the rest immediately. *De gustibus non est disputandum.*

Handling Outlook.com Failures

Although any computer system in general — and any online system in particular — has failures, Outlook.com, and Hotmail before it, seems (at least to me) to be more susceptible than Gmail.

I recall one particular incident in January 2011, when Hotmail went down and took all the mail from 17,000 users with it. In the grand Hotmail scheme of things, 17,000 users is a very tiny drop in the 300-million-plus subscribers bucket. But if you're one of the 17,000, your opinion may well vary. Ultimately, all those customers got their mail back, but it took up to three days to restore from tape backups (yes, tape!).

If Outlook.com starts acting up on you, here are two websites you should consult:

>> **The Microsoft Hotmail, er, Outlook.com, uh Microsoft 365 Service Status site** (see Figure 4-5) gives you the latest information about Outlook.com's current health — from Microsoft's point of view. Unfortunately, in the past, the site has been criticized for being very slow to recognize reality. In the past few years, Microsoft's network going down has, at times, also taken the status reporting sites down (`https://portal.office.com/servicestatus`).

>> **Downrightnow,** which isn't aligned with Microsoft, gives you a crowdsourced consensus view of what's really happening with Outlook.com/Hotmail. Downrightnow (shown in Figure 4-6) not only actively solicits comments from people who visit the site but also has a Twitter monitoring program that finds some of the tweeted complaints in real time (`www.downrightnow.com/hotmail`). Yes, it's Hotmail, not Outlook.

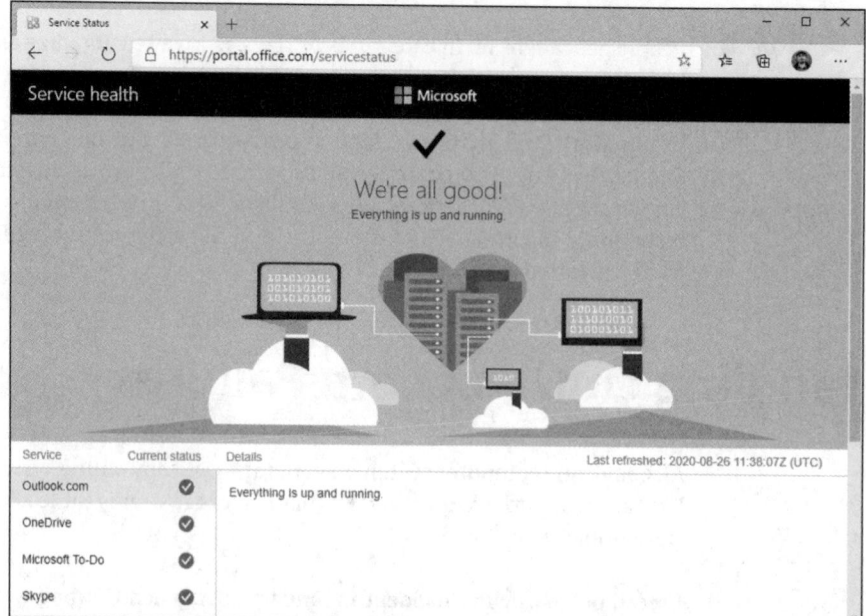

FIGURE 4-5:
Microsoft's
Outlook Service
Status site gives
a broad overview
of the current
Outlook.com
status.

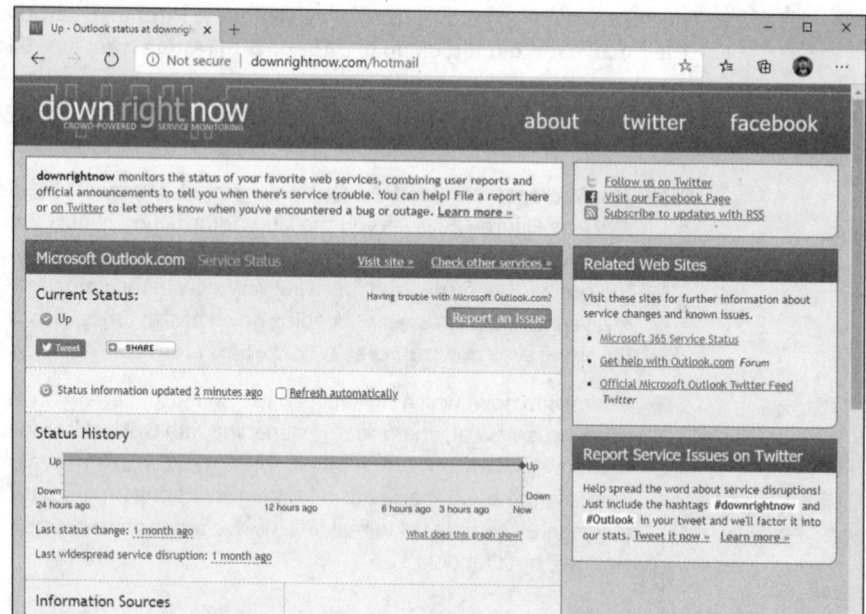

FIGURE 4-6:
Compare the
Microsoft party
line with the
crowdsourced
Downrightnow.

Importing Outlook.com Messages into Gmail

If you find that you prefer Gmail to Outlook.com, you don't have to give up your @hotmail.com, @live.com, or @outlook.com email address. Gmail gladly — I'm tempted to write *gleefully* — takes your Outlook.com mail, pulls it into Gmail and, if you reply to a message, tacks your @hotmail.com, @live.com, or @outlook.com address onto it. Your correspondents won't know that you've switched email providers.

I know. I've been doing it for years.

TIP

Assuming you have both a Gmail and an Outlook.com email address, here's how to set up Gmail so you can read and respond to your Outlook.com mail via the Gmail interface:

1. **Fire up Gmail, and log in with your account.**

2. **In Gmail, tap or click the gear icon and choose See All Settings.**

   The Settings page appears, as shown in Figure 4-7.

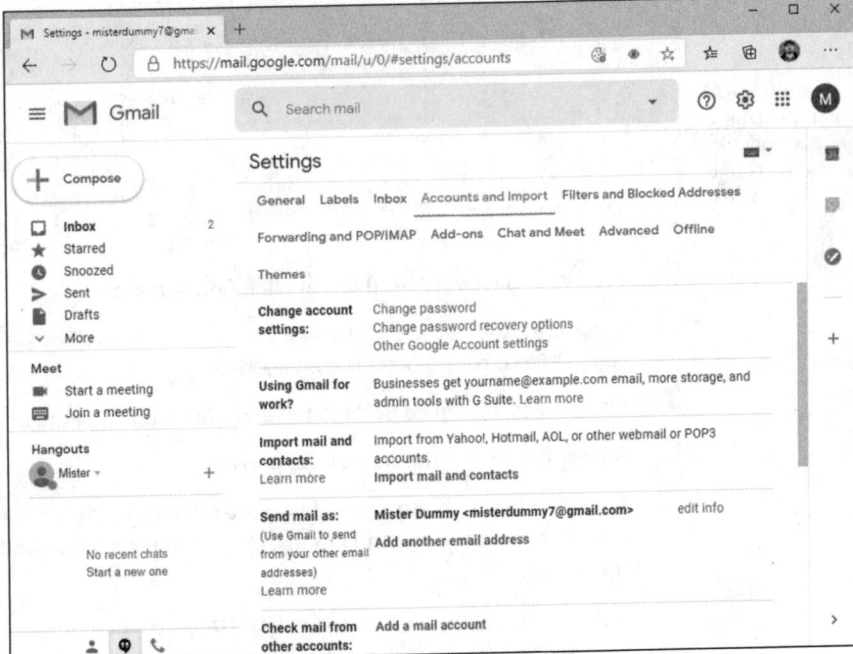

FIGURE 4-7:
The Gmail
Settings page.

3. **At the top, tap or click Accounts and Import. Under the Check Mail from Other Accounts heading, tap or click Add a Mail Account.**

   Gmail asks for the email address.

4. **Type your @hotmail.com, @live.com, or @outlook.com email address — the full address — and then tap or click Next.**

5. **Choose Import Emails from My Other Account (POP3) and then press Next.**

   Gmail fills in all the details for hooking into an Outlook.com (or Hotmail) account, as shown in Figure 4-8, and asks for your password.

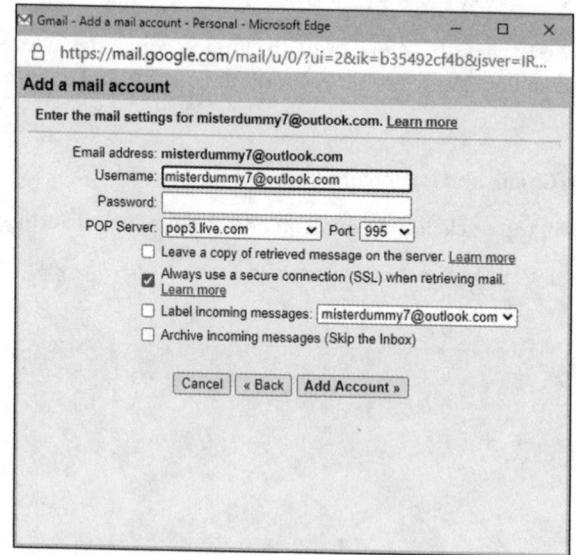

6. **Type your password, and tap or click Add Account.**

   Gmail asks whether you want to be able to send email using your @hotmail. com, @live.com, or @outlook.com address.

7. **Choose Yes, I Want to Be Able to Send Mail, and then tap or click Next.**

8. **Accept the rest of the default responses.**

9. **If you are asked to enter the password once again, this time for the SMTP server that is used to send emails, do so and then press Add Account one more time.**

   Gmail sends a message to your Outlook.com account to make sure you own it, with a verification code.

10. **Manually enter the verification code you received on Outlook.com, and press Verify or tap or click the link in that email message.**

You're all set up.

Gmail's melding onto your Outlook.com account doesn't change anything inside Outlook.com: You still get your mail in Outlook.com and can respond to it there, if you don't want to use Gmail.

Questions about Outlook.com? Go to `http://answers.microsoft.com`.

Weighing the Alternatives

ASK
WOODY.COM

In Book 4, Chapter 1, I talk about choosing an email program. Hotmail, Outlook. com is just one of many, many email programs. At this moment, Microsoft offers about a dozen different email programs:

>> The tiled, Windows 10 Mail app (see Book 4, Chapter 1)

>> Outlook.com, formerly Hotmail (this chapter)

>> Outlook inside Office (many flavors in various versions of Office or Microsoft 365, some of them Exchange Server-based, some on Windows, Mac, iPad, iPhone, or Android tablet or phone)

>> The Outlook Web App, a feature in Exchange Server, which is looking more and more like Outlook.com

Apart from Outlook Express and Windows Mail, and to a lesser extent the various versions of Office Outlook, no two Microsoft email programs look even vaguely similar. In particular, Outlook.com doesn't act anything at all like Outlook inside Office.

Microsoft isn't the only email game in town, of course. Yahoo! Mail still has lots of users, especially in the United States. Gmail's in the same league, although its appeal reaches worldwide. Microsoft's been trying to catch up with Gmail for years, and its switch to Outlook.com is widely viewed as an attempt to shore up Hotmail's rapidly declining market share.

In Chapter 3 of this minibook, I cover Gmail in some depth and branch out to show you how Gmail, Google Drive, and Google Apps cooperate.

REMEMBER

Outlook.com doesn't tie in with the other Microsoft apps the same way that Gmail ties in with G Suite. Microsoft's approach to an all-encompassing application solution, Microsoft 365, uses Outlook and its variants for managing mail, not Outlook.com. (Confusing, yes, I know.) Although you can get your Outlook.com messages fed into Outlook, and you can coerce the Windows 10 tiled Mail program to grab your Outlook.com messages, Outlook.com isn't integrated into Microsoft's Grand Scheme. Yet.

I go through a metric ton of email. I use Gmail, and rely heavily on its automatic Important/Everything else scanning and management. If you're shopping for an email program, make sure you check it out.

Chapter **5**

Best Free Windows Add-Ons

**ASK
WOODY.COM**

Much as I love — and hate, and love to hate — Windows 10, it has a few glaring holes that can be fixed by only non–Microsoft software.

In this chapter, I step you through two different kinds of software. First come the (few) programs that you need to fix holes in Windows. Second is a much larger group of programs that just make Windows work better. Both of the collections have two things in common: They're absolutely free for personal use, with one exception (which costs $5).

Windows 10 apps are still in their infancy, but I've found one that you'll like. A year or two from now, I hope to include many more in this Hall of Cheap Charlie Honor.

At the end of this chapter, I turn to one of my favorite topics: Software that you *don't* need and should never pay one cent to acquire. There are lots of snake oil salesmen out there. This chapter tells you why they're just blowing smoke.

Windows Apps You Absolutely Must Have

Depending on what kind of Windows 10 machine you have, there's a short and sweet list of free software that you definitely need.

File History

It isn't an add-on. There, I fooled you to get your attention. I don't know how Windows 10 users miss this one, but File History (see Figure 5-1) is a fantastic backup application; it's easy to use, and it is part of Windows. You already own it.

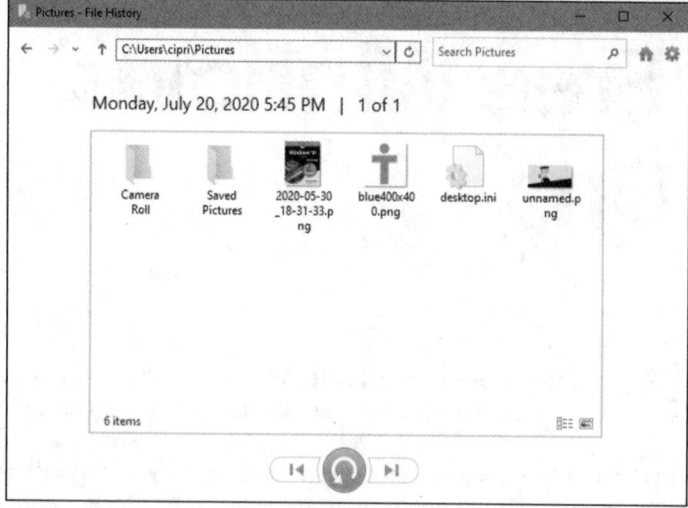

FIGURE 5-1:
File History, the
Windows 10
version of Time
Machine.

Microsoft's telemetry says that more than 80 percent of all Windows 7 users missed the analogous feature in the older version of Windows. Now you have no excuse. All it takes is a USB drive or a hard drive.

Think of File History as the Windows version of Apple's long-admired Time Machine. You get full backups, automatically, and it's easy to retrieve all the earlier copies of a file.

Granted, if you store all your data on OneDrive, Dropbox, Google Drive, or any of the other cloud services, your need for File History goes way down because its functions are built into the online package. But for most of us who still stick things on our PCs, it's a godsend.

If you haven't yet turned on File History, drop everything, head over to Book 8, Chapter 1, and turn it on. Also, keep the backup drive that you use for File History plugged in so that it can do its job daily.

I apologize for the deception. From this point on, I turn to add-ons.

VLC Media Player

Although Microsoft made a few minor improvements to its media handling in Windows 10 — adding the ability to play FLAC lossless audio, MKV video, and a handful of less interesting media formats — it remains woefully underpowered in its ability to work with common media files.

TIP

Find a DVD movie somewhere — if you don't have one, rent one . . . if you can find a place to rent them now — and stick the DVD in your PC. A Windows notification appears, and you can tap or click that notification and play the DVD. It ought to be like falling off a log.

Unfortunately, some Windows 10 PCs — brand-spanking new machines — won't play DVD movies. Why? Microsoft decided that, even though it shipped the DVD-playing capability in previous versions of Windows, putting that capability in Windows 8 and later just cost too much. You can read the details on my blog at www.infoworld.com/article/2616896/microsoft-windows/update--windows-8-won-t-be-able-to-play-dvds.html.

Most PC makers step in and provide the DVD movie-playing software with their new machines, but they're under no obligation to do so. That's why I suggest you get a DVD movie and see whether it'll play.

If it won't play, a simple solution is the free VLC Media Player program. In fact, VLC is so good that I use it and recommend it for all media playing — music and movies. VLC includes the small translation programs (called *codecs)* that let you play just about any kind of music or video on your Windows 10 PC.

**ASK
WOODY.COM**

Another poster child for open source, VLC Media Player plays just about anything — including YouTube Flash FLV files — with no additional software, downloads, or headaches.

Unlike other media players, VLC sports simple controls; built-in codecs for almost every file type imaginable; and a large, vocal online support community. VLC plays Internet streaming media with a click, records played media, converts between file types, and even supports individual-frame screenshots. VLC is well-known for tolerating incomplete or damaged media files. It will even start to play downloaded media before the download is finished.

Hop over to VLC (www.videolan.org) and install it (see Figure 5-2). Yeah, it's ugly. But it works very well indeed.

FIGURE 5-2:
VLC Media Player
plays every song
and video type
imaginable, even
your video DVDs.

VLC has played with the idea of shipping a Windows 10 app version of VLC but results as of this writing are disappointing. Unless you're running Windows 10 in S mode — and thus can't run programs on the desktop — I'd give the Microsoft Store version a pass.

LastPass

In Book 9, Chapter 4, I talk about two password managers, LastPass and RoboForm. Both are excellent choices. Most people, in my experience, prefer LastPass, but you ought to look at Book 9, Chapter 4 and see if your circumstances are different. I use LastPass religiously, in all my browsers, on all my computers: Windows, Android, iOS, Chrome OS, Mac, you name it.

LastPass (shown in Figure 5-3) keeps track of your user IDs, passwords, and other settings; stores them in the cloud; and offers them to you with a click. LastPass does its AES-256 encrypting and decrypting on your PC, using a master password that you have to remember. The data that gets stored in the cloud is encrypted, and without the key, the stored passwords can't be broken, unless you know somebody who can crack AES-256 encryption.

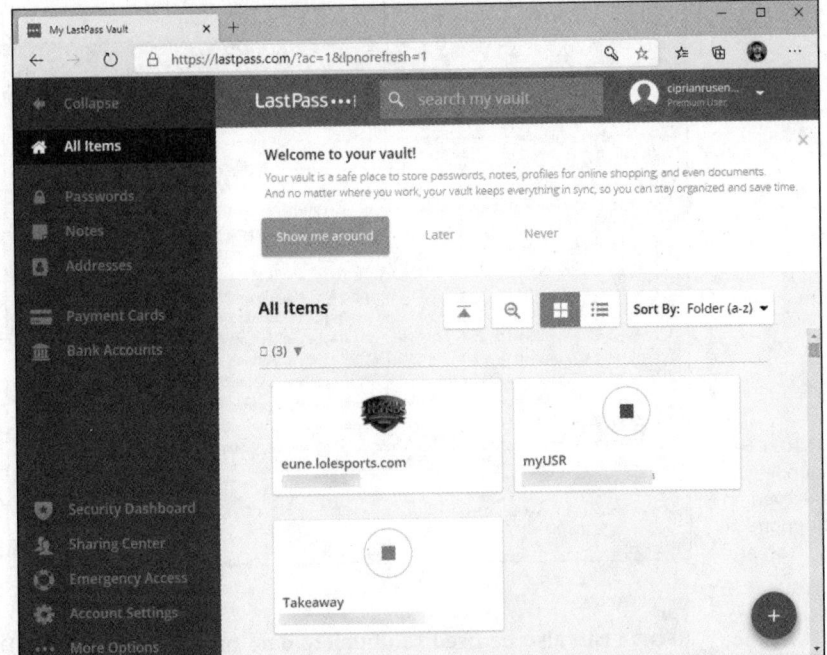

FIGURE 5-3:
LastPass gives you one place to store all your passwords — and all the encrypting is done on your computer.

LastPass works as a browser add-on for Edge (see Figure 5-3), IE, Firefox, or Chrome, so all your passwords are stored in one place, accessible to any PC you happen to be using — if you have the master password.

LastPass is free for personal use on PCs. The Premium version, which works on all sorts of mobile devices, costs $3 a month.

Recuva

ASK WOODY.COM

File undelete has been a mainstay PC utility since DOS. But there's never been an undeleter better than Recuva (pronounced "recover"), which is fast, thorough, and free. See Figure 5-4.

When you throw out the Windows Recycle Bin trash, the files aren't destroyed; rather, the space they occupy is earmarked for new data. Undelete routines scan the flotsam and jetsam and put the pieces back together.

As long as you haven't added new data to a drive, undelete (almost) always works; if you've added some data, there's still a good chance you can get most of the deleted stuff back.

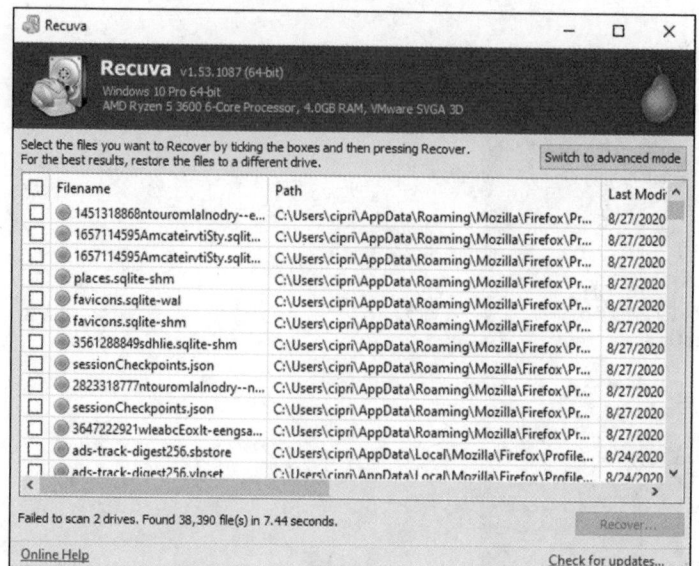

Recuva can also be used to undelete data on a USB drive, an SD card, and many phones and cameras that can be attached to your PC.

Powerful stuff. For more advanced features, there's a Pro version for $20.

The Best of the Rest — All Free

Here are my recommendations for useful software that you may or may not want, depending on your circumstances.

Hey, the price is right.

Nextpad (Notepad replacement)

After years of using the old Notepad from Windows, and more years working with Notepad ++ (which is still my favorite editor for writing code or hand-writing HTML), I finally found a quick, light, simple Notepad replacement that doesn't bend my brain.

Wonder of wonders, it's a Windows 10 app, which is to say it's a tiled (formerly Metro) app that you get from the Microsoft Store.

Nextpad (see Figure 5-5) has tabs for working on multiple documents, it saves your changes automatically, and it doesn't store your stuff in the cloud. A straight-up text editor with few frills and lots of moxie. And it's free. Get it from the Microsoft Store.

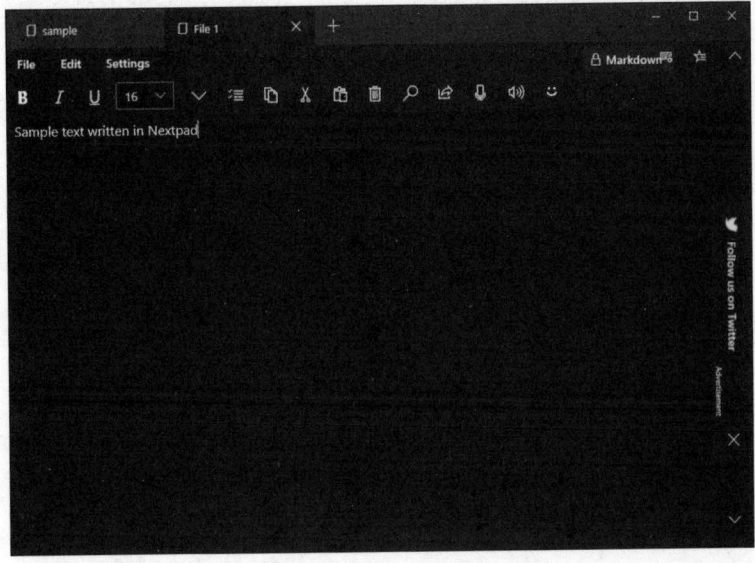

FIGURE 5-5:
Nextpad runs rings around Notepad, and it's simple, fast, and easy.

Ninite

Leading the list of traditional desktop programs is a program that helps you install (and update) other programs. Actually, it isn't a program. It's a website. I talk about it in Book 4, Chapter 4.

When you start looking at desktop applications, your first stop should be to `https://ninite.com/` (see Figure 5-6). Simply click the applications you want and Ninite will download the latest versions free of crapware, install them, and leave you in the driver's seat.

Need to update your apps? Run Ninite again. Everything's brought up to date, no junkware, no hassle. For the full royal treatment — where Ninite notifies you of changes to programs that you've installed —Ninite Updater ($9.99/year) works like a champ.

The beauty of the Ninite approach is that all these apps are a click away — no fuss, no nags, no charge. It's the best way I know to install a bunch of good programs on a new machine in just a few minutes. The downside? It misses a few of my favorite desktop apps. More details in Book 9, Chapter 4.

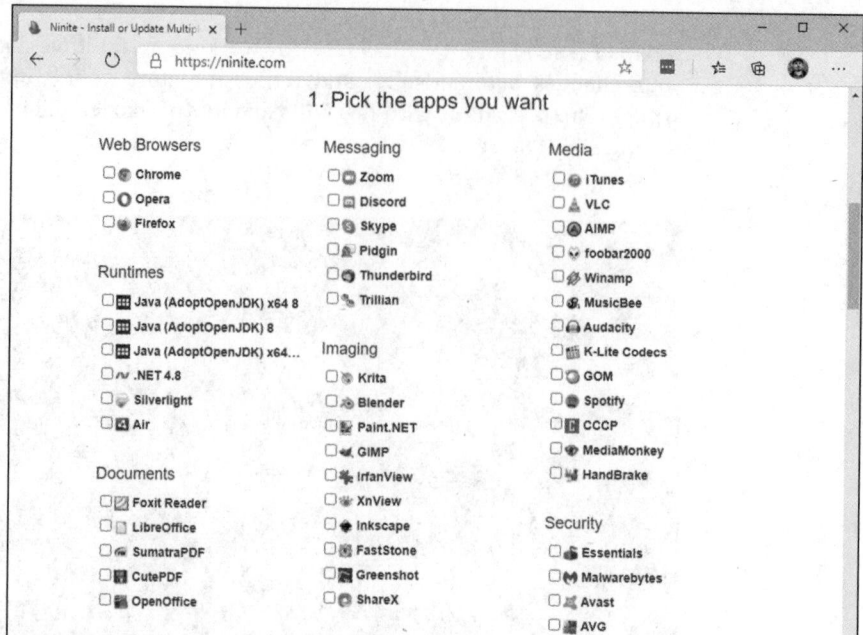

FIGURE 5-6:
Install or update almost any desktop app, any time, with Ninite.

Revo Uninstaller

Revo Uninstaller (`www.revouninstaller.com/revo_uninstaller_free_download.html`) well and truly uninstalls programs, and it does so in an unexpected way.

When you use Revo, it runs the program's uninstaller and watches while the uninstaller works, looking for the location of program files and for Registry keys that the uninstaller zaps. It then goes in and removes leftover pieces, based on the locations and keys that the program's uninstaller took out. Revo also consults its own internal database for commonly left-behind bits and roots those out as well.

Revo gives you a great deal of flexibility in deciding just how much you want to clean and what you want to save. For most programs, the recommended Moderate setting strikes a good balance between zapping problematic pieces and deleting things that really shouldn't be deleted.

TIP

The not-free Pro version monitors your system when you install a program, making removal easier and more complete. Pro will also uninstall remnants of programs that have already been uninstalled.

If you uninstall programs — whether to tidy up your system or to get rid of something that's bothering you — it's worth its weight in gold.

Paint.net

In Book 7, Chapter 6, I talk about the Microsoft Paint program, which can help you put together graphics in a pinch. For powerful, easy-to-use photo editing, with layers, plugins, and all sorts of special effects, along with a compact and easily understood interface, I stick with Paint.net.

With dozens of good — even great — free image editors around, it's hard to choose one above the others. Irfanview, for example, has tremendous viewing, organizing, and resizing capabilities.

Although Paint.net requires the Windows .Net Framework, the program puts all the editing tools a nonprofessional might reasonably expect into a remarkably intuitive package. Download it at www.getpaint.net and give it a try.

7-Zip

Another venerable Windows utility, 7-Zip (www.7-zip.org) still rates as a must-have, even though Windows supports the Zip format natively. Why? Because some people of the Apple persuasion will send you RAR files from time to time, and 7-Zip is the fast, easy, completely free way to handle them.

ASK
WOODY.COM

7-Zip also creates self-extracting EXE files, which can come in handy (although heaven help you if you ever try to email one — most email scanners won't let an EXE file through). And it supports AES-256 bit encryption. The interface rates as clunky by modern standards (see Figure 5-7), but it gets the job done with Zip, RAR, CAB, ARJ, TAR, 7z, and many lesser-known formats. It even lets you extract files from ISO CD images.

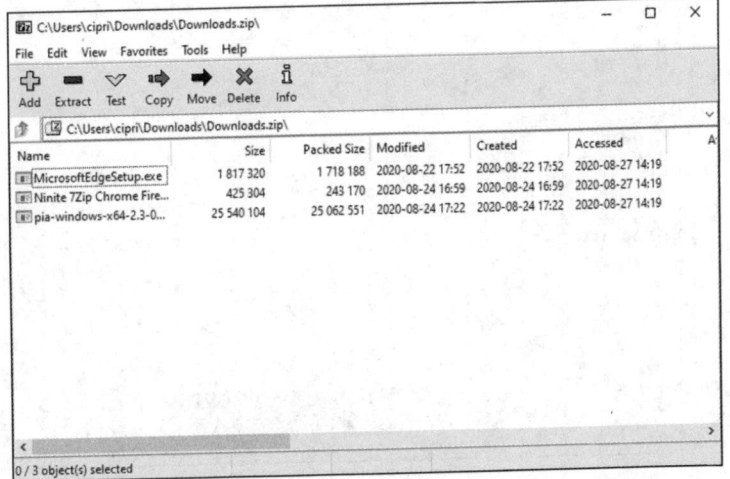

FIGURE 5-7:
7-Zip may not have the greatest interface, but it's a workhorse.

Another poster boy for the open-source community, 7-Zip goes in easily, never nags, and wouldn't dream of dropping an unwanted toolbar on your system. Enlightened.

You don't need to register or pay for 7-Zip. Don't fall for a website with a similar name. To get the real, original, one and only free 7-Zip, with a crapware-free installer, go to 7-zip.org. Download it from www.7-zip.org/download.html.

qBittorrent

If you aren't yet using torrents, now's the time to start. Torrents have taken a bad rap for spreading illegal, pirate software. Although that reputation is entirely deserved, it's also true that many torrents are absolutely legitimate. Torrents are the single most efficient way to distribute files that exists.

For years I've used and recommended uTorrent, but the current version's installer includes crapware — and in previous versions, it's installed some really obnoxious crapware. Worse, the uTorrent "Date hot Russians" ads and their ilk make it tough to torrent in mixed company.

Instead, try qBittorrent, shown in Figure 5-8. It's simple (no Java, no .NET), fast, and easy to use, and it supports magnet links (which simplify downloads), with extensive bandwidth reporting and management. Download it from www.qbittorrent.org

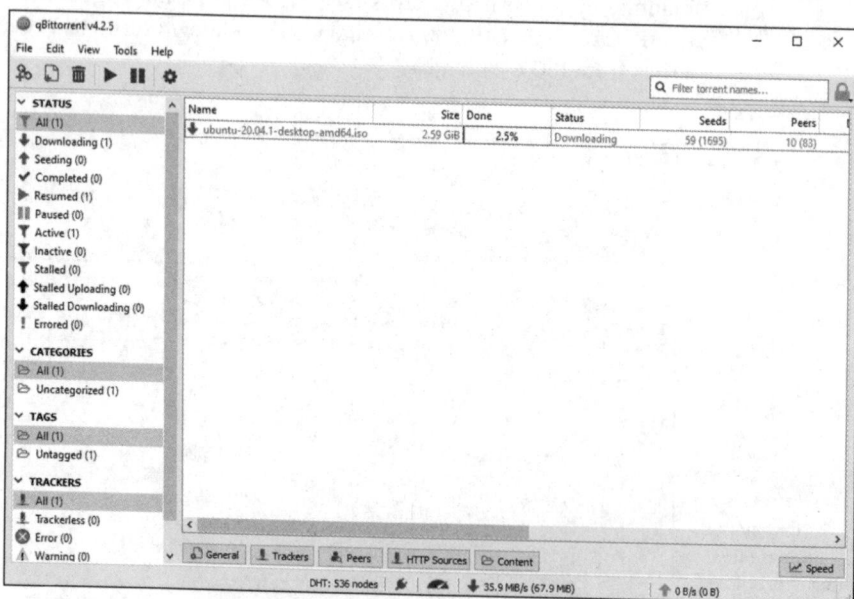

FIGURE 5-8:
qBittorrent
doesn't have
uTorrent's
baggage.

Dropbox, Google Drive, OneDrive, or . . .

Even if the thought of putting your data on the Internet drives you nuts, sooner or later you're going to want a way to store data away from your main machine, and you're going to want an easy way to share data either with other people, or with other computers (desktops, laptops, tablets, and phones).

I give you an overview of the options available in Book 7, Chapter 4. There's no obvious winner — no cloud storage that's inherently better than any of the others. Just pick one and get it set up. Someday, it'll save your tail. I use Dropbox for my important files (including the files for this book). However, I also use OneDrive for files that I might need on a mobile version of Office, and I use Google Drive for everything else, especially the amazing Google Photos app, which I discuss in Book 4, Chapter 3.

They all have free introductory options, and some give you an enormous amount of storage for free, for a small fee, or even a very large amount of storage if you subscribe to a related service – that's the Microsoft 365 shtick, formerly known as Office 365.

Other interesting free software

If you connect over public Wi-Fi, such as in a coffee shop, you really should use a Virtual Private Network (VPN). I talk about VPNs in Book 9, Chapter 4. Session hijacking, pioneered by the program Firesheep, can let others pose as you, even while your session is in progress. Using secure sockets (SSL) helps, but even those can be subverted in certain circumstances. Best bet is to stick with VPN.

Need to rip a DVD? Forget trying to use Windows. Get the open-source, free and junk-free HandBrake, `https://handbrake.fr`. Works like a champ on any DVD.

Wonder what programs run whenever you start Windows? Look at Microsoft's venerable and free-as-a-breeze Autoruns, `https://technet.microsoft.com/en-us/sysinternals`. Autoruns finds more autostarting programs (add-ins, drivers, codecs, gadgets, shell extensions, whatever) in more obscure places than any other program, anywhere. Autoruns not only lists the autorunning programs but also lets you turn off individual programs. It has many minor features, including the capability to filter out Microsoft-signed programs, a quick way to jump to folders holding autostarting programs, and a command-line version that lets you display file hashes. Autoruns doesn't require installation. It's a program that runs and collects its information, displays it (with a rather rudimentary user interface), lets you wrangle with your system, and then fades away.

TIP

Want to know what hardware you have? It's a common question that's easily answered with a nifty, free utility called HWiNFO, available at www.hwinfo.com. HWiNFO delves into every nook and cranny. From the summary to detailed Device Manager-style trees of information — entire forests of information — HWiNFO can tell you everything anyone could want to know about your machine. There's a separate real-time monitoring panel that tells you the current status of every-thing under the sun: temperatures, speeds, usage, clocks, voltages, wattages, hard drive SMART stats, read rates, write rates, GPU load, network throughput, and on and on.

You may not need to buy Microsoft Office

Maybe.

If your needs are simple and you don't have to edit fancy documents created in Word, Excel, or PowerPoint, you may be able to get by with the Google apps (which I discuss in Chapter 3 of this minibook) or LibreOffice. The web-based Office Online apps are also good — and free for personal use. If you're moonlighting with a Mac, the iWork apps might do, too.

TIP

Do the math: LibreOffice, free. Google apps, free for personal use. iWork apps, free. Microsoft 365 Home (which includes five licenses) $100/year, forever. Office 2019 Home & Student (for personal use only, no Outlook), $150. Home & Business, $220. Office Online, free for personal use.

The big advantage to Microsoft 365: You get not only five licenses of the latest versions of the Office programs — Word, Excel, PowerPoint, Outlook, OneNote, Access, Publisher — for PCs or Macs, but also licenses for five tablets (includ-ing Office for iPad, which is a tremendous product), and five licenses for phones (largely forgettable). In addition, you get 1TB of OneDrive online storage per user for up to five users. Unless you have a visceral reaction to renting Office — I can sympathize — Microsoft 365 at $100/year or less comes across as quite a bargain.

Whenever somebody asks me, "Why do you recommend Office when OpenOffice/ LibreOffice does everything for free?" I have to cringe. It's true that Microsoft Office is expensive, and with Microsoft 365 you're locked into the annual fee. It's also true that good, but not great, alternatives exist — including Google's G Suite (which I discuss in Chapter 3 of this minibook), among many others.

Here are two substantial problems:

>> As much as I would love to recommend a free replacement for Word, Excel, or PowerPoint, the simple fact is that the free alternatives (other than Office

Online) aren't 100-percent compatible. In fact, for anything except the simplest formatting and most basic features, they aren't compatible at all. Even Microsoft's free Office Online Apps aren't as full-featured as the real Word, Excel, or PowerPoint. If your needs are modest, by all means, explore the alternatives. But if you have to edit a document that somebody else is going to use and it has any unusual formatting, you may end up with an unusable mess.

>> Many people don't realize it, but OpenOffice.org isn't the same organization it used to be. In fact, there's an ongoing debate about the superiority of the new OpenOffice.org (which now belongs to Apache) and the renegade offshoot LibreOffice (www.libreoffice.org). Basically, some feel that OpenOffice.org moved away from its open-source roots when Oracle owned it, so a new organization, LibreOffice, forked the code and has released several new versions that are not associated with OpenOffice.org or Oracle. So you're left with two organizations, slightly different products, and no clear indication of which version (if either) will be around for the long term.

ASK WOODY.COM

If you can get by with G Suite — that's what I use for everything except books — go through the steps in Chapter 3 of this minibook. If you have to use Office, do yourself a favor and first try the free-for-personal-use Office Online, at https://products.office.com/en-us/office-online/documents-spreadsheets-presentations-office-online.

Don't Pay for Software You Don't Need!

If you've moved to Windows 10, there's a raft of software — entire *categories* of software — that you simply don't need. Why pay for it?

ASK WOODY.COM

Many people write to ask me for recommendations about antivirus software, utility programs, Registry cleaners, or backup programs. They cite comparative reviews — even articles that I wrote a few years ago — debating the merits and flaws of various packages.

Time and again, I have to tell them that all the information they know is wrong. On second thought, I guess the accumulated knowledge isn't so much wrong as obsolete.

The simple fact is, if you moved up to Windows 10, you wouldn't need lots of that stuff — and the old reviews are just that. Old reviews.

In this, the last section of the last chapter of this book, I'm going to lay it on the line — point out what you don't need, in my considered opinion — and try to save you a bunch of money. With any luck at all, this handful of tips will save you the price of the book.

Windows 10 has all the antivirus software you need

Windows Security, formerly known as Windows Defender, works great. And it doesn't cost a cent. I've railed against the big antivirus companies for years. And I'll rail once again. You don't need to pay a penny for antivirus, antispyware, anti-anything software, and you don't need a fancy outbound firewall, either. (I talk about Windows Security and the Windows Defender Firewall in Book 9, Chapter 3.)

You do need *other* security programs, however. I list those in Book 9, Chapter 4. They're free.

Windows 10 doesn't need a disk defragger

Because of the way Windows stores data on a hard drive and reclaims the areas left behind when deleting data, your drives can start to look like a patchwork quilt, with data scattered all over the place. *Defragmentation* reorganizes the data, plucking data off the drive and putting files back together again, ostensibly to speed up hard drive access.

Although it's true that horribly fragmented hard drives — many of them hand-crafted by defrag software companies trying to prove their worth — run slower than defragged drives, in practice the differences aren't that remarkable, particularly if you defrag your hard drives every month or two or six. (Note that you should never defrag a solid-state drive.) In practice, even moderately bad fragmentation doesn't make a noticeable difference in performance, although running a defrag every now and again helps.

With Windows 10, you don't need to run a defrag. Ever. If you have a solid-state drive, you don't need (or want) a defrag — it wears out your drive and doesn't improve anything. If you have a whirling-platter hard drive, Windows runs a defrag for you, by default, one day every week.

Windows 10 doesn't need a disk partitioner

I personally hate disk partitioning, but rather than get into a technical argument (yes, I know that dual-boot systems with a single hard drive need multiple partitions), I limit myself to extolling the virtues of Windows 10's partition manager.

No, Windows 10 doesn't have a full-fledged, disk partition manager. But it does everything with partitions that most people need — and it gets the job done without messing up your hard drive. Which is more than I can say for some third-party disk partition managers.

To run Windows 10's built-in disk partitioner, type **partition** in the Windows 10 search box. Click the first link, which should be Create and Format Hard Disk Partitions. That puts you in the Disk Management program, where it's right-click easy to see, delete, expand, and change your partitions. If you want to create a new partition, right-click in any empty area and choose Create Volume. If you want to make a new partition on a volume that's full, right-click on the volume and choose Shrink Volume.

Windows 10 doesn't need a Registry cleaner

I've never seen a real-world example of a Windows 10 machine that improved in any significant way after running a Registry cleaner. As with defraggers, Registry cleaners may have served a useful purpose for Windows XP, but nowadays, I think they're useless (correction: worse than useless). I've never found a single run of a single Registry cleaner that caused anything but grief.

There's a great quote that (as best I can tell) originated on the DSLReports forum in March 2005. A poster who goes by the handle Jabarnut states, "The Registry is an enormous database, and all this cleaning really doesn't amount to much . . . I've said this before, but I liken it to sweeping out one parking space in a parking lot the size of Montana." And that's the long and short of it.

Jabarnut is correct: The Registry is a giant database — a particularly simple one. As with all big databases, sooner or later some of the entries get stale; they refer to programs that have been deleted from the system or to settings for obsolete versions of programs. Sure, you can go in and clean up the pointers that lead nowhere, but why bother? Registry cleaners are notorious for messing up systems by cleaning things that shouldn't be touched.

Windows 10 doesn't need a backup program

The built-in backup options, which I discuss at length in Book 8, Chapter 1, work very well.

The only possible exception is if you're paranoid enough to want a full ghost backup of your hard drive. In that case, yes, you have to acquire (possibly buy) a backup program. But why bother? Windows 10's Restore works very well indeed.

Don't turn off services or hack your Registry

I just love it when someone writes to me, all excited because he's found a Windows service that he can turn off, with no apparent ill effect. Other people tell me about this really neat Windows pre-fetch hack they've found, in which a couple of flipped bits in the Registry can significantly speed up your computer. Before they changed, Windows boot times were sooooo slow. Now, with the hack, it's like having a new PC all over again!

Meh.

I call it the Registry Placebo Effect. If you find an article or a book or a YouTube video that shows you how to reach into the bowels of Windows to change something, and the article (or book or video) says that this change makes your machine run faster, well — by golly — when you try it, your machine runs faster! I mean, just try it for yourself: Your machine will run *so* much better.

Yeah. Sure. Once upon a time, when dinosaurs walked the earth, it's possible that turning off a few Windows *services* (little Windows subprograms that run automatically every time you boot) may have added a minuscule performance boost to your daily Windows ME routine. Bob may have jumped up faster, or Clippy could have offered his helpful admonitions a fraction of a millisecond more quickly. But these days, turning off Windows services is just plain stupid. Why? The service you turn off may be needed, oh, once every year. If the service isn't there, your PC may crash or lock up or behave in some strange way. Services are tiny, low-overhead critters. Let them be.

That covers the high points. I hope this chapter alone paid for the book — and the rest is just gravy!

Index

Symbols and Numerics

H

I

R

About the Authors

Woody Leonhard's dozens of books about Windows and Office cover everything you need to know, without putting you to sleep. He's taken home eight Computer Press Association awards and two Jesse H. Neal awards. His Erdős number is 3.

Woody writes an ongoing for column for www.Computerworld.com called Woody on Windows. His own website, www.AskWoody.com, draws almost a million visits a month.

Woody specializes in telling the truth about Windows and Office — hold the sugar coating — whether Microsoft likes it or not.

Ciprian Adrian Rusen is a tech blogger and author of several titles about Windows and Office. He has been recognized by Microsoft as a Windows Insider MVP, an honorary title given for his public contribution and expertise to the Windows ecosystem. This book is one of the many ways in which he helps Windows users worldwide.

Ciprian leads the team at www.digitalcitizen.life, a website that provides useful how-to content for Windows, Android, iOS, and Mac. If you want to learn how to tame the computers, smartphones, and gadgets that you use daily, subscribe to his blog.

Dedications

Woody: To Addie and Andy, who have made it all possible.

Ciprian: To Cristina and her gorgeous smile, which was there during the tough lockdown days.

Author's Acknowledgments

Many thanks to everyone involved in the process of bringing this book to light in spite of horrendous deadlines — and those who worked on previous Windows All-in-One books, refining the format and tightening the approach. It's been quite a road from Windows XP to Windows 10, and you all deserve enormous praise.

Thanks to Dell, for loaning me a gorgeous XPS-15 with all the fixings so I could test Windows 10's more exotic new features. Also, thanks to ASUS for loaning a cutting-edge ASUS ZenBook Duo that we used to update this book to its 4th edition and make it the best version yet.

And thanks to Susan Pink, Ryan Williams, and Steve Hayes for bringing this massive tome together in record time.

Particular thanks to the folks on www.AskWoody.com who keep on top of all the problems — and answers — that make this book and the site tick, especially Abbodi86, Kirsty, MrBrian, and PKCano, along with all the other AskWoody MVPs.

In addition, the following select group of kind people have helped keep the Ask-Woody.com question-and-answer site going. Our AskWoody benefactor hall of fame:

BillC, BJM, Cosmo, CyGuy, David F., dencorso, Gail, Ikester, jim, Jim in Yakima, Marie, Marty, Mary F, Mike D, Mike in Texas, Mr. D, Northwest Rick, Pedro, PKCano, Q, samak, Schnarph, VFRJohn, Allen Adolfsen, Stewart Agreen, Neil Ames, Robert Apted, Fabio Araujo, AJ Averett, George Barclay, Peter Barker, Kathleen Barron, Stan Beben, William Becker, Ken Beebe, Gene Beeman, Kenneth Beers, Robert Bellanca, Anita Benike, Stan Bennett, Marshall Blatz, Jakob Bohm, Scott Bostwick, Susan Brocklebank, David Brotherton, Michael Brown, Gunnar Brundin, Alexander Bryant, Ed Buckley, Janet Burk, Elizabeth Burke, William Burt, Karen Burton, Gary Cahn, Susan Campbell, Noel Carboni, Sam Carson, Francis Cerra, Edward Chambers, Timothy Chan, Eidens Christoph, Brian Citrine, Andrea Clarkson, George Clarkson, Shillest Clayton, Mark Cohen, Dan Collison, Oscar Colombo, Joseph Conway, Michael Crisp, R. Lee Cummings, Robert Cummins, Harold Cunningham, Steve DeRose, Tim Downey, William Drummond, Marty Eggers, J. W. Evans III, David F. Franklin, David Fox, Lori Gallagher, Elizabeth Gattone, William Gilbert, Sarah Grafton, Kenneth Greaves, Robert Griewahn, Tim Griffiths, Gordon Griswold, Patrick Groleau, Donald Haddad, Richard Hall, Evelyn Hamby, Lynn Hancock, Gene Harmon, Jeremy Harpur, Adrian Hecker, Sherry K. Heinze, Joe Hendrickson, Connie Hester, Christopher Hill, Cliff Hogan, Dave Holt, Kevin James, Susan Johnston, Martin Joyce, Ken Kennedy, Jane Kerber, John Kiefer, Robert King, Ax Kramer, Chris Kubas, Terry Kukral, Jan Kyster, P. Lane, Frank Larimer, James Lavery, Keith Layton, Eric Levin, Gareth Lewis, Laurie Lindsay, Jay Linn, Loretta Linstad, John Lockhart, Mark Logsdon, George Losoncy, Stephen Mackinder, Layne Marshal, Raymond Martin, Brad Matthews, Willie McClure, Jane McDaniel, Gary McLerran, Matthew Medeiros, Robert Meijer, Bob Miller, James G. Miller, John Moffett, Charlotte Morrill, Alejandro Perez Muñoz, Bernie Nelson, Jim North, Carole Norton, Lawrence Nowak, Maryann Ojala, Dale Osland, India Overland, Frank Pajerski, Allan Papkin, Daniel Pareja, Vernon Parenton, Adrian Pawsey, Jim Perlman, Jim Phelps, Maxim Pimenov, Francesco Lo

Polito, Wayne Powell, Dina Preece, Robert Primak, ProDigital Software, Provider Marketing, Shelly Pyles, William Rackley, Gary Rash, Karl Rasmusson, Chris Ratkowski, Joseph Raynoha, Patrick Reagan, ReConnect-IT Consulting, Vincent Reed, Karla Rizzo, Peter Rothschild, Robert Rottman, John Ruisch, Bill Ruppert, Robert Saddington, Gay Schierholz, Morty Schiller, Bob Schmidt, Heinz-Dieter Schulze, John Scott, Anthony Shardlow, Evelyn Sheldon, Susan Sims, Elaine Slansky, Malcolm Slater, Dr. Ron Slovikoski, Charles Smart, Dave Smith, Don Smith, Gregory Smith, Mike Smith, R. Smith, Robert Smith, Sue Smith, D Sobertson, Richard Solo, Bob Souer, Clive Spencer, Steven Spencer, Philip Spohn, Todd Starcher, C Stark, Betty Staton, David Steffens, Neil Stober, Christopher Stone, Eric Strite, Arianna Sunbear, Lesley Swift, Lewis L Szerecz, Theodore Tanalski, Canadian Tech, Nick Teti, Edward Tobin, P. Toblerone, David Todd, Kris Trimmer, Tietopalvelu Tuomi, Debra Vandenbroucke, James VanSickle, Kevin Webb, Wellington Webb, Sue Weiss, Olivier Wenger, Marc Whiston, Janet White, Thomas Will, Dawn Williams, David Windstrom, J. A. Wolters, Patrick Woods, Mark D Worthen PsyD, Mark Wyatt, and Paul Zammit.

Many thanks to all of you!

Publisher's Acknowledgments

Executive Editor: Steve Hayes

Project and Copy Editor: Susan Pink

Technical Editor: Ryan Williams

Production Editor: Siddique Shaik

Cover Image: © Sergey Parantaev/Shutterstock

Take dummies with you everywhere you go!

Whether you are excited about e-books, want more from the web, must have your mobile apps, or are swept up in social media, dummies makes everything easier.

Find us online!

Leverage the power

Dummies is the global leader in the reference category and one of the most trusted and highly regarded brands in the world. No longer just focused on books, customers now have access to the dummies content they need in the format they want. Together we'll craft a solution that engages your customers, stands out from the competition, and helps you meet your goals.

Advertising & Sponsorships

Connect with an engaged audience on a powerful multimedia site, and position your message alongside expert how-to content. Dummies.com is a one-stop shop for free, online information and know-how curated by a team of experts.

- Targeted ads
- Video
- Email Marketing
- Microsites
- Sweepstakes sponsorship

20 MILLION PAGE VIEWS EVERY SINGLE MONTH

15 MILLION UNIQUE VISITORS PER MONTH

43% OF ALL VISITORS ACCESS THE SITE VIA THEIR MOBILE DEVICES

700,000 NEWSLETTER SUBSCRIPTIONS TO THE INBOXES OF *300,000* UNIQUE INDIVIDUALS EVERY WEEK

# of dummies

Custom Publishing

Reach a global audience in any language by creating a solution that will differentiate you from competitors, amplify your message, and encourage customers to make a buying decision.

- Apps
- Books
- eBooks
- Video
- Audio
- Webinars

Brand Licensing & Content

Leverage the strength of the world's most popular reference brand to reach new audiences and channels of distribution.

For more information, visit **dummies.com/biz**

PERSONAL ENRICHMENT

9781119187790	9781119179030	9781119293354	9781119293347	9781119310068	9781119235606
USA $26.00	USA $21.99	USA $24.99	USA $22.99	USA $22.99	USA $24.99
CAN $31.99	CAN $25.99	CAN $29.99	CAN $27.99	CAN $27.99	CAN $29.99
UK £19.99	UK £16.99	UK £17.99	UK £16.99	UK £16.99	UK £17.99

9781119251163	9781119235491	9781119279952	9781119283133	9781119287117	9781119130246
USA $24.99	USA $26.99	USA $24.99	USA $24.99	USA $24.99	USA $22.99
CAN $29.99	CAN $31.99	CAN $29.99	CAN $29.99	CAN $29.99	CAN $27.99
UK £17.99	UK £19.99	UK £17.99	UK £17.99	UK £16.99	UK £16.99

PROFESSIONAL DEVELOPMENT

9781119311041	9781119255796	9781119293439	9781119281467	9781119280651	9781119251132	9781119310563
USA $24.99	USA $39.99	USA $26.99	USA $26.99	USA $29.99	USA $24.99	USA $34.00
CAN $29.99	CAN $47.99	CAN $31.99	CAN $31.99	CAN $35.99	CAN $29.99	CAN $41.99
UK £17.99	UK £27.99	UK £19.99	UK £19.99	UK £21.99	UK £17.99	UK £24.99

9781119181705	9781119263593	9781119257769	9781119293477	9781119265313	9781119239314	9781119293323
USA $29.99	USA $26.99	USA $29.99	USA $26.99	USA $24.99	USA $29.99	USA $29.99
CAN $35.99	CAN $31.99	CAN $35.99	CAN $31.99	CAN $29.99	CAN $35.99	CAN $35.99
UK £21.99	UK £19.99	UK £21.99	UK £19.99	UK £17.99	UK £21.99	UK £21.99

dummies.com

dummies®
A Wiley Brand

Learning Made Easy

ACADEMIC

9781119293576
USA $19.99
CAN $23.99
UK £15.99

9781119293637
USA $19.99
CAN $23.99
UK £15.99

9781119293491
USA $19.99
CAN $23.99
UK £15.99

9781119293460
USA $19.99
CAN $23.99
UK £15.99

9781119293590
USA $19.99
CAN $23.99
UK £15.99

9781119215844
USA $26.99
CAN $31.99
UK £19.99

9781119293378
USA $22.99
CAN $27.99
UK £16.99

9781119293521
USA $19.99
CAN $23.99
UK £15.99

9781119239178
USA $18.99
CAN $22.99
UK £14.99

9781119263883
USA $26.99
CAN $31.99
UK £19.99

Available Everywhere Books Are Sold

dummies.com

dummies®
A Wiley Brand